WILKS · Mathematical Statistics

WILLIAMS · Diffusion, Markov Processes, and Martingales, Volume I: Foundations

ZACKS · The Theory of Statistical Inference

Applied Probability and Statistics

BAILEY · The Elements of Stochastic Processes with Applications to the Natural Sciences

BAILEY · Mathematics, Statistics and Systems for Health

BARTHOLOMEW · Stochastic Models for Social Processes, *Second Edition*

BARTHOLOMEW and FORBES · Statistical Techniques in Manpower Planning

BECK and ARNOLD · Parameter Estimation in Engineering and Science

BENNETT and FRANKLIN · Statistical Analysis in Chemistry and the Chemical Industry

BHAT · Elements of Applied Stochastic Processes

BLOOMFIELD · Fourier Analysis of Time Series: An Introduction

BOX · R. A. Fisher, The Life of a Scientist

BOX and DRAPER · Evolutionary Operation: A Statistical Method for Process Improvement

BOX, HUNTER, and HUNTER · Statistics for Experimenters: An Introduction to Design, Data Analysis, and Model Building

BROWN and HOLLANDER · Statistics: A Biomedical Introduction

BROWNLEE · Statistical Theory and Methodology in Science and Engineering, *Second Edition*

BURY · Statistical Models in Applied Science

CHAMBERS · Computational Methods for Data Analysis

CHATTERJEE and PRICE · Regression Analysis by Example

CHERNOFF and MOSES · Elementary Decision Theory

CHOW · Analysis and Control of Dynamic Economic Systems

CLELLAND, DECANI, BROWN, BURSK, and MURRAY · Basic Statistics with Business Applications, *Second Edition*

COCHRAN · Sampling Techniques, *Third Edition*

COCHRAN and COX · Experimental Designs, *Second Edition*

CO ′ · Planning of Experiments

COX and MILLER · The Theory of Stochastic Processes, *Second Edition*

DANIEL · Biostatistics: A Foundation for Analysis in the Health Sciences, *Second Edition*

DANIEL · Application of Statistics to Industrial Experimentation

DANIEL and WOOD · Fitting Equations to Data

DAVID · Order Statistics

DEMING · Sample Design in Business Research

DODGE and ROMIG · Sampling Inspection Tables, *Second Edition*

DRAPER and SMITH · Applied Regression Analysis

DUNN · Basic Statistics: A Primer for the Biomedical Sciences, *Second Edition*

DUNN and CLARK · Applied Statistics: Analysis of Variance and Regression

ELANDT-JOHNSON · Probability Models and Statistical Methods in Genetics

FLEISS · Statistical Methods for Rates and Proportions

GALAMBOS · The Asymptotic Theory of Extreme Order Statistics

GIBBONS, OLKIN, and SOBEL · Selecting and Ordering Populations: A New Statistical Methodology

GNANADESIKAN · Met[illegible] Statistical Data Analysis of Multivariate Observations

ontinued on back

Probability Theory

Probability Theory

R. G. LAHA and V. K. ROHATGI
Bowling Green State University

JOHN WILEY & SONS
New York Chichester Brisbane Toronto

Library of Congress Cataloging in Publication Data:

Laha, R. G.
Probability theory.

(Wiley series in probability and mathematical statistics)
Includes bibliographical references and indexes.
1. Probabilities. 2. Measure theory. I. Rohatgi, V. K., 1939– joint author. II. Title.

QA273.L18 519.2 78-24431
ISBN 0-471-03262-X

Printed in the United States of America

10 9 8 7 6 5 4 3 2 1

Preface

This book is designed for a beginning or an intermediate graduate course in probability theory. It is intended for a serious student of probability, whether a mathematics and/or a statistics major. A working knowledge of real analysis at the level of Royden, complex analysis at the level of Hille, and abstract measure theory at the level of Halmos is assumed. For students who wish to specialize in probability theory it is expected that this course will be followed by one in stochastic processes and/or one in probability measures on abstract spaces for research preparation. For others, a course based on this book will accord adequate preparation.

The objective of this book is to make the basic concepts of probability theory easily accessible to both students and research workers in a comprehensive manner. We believe that a student of probability theory at graduate level should have a comprehensive knowledge of both the measure-theoretic foundations and the analytical tools of probability theory. This should include, for example, the axiomatic foundations of probability theory as developed by Kolmogorov and the analytical tools as developed, amongst others, by Lévy, Cramér, Feller, and Prokhorov. Some of the existing books on the subject do not concentrate long enough on the central topics such as laws of large numbers, the law of the iterated logarithms, infinitely divisible laws, and the central limit theory, while others do not treat the analytical tools in a comprehensive manner. This book is written to overcome not only these drawbacks but also to meet the requirements set out above.

Some special features of this book are as follows:

1. The strong interrelationship between probability theory and mathematical analysis is emphasized. For example, a detailed discussion of the properties of characteristic functions and $\mathscr{L}_p$ spaces, and a chapter on random variables taking values in normed linear spaces, are included. We emphasize classical as well as modern methods.

2. Special stress is placed on probability that is applicable rather than probability as analysis. Applications of probability, in particular, to statistics and analysis are emphasized.
3. Some recent developments in probability theory are included. For example, a detailed proof of Prokhorov's theorem and its applications (Section 3.9) are given. Section 3.8 deals with semigroups of probability distributions and their infinitesimal generators. In Chapter 7 we prove the Minlos–Sazonov theorem and derive an analogue of the Lévy–Khintchine representation of the Fourier transform of an infinitely divisible probability measure on a Hilbert space. We also discuss in detail the general central limit problem in a Hilbert space.
4. Every attempt has been made to make the book self-contained. Only well-known results from analysis and measure theory have been used. We have avoided using results by quotation from sources such as monographs or research papers.
5. A large number of examples and remarks elucidate the text.
6. An adequate number of problems at the end of each chapter (subdivided by sections) supplement the text.
7. Notes and comments at the end of each chapter include references to sources and to additional reading material.
8. An extensive list of references is included.

A few words are in order about the selection of topics and applications. The choice of topics is somewhat traditional, but the reader will find here some material that is available only in specialized monographs. The ordering of the chapters is for ease in presentation. However, the reader need not follow this order. For example, most of Chapter 6 can easily be understood after Chapter 3. Similarly parts of Chapter 7 can also be followed after Chapter 3 has been read. As for the choice of applications, only those are included for which little or no preparatory work is needed. In view of our intended audience, applications to statistics and analysis figure prominently in our selection, whereas applications to number theory, stochastic processes, and the like are given less attention.

We do not claim any originality in methods of proofs or their presentation. However, a special attempt has been made to present complete proofs in a lucid and precise manner. Most of the results that are included in this book are fairly well known. For this reason we have avoided overburdening the text with credit references. Rather we have cited, wherever possible, monographs and books where such references can be found. Similarly we have referred to only the sources where the results are stated or proved, in the formulation best suited to our needs. For the organization and presentation of the material we have relied heavily on the well known works of Loève,

Lukacs, and Gnedenko and Kolmogorov. To these authors we express our indebtedness.

The numbering of chapters, sections, subsections, theorems, and so forth is traditional. Each chapter has been subdivided into several sections. Each section has been further subdivided into subsections wherever necessary. Definitions, theorems, equations, and so on are numbered consecutively within each section. Thus equation $(i.j.k)$ stands for the kth equation in Section j of Chapter i. Section $i.j$ stands for Section j in Chapter i; Section $i.j.k$ stands for Subsection k of Section j in Chapter i; and so on.

References are given at the end of the book and are denoted by numbers enclosed in brackets: [ ].

The set of lecture notes on which this book is based has been used by both of us over the last 10 years. The first-named author class-tested a major portion of the present version during the year 1977–1978.

We are indebted to Professors J. Sethuraman of Florida State University, A. J. Terzuoli of the Polytechnic Institute of New York, and H. Braun of Princeton University who read the first draft and made many suggestions. We are specially grateful to Professors R. J. Tompkins and C. C. Heyde for making numerous suggestions. Thanks are also due to D. Borowiak for his assistance in proofreading.

Kae Lea Main deserves our special thanks for her efficient and error-free typing. Finally we express our indebtedness to Ms. Beatrice Shube of John Wiley & Sons for her efficient editorial guidance.

R. G. LAHA
V. K. ROHATGI

Bowling Green, Ohio
January 1979

Contents

Probability Theory

CHAPTER 1

Basic Concepts of Probability Theory

The fundamental concepts of probability theory have roots in measure theory. Like any branch of mathematics, probability theory has its own terminology and its own tools. In this chapter we introduce some of this terminology and study some basic concepts of probability theory. It is assumed that the reader has a working knowledge of measure theory at the level of Halmos [31], and of real analysis at the level of Royden [71]. We will frequently use results from measure theory and real analysis without reference to their source. Results which are of particular importance in probability theory, however, are proved in some detail. These include for example, Theorems 1.1.2 and 1.1.5.

1.1 PROBABILITY SPACES AND RANDOM VARIABLES

1.1.1 Notation and Probability Terminology

We denote by Ω a nonempty set. The elements of Ω will be called *points* and be denoted generically by ω. The following set-theoretic notation will be used:

Points	ω
Sets	capital letters E, F, G, etc.
Union	$E \cup F$, $\bigcup_\alpha E_\alpha$
Intersection	$E \cap F$, $\bigcap_\alpha E_\alpha$
Complement	E^c
Difference	$E - F = E \cap F^c$
Singleton set	$\{\omega\}$
Set inclusion	$E \subset F$ (not excluding $E = F$)

Classes	script capital letters $\mathscr{A}$, $\mathscr{B}$, $\mathscr{S}$, etc.
Inclusion	$\mathscr{A} \subset \mathscr{B}$ (not excluding $\mathscr{A} = \mathscr{B}$)
Belonging to	$\omega \in E$, $E \in \mathscr{S}$
Empty set	$\varnothing$

In the following list we give the correspondences between the probability and measure-theoretic terms which are frequently used:

Sample space	Measurable space
Probability	Normed measure
Probability space	Normed measure space
Elementary event	Singleton set
Event	Measurable set
Sure event	Whole space Ω
Impossible event	Empty set $\varnothing$
Almost sure, almost surely (a.s.) (with probability 1)	Almost everywhere (a.e.)
Random variable	(Finite-valued numerical) measurable function
Expectation	Integral

We summarize below in probability language some results which are specializations of corresponding results in measure theory.

1.1.2 Probability Space

Let Ω be a nonempty set. Let $\mathscr{S}$ be a σ-field of subsets of Ω, that is, a nonempty class of subsets of Ω which contains Ω and is closed under countable union and complementation.

Let P be a measure defined on $\mathscr{S}$ satisfying $P(\Omega) = 1$. Then the triple $(\Omega, \mathscr{S}, P)$ is called a *probability space*, and P, a *probability measure*. The set Ω is the *sure event*, and elements of $\mathscr{S}$ are called *events*. Singleton sets $\{\omega\}$ are called *elementary events*. The symbol $\varnothing$ denotes the empty set and is known as the *null* or *impossible event*. Unless otherwise stated, the probability space $(\Omega, \mathscr{S}, P)$ is fixed, and $A, B, C, \ldots,$ with or without subscripts, represent events.

We note that, if $A_n \in \mathscr{S}$, $n = 1, 2, \ldots,$ then A_n^c, $\bigcup_{n=1}^{\infty} A_n$, $\bigcap_{n=1}^{\infty} A_n$, $\liminf_{n\to\infty} A_n$, $\limsup_{n\to\infty} A_n$, and $\lim_{n\to\infty} A_n$ (if it exists) are events. Also, the probability measure P is defined on $\mathscr{S}$, and for all events A, A_n

$$P(A) \geq 0, \qquad P\left(\bigcup_{n=1}^{\infty} A_n\right) = \sum_{n=1}^{\infty} P(A_n)(A_n\text{'s disjoint}), \qquad P(\Omega) = 1.$$

It follows that

$$P(\emptyset) = 0, \qquad P(A) \le P(B) \quad \text{for } A \subset B, \qquad P\left(\bigcup_{n=1}^{\infty} A_n\right) \le \sum_{n=1}^{\infty} P(A_n).$$

Moreover,

$$P\left(\liminf_{n\to\infty} A_n\right) \le \liminf_{n\to\infty} P(A_n) \le \limsup_{n\to\infty} P(A_n) \le P\left(\limsup_{n\to\infty} A_n\right),$$

and, if $\lim_{n\to\infty} A_n$ exists, then

$$P\left(\lim_{n\to\infty} A_n\right) = \lim_{n\to\infty} P(A_n).$$

The last result is known as the continuity property of probability measures.

Example 1.1.1. Let $\Omega = \{\omega_j : j \ge 1\}$, and let $\mathscr{S}$ be the σ-field of all subsets of Ω. Let $\{p_j, j \ge 1\}$ be any sequence of nonnegative real numbers satisfying $\sum_{j=1}^{\infty} p_j = 1$. Define P on $\mathscr{S}$ by

$$P(E) = \sum_{\omega_j \in E} p_j, \qquad E \in \mathscr{S}.$$

Then P defines a probability measure on $(\Omega, \mathscr{S})$, and $(\Omega, \mathscr{S}, P)$ is a probability space.

Example 1.1.2. Let $\Omega = (0, 1]$ and $\mathscr{S} = \mathscr{B}$ be the σ-field of Borel sets on Ω. Let λ be the Lebesgue measure on $\mathscr{B}$. Then $(\Omega, \mathscr{S}, \lambda)$ is a probability space.

Definition 1.1.1. Let $(\Omega, \mathscr{S}, P)$ be a probability space. A real-valued function X defined on Ω is said to be a random variable if

$$X^{-1}(E) = \{\omega \in \Omega : X(\omega) \in E\} \in \mathscr{S} \qquad \text{for all } E \in \mathscr{B},$$

where $\mathscr{B}$ is the σ-field of Borel sets in $\mathbb{R} = (-\infty, \infty)$; that is, a random variable X is a measurable transformation of $(\Omega, \mathscr{S}, P)$ into $(\mathbb{R}, \mathscr{B})$.

We note that it suffices to require that $X^{-1}(I) \in \mathscr{S}$ for all intervals I in $\mathbb{R}$, or for all semiclosed intervals $I = (a, b]$, or for all intervals $I = (-\infty, b]$, and so on. Unless otherwise specified, $X, Y, \ldots$, with or without subscripts, will represent random variables.

We note that a random variable X defined on $(\Omega, \mathscr{S}, P)$ induces a measure P_X on $\mathscr{B}$ defined by the relation

$$P_X(E) = P\{X^{-1}(E)\} \qquad (E \in \mathscr{B}).$$

Clearly P_X is a probability measure on $\mathscr{B}$ and is called the *probability distribution* or, simply, the *distribution* of X. We note that P_X is a Lebesgue–Stieltjes measure on $\mathscr{B}$.

Definition 1.1.2. For every $x \in \mathbb{R}$ set

$$F_X(x) = P_X(-\infty, x] = P\{\omega \in \Omega: X(\omega) \leq x\}. \tag{1.1.1}$$

We call $F_X = F$ the distribution function of the random variable X.

In the following we write $\{X \leq x\}$ for the event $\{\omega \in \Omega: X(\omega) \leq x\}$. We first prove the following elementary property of a distribution function.

Theorem 1.1.1. The distribution function F of a random variable X is a nondecreasing, right-continuous function on $\mathbb{R}$ which satisfies

$$F(-\infty) = \lim_{x \to -\infty} F(x) = 0$$

and

$$F(+\infty) = \lim_{x \to \infty} F(x) = 1.$$

Proof. Note that for every $x \in \mathbb{R}$ and $h > 0$

$$F(x + h) - F(x) = P\{x < X \leq x + h\} \geq 0,$$

so that F is nondecreasing.

Next, let $\{h_n\}$ be a sequence of real numbers such that $0 < h_n \downarrow 0$, as $n \to \infty$. Then, for every $n \geq 1$,

$$F(x + h_n) - F(x) = P\{x < X \leq x + h_n\}.$$

It follows from the continuity property of P that

$$\lim_{n \to \infty} [F(x + h_n) - F(x)] = 0,$$

and hence that F is right-continuous.

Finally, for every $N \geq 1$ we have

$$F(N) - F(-N) = P\{-N < X \leq N\}.$$

Taking the limit of both sides as $N \to \infty$ and using the continuity property once again, we conclude that

$$F(+\infty) - F(-\infty) = 1.$$

But since $0 \leq F(x) \leq 1$ for every $x \in \mathbb{R}$, it follows that $F(-\infty) = 0$ and $F(+\infty) = 1$. ■

Corollary. A distribution function F is continuous at $x \in \mathbb{R}$ if and only if $P\{\omega: X(\omega) = x\} = 0$.

Proof. The proof of the corollary is an immediate consequence of the fact that

$$P\{X = x\} = F(x) - F(x - 0). \tag{1.1.2}$$

Remark 1.1.1. Let X be a random variable, and let g be a Borel-measurable function defined on $\mathbb{R}$. Then $g(X)$ is also a random variable whose distribution is determined by that of X.

We now show that a function F on $\mathbb{R}$ with the properties stated in Theorem 1.1.1 determines uniquely a probability measure P_F on $\mathscr{B}$.

Theorem 1.1.2. Let F be a nondecreasing, right-continuous function defined on $\mathbb{R}$ and satisfying

$$F(-\infty) = 0 \qquad \text{and} \qquad F(+\infty) = 1.$$

Then there exists a probability measure $P = P_F$ on $\mathscr{B}$ determined uniquely by the relation

$$P_F(-\infty, x] = F(x) \qquad \text{for every } x \in \mathbb{R}. \tag{1.1.3}$$

Proof. Let $\mathscr{P}$ be the class of all bounded left-open, right-closed intervals of the form $(a, b]$, $-\infty < a < b < \infty$. Define a set function P_F on $\mathscr{P}$ by the relation

$$P_F(a, b] = F(b) - F(a).$$

We write $P = P_F$ and note that $0 \le P(E) \le 1$ for all $E \in \mathscr{P}$. The proof of Theorem 1.1.2 is based on the following steps.

STEP 1. Let $E_0 \in \mathscr{P}$, and $E_n \in \mathscr{P}$, $n = 1, 2, \ldots$ be a sequence of disjoint sets such that $E_n \subset E_0$ for every n. We show that

$$\sum_{n=1}^{\infty} P(E_n) \le P(E_0) \tag{1.1.4}$$

holds.

Let us first consider the case where the sequence E_n consists of only a finite number of sets, say $E_1, E_2, \ldots, E_N$. Set $E_n = (a_n, b_n], 0 \le n \le N$. Without loss of generality, we may assume that $a_1 \le a_2 \le \cdots \le a_N$. Since the E_n are disjoint and $E_n \subset E_0$ for each n, it follows that

$$a_0 \le a_1 < b_1 \le a_2 < b_2 \le \cdots \le a_N < b_N \le b_0.$$

Then

$$\sum_{n=1}^{N} P(E_n) = \sum_{n=1}^{N} [F(b_n) - F(a_n)]$$
$$\leq \sum_{n=1}^{N} [F(b_n) - F(a_n)] + \sum_{n=1}^{N-1} [F(a_{n+1}) - F(b_n)]$$
$$= F(b_N) - F(a_1)$$
$$\leq P(E_0).$$

Next, let the sequence $\{E_n\}$ be countably infinite. Since

$$\sum_{n=1}^{N} P(E_n) \leq P(E_0) \qquad \text{for every } N,$$

letting $N \to \infty$ yields (1.1.4).

STEP 2. Let $K_0 = [a_0, b_0]$, $-\infty < a_0 < b_0 < \infty$, and let $V_n = (a_n, b_n)$, $-\infty < a_n < b_n < \infty, n = 1, 2, \ldots, N$, be such that $K_0 \subset \bigcup_{n=1}^{N} V_n$. Then we show that

$$F(b_0) - F(a_0) \leq \sum_{n=1}^{N} [F(b_n) - F(a_n)]. \tag{1.1.5}$$

Since $K_0 \subset \bigcup_{n=1}^{N} V_n$, there exists an integer k_1, $1 \leq k_1 \leq N$, such that $a_0 \in V_{k_1}$. If $b_0 \in V_{k_1}$, then clearly (1.1.5) holds. Otherwise, $b_{k_1} < b_0$. In this case there exists an integer k_2, $1 \leq k_2 \leq N$, such that $b_{k_1} \in V_{k_2}$. If $b_{k_2} < b_0$, there exists an integer k_3, $1 \leq k_3 \leq N$, such that $b_{k_2} \in V_{k_3}$, and so on. Clearly this process must terminate after a finite number of steps when we have obtained a set V_{k_m} from the sequence $\{V_n\}$ such that $b_0 \in V_{k_m}$. We may assume, without loss of generality, that $m = N$ and that $V_{k_n} = V_n$ for $1 \leq n \leq N$. We have the set of inequalities

$$a_1 < a_0 < b_1, \qquad a_2 < b_1 < b_2, \ldots, a_N < b_{N-1} < b_N, \qquad a_N < b_0 < b_N.$$

Hence

$$F(b_0) - F(a_0) \leq F(b_N) - F(a_1)$$
$$= F(b_1) - F(a_1) + \sum_{j=1}^{N-1} [F(b_{j+1}) - F(b_j)]$$
$$\leq \sum_{n=1}^{N} [F(b_n) - F(a_n)].$$

STEP 3. Let $E_0 \in \mathscr{P}$ and let $E_n \in \mathscr{P}, n = 1, 2, \ldots$, be such that $E_0 \subset \bigcup_{n=1}^{\infty} E_n$. Then we show that

$$P(E_0) \leq \sum_{n=1}^{\infty} P(E_n). \tag{1.1.6}$$

Set $E_n = (a_n, b_n]$, $n = 0, 1, 2, \ldots$. Let $\epsilon > 0$ so that $0 < \epsilon < b_0 - a_0$. Write $K_0 = [a_0 + \epsilon, b_0]$ and $V_n = (a_n, b_n + \epsilon_n)$, $n = 1, 2, \ldots$, where $\epsilon_n > 0$ for each n. Then $K_0 \subset E_0$ and $E_n \subset V_n$ for every n so that

$$K_0 \subset E_0 \subset \bigcup_{n=1}^{\infty} E_n \subset \bigcup_{n=1}^{\infty} V_n.$$

Since K_0 is a compact set in $\mathbb{R}$ and the sequence $\{V_n\}$ forms an open covering of K_0, it follows from the Heine–Borel theorem that $\{V_n\}$ contains a finite subsequence, say $V_1, V_2, \ldots, V_N$, such that $K_0 \subset \bigcup_{n=1}^{N} V_n$. It follows from (1.1.5) that

$$\begin{aligned} F(b_0) - F(a_0 + \epsilon) &\leq \sum_{n=1}^{N} [F(b_n + \epsilon_n) - F(a_n)] \\ &\leq \sum_{n=1}^{\infty} [F(b_n + \epsilon_n) - F(a_n)]. \end{aligned}$$

Since F is right-continuous, we can choose ϵ_n such that

$$F(b_n + \epsilon_n) \leq F(b_n) + \frac{\epsilon}{2^n}$$

holds for each n. Then we get

$$F(b_0) - F(a_0 + \epsilon) \leq \sum_{n=1}^{\infty} [F(b_n) - F(a_n)] + \epsilon,$$

which yields (1.1.6) on letting $\epsilon \downarrow 0$.

STEP 4. The set function P is a measure on $\mathscr{P}$. It suffices to prove that P is countably additive. Let $\{E_n\}$ be a sequence of disjoint sets from $\mathscr{P}$ such that $\bigcup_{n=1}^{\infty} E_n = E \in \mathscr{P}$. From (1.1.4) and (1.1.6) we get $P(E) = \sum_{n=1}^{\infty} P(E_n)$.

STEP 5. Let $\mathscr{R}$ be the class of all finite unions of disjoint sets belonging to $\mathscr{P}$. We show that there exists a unique measure $\bar{P}$ on $\mathscr{R}$ such that the restriction of $\bar{P}$ to $\mathscr{P}$ coincides with P.

Let $E \in \mathscr{R}$ be arbitrary. Then there exists a finite number of disjoint sets $E_j \in \mathscr{P}, j = 1, 2, \ldots, N$, such that $E = \bigcup_{j=1}^{N} E_j$. Set

$$\bar{P}(E) = \sum_{j=1}^{N} P(E_j). \tag{1.1.7}$$

First note that $\bar{P}$ is uniquely defined by (1.1.7). In fact, let $E = \bigcup_{k=1}^{M} F_k$ be another representation of E such that $F_k \in \mathscr{P}$, $k = 1, 2, \ldots, M$, and the F_k's are disjoint. Clearly $E_j = \bigcup_{k=1}^{M} (E_j \cap F_k)$, $j = 1, 2, \ldots, N$, and $F_k = \bigcup_{j=1}^{N} (E_j \cap F_k)$, $k = 1, 2, \ldots, M$. The sets $E_j \cap F_k$ are disjoint and belong to $\mathscr{P}$ for all $j = 1, 2, \ldots, N$, $k = 1, 2, \ldots, M$. It follows that

$$\sum_{j=1}^{N} P(E_j) = \sum_{j=1}^{N} \sum_{k=1}^{M} P(E_j \cap F_k)$$

$$= \sum_{k=1}^{M} P(F_k),$$

and hence that $\bar{P}$ is uniquely defined. Clearly $\bar{P}(E) = P(E)$ for every $E \in \mathscr{P}$.

Next, we note that we need only prove that $\bar{P}$ is countably additive. Clearly $\bar{P}$ is finitely additive. Let $E_n \in \mathscr{R}$ be a sequence of disjoint sets such that $E = \bigcup_{n=1}^{\infty} E_n \in \mathscr{R}$. Since $E_n \in \mathscr{R}$, we can write $E_n = \bigcup_k E_{nk}$, where $E_{nk} \in \mathscr{P}$ is a finite sequence of disjoint sets. Also,

$$\bar{P}(E_n) = \sum_k P(E_{nk}) \qquad \text{for } n \geq 1.$$

Let us first consider the case where $E \in \mathscr{P}$. Then $E = \bigcup_{n,k} E_{nk}$, where $E_{nk} \in \mathscr{P}$ is a sequence of disjoint sets. Since P is a measure on $\mathscr{P}$ and $\bar{P}$ coincides with P on $\mathscr{P}$, we have

$$\bar{P}(E) = P(E) = \sum_{n,k} P(E_{nk}) = \sum_n \bar{P}(E_n).$$

Next, let $E \in \mathscr{R}$. Then there exists a finite number of disjoint sets $F_j \in \mathscr{P}$ such that $E = \bigcup_j F_j$. Since $F_j \in \mathscr{P}$, it follows from the above that

$$\bar{P}(F_j) = \sum_n \bar{P}(F_j \cap E_n).$$

Since $\bar{P}$ is finitely additive on $\mathscr{R}$, we have

$$\bar{P}(E) = \sum_j \bar{P}(F_j) = \sum_j \sum_n \bar{P}(F_j \cap E_n)$$

$$= \sum_n \bar{P}(E_n),$$

which proves the countable additivity of $\bar{P}$.

Finally we show that $\bar{P}$ is the unique extension of P to $\mathscr{R}$. Indeed, let $\bar{P}_1$

and $\bar{P}_2$ be two such extensions. Let $E \in \mathscr{R}$. Then $E = \bigcup_j E_j$, where $E_j \in \mathscr{P}$ is a finite sequence of disjoint sets. Thus

$$\begin{aligned}\bar{P}_1(E) &= \sum_j \bar{P}_1(E_j)\\ &= \sum_j P(E_j)\\ &= \sum_j \bar{P}_2(E_j)\\ &= \bar{P}_2(E).\end{aligned}$$

In what follows we write P for $\bar{P}$. To complete the proof of the theorem we need only note that $\mathscr{R}$ is a ring and P is a finite measure on $\mathscr{R}$. It follows from the Carathéodory extension theorem (Halmos [31], p. 54) that P can be uniquely extended to a measure on the σ-ring $\mathscr{S}(\mathscr{R})$ generated by $\mathscr{R}$. We note that $\mathscr{B} = \mathscr{S}(\mathscr{R})$ is precisely the σ-field of Borel sets in $\mathbb{R}$. Clearly $P(\mathbb{R}) = 1$, so that P is a probability measure. This completes the proof of Theorem 1.1.2. ∎

Remark 1.1.2. Let F be a bounded nondecreasing, right-continuous function defined on $\mathbb{R}$ satisfying $F(-\infty) = 0$. Then it is clear from the proof of Theorem 1.1.2 that there exists a finite measure $\mu = \mu_F$ on $\mathscr{B}$ determined uniquely by $\mu_F(-\infty, x] = F(x)$, $x \in \mathbb{R}$.

Remark 1.1.3. Let F on $\mathbb{R}$ satisfy the conditions of Theorem 1.1.2. Then there exists a random variable X on some probability space such that F is the distribution function of X. In fact, consider the probability space $(\mathbb{R}, \mathscr{B}, P)$, where P is the probability measure as constructed in Theorem 1.1.2. Let $X(\omega) = \omega$, for all $\omega \in \mathbb{R}$. It is easy to see that F is the distribution function of the random variable X.

Let F be a distribution function, and let $x \in \mathbb{R}$ be a discontinuity point of F. Then $p(x) = F(x) - F(x - 0)$ is called the *jump* of F at x. A point is said to be a *point of increase* of F if, for every $\epsilon > 0$, $F(x + \epsilon) - F(x - \epsilon) > 0$.

Some other elementary properties of a distribution function are obtained in the following propositions.

Proposition 1.1.1. Let F_1 and F_2 be two distribution functions such that

$$F_1(x) = F_2(x) \qquad \text{for all } x \in D,$$

where D is everywhere dense in $\mathbb{R}$. Then $F_1(x) = F_2(x)$ for every $x \in \mathbb{R}$.

Proof. Let $x \in \mathbb{R}$. Then there exists a sequence $x_n \in D$ such that $x_n \downarrow x$ as $n \to \infty$. Also $F_1(x_n) = F_2(x_n)$ for every n. The result follows using the right continuity of F_1 and F_2. ■

Proposition 1.1.2. The set of discontinuity points of a distribution function F is countable.

Proof. Let D be the set of all discontinuity points of F. Let n be a positive integer, and let

$$D_n = \left\{x \in D: \frac{1}{n+1} < p(x) \leq \frac{1}{n}\right\}.$$

Since $F(+\infty) - F(-\infty) = 1$, the number of elements in D_n cannot exceed n. Clearly $D = \bigcup_{n=1}^{\infty} D_n$, and it follows that D is countable. ■

Proposition 1.1.3. Let F_1 and F_2 be two distribution functions, and let C_1 and C_2, respectively, be their sets of continuity points. If

$$F_1(x) = F_2(x) \qquad \text{for } x \in C_1 \cap C_2,$$

then $F_1(x) = F_2(x)$ for all $x \in \mathbb{R}$.

Proof. We need only note that the set

$$\mathbb{R} - (C_1 \cap C_2) = (\mathbb{R} - C_1) \cup (\mathbb{R} - C_2)$$

is countable in view of Proposition 1.1.2. The proof is now an immediate consequence of Proposition 1.1.1. ■

1.1.3 Decomposition of a Distribution Function

We now prove an important property of a distribution function. Let F be a distribution function, and let $\{x_n\}$ be the countable set of all discontinuity points of F. Set

$$G(x) = \sum_{x_n \leq x} p(x_n) \qquad \text{for } x \in \mathbb{R}, \tag{1.1.8}$$

where $p(x_n)$ is the jump of F at x_n. We note that G is a step function which is nondecreasing, is right-continuous, and satisfies $G(-\infty) = 0$ and $G(+\infty) = \alpha$, $0 \leq \alpha \leq 1$. Define the function H by

$$H(x) = F(x) - G(x) \qquad \text{for } x \in \mathbb{R}. \tag{1.1.9}$$

Proposition 1.1.4. The function H defined in (1.1.9) is nondecreasing and continuous on $\mathbb{R}$.

Proof. Clearly H is right-continuous. Let $x' < x$. Then it follows from (1.1.9) that

$$H(x) - H(x') = [F(x) - F(x')] - [G(x) - G(x')].$$

Taking the limit on both sides as $x' \uparrow x$, we see that H is also left-continuous.

We next show that H is nondecreasing. We first prove that for $x' < x$ the inequality

$$\sum_{x' < x_n \leq x} p(x_n) \leq F(x) - F(x') \tag{1.1.10}$$

holds. Suppose that F has only a finite number, say N, of discontinuity points x_n in the interval $(x', x]$. Without loss of generality we may assume that

$$x' < x_1 < x_2 < \cdots < x_N \leq x.$$

By the monotone property of F it follows that

$$F(x') \leq F(x_1 - 0) < F(x_1) \leq F(x_2 - 0) < F(x_2) \leq \cdots$$
$$\leq F(x_N - 0) < F(x_N) \leq F(x).$$

Since $p(x_n) = F(x_n) - F(x_n - 0)$, we have

$$\sum_{n=1}^{N} p(x_n) = F(x_N) - F(x_1 - 0) + \sum_{j=1}^{N-1} [F(x_j) - F(x_{j+1} - 0)]$$
$$\leq F(x) - F(x'). \tag{1.1.11}$$

Next suppose that F has a countably infinite set of discontinuity points in $(x', x]$. For every $N \geq 1$ (1.1.11) holds. Taking the limit as $N \to \infty$, we obtain (1.1.10). Consequently

$$H(x) - H(x') = F(x) - F(x') - \sum_{x' < x_n \leq x} p(x_n) \geq 0,$$

and H is nondecreasing. This completes the proof of the assertion. ∎

We note that $H(-\infty) = 0$ and $H(+\infty) = 1 - \alpha$. Thus we have the decomposition

$$F(x) = G(x) + H(x), \tag{1.1.12}$$

which is unique in view of the fact that G is a step function and H is continuous.

Theorem 1.1.3. Every distribution function F admits the decomposition

$$F(x) = \alpha F_d(x) + (1 - \alpha)F_c(x) \qquad (0 \leq \alpha \leq 1), \tag{1.1.13}$$

where F_d and F_c are both distribution functions such that F_d is a step function and F_c is continuous. Moreover, decomposition (1.1.13) is unique.

Proof. Let $0 < \alpha < 1$. Set

$$F_d(x) = \frac{1}{\alpha} G(x) \qquad \text{and} \qquad F_c(x) = \frac{1}{1-\alpha} H(x),$$

where G and H are as defined in (1.1.8) and (1.1.9), respectively. If $\alpha = 1$, set $F(x) = G(x)$; if $\alpha = 0$, set $F(x) = H(x)$. The result now follows from (1.1.12). ∎

Remark 1.1.4. A distribution function F is said to be *discrete* if $F(x) = G(x)$ and *continuous* if $F(x) = H(x)$, where G and H are as defined in (1.1.8) and (1.1.9), respectively.

We now give examples of some standard discrete distributions. Examples of some standard continuous distributions will be taken up later.

Example 1.1.3. For every $x \in \mathbb{R}$, define

$$\epsilon(x) = \begin{cases} 0 & \text{if } x < 0, \\ 1 & \text{if } x \geq 0. \end{cases}$$

Clearly $\epsilon(x)$ defines a distribution function. A random variable X with distribution function $F(x) = \epsilon(x - c)$, where c is a fixed constant, is known as the random variable degenerate at c. Here F has exactly one jump point at $x = c$, and the jump of F at c is 1.

Example 1.1.4. Let $0 < p < 1$. Consider the distribution function $F(x) = p\epsilon(x - 1) + (1 - p)\epsilon(x)$. It has two discontinuity points, $x = 0$ and $x = 1$, with jumps $(1 - p)$ and p, respectively. A random variable X with distribution function F is known as a *Bernoulli* random variable.

Example 1.1.5. Let $0 < p < 1$, and consider the distribution function

$$F(x) = \sum_{k=0}^{n} \binom{n}{k} p^k (1 - p)^{n-k} \epsilon(x - k).$$

The distribution function F has $n + 1$ discontinuity points with jump $\binom{n}{k} p^k (1 - p)^{n-k}$ at k, $k = 0, 1, 2, \ldots, n$. A random variable X with distribution function F defined above is known as a *binomial* random variable.

Example 1.1.6. A random variable X with distribution function F defined by

$$F(x) = \sum_{k=0}^{\infty} e^{-\lambda} \frac{\lambda^k}{k!} \epsilon(x - k) \qquad (\lambda > 0)$$

is called a *Poisson* random variable.

We next show that the continuous distribution function F_c in (1.1.13) has a further decomposition. In view of Theorem 1.1.2 F_c determines uniquely a probability measure P_c on $\mathscr{B}$. In the following λ denotes the Lebesgue measure on $\mathscr{B}$. It follows from the Lebesgue decomposition theorem that we can write

$$(1.1.14) \qquad P_c(E) = P_s(E) + P_{ac}(E) \qquad (E \in \mathscr{B})$$

uniquely, where P_s is singular and P_{ac} is absolutely continuous with respect to λ. Thus there exist two disjoint Borel sets A and B with $A \cup B = \mathbb{R}$, such that, for every $E \in \mathscr{B}$, the relation

$$P_s(E \cap A) = \lambda(E \cap B) = 0$$

holds. On the other hand, the Radon–Nikodym theorem implies that there exists a nonnegative Borel-measurable function f on $\mathbb{R}$ such that

$$(1.1.15) \qquad P_{ac}(E) = \int_E f\, d\lambda \qquad \text{for } E \in \mathscr{B}.$$

Here f is the Radon–Nikodym derivative $dP_{ac}/d\lambda$. Now set

$$(1.1.16) \quad G_s(x) = P_s(-\infty, x] \qquad \text{and} \qquad G_{ac}(x) = P_{ac}(-\infty, x] \quad \text{for } x \in \mathbb{R}.$$

Clearly G_s and G_{ac} are nondecreasing, right-continuous functions determined uniquely by P_s and P_{ac}, respectively. Here

$$G_s(-\infty) = G_{ac}(-\infty) = 0, \qquad G_s(+\infty) = \beta,$$

and

$$G_{ac}(+\infty) = 1 - \beta,$$

where $0 \leq \beta \leq 1$.

Theorem 1.1.4. Every continuous distribution function F_c admits the decomposition

$$(1.1.17) \qquad F_c(x) = \beta F_s(x) + (1 - \beta)F_{ac}(x) \qquad (0 \leq \beta \leq 1),$$

where F_s and F_{ac} are both distribution functions determined uniquely by F_c.

Proof. For $0 < \beta < 1$, set

$$F_s(x) = \frac{1}{\beta} G_s(x) \qquad \text{and} \qquad F_{ac}(x) = \frac{1}{1-\beta} G_{ac}(x),$$

where G_s and G_{ac} are defined in (1.1.16). For $\beta = 1$, set $F_c(x) = G_s(x)$; and for $\beta = 0$, set $F_c(x) = G_{ac}(x)$. ■

Theorems 1.1.3 and 1.1.4 together yield the following result immediately.

Theorem 1.1.5. Every distribution function F admits a unique decomposition

(1.1.18) $$F(x) = \alpha_1 F_d(x) + \alpha_2 F_s(x) + \alpha_3 F_{ac}(x) \qquad (x \in \mathbb{R}),$$

where $\alpha_1, \alpha_2, \alpha_3 \geq 0$ and $\sum_{j=1}^{3} \alpha_j = 1$. Here F_d, F_s, and F_{ac} are distribution functions.

In Theorem 1.1.5 we note that F_d is a step function. We now show that F_s is continuous singular, whereas F_{ac} is absolutely continuous.

Proposition 1.1.5. In the correspondence $F \leftrightarrow P_F$ in Theorem 1.1.2, F is absolutely continuous on $\mathbb{R}$ if and only if P_F is absolutely continuous with respect to the Lebesgue measure λ on $\mathscr{B}$.

Proof. Suppose that $P_F \ll \lambda$. Let $\epsilon > 0$. Then there exists a $\delta = \delta_\varepsilon > 0$ such that $P_F(E) < \epsilon$ for $E \in \mathscr{B}$ for which $\lambda(E) < \delta$. Let $\{(a_j, b_j), j = 1, 2, \ldots, N\}$ be a finite disjoint class of bounded open intervals in $\mathbb{R}$ with

$$\sum_{j=1}^{N} (b_j - a_j) < \delta.$$

Then clearly

$$\lambda\left(\bigcup_{j=1}^{N} (a_j, b_j]\right) < \delta$$

so that $P_F(\bigcup_{j=1}^{N} (a_j, b_j]) < \epsilon$. Therefore

$$\sum_{j=1}^{N} [F(b_j) - F(a_j)] = \sum_{j=1}^{N} P_F(a_j, b_j] < \epsilon,$$

which proves that F is absolutely continuous.

Conversely, suppose that F is absolutely continuous on $\mathbb{R}$. This means that to every $\epsilon > 0$ there corresponds a $\delta > 0$ such that

$$\sum_{j=1}^{N} [F(b_j) - F(a_j)] < \epsilon$$

for every finite disjoint class $\{(a_j, b_j): j = 1, 2, \ldots, N\}$ of bounded open intervals for which $\sum_{j=1}^{N} (b_j - a_j) < \delta$. Let $E \in \mathscr{B}$ be such that $\lambda(E) = 0$. Then there exists a sequence of disjoint intervals $\{(a_j, b_j], j = 1, 2, \ldots\}$ such that

$$E \subset \bigcup_{j=1}^{\infty} (a_j, b_j] \qquad \text{and} \qquad \sum_{j=1}^{\infty} (b_j - a_j) < \delta.$$

It follows that $\sum_{j=1}^{N} [F(b_j) - F(a_j)] < \epsilon$ for every $N \geq 1$. Hence

$$P_F(E) \leq \sum_{j=1}^{\infty} [F(b_j) - F(a_j)] \leq \epsilon.$$

Since $\epsilon > 0$ is arbitrary, we see that $P_F \ll \lambda$. ■

Corollary. In the notation of Proposition 1.1.5 let f be the Radon-Nikodym derivative $dP_F/d\lambda$. Then the integral representation

$$F(x) = \int_{-\infty}^{x} f(y)\, d\lambda(y) \qquad (x \in \mathbb{R}) \tag{1.1.19}$$

holds.

Proof. The proof of the corollary is an immediate consequence of the Radon–Nikodym theorem. Here f is known as the *probability density function* of the distribution function F. Clearly $f(x) \geq 0$ for all $x \in \mathbb{R}$, and $\int_{-\infty}^{\infty} f(x)\, d\lambda(x) = 1$.

Proposition 1.1.6. In the correspondence $F \leftrightarrow P_F$ of Theorem 1.1.2, F is continuous singular if and only if P_F is singular with respect to λ and $P_F\{x\} = 0$ for every $x \in \mathbb{R}$.

Proof. We recall that a continuous, nondecreasing function F is singular if F is not a constant and its derivative $F' = 0$ a.e. (λ), where λ is the Lebesgue measure on $\mathbb{R}$.

First suppose that F is not singular. Then F has a unique decomposition $F = F_1 + F_2$ (Natanson [62], pp. 263–264), where F_1 and F_2 are nondecreasing, F_1 is singular, and F_2 is absolutely continuous on $\mathbb{R}$. Clearly $F_2 \neq 0$. Let μ_1 and μ_2 be the finite measures on $\mathscr{B}$ corresponding to F_1 and F_2, respectively. It follows from Proposition 1.1.5 that $\mu_2 \not\equiv 0$ and, moreover, $\mu_2 \ll \lambda$. Let P_F be the probability measure on $\mathscr{B}$ defined by

$$P_F(E) = \mu_1(E) + \mu_2(E) \qquad (E \in \mathscr{B}).$$

Clearly P_F is the measure corresponding to F. We show that P_F is not singular with respect to λ. If P_F is singular with respect to λ, there exist disjoint sets $A, A^c \in \mathscr{B}$ such that, for every $E \in \mathscr{B}$, $P_F(E \cap A) = \lambda(E \cap A^c) = 0$. It follows that $\mu_2(E \cap A) = 0$. On the other hand, since $\mu_2 \ll \lambda$, $\lambda(E \cap A^c) = 0$ implies $\mu_2(E \cap A^c) = 0$ and hence $\mu_2(E) = 0$ for all $E \in \mathscr{B}$. This contradiction proves that F singular implies P_F singular.

Conversely, we show that P_F not singular with $P_F\{x\} = 0$ for all $x \in \mathbb{R}$ implies that F is not singular. According to the Lebesgue decomposition theorem, we have the unique decomposition $P_F = \mu_1 + \mu_2$, where μ_1 and μ_2

are finite measures on $\mathscr{B}$ such that μ_1 is singular with respect to λ and $\mu_2 \ll \lambda$. Clearly $\mu_2 \not\equiv 0$. Let F_1 and F_2 be nondecreasing functions which correspond to μ_1 and μ_2, respectively (as in Remark 1.1.2). Then F_1 and F_2 are both continuous, nondecreasing functions on $\mathbb{R}$ such that $F_1(-\infty) = F_2(-\infty) = 0$. Clearly $F(x) = F_1(x) + F_2(x)$ for $x \in \mathbb{R}$. From Proposition 1.1.5, F_2 is absolutely continuous and $F_2 \not\equiv 0$. Suppose that F is singular. Clearly $F' = F_1' + F_2'$ and $F_1' \geq 0$, $F_2' \geq 0$ a.e. (λ) on $\mathbb{R}$. Since $F' = 0$ a.e. (λ), it follows that $F_2' = 0$ a.e. (λ). Since F_2 is absolutely continuous, $F_2(x) \equiv c$, $x \in \mathbb{R}$, where c is a constant. Since $F_2(-\infty) = 0$, $F_2 \equiv 0$, which is a contradiction. ■

Remark 1.1.5. A continuous distribution function F_c is said to be *singular* if $F_c(x) = F_s(x)$, and is said to be *absolutely continuous* if $F_c(x) = F_{ac}(x)$ in (1.1.18).

We now present examples of some well-known absolutely continuous and singular distributions.

Example 1.1.7. Let $a < b$ be two real numbers. Clearly the function

$$f(x) = \begin{cases} \dfrac{1}{b-a}, & a < x < b, \\ 0, & \text{otherwise,} \end{cases}$$

defines a probability density function. Also, the function

$$F(x) = \begin{cases} 0, & x \leq a, \\ \dfrac{x-a}{b-a}, & a < x < b, \\ 1, & x \geq b, \end{cases}$$

defines the distribution function that corresponds to f. The absolutely continuous distribution function F is known as the *uniform distribution* over (a, b).

Example 1.1.8. The function

$$f(x) = \frac{1}{\sigma\sqrt{2\pi}} e^{-(x-\mu)^2/2\sigma^2}, \qquad x \in \mathbb{R},$$

where $\mu \in \mathbb{R}$, and $\sigma > 0$ defines a probability density function known as the *normal probability density* function.

Example 1.1.9. Consider the function

$$f(x) = \begin{cases} \dfrac{1}{\Gamma(\alpha)\beta^\alpha} x^{\alpha-1} e^{-x/\beta}, & x > 0, \\ 0, & x \leq 0, \end{cases}$$

where $\alpha > 0$, $\beta > 0$. Clearly f defines a probability density function which is called a *gamma probability density* function.

Example 1.1.10. The probability density function

$$f(x) = \frac{\lambda}{\pi} \frac{1}{\lambda^2 + (x - \theta)^2}, \qquad x \in \mathbb{R},$$

defines a *Cauchy distribution*. Here $\lambda > 0$ and $\theta \in \mathbb{R}$ are fixed real numbers.

Example 1.1.11. In this example we construct a singular distribution function. First we construct *Cantor's (ternary) set*. We divide the interval $[0, 1]$ into three equal intervals from which we choose the middle open interval $(\frac{1}{3}, \frac{2}{3})$. Next we subdivide each of the remaining two intervals into three equal intervals, and choose the middle open interval in each case. This process is continued, that is, we define open sets $D_1, D_2, \ldots$ as follows:

$$D_1 = \left(\frac{1}{3}, \frac{2}{3}\right)$$

$$D_2 = \left(\frac{1}{3^2}, \frac{2}{3^2}\right) \cup \left(\frac{2}{3} + \frac{1}{3^2}, \frac{2}{3} + \frac{2}{3^2}\right)$$

$$D_3 = \left(\frac{1}{3^3}, \frac{2}{3^3}\right) \cup \left(\frac{2}{3^2} + \frac{1}{3^3}, \frac{2}{3^2} + \frac{2}{3^3}\right) \cup \left(\frac{2}{3} + \frac{1}{3^3}, \frac{2}{3} + \frac{2}{3^3}\right)$$
$$\cup \left(\frac{2}{3} + \frac{1}{3^2} + \frac{1}{3^3}, \frac{2}{3} + \frac{2}{3^2} + \frac{2}{3^3}\right)$$

$$\vdots$$

$$D_n = \left(\frac{1}{3^n}, \frac{2}{3^n}\right) \cup \left(\frac{7}{3^n}, \frac{8}{3^n}\right) \cup \cdots \cup \left(\frac{3^n - 2}{3^n}, \frac{3^n - 1}{3^n}\right)$$
$$\equiv \bigcup_{r=1}^{2^{n-1}} D_{n,r}.$$

Consider the set $D = \bigcup_{n=1}^{\infty} D_n$. Clearly D is an open set. Moreover $\lambda(D_n) = 2^{n-1}/3^n$, so that $\lambda(D) = \sum_{n=1}^{\infty} \lambda(D_n) = 1$, where λ is the Lebesgue measure. The set $C = [0, 1] \cap D^c$, where D^c is the complement of set D, is a closed set and is called the *Cantor set*. Clearly $\lambda(C) = 0$.

Let $x \in [0, 1]$. Then we can express x by its ternary expansion as $x = \sum_{i=1}^{\infty} (a_i/3^i)$, where $a_i = 0, 1$, or 2, $i = 1, 2, \ldots$. In that case, we write $x = 0 \cdot a_1 a_2 \ldots a_n \ldots$. It is easy to see that C contains those and only those points $x \in [0, 1]$ which have at least one ternary expansion $x = 0 \cdot a_1 a_2, \ldots$, where $a_1, a_2, \ldots$ take the value 0 or 2 only. Now set $b_n = \frac{1}{2}a_n$ so that b_n is either 0 or 1. Define F on C by

$$F(x) \equiv 0 \cdot b_1 b_2 b_3 \ldots,$$

where the expansion on the right is now interpreted as a binary expansion in terms of the digits 0 and 1. Clearly F maps $[0, 1]$ onto a proper subset of C. Moreover F is not one-to-one. Indeed, $F(x_1) = F(x_2)$ if and only if x_1 and x_2 are end points of one of the open intervals deleted from $[0, 1]$ in the construction of the set C. Thus F is increasing on C, and strictly increasing except for such pairs of end points.

We next extend F so that its domain is $[0, 1]$ as follows. If $x \in D$, then $x \in D_{n,r}$ for some r and some n, $r = 1, 2, \ldots, 2^{n-1}$, $n = 1, 2, \ldots$. Define $F(x)$ by its common value at the end points of $D_{n,r}$. Thus

$$F(x) = \frac{1}{2} \quad \text{on} \quad \left[\frac{1}{3}, \frac{2}{3}\right], \qquad F(x) = \frac{1}{4} \quad \text{on} \quad \left[\frac{1}{3^2}, \frac{2}{3^2}\right],$$

$$F(x) = \frac{3}{4} \quad \text{on} \quad \left[\frac{7}{9}, \frac{8}{9}\right],$$

and so on. The function F is now defined on the closed bounded interval $[0, 1]$. Clearly F is nondecreasing on $[0, 1]$. Since the range of F is the entire interval $[0, 1]$, F has no jump discontinuities. It follows that F is continuous. Since F is (locally) constant on the open subset D, which has measure 1, $F'(x) = 0$ a.e. in $[0, 1]$. It follows that $\int_0^1 F'(x)\,dx = 0$. Since $F(0) = 0$ and $F(1) = 1$,

$$\int_0^1 F'(x)\,dx = 0 < F(1) - F(0),$$

and we conclude that F is not absolutely continuous. Hence F is a continuous singular function on $[0, 1]$. Defining $G(x) = 0$ for $x < 0$, $G(x) = 1$ for $x > 1$, and $G(x) = F(x)$ for $x \in [0, 1]$, we conclude that G is a continuous singular distribution function. We note that $P_G(C) = 1$ although $\lambda(C) = 0$.

1.1.4 Random Vectors

We next introduce the concept of a *random vector*. Let $\mathbb{R}_n$ denote the n-dimensional Euclidean space and let $\mathscr{B}_n$ denote the σ-field of Borel subsets of $\mathbb{R}_n$.†

Definition 1.1.3. Let $(\Omega, \mathscr{S}, P)$ be a probability space. A mapping $\mathbf{X}: \Omega \to \mathbb{R}_n$ is said to be a random vector if, for every $E \in \mathscr{B}_n$,

$$\mathbf{X}^{-1}(E) = \{\omega \in \Omega: \mathbf{X}(\omega) \in E\} \in \mathscr{S},$$

that is, $\mathbf{X}$ is a measurable transformation of $(\Omega, \mathscr{S}, P)$ into $(\mathbb{R}_n, \mathscr{B}_n)$.

Let $\mathbf{X}$ be an arbitrary mapping of $\Omega \to \mathbb{R}_n$. Then for every $\omega \in \Omega$ we can write

$$\mathbf{X}(\omega) = (X_1(\omega), X_2(\omega), \ldots, X_n(\omega))$$

where each X_j is a mapping of $\Omega \to \mathbb{R}$.

Proposition 1.1.7. $\mathbf{X}$ is a random vector if and only if each X_j, $1 \le j \le n$, is a random variable.

Proof. Let $E \subset \mathbb{R}_n$ be such that

$$E = \prod_{j=1}^{n} E_j, \qquad E_j \subset \mathbb{R}, \quad j = 1, 2, \ldots, n.$$

Then we clearly have the relation

$$\mathbf{X}^{-1}(E) = \bigcap_{j=1}^{n} X_j^{-1}(E_j). \tag{1.1.20}$$

Suppose that $\mathbf{X}$ is a random vector. Then, for every $E \in \mathscr{B}_n$, $\mathbf{X}^{-1}(E) \in \mathscr{S}$. In particular, let $E_j \in \mathscr{B}$, $j = 1, 2, \ldots, n$. Then it follows from (1.1.20) that $\bigcap_{j=1}^{n} X_j^{-1}(E_j) \in \mathscr{S}$. Setting $E_j = \mathbb{R}$ for $j = 2, 3, \ldots, n$, we see at once that X_1 is a random variable. Similarly $X_2, X_3, \ldots, X_n$ are also random variables.

Conversely, suppose that each X_j is a random variable. Then, for every $E \subset \mathbb{R}_n$ which is of the form $E = \prod_{j=1}^{n} E_j$, $E_j \in \mathscr{B}, j = 1, 2, \ldots, n$, $\mathbf{X}^{-1}(E) \in \mathscr{S}$. Let $\mathscr{E}$ be the class of all subsets E of $\mathbb{R}_n$ for which $\mathbf{X}^{-1}(E) \in \mathscr{S}$. It is easy to verify that $\mathscr{E}$ is a σ-field so that $\mathscr{B}_n \subset \mathscr{E}$. Hence $\mathbf{X}$ is a random vector. ∎

† For convenience we write $\mathbb{R} = \mathbb{R}_1$ and $\mathscr{B} = \mathscr{B}_1$.

Clearly $\mathbf{X}$ induces a probability measure $P_{\mathbf{X}}$ on $\mathscr{B}_n$, defined by

$$P_{\mathbf{X}}(E) = P\{\mathbf{X}^{-1}(E)\} \qquad \text{for every } E \in \mathscr{B}_n. \tag{1.1.21}$$

Here $P_{\mathbf{X}}$ is known as the *probability distribution of* $\mathbf{X}$.

Definition 1.1.4. Let $\mathbf{X} = (X_1, X_2, \ldots, X_n)$ be a random vector defined on a probability space $(\Omega, \mathscr{S}, P)$. Define a function F on $\mathbb{R}_n$ by the relation

$$F(x_1, x_2, \ldots, x_n) = P\{\omega: X_1(\omega) \le x_1, X_2(\omega) \le x_2, \ldots, X_n(\omega) \le x_n\},$$

$$x_i \in \mathbb{R}, 1 \le i \le n. \tag{1.1.22}$$

Here F is known as the joint distribution function of the random vector $\mathbf{X}$.

Remark 1.1.6. Let $\mathbf{X}: (\Omega, \mathscr{S}, P) \to (\mathbb{R}_n, \mathscr{B}_n)$ be a random vector, and let $\mathbf{g}: \mathbb{R}_n \to \mathbb{R}_k$ be a measurable mapping. Then $\mathbf{g}(\mathbf{X}): \Omega \to \mathbb{R}_k$ is a random vector.

Remark 1.1.7. Let $F = F_{X_1, X_2, \ldots, X_n}$ be the distribution function of a random vector $\mathbf{X} = (X_1, X_2, \ldots, X_n)$, $n \ge 2$. Set

$$\lim_{x_n \to \infty} F_{X_1, X_2, \ldots, X_n}(x_1, x_2, \ldots, x_n) = F_{X_1, X_2, \ldots, X_{n-1}}(x_1, x_2, \ldots, x_{n-1}).$$

Then $F_{X_1, X_2, \ldots, X_{n-1}}$ is a distribution function which is known as the *marginal distribution function* of the random vector $(X_1, X_2, \ldots, X_{n-1})$. In general the joint distribution of any subset of $X_1, X_2, \ldots, X_n$ will be referred to as a marginal distribution function.

The following result can be proved along the lines of the proofs of Theorems 1.1.1 and 1.1.2.

Theorem 1.1.6. Let $\mathbf{X}$ be a random vector with distribution function F. Then

(i) $F(x_1, x_2, \ldots, x_n) \to 0$ as $x_i \to -\infty$ for at least one i.

$F(x_1, x_2, \ldots, x_n) \to 1$ as $x_i \to +\infty$ for each i, $1 \le i \le n$.

(ii) F is right-continuous in each argument.

(iii) For all $h_i \geq 0$, $i = 1, 2, \ldots, n$, and $x_i \in \mathbb{R}$, $1 \leq i \leq n$, the following inequality holds:

$$\begin{aligned} &F(x_1 + h_1, x_2 + h_2, \ldots, x_n + h_n) \\ &\quad - [F(x_1, x_2 + h_2, \ldots, x_n + h_n) + \cdots \\ &\quad + F(x_1 + h_1, x_2 + h_2, \ldots, x_{n-1} + h_{n-1}, x_n)] \\ &\quad + [F(x_1, x_2, x_3 + h_3, \ldots, x_n + h_n) + \cdots \\ &\quad + F(x_1 + h_1, x_2 + h_2, \ldots, x_{n-2} + h_{n-2}, x_{n-1}, x_n)] - \cdots \\ &\quad + (-1)^{n-1} F(x_1, x_2, \ldots, x_n) \geq 0. \end{aligned} \tag{1.1.23}$$

Conversely, any function F defined on $\mathbb{R}_n$ and satisfying conditions (i), (ii), and (iii) above determines uniquely a probability measure P_F on $\mathscr{B}_n$, the σ-field of Borel sets on $\mathbb{R}_n$.

Remark 1.1.8. Condition (iii) in Theorem 1.1.6 implies that F is nondecreasing in each argument. However, a function F which is nondecreasing in each argument does not necessarily satisfy (1.1.23). For example, consider the function

$$f(x, y) = \begin{cases} 0 & \text{if } x < 0 \text{ or } x + y < 1 \text{ or } y < 0, \\ 1 & \text{otherwise.} \end{cases}$$

Clearly f is nondecreasing in each argument, but

$$f(1, 1) - f(1, \tfrac{1}{3}) - f(\tfrac{1}{3}, 1) + f(\tfrac{1}{3}, \tfrac{1}{3}) = -1,$$

and hence (1.1.23) does not hold.

Remark 1.1.9. As in the univariate case, we may consider the three types of joint distribution functions: discrete, absolutely continuous, and singular. We say that a joint distribution function F of n random variables is *discrete* if there exists a countable subset E of $\mathbb{R}_n$ such that $P_F(E) = 1$, where P_F is the measure over $\mathbb{R}_n$ determined by F. It is said to be *absolutely continuous* if there exists a Borel-measurable function f defined on $\mathbb{R}_n$ such that

$$F(x_1, x_2, \ldots, x_n) = \int_{-\infty}^{x_1} \cdots \int_{-\infty}^{x_n} f(t_1, \ldots, t_n)\, dt_1 \ldots dt_n$$

for all $x_i \in \mathbb{R}$, $i = 1, 2, \ldots, n$. Here f is known as the *joint probability density function* of F (or of the corresponding random vector $\mathbf{X}$). Finally we say that F is *continuous singular* if it is continuous and there exists a Borel set E in $\mathbb{R}_n$ with Lebesgue measure zero such that $P_F(E) = 1$.

Remark 1.1.10. We know that the distribution function of a random variable has at most a countable number of discontinuity points. Let F be the joint distribution function of a random vector $\mathbf{X} = (X_1, X_2, \ldots, X_n)$. Then F is continuous everywhere except over the union of a countable set of hyperplanes of the form $x_k = c$, $1 \le k \le n$ (Problem 1.5.9).

Example 1.1.12. Let $\boldsymbol{\mu} = (\mu_1, \mu_2, \ldots, \mu_n)$ be a real n-vector, and $\mathbf{M}$ be a real positive definite symmetric matrix of order n, $n \ge 1$. Writing $|\mathbf{M}|$ for the determinant of $\mathbf{M}$, we note that $|\mathbf{M}| > 0$, The inverse $\mathbf{M}^{-1}$ exists and is also positive definite. It follows that the quadratic form

$$Q(\mathbf{x}, \boldsymbol{\mu}) = (\mathbf{x} - \boldsymbol{\mu})\mathbf{M}^{-1}(\mathbf{x} - \boldsymbol{\mu})' = \sum_{i=1}^{n} \sum_{j=1}^{n} (\mathbf{M}^{-1})_{ij}(x_i - \mu_i)(x_j - \mu_j)$$

is nonnegative and vanishes only for $\mathbf{x} = \boldsymbol{\mu}$. The function f defined for all real n-vectors $\mathbf{x} = (x_1, x_2, \ldots, x_n)$ by

$$f(\mathbf{x}) = (2\pi)^{-n/2} |\mathbf{M}|^{-1/2} \exp\{-\tfrac{1}{2}Q(\mathbf{x}, \boldsymbol{\mu})\},$$

is known as the *multivariate normal density*. We leave it to the reader to check that f is a joint probability density function (Problem 1.5.10). One can also show, for example, that the marginal distribution of any subset of the random vector $\mathbf{X} = (X_1, X_2, \ldots, X_n)$ that has a multivariate normal distribution is once again a multivariate normal distribution.

1.1.5 Infinite-Dimensional Random Variables—Consistency Theorem

Suppose we observe n random variables $X_1, X_2, \ldots, X_n$. In many applications of probability we are interested in the limiting behavior of sequences such as $\sum_{k=1}^{n} X_k/n$ as $n \to \infty$. Such limiting events cannot be defined in any of the spaces $\{(\mathbb{R}_n, \mathscr{B}_n)\}$, since they involve infinitely many coordinates. It is therefore desirable to find a probability space which contains such events and at the same time contains all the information contained in every $(\mathbb{R}_n, \mathscr{B}_n, P_n)$, where P_n is the probability distribution of $X_1, X_2, \ldots, X_n$. This is what we plan to do in the following. For this purpose we introduce some new concepts.

Let T be an arbitrary index set, and consider the collection $\mathbb{R}_T$ of all sets of the form $\omega = \{x_t\}$, where $x_t \in \mathbb{R}$ and $t \in T$. Clearly

$$\mathbb{R}_T = \prod_{t \in T} \mathbb{R}.$$

In particular, if $T = \{1, 2, \ldots, n\}$, where $n \ge 1$ is an integer, then $\mathbb{R}_T = \mathbb{R}_n$ is the n-dimensional Euclidean space. If $T = \mathbb{R}$, then $\mathbb{R}_T$ is the set of all real-valued functions defined on $\mathbb{R}$.

Let $\omega = \{x_t\} \in \mathbb{R}_T$, and let $\{t_1, t_2, \ldots, t_n\}$ be a finite subset of T. We set

$$\pi_{t_1, t_2, \ldots, t_n}(\omega) = (x_{t_1}, x_{t_2}, \ldots, x_{t_n}) \in \mathbb{R}_n.$$

Here $\pi_{t_1, t_2, \ldots, t_n}$ is called a *projection* of ω onto $\mathbb{R}_n$.

Definition 1.1.5. A subset A of $\mathbb{R}_T$ is said to be a cylinder set if there exists a nonempty finite subset $\{t_1, t_2, \ldots, t_n\}$ of T, and a subset M of $\mathbb{R}_n$ such that A can be represented in the form

$$A = \pi^{-1}_{t_1, t_2, \ldots, t_n}(M) = \{\omega \in \mathbb{R}_T : (x_{t_1}, x_{t_2}, \ldots, x_{t_n}) \in M\}. \tag{1.1.24}$$

A cylinder set A is said to be a Borel cylinder set if $M \in \mathscr{B}_n$.

We see easily that the class of all Borel cylinder sets in $\mathbb{R}_T$ is a field which we denote by $\mathscr{F}_T$. Let $\mathscr{B}_T$ be the σ-field generated by $\mathscr{F}_T$. Then $\mathscr{B}_T$ is called the *Borel σ-field* of sets in $\mathbb{R}_T$, and its elements are called *Borel sets* of $\mathbb{R}_T$. We first consider the case where a probability measure P_T is defined on $\mathscr{F}_T$. In view of the extension theorem P_T can be uniquely extended to a probability measure on $\mathscr{B}_T$ so that $(\mathbb{R}_T, \mathscr{B}_T, P_T)$ is a probability space.

For every $t_0 \in T$ we set $X_{t_0}(\omega) = x_{t_0}$, where $\omega = \{x_t\}$. It follows at once from (1.1.24) that $\{X_t, t \in T\}$ is a collection of random variables. Moreover, every finite subset $\{X_{t_1}, \ldots, X_{t_n}\} \subset \{X_t, t \in T\}$ is an n-dimensional random vector and hence determines uniquely an n-dimensional probability distribution $P_{t_1, t_2, \ldots, t_n}$ on $\mathscr{B}_n$, defined by the relation

$$P_T(A) = P_{t_1, t_2, \ldots, t_n}(M),$$

whenever A is a Borel cylinder set related to M by (1.1.24). In this case $P_{t_1, t_2, \ldots, t_n}$ determines uniquely the corresponding distribution function $F_{t_1, t_2, \ldots, t_n}$ of $(X_{t_1}, X_{t_2}, \ldots, X_{t_n})$ on $\mathbb{R}_n$. We have thus obtained the following result.

Proposition 1.1.8. Let P_T be a probability measure on $\mathscr{F}_T$. Then P_T is determined uniquely on $\mathscr{B}_T$ by the set of all finite-dimensional distribution functions $\{F_{t_1, t_2, \ldots, t_n}\}$.

Next we consider the case where a given set of finite-dimensional distribution functions $\{F_{t_1, t_2, \ldots, t_n}\}$ determines uniquely a probability measure on $\mathscr{F}_T$ and consequently on $\mathscr{B}_T$.

Theorem 1.1.7 (Kolmogorov). Let $\{F_{t_1, t_2, \ldots, t_n}\}$ be a given set of finite-dimensional distribution functions which satisfy the following consistency conditions:

(i) $F_{t_{i_1}, t_{i_2}, \ldots, t_{i_n}}(a_{i_1}, a_{i_2}, \ldots, a_{i_n}) = F_{t_1, t_2, \ldots, t_n}(a_1, a_2, \ldots, a_n)$

where $\{i_1, i_2, \ldots, i_n\}$ is a permutation of $\{1, 2, \ldots, n\}$.

(ii) $F_{t_1, t_2, \ldots, t_m}(a_1, a_2, \ldots, a_m) = F_{t_1, t_2, \ldots, t_n}(a_1, a_2, \ldots, a_m, +\infty, \ldots, +\infty)$

for $m < n$.

Then the set $\{F_{t_1, t_2, \ldots, t_n}\}$ determines uniquely a probability measure P_T on $\mathscr{F}_T$ and hence on $\mathscr{B}_T$.

Proof. First we note that every n-dimensional distribution function $F_{t_1, t_2, \ldots, t_n}$ determines uniquely the corresponding probability distribution $P_{t_1, t_2, \ldots, t_n}$ on $\mathscr{B}_n$. For every $A \in \mathscr{F}_T$ we set

$$P_T(A) = P_{t_1, t_2, \ldots, t_n}(M) \tag{1.1.25}$$

where $M \in \mathscr{B}_n$ and is given by (1.1.24). Clearly $0 \leq P_T(A) \leq 1$ for $A \in \mathscr{F}_T$. We show that P_T on $\mathscr{F}_T$ defined by (1.1.25) is independent of the representation of A given by (1.1.24). In fact, suppose A has two different representations:

$$\begin{aligned} A &= \pi^{-1}_{t_{i_1}, \ldots, t_{i_n}}(M_1), \\ &= \pi^{-1}_{t_{j_1}, \ldots, t_{j_m}}(M_2), \end{aligned}$$

where $M_1 \in \mathscr{B}_n$ and $M_2 \in \mathscr{B}_m$. Let $\{X_t, t \in T\}$ be the collection of random variables as defined in the paragraph preceding Proposition 1.1.8. Without loss of generality we assume that both the sets

$$\{X_{t_{i_1}}, \ldots, X_{t_{i_n}}\} \qquad \text{and} \qquad \{X_{t_{j_1}}, \ldots, X_{t_{j_m}}\}$$

of random variables are some subsets of a finite set of random variables, say $\{X_{t_1}, \ldots, X_{t_N}\}$. Using consistency conditions (i) and (ii) and the properties of probability distributions, we see that

$$\begin{aligned} P_{t_{i_1}, \ldots, t_{i_n}}(M_1) &= P_{t_1, t_2, \ldots, t_N}\{(x_{t_1}, \ldots, x_{t_N}) : (x_{t_{i_1}}, \ldots, x_{t_{i_n}}) \in M_1\} \\ &= P_{t_1, t_2, \ldots, t_N}\{(x_{t_1}, \ldots, x_{t_N}) : (x_{t_{j_1}}, \ldots, x_{t_{j_m}}) \in M_2\} \\ &= P_{t_{j_1}, \ldots, t_{j_m}}(M_2), \end{aligned}$$

so that P_T defined on $\mathscr{F}_T$ by (1.1.25) does not depend on the representation of A.

Next we show that P is finitely additive on $\mathscr{F}_T$. Let

$$A = \pi^{-1}_{t_{i_1}, \ldots, t_{i_n}}(M_1)$$

and

$$B = \pi^{-1}_{t_{j_1}, \ldots, t_{j_m}}(M_2)$$

be two disjoint sets in $\mathscr{F}_T$. As before, we assume that both the sets

$$\{X_{t_{i_1}}, \ldots, X_{t_{i_n}}\} \qquad \text{and} \qquad \{X_{t_{j_1}}, \ldots, X_{t_{j_m}}\}$$

of random variables are subsets of a finite set, say $\{X_{t_1}, \ldots, X_{t_N}\}$. Since $A \cap B = \varnothing$, we have

$$\begin{aligned} P_T(A \cup B) &= P_{t_1, t_2, \ldots, t_N}\{(x_{t_1}, \ldots, x_{t_N}) : (x_{t_{i_1}}, \ldots, x_{t_{i_n}}) \in M_1 \\ &\qquad\qquad \text{or} \quad (x_{t_{j_1}}, \ldots, x_{t_{j_m}}) \in M_2\} \\ &= P_{t_1, t_2, \ldots, t_N}\{(x_{t_1}, \ldots, x_{t_N}) : (x_{t_{i_1}}, \ldots, x_{t_{i_n}}) \in M_1\} \\ &\quad + P_{t_1, t_2, \ldots, t_N}\{(x_{t_1}, \ldots, x_{t_N}) : (x_{t_{j_1}}, \ldots, x_{t_{j_m}}) \in M_2\} \\ &= P_{t_{i_1}, \ldots, t_{i_n}}(M_1) + P_{t_{j_1}, \ldots, t_{j_m}}(M_2) \\ &= P_T(A) + P_T(B). \end{aligned}$$

Finally we show that P_T on $\mathscr{F}_T$ is countably additive. Since P_T is finite, nonnegative, and finitely additive on $\mathscr{F}_T$, this will follow if we can show that P_T is continuous from above at $\varnothing$. Let $A_n \in \mathscr{F}_T$, $n \geq 1$, such that $A_n \downarrow \varnothing$. We show that $P_T(A_n) \downarrow 0$ as $n \to \infty$.

Suppose there exists a decreasing sequence $A_n \in \mathscr{F}_T$ such that

$$\lim_{n \to \infty} P_T(A_n) = \epsilon_0 > 0.$$

We show that $\bigcap_{n=1}^{\infty} A_n \neq \varnothing$. We first note that, since P_T is finitely additive and nonnegative, P_T is monotone so that $P_T(A_N) \downarrow \epsilon_0$ as $n \to \infty$. For every $n \geq 1$ we can write

$$A_n = \pi_{t_1, t_2, \ldots, t_n}^{-1}(M_n),$$

so that

$$P_T(A_n) = P_{t_1, t_2, \ldots, t_n}(M_n) = P_n(M_n),$$

say. Then we have

$$P_n(M_n) = P_T(A_n) \geq \epsilon_0 > 0, \qquad n \geq 1.$$

Since $M_n \in \mathscr{B}_n$, we can choose a compact set $K_n(\epsilon_0) = K_n \in \mathscr{B}_n$ (see, for example, Proposition 3.9.1) such that $K_n \subset M_n$ and

$$P_n(M_n - K_n) < \frac{\epsilon_0}{2^{n+1}}, \qquad n \geq 1.$$

Clearly K_n is a closed, bounded set in $\mathbb{R}_n$. Let $B_n \in \mathscr{F}_T$ be such that

$$B_n = \pi_{t_1, t_2, \ldots, t_n}^{-1}(K_n).$$

We see easily that $B_n \subset A_n$ and, moreover,

$$\begin{aligned} P_T(A_n - B_n) &= P_n(M_n - K_n) \\ &< \frac{\epsilon_0}{2^{n+1}}, \qquad n \geq 1. \end{aligned}$$

Now for every $n \geq 1$ we set

$$C_n = B_1 \cap B_2 \cap \cdots \cap B_n.$$

Clearly $C_n \in \mathscr{F}_T$ and $C_n \subset B_n \subset A_n$. From the finite subadditivity of P_T we obtain

$$\begin{aligned} P_T(A_n - C_n) &= P_T\left(\bigcup_{k=1}^{n} (A_n - B_k)\right) \\ &\leq P_T\left(\bigcup_{k=1}^{n} (A_k - B_k)\right) \\ &\leq \sum_{k=1}^{n} P_T(A_k - B_k) \\ &< \epsilon_0 \sum_{k=1}^{n} \frac{1}{2^{k+1}} < \frac{\epsilon_0}{2}. \end{aligned}$$

Consequently

$$P_T(A_n) - P_T(C_n) = P_T(A_n - C_n) < \frac{\epsilon_0}{2},$$

which implies that

$$\begin{aligned} P_T(C_n) &> P_T(A_n) - \frac{\epsilon_0}{2} \\ &> \frac{\epsilon_0}{2}, \qquad n \geq 1. \end{aligned}$$

In particular, $C_n \neq \varnothing$ for every $n \geq 1$. Hence for every n we can choose a point $\omega^{(n)} = \{x_t^{(n)}\}$ in C_n. Clearly the point $\omega^{(n+v)} \in B_n$ for every $v \geq 0$. Consequently

$$(x_{t_1}^{(n+v)}, x_{t_2}^{(n+v)}, \ldots, x_{t_n}^{(n+v)}) \in K_n$$

for every $v \geq 0$ and $n \geq 1$. Since $K_n \subset \mathbb{R}_n$ is compact, it is also sequentially compact. From the Bolzano–Weirstrass theorem and the usual method of diagonalization, it follows that the sequence $\{\omega^{(n)}\}$ contains an infinite subsequence

$$\omega^{(n_1)}, \omega^{(n_2)}, \ldots, \omega^{(n_k)}, \ldots,$$

where $n_1 < n_2 < \cdots < n_k$ ($n_k \to \infty$ as $k \to \infty$), such that every coordinate $x_{t_j}^{(n_k)}$ of $\omega^{(n_k)}$ converges to a finite limit, say x_j for $j = 1, 2, \ldots, k$ and $k \geq 1$. Moreover, since K_k is closed, we conclude that $(x_1, x_2, \ldots, x_k) \in K_k$ for $k \geq 1$.

Finally, let $\omega \in \mathbb{R}_T$ be such that

$$\pi_{t_1, t_2, \ldots, t_n}(\omega) = (x_{t_1}, x_{t_2}, \ldots, x_{t_n}) = (x_1, x_2, \ldots, x_n)$$

for $n \geq 1$. It follows that $\omega \in B_n$ for every n, and consequently $\omega \in A_n$ for $n \geq 1$. Hence $\omega \in \bigcap_{n=1}^{\infty} A_n$, and $\bigcap_{n=1}^{\infty} A_n \neq \varnothing$. This completes the proof of the theorem. ■

As an application of Theorem 1.1.7. we consider the following example.

Example 1.1.13. Let $(\Omega, \mathscr{S}, P)$ be a probability space. Let T be an arbitrary index set. Let $\mathfrak{X} = \{X_t, t \in T\}$ be a collection of random variables defined on $(\Omega, \mathscr{S}, P)$. We now consider $\mathfrak{X}$ as a mapping of Ω into $\mathbb{R}_T = \prod_{t \in T} \mathbb{R}$. The problem that naturally arises is how to define a probability measure P_T on a suitable σ-field $\mathscr{B}_T$ of subsets in $\mathbb{R}_T$ so that $(\mathbb{R}_T, \mathscr{B}_T, P_T)$ is a probability space and, moreover, $\mathfrak{X}$ is a measurable mapping of $(\Omega, \mathscr{S}, P)$ into $(\mathbb{R}_T, \mathscr{B}_T, P_T)$. This can be done with the help of Theorem 1.1.7. For this purpose let $\mathscr{B}_T$ be the σ-field of subsets of $\mathbb{R}_T$ generated by the field of all Borel cylinder sets. Let $\{F_{t_1, t_2, \ldots, t_n}\}$ be the collection of all finite-dimensional distribution functions determined by the finite subsets $\{X_{t_1}, X_{t_2}, \ldots, X_{t_n}\}$ of $\mathfrak{X}$. Then clearly $\{F_{t_1, t_2, \ldots, t_n}\}$ satisfies the consistency conditions of Theorem 1.1.7, and hence it follows that $\{F_{t_1, t_2, \ldots, t_n}\}$ determines uniquely a probability measure P_T on $\mathscr{B}_T$ so that $(\mathbb{R}_T, \mathscr{B}_T, P_T)$ is a probability space. Moreover, for every $A \in \mathscr{B}_T$, $\mathfrak{X}^{-1}(A) \in \mathscr{S}$, so that $\mathfrak{X}$ is a measurable transformation from $(\Omega, \mathscr{S}, P)$ to $(\mathbb{R}_T, \mathscr{B}_T, P_T)$. In fact, if $A \in \mathbb{R}_T$, let

$$A = \pi_{t_1, t_2, \ldots, t_n}^{-1}(M)$$

for some $M \in \mathscr{B}_n$ and some $\{t_1, t_2, \ldots, t_n\} \subset T$. Then

$$\mathfrak{X} = \{X_t\} \in A \Leftrightarrow (X_{t_1}, X_{t_2}, \ldots, X_{t_n}) \in M,$$

so that

$$\begin{aligned}\mathfrak{X}^{-1}(A) &= \{\omega \in \Omega: \mathfrak{X}(\omega) \in A\} \\ &= \{\omega \in \Omega: (X_{t_1}(\omega), X_{t_2}(\omega), \ldots, X_{t_n}(\omega)) \in M\} \in \mathscr{S}.\end{aligned}$$

Remark 1.1.11. Theorem 1.1.7 is used frequently in probability theory. It is what distinguishes probability theory from abstract measure theory. As shown in Example 1.1.13, Theorem 1.1.7 allows us to make statements such as the following. Let $\mathfrak{X}$ be a collection of random variables on a probability space $(\Omega, \mathscr{S}, P)$. Then the probability $P\{\mathfrak{X} \in A\}$ is determined for all $A \in \mathscr{B}_T$ by the class $\{F_{t_1, t_2, \ldots, t_n}\}$ of all finite-dimensional distribution functions of finite subsets $\{X_{t_1}, X_{t_2}, \ldots, X_{t_n}\}$ of $\mathfrak{X}$.

1.2 MATHEMATICAL EXPECTATION

We now study some characteristics of a random variable (or of its distribution function). These play an important role in the study of probability theory.

1.2.1 Definitions

Let $(\Omega, \mathscr{S}, P)$ be a probability space, and X be a random variable defined on it. Let g be a real-valued Borel-measurable function on $\mathbb{R}$. Then $g(X)$ is also a random variable.

Definition 1.2.1. We say that the mathematical expectation (or, simply, the expectation) of $g(X)$ exists if $g(X)$ is integrable over Ω with respect to P. In this case we define the expectation $\mathscr{E}g(X)$ of the random variable $g(X)$ by

$$\mathscr{E}g(X) = \int_{\Omega} g(X(\omega))\, dP(\omega) = \int_{\Omega} g(X)\, dP. \tag{1.2.1}$$

Remark 1.2.1. Since integrability is equivalent to absolute integrability, it follows that $\mathscr{E}g(X)$ exists if and only if $\mathscr{E}|g(X)|$ exists.

Let P_X be the probability distribution of X. Suppose that $\mathscr{E}g(X)$ exists. Then it follows (Halmost [31], p. 163) that g is also integrable over $\mathbb{R}$ with respect to P_X. Moreover, the relation

$$\int_{\Omega} g(X)\, dP = \int_{\mathbb{R}} g(t)\, dP_X(t) \tag{1.2.2}$$

holds. We note that the integral on the right-hand side of (1.2.2) is the Lebesgue–Stieltjes integral of g with respect to P_X.

In particular, if g is continuous on $\mathbb{R}$ and $\mathscr{E}g(X)$ exists, we can rewrite (1.2.2) as follows:

$$\int_{\Omega} g(X)\, dP = \int_{\mathbb{R}} g\, dP_X = \int_{-\infty}^{\infty} g(x)\, dF(x), \tag{1.2.3}$$

where F is the distribution function corresponding to P_X, and the last integral is a Riemann–Stieltjes integral. Two important special cases of (1.2.3) are as follows.

CASE 1. Let F be discrete with the set of discontinuity points $\{x_n, n = 1, 2, \ldots\}$. Let $p(x_n)$ be the jump of F at x_n, $n = 1, 2, \ldots$. Then $\mathscr{E}g(X)$ exists if and only if $\sum_{n=1}^{\infty} |g(x_n)|p(x_n) < \infty$, and in that case we have

$$\mathscr{E}g(X) = \sum_{n=1}^{\infty} g(x_n)p(x_n). \tag{1.2.3a}$$

CASE 2. Let F be absolutely continuous on $\mathbb{R}$ with probability density function $f(x) = F'(x)$. Then $\mathscr{E}g(X)$ exists if and only if $\int_{-\infty}^{\infty} |g(x)| f(x)\, dx < \infty$, and in that case we have

$$\mathscr{E}g(X) = \int_{-\infty}^{\infty} g(x) f(x)\, dx. \tag{1.2.3b}$$

We now state some elementary properties of random variables with finite expectations which follow as immediate consequences of the properties of integrable functions. Denote by $\mathscr{L}_1 = \mathscr{L}_1(\Omega, \mathscr{S}, P)$ the set of all random variables with finite expectations. In the following we write a.s. to abbreviate "almost surely (everywhere) with respect to the probability distribution of X on $(\mathbb{R}, \mathscr{B})$." As before, a.e. without qualification refers to the Lebesgue measure λ on $\mathbb{R}$.

(a) $X, Y \in \mathscr{L}_1$ and $\alpha, \beta \in \mathbb{R} \Rightarrow \alpha X + \beta Y \in \mathscr{L}_1$ and $\mathscr{E}(\alpha X + \beta Y) = \alpha \mathscr{E} X + \beta \mathscr{E} Y$.

(b) $X \in \mathscr{L}_1 \Rightarrow |\mathscr{E} X| \le \mathscr{E}|X|$.

(c) $X \in \mathscr{L}_1$, $X \ge 0$ a.s. $\Rightarrow \mathscr{E} X \ge 0$.

(d) Let $X \in \mathscr{L}_1$. Then $\mathscr{E}|X| = 0 \Leftrightarrow X = 0$ a.s.

(e) For $E \in \mathscr{S}$, write χ_E for the indicator function of the set E, that is, $\chi_E = 1$ on E and $=0$ otherwise. Then $X \in \mathscr{L}_1 \Rightarrow X\chi_E \in \mathscr{L}_1$, and we write

$$\int_E X\, dP = \mathscr{E}(X\chi_E).$$

Also, $\mathscr{E}(|X|\chi_E) = 0 \Leftrightarrow$ either $P(E) = 0$ or $X = 0$ a.s. on E.

(f) If $X \in \mathscr{L}_1$, then $X = 0$ a.s. $\Leftrightarrow \mathscr{E}(X\chi_E) = 0$ for all $E \in \mathscr{S}$.

(g) Let $X \in \mathscr{L}_1$, and define the set function Q_X on $\mathscr{S}$ by

$$Q_X(E) = \int_E X\, dP, \qquad E \in \mathscr{S}.$$

Then Q_X is countably additive on $\mathscr{S}$, and $Q_X \ll P$. In particular, Q_X is a finite measure on $\mathscr{S}$ if $X \ge 0$ a.s.

(h) Let $X \in \mathscr{L}_1$ and $E \in \mathscr{S}$. If $\alpha \le X \le \beta$ a.s. on E for $\alpha, \beta \in \mathbb{R}$, then the inequality

$$\alpha P(E) \le \int_E X\, dP \le \beta P(E)$$

holds.

(i) Let $Y \in \mathscr{L}_1$, and X be a random variable such that $|X| \le |Y|$ a.s. Then

$X \in \mathscr{L}_1$ and $\mathscr{E}|X| \le \mathscr{E}|Y|$. In particular, if X is bounded a.s., then $X \in \mathscr{L}_1$.

Remark 1.2.2. It is easy to see that $\mathscr{L}_1$ is a vector space over $\mathbb{R}$ if we define the null vector $X = 0$ if $X = 0$ a.s. We will see later that $\mathscr{L}_1$ is a Banach space with respect to the norm $\|X\|_1 = \mathscr{E}|X|$, $X \in \mathscr{L}_1$.

Let X be a random variable on $(\Omega, \mathscr{S}, P)$, and let $g(X)$ be integrable over Ω with respect to P. We next study some particular forms of g which will be used subsequently.

(a) Let $g(x) = x^n$, where n is a positive integer. Then $\alpha_n = \mathscr{E}X^n$, if it exists, is called the *moment of order n* of the random variable X.
(b) Let $g(x) = |x|^\lambda$, where λ is a positive real number. Then $\beta_\lambda = \mathscr{E}|X|^\lambda$, if it exists, is called the *absolute moment of order λ* of the random variable X.
(c) Let $g(x) = (x - \gamma)^n$, where γ is a real number and n is a positive integer. Then $\mathscr{E}(X - \gamma)^n$, if it exists, is known as the *moment of order n about the point γ*. In particular, if $\gamma = \mathscr{E}X$, then $\mathscr{E}(X - \mathscr{E}X)^n$ is called the *central moment of order n* and is denoted by μ_n. Clearly $\mu_1 = 0$. For $n = 2$,

(1.2.4) $$\mu_2 = \mathscr{E}(X - \mathscr{E}X)^2 = \mathscr{E}X^2 - (\mathscr{E}X)^2$$

is called the *variance* of X and is denoted by $\operatorname{var}(X)$. The positive square root of $\operatorname{var}(X)$ is called the *standard deviation* of X. We note that

(1.2.5) $$\begin{cases} & \operatorname{var}(X) \ge 0 \\ \text{and} & \\ & \operatorname{var}(X) = 0 \Leftrightarrow X = c \text{ a.s., } c \text{ constant.} \end{cases}$$

(d) Let $g(x) = e^{tx}$, $t \in (-\delta, \delta)$ for some positive real number δ. Then we write $M(t) = \mathscr{E}e^{tX}$, if it exists, and call it the *moment generating function* of the random variable X.
(e) Let $g(x) = e^{itx} = \cos tx + i \sin tx$, where $t \in \mathbb{R}$ and $i = \sqrt{-1}$. Then we write

(1.2.6) $$\varphi(t) = \mathscr{E}e^{itX} = \mathscr{E}\cos tX + i\mathscr{E}\sin tX,$$

and call φ the *characteristic function* of X. We note that the characteristic function φ of a random variable X always exists.

In the following examples we briefly introduce some standard distributions. The reader is asked to check the results stated in each example.

Example 1.2.1. Let X be a random variable with binomial distribution given by

$$p(k) = P\{X = k\} = \binom{n}{k}p^k(1 - p)^{n-k}, \qquad k = 0, 1, \dots, n,$$

where $0 < p < 1$. Then moments of all order exist. We have

$$\mathscr{E}X = np,$$

$$\text{var}(X) = np(1 - p).$$

Also,

$$\varphi(t) = \mathscr{E}e^{itX} = [(1 - p) + pe^{it}]^n.$$

Example 1.2.2. Let X be a random variable with Poisson distribution

$$p(k) = P\{X = k\} = \frac{e^{-\lambda}\lambda^k}{k!}, \qquad k = 0, 1, 2, \ldots,$$

where $\lambda > 0$. In this case

$$\mathscr{E}X = \sum \frac{ke^{-\lambda}\lambda^k}{k!} = \lambda,$$

$$\text{var}(X) = \lambda,$$

and

$$\varphi(t) = \mathscr{E}e^{itX} = \exp\{\lambda(e^{it} - 1)\}.$$

Example 1.2.3. If X has a uniform distribution on (a, b), then

$$\mathscr{E}X = \frac{a + b}{2},$$

$$\text{var}(X) = \frac{(b - a)^2}{12},$$

and, if $t \neq 0$,

$$\varphi(t) = \frac{1}{t(b - a)}(e^{itb} - e^{ita}).$$

Example 1.2.4. Let the random variable X have the normal distribution given by the probability density function

$$f(x) = (2\pi\sigma^2)^{-1/2}e^{-(x-\mu)^2/2\sigma^2}, \qquad x \in \mathbb{R},$$

where $\sigma^2 > 0$ and $\mu \in \mathbb{R}$. Then $\mathscr{E}|X|^\gamma < \infty$ for every $\gamma > 0$, and we have

$$\mathscr{E}X = \mu$$

and

$$\text{var}(X) = \sigma^2.$$

Also,

$$\begin{aligned}\varphi(t) &= \mathscr{E}e^{itX}\\ &= e^{it\mu}e^{-t^2\sigma^2/2}, \qquad t \in \mathbb{R}.\end{aligned}$$

Example 1.2.5. Let X have a Cauchy probability density function

$$f(x) = \frac{1}{\pi(1 + x^2)}, \qquad x \in \mathbb{R}.$$

In this case $\int_{-\infty}^{\infty} |x| f(x)\, dx = \infty$, so that $\mathscr{E}X$ does not exist. The characteristic function of X is given by

$$\varphi(t) = \frac{1}{\pi}\int_{-\infty}^{\infty} \frac{e^{itx}}{1 + x^2}\, dx = e^{-|t|}, \qquad t \in \mathbb{R}.$$

Writing $Y = \mu X + \theta$, where $\mu > 0$ and $\theta \in \mathbb{R}$, we see that the random variable Y has the probability density function

$$f_Y(y) = \frac{\mu}{\pi}\frac{1}{\mu^2 + (y - \theta)^2}, \qquad y \in \mathbb{R}.$$

The characteristic function of Y is given by

$$\begin{aligned}\varphi_Y(t) &= e^{it\theta}\mathscr{E}e^{i\mu tX}\\ &= e^{it\theta - \mu|t|}, \qquad t \in \mathbb{R}.\end{aligned}$$

Example 1.2.6. Let X be a random variable with gamma probability density function given by

$$f(x) = \begin{cases} \dfrac{1}{\Gamma(\alpha)\beta^\alpha} x^{\alpha-1}e^{-x/\beta}, & x > 0,\\ 0, & x \le 0,\end{cases}$$

where $\alpha > 0$, $\beta > 0$. Then

$$\mathscr{E}X = \alpha\beta$$

and

$$\operatorname{var}(X) = \alpha\beta^2.$$

Also

$$\varphi(t) = (1 - i\beta t)^{-\alpha}, \qquad t \in \mathbb{R}.$$

1.2.2 Some Important Properties

We first consider some inequalities concerning moments. The Markov and the Cauchy–Schwartz inequalities, in particular, are among the most commonly used inequalities in probability theory.

Proposition 1.2.1. Suppose that $\mathscr{E}|X|^\lambda < \infty$ for some $\lambda > 0$. Then $\mathscr{E}|X|^\nu < \infty$ for $0 \leq \nu \leq \lambda$.

Proof. Clearly $|x| \leq 1 \Rightarrow |x|^\nu \leq 1$, and $|x| > 1 \Rightarrow |x|^\nu \leq |x|^\lambda$ for $\nu \leq \lambda$. It follows that $|X|^\nu \leq 1 + |X|^\lambda$ a.s., so that $\mathscr{E}|X|^\nu < \infty$ from Property (i) above. ∎

Proposition 1.2.2. Let X be a random variable, and let g be a nonnegative Borel-measurable function such that $\mathscr{E}g(X) < \infty$. Suppose that g is even and nondecreasing on $[0, \infty)$. Then, for every $\epsilon > 0$,

$$P\{|X| \geq \epsilon\} \leq \frac{\mathscr{E}g(X)}{g(\epsilon)}. \tag{1.2.7}$$

Proof. Let $E = \{|X| \geq \epsilon\}$. Then

$$\begin{aligned}\mathscr{E}g(X) &= \int_E g(X)\,dP + \int_{E^c} g(X)\,dP \\ &\geq g(\epsilon)P(E).\end{aligned}$$ ∎

Corollary (Markov's Inequality). If $\mathscr{E}|X|^\lambda < \infty$ for some $\lambda > 0$, then

$$P\{|X| \geq \epsilon\} \leq \frac{\mathscr{E}|X|^\lambda}{\epsilon^\lambda} \qquad (\epsilon > 0). \tag{1.2.8}$$

In particular, if $\lambda = 2$, we get

$$P\{|X| \geq \epsilon\} \leq \frac{\mathscr{E}X^2}{\epsilon^2} \qquad (\epsilon > 0), \tag{1.2.9}$$

which is known as the *Chebyshev inequality.*

Proposition 1.2.3. Let X and g be as in Proposition 1.2.2, and suppose further that a.s. sup $g(X) < \infty$. Then the inequality

$$P\{|X| \geq \epsilon\} \geq \frac{\mathscr{E}g(X) - g(\epsilon)}{\text{a.s. sup } g(X)} \tag{1.2.10}$$

holds for every $\epsilon > 0$.

Proof. Let $E = \{|X| \geq \epsilon\}$. Then

$$\mathscr{E}g(X) = \int_E g(X)\, dP + \int_{E^c} g(X)\, dP$$
$$\leq \text{a.s. sup } g(X)P(E) + g(\epsilon)P(E^c)$$
$$\leq \text{a.s. sup } g(X)P(E) + g(\epsilon). \quad \blacksquare$$

Corollary. For $\lambda > 0$ and every $\epsilon > 0$ the following inequality holds:

$$\mathscr{E}\frac{|X|^\lambda}{1 + |X|^\lambda} - \frac{\epsilon^\lambda}{1 + \epsilon^\lambda} \leq P\{|X| \geq \epsilon\} \leq \frac{1 + \epsilon^\lambda}{\epsilon^\lambda}\,\mathscr{E}\frac{|X|^\lambda}{1 + |X|^\lambda}. \tag{1.2.11}$$

Proposition 1.2.4 (Cauchy–Schwartz Inequality). Let X and Y be any two random variables with $\mathscr{E}X^2 < \infty$ and $\mathscr{E}Y^2 < \infty$. Then $\mathscr{E}|XY| < \infty$, and the following inequality holds:

$$[\mathscr{E}|XY|]^2 \leq \mathscr{E}X^2 \cdot \mathscr{E}Y^2. \tag{1.2.12}$$

Equality (1.2.12) holds if and only if Y is a linear function of X a.s.

Proof. Inequality (1.2.12) holds trivially if at least one of X or Y is zero a.s. Thus we assume that $\mathscr{E}X^2 > 0$ and $\mathscr{E}Y^2 > 0$. Setting

$$a = \frac{X}{[\mathscr{E}X^2]^{1/2}} \quad \text{and} \quad b = \frac{Y}{[\mathscr{E}Y^2]^{1/2}}$$

in the inequality $2|ab| \leq a^2 + b^2$, we obtain

$$|XY| \leq \tfrac{1}{2}[\mathscr{E}X^2\mathscr{E}Y^2]^{1/2}\left[\frac{X^2}{\mathscr{E}X^2} + \frac{Y^2}{\mathscr{E}Y^2}\right] \text{ a.s.}$$

It follows that $\mathscr{E}|XY| < \infty$. Taking expectations on both sides, we obtain (1.2.12). ■

Corollary (Lyapounov's Inequality). Let X be a random variable with $\beta_n < \infty$ for some positive integer n. Then

$$\beta_{k-1}^{1/(k-1)} \leq \beta_k^{1/k}, \qquad k = 2, 3, \ldots, n. \tag{1.2.13}$$

Proof. Replacing X by $|X|^{(k-1)/2}$ and Y by $|X|^{(k+1)/2}$ in (1.2.12), we get $\beta_k^2 \leq \beta_{k-1}\beta_{k+1}$, $k = 1, 2, \ldots, n-1$, where $\beta_0 = 1$. Inequality (1.2.13) follows by induction.

Proposition 1.2.5 (Hölder's Inequality). Let p and q be two positive real numbers satisfying $1 < p$, $q < \infty$, and $p^{-1} + q^{-1} = 1$. Let X and Y

be two random variables such that $\mathscr{E}|X|^p < \infty$ and $\mathscr{E}|Y|^q < \infty$. Then $\mathscr{E}|XY| < \infty$, and the inequality

$$\mathscr{E}|XY| \leq [\mathscr{E}|X|^p]^{1/p}[\mathscr{E}|Y|^q]^{1/q} \tag{1.2.14}$$

holds.

Proof. We first prove the following result. Let p and q be as defined above, and let $\alpha \geq 0$, $\beta \geq 0$. Then the inequality

$$0 \leq \alpha\beta \leq \frac{\alpha^p}{p} + \frac{\beta^q}{q} \tag{1.2.15}$$

holds. In fact we note that (1.2.15) holds trivially if at least one of α and β equals zero. We assume, therefore, that $\alpha > 0$, $\beta > 0$. We define the function φ on $(0, \infty)$ by setting

$$\varphi(t) = \frac{t^p}{p} + \frac{t^{-q}}{q}.$$

Then $\varphi'(t) = t^{p-1} - t^{-q-1}$. We let t_0 (>0) be a solution of $\varphi'(t) = 0$. Then $t_0^{p+q} = 1$, which implies that $t_0 = 1$. On the other hand, we note that $\varphi'(t) < 0$ for $0 < t < 1$, and $\varphi'(t) > 0$ for $t > 1$. Moreover,

$$\lim_{t \downarrow 0} \varphi(t) = \lim_{t \to \infty} \varphi(t) = +\infty.$$

It follows that $t = 1$ is the unique minimum of φ on $(0, \infty)$. Thus we have $\varphi(t) \geq \varphi(1)$ for $t \in (0, \infty)$, which implies

$$\frac{t^p}{p} + \frac{t^{-q}}{q} \geq 1, \qquad t > 0. \tag{1.2.16}$$

Setting $t = \alpha^{1/q}/\beta^{1/p}$ in (1.2.16), we obtain (1.2.15).

Returning to the proof of the proposition, we note that (1.2.14) holds trivially if at least one of $\mathscr{E}|X|^p$ or $\mathscr{E}|Y|^q$ is zero. We assume therefore that $\mathscr{E}|X|^p > 0$ and $\mathscr{E}|Y|^q > 0$. For $\omega \in \Omega$ we set

$$\alpha = \frac{|X(\omega)|}{[\mathscr{E}|X|^p]^{1/p}}, \qquad \beta = \frac{|Y(\omega)|}{[\mathscr{E}|Y|^q]^{1/q}}$$

in (1.2.15) to obtain the inequality

$$0 < |XY| \leq [\mathscr{E}|X|^p]^{1/p}[\mathscr{E}|Y|^q]^{1/q}\left[\frac{1}{p}\frac{|X|^p}{\mathscr{E}|X|^p} + \frac{1}{q}\frac{|Y|^q}{\mathscr{E}|Y|^q}\right],$$

which holds a.s. Clearly $\mathscr{E}|XY| < \infty$; and, taking the expectation on both sides, we verify that (1.2.14) holds. ■

Remark 1.2.3. Proposition 1.2.4 is a special case of Proposition 1.2.5 where $p = q = 2$.

Proposition 1.2.6 (Minkowski Inequality). Let $1 \leq p < \infty$. Let X and Y be two random variables such that $\mathscr{E}|X|^p < \infty$, $\mathscr{E}|Y|^p < \infty$. Then $\mathscr{E}|X + Y|^p < \infty$, and the inequality

$$[\mathscr{E}|X + Y|^p]^{1/p} \leq [\mathscr{E}|X|^p]^{1/p} + [\mathscr{E}|Y|^p]^{1/p} \tag{1.2.17}$$

holds.

Proof. It suffices to prove (1.2.17) when $1 < p < \infty$. We choose q such that $1/p + 1/q = 1$ holds. Then $1 < q < \infty$. Proceeding as in Proposition 1.2.5, we can show that, if $\alpha_1, \alpha_2, \beta_1, \beta_2$ are arbitrary real numbers, the inequality

$$|\alpha_1\beta_1 + \alpha_2\beta_2| \leq (|\alpha_1|^p + |\alpha_2|^p)^{1/p}(|\beta_1|^q + |\beta_2|^q)^{1/q} \tag{1.2.18}$$

holds. We let $\omega \in \Omega$, and set

$$\alpha_1 = |X(\omega)|, \qquad \alpha_2 = |Y(\omega)|, \qquad \beta_1 = \beta_2 = |(X + Y)(\omega)|^{p-1}$$

in (1.2.18). We obtain

$$|X + Y|^p \leq 2^{p/q}(|X|^p + |Y|^p) \text{ a.s.} \tag{1.2.19}$$

It follows immediately that $\mathscr{E}|X + Y|^p < \infty$. Next we note that

$$|X + Y|^p \leq |X| \cdot |X + Y|^{p-1} + |Y| \cdot |X + Y|^{q-1} \text{ a.s.}$$

Taking the expectation on both sides and making some elementary computations, we obtain (1.2.17). ∎

Remark 1.2.4. Let $1 \leq p < \infty$, and $\mathscr{L}_p = \mathscr{L}_p(\Omega, \mathscr{S}, P)$ be the set of all random variables X on Ω with $\mathscr{E}|X|^p < \infty$. Then it follows immediately from Proposition 1.2.6 that the set $\mathscr{L}_p$ is a vector space over $\mathbb{R}$ if we define the null vector $X = 0$ if $X = 0$ a.s. We will show later that $\mathscr{L}_p$ is a Banach space with respect to the norm $\|X\|_p = [\mathscr{E}|X|^p]^{1/p}$, $X \in \mathscr{L}_p$.

We next derive some conditions for the existence of the moments of a random variable in terms of its distribution function.

Proposition 1.2.7. Let X be a random variable defined on $(\Omega, \mathscr{S}, P)$. Then

$$\mathscr{E}|X|^p < \infty \Rightarrow x^p P\{|X| \geq x\} \to 0 \qquad \text{as } x \to \infty.$$

Proof. Let F be the distribution function of X. Then

$$\mathscr{E}|X|^p = \int_{|t| \geq x} |t|^p \, dF(t) + \int_{|t| < x} |t|^p \, dF(t),$$

and since $\mathscr{E}|X|^p < \infty$ it follows that $\lim_{x \to \infty} \int_{|t| \geq x} |t|^p \, dF(t) = 0$. But

$$\int_{|t| \geq x} |t|^p \, dF(t) \geq x^p P\{|X| \geq x\},$$

and the result follows on taking limits as $x \to \infty$ on both sides. ∎

Proposition 1.2.8. Let $X \geq 0$ a.s., and let F be its distribution function. Then

$$\mathscr{E}X < \infty \Leftrightarrow \int_0^\infty [1 - F(x)] \, dx < \infty.$$

In this case the relation

$$\mathscr{E}X = \int_0^\infty [1 - F(x)] \, dx \tag{1.2.20}$$

holds.

Proof. Suppose that $\mathscr{E}X < \infty$. Then

$$\mathscr{E}X = \int_0^\infty x \, dF(x) = \lim_{n \to \infty} \int_0^n x \, dF(x).$$

On integration by parts, we obtain

$$\int_0^n x \, dF(x) = nF(n) - \int_0^n F(x) \, dx$$

$$= -n[1 - F(n)] + \int_0^n [1 - F(x)] \, dx. \tag{1.2.21}$$

Since $\mathscr{E}X < \infty$, $n[1 - F(n)] \to 0$ as $n \to \infty$ by Proposition 1.2.7. Hence (1.2.20) holds.

Conversely, suppose $\int_0^\infty [1 - F(x)] \, dx < \infty$. Clearly for every n

$$\int_0^n x \, dF(x) \leq \int_0^n [1 - F(x)] \, dx \leq \int_0^\infty [1 - F(x)] \, dx,$$

and it follows that $\mathscr{E}X < \infty$. In view of (1.2.21) relation (1.2.20) holds. ∎

Corollary 1. Let X be a random variable with distribution function F. Then $\mathscr{E}|X| < \infty$ if and only if both the integrals $\int_{-\infty}^{0} F(x)\,dx$ and $\int_{0}^{\infty} [1 - F(x)]\,dx$ are finite. In this case

$$\mathscr{E}X = \int_{0}^{\infty} [1 - F(x)]\,dx - \int_{-\infty}^{0} F(x)\,dx. \tag{1.2.22}$$

Corollary 2 (Moments Lemma). Let X be a random variable, and let $0 < p < \infty$. Then $\mathscr{E}|X|^p < \infty \Leftrightarrow \sum_{n=1}^{\infty} P\{|X| \geq n^{1/p}\} < \infty$.

Proof. From Proposition 1.2.8

$$\mathscr{E}|X|^p = \int_{0}^{\infty} P\{|X|^p \geq x\}\,dx$$

$$= p \int_{0}^{\infty} x^{p-1} P\{|X| \geq x\}\,dx.$$

It follows that

$$\mathscr{E}|X|^p < \infty \Leftrightarrow \int_{0}^{\infty} x^{p-1} P\{|X| \geq x\}\,dx < \infty.$$

A simple application of the integral test now completes the proof of the corollary.

Corollary 3. Let X be a random variable satisfying $n^p P\{|X| \geq n\} \to 0$ as $n \to \infty$ for some $p > 0$. Then $\mathscr{E}|X|^q < \infty$ for $0 \leq q < p$.

1.2.3 Mathematical Expectation of a Random Vector

Let $(\Omega, \mathscr{S}, P)$ be a probability space, and let $\mathbf{X} = (X_1, X_2, \ldots, X_n)$ be a random vector defined on $(\Omega, \mathscr{L}, P)$. Let $g: \mathbb{R}_n \to \mathbb{R}_1$ be a Borel-measurable function. Then $g(\mathbf{X})$ is a random variable.

Definition 1.2.2. We say that the mathematical expectation of $g(\mathbf{X})$ exists if $g(\mathbf{X})$ is integrable over Ω with respect to P. In this case we write

$$\mathscr{E}g(\mathbf{X}) = \int_{\Omega} g(\mathbf{X})\,dP = \int_{\mathbb{R}_n} g(t_1, t_2, \ldots, t_n)\,dP_{\mathbf{X}}(t_1, t_2, \ldots, t_n), \tag{1.2.23}$$

where $P_{\mathbf{X}}$ is the probability distribution of $\mathbf{X}$.

Suppose that g is continuous on $\mathbb{R}_n$ and $\mathscr{E}g(\mathbf{X})$ exists. Then we can write (1.2.23) as

$$\mathscr{E}g(\mathbf{X}) = \int_{\mathbb{R}_n} g(x_1, x_2, \ldots, x_n)\,dF(x_1, x_2, \ldots, x_n), \tag{1.2.24}$$

where F is the joint distribution function corresponding to $P_{\mathbf{X}}$, and the last integral in (1.2.24) is a Riemann–Stieltjes integral.

Some useful forms of g are discussed below.

(a) Let

$$g(x_1, x_2, \ldots, x_n) = \prod_{j=1}^{n} x_j^{v_j},$$

where $v_j, j = 1, 2, \ldots, n$, are nonnegative integers. Then

$$\alpha_{v_1, v_2, \ldots, v_n} = \mathscr{E}(X_1^{v_1} X_2^{v_2} \cdots X_n^{v_n})$$

is known, if it exists, as the *moment of order* v_1 *in* X_1, v_2 *in* $X_2, \ldots,$ *and* v_n *in* X_n. In a similar manner one can define absolute moments and central order moments of $g(\mathbf{X})$. Clearly

$$\alpha_{0, 0, \ldots, 0, v_j, 0, \ldots, 0} = \mathscr{E}X_j^{v_j}$$

is the moment of order v_j of the random variable X_j, and

$$\alpha_{1, 0, 0, \ldots, 0} = \mathscr{E}X_1,$$

$$\alpha_{2, 0, 0, \ldots, 0} - (\alpha_{1, 0, 0, \ldots, 0})^2 = \operatorname{var}(X_1),$$

and so on.

Another special case of importance occurs when $v_i = v_j = 1$ for $i \neq j$, and $v_k = 0$ for $k = 1, 2, \ldots, n$, $k \neq i$, $k \neq j$. In this case we write

$$\sigma_{jj} = \operatorname{var}(X_j), \qquad j = 1, 2, \ldots, n,$$

and, for $j \neq k$,

$$\begin{aligned}\sigma_{jk} &= \mathscr{E}\{(X_j - \mathscr{E}X_j)(X_k - \mathscr{E}X_k)\} \\ &= \mathscr{E}(X_j X_k) - \mathscr{E}X_j \mathscr{E}X_k.\end{aligned}$$

We call σ_{jk}, if it exists, the *covariance* between X_j and X_k, $j, k = 1, 2, \ldots, n$, $j \neq k$. It follows from the Cauchy–Schwartz inequality (Proposition 1.2.4) that σ_{jk} exists whenever $\sigma_{jj} < \infty$ and $\sigma_{kk} < \infty$ and the inequality $|\sigma_{jk}| \leq \{\sigma_{jj}\sigma_{kk}\}^{1/2}$ holds. The $n \times n$ matrix $\mathbf{\Sigma} = ((\sigma_{jk}))_{j,k=1,2,\ldots,n}$ is called the *variance-covariance* (or *dispersion*) matrix of the random vector $\mathbf{X}$. We note that $\mathbf{\Sigma}$ is a real symmetric, positive semidefinite matrix.

Next, we write $\rho_{jk} = \sigma_{jk}/\{\sigma_{jj}\sigma_{kk}\}^{1/2}$, $j \neq k$, and call it the *coefficient of correlation between* X_j *and* X_k. It follows again from Proposition 1.2.4 that $|\rho_{jk}| \leq 1$ and $\rho_{jk} = \pm 1$ if and only if X_j is a linear function of X_k a.s.

(b) Let

$$g(x_1, x_2, \ldots, x_n) = \exp\left(\sum_{j=1}^{n} t_j x_j\right),$$

where $t_k \in (-\delta, \delta)$, $k = 1, 2, \ldots, n$, for some $\delta > 0$. Then we write

$$M(t_1, t_2, \ldots, t_n) = \mathscr{E} \exp\left(\sum_{j=1}^{n} t_j X_j\right)$$

and call it the *moment generating function of* $\mathbf{X}$, if $\mathscr{E} \exp(\sum_{j=1}^{n} t_j X_j)$ exists.

(c) Let

$$g(x_1, x_2, \ldots, x_n) = \exp\left(\sum_{j=1}^{n} i t_j x_j\right),$$

where $t_k \in \mathbb{R}$, $k = 1, 2, \ldots, n$, and $i = \sqrt{-1}$. Then write

$$\varphi(t_1, t_2, \ldots, t_n) = \mathscr{E} \exp\left(i \sum_{j=1}^{n} t_j X_j\right), \tag{1.2.25}$$

and call φ the *characteristic function of* $\mathbf{X}$. We note that the characteristic function φ of $\mathbf{X}$ always exists. Clearly

$$\varphi(0, 0, \ldots, t_j, 0, \ldots, 0) = \mathscr{E} e^{it_j X_j},$$

which is the characteristic function of X_j. Similarly

$$\varphi(0, 0, \ldots, 0, t_j, 0, \ldots, 0, t_k, 0, \ldots, 0) = \mathscr{E} e^{i(t_j X_j + t_k X_k)}$$

is the characteristic function of the random vector (X_j, X_k), and so on.

Example 1.2.7. Consider the random vector $(X_1, X_2, \ldots, X_{k-1})$, $k \geq 3$, which has the *multinomial distribution* given by

$$P\{X_1 = x_1, X_2 = x_2, \ldots, X_{k-1} = x_{k-1}\}$$

$$= \begin{cases} \dfrac{n!}{x_1!\, x_2! \ldots (n - x_1 - \ldots - x_{k-1})!} p_1^{x_1} p_2^{x_2} \cdots p_k^{n - x_1 - \cdots - x_{k-1}} & \text{if } \sum_1^{k-1} x_i \leq n, \\ 0 & \text{otherwise}, \end{cases} \tag{1.2.26}$$

where $p_i \geq 0$, $i = 1, 2, \ldots, k - 1$, and $\sum_{i=1}^{k} p_i = 1$ and $x_1, x_2, \ldots, x_{k-1}$ are nonnegative integers. Let us write $x_k = n - \sum_1^{k-1} x_i$. Then the characteristic function of $(X_1, X_2, \ldots, X_{k-1})$ is given by

$$\varphi(t_1, t_2, \ldots, t_{k-1}) = (p_1 e^{it_1} + \cdots + p_{k-1} e^{it_{k-1}} + p_k)^n \tag{1.2.27}$$

for all $t_1, t_2, \ldots, t_{k-1} \in \mathbb{R}$. Clearly

$$\varphi(0, 0, \ldots, 0, t_j, 0, \ldots, 0) = (p_j e^{it_j} + 1 - p_j)^n,$$
$$\varphi(0, 0, \ldots, 0, t_l, 0, \ldots, 0, t_m, 0, \ldots, 0) = (p_l e^{it_l} + p_m e^{it_m} + 1 - p_l - p_m)^n,$$

and so on. One can show easily from (1.2.26) that each X_j is a binomial random variable so that

$$\mathscr{E}X_j = np_j, \qquad \sigma_{jj} = \text{var}(X_j) = np_j(1 - p_j), \qquad j = 1, 2, \ldots, k-1.$$

Moreover, for $l \neq m$,

$$\sigma_{lm} = \mathscr{E}\{(X_l - \mathscr{E}X_l)(X_m - \mathscr{E}X_m)\} = -np_l p_m,$$

so that

$$\rho_{lm} = -\left[\frac{p_l p_m}{(1 - p_l)(1 - p_m)}\right]^{1/2}, \qquad l, m = 1, 2, \ldots, k-1 \quad (l \neq m).$$

Example 1.2.8. Let $\mathbf{X} = (X_1, X_2, \ldots, X_n)$ be a random vector having multivariate normal density given by

$$f(x_1, x_2, \ldots, x_n) = \frac{|\mathbf{\Sigma}|^{-1/2}}{(2\pi)^{k/2}} \exp\{-\tfrac{1}{2}(\mathbf{x} - \boldsymbol{\mu})\mathbf{\Sigma}^{-1}(\mathbf{x} - \boldsymbol{\mu})'\},$$

where $\mathbf{\Sigma}$ is a positive definite matrix. Let us introduce new variables $\mathbf{y} = (y_1, y_2, \ldots, y_n)$, defined by the nonsingular transformation

$$\mathbf{x} = \boldsymbol{\mu} + \mathbf{y}\mathbf{A},$$

where $\mathbf{A}$ is an $n \times n$ matrix. Clearly it is possible to choose $\mathbf{A}$ so that $\mathbf{A}\mathbf{\Sigma}^{-1}\mathbf{A}' = \mathbf{I}$, where $\mathbf{I}$ is the $n \times n$ identity matrix. We have

$$i\mathbf{x}\mathbf{t}' - \tfrac{1}{2}(\mathbf{x} - \boldsymbol{\mu})\mathbf{\Sigma}^{-1}(\mathbf{x} - \boldsymbol{\mu})' = i\boldsymbol{\mu}\mathbf{t}' + i\mathbf{y}\mathbf{A}\mathbf{t}' - \tfrac{1}{2}\mathbf{y}\mathbf{y}',$$

where $\mathbf{t} = (t_1, t_2, \ldots, t_n)$, and it follows that

$$\begin{aligned}
\varphi(\mathbf{t}) &= \mathscr{E} \exp(i\mathbf{X}\mathbf{t}') \\
&= |\mathbf{\Sigma}|^{-1/2}(2\pi)^{-n/2} \int_{\mathbb{R}_n} \exp\{-\tfrac{1}{2}(\mathbf{x} - \boldsymbol{\mu})\mathbf{\Sigma}^{-1}(\mathbf{x} - \boldsymbol{\mu})'\}\, d\mathbf{x} \\
&= \frac{\exp(i\boldsymbol{\mu}\mathbf{t}')}{(2\pi)^{n/2}} \int_{\mathbb{R}_n} \exp(i\mathbf{y}\mathbf{A}\mathbf{t}' - \tfrac{1}{2}\mathbf{y}\mathbf{y}')\, d\mathbf{y}.
\end{aligned}$$

Next, we write $\mathbf{u} = (u_1, u_2, \ldots, u_n) = \mathbf{tA}'$. Then we have

$$i\mathbf{yAt}' - \tfrac{1}{2}\mathbf{yy}' = i\mathbf{yu}' - \tfrac{1}{2}\mathbf{yy}' = \sum_{j=1}^{n} (iy_j u_j - \tfrac{1}{2}y_j^2),$$

and hence

$$\begin{aligned} \varphi(\mathbf{t}) &= \exp(i\boldsymbol{\mu}\mathbf{t}' - \tfrac{1}{2}\mathbf{uu}') \\ &= \exp(i\boldsymbol{\mu}\mathbf{t}' - \tfrac{1}{2}\mathbf{tA}'\mathbf{At}') \\ &= \exp(i\boldsymbol{\mu}\mathbf{t}' - \tfrac{1}{2}\mathbf{t}\boldsymbol{\Sigma}\mathbf{t}'). \end{aligned} \tag{1.2.28}$$

Clearly

$$\varphi(0, 0, \ldots, 0, t_j, 0, \ldots, 0) = \exp(i\mu_j t_j - \tfrac{1}{2}\sigma_{jj} t_j^2),$$

and

$$\varphi(0, 0, \ldots, 0, t_l, 0, \ldots, 0, t_m, 0, \ldots, 0) = \exp\left\{ i(\mu_l t_l + \mu_m t_m) - \frac{\sigma_{ll} t_l^2 + 2\sigma_{lm} t_l t_m + \sigma_{mm} t_m^2}{2} \right\},$$

where $\boldsymbol{\Sigma} = ((\sigma_{lm}))_{l, m = 1, 2, \ldots, n}$.

1.3 CONVERGENCE CONCEPTS

Let $(\Omega, \mathscr{S}, P)$ be a probability space, and let $\{X_n, n \geq 1\}$ be a sequence of random variables defined on it. In this section we consider some useful concepts of convergence of the sequence $\{X_n\}$. It is expected that the reader will have been exposed to some or all of the concepts introduced here in his or her measure theory course. Nevertheless we go into some important details for the sake of completeness.

1.3.1 Almost Sure Convergence

The concept of almost sure (a.s.) convergence in probability theory is identical with the concept of almost everywhere (a.e.) convergence in measure theory. For completeness we give the following definitions and prove some basic results.

Definition 1.3.1. The sequence of random variables $\{X_n\}$ is said to converge a.s. to the random variable X if and only if there exists a set $E \in \mathscr{S}$ with

$P(E) = 0$ such that, for every $\omega \in E^c$, $|X_n(\omega) - X(\omega)| \to 0$ as $n \to \infty$. In that case we write $X_n \xrightarrow{\text{a.s.}} X$.

Definition 1.3.2. The sequence $\{X_n\}$ is said to be Cauchy (fundamental) a.s. if there exists a set $E \in \mathscr{S}$ with $P(E) = 0$ such that, for every $\omega \in E^c$, $|X_n(\omega) - X_m(\omega)| \to 0$ as $m, n \to \infty$.

We now show that the two concepts defined above are equivalent.

Proposition 1.3.1. Let $\{X_n\}$ be a sequence of random variables defined on $(\Omega, \mathscr{S}, P)$. Then

$$X_n \xrightarrow{\text{a.s.}} X \Leftrightarrow \{X_n\} \text{ is Cauchy a.s.}$$

Proof. If $X_n \xrightarrow{\text{a.s.}} X$, there exists a null set $E \in \mathscr{S}$ such that $X_n(\omega) \to X(\omega)$ for every $\omega \in E^c$ as $n \to \infty$. Hence, for $\omega \in E^c$ and for m, n arbitrary positive integers,

$$|X_m(\omega) - X_n(\omega)| \le |X_m(\omega) - X(\omega)| + |X_n(\omega) - X(\omega)| \to 0 \qquad \text{as } m, n \to \infty.$$

Consequently $\{X_n\}$ is Cauchy a.s.

Conversely, if $\{X_n\}$ is Cauchy a.s., there exists a null set $E \in \mathscr{S}$ such that, for every $\omega \in E^c$, $|X_n(\omega) - X_m(\omega)| \to 0$ as $m, n \to \infty$. Thus the sequence $\{X_n(\omega)\}$ is a Cauchy sequence of real numbers. Consequently there exists a unique real number, say $X(\omega)$, such that $X(\omega) = \lim_{n\to\infty} X_n(\omega)$ for every $\omega \in E^c$, that is, $X_n \xrightarrow{\text{a.s.}} X$. ■

We note that, if $X_n \xrightarrow{\text{a.s.}} X$, then X is also a random variable which is unique a.s. We next derive the following criterion for a.s. convergence.

Proposition 1.3.2.

(a) The sequence $\{X_n\}$ of random variables converges to a random variable X a.s. if and only if

$$\text{(1.3.1)} \quad \lim_{n\to\infty} P\left\{\bigcup_{m=n}^{\infty} (|X_m - X| \ge \epsilon)\right\} = 0 \qquad \text{for every } \epsilon > 0.$$

(b) $\{X_n\}$ is Cauchy a.s. if and only if

$$\text{(1.3.2)} \quad \lim_{n\to\infty} P\left\{\bigcup_{m=1}^{\infty} (|X_{m+n} - X_n| \ge \epsilon)\right\} = 0 \qquad \text{for every } \epsilon > 0.$$

Proof. (a) For every $\epsilon > 0$ set

$$E_n(\epsilon) = (|X_n - X| \geq \epsilon), \qquad n \geq 1.$$

Let

$$D = \{\omega \in \Omega: X_n(\omega) \nrightarrow X(\omega) \text{ as } n \to \infty\}.$$

Clearly

$$\begin{aligned} D &= \bigcup_{\epsilon > 0} \limsup_n E_n(\epsilon) \\ &= \bigcup_{\epsilon > 0} \left(\bigcap_{n=1}^{\infty} \bigcup_{m=n}^{\infty} E_m(\epsilon) \right) \\ &= \bigcup_{k=1}^{\infty} \left(\bigcap_{n=1}^{\infty} \bigcup_{m=n}^{\infty} E_m\left(\frac{1}{k}\right) \right). \end{aligned}$$

It follows that

$$\begin{aligned} X_n \overset{\text{a.s.}}{\to} X &\Leftrightarrow P(D) = 0 \\ &\Leftrightarrow P\left(\limsup_n E_n(\epsilon)\right) = 0 \qquad \text{for every } \epsilon > 0. \end{aligned}$$

Since $\bigcup_{m=n}^{\infty} E_m(\epsilon) \downarrow \limsup_n E_n(\epsilon)$, we see that

$$P\left(\limsup_n E_n(\epsilon)\right) = \lim_n P\left(\bigcup_{m=n}^{\infty} E_m(\epsilon) \right),$$

so that

$$X_n \overset{\text{a.s.}}{\to} X \Leftrightarrow \lim_{n \to \infty} P\left\{ \bigcup_{m=n}^{\infty} (|X_m - X| \geq \epsilon) \right\} = 0 \qquad \text{for every } \epsilon > 0,$$

and the proof of (a) is complete.

(b) Note that

$$\bigcup_{m=n}^{\infty} \left(|X_m - X_n| \geq \frac{\epsilon}{2} \right) \supset \left(\sup_{m \geq n} |X_m - X_n| \geq \epsilon \right) \supset \bigcup_{m=n}^{\infty} (|X_m - X_n| \geq \epsilon).$$

It follows from (a) that $\{X_n\}$ is Cauchy a.s.

$$\begin{aligned} &\Leftrightarrow \lim_{n \to \infty} P\left\{ \bigcup_{m=n}^{\infty} (|X_m - X_n| \geq \epsilon) \right\} = 0 \qquad \text{for every } \epsilon > 0, \\ &\Leftrightarrow \lim_{n \to \infty} P\left(\sup_{m \geq n} |X_m - X_n| \geq \epsilon \right) = 0 \qquad \text{for every } \epsilon > 0, \\ &\Leftrightarrow \lim_{n \to \infty} P\left\{ \bigcup_{m=1}^{\infty} (|X_{m+n} - X_n| \geq \epsilon) \right\} = 0 \qquad \text{for every } \epsilon > 0. \end{aligned}$$

■

Corollary 1. The sequence $\{X_n\}$ converges a.s. to the random variable X if $\sum_{n=1}^{\infty} P\{|X_n - X| \geq \epsilon\} < \infty$ for all $\epsilon > 0$.

Proof. Since

$$P\left\{\bigcup_{m=n}^{\infty} (|X_m - X| \geq \epsilon)\right\} \leq \sum_{m=n}^{\infty} P\{|X_m - X| \geq \epsilon\},$$

the result follows from part (a) of Proposition 1.3.2.

We remark that a sequence $\{X_n\}$ is said to converge *completely* to a random variable X if $\sum_{n=1}^{\infty} P\{|X_n - X| \geq \epsilon\} < \infty$ for every $\epsilon > 0$. This concept of convergence was introduced by Hsu and Robbins [38]. According to Corollary 1, complete convergence $\Rightarrow$ a.s. convergence.

Corollary 2. Let $\{X_n\}$ be a sequence of random variables, and let X be a random variable with $\mathscr{E}(X_n - X)^2 < \infty$, $n \geq 1$, such that $\sum_{n=1}^{\infty} \mathscr{E}(X_n - X)^2 < \infty$. Then $X_n \xrightarrow{\text{a.s.}} X$.

Example 1.3.1. Let $\{X_n\}$ be a sequence of random variables with $P\{X_n = \pm 1/n\} = \frac{1}{2}$. Then, for $j < k$, $|X_j| > |X_k|$ a.s., so that $\{|X_j| \geq \epsilon\} \supset \{|X_k| \geq \epsilon\}$, and it follows that

$$\bigcup_{j=n}^{\infty} (|X_j| \geq \epsilon) = \{|X_n| \geq \epsilon\}.$$

Choosing $n > 1/\epsilon$, we see that

$$P\left\{\bigcup_{j=n}^{\infty} [|X_j| \geq \epsilon]\right\} = P\{|X_n| \geq \epsilon\} \leq P\left\{|X_n| > \frac{1}{n}\right\} = 0,$$

and it follows from Proposition 1.3.2 that $X_n \xrightarrow{\text{a.s.}} 0$. Alternatively, one can apply Corollary 1 or Corollary 2 to arrive at the same conclusion.

1.3.2 Convergence in Probability

A somewhat weaker concept of convergence is that of convergence in probability, which is identical with the concept of convergence in measure.

Definition 1.3.3. Let $\{X_n, n \geq 1\}$ be a sequence of random variables defined on a probability space $(\Omega, \mathscr{S}, P)$. We say that X_n converges in probability to the random variable X if for every $\epsilon > 0$

$$P\{\omega \in \Omega: |X_n(\omega) - X(\omega)| \geq \epsilon\} \to 0$$

as $n \to \infty$. In this case we write $X_n \xrightarrow{P} X$.

Definition 1.3.4. We say that $\{X_n\}$ is Cauchy in probability (or fundamental in probability) if for every $\epsilon > 0$

$$P\{|X_n - X_m| \geq \epsilon\} \to 0$$

as $m, n \to \infty$.

Remark 1.3.1. If $X_n \xrightarrow{P} X$, then X is unique a.s. in the sense that if $X_n \xrightarrow{P} X$ and $X_n \xrightarrow{P} Y$, then $X = Y$ a.s.

Remark 1.3.2. Let $X_n \xrightarrow{P} X$; then $\{X_n\}$ is Cauchy in probability. The converse statement is also true and is proved in Proposition 1.3.5.

Remark 1.3.3. Let $X_n \xrightarrow{P} X$, and let $\{X_{n_k}\}$ be any infinite subsequence of $\{X_n\}$. Then $X_{n_k} \xrightarrow{P} X$ as $k \to \infty$.

We next show that convergence in probability is weaker than convergence a.s.

Proposition 1.3.3. We have

$$X_n \xrightarrow{\text{a.s.}} X \Rightarrow X_n \xrightarrow{P} X.$$

Proof. From Proposition 1.3.2, $X_n \xrightarrow{\text{a.s.}} X$ implies that for every $\epsilon > 0$

$$\lim_{n\to\infty} P\left\{\bigcup_{m=n}^{\infty} (|X_m - X| \geq \epsilon)\right\} = 0.$$

Since for every $n \geq 1$

$$(|X_n - X| \geq \epsilon) \subset \bigcup_{m=n}^{\infty} (|X_m - X| \geq \epsilon),$$

the result follows immediately. ■

Remark 1.3.4. That the converse of Proposition 1.3.3 is not true in general is demonstrated in Example 1.3.2 below. In the following special case, however, convergence in probability implies a.s. convergence. Let $(\Omega, \mathscr{S}, P)$ be a probability space. An event A of positive probability is said to be an *atom* if $B \in \mathscr{S}$ with $B \subset A \Rightarrow P(B) = 0$ or $P(B) = P(A)$. Let $\mathfrak{X}$ be the space of all random variables defined on $(\Omega, \mathscr{S}, P)$ such that, if $X, Y \in \mathfrak{X}$, then $P\{X \neq Y\} > 0$. Then convergence in probability implies a.s. convergence for all sequences from $\mathfrak{X}$ if and only if Ω is the union of a countable number of disjoint atoms. We ask the reader to construct a proof of this result in Problem 1.5.27. (See also Problem 1.5.6.)

Example 1.3.2. Let $\Omega = [0, 1]$, and let $\mathscr{S}$ be the class of all Borel sets on Ω. Let P be the Lebesgue measure. For any positive integer n, choose integer m with $2^m \le n < 2^{m+1}$. Clearly $n \to \infty \Leftrightarrow m \to \infty$. We can write $n \ge 1$ as $n = 2^m + k$, $k = 0, 1, \ldots, 2^{m-1}$. Let us define X_n on Ω by

$$X_n(\omega) = \begin{cases} 1 & \text{if } \omega \in \left[\dfrac{k}{2^m}, \dfrac{k+1}{2^m}\right], \\ 0 & \text{otherwise,} \end{cases}$$

if $n = 2^m + k$. Then X_n is a random variable which satisfies

$$P\{|X_n| \ge \epsilon\} = \begin{cases} \dfrac{1}{2^m} & \text{if } 0 < \epsilon < 1, \\ 0 & \text{if } \epsilon \ge 1, \end{cases}$$

so that $X_n \xrightarrow{P} 0$. However $X_n \overset{\text{a.s.}}{\nrightarrow} 0$. In fact, for any $\omega \in [0, 1]$, there are an infinite number of intervals of the form $[k/2^m, (k+1)/2^m]$ which contain ω. Such a sequence of intervals depends on ω. Let us denote it by

$$\left\{\left[\frac{k_m}{2^m}, \frac{k_m + 1}{2^m}\right], m = 1, 2, \ldots\right\},$$

and let $n_m = 2^m + k_m$. Then $X_{n_m}(\omega) = 1$, but $X_n(\omega) = 0$, if $n \ne n_m$. If follows that $\{X_n\}$ does not converge at ω. Since ω is arbitrary, X_n does not converge a.s. to any random variable.

Proposition 1.3.4. Let $\{X_n\}$ be a sequence of random variables which converges in probability to the random variable X. Then there exists a subsequence $\{X_{n_k}\} \subset \{X_n\}$ which converges a.s. to X.

Proof. We note that $\{X_n\}$ is Cauchy in probability. In fact, for every $\epsilon > 0$ and for every $m, n \ge 1$,

$$P\{|X_m - X_n| \ge \epsilon\} \le P\left\{|X_m - X| \ge \frac{\epsilon}{2}\right\} + P\left\{|X_n - X| \ge \frac{\epsilon}{2}\right\}.$$

Therefore for every integer $k \ge 1$ there exists an integer $N(k)$ such that, for all $n \ge N(k)$ and $m \ge N(k)$, the inequality

$$P\left\{|X_n - X_m| \ge \frac{1}{2^k}\right\} < \frac{1}{2^k}$$

holds. Set

$$n_1 = N(1), \qquad n_2 = \max(n_1 + 1, N(2)), \ldots, n_k = \max(n_{k-1} + 1, N(k)), \ldots.$$

Clearly $n_k \uparrow \infty$ as $k \to \infty$, so that $\{X_{n_k}\}$ is an infinite subsequence of $\{X_n\}$. We now show that $\{X_{n_k}\}$ has the required property.

For convenience, we write

$$Y_k = X_{n_k} \quad \text{and} \quad E_k = \left\{ |Y_k - Y_{k+1}| \geq \frac{1}{2^k} \right\}, \quad k = 1, 2, \ldots .$$

Then $P(E_k) < 1/2^k$; moreover, on the set E_k^c, $|Y_k - Y_{k+1}| < 1/2^k$. Let $n \geq 1$, and write $F_n = \bigcup_{k=n}^{\infty} E_k$. Then

$$P(F_n) \leq \sum_{k=n}^{\infty} P(E_k) < \frac{1}{2^{n-1}}.$$

Also, on $F_n^c = \bigcap_{k=n}^{\infty} E_k^c$ we have

$$|Y_k - Y_{k+1}| < \frac{1}{2^k}, \quad k \geq n.$$

Let $n \geq 1$ be fixed, and let N_0 be a positive integer, $N_0 \geq n$. Then, for every $m \geq N_0$, $v \geq 1$, and $\omega \in F_n^c$, we have

$$|Y_{m+v}(\omega) - Y_m(\omega)| \leq \sum_{k=N_0}^{\infty} |Y_k(\omega) - Y_{k+1}(\omega)| < \frac{1}{2^{N_0 - 1}}.$$

Let $\epsilon > 0$, and choose $N_0 = N_0(\epsilon)$ such that $N_0 \geq n$ and $1/2^{N_0-1} < \epsilon$. Then, for every $m \geq N_0$ and $v \geq 1$, the inequality

$$\sup_{\omega \in F_n^c} |Y_{m+v}(\omega) - Y_m(\omega)| < \epsilon$$

holds for every $n \geq 1$. This implies (Problem 1.5.28) that the sequence $\{Y_k\}$ is uniformly Cauchy on F_n^c for every $n \geq 1$. We now set $F = \bigcap_{n=1}^{\infty} F_n$. Then $F \subset F_n$ for every n, and $P(F) < 1/2^{n-1}$ for every n. Thus $P(F) = 0$. Clearly $\{Y_k\}$ is Cauchy on F^c, so that $\{Y_k\}$ is Cauchy a.s. It follows that there exists a random variable Y such that $Y_k \xrightarrow{\text{a.s.}} Y$. Then $Y_k \xrightarrow{P} Y$, and it follows from Remark 1.3.3 that $X = Y$ a.s. ∎

Proposition 1.3.5. The sequence $\{X_n\}$ converges in probability to a random variable X if and only if it is Cauchy in probability.

Proof. We need only to prove that, if $\{X_n\}$ is Cauchy in probability, it converges in probability to some random variable X. From Proposition 1.3.4 it follows that $\{X_n\}$ has a subsequence $\{X_{n_k}\}$ which converges a.s. to a random variable X. Consequently $X_{n_k} \xrightarrow{P} X$ as $n_k \to \infty$. Clearly for every $\epsilon > 0$

$$P\{|X_n - X| \geq \epsilon\} \leq P\left\{|X_n - X_{n_k}| \geq \frac{\epsilon}{2}\right\} + P\left\{|X_{n_k} - X| \geq \frac{\epsilon}{2}\right\}.$$

Since $\{X_n\}$ is Cauchy in probability, the result follows. ∎

1.3.3 Convergence in Mean

Let $(\Omega, \mathscr{S}, P)$ be a probability space, and let $\mathscr{L}_1 = \mathscr{L}_1(\Omega, \mathscr{S}, P)$ be the set of all random variables X having finite expectations.

Definition 1.3.5. Let $\{X_n\}$ be a sequence of random variables such that $X_n \in \mathscr{L}_1$, $n \geq 1$. We say that $\{X_n\}$ converges in mean to a random variable $X \in \mathscr{L}_1$ if $\mathscr{E}|X_n - X| \to 0$ as $n \to \infty$. In this case we write $X_n \overset{\mathscr{L}_1}{\to} X$ and note that the limiting random variable X, if it exists, is unique a.s.

Definition 1.3.6. The sequence $\{X_n\}$ is said to be Cauchy in mean if $\mathscr{E}|X_n - X_m| \to 0$ as $m, n \to \infty$.

First we recall from measure theory the following three important results on convergence in mean.

(a) (Lebesgue Dominated Convergence Theorem). Let $\{X_n\}$ be a sequence of random variables such that X_n converges in probability (or a.s.) to a random variable X. Suppose there exists a random variable $Y \in \mathscr{L}_1$ such that $|X_n| \leq |Y|$ a.s. Then $X_n, X \in \mathscr{L}_1$, and, moreover, $X_n \overset{\mathscr{L}_1}{\to} X$. In particular, $\mathscr{E}X_n \to \mathscr{E}X$ as $n \to \infty$.

(b) (Monotone Convergence Theorem). Let $\{X_n\}$ be a nondecreasing (a.s.) sequence of nonnegative (a.s.) random variables which converges a.s. to a random variable X.
   (1) Suppose that $X_n \in \mathscr{L}_1$ for all n and $\lim_{n\to\infty} \mathscr{E}X_n < \infty$. Then $X \in \mathscr{L}_1$ and $\lim_{n\to\infty} \mathscr{E}X_n = \mathscr{E}X$.
   (2) Conversely, if $X \in \mathscr{L}_1$, then each $X_n \in \mathscr{L}_1$ and $\lim_{n\to\infty} \mathscr{E}X_n = \mathscr{E}X$.

(c) (Fatou's Lemma). Let $X_n \in \mathscr{L}_1$, $n \geq 1$, be a sequence of (a.s.) nonnegative random variables such that $\liminf_{n\to\infty} \mathscr{E}X_n < \infty$. Then $X_* = \liminf_{n\to\infty} X_n \in \mathscr{L}_1$, and the inequality

$$\mathscr{E}X_* \leq \liminf_{n\to\infty} \mathscr{E}X_n$$

holds.

We next study some properties of convergence in mean.

Proposition 1.3.6. We have the following:

(a) $X_n \overset{\mathscr{L}_1}{\to} X \Rightarrow X_n \overset{P}{\to} X$.
(b) $\{X_n\}$ Cauchy in mean $\Rightarrow \{X_n\}$ Cauchy in probability.

Proof. The proofs of both parts follow immediately from Markov's inequality. ∎

Proposition 1.3.7. The sequence $\{X_n\}$, $X_n \in \mathscr{L}_1$, is Cauchy in mean if and only if it converges in mean to some random variable $X \in \mathscr{L}_1$.

Proof. Since for every $m, n \geq 1$

$$|X_n - X_m| \leq |X_n - X| + |X_m - X|,$$

it follows that $X_n \xrightarrow{\mathscr{L}_1} X \Rightarrow \{X_n\}$ Cauchy in mean.

To prove the converse, we first note that $\lim_{n\to\infty} \mathscr{E}|X_n|$ exists and is finite. In fact,

$$|\mathscr{E}|X_n| - \mathscr{E}|X_m|| \leq \mathscr{E}|X_n - X_m| \to 0 \qquad \text{as } m, n \to \infty,$$

so that the sequence of real numbers $\{\mathscr{E}|X_n|\}$ is a Cauchy sequence and hence converges to a finite real number. Next we show that there exists an $X \in \mathscr{L}_1$ such

$$\mathscr{E}|X| \leq \lim_{n\to\infty} \mathscr{E}|X_n|.$$

Indeed, since $\{X_n\}$ is Cauchy in probability, in view of Propositions 1.3.5 and 1.3.4, it follows that it contains a subsequence $\{X_{n_k}\}$ which converges a.s. to some random variable X. Clearly

$$\liminf_k \mathscr{E}|X_{n_k}| = \lim_{k\to\infty} \mathscr{E}|X_{n_k}| = \lim_{n\to\infty} \mathscr{E}|X_n| < \infty,$$

so that we conclude from Fatou's lemma that $X \in \mathscr{L}_1$. Moreover, the inequality

$$\mathscr{E}|X| \leq \lim_{n\to\infty} \mathscr{E}|X_n|$$

holds. Finally, we show that the sequence $\{X_n\}$ converges in mean to X. Let $m \geq 1$ be a fixed positive integer. Then the sequence $\{X_n - X_m, n \geq 1\}$ is Cauchy in mean. Proceeding as above with $\{X_n\}$ replaced by $\{X_n - X_m\}$, we obtain the inequality

$$\mathscr{E}|X - X_m| \leq \lim_{n\to\infty} \mathscr{E}|X_n - X_m| \qquad \text{for all } m \geq 1.$$

Letting $m \to \infty$, we obtain the required result. ∎

We have essentially proved the following result.

Theorem 1.3.1. Let $\mathscr{L}_1$ be the set of all random variables X with $\mathscr{E}|X| < \infty$. In this set, define $X = 0$ if and only if $X = 0$ a.s. Then $\mathscr{L}_1$ is a (real) Banach space with respect to the norm $\|X\|_1 = \mathscr{E}|X|$ for $X \in \mathscr{L}_1$.

Proof. It follows immediately from Proposition 1.3.7 that $\mathscr{L}_1$ is a complete metric space with respect to the metric ρ_1, defined by

$$\rho_1(X, Y) = \mathscr{E}|X - Y|, \qquad X, Y \in \mathscr{L}_1.$$

This completes the proof. ■

Example 1.3.3. In Example 1.3.1, $\mathscr{E}|X_n| = 0$ for every n, so that $X_n \xrightarrow{\mathscr{L}_1} 0$. Thus $X_n \xrightarrow{\mathscr{L}_1} X$ and $X_n \xrightarrow{\text{a.s.}} X$ are both possible. In Example 1.3.2, $\mathscr{E}|X_n| = 1/2^m \to 0$ as n, and hence m, $\to \infty$, so that $X_n \xrightarrow{\mathscr{L}_1} 0$ but $X_n \not\xrightarrow{\text{a.s.}} 0$.

Remark 1.3.5. Let X be a random variable. Then $\mathscr{E}|X| < \infty$ if and only if there exists a sequence $\{X_n\}$ of random variables in $\mathscr{L}_1$ which is Cauchy in mean and which converges to X in probability. Moreover, in that case $X_n \xrightarrow{\mathscr{L}_1} X$.

Remark 1.3.6. Let $X \in \mathscr{L}_1$. Then for every $\epsilon > 0$ there exists a simple random variable (simple function) $Y \in \mathscr{L}_1$ such that $\mathscr{E}|X - Y| < \epsilon$.

Finally we obtain the following necessary and sufficient condition for convergence in mean.

Proposition 1.3.8. Let $X_n \in \mathscr{L}_1$, $n = 1, 2, \ldots$. Then the sequence $\{X_n\}$ converges in mean to some random variable $X \in \mathscr{L}_1$ if and only if the following two conditions are satisfied:

(i) $X_n \xrightarrow{P} X$.
(ii) For every $\epsilon > 0$, there exists a $\delta = \delta(\epsilon) > 0$ (independent of n) such that the inequality

$$\int_E |X_n|\, dP < \epsilon$$

holds for every event E for which $P(E) < \delta$ and for all n.

Proof. First suppose that there exists an $X \in \mathscr{L}_1$ such that $X_n \xrightarrow{\mathscr{L}_1} X$. Clearly (i) holds. To prove (ii), we note that for $\epsilon > 0$ there exists an $n_0 = n_0(\epsilon) \geq 1$ such that, for all $n \geq n_0$, $\mathscr{E}|X_n - X_{n_0}| < \epsilon/2$. On the other hand, in view of the absolute continuity of the indefinite integral with respect to

P it follows that there exists a $\delta = \delta(\epsilon) > 0$ such that $\int_E |X_n|\, dP < \epsilon/2 < \epsilon$ for every $E \in \mathscr{S}$ with $P(E) < \delta$ and for $n \le n_0$. Next let $n > n_0$. Then

$$\begin{aligned}\int_E |X_n|\, dP &\le \int_E |X_n - X_{n_0}|\, dP + \int_E |X_{n_0}|\, dP \\ &< \mathscr{E}|X_n - X_{n_0}| + \int_E |X_{n_0}|\, dP \\ &< \epsilon,\end{aligned}$$

which proves (ii).

Conversely, suppose that (i) and (ii) hold. Let $\epsilon > 0$ and let $m, n \ge 1$. Let

$$E_{mn} = \{|X_n - X_m| \ge \epsilon\}.$$

Since $X_n \xrightarrow{P} X$, $P(E_{mn}) \to 0$ as $m, n \to \infty$. Therefore

$$\begin{aligned}\mathscr{E}|X_n - X_m| &= \int_{E_{mn}^c} |X_n - X_m|\, dP + \int_{E_{mn}} |X_n - X_m|\, dP \\ &< \epsilon + \sup_n \int_{E_{mn}} |X_n|\, dP + \sup_m \int_{E_{mn}} |X_m|\, dP. \qquad (1.3.3)\end{aligned}$$

For $\epsilon > 0$ let $\delta = \delta(\epsilon)$ be determined from (ii). Then for this δ there exists an $N = N(\delta)$ such that, for all $m, n \ge N$, $P(E_{mn}) < \delta$. It is now easily verified that $X_n \xrightarrow{\mathscr{L}_1} X$. ■

1.3.4 Convergence in the *p*th Mean

Let $(\Omega, \mathscr{S}, P)$ be a probability space, and let $1 \le p < \infty$. Denote by $\mathscr{L}_p = \mathscr{L}_p(\Omega, \mathscr{S}, P)$ the set of all random variables X on Ω such that $\mathscr{E}|X|^p < \infty$.

Definition 1.3.7. Let $X_n \in \mathscr{L}_p$, $n = 1, 2, \ldots$. We say that the sequence $\{X_n\}$ converges in the pth mean to a random variable $X \in \mathscr{L}_p$ if $\mathscr{E}|X_n - X|^p \to 0$ as $n \to \infty$. In this case write $X_n \xrightarrow{\mathscr{L}_p} X$, and note that the limiting random variable X, if it exists, is unique a.s.

Definition 1.3.8. The sequence $\{X_n\}$ is said to be Cauchy in the pth mean if $\mathscr{E}|X_n - X_m|^p \to 0$ as $m, n \to \infty$.

Clearly $X_n \xrightarrow{\mathscr{L}_p} X \Rightarrow X_n \xrightarrow{P} X$ and $\{X_n\}$ Cauchy in the pth mean $\Rightarrow \{X_n\}$ Cauchy in probability.

Example 1.3.4. In Example 1.3.1 the sequence $\{X_n\}$ converges in the pth mean to zero for any $p \ge 1$. It also converges a.s. to zero.

Proposition 1.3.9. The sequence $\{X_n\}$, $X_n \in \mathscr{L}_p$, is Cauchy in the pth mean if and only if it converges in the pth mean to some random variable $X \in \mathscr{L}_p$.

Proof. Clearly convergence in the pth mean implies Cauchy convergence in the pth mean.

To prove the converse we note that the space $\mathscr{L}_p$ is a normed linear space with respect to the norm $\|X\|_p = \{\mathscr{E}|X|^p\}^{1/p}$, where we set the null vector $X = 0$ if and only if $X = 0$ a.s. To establish that $\mathscr{L}_p$ is complete with respect to the metric ρ_p, defined by

$$\rho_p(X, Y) = \mathscr{E}\|X - Y\|_p, \qquad X, Y \in \mathscr{L}_p,$$

it is sufficient to show that every absolutely convergent series of elements in $\mathscr{L}_p$ converges to some element in $\mathscr{L}_p$. Let $X_n \in \mathscr{L}_p$ be such that $\sum_{n=1}^{\infty} \|X_n\|_p = \gamma < \infty$. For every $N \geq 1$ define the random variable Y_N on Ω by

$$Y_N = \sum_{n=1}^{N} |X_n|.$$

In view of Minkowski's inequality $Y_N \in \mathscr{L}_p$, and, moreover,

$$\|Y_N\|_p \leq \sum_{n=1}^{N} \|X_n\|_p \leq \gamma \qquad \text{for every } N \geq 1,$$

so that

$$\mathscr{E} Y_N^p \leq \gamma^p.$$

Hence

$$\lim_{N\to\infty} \mathscr{E} Y_N^p \leq \gamma^p.$$

Set $Y(\omega) = \lim_{N\to\infty} Y_N(\omega)$, $\omega \in \Omega$. It follows at once from Fatou's lemma that $Y^p \in \mathscr{L}_1$ and $\mathscr{E}Y^p \leq \gamma^p$. Thus $Y \in \mathscr{L}^p$, and, moreover, $\|Y\|_p < \gamma$. Now set $E = \{\omega \in \Omega: Y(\omega) = +\infty\}$. Then clearly $P(E) = 0$. For every $N \geq 1$ and for every $\omega \in E^c$ define $Z_N(\omega) = \sum_{n=1}^{N} X_n(\omega)$. Then $|Z_N(\omega)| \leq Y(\omega)$, $\omega \in E^c$, and for every $\omega \in E^c$ the series $\sum_{n=1}^{\infty} X_n(\omega)$ converges absolutely to $Y(\omega)$ and hence converges (pointwise) to some real number $Z(\omega)$. For every $\omega \in E$ define $Z(\omega) = 0$. It follows that the sequence $\{Z_n\}$ converges a.s. to Z on Ω. Moreover, $|Z| \leq Y$ a.s. Therefore $|Z|^p \in \mathscr{L}_1$, and hence $Z \in \mathscr{L}_p$.

On the other hand, for every $N \geq 1$ we have

$$|Z_N - Z|^p \leq 2^p Y^p \text{ a.s.}$$

It follows from the Lebesgue dominated convergence theorem that $\mathscr{E}|Z_N - Z|^p \to 0$ as $N \to \infty$, which, in turn, implies that $\|Z_N - Z\|_p \to 0$.

Hence we conclude that the series $\sum_{n=1}^{\infty} X_n$ converges to the element $Z \in \mathscr{L}_p$. This completes the proof of Proposition 1.3.9. ∎

We have essentially established the following result.

Theorem 1.3.2. Let $1 \le p < \infty$, and let $\mathscr{L}_p$ be the set of all random variables X with $\mathscr{E}|X|^p < \infty$. In this set, define the null vector $X = 0$ if and only if $X = 0$ a.s. Then $\mathscr{L}_p$ is a (real) Banach space with respect to the norm $\|X\|_p = [\mathscr{E}|X|^p]^{1/p}$ for $X \in \mathscr{L}_p$.

Remark 1.3.7. For the case $p = 1$, Proposition 1.3.9 provides an alternative proof of Proposition 1.3.7.

Remark 1.3.8. For the case $p = 2$, $\mathscr{L}_2$ is a real Hilbert space with respect to the inner product $\langle X, Y \rangle = \mathscr{E}(XY)$, $X, Y \in \mathscr{L}_2$.

Remark 1.3.9. Let $1 \le p < \infty$, and let $X \in \mathscr{L}_p$. Let $\epsilon > 0$. Then there exists a simple random variable $Y \in \mathscr{L}_p$ such that $\mathscr{E}|X - Y|^p < \epsilon$.

1.4 INDEPENDENCE

In this section we study a fundamental concept that is peculiar to probability theory, namely, the concept of independence. Let $(\Omega, \mathscr{S}, P)$ be a probability space which is fixed but otherwise arbitrary. Unless specified otherwise, all random variables and random vectors are defined on this space. Let T be an index set. Indices t vary on T, and events of a class have the index of the class.

Definition 1.4.1. Let $\mathscr{E}_t = \{E_t\}$ be a family of events. We say that $\mathscr{E}_t$ is a family of independent events or, simply, that events $\{E_t\}$ are independent if for every finite subset $\{t_1, t_2, \ldots, t_n\}$ of T

$$P\left(\bigcap_{j=1}^{n} E_{t_j}\right) = \prod_{j=1}^{n} P(E_{t_j}). \tag{1.4.1}$$

In a similar manner we speak of *independent classes of events.* Classes $\{\mathscr{E}_t\}$ of events are said to be independent if events chosen arbitrarily, one from each class, form an independent family of events. Thus $\mathscr{E}_t$ and $\mathscr{E}_{t'}$ are independent if, for every $E_t \in \mathscr{E}_t$ and every $E_{t'} \in \mathscr{E}_{t'}$, we have

$$P(E_t \cap E_{t'}) = P(E_t)P(E_{t'}).$$

Remark 1.4.1. Let $\mathscr{E}_t$ be a family of independent events. Then it is clear that, if $\mathscr{E}'_{t'} \subset \mathscr{E}_t$, $t' \in T' \subset T$, then $\{\mathscr{E}'_{t'}\}$ are also independent, that is, subclasses of independent classes are independent.

We next consider the concept of independence of random variables. Let $\{X_t\}$ be random variables, and let $X_t^{-1}(\mathscr{B})$ be the sub-σ-field of events induced by X_t in the space $(\Omega, \mathscr{S}, P)$.

Definition 1.4.2. The family of random variables $\{X_t, t \in T\}$ is said to be independent if $\{X_t^{-1}(\mathscr{B}), t \in T\}$ forms a family of independent classes of events.

Since subclasses of independent classes are independent and since a Borel-measurable function of X_t induces a sub-σ-field of events contained in $X_t^{-1}(\mathscr{B})$, the following result is immediate.

Theorem 1.4.1. Let $\{X_t, t \in T\}$ be independent random variables. For every $t \in T$, let g_t be a Borel-measurable function defined on $\mathbb{R}$. Then $\{g_t(X_t), t \in T\}$ is also a family of independent random variables.

Remark 1.4.2. The independence of a family of random vectors can be defined in a manner analogous to Definition 1.4.2. A random vector version of Theorem 1.4.1 holds. We leave the reader to furnish the details.

As an immediate consequence of Definition 1.4.2 we obtain the following criterion for independence of random variables.

Theorem 1.4.2. The family $\{X_t, t \in T\}$ of random variables is independent if and only if, for every finite collection $\{E_{t_1}, E_{t_2}, \ldots, E_{t_n}\}$ of Borel sets in $\mathbb{R}$,

$$P\left(\bigcap_{j=1}^{n} \{X_{t_j} \in E_{t_j}\}\right) = \prod_{j=1}^{n} P\{X_{t_j} \in E_{t_j}\}. \tag{1.4.2}$$

Theorem 1.4.3. Let X and Y be two independent random variables. If $\mathscr{E}|X| < \infty$ and $\mathscr{E}|Y| < \infty$, then $\mathscr{E}|XY| < \infty$, and

$$\mathscr{E}(XY) = \mathscr{E}X\mathscr{E}Y. \tag{1.4.3}$$

Conversely, if $\mathscr{E}|XY| < \infty$ and neither X nor Y is degenerate at zero, then $\mathscr{E}|X| < \infty$ and $\mathscr{E}|Y| < \infty$.

Proof. We first consider the case where both X and Y are (a.s.) nonnegative simple random variables. Then we can write

$$X = \sum_{i=1}^{n} \alpha_i \chi_{E_i} \qquad \text{and} \qquad Y = \sum_{j=1}^{m} \beta_j \chi_{F_j},$$

where the E_i's are disjoint events and so are the F_j's. Also, we may assume that the α_i's and β_j's are all distinct, so that

$$E_i = \{\omega \in \Omega: X(\omega) = \alpha_i\}, \qquad i = 1, 2, \ldots, n,$$

and

$$F_j = \{\omega \in \Omega: Y(\omega) = \beta_j\}, \qquad j = 1, 2, \ldots, m.$$

Since X and Y are independent, the classes $\{E_1, E_2, \ldots, E_n\}$ and $\{F_1, F_2, \ldots, F_m\}$ are independent, so that

$$P(E_i \cap F_j) = P(E_i)P(F_j)$$

for $i = 1, 2, \ldots, n$, and $j = 1, 2, \ldots, m$. In this case XY is also a simple random variable which can be written as

$$XY = \sum_{i=1}^{n} \sum_{j=1}^{m} \alpha_i \beta_j \chi_{E_i \cap F_j}.$$

Hence

$$\begin{aligned}
\mathscr{E}XY &= \int_{\Omega} XY \, dP \\
&= \sum_{i=1}^{n} \sum_{j=1}^{m} \alpha_i \beta_j P(E_i \cap F_j) \\
&= \left[\sum_{i=1}^{n} \alpha_i P(E_i)\right]\left[\sum_{j=1}^{m} \beta_j P(F_j)\right] \\
&= \mathscr{E}X\mathscr{E}Y.
\end{aligned}$$

Next we consider the case where X and Y are nonnegative random variables. For every $n \geq 1$ and every $\omega \in \Omega$, we define

$$X_n(\omega) = \begin{cases} \dfrac{i-1}{2^n} & \text{if } \dfrac{i-1}{2^n} \leq X(\omega) < \dfrac{i}{2^n}, \quad i = 1, 2, \ldots, n2^n, \\ n & \text{if } X(\omega) \geq n, \end{cases}$$

and

$$Y_n(\omega) = \begin{cases} \dfrac{j-1}{2^n} & \text{if } \dfrac{j-1}{2^n} \leq Y(\omega) < \dfrac{j}{2^n}, \quad j = 1, 2, \ldots, n2^n, \\ n & \text{if } Y(\omega) \geq n. \end{cases}$$

Then $\{X_n\}$ and $\{Y_n\}$ are sequences of simple random variables such that $0 \le X_n \uparrow X$ and $0 \le Y_n \uparrow Y$. Clearly $0 \le X_n Y_n \uparrow XY$. Since X and Y are independent, we can verify easily that, for every $n \ge 1$, X_n and Y_n are also independent, and it follows that

$$\mathscr{E}(X_n Y_n) = \mathscr{E}X_n \mathscr{E}Y_n \text{ for every } n.$$

On the other hand, in view of the monotone convergence theorem we have

$$\lim_{n\to\infty} \mathscr{E}X_n = \mathscr{E}X, \qquad \lim_{n\to\infty} \mathscr{E}Y_n = \mathscr{E}Y, \qquad \lim_{n\to\infty} \mathscr{E}(X_n Y_n) = \mathscr{E}(XY).$$

Therefore

$$\mathscr{E}XY = \mathscr{E}X\mathscr{E}Y.$$

Finally, suppose that X and Y are arbitrary independent random variables. Then the nonnegative random variables X^+ and X^- are independent of Y^+ and Y^-, and, moreover,

$$\mathscr{E}(X^+Y^+) = \mathscr{E}X^+\mathscr{E}Y^+, \qquad \mathscr{E}(X^+Y^-) = \mathscr{E}X^+\mathscr{E}Y^-,$$
$$\mathscr{E}(X^-Y^+) = \mathscr{E}X^-\mathscr{E}Y^+, \qquad \mathscr{E}(X^-Y^-) = \mathscr{E}X^-\mathscr{E}Y^-.$$

Now, if $\mathscr{E}X$, $\mathscr{E}Y$ exist, so do $\mathscr{E}X^+$, $\mathscr{E}X^-$, $\mathscr{E}Y^+$, $\mathscr{E}Y^-$ and hence also $\mathscr{E}X^+Y^+$, $\mathscr{E}X^+Y^-$, etc. Therefore $\mathscr{E}|XY| < \infty$, and we have

$$\begin{aligned}\mathscr{E}XY &= \mathscr{E}(X^+ - X^-)(Y^+ - Y^-)\\ &= \mathscr{E}(X^+ - X^-)\mathscr{E}(Y^+ - Y^-)\\ &= \mathscr{E}X\mathscr{E}Y.\end{aligned}$$

Conversely, suppose that $\mathscr{E}|XY| < \infty$. Clearly $|X| \ge 0$ a.s., $|Y| \ge 0$ a.s., and $|X|$ and $|Y|$ are independent. Hence

$$\mathscr{E}|X|\mathscr{E}|Y| = \mathscr{E}|XY| < \infty.$$

Since neither X nor Y is degenerate at zero, it follows that $\mathscr{E}|X| < \infty$ and $\mathscr{E}|Y| < \infty$. This completes the proof of Theorem 1.4.3. ∎

The above result can easily be extended as follows.

Theorem 1.4.4. Let $X_1, X_2, \ldots, X_n$ be independent random variables, and let $g_1, g_2, \ldots, g_n$ be Borel-measurable functions on $\mathbb{R}$ such that $\mathscr{E}|g_j(X_j)| < \infty$ for $j = 1, 2, \ldots, n$. Then $\mathscr{E}\prod_{j=1}^n |g_j(X_j)| < \infty$, and, moreover, the relation

$$\mathscr{E}\left[\prod_{j=1}^n g_j(X_j)\right] = \prod_{j=1}^n \mathscr{E}g_j(X_j) \tag{1.4.4}$$

holds. Conversely, if $\mathscr{E}\prod_{j=1}^n |g_j(X_j)| < \infty$, then $\mathscr{E}|g_j(X_j)| < \infty$ for $j = 1, 2, \ldots, n$ [provided that none of the $g_j(X_j)$'s is degenerate at zero], and relation (1.4.4) holds.

Corollary 1. Let X and Y be two independent random variables such that $\text{var}(X) < \infty$, $\text{var}(Y) < \infty$. Then the following conditions hold:

(i) $\text{cov}(X, Y) = 0$
(ii) $\text{var}(X + Y) = \text{var}(X) + \text{var}(Y)$.

Corollary 2. Let X and Y be two independent random variables, and let φ, φ_1, and φ_2 be the characteristic functions of $X + Y$, X, and Y, respectively. Then

$$\varphi(t) = \varphi_1(t)\varphi_2(t), \qquad t \in \mathbb{R}. \tag{1.4.5}$$

Corollary 3. Let X and Y be two independent random variables, and let φ, φ_1, and φ_2 be the characteristic functions of (X, Y), X, and Y, respectively. Then

$$\varphi(t_1, t_2) = \varphi_1(t_1)\varphi_2(t_2), \qquad t_1, t_2 \in \mathbb{R}. \tag{1.4.6}$$

Remark 1.4.3. In Corollary 1 it is sufficient to assume that $\text{cov}(X, Y) = 0$ for (ii) to hold. It is clear that $\text{cov}(X, Y) = 0$, in general, does not imply independence of X and Y. The converse of Corollary 2 does not hold, but that of Corollary 3 does. This result will be obtained in Chapter 3.

Finally we obtain the following useful criterion for independence of random variables in terms of their distribution functions.

Theorem 1.4.5. Let X and Y be two random variables, and let F, F_1, and F_2 be the distribution functions of (X, Y), X, and Y, respectively. Then X and Y are independent if and only if

$$F(x, y) = F_1(x)F_2(y), \qquad x \in \mathbb{R}, \quad y \in \mathbb{R}. \tag{1.4.7}$$

Proof. Suppose that X and Y are independent. Then it follows from (1.4.2) that

$$P\{X \in E_1, Y \in E_2\} = P\{X \in E_1\}P\{Y \in E_2\} \tag{1.4.8}$$

holds for all $E_1, E_2 \in \mathscr{B}$ and, in particular, for $E_1 = (-\infty, x]$ and $E_2 = (-\infty, y]$. Hence (1.4.7) holds.

Next suppose that (1.4.7) holds. Let $y \in \mathbb{R}$ be fixed, and let $\mathfrak{M}$ be the class of all subsets E of $\mathbb{R}$ with $X^{-1}(E) \in \mathscr{S}$ for which the relation

$$P\{X \in E, Y \le y\} = P\{X \in E\}P\{Y \le y\} \tag{1.4.9}$$

holds. We will show that $\mathscr{B} \subset \mathfrak{M}$. Clearly, $\mathfrak{M}$ contains all intervals of the form $(x_1, x_2]$, $(-\infty, x_1]$, (x_2, ∞), and $\mathbb{R}$. Also, $\mathfrak{M}$ is closed under countable union of disjoint sets. Indeed, let $E_n \in \mathfrak{M}$ be a countable sequence of disjoint sets. Let $E = \bigcup_{n=1}^{\infty} E_n$. Then

$$P\{X \in E, Y \le y\} = \sum_{n=1}^{\infty} P\{X \in E_n, Y \le y\}$$
$$= P(E)P\{Y \le y\},$$

so that $E \in \mathfrak{M}$. Now we note that $\mathfrak{M}$ is a monotone class in view of the continuity of probability measures and relation (1.4.9). Let $\mathscr{F}$ be the class of all finite unions of disjoint sets of the form $(x_1, x_2]$, $(-\infty, x_1]$, (x_2, ∞) and $\mathbb{R}$. Then $\mathscr{F}$ is a field of subsets of $\mathbb{R}$; moreover, $\mathscr{F} \subset \mathfrak{M}$. Let $\mathfrak{M}(\mathscr{F})$ be the monotone class generated by $\mathscr{F}$. Clearly $\mathfrak{M}(\mathscr{F}) \subset \mathfrak{M}$. Let $\mathscr{S}(\mathscr{F})$ be the σ-field generated by $\mathscr{F}$. Then $\mathfrak{M}(\mathscr{F}) = \mathscr{S}(\mathscr{F}) = \mathscr{B}$, and it follows that $\mathscr{B} \subset \mathfrak{M}$.

Next, we let $E \in \mathscr{B}$ be fixed. We repeat the same argument with Y, and using relation (1.4.9) we get (1.4.8). This completes the proof. ■

Corollary 1. Let X and Y be two discrete random variables with mass concentrated, respectively, in $\{x_i\}$ and $\{y_j\}$. Then X and Y are independent if and only if the relation

$$P\{X = x_i, Y = y_j\} = P\{X = x_i\}P\{Y = y_j\} \tag{1.4.10}$$

holds for all x_i, y_j.

Corollary 2. Let (X, Y) be a random vector with an absolutely continuous joint distribution function that has absolutely continuous marginal distribution functions. Let f, f_1, and f_2, respectively, be the probability density functions of (X, Y), X, and Y. Then X and Y are independent if and only if

$$f(x, y) = f_1(x) f_2(y), \qquad x, y \in \mathbb{R}. \tag{1.4.11}$$

Remark 1.4.4. Theorem 1.4.5 and its corollaries can easily be extended to any finite number of random variables or random vectors.

1.5 PROBLEMS

SECTION 1.1

1. (a) Let $\mathscr{S}_1 \subset \mathscr{S}_2 \subset \ldots$ be a denumerable sequence of σ-fields of subsets of Ω. (Here the inclusion is proper.) Construct an example to show that $\bigcup_{n=1}^{\infty} \mathscr{S}_n$ need not be a σ-field.

   (b) Prove the following much stronger result. Let $\mathscr{S}_1 \subset \mathscr{S}_2 \subset \ldots$ be a denumerable sequence of σ-fields. Then $\bigcup_{n=1}^{\infty} \mathscr{S}_n$ cannot be a σ-field.

   (Broughton and Huff [10])

2. Let $\mathcal{S}$ be the σ-field generated by a countable collection of disjoint sets $\{E_n\}$. Show that any $E \in \mathcal{S}$ can be represented as a union of a countable subcollection of the E_n.

3. Let Ω be a nonempty set, and $\mathcal{S}$ be a σ-field of subsets of Ω. Let P be a finite non-negative set function defined on $\mathcal{S}$. Show that P is countably additive $\Leftrightarrow$ P is finitely additive and, if $E_n \downarrow \emptyset$, then $P(E_n) \to 0$.

4. Let $(\Omega, \mathcal{S}, P)$ be a probability space. Establish the following results.

(a) (Principle of Inclusion-Exclusion) If $A_1, A_2, \ldots, A_n \in \mathcal{S}$, then

$$P\left(\bigcup_{k=1}^{n} A_k\right) = \sum_{k=1}^{n} P(A_k) - \sum_{k_1<k_2}^{n} P(A_{k_1} \cap A_{k_2}) + \cdots + (-1)^{n+1} P\left(\bigcap_{k=1}^{n} A_k\right).$$

(b) (Bonferroni's Inequality) Given $n(>1)$ events, $A_1, A_2, \ldots, A_n$,

$$\sum_{i=1}^{n} P(A_i) - \sum_{i<j} P(A_i \cap A_j) \le P\left(\bigcup_{i=1}^{n} A_i\right) \le \sum_{i=1}^{n} P(A_i).$$

5. Let X be a random variable defined on $(\Omega, \mathcal{S})$. Then the σ-field induced by X, denoted by $\sigma(X)$, is defined to be the σ-field generated by the collection of sets

$$\{[\omega \in \Omega: X(\omega) \le x], x \in \mathbb{R}\}.$$

Show that, if g is a Borel-measurable function on $\mathbb{R}$ and $Y = g(X)$, then $\sigma(Y) \subset \sigma(X)$.

6. Let $(\Omega, \mathcal{S}, P)$ be a probability space. An event $A \in \mathcal{S}$ is said to be an atom if $P(A) > 0$ and, for $B \in \mathcal{S}$ with $B \subset A$, either $P(B) = 0$ or $P(B) = P(A)$. Show that Ω can be written as $\Omega = A \cup \bigcup_{k=1}^{\infty} A_k$, where $A, A_1, A_2, \ldots$ are disjoint events, each A_k is either an atom or empty, and A has the property that, given $B \in \mathcal{S}$, $B \subset A$, and any ϵ, $0 < \epsilon < P(B)$, there exists a $C \in \mathcal{S}$ with $P(C) = \epsilon$. This decomposition of Ω is unique except for P-null sets. Show also that random variables are constant on atoms.

7. Construct an example of a discrete distribution function F such that every real number is a point of increase of F.

8. Let X be a random variable with a continuous distribution function F. Show that the random variable $F(X)$ has a uniform distribution on $[0, 1]$. Conversely, let F be a distribution function, and let X be a random variable whose distribution function is uniform on $[0, 1]$. Show that there exists a Borel-measurable function h such that F is the distribution function of $h(X)$.

9. Let F be the joint distribution function of a random vector $\mathbf{X} = (X_1, X_2, \ldots, X_n)$. Show that F is continuous everywhere except over the union of a countable set of hyperplanes $x_j = c$, $j = 1, 2, \ldots, n$.

10. Show that the function f defined in Example 1.1.12 is a joint probability density function. Find the marginal density function of $(X_1, X_2, \ldots, X_{n-1})$.

11. Let f_1, f_2, f_3 be three probability density functions with corresponding distribution functions F_1, F_2, F_3, and let α be a constant, $|\alpha| \le 1$. Show that the function

$$f_\alpha(x_1, x_2, x_3) = \left[\prod_{i=1}^{3} f_i(x_i)\right]\left\{1 + \alpha \prod_{i=1}^{3}[2F_i(x_i) + 1]\right\}$$

defines a joint probability density function for each $\alpha \in [-1, 1]$. Moreover, f_1, f_2, and f_3 are the marginal probability density functions of f_α for every α.

SECTION 1.2

12. For the random variables with the following distributions compute the means, the variances, and the characteristic functions.

(a) $P\{X = k\} = \binom{n}{k} p^k (1-p)^{n-k}$, $k = 0, 1, \ldots, n$, $0 \le p \le 1$.
(b) $P\{X = k\} = e^{-\lambda}\lambda^k/k!$, $k = 0, 1, 2, \ldots$, $\lambda > 0$.
(c) $f(x) = 1/(b-a)$ if $x \in (a, b)$, and $= 0$ otherwise.
(d) $f(x) = (2\pi\sigma^2)^{-1/2} e^{-(x-\mu)^2/2\sigma^2}$, $x \in \mathbb{R}$, $\sigma^2 > 0$, and $\mu \in \mathbb{R}$.
(e) $f(x) = [\Gamma(\alpha)\beta^\alpha]^{-1} x^{\alpha-1} e^{-x/\beta}$ if $x > 0$, and $= 0$ otherwise, where $\alpha > 0$, $\beta > 0$.
(f) $f(x) = [B(\alpha, \beta)]^{-1} x^{\alpha-1}(1-x)^{\beta-1}$ if $x \in (0, 1)$, and $= 0$ otherwise, where $\alpha > 0$, $\beta > 0$.

[In (c) through (f) f defines the probability density function of the random variable X.]

13. In Problem 12 parts (d), (e), and (f) show that moments of all order exist. In each case compute $\mathscr{E}|X|^\gamma$, $\gamma > 0$.

14. For the Cauchy law given by the probability density function

$$f(x) = \frac{\mu}{\pi}[\mu^2 + (x - \theta)^2]^{-1}, \qquad x \in \mathbb{R},$$

where $\mu > 0$ and $\theta \in \mathbb{R}$ are constants, show that moments of order $\alpha \ge 1$ do not exist but those of order $\alpha < 1$ do. Compute the corresponding characteristic function.

15. Show that the Markov's inequality cannot be improved unless some additional restrictions are imposed.

16. (a) Show that all members of the family of distribution functions defined by

$$F_\lambda(x) = \left(\int_0^\infty x^{-\ln x}\,dx\right)^{-1} \int_0^\infty x^{-\ln x}[1 - \lambda \sin(4\pi \ln x)]\,dx$$

have the same set of moments.

(b) Do the same for probability density functions

$$f_\lambda(x) = ke^{-|x|^t}\{1 + \lambda \sin[|x|^t \tan(\pi t)]\}, \qquad x \in \mathbb{R}, \quad 0 < t < 1, \quad |\lambda| \le 1.$$

Here k is the appropriate constant.

17. Is the Minkowski inequality valid for $p < 1$?

18. When does equality hold in the Lyapounov inequality?

19. (a) (Riesz) Let X be a random variable with distribution function F, and let g be a strictly convex function on the range of F. Let φ be a Borel-measurable function. Suppose that $\mathscr{E}X$, $\mathscr{E}\varphi(X)$, and $\mathscr{E}g(X)$ all exist, and write $\mathscr{E}X = \mu$. Let $t(x) = g(\mu) + k(x - \mu)$ be a line of support for g at $x = \mu$. Also, let $h(x) = g(x) - t(x)$. Then for every $\epsilon > 0$ the inequality

$$\mathscr{E}|\varphi(X)| \leq \sup_{|x-\mu|<\epsilon} |\varphi(x)| + \left[\sup_{|x-\mu|\geq\epsilon} \frac{|\varphi(x)|}{h(x)}\right]\mathscr{E}h(X)$$

holds.

(b) Derive Markov's inequality from Riesz's inequality.

(c) Show that for $a > 0$ and $\epsilon > 0$

$$P\{|X - \mu| > \epsilon\} \leq Me^{-a\mu}(\mathscr{E}e^{aX} - e^{a\mu}),$$

where

$$M = (e^{-a\epsilon} + a\epsilon - 1)^{-1}.$$

20. Let X be a random variable satisfying

$$\frac{P(|X| > \alpha k)}{P(|X| > k)} \to 0 \qquad \text{as } k \to \infty \quad \text{for } \alpha > 1.$$

Show that $\mathscr{E}|X|^\nu < \infty$ for all $\nu > 0$,

21. Show that the function $\ln \mathscr{E}|X|^r$ is a convex function of r provided that $\mathscr{E}|X|^r < \infty$.

22. Show that a random variable X possesses moments of all orders if and only if

$$\limsup_{n\to\infty} [P\{|X| > \alpha^n\}]^{1/n} = 0 \qquad \text{for } \alpha > 1.$$

23. Let X be a random variable and $0 < r < 1$. Suppose that $nP\{|X| > n^{1/r}\} \to 0$ as $n \to \infty$. Show that

$$n^{1-1/r} \int_{|x|<n^{1/r}} x \, dP\{X \leq x\} \to 0 \qquad \text{as } n \to \infty.$$

24. (a) Let X be a random variable with zero mean and finite variance σ^2. Show that

$$P\{X > x\} \leq \frac{\sigma^2}{\sigma^2 + x^2} \qquad \text{if } x > 0$$

and

$$P\{X > x\} \leq \frac{x^2}{\sigma^2 + x^2} \qquad \text{if } x < 0.$$

(b) Let $\mathscr{E}|X|^4 < \infty$, and let $\mathscr{E}X = 0$, $\mathscr{E}X^2 = \sigma^2$. Then for $k > 1$ show that

$$P\{|X| \geq k\sigma\} \geq \frac{\mu_4 - \sigma^4}{\mu_4 + \sigma^4 k^4 - 2k^2\sigma^4},$$

where $\mu_4 = \mathscr{E}X^4$.

(c) Show that the inequalities in (a) and (b) cannot be improved.

SECTION 1.3

25. Let $\{\mathbf{X}^{(n)} = (X_{1n}, X_{2n}, \ldots, X_{kn}), n \geq 1\}$ be a sequence of k-dimensional random vectors. We say that the sequence $\{\mathbf{X}^{(n)}\}$ converges a.s. to a k-dimensional random vector $\mathbf{X}$ if $P\{\mathbf{X}^{(n)} \to \mathbf{X} \text{ as } n \to \infty\} = 1$, and in that case we write $\mathbf{X}^{(n)} \xrightarrow{\text{a.s.}} \mathbf{X}$. Writing $|\mathbf{x}| = (\sum_{j=1}^{k} x_j^2)^{1/2}$, we say that the sequence $\{\mathbf{X}^{(n)}\}$ converges in probability to $\mathbf{X}$ if $P\{|\mathbf{X}^{(n)} - \mathbf{X}| \geq \epsilon\} \to 0$ as $n \to \infty$ for every $\epsilon > 0$, and we write $\mathbf{X}^{(n)} \xrightarrow{P} \mathbf{X}$. One can obtain analogues of the one-dimensional results proved in the text. In particular, show the following:

(a) $\mathbf{X}^{(n)} \xrightarrow{\text{a.s.}} \mathbf{X} \Rightarrow \mathbf{X}^{(n)} \xrightarrow{P} \mathbf{X}$.

(b) $\mathbf{X}^{(n)} \xrightarrow{P} \mathbf{X} \Rightarrow$ there exists a subsequence $\{\mathbf{X}^{(n_k)}\}$ of $\{\mathbf{X}^{(n)}\}$ such that $\mathbf{X}^{(n_k)} \xrightarrow{\text{a.s.}} \mathbf{X}$.

(c) $\mathbf{X}^{(n)} \xrightarrow{P} \mathbf{X} \Leftrightarrow$ every $\{\mathbf{X}^{(n_k)}\} \subset \{\mathbf{X}^{(n)}\}$ contains a subsequence $\{X^{(n_k')}\}$ such that $\mathbf{X}^{(n_k')} \xrightarrow{\text{a.s.}} \mathbf{X}$.

(d) $\mathbf{X}^{(n)} \xrightarrow{P} \mathbf{X}, \boldsymbol{g}: \mathbb{R}_k \to \mathbb{R}_l, 1 \leq l \leq k$, continuous $\Rightarrow \boldsymbol{g}(\mathbf{X}^n) \xrightarrow{P} \boldsymbol{g}(\mathbf{X})$.

(e) Construct an example to show that in part (*d*) one cannot relax the continuity of g to Borel measurability.

26. Show that $X_n \xrightarrow{\text{a.s.}} X$ implies that there exists a subsequence $\{X_{n_k}\}$ of $\{X_n\}$ such that $X_{n_k} \to X$ completely.

27. (a) Show that the following statements are equivalent:

(1) Ω is the finite union of disjoint atoms.
(2) $X_n \xrightarrow{P} X \Leftrightarrow X_n \xrightarrow{\mathscr{L}_p} X$ for some (all) $p > 0$.
(3) $X_n \xrightarrow{\text{a.s.}} X \Leftrightarrow X_n \xrightarrow{\mathscr{L}_p} X$ for all $p > 0$.
(4) $X_n \to X$ completely $\Leftrightarrow X_n \xrightarrow{\mathscr{L}_p} X$ for all $p > 0$.

[Here $X, X_n \in \mathfrak{X}$, where $\mathfrak{X}$ is the space of all random variables defined on $(\Omega, \mathscr{S}, P)$ such that $X, Y \in \mathfrak{X} \Rightarrow P\{X \neq Y\} > 0$.]

(b) Prove the result stated in Remark 1.3.4.

28. Let $\{X_n\}$ be a sequence of random variables defined on $(\Omega, \mathscr{S}, P)$. We say that $\{X_n\}$ converges to a random variable X uniformly a.s. if there exists a P-null event E such that, for every $\epsilon > 0$, one can find an integer $N = N(\epsilon)$ with the property that

$$|X_n(\omega) - X(\omega)| < \epsilon \qquad \text{for } n \geq N, \quad \omega \notin E.$$

In a similar manner we can define a uniformly Cauchy (fundamental) a.s. sequence of random variables. Show that a sequence $\{X_n\}$ of random variables converges uniformly a.s. to some random variable X if and only if $\{X_n\}$ is uniformly Cauchy a.s.

29. (a) Let $\{X_n\}$ be a sequence of random variables defined on $(\Omega, \mathscr{S}, P)$ and $X_n \xrightarrow{\text{a.s.}} X$. According to Egoroff's theorem, for every $\epsilon > 0$ there exists an $E \in \mathscr{S}$ with $P(E^c) < \epsilon$ such that $\{X_n\}$ converges uniformly on E. Motivated by this result, we say that $\{X_n\}$ converges almost uniformly (a.u.) to X and write $X_n \xrightarrow{\text{a.u.}} X$ if, for each $\epsilon > 0$, there is an $E \in \mathscr{S}$ with $P(E^c) < \epsilon$ such that $X_n \xrightarrow{\text{a.u.}} X$ on E. Clearly uniform convergence a.s. implies a.u. convergence. Show that the converse is not true.

(b) Show that a.u. convergence $\Rightarrow$ convergence in probability.

(c) Show that, if $\{X_n\}$ is Cauchy in probability, there is a subsequence $\{X_{n_k}\}$ that is a.u. convergent.

(d) Show that a.u. convergence $\Rightarrow$ a.s. convergence.

30. Let $\{p_n\}$ be a sequence of probability density functions. Suppose that $p_n(x) \to p(x)$ as $n \to \infty$. If p is a probability density function, show that

$$\int_{-\infty}^{\infty} |p_n(x) - p(x)| dx \to 0 \qquad \text{as } n \to \infty.$$

31. Let $0 < p < 1$, and let $\mathscr{L}_p \subset \mathfrak{X}$ ($\mathfrak{X}$ as defined in Problem 27) be the subspace of all random variables with $\mathscr{E}|X|^p < \infty$. Show that $\rho(X, Y) = \mathscr{E}|X - Y|^p$, $X, Y \in \mathscr{L}_p$, defines a metric and $\mathscr{L}_p$, $0 < p < 1$, is a metric space.

32. Prove the result stated in Remark 1.3.3.

33. Prove the statement in Remark 1.3.9.

34. Let $\{X_n\}$ be a sequence of random variables satisfying

$$X_1 > X_2 > X_3 > \cdots > 0 \quad \text{a.s.}$$

Then show that $X_n \xrightarrow{P} 0 \Rightarrow X_n \xrightarrow{\text{a.s.}} 0$.

35. Let $\{X_n\}$ be a sequence of random variables such that $X_n \xrightarrow{\text{a.s.}} X$. Let P_n and P be the probability distributions of X_n and X, respectively, $n \geq 1$. Construct an example to show that $P_n(A) \not\to P(A)$ as $n \to \infty$ for all $A \in \mathscr{B}$.

36. Let $\mathscr{L}$ be a real normed linear space with norm $\|\cdot\|$ Then a mapping $l: x \to l(x)$ of $\mathscr{L}$ into $\mathbb{R}$ is said to be a functional on $\mathscr{L}$. A function l is said to be linear if it satisfies $l(\alpha x + \beta y) = \alpha l(x) + \beta l(y)$ for all $x, y \in \mathscr{L}$ and $\alpha, \beta \in \mathbb{R}$. A functional l is said to be bounded if there exists a $\gamma > 0$ such that $|l(x)| \leq \gamma \|x\|$, $x \in \mathscr{L}$. In this case we define the norm of l by

$$\|l\| = \sup_{x \in \mathscr{L}} \frac{|l(x)|}{\|x\|}.$$

Let $1 < p < \infty$, and let l be a bounded linear functional on $\mathscr{L}_p = \mathscr{L}_p(\Omega, \mathscr{S}, P)$, the space of all random variables X on $(\Omega, \mathscr{S}, P)$ with $\mathscr{E}|X|^p < \infty$. Show that there exists a random variable $Y \in \mathscr{L}_q$, where $q = p/(p-1)$ (determined uniquely by l), such that $l(X) = \mathscr{E}(XY)$ for all $X \in \mathscr{L}_p$. Moreover, in this case $\|l\| = \|Y\|_q$. The result also holds for $p = 1$ if we define $\mathscr{L}_\infty = \mathscr{L}_\infty(\Omega, \mathscr{S}, P)$ as the space of all a.s. bounded random variables defined on Ω. In this case $\mathscr{L}_\infty$ is a real Banach space with norm $\|X\| = \text{ess. sup}|X|$.

37. Let $X_n \in \mathscr{L}_p$, $n \geq 1$. Then show that

$$X_n \xrightarrow{\mathscr{L}_p} X \Leftrightarrow X_n \xrightarrow{P} X \qquad \text{and} \qquad \mathscr{E}|X_n|^p \to \mathscr{E}|X|^p < \infty.$$

38. If the X_n are a.s. uniformly bounded and $X_n \xrightarrow{P} X$, show that

$$X_n \xrightarrow{\mathscr{L}_p} X \text{ for any } p.$$

39. If $X_n \xrightarrow{\mathscr{L}_p} X$, show that $X_n \xrightarrow{\mathscr{L}_s} X$ for $s \leq p$.

SECTION 1.4

40. Let X and Y be two jointly distributed random variables such that

$$\varphi(t) = \varphi_1(t)\varphi_2(t), \qquad t \in \mathbb{R},$$

where φ, φ_1, and φ_2 are the characteristic functions of $X + Y$, X, and Y, respectively. Construct an example to show that X and Y need not be independent.

41. Let X and Y be independent with distribution functions F and G, respectively. Find the distribution functions of $X + Y$, XY, and X/Y (provided that $P\{Y = 0\} = 0$) in terms of F and G.

42. Let X and Y be independent random variables, and suppose that $\mathscr{E}|X + Y|^r < \infty$ for some $r > 0$. Then show that $\mathscr{E}|X|^r < \infty$, $\mathscr{E}|Y|^r < \infty$.

43. Let X and Y be independent random variables such that X and $X - Y$ are independent. Then show that X must be degenerate.

44. If X and Y are independent so that XY is degenerate at $c \neq 0$, show that X and Y must also be degenerate.

45. Let (X, Y) be a bivariate normal random variable with means 0, variances 1, and covariance ρ. Show that X^2 and Y^2 are independent if and only if $\rho = 0$.

46. Let X and Y be independent, and suppose that at least one of X and Y has a probability density function. Show that $X + Y$ also has a probability density function, that is, show that the distribution function of $X + Y$ is absolutely continuous. What if X and Y are not independent?

47. Let X be a random variable defined on $(\Omega, \mathscr{S}, P)$. The concentration function $Q(X; l)$ of X is defined by the relation

$$Q(X; l) = \sup_{x \in \mathbb{R}} P\{x \leq X \leq x + l\}$$

for every $l \geq 0$. Clearly Q is a nondecreasing function of l satisfying $0 \leq Q(x; l) \leq 1$ for all $l \geq 0$.

(a) Show that there exists an x_l such that

$$Q(X; l) = P\{x_l \leq X \leq x_l + l\}$$

and that the distribution function of X is continuous if and only if $Q(X; 0) = 0$.

(b) Let X and Y be independent random variables. Show that

$$Q(X + Y; l) \leq \min\{Q(X; l), Q(Y; l)\} \text{ for all } l \geq 0.$$

48. Let $\Omega = [0, 1]$, $\mathscr{S}$ be the Borel σ-field of subsets of Ω, and P be the Lebesgue measure on $(\Omega, \mathscr{S})$. Define $R_0(\omega) = 1$ if $0 \leq \omega \leq \frac{1}{2}$, and $= -1$ if $\frac{1}{2} < \omega \leq 1$. For $1 < \omega < \infty$ define R_n to be periodic with period 1. Thus

$$R_n(\omega) = R_0(2^n\omega), \qquad \omega \in [0, 1], \quad n \geq 1.$$

The sequence $\{R_n\}$ is called a sequence of Rademacher functions. Another representation of R_n is the following:

$$R_n(\omega) = \operatorname{sgn} \sin(2^{n+1}\pi\omega), \qquad \omega \in [0, 1], \quad n \geq 0,$$

where sgn $x = 1$ if $x > 0$, $= 0$ if $x = 0$, and $= -1$ if $x < 0$. Clearly $\operatorname{sgn} \sin(2\pi\omega) = R_0(\omega)$, $0 \leq \omega < \infty$, so that $R_n(\omega) = R_0(2^n\omega)$. Yet another representation can be given in terms of the binary expansion of numbers in [0, 1].

Show that $\{R_n\}$ is a sequence of independent random variables all with common distribution function F, given by

$$F(x) = 0 \quad \text{if } x < -1, \qquad = \tfrac{1}{2} \text{ if } -1 \leq x < 1, \qquad \text{and} \qquad = 1 \text{ if } 1 \leq x < \infty.$$

NOTES AND COMMENTS

Throughout this book we follow the axiomatic approach to probability theory which was systematically developed by Kolmogorov [48]. A key feature of this approach is the countable additivity or, equivalently, the continuity property of the probability measure. A somewhat less successful approach which uses only the concept of finite additivity has also been followed by various authors. In particular, we refer to Dunford and Schwartz [19], Chapter III, Sections 1 through 3, Dubins and Savage [18], and De Finetti [16]. A different approach based on the concept of conditional probability is followed by Rényi [68]. Yet another approach to the subject would be to start with the axiomatic definition of an expectation operator. See, for example, Whittle [87], Chapter 2. In this development functional analysis plays a more dominant role than measure theory. Finally we note that the classical approach to probability through relative frequency is due to Von Mises [86].

For a detailed treatment of the measure-theoretic concepts used in this book we refer to Halmos [31]. For related results in real analysis we recommend Royden [71], Hewitt and Stromberg [33], and Kolmogorov and Fomin [49, 50]. It is desirable, though not necessary, that the student be already familiar with the elementary concepts of probability and statistics at the level of Feller [24], Fisz [26], and Rohatgi [70].

CHAPTER 2

The Laws of Large Numbers

In probability theory we study limit theorems which are of the following two types:

(a) Strong limit theorems. These deal with the a.s. convergence of a sequence of random variables.
(b) Weak limit theorems. These deal with the convergence in probability of a sequence of random variables, as well as convergence of a sequence of distribution functions.

In this chapter, for the most part, we shall study some important strong limit theorems of probability theory.

We begin with the following example. Consider n independent tossings of a coin with constant probability p of success (heads). Let r be the frequency of heads, so that r/n is the relative frequency of heads in n trials. For sufficiently large n it is reasonable to expect that r/n will be close to p. The laws of large numbers make this statement mathematically precise. According to the weak law of large numbers, $r/n \xrightarrow{P} p$; that is, if the number of trials is very large, the probability that r/n differs from p is very small. A stronger assertion is made by the strong law of large numbers, namely, that $r/n \xrightarrow{\text{a.s.}} p$. The law of the iterated logarithm, on the other hand, deals with the growth rate of r as $n \to \infty$.

2.1 THE WEAK LAW OF LARGE NUMBERS

Let $(\Omega, \mathscr{S}, P)$ be a probability space, and let $\{X_n\}$ be a sequence of random variables defined on it. We now study the weak law of large numbers, which deals with conditions of convergence in probability of the sequence of partial sums $S_n = \sum_{k=1}^{n} X_k$, $n = 1, 2, \ldots$.

Theorem 2.1.1 (Chebyshev's Weak Law of Large Numbers). Let $\{X_n\}$ be a sequence of independent random variables with $\mathscr{E}X_n^2 < \infty$, $n \geq 1$. Suppose there exists a $\gamma > 0$ such that $\text{var}(X_n) \leq \gamma$ for all n. Set $S_n = \sum_{k=1}^{n} X_k$, $n \geq 1$. Then

$$\frac{S_n - \mathscr{E}S_n}{n} \xrightarrow{P} 0 \qquad \text{as } n \to \infty.$$

Proof. Let $\epsilon > 0$. Then from Chebyshev's inequality we obtain

$$P\left\{\left|\frac{S_n - \mathscr{E}S_n}{n}\right| \geq \epsilon\right\} \leq \frac{\text{var}(S_n)}{n^2\epsilon^2}$$

$$\leq \frac{\gamma}{n\epsilon^2} \to 0 \qquad \text{as } n \to \infty.$$

This proves Theorem 2.1.1. ■

Remark 2.1.1. Let $\{X_n\}$ be a sequence of random variables, and $\{A_n\}$ be a sequence of constants. A statement of the type

$$\frac{S_n - A_n}{n} \xrightarrow{P} 0 \qquad \text{as } n \to \infty$$

is called a *weak law of large numbers.* Sometimes one seeks sequences of constants $\{A_n\}$ and $\{B_n\}$, $B_n > 0$ and $B_n \to \infty$ as $n \to \infty$, such that

$$B_n^{-1}(S_n - A_n) \xrightarrow{P} 0 \qquad \text{as } n \to \infty.$$

Here we consider only the case $B_n = n$. In Theorem 2.1.3 we will show that, when the X_n are independent and identically distributed with common mean μ, the weak law of large numbers holds if we choose $A_n = n\mu$.

We next consider an important particular case of Theorem 2.1.1. For this purpose we need the following definition.

Definition 2.1.1. Two random variables X and Y are said to be identically distributed if their probability distributions P_X and P_Y coincide on $\mathscr{B}$.

Clearly X and Y are identically distributed if and only if their distribution functions coincide on $\mathbb{R}$. Moreover, in this case, for any real-valued Borel-measureable function g on $\mathbb{R}$ which is integrable with respect to P_X the relation

$$\mathscr{E}g(X) = \mathscr{E}g(Y)$$

holds.

Corollary to Theorem 2.1.1. Let $\{X_n\}$ be a sequence of independent and identically distributed random variables with common mean μ and variance $\sigma^2 < \infty$. Then

$$\frac{S_n}{n} \xrightarrow{P} \mu \qquad \text{as } n \to \infty.$$

Remark 2.1.2. It is clear from the proof of Theorem 2.1.1 that all we need is the condition that $\operatorname{var}(S_n)/n^2 \to 0$ as $n \to \infty$. The condition of independence is not necessary.

We next give a necessary and sufficient condition for convergence in probability.

Theorem 2.1.2. Let $\{Y_n\}$ be an arbitrary sequence of random variables. Then $Y_n \xrightarrow{P} 0$ if and only if

$$\mathscr{E}\frac{|Y_n|^r}{1 + |Y_n|^r} \to 0 \qquad \text{for some } r > 0.$$

Proof. Let $\epsilon > 0$. According to (1.2.11), we have

$$\mathscr{E}\frac{|Y_n|^r}{1 + |Y_n|^r} - \frac{\epsilon^r}{1 + \epsilon^r} \leq P\{|Y_n| \geq \epsilon\} \leq \frac{1 + \epsilon^r}{\epsilon^r}\,\mathscr{E}\frac{|Y_n|^r}{1 + |Y_n|^r}.$$

The proof follows immediately on letting $n \to \infty$. ■

Corollary. Let $\{X_n\}$ be an arbitrary sequence of random variables. Then

$$\frac{S_n - \mathscr{E}S_n}{n} \xrightarrow{P} 0 \qquad \text{as } n \to \infty$$

if and only if

$$\mathscr{E}\frac{(S_n - \mathscr{E}S_n)^2}{n^2 + (S_n - \mathscr{E}S_n)^2} \to 0 \qquad \text{as } n \to \infty.$$

We note that Theorem 2.1.1, its corollary, and the result in Remark 2.1.2 all follow from Theorem 2.1.2.

Finally we prove the following stronger result for the case of independent and identically distributed random variables, which is due to Khintchine.

Theorem 2.1.3. Let $\{X_n\}$ be a sequence of independent and identically distributed random variables with common mean μ. Then

$$\frac{S_n}{n} \xrightarrow{P} \mu \qquad \text{as } n \to \infty.$$

Proof. For the proof we use the method of truncation, which we shall discuss in detail in the next section. Let $\delta > 0$, and let $n \geq 1$ be a fixed integer. Then for $1 \leq k \leq n$ define

$$X_k^* = \begin{cases} X_k & \text{if } |X_k| < \delta n, \\ 0 & \text{otherwise.} \end{cases}$$

Then the X_k^* are also independent and identically distributed. Let X be a random variable which has the same distribution as each X_n. Then $\mathscr{E}X = \mu$. We write $\mathscr{E}|X| = \beta < \infty$. Define

$$E_n = \{\omega \in \Omega : |X(\omega)| < \delta n\}, \qquad n \geq 1.$$

Then for $1 \leq k \leq n$

$$\mu_n^* = \mathscr{E}X_k^* = \int_\Omega X_k^* \, dP = \int_{E_n} X \, dP = \mathscr{E}(X\chi_{E_n}).$$

Also,

$$\begin{aligned} \operatorname{var}(X_k^*) &= \mathscr{E}X_k^{*2} - (\mathscr{E}X_k^*)^2 \\ &\leq \mathscr{E}X_k^{*2} \\ &= \int_{E_n} X^2 \, dP \\ &\leq \delta n \int_{E_n} |X| \, dP \\ &\leq \delta\beta n. \end{aligned}$$

Write $S_n^* = \sum_{k=1}^n X_k^*$. Then $\mathscr{E}S_n^* = n\mu_n^*$, and

$$\operatorname{var}(S_n^*) \leq n(\delta\beta n) = \beta\delta n^2,$$

so that

$$\frac{\mathscr{E}S_n^*}{n} = \mu_n^* \qquad \text{and} \qquad \operatorname{var}\left(\frac{S_n^*}{n}\right) \leq \beta\delta.$$

Applying Chebyshev's inequality to S_n^*/n, we conclude that

(2.1.1) $$P\left\{\left|\frac{S_n^*}{n} - \mu_n^*\right| \geq \epsilon\right\} \leq \frac{\beta\delta}{\epsilon^2} \qquad \text{for every } \epsilon > 0.$$

Also, using the Lebesgue dominated convergence theorem, we conclude that $\mu_n^* \to \mu$ as $n \to \infty$. For given $\epsilon > 0$ there exists a positive integer $N^*(\epsilon) = N^*$ such that for all $n \geq N^*$

$$|\mu_n^* - \mu| < \epsilon.$$

This, together with (2.1.1), implies that for all $n \geq N^*$ we have

$$P\left\{\left|\frac{S_n^*}{n} - \mu\right| \geq 2\epsilon\right\} \leq \frac{\beta\delta}{\epsilon^2}. \tag{2.1.2}$$

For $1 \leq k \leq n$ we write $Y_k = X_k - X_k^*$. Then

$$\begin{aligned} P\{Y_k \neq 0\} &= P\{|X_k| \geq \delta n\} = P\{|X| \geq \delta n\} \\ &\leq \frac{1}{\delta n}\int_{E_n^c} |X|\, dP, \end{aligned}$$

so that

$$\begin{aligned} P\left\{\sum_{k=1}^{n} Y_k \neq 0\right\} &\leq \sum_{k=1}^{n} P\{Y_k \neq 0\} \\ &\leq \frac{1}{\delta}\int_{E_n^c} |X|\, dP. \end{aligned}$$

It follows that

$$P\left\{\frac{S_n - S_n^*}{n} \neq 0\right\} \leq \frac{1}{\delta}\int_{E_n^c} |X|\, dP. \tag{2.1.3}$$

Combining (2.1.2) and (2.1.3), we obtain for all $n \geq N^*$

$$P\left\{\left|\frac{S_n}{n} - \mu\right| \geq 2\epsilon\right\} \leq \frac{\beta\delta}{\epsilon^2} + \frac{1}{\delta}\int_{E_n^c} |X|\, dP.$$

We note that $\int_{E_n^c} |X|\, dP \to 0$ as $n \to \infty$. Hence for a given $\delta > 0$ there exists an $N^{**} > 0$ such that for all $n \geq N^{**}$

$$\int_{E_n^c} |X|\, dP < \delta^2.$$

We set $N = \max(N^*, N^{**})$. Then for all $n \geq N$ we have

$$P\left\{\left|\frac{S_n}{n} - \mu\right| \geq 2\epsilon\right\} \leq \frac{\beta\delta}{\epsilon^2} + \delta.$$

Since $\delta > 0$ and $\epsilon > 0$ are arbitrary, it follows that $S_n/n \xrightarrow{P} \mu$ as $n \to \infty$ and the proof is complete. ■

2.2 BASIC TECHNIQUES

Here we discuss some important tools of probability (limit) theory which we shall use in our subsequent investigation.

2.2.1 Borel–Cantelli Lemma

We begin with the proof of the Borel–Cantelli lemma.

Proposition 2.2.1 (Borel–Cantelli Lemma). Let $(\Omega, \mathscr{S}, P)$ be a probability space, and let $\{E_n\}$ be a sequence of events.

(a) If $\sum_{n=1}^{\infty} P(E_n) < \infty$, then $P(\limsup_n E_n) = 0$.
(b) If, in addition, the E_n are independent, then $P(\limsup_n E_n) = 0$ or $= 1$ according as the series $\sum_{n=1}^{\infty} P(E_n)$ converges or diverges.

Proof. (a) Set $E = \limsup_n E_n = \bigcap_{n=1}^{\infty} \bigcup_{m=n}^{\infty} E_m$. Then we can write $E = \bigcap_{n=1}^{\infty} F_n$, where $F_n = \bigcup_{m=n}^{\infty} E_m$. Now for every positive integer n

$$P(F_n) \leq \sum_{m=n}^{\infty} P(E_m) \to 0 \qquad \text{as } n \to \infty.$$

Moreover, $F_n \downarrow E$ as $n \to \infty$, so that $P(E) = \lim_{n \to \infty} P(F_n) = 0$, as asserted.

(b) Suppose that the E_n are independent. From part (a) it follows that $\sum_{n=1}^{\infty} P(E_n) < \infty \Rightarrow P(\limsup_n E_n) = 0$. It remains to show that $\sum_{n=1}^{\infty} P(E_n) = \infty \Rightarrow P(\limsup_n E_n) = 1$. As before, set $E = \limsup_n E_n$. Then

$$E^c = \liminf_n E_n^c = \bigcup_{n=1}^{\infty} \bigcap_{m=n}^{\infty} E_m^c.$$

Note that the E_m^c are also independent. Let $N, n (N > n)$ be two positive integers. Then

$$\begin{aligned} P\left(\bigcap_{m=n}^{\infty} E_m^c\right) &\leq P\left(\bigcap_{m=n}^{N} E_m^c\right) \\ &= \prod_{m=n}^{N} [1 - P(E_m)] \\ &\leq \exp\left\{-\sum_{m=n}^{N} P(E_m)\right\}. \end{aligned}$$

Taking the limit as $N \to \infty$, we have

$$P\left(\bigcap_{m=n}^{\infty} E_m^c\right) \leq \exp\left\{-\sum_{m=n}^{\infty} P(E_m)\right\} \qquad \text{for every } n.$$

Since $\sum_{n=1}^{\infty} P(E_n) = \infty$, it follows that $P(\bigcap_{m=n}^{\infty} E_m^c) = 0$ for every n. This implies that $P(E^c) = 0$, and hence $P(E) = 1$. ■

Corollary. Let $\{E_n\}$ be an independent sequence of events. Then

$$P\left(\limsup_n E_n\right) = 0 \Leftrightarrow \sum_{n=1}^{\infty} P(E_n) < \infty.$$

Corollary 2. Let $\{X_n\}$ be a sequence of independent random variables. Then $X_n \xrightarrow{\text{a.s.}} 0$ if and only if $\sum_{n=1}^{\infty} P\{|X_n| \geq \epsilon\} < \infty$ for every $\epsilon > 0$.

Proof. The proof of Corollary 2 follows immediately from Proposition 2.2.1 by setting $E_n = \{|X_n| \geq \epsilon\}$ and from Corollary 1 to Proposition 1.3.2.

Remark 2.2.1. In view of the definition of the limit superior of a sequence of events, it is customary to write

$$\limsup_n E_n = E_n \text{ i.o.}$$

where i.o. is an abbreviation for "infinitely often." Thus, if the E_n are independent events, $P(E_n \text{ i.o.}) = 0$ if and only if $\sum_{n=1}^{\infty} P(E_n) < \infty$. Similarly, if $\{X_n\}$ is a sequence of independent random variables, $X_n \xrightarrow{\text{a.s.}} 0$ if and only if $P\{|X_n| \geq \epsilon \text{ i.o.}\} = 0$ for every $\epsilon > 0$.

Example 2.2.1. Let A_n denote the event that a head turns up on both the nth and $(n + 1)$st toss of a fair coin. Let $A = \limsup_n A_n$. Then A is the event that two successive heads will appear i.o. in repeated tossings of a fair coin. Clearly $P(A_n) = \frac{1}{4}$, so that $\sum_{n=1}^{\infty} P(A_{2n}) = \infty$. Since, however, $\{A_{2n}\}$ is an independent sequence of events, it follows that $P(A) = 1$. Note that the A_n are not independent.

Example 2.2.2. Let $\{X_n\}$ be a sequence of random variables, and suppose that $X_n \xrightarrow{P} 0$ as $n \to \infty$. We now obtain an alternative proof of Proposition 1.3.4 as a simple application of the Borel–Cantelli lemma, that is, we show that $\{X_n\}$ contains a subsequence $\{X_{n_k}\}$ such that $X_{n_k} \xrightarrow{\text{a.s.}} 0$ as $k \to \infty$. Let $\epsilon > 0$. Then $P\{|X_n| \geq \epsilon\} \to 0$, and we can choose positive integers $n_k \to \infty$ such that $P\{|X_{n_k}| \geq 2^{-k}\} \leq 2^{-k}$ for each $k \geq 1$. Hence $\sum_{k=1}^{\infty} P\{|X_{n_k}| \geq 2^{-k}\} < \infty$. Choosing $E_k = P\{|X_{n_k}| \geq 2^{-k}\}$ in Proposition 2.2.1(a), we see immediately (Remark 2.2.1) that $X_{n_k} \xrightarrow{\text{a.s.}} 0$ as $k \to \infty$.

Remark 2.2.2. The converse of Proposition 2.2.1(a) does not hold if we drop the assumption of independence. For example, let $\Omega = [0, 1]$, $\mathscr{S}$ be the Borel σ-field of subsets of Ω, and P be the Lebesgue measure. Let $E_n = (0, 1/n)$, $n = 1, 2, \ldots$. Then $E_n \downarrow$ and $\limsup_n E_n = \bigcap_{n=1}^{\infty} E_n = \varnothing$, so that $P(E_n \text{ i.o.}) = 0$ but $\sum_{n=1}^{\infty} P(E_n) = \sum_{n=1}^{\infty} (1/n) = \infty$. Here the E_n are clearly dependent.

Part (b) of Proposition 2.2.1 can be extended in several directions when the assumption of independence is dropped. Needless to say, some additional condition(s) will then have to be imposed. We consider one such extension in the following proposition.

Proposition 2.2.2. Let $\{E_n\}$ be any sequence of events such that $\sum_{n=1}^{\infty} P(E_n) = \infty$ and

$$\liminf_{n\to\infty} \frac{\sum_{j=1}^{n}\sum_{k=1}^{n} P(E_j \cap E_k)}{[\sum_{j=1}^{n} P(E_j)]^2} = 1.$$

Then $P(\limsup_n E_n) = 1$.

Proof. Let χ_{E_n} be the indicator function of E_n. Then $\mathscr{E}\chi_{E_n} = P(E_n)$, and by Chebyshev's inequality we have

$$P\left\{\left|\sum_{k=1}^{n} \chi_{E_k} - \sum_{k=1}^{n} P(E_k)\right| \geq \epsilon \sum_{k=1}^{n} P(E_k)\right\} \leq \frac{\operatorname{var}(\sum_{k=1}^{n} \chi_{E_k})}{\epsilon^2[\sum_{j=1}^{n} P(E_j)]^2}.$$

Now

$$\operatorname{var}\left(\sum_{k=1}^{n} \chi_{E_k}\right) = \sum_{j=1}^{n}\sum_{k=1}^{n} P(E_j \cap E_k) - \left[\sum_{j=1}^{n} P(E_j)\right]^2,$$

and it follows from the hypothesis that

$$\liminf_{n\to\infty} P\left\{\left|\sum_{k=1}^{n} \chi_{E_k} - \sum_{k=1}^{n} P(E_k)\right| \geq \frac{1}{2}\sum_{k=1}^{n} P(E_k)\right\} = 0.$$

Hence

$$\liminf_{n\to\infty} P\left\{\sum_{k=1}^{n} \chi_{E_k} < \frac{1}{2}\sum_{k=1}^{n} P(E_k)\right\} = 0.$$

It follows that there exists a sequence of integers $n_1 < n_2 < \cdots$ such that

$$\sum_{j=1}^{\infty} P\left\{\sum_{k=1}^{n_j} \chi_{E_k} < \frac{1}{2}\sum_{k=1}^{n_j} P(E_k)\right\} < \infty.$$

From part (a) of the Borel-Cantelli lemma we conclude that with probability 1

$$\sum_{k=1}^{n_j} \chi_{E_k} \geq \frac{1}{2}\sum_{k=1}^{n_j} P(E_k)$$

except for a finite number of values of j. In view of our hypothesis that $\sum_{k=1}^{\infty} P(E_k) = \infty$, it follows that $\sum_{k=1}^{\infty} \chi_{E_k}$ diverges with probability 1. Thus $P(\limsup_n E_n) = 1$ as asserted. ∎

Corollary. Let $\{E_n\}$ be pairwise independent, that is, suppose that $P(E_j \cap E_k) = P(E_j)P(E_k)$ for all $j, k, j \neq k$. In addition, assume that $\sum_{n=1}^{\infty} P(E_n) = \infty$. Then $P(\limsup_n E_n) = 1$.

Proof. In view of pairwise independence

$$\frac{\sum_{j=1}^n \sum_{k=1}^n P(E_j \cap E_k)}{[\sum_{k=1}^n P(E_k)]^2}$$

$$= \frac{\sum_{j=1}^n \sum_{k=1}^n P(E_j)P(E_k) + \sum_{j=1}^n P(E_j)}{[\sum_{k=1}^n P(E_k)]^2}$$

$$= \frac{\sum_{j=1}^n \sum_{k=1, j \neq k}^n P(E_j)P(E_k) + \sum_{j=1}^n P(E_j)[1 - P(E_j)]}{[\sum_{k=1}^n P(E_k)]^2}.$$

Since, however,

$$0 \le \sum_{j=1}^n P(E_j)[1 - P(E_j)] \le \sum_{j=1}^n P(E_j),$$

it follows that

$$\liminf_{n\to\infty} \frac{\sum_{j=1}^n \sum_{k=1}^n P(E_j \cap E_k)}{[\sum_{j=1}^n P(E_j)]^2} = 1.$$

2.2.2 Tail Events, Tail σ-Fields, and Tail Functions

Let $\{X_n\}$ be a sequence of independent random variables defined on $(\Omega, \mathscr{S}, P)$. An interesting problem in probability is the following: What is the probability that the series $\sum_{i=1}^\infty X_i$ converges? A remarkable result that we shall prove is that this probability must be either 0 or 1. The same property is shared by many other events associated with independent random variables. For a precise formulation of this result, known as the zero-one law, we need some preliminaries.

Let X be a random variable on $(\Omega, \mathscr{S}, P)$, and let $\sigma(X)$ be the smallest σ-field induced by X, that is, $\sigma(X)$ is the σ-field of subsets of Ω generated by sets of the form $\{\omega: X(\omega) \le x\}$, $x \in \mathbb{R}$. Clearly $\sigma(X) \subset \mathscr{S}$. Next, let $X_1, X_2, \ldots, X_n$ be a finite number of random variables defined on $(\Omega, \mathscr{S}, P)$. Write $\sigma(X_1, X_2, \ldots, X_n)$ for the σ-field generated by the class of sets of the form $\{\omega \in \Omega: X_1(\omega) \le x_1, X_2(\omega) \le x_2, \ldots, X_n(\omega) \le x_n\}$, where $x_i \in \mathbb{R}$, $i = 1, 2, \ldots, n$. Clearly

$$\sigma(X_{i_1}, X_{i_2}, \ldots, X_{i_k}) \subset \sigma(X_1, X_2, \ldots, X_n)$$

whenever $\{i_1, i_2, \ldots, i_k\} \subset \{1, 2, \ldots, n\}$. It is easy to verify that

$$\sigma(X_1, X_2, \ldots, X_n) = \sigma\left(\bigcup_{k=1}^n \sigma(X_k)\right).$$

Finally, suppose $\{X_n\}$ is an infinite sequence of random variables defined on $(\Omega, \mathscr{S}, P)$. Let $n \geq 1$, and consider the sequence of σ-fields

$$\sigma(X_n) \subset \sigma(X_n, X_{n+1}) \subset \sigma(X_n, X_{n+1}, X_{n+2}) \subset \cdots.$$

Clearly $\bigcup_{k=1}^{\infty} \sigma(X_n, X_{n+1}, \ldots, X_{n+k})$ is a field. Let $\sigma(X_n, X_{n+1}, \ldots)$ denote the σ-field generated by $\bigcup_{k=1}^{\infty} \sigma(X_n, X_{n+1}, \ldots, X_{n+k})$. Since the sequence of σ-fields

$$\sigma(X_n, X_{n+1}, \ldots), \sigma(X_{n+1}, X_{n+2}, \ldots), \ldots$$

is nonincreasing, the limit of this sequence exists and is a σ-field. We write

$$\mathscr{T} = \bigcap_{n=1}^{\infty} \sigma(X_n, X_{n+1}, \ldots).$$

Then $\mathscr{T}$ is known as the *tail σ-field* of the sequence $\{X_n\}$. An event $E \in \mathscr{T}$ is called a *tail event*, and any function on Ω measurable with respect to $\mathscr{T}$ is known as a *tail function*.

Remark 2.2.3. Intuitively, a tail function does not depend on any finite segment of the sequence $\{X_n\}$, that is, its value is not affected by changing a finite number of the X_n. Similarly, a tail event is one whose probability remains unaltered when a finite number of the X_n are changed.

Some important examples of tail functions are the following:

$$\liminf_{n\to\infty} X_n, \qquad \limsup_{n\to\infty} X_n, \qquad \liminf_{n\to\infty} \frac{1}{n}\sum_{k=1}^{n} X_k, \qquad \limsup_{n\to\infty} \frac{1}{n}\sum_{k=1}^{n} X_k.$$

Proposition 2.2.3. Let $\{X_n\}$ be a sequence of independent random variables. Then, for any $n \geq 1$, $\sigma(X_1, X_2, \ldots, X_n)$ and $\sigma(X_{n+1}, X_{n+2}, \ldots)$ are independent classes of events.

Proof. We set $E_n = \{X_1 \leq x_1, X_2 \leq x_2, \ldots, X_n \leq x_n\}$ and

$$F_{n,m} = \{X_{n+1} \leq x_{n+1}, X_{n+2} \leq x_{n+2}, \ldots, X_{n+m} \leq x_{n+m}\},$$

where $x_1, x_2, \ldots, x_{n+m}$ are real numbers, $n \geq 1$, $m \geq 1$. Since the X_n are independent,

$$P(E_n \cap F_{n,m}) = P(E_n)P(F_{n,m}). \tag{2.2.1}$$

For a fixed E_n, let $\mathscr{C}_{n,m}$ be the class of all sets G for which

$$P(E_n \cap G) = P(E_n)P(G).$$

Clearly $\mathscr{C}_{n,m}$ is a σ-field which contains $\sigma(X_{n+1}, X_{n+2}, \ldots, X_{n+m})$. It follows that (2.2.1) holds for all

$$G_{n,m} \in \sigma(X_{n+1}, X_{n+2}, \ldots, X_{n+m}).$$

Since $\sigma(X_{n+1}, X_{n+2}, \ldots)$ is generated by $\bigcup_{m=1}^{\infty} \sigma(X_{n+1}, X_{n+2}, \ldots, X_{n+m})$, it follows that

$$P(E_n \cap F_n) = P(E_n)P(F_n) \tag{2.2.2}$$

for all $F_n \in \sigma(X_{n+1}, X_{n+2}, \ldots)$. This is accomplished by taking an increasing sequence of sets $\{G_{n,m}, m = 1, 2, \ldots\}$. Thus (2.2.2) holds for each fixed E_n of the form $\{X_1 \le x_1, X_2 \le x_2, \ldots, X_n \le x_n\}$ and every

$$F_n \in \sigma(X_{n+1}, X_{n+2}, \ldots).$$

By a similar argument it is easy to show that (2.2.2) holds for every set $E_n \in \sigma(X_1, X_2, \ldots, X_n)$ and $F_n \in \sigma(X_{n+1}, X_{n+2}, \ldots)$, that is, the σ-fields $\sigma(X_1, X_2, \ldots, X_n)$ and $\sigma(X_{n+1}, X_{n+2}, \ldots)$ are independent. ■

We are now ready to prove the following important result due to Kolmogorov.

Theorem 2.2.1 (Kolmogorov's Zero-One Law). Let $\{X_n\}$ be a sequence of independent random variables. Then the probability of any tail event is either 0 or 1. Moreover, any tail function is a.s. constant, that is, if Y is a random variable such that $\sigma(Y) \subset \mathscr{T}$, then $Y = c$ a.s., where c is a constant.

Proof. Let $\mathscr{T}$ be the tail σ-field of $\{X_n\}$, and let $A \in \mathscr{T}$. Then

$$A \in \sigma(X_{n+1}, X_{n+2}, \ldots), \qquad n = 0, 1, 2, \ldots,$$

and it follows from Proposition 2.2.3 that

$$P(A \cap E_n) = P(A)P(E_n)$$

for every $E_n \in \sigma(X_1, X_2, \ldots, X_n)$ and all n. Hence

$$P(A \cap E) = P(A)P(E)$$

for every $E \in \sigma(X_1, X_2, \ldots)$. Since $E \in \mathscr{T}$, $E \in \sigma(X_1, X_2, \ldots)$, and by taking $E = A$ we see that

$$P(A) = [P(A)]^2,$$

that is, $P(A) = 0$ or 1.

To complete the proof of the theorem, let Y be a tail function. Then $\{Y \le y\}$ is a tail event for each $y \in \mathbb{R}$ and $F(y) = P\{Y \le y\}$ equals either 0 or 1. Since F is nondecreasing,

$$F(y) = 0 \qquad \text{for some } y = y_1 \Rightarrow F(y) = 0 \quad \text{for } y < y_1$$

and

$$F(y) = 1 \qquad \text{for some } y = y_2 \Rightarrow F(y) = 1 \quad \text{for } y > y_2.$$

It follows that there exists a constant c such that $F(y) = 0$ for $y < c$, and $F(y) = 1$ for $y \geq c$. ∎

Corollary. Let $\{X_n\}$ be a sequence of independent random variables. Then the following assertions hold:

(a) $\{X_n\}$ either converges a.s. to a finite limit or diverges a.s.
(b) $\sum X_n$ either converges a.s. to a finite limit or diverges a.s.
(c) If $b_n \uparrow \infty$ as $n \to \infty$, the sequence $\{b_n^{-1} \sum_{k=1}^{n} X_k\}$ either converges a.s. to a finite limit or diverges a.s.

Moreover, if the sequence in part (a) or (c) converges a.s., the limit is a constant a.s.

Proof. We will prove part (a) and leave completion of the proofs of (b) and (c) to the reader. Let $A = \{\limsup_n X_n(\omega) = \infty\}$. Then, for each $\omega \in A$, $\limsup_n X_n(\omega) > x$ for all $x \in \mathbb{R}$. Since $A \in \mathscr{T}$, $P(A) = 0$ or 1. If $P(A) = 1$, then X_n diverges a.s. If, on the other hand, $P(A) = 0$, then $\limsup_n X_n(\omega) < \infty$, and in that case let $B = \{\liminf_n X_n(\omega) = -\infty\}$. Clearly $P(B) = 0$ or 1. If $P(B) = 1$, then X_n diverges a.s. to $-\infty$. If $P(A) = 0$ and $P(B) = 0$, we must have

$$-\infty < \liminf_n X_n(\omega) < \limsup_n X_n(\omega) < \infty \qquad \text{a.s.}$$

Since, however, $\liminf_n X_n$ and $\limsup_n X_n$ are tail functions, each of them is a constant a.s. Also, $\{\liminf_n X_n(\omega) \neq \limsup_n X_n(\omega)\} \in \mathscr{T}$, so that

$$P\left\{\liminf_n X_n(\omega) \neq \limsup_n X_n(\omega)\right\} = 0 \qquad \text{or} \qquad 1$$

It follows that, if this last probability is zero, the X_n converge a.s. to a finite limit; if not, the X_n diverge a.s.

The Kolmogorov zero-one law applies to any sequence of independent random variables $\{X_n\}$ but not to the sequence of its partial sums $\{S_n\}$, $S_n = \sum_{k=1}^{n} X_k$, $n \geq 1$. In general, tail events (and tail functions) on $\{S_n\}$ are not tail events (and tail functions) on $\{X_n\}$, and the Kolmogorov zero-one law does not apply. If, however, the X_n are, in addition, identically distributed, then the S_n are *exchangeable* and a much wider class of events has probability 0 or 1.

Definition 2.2.1. Let $X_1, X_2, \ldots, X_n$ be random variables defined on a probability space $(\Omega, \mathscr{S}, P)$. We say that $X_1, X_2, \ldots, X_n$ are exchangeable or symmetrically dependent if, for every one of the $n!$ permutations $\{i_1, i_2, \ldots, i_n\}$ of $\{1, 2, \ldots, n\}$, the random vectors $(X_1, X_2, \ldots, X_n)$ and $(X_{i_1}, X_{i_2}, \ldots, X_{i_n})$ have the same distribution. An infinite sequence of random variables $\{X_n\}$ is said to be exchangeable if for each $n \geq 1$ the random variables $X_1, X_2, \ldots, X_n$ are exchangeable. A measurable function $g(X_1, X_2, \ldots)$ is exchangeable if it is invariant under all finite permutations of its arguments:

$$g(X_1, X_2, \ldots, X_n, \ldots) = g(X_{i_1}, X_{i_2}, \ldots, X_{i_n}, X_{n+1}, \ldots).$$

In particular, an event based on $(X_1, X_2, \ldots)$ is symmetric or exchangeable if its indicator function is exchangeable.

Let $\{X_n\}$ be a sequence of random variables, and let $\mathbb{R}_\infty$ be the infinite-dimensional Euclidean space. Let $A \in \sigma(X_1, X_2, \ldots)$. Clearly A is an exchangeable event if and only if there exists a Borel set $B_\infty \subset \mathbb{R}_\infty$ such that, for each $n \geq 1$ and each permutation $\{i_1, i_2, \ldots, i_n\}$ of $\{1, 2, \ldots, n\}$,

$$A = \{(X_1, X_2, \ldots) \in B_\infty\} = \{(X_{i_1}, X_{i_2}, \ldots, X_{i_n}, X_{n+1}, \ldots) \in B_\infty\}.$$

Example 2.2.3. Let $X_1, X_2, \ldots, X_n$ be independent and identically distributed. Set $S_n = \sum_{k=1}^{n} X_k, n \geq 1$, and $Y_k = X_k - n^{-1}S_n, k \geq 1$. Then $Y_1, Y_2, \ldots, Y_{n-1}$ are exchangeable random variables.

Example 2.2.4. Let $X_1, X_2, \ldots, X_n$ be the abscissas of n points chosen independently and randomly according to some common density function on the interval $(0, 1)$. Let $Y_1, Y_2, \ldots, Y_{n+1}$ be the lengths of the $n + 1$ subintervals into which $(0, 1)$ is subdivided by the points $X_1, X_2, \ldots, X_n$. Then the random variables $Y_1, Y_2, \ldots, Y_{n+1}$ are exchangeable.

It follows from Definition 2.2.1 that any tail event is exchangeable. In fact, if $A \in \mathscr{T}$ and $\{i_1, i_2, \ldots, i_n\}$ is a permutation of $\{1, 2, \ldots, n\}$, we may write A in the form $\{(X_{n+1}, X_{n+2}, \ldots) \in A_{n+1}\}$ where A_{n+1} is a Borel set in $\mathbb{R}_\infty$. This follows since

$$A \in \sigma(X_{n+1}, X_{n+2}, \ldots).$$

Thus

$$\begin{aligned} A &= \{(X_1, X_2, \ldots) \in \mathbb{R}_n \times A_{n+1}\} \\ &= \{(X_{i_1}, X_{i_2}, \ldots, X_{i_n}, X_{n+1}, \ldots) \in \mathbb{R}_n \times A_{n+1}\}. \end{aligned}$$

There are exchangeable events that are not tail events. For example, the event $\{X_n = 0 \text{ for all } n\}$ is exchangeable but is not a tail event.

It is easily verified that exchangeable events on $(X_1, X_2, \ldots)$ form a σ-field $\mathscr{E}$ and exchangeable functions are $\mathscr{E}$-measurable. Clearly $\mathscr{T} \subset \mathscr{E}$. If, in particular, the X_i are independent and identically distributed, the tail events and tail functions on $\{S_n\}$ are exchangeable.

Theorem 2.2.2 (Hewitt–Savage Zero-One Law). Let $\{X_n\}$ be a sequence of independent, indentically distributed random variables. Then every exchangeable event in $\sigma(X_1, X_2, \ldots)$ has probability 0 or 1.

Proof. Let $A \in \sigma(X_1, X_2, \ldots)$ be exchangeable. It follows from Theorem D, page 56, of Halmos [31] that there exists $A_n \in \sigma(X_1, X_2, \ldots, X_n)$ such that

$$P(A_n \triangle A) = P((A_n - A) \cup (A - A_n)) \to 0.$$

Let $B_n \in \mathscr{B}_n$, and B_∞ be a Borel set in $\mathbb{R}_\infty$ (see Section 1.1) such that

$$A_n = \{(X_1, X_2, \ldots, X_n) \in B_n\} \qquad \text{and} \qquad A = \{(X_1, X_2, \ldots) \in B_\infty\}.$$

Write

$$A_n^* = \{(X_{n+1}, X_{n+2}, \ldots, X_{2n}) \in B_n\}$$

and

$$A^* = \{(X_{n+1}, X_{n+2}, \ldots, X_{2n}, X_1, X_2, \ldots, X_n, X_{2n+1}, \ldots) \in B_\infty\}.$$

Then

$$P(A_n \cap A_n^*) = P(A_n)P(A_n^*),$$

since A_n and A_n^* are independent. Since the X_n are identically distributed, $P(A_n) = P(A_n^*)$. Now $P(A_n \triangle A) \to 0$ implies $P(A_n) \to P(A)$, so that

$$P(A_n \cap A_n^*) \to [P(A)]^2 \qquad \text{as } n \to \infty.$$

Also, A is exchangeable, so that

$$P(A_n^* \triangle A) = P(A_n^* \triangle A^*) \qquad \text{for all } n \geq 1.$$

Once again, since the X_n are identically distributed,

$$P(A_n^* \triangle A^*) = P(A_n \triangle A),$$

and it follows that $P(A_n^* \triangle A^*) \to 0$ as $n \to \infty$. Thus

$$P(A_n \cap A_n^*) \to P(A) \qquad \text{as } n \to \infty.$$

Therefore $[P(A)]^2 = P(A)$, and $P(A) = 0$ or $= 1$. ∎

Corollary. Let $X_1, X_2, \ldots$ be independent and identically distributed. Write $S_n = \sum_{k=1}^n X_k$, and consider the sequence $\{S_n\}$. Every tail event on the sequence $\{S_n\}$ has probability 0 or 1.

Proof. Let A be a tail event on $S_1, S_2, \ldots$. It suffices to prove that A is an exchangeable event. Let $\{i_1, i_2, \ldots, i_n\}$ be a permutation of $\{1, 2, \ldots, n\}$. Then we may write A in the form $\{(S_{n+1}, S_{n+2}, \ldots) \in A_{n+1}\}$. Thus

$$\begin{aligned} A &= \{(S_1, S_2, \ldots, S_n, S_{n+1}, \ldots) \in \mathbb{R}_n \times A_{n+1}\} \\ &= \{(X_{i_1}, X_{i_2}, \ldots, X_{i_n}, S_{n+1}, \ldots) \in \mathbb{R}_n \times A_{n+1}\} \\ &= \{(X_{i_1}, X_{i_2}, \ldots, X_{i_n}, X_{n+1}, \ldots) \in \mathbb{R}_n \times A_{n+1}\}, \end{aligned}$$

and it follows that A is an exchangeable event.

As an immediate consequence we see that $P(S_n \in A_n \text{ i.o.}) = 0$ or $=1$ for any set $A_n \in \mathscr{B}$. Similarly, $\liminf_n S_n$ and $\limsup_n S_n$ are degenerate random variables.

2.2.3 Centering and Truncation

We say that a random variable X is *centered* at a point c if we replace X by $X - c$. The choice of the constant c plays an important role in some problems of probability theory. One such centering constant is a *median*, which we now define. Let X be a random variable. A finite real number $\text{med}(X)$ is called a *median* of X if

$$(2.2.3) \qquad P\{X \geq \text{med}(X)\} \geq \tfrac{1}{2} \quad \text{and} \quad P\{X \leq \text{med}(X)\} \geq \tfrac{1}{2}.$$

Equivalently, $\text{med}(X)$ is a median if

$$(2.2.4) \qquad F(\text{med}(X) - 0) \leq \tfrac{1}{2} \leq F(\text{med}(X)),$$

where F is the distribution function of X.

If X is integrable, we may choose $c = \mathscr{E}X$ and say that a random variable X is *centered at its expectation if* $\mathscr{E}X = 0$.

Let $c > 0$ be a constant. Set

$$X^c = \begin{cases} X & \text{if } |X| < c, \\ 0 & \text{if } |X| \geq c. \end{cases}$$

The random variable X^c defined above is known as *X truncated at c*. Since X^c is a bounded random variable, moments of all orders of X^c exist and are finite.

Let $\{X_n\}$ be a sequence of random variables. For $\epsilon > 0$, choose c_n such that

$$P\{|X_n| \geq c_n\} < \frac{\epsilon}{2^n}, \qquad n = 1, 2, \ldots.$$

Then

$$P\left\{\bigcup_{n=1}^{\infty} (X_n \neq X_n^{c_n})\right\} \leq \sum_{n=1}^{\infty} P\{|X_n| \geq c_n\} < \epsilon.$$

Two sequences of random variables $\{X_n\}$ and $\{X'_n\}$ are said to be *tail equivalent* if they differ a.s. only by a finite number of terms; that is, for almost all $\omega \in \Omega$, there exists a finite positive integer $n(\omega)$ such that for all $n \geq n(\omega)$ the two sequences $\{X_n\}$ and $\{X'_n\}$ are identical. In other words,

$$P\{X_n \neq X'_n \text{ i.o.}\} = 0.$$

We say that $\{X_n\}$ and $\{X'_n\}$ are *convergence equivalent* if they converge on the same event except for a null event.

Proposition 2.2.4. Let $\{X_n\}$ and $\{X'_n\}$ be two sequences of random variables such that $\sum_{n=1}^{\infty} P\{X_n \neq X'_n\} < \infty$. Then the following assertions hold:

(a) The sequences $\{X_n\}$ and $\{X'_n\}$ are tail equivalent.
(b) The series $\sum X_n$ and $\sum X'_n$ are convergence equivalent.
(c) The sequences $\{b_n^{-1} \sum_{k=1}^{n} X_k\}$ and $\{b_n^{-1} \sum_{k=1}^{n} X'_k\}$, where $b_n \uparrow \infty$ as $n \to \infty$, converge on the same event and to the same limit, except possibly for a null event.

Proof. Clearly

$$\begin{aligned} P\{X_n \neq X'_n \text{ i.o.}\} &= \lim_n P\left\{\bigcup_{k=n}^{\infty} [X_k \neq X'_k]\right\} \\ &\leq \lim_n \sum_{k=n}^{\infty} P\{X_k \neq X'_k\} \\ &= 0 \end{aligned}$$

so that (a) follows. The proofs of (b) and (c) are immediate. ∎

2.3 THE STRONG LAW OF LARGE NUMBERS

Let $\{X_n\}$ be a sequence of random variables defined on $(\Omega, \mathscr{S}, P)$. The strong law of large numbers deals with a.s. convergence of the sequence of partial sums $\{\sum_{k=1}^{n} X_k\}$.

2.3.1 Kolmogorov's Inequalities

We first prove two inequalities due to Kolmogorov which play a key role in our investigation.

Proposition 2.3.1. Let $X_1, X_2, \ldots, X_n$ be independent random variables such that $\mathscr{E}X_k = 0$ and $\operatorname{var}(X_k) = \mathscr{E}X_k^2 = \sigma_k^2 < \infty$, $k = 1, 2, \ldots, n$. Let

$S_k = \sum_{j=1}^{k} X_j$, $k = 1, 2, \ldots, n$. Then for every $\epsilon > 0$ the inequality

$$P\left\{\max_{1 \le k \le n} |S_k| \ge \epsilon\right\} \le \frac{1}{\epsilon^2} \sum_{k=1}^{n} \sigma_k^2 \tag{2.3.1}$$

holds.

Proof. Let $\epsilon > 0$, and define the sets

$$E_1 = \{|S_1| \ge \epsilon\}$$

and

$$E_k = \{|S_k| \ge \epsilon\} \cap \bigcap_{j=1}^{k-1} \{|S_j| < \epsilon\}$$

for $k = 2, 3, \ldots, n$. Set

$$E = \left\{\max_{1 \le k \le n} |S_k| \ge \epsilon\right\}.$$

Clearly the E_k are disjoint events, and $E = \bigcup_{k=1}^{n} E_k$. Since the X_k are independent and $\mathscr{E}X_k = 0$ for all k,

$$\begin{aligned} \sum_{k=1}^{n} \sigma_k^2 &= \operatorname{var}(S_n) \\ &= \int_{\Omega} S_n^2 \, dP \\ &\ge \int_{E} S_n^2 \, dP = \sum_{k=1}^{n} \int_{E_k} S_n^2 \, dP. \end{aligned} \tag{2.3.2}$$

Now for $1 \le k \le n$

$$\begin{aligned} \int_{E_k} S_n^2 \, dP &= \int_{E_k} (S_k + X_{k+1} + \cdots + X_n)^2 \, dP \\ &= \int_{E_k} S_k^2 \, dP + \sum_{j=k+1}^{n} \int_{E_k} X_j^2 \, dP \\ &\quad + 2 \sum_{j=k+1}^{n} \int_{E_k} S_k X_j \, dP \\ &\quad + \sum_{\substack{j \ne j' \\ j, j' = k+1}}^{n} \int_{E_k} X_j X_{j'} \, dP. \end{aligned} \tag{2.3.3}$$

Note that for $j = 1, 2, \ldots, n - k$

$$\int_{E_k} S_k X_{k+j}\, dP = \mathscr{E}(\chi_{E_k} S_k)\mathscr{E} X_{k+j} = 0,$$

and for $j \neq j'$

$$\int_{E_k} X_{k+j} X_{k+j'}\, dP = P(E_k)\mathscr{E} X_{k+j}\mathscr{E} X_{k+j'} = 0.$$

It therefore follows from (2.3.3) that

$$\int_{E_k} S_n^2\, dP \geq \int_{E_k} S_k^2\, dP \geq \epsilon^2 P(E_k),$$

so that from (2.3.2) we obtain

$$\sum_{k=1}^{n} \sigma_k^2 \geq \epsilon^2 \sum_{k=1}^{n} P(E_k) = \epsilon^2 P(E).$$

This completes the proof of our assertion. ■

Corollary. Let $X_1, X_2, \ldots, X_n$ be independent random variables with $\operatorname{var}(X_k) = \sigma_k^2 < \infty$, $k = 1, 2, \ldots, n$. Then for every $\epsilon > 0$

$$P\left\{\max_{1 \leq k \leq n} |S_k - \mathscr{E}(S_k)| \geq \epsilon\right\} \leq \frac{1}{\epsilon^2} \sum_{k=1}^{n} \sigma_k^2.$$

In particular, if $n = 1$, we get Chebyshev's inequality.

Proposition 2.3.2. Let $X_1, X_2, \ldots, X_n$ be independent random variables, and let $\gamma > 0$ be a constant such that $P\{|X_k| \leq \gamma\} = 1$ for $1 \leq k \leq n$. Then for every $\epsilon > 0$

$$P\left\{\max_{1 \leq k \leq n} |S_k - \mathscr{E}(S_k)| \geq \epsilon\right\} \geq 1 - \frac{(\epsilon + 2\gamma)^2}{\operatorname{var}(S_n)}. \tag{2.3.4}$$

Proof. Clearly $|\mathscr{E} X_k| \leq \gamma$, and $P\{|X_k - \mathscr{E} X_k| \leq 2\gamma\} = 1$. We may therefore assume that the X_k are centered at $\mathscr{E} X_k$ and satisfy $P\{|X_k| \leq 2\gamma\} = 1$.

Let $\epsilon > 0$. Write $S_0 = 0$, and set

$$F_0 = \Omega \qquad \text{and} \qquad F_k = \bigcap_{i=1}^{k} \{|S_i| < \epsilon\}, \qquad k = 1, 2, \ldots, n.$$

Clearly

$$F_n \subset F_{n-1} \subset \cdots \subset F_1 \subset F_0 = \Omega.$$

Moreover,

$$F_n^c = \left\{ \max_{1 \le k \le n} |S_k| \ge \epsilon \right\}.$$

Let the sets E_k and E be defined as in Proposition 2.3.1. It is easy to see that $E_k \cup F_k = F_{k-1}$ and $E_k \cap F_k = \varnothing$, $k = 1, 2, \ldots, n$. Moreover, $E = F_n^c$. Now we compute

$$I_k = \int_{F_k} S_k^2 \, dP - \int_{F_{k-1}} S_{k-1}^2 \, dP.$$

We note that

$$\int_{F_k} S_k^2 \, dP = \int_{F_{k-1}} S_k^2 \, dP - \int_{E_k} S_k^2 \, dP,$$

so that

$$\begin{aligned} I_k &= \int_{F_{k-1}} (S_k^2 - S_{k-1}^2) \, dP - \int_{E_k} S_k^2 \, dP \\ &= \int_{F_{k-1}} X_k^2 \, dP + 2 \int_{F_{k-1}} S_{k-1} X_k \, dP - \int_{E_k} S_k^2 \, dP. \end{aligned}$$

Clearly

$$\begin{aligned} \int_{F_{k-1}} X_k^2 \, dP &= \mathscr{E}(\chi_{F_{k-1}}) \mathscr{E} X_k^2 \\ &= P(F_{k-1}) \mathscr{E} X_k^2 \\ &\ge P(F_n) \mathscr{E} X_k^2 \end{aligned}$$

and

$$\int_{F_{k-1}} S_{k-1} X_k \, dP = \mathscr{E}(\chi_{F_{k-1}} S_{k-1}) \mathscr{E} X_k = 0.$$

On the set E_k we have

$$|S_k| \le |S_{k-1}| + |X_k| \le \epsilon + 2\gamma,$$

so that

$$\int_{E_k} S_k^2 \, dP \le (\epsilon + 2\gamma)^2 P(E_k).$$

Thus we obtain

$$I_k \ge P(F_n) \mathscr{E} X_k^2 - (\epsilon + 2\gamma)^2 P(E_k) \qquad \text{for } k = 1, 2, \ldots, n.$$

Summing both sides of this last inequality for $k = 1, 2, \ldots, n$, we get

$$\int_{F_n} S_n^2 \, dP \geq P(F_n) \sum_{k=1}^{n} \mathscr{E} X_k^2 - (\epsilon + 2\gamma)^2 P(E)$$
$$= P(F_n)\operatorname{var}(S_n) - (\epsilon + 2\gamma)^2 P(E).$$

On the other hand,

$$\int_{F_n} S_n^2 \, dP \leq \epsilon^2 P(F_n),$$

and hence

$$\epsilon^2 P(F_n) \geq \int_{F_n} S_n^2 \, dP \geq P(F_n)\operatorname{var}(S_n) - (\epsilon + 2\gamma)^2 P(E).$$

It follows that

$$P(F_n) \leq \frac{(\epsilon + 2\gamma)^2}{\operatorname{var}(S_n)},$$

which yields (2.3.4) immediately. ∎

2.3.2 Series of Random Variables

We shall now investigate a.s. convergence of a series of independent random variables.

Proposition 2.3.3. Let $\{X_n\}$ be a sequence of independent random variables with

$$\sigma_n^2 = \operatorname{var}(X_n) < \infty, \qquad n = 1, 2, \ldots.$$

If $\sum_{n=1}^{\infty} \sigma_n^2 < \infty$, then $\sum_{n=1}^{\infty} (X_n - \mathscr{E} X_n)$ converges a.s.

Proof. We set $S_n = \sum_{k=1}^{n} X_k$, and apply Proposition 2.3.1 to the set of independent random variables $X_{n+1} - \mathscr{E} X_{n+1}, X_{n+2} - \mathscr{E} X_{n+2}, \ldots, X_{n+m} - \mathscr{E} X_{n+m}$, where n and m are any two positive integers. Then for every $\epsilon > 0$ we obtain

$$P\left\{\max_{1 \leq k \leq m} |S_{n+k} - S_n - \mathscr{E}(S_{n+k}) + \mathscr{E}(S_n)| \geq \epsilon\right\} \leq \frac{1}{\epsilon^2} \sum_{k=1}^{m} \sigma_{n+k}^2. \tag{2.3.5}$$

Let us write

$$T_k = S_k - \mathscr{E}(S_k),$$
$$\Delta_k = \sup_{\nu \geq 1}\{|T_{k+\nu} - T_k|\},$$

and

$$\Delta = \inf_{k \geq 1} \Delta_k.$$

Then we take the limit of both sides in (2.3.5) as $m \to \infty$ to obtain

$$P\{\Delta_n \geq \epsilon\} \leq \frac{1}{\epsilon^2} \sum_{k=n+1}^{\infty} \sigma_k^2,$$

so that

$$P\{\Delta \geq \epsilon\} \leq \frac{1}{\epsilon^2} \sum_{k=n+1}^{\infty} \sigma_k^2 \qquad \text{for every } n \geq 1.$$

Since $\sum_{k=1}^{\infty} \sigma_k^2 < \infty$, it follows that $P\{\Delta \geq \epsilon\} = 0$. This implies that the sequence $\{S_n\}$ is Cauchy a.s. and hence converges a.s. ■

Proposition 2.3.4. Let $\{X_n\}$ be a sequence of independent random variables such that $|X_n| \leq \gamma$ a.s. for some $\gamma > 0$. Let $\text{var}(X_n) = \sigma_n^2$, $n = 1, 2, \ldots$. If $\sum_{n=1}^{\infty} (X_n - \mathscr{E}X_n)$ converges a.s., then $\sum_{n=1}^{\infty} \sigma_n^2 < \infty$.

Proof. We use the same notation as in Proposition 2.3.3. It follows from Proposition 2.3.2 that

$$P\{\Delta_n \geq \epsilon\} \geq 1 - \frac{(\epsilon + 2\gamma)^2}{\sum_{k=n+1}^{\infty} \sigma_k^2}.$$

Suppose that $\sum_{k=1}^{\infty} \sigma_k^2 = \infty$. Then for every $n \geq 1$ we obtain

$$P\{\Delta_n \geq \epsilon\} = 1,$$

so that $P\{\Delta \geq \epsilon\} = 1$, and hence the series $\sum_{n=1}^{\infty} (X_n - \mathscr{E}X_n)$ diverges a.s. This completes the proof. ■

Remark 2.3.1. Let $\{X_n\}$ be a sequence of independent random variables such that $|X_n| \leq \gamma$ a.s. for some $\gamma > 0$, $n = 1, 2, \ldots$. Let $\text{var}(X_n) = \sigma_n^2$, $n = 1, 2, \ldots$. Then the series $\sum_{n=1}^{\infty} (X_n - \mathscr{E}X_n)$ converges or diverges a.s. according as $\sum_{n=1}^{\infty} \sigma_n^2 < \infty$ or $= \infty$.

Proposition 2.3.5. Let $\{X_n\}$ be independent so that $|X_n| \leq \gamma$ a.s., $n = 1, 2, \ldots$. Then the series $\sum_{n=1}^{\infty} X_n$ converges a.s. if and only if the following conditions hold:

(i) $|\sum_{n=1}^{\infty} \mathscr{E}X_n| < \infty$.
(ii) $\sum_{n=1}^{\infty} \text{var}(X_n) < \infty$.

Proof. It is sufficient to show that the a.s. convergence of $\sum X_n$ implies (i) and (ii). For this purpose we use the *method of symmetrization*. We introduce a sequence $\{X'_n\}$ of independent random variables with the following properties:

(a) For every n, X_n and X'_n are identically distributed.
(b) The combined sequence $\{X_n, X'_n\}_{n=1}^{\infty}$ is a sequence of independent random variables.

Next we form the symmetrized sequence $X_n^s = X_n - X'_n$, $n = 1, 2, \ldots$. Clearly, $\{X_n^s\}$ is a sequence of independent random variables; moreover,

$$|X_n^s| \le |X_n| + |X'_n| \le 2\gamma \text{ a.s.},$$

$$\mathscr{E}X_n^s = \mathscr{E}X_n - \mathscr{E}X'_n = 0,$$

and

$$\operatorname{var}(X_n^s) = 2 \operatorname{var}(X_n)$$

for each $n = 1, 2, \ldots$. Since $\sum X_n$ converges a.s., so does $\sum X'_n$ and hence also the series $\sum_{n=1}^{\infty} X_n^s$. It follows from Proposition 2.3.4 that $\sum_{n=1}^{\infty} \operatorname{var}(X_n^s) < \infty$, and consequently $\sum_{n=1}^{\infty} \operatorname{var}(X_n) < \infty$. Therefore from Proposition 2.3.3 $\sum_{n=1}^{\infty} (X_n - \mathscr{E}X_n)$ converges a.s., and hence

$$\sum_{n=1}^{\infty} \mathscr{E}X_n = \sum_{n=1}^{\infty} X_n - \sum_{n=1}^{\infty} (X_n - \mathscr{E}X_n)$$

also converges. ∎

Proposition 2.3.6 (Three-Series Criterion). For $c > 0$, let X^c denote the random variable X truncated at c as defined earlier. Let $\{X_n\}$ be a sequence of independent random variables. Then the series $\sum X_n$ converges a.s. if and only if, for a fixed $c > 0$, the following three series converge:

(a) $\sum_{n=1}^{\infty} P\{|X_n| \ge c\}$.
(b) $\sum_{n=1}^{\infty} \mathscr{E}X_n^c$.
(c) $\sum_{n=1}^{\infty} \operatorname{var}(X_n^c)$.

Moreover, if (a), (b), (c) converge for some $c > 0$, they converge for all $c > 0$.

Proof. Suppose $\sum X_n$ converges a.s. Then $X_n \xrightarrow{\text{a.s.}} 0$, and hence, for any $c > 0$, $P\{|X_n| \ge c \text{ i.o.}\} = 0$. By the Borel–Cantelli lemma, $\sum_{n=1}^{\infty} P\{|X_n| \ge c\} < \infty$. In view of the equivalence lemma (Proposition 2.2.4) $\sum X_n$ and $\sum X_n^c$ are convergence equivalent. Then (b) and (c) are immediate consequences of Proposition 2.3.5.

Conversely, if (a) converges, $\sum X_n$ and $\sum X_n^c$ are convergence equivalent. If, in addition, (b) and (c) converge, then from Proposition 2.3.5 we conclude that the series $\sum_{n=1}^{\infty} X_n^c$ converges a.s., and hence so does $\sum_{n=1}^{\infty} X_n$. ■

2.3.3 The Strong Law of Large Numbers

We first prove two elementary results on convergence of sequences of real numbers.

Lemma 2.3.1 (Toeplitz Lemma). Let $\{a_n\}$ be a sequence of real numbers such that $a_n \to a$ as $n \to \infty$. Then $n^{-1} \sum_{k=1}^{n} a_k \to a$ as $n \to \infty$.

Proof. Since $a_n \to a$, for $\epsilon > 0$ there exists $n_0(\epsilon) = n_0$ such that for $n \geq n_0$

$$|a_n - a| < \frac{\epsilon}{2}.$$

Next choose an integer $n_0^* > n_0$ such that

$$\frac{1}{n_0^*} \sum_{k=1}^{n_0} |a_k - a| < \frac{\epsilon}{2}.$$

Let $n > n_0^*$. Then

$$\begin{aligned}\left|\frac{1}{n} \sum_{k=1}^{n} a_k - a\right| & \\ &\leq \frac{1}{n_0^*} \sum_{k=1}^{n_0} |a_k - a| + \frac{1}{n} \sum_{k=n_0+1}^{n} |a_k - a| \\ &< \frac{\epsilon}{2} + \frac{n - n_0}{n} \frac{\epsilon}{2} < \epsilon.\end{aligned}$$

■

Lemma 2.3.2 (Kronecker Lemma). Let $\{a_n\}$ be a sequence of real numbers such that $\sum_{n=1}^{\infty} a_n$ converges. Then $n^{-1} \sum_{k=1}^{n} k a_k \to 0$ as $n \to \infty$.

Proof. We set $s_0 = 0$ and $s_n = \sum_1^n a_k$, $n = 1, 2, \ldots$. Clearly

$$\begin{aligned}\frac{1}{n} \sum_{k=1}^{n} k a_k &= \frac{1}{n} \sum_{k=1}^{n} k(s_k - s_{k-1}) \\ &= s_n - \frac{1}{n} \sum_{k=1}^{n-1} s_k.\end{aligned}$$

The sequence $\{s_n\}$ converges to a finite limit, say s. In view of the Toeplitz lemma, $n^{-1} \sum_{k=1}^{n-1} s_k \to s$ as $n \to \infty$. Hence we have the result. ■

We next prove the following result, due to Kolmogorov.

Proposition 2.3.7. Let $\{X_n\}$ be a sequence of independent random variables with $\operatorname{var}(X_n) = \sigma_n^2 < \infty$, $n = 1, 2, \ldots$. Suppose that $\sum_{n=1}^{\infty} \sigma_n^2/n^2 < \infty$. Let $S_n = \sum_{k=1}^{n} X_k$, $n = 1, 2, \ldots$. Then the sequence $\{n^{-1}(S_n - \mathscr{E}S_n)\}$ converges a.s. to zero.

Proof. We set $Y_n = (X_n - \mathscr{E}X_n)/n$, $n = 1, 2, \ldots$. Then $\mathscr{E}Y_n = 0$ and $\operatorname{var}(Y_n) = \sigma_n^2/n^2$, so that $\sum_{n=1}^{\infty} \operatorname{var}(Y_n) < \infty$. It follows from Proposition 2.3.3 that $\sum_{n=1}^{\infty} Y_n$ converges a.s. We now apply the Kronecker lemma to the sequence $\{Y_n\}$ to obtain the required result. ■

Remark 2.3.2. Let $\{X_n\}$ be a sequence of random variables, and $\{A_n\}$ be a sequence of constants. A statement of the type

$$\frac{S_n - A_n}{n} \xrightarrow{\text{a.s.}} 0 \qquad \text{as } n \to \infty$$

is known as a *strong law of large numbers*. Sometimes one seeks sequences $\{A_n\}$ and $\{B_n\}$, $B_n > 0$ and $B_n \to \infty$, such that

$$B_n^{-1}(S_n - A_n) \xrightarrow{\text{a.s.}} 0 \qquad \text{as } n \to \infty,$$

but we do not propose to do so here. The main result is the Kolmogorov strong law of large numbers, which is proved below.

We are now in a position to prove the following fundamental result, known as the strong law of large numbers, for sequences of independent, identically distributed random variables.

Theorem 2.3.1 (Kolmogorov). Let $\{X_n\}$ be a sequence of independent, identically distributed random variables. Let $S_n = \sum_{k=1}^{n} X_k$, $n = 1, 2, \ldots$. Then the sequence $\{n^{-1}S_n\}$ converges a.s. to a finite limit α if and only if $\mathscr{E}|X_n| < \infty$. Moreover, in this case $\mathscr{E}X_n = \alpha$.

Proof. Suppose that $\mathscr{E}|X_1| < \infty$. We show that $n^{-1}S_n \xrightarrow{\text{a.s.}} \mathscr{E}X_1$ as $n \to \infty$. Introduce a random variable X which is identically distributed as X_1. Set

$$E_0 = \Omega \qquad \text{and} \qquad E_n = \{|X| \geq n\}, \qquad n = 1, 2, \ldots.$$

Clearly $E_n \downarrow \varnothing$. Moreover, for every $n \geq 0$, $E_n = \bigcup_{k=n}^{\infty} (E_k - E_{k+1})$ is a union of disjoint events. We note that

$$\sum_{n=1}^{\infty} P(E_n) = \sum_{n=1}^{\infty} nP(E_n - E_{n+1})$$

and

$$1 + \sum_{n=1}^{\infty} P(E_n) = \sum_{n=0}^{\infty} P(E_n)$$

$$= \sum_{n=1}^{\infty} nP(E_{n-1} - E_n).$$

By Proposition 1.2.8 we see that

$$\sum_{n=1}^{\infty} P(E_n) \leq E|X| \leq 1 + \sum_{n=1}^{\infty} P(E_n),$$

and it follows that $\sum_{n=1}^{\infty} P(E_n) < \infty$.

For every $n \geq 1$ let X_n^* be X_n truncated at n, so that

$$X_n^* = \begin{cases} X_n & \text{if } |X_n| < n, \\ 0 & \text{if } |X_n| \geq n. \end{cases}$$

Then

$$\operatorname{var}(X_n^*) \leq \mathscr{E} X_n^{*2}$$

$$= \int_{\{|X_n| < n\}} X_n^2 \, dP$$

$$= \int_{E_n^c} X^2 \, dP$$

$$\leq \sum_{k=1}^{n} k^2 P(E_{k-1} - E_k),$$

so that

$$\sum_{n=1}^{\infty} \frac{\operatorname{var}(X_n^*)}{n^2} \leq \sum_{n=1}^{\infty} \sum_{k=1}^{n} \frac{k^2}{n^2} P(E_{k-1} - E_k)$$

$$= \sum_{k=1}^{\infty} k^2 P(E_{k-1} - E_k) \sum_{n=k}^{\infty} \frac{1}{n^2}.$$

But we note that for $k \geq 1$

$$\sum_{n=k}^{\infty} \frac{1}{n^2} \leq \frac{1}{k^2} + \int_k^{\infty} \frac{dx}{x^2}$$

$$= \frac{1}{k^2} + \frac{1}{k} \leq \frac{2}{k},$$

so that

$$\sum_{n=1}^{\infty} \frac{\operatorname{var}(X_n^*)}{n^2} \le 2 \sum_{k=1}^{\infty} kP(E_{k-1} - E_k)$$
$$= 2\left[1 + \sum_{n=1}^{\infty} P(E_n)\right]$$
$$< \infty.$$

We now set $S_n^* = \sum_{k=1}^n X_k^*$, $n = 1, 2, \ldots$. From Proposition 2.3.7 applied to $\{X_n^*\}$, we conclude that $n^{-1}(S_n^* - \mathscr{E}S_n^*) \xrightarrow{\text{a.s.}} 0$ as $n \to \infty$. We first show that $n^{-1}\mathscr{E}S_n^* \to \mathscr{E}X$.

For every n

$$\mathscr{E}X_n^* = \int_{\{|X_n| < n\}} X_n \, dP = \int_{E_n^c} X \, dP$$
$$= \mathscr{E}(X\chi_{E_n^c}).$$

Since $E_n^c \uparrow \Omega$, we conclude that $X\chi_{E_n^c} \xrightarrow{\text{a.s.}} X$ as $n \to \infty$. Using the Lebesgue dominated convergence theorem, we conclude that $\mathscr{E}X_n^* \to \mathscr{E}X$. Consequently $n^{-1}\mathscr{E}S_n^* \to \mathscr{E}X$ by the Toeplitz lemma.

Finally we note that $\{X_n\}$ and $\{X_n^*\}$ are tail-equivalent, so that

$$n^{-1}(S_n - S_n^*) \xrightarrow{\text{a.s.}} 0.$$

Hence $n^{-1}S_n \xrightarrow{\text{a.s.}} \mathscr{E}X$ as $n \to \infty$.

Conversely, suppose that the sequence $\{n^{-1}S_n\}$ converges a.s. to a finite limit α. We show that $\mathscr{E}|X| < \infty$, and, moreover, $\mathscr{E}X = \alpha$. Since

$$\frac{X_n}{n} = \frac{S_n}{n} - \frac{n-1}{n}\frac{S_{n-1}}{n-1},$$

we conclude that $X_n/n \xrightarrow{\text{a.s.}} 0$ as $n \to \infty$. In view of Corollary 2 to Proposition 2.2.1, we see that $\sum_{n=1}^{\infty} P\{|X_n| \ge n\epsilon\} < \infty$ for every $\epsilon > 0$. In particular, $\sum_{n=1}^{\infty} P(E_n) < \infty$, and hence $\mathscr{E}|X| < \infty$. It now follows from the first part of the theorem that $\alpha = \mathscr{E}X$. This completes the proof of Theorem 2.3.1. ■

2.4 THE LAW OF THE ITERATED LOGARITHM

Let $\{X_n\}$ be a sequence of independent random variables, and let $S_n = \sum_{k=1}^n X_k$, $n \ge 1$. In this section we investigate the rate of growth of partial sums $\{S_n\}$. If the X_n are assumed, in addition, to be identically distributed with $\mathscr{E}X_i = 0$, the strong law of large numbers tells us that $|S_n| = o(n)$ a.s., that is, whatever $\epsilon > 0$

$$P\{|S_n| > \epsilon n \text{ i.o.}\} = 0,$$

so that the growth rate of S_n is smaller than n. If we assume that $\mathscr{E}X_i^2$ is finite and equals 1, a simple application of the Lévy central limit theorem (Corollary 1 to Theorem 5.1.1) shows that the growth rate of S_n is larger than $\sqrt{n}$. It follows that the growth rate of S_n is larger than $\sqrt{n}$ and smaller than n. We note that $\limsup_n\{S_n/(2n \ln \ln n)^{1/2}\}$ is a tail function, so by Kolmogorov's zero-one law it follows that

$$\limsup_{n\to\infty} \frac{S_n}{(2n \ln \ln n)^{1/2}} = c \qquad \text{a.s.,}$$

where c is a constant. The law of the iterated logarithm says that $c = 1$. This section is devoted to a proof of the law of the iterated logarithm.

2.4.1 Exponential Bounds

Let $\{X_n\}$ be a sequence of independent random variables with $\mathscr{E}X_n = 0$ and $\mathscr{E}X_n^2 < \infty$ for all $n \geq 1$. Set $S_n = \sum_{k=1}^n X_k$ and $s_n^2 = \sum_{k=1}^n \mathscr{E}X_k^2$ for all $n = 1, 2, \ldots$.

Proposition 2.4.1 (Kolmogorov). Let $|X_k| \leq cs_n$ a.s. (where $c > 0$ is a constant) for each k, $1 \leq k \leq n$, $n \geq 1$, and suppose that $\epsilon > 0$.

(a) If $\epsilon c \leq 1$, then for every $n \geq 1$ we have

$$P\{S_n > \epsilon s_n\} < \exp\left\{-\frac{\epsilon^2}{2}\left(1 - \frac{\epsilon c}{2}\right)\right\}. \tag{2.4.1}$$

If $\epsilon c \geq 1$, then for every $n \geq 1$ we have

$$P\{S_n > \epsilon s_n\} < \exp\left(-\frac{\epsilon}{4c}\right). \tag{2.4.2}$$

(b) For any given $\gamma > 0$, there exist constants $\epsilon(\gamma)$ (sufficiently large) and $\pi(\gamma)$ (sufficiently small) such that for $\epsilon \geq \epsilon(\gamma)$ and $\epsilon c \leq \pi(\gamma)$

$$P\{S_n > \epsilon s_n\} > \exp\left\{-\frac{\epsilon^2}{2}(1 + \gamma)\right\}. \tag{2.4.3}$$

Proof. (a) Let $t > 0$, and let X be a random variable with $|X| \leq c < \infty$ a.s. and $\mathscr{E}X = 0$. Then $\mathscr{E}|X|^n \leq c^n$ and $\mathscr{E}|X|^r \leq c^{r-2}\mathscr{E}X^2$ for $r \geq 2$. We have

$$\begin{aligned} \mathscr{E}\exp(tX) &= \sum_{r=0}^{\infty} t^r \frac{\mathscr{E}X^r}{r!} \\ &\leq 1 + \frac{t^2}{2}\mathscr{E}X^2\left(1 + \frac{tc}{3} + \frac{t^2c^2}{3\cdot 4} + \cdots\right). \end{aligned}$$

If $tc \le 1$, then

$$\mathscr{E}\exp(tX) \le 1 + \frac{t^2}{2}\mathscr{E}X^2\left[1 + \frac{tc}{3}\left(1 + \frac{tc}{4} + \cdots\right)\right] < 1 + \frac{t^2}{2}\mathscr{E}X^2\left(1 + \frac{tc}{2}\right)$$

$$\le \exp\left\{\frac{t^2}{2}\mathscr{E}X^2\left(1 + \frac{tc}{2}\right)\right\},$$

since $1 + x \le \exp(x)$ for all $x \in \mathbb{R}$. Therefore, for fixed $n \ge 1$ and $tc \le 1$,

$$P\{S_n > \epsilon s_n\} \le \exp(-t\epsilon)\mathscr{E}\exp\left(t\frac{S_n}{s_n}\right)$$

$$< \exp(-t\epsilon)\exp\left\{\frac{t^2}{2}\left(1 + \frac{tc}{2}\right)\right\}. \tag{2.4.4}$$

We put $t = \epsilon$ in (2.4.4) to get (2.4.1). If, on the other hand, $\epsilon c \ge 1$, we put $t = 1/c$, so that $tc = 1$ and (2.4.4) still holds. We get

$$P\{S_n > \epsilon s_n\} < \exp\left(-\frac{\epsilon}{c}\right)\exp\left\{\frac{1}{2c^2}\left(1 + \tfrac{1}{2}\right)\right\}$$

$$\le \exp\left(-\frac{\epsilon}{c} + \frac{3}{4}\frac{\epsilon}{c}\right),$$

since $\epsilon c \ge 1$.

(b) It suffices to prove (2.4.3) for ϵ sufficiently large and ϵc sufficiently small. Let $t > 0$ be fixed (to be specified later), and choose $c > 0$ such that $tc \le 1$. Let X be a random variable with $|X| \le c < \infty$ a.s. and $\mathscr{E}X = 0$. Then $\mathscr{E}X^r \ge -c^{-r-2}\mathscr{E}X^2$ for $r \ge 2$, and we have

$$\mathscr{E}\exp(tX) \ge 1 + \frac{t^2\mathscr{E}X^2}{2}\left(1 - \frac{tc}{3} - \frac{t^2c^2}{3\cdot 4} - \cdots\right)$$

$$> 1 + \frac{t^2}{2}\mathscr{E}X^2\left(1 - \frac{tc}{2}\right)$$

$$\ge \exp\left\{\frac{t^2}{2}(1 - tc)\mathscr{E}X^2\right\},$$

since $1 + x \ge \exp(x - x^2)$ for all $x > 0$. Therefore

$$\mathscr{E}\exp\left(t\frac{S_n}{s_n}\right) > \exp\left\{\frac{t^2}{2}(1 - tc)\right\}.$$

Let $0 < \beta < 1$ be fixed, and choose c sufficiently small so that $tc \le \beta^2/4$. Then

$$\mathscr{E} \exp\left(t \frac{S_n}{s_n}\right) \ge \exp\left\{\frac{t^2}{2}\left(1 - \frac{\beta^2}{4}\right)\right\}. \tag{2.4.5}$$

Set $q(x) = P\{S_n > xs_n\}$, $x \in \mathbb{R}$. On integration by parts we have

$$\mathscr{E} \exp\left(t \frac{S_n}{s_n}\right) = -\int_{-\infty}^{\infty} \exp(tx)\, dq(x) = t \int_{-\infty}^{\infty} \exp(tx) q(x)\, dx$$

$$= t \sum_{k=1}^{5} J_k, \tag{2.4.6}$$

where

$$J_1 = \int_{A_1} \exp(tx)q(x)\, dx, \quad A_1 = (-\infty, 0],$$

$$J_2 = \int_{A_2} \exp(tx)q(x)dx, \quad A_2 = (0, t(1 - \beta)],$$

$$J_3 = \int_{A_3} \exp(tx)q(x)\, dx, \quad A_3 = (t(1 - \beta), t(1 + \beta)],$$

$$J_4 = \int_{A_4} \exp(tx)q(x)\, dx, \quad A_4 = (t(1 + \beta), 8t],$$

and

$$J_5 = \int_{A_5} \exp(tx)q(x)\, dx, \quad A_5 = (8t, \infty).$$

Let us first find upper bounds for J_1 and J_5. We have

$$tJ_1 = \int_{-\infty}^{0} t \exp(tx)q(x)\, dx < \int_{-\infty}^{0} t \exp(tx)\, dx = 1.$$

Let $x \in A_5$. Let $8tc < 1$. If $cx \le 1$, then from (2.4.1) we get

$$q(x) < \exp\left\{-\frac{x^2}{2}\left(1 - \frac{xc}{2}\right)\right\} < \exp\left(-\frac{x^2}{4}\right) \le \exp(-2tx),$$

since $x > 8t$. If $cx > 1$, then from (2.4.2) we get

$$q(x) < \exp\left(-\frac{x}{4c}\right) < \exp(-2tx).$$

Therefore, for all $x \in A_5$, $q(x) < \exp(-2tx)$ for $8tc < 1$, and we have

$$tJ_5 = \int_{8t}^{\infty} t \exp(tx)q(x)\,dx < \int_{8t}^{\infty} t \exp(tx)\exp(-2tx)\,dx$$
$$< \int_0^{\infty} t \exp(-tx)\,dx = 1.$$

It follows that

$$t(J_1 + J_5) < 2. \tag{2.4.7}$$

Let us next consider A_2 and A_4. Choose c so that $8tc < 1$, and note that $xc < 1$ for $x \in A_2 \cup A_4$. It follows from (2.4.1) that

$$\exp(tx)q(x) < \exp\left\{tx - \frac{x^2}{2}\left(1 - \frac{xc}{2}\right)\right\}$$
$$\leq \exp\left\{tx - \frac{x^2}{2}(1 - 4tc)\right\}.$$

Set $g(x) = tx - (x^2/2)(1 - 4tc)$ for $x \in \mathbb{R}$. Then the quadratic function g is maximized at $x = x_0$, where

$$x_0 = \frac{t}{1 - 4tc}.$$

Choosing c sufficiently small so that $4tc(1 - \beta)^2 < \beta^2/2$, we see that $x_0 \in A_3$. Therefore, for $x \in A_2$ and c sufficiently small,

$$g(x) \leq g(t(1 - \beta)) = \frac{t^2}{2}(1 - \beta)(1 + \beta + 4tc - 4tc\beta)$$
$$< \frac{t^2}{2}\left(1 - \frac{\beta^2}{2}\right).$$

It follows that

$$tJ_2 = \int_0^{t(1-\beta)} t \exp(tx)q(x)\,dx < t\int_0^{t(1-\beta)} \exp(g(x))\,dx$$
$$< t^2 \exp\left\{\frac{t^2}{2}\left(1 - \frac{\beta^2}{2}\right)\right\}.$$

Similarly, since $x_0 \in A_3$, we obtain

$$tJ_4 = t\int_{t(1+\beta)}^{8t} \exp(tx)q(x)\,dx \leq t\int_{t(1+\beta)}^{8t} \exp(g(x))\,dx$$
$$< 8t^2 \exp\left\{\frac{t^2}{2}\left(1 - \frac{\beta^2}{2}\right)\right\}.$$

Combining the two, we get

$$t(J_2 + J_4) < 9t^2 \exp\left\{\frac{t^2}{2}\left(1 - \frac{\beta^2}{2}\right)\right\}. \tag{2.4.8}$$

Let $\epsilon(\gamma)$ be fixed, and let $\epsilon \geq \epsilon(\gamma)$. Set $t = \epsilon/(1 - \beta)$. From (2.4.5) we see that

$$9t^2 \exp\left\{\frac{t^2}{2}\left(1 - \frac{\beta^2}{2}\right)\right\} \leq 9t^2 \exp\left(-\frac{t^2\beta^2}{2}\right)\mathscr{E} \exp\left(\frac{tS_n}{s_n}\right). \tag{2.4.9}$$

Now choose $\epsilon(\gamma)$ sufficiently large (hence t large) so that for $t(1 - \beta) = \epsilon \geq \epsilon(\gamma)$

$$\mathscr{E} \exp\left(t\frac{S_n}{s_n}\right) > 8 \tag{2.4.10}$$

and

$$t(J_2 + J_4) < \tfrac{1}{4}\mathscr{E} \exp\left(t\frac{S_n}{s_n}\right). \tag{2.4.11}$$

This is always possible in view of (2.4.9) and (2.4.10), since

$$9t^2 \exp\left(-\frac{t^2\beta^2}{2}\right) \to 0 \qquad \text{as } t \to \infty.$$

It follows from (2.4.10) and (2.4.7) that

$$t(J_1 + J_5) < \tfrac{1}{4}\mathscr{E} \exp\left(t\frac{S_n}{s_n}\right). \tag{2.4.12}$$

In view of (2.4.11), (2.4.12), and (2.4.6) we have

$$\mathscr{E} \exp\left(t\frac{S_n}{s_n}\right) = t \sum_{k=1}^{5} J_k < tJ_3 + \tfrac{1}{2}\mathscr{E} \exp\left(t\frac{S_n}{s_n}\right),$$

so that

$$\begin{aligned} tJ_3 &> \tfrac{1}{2}\mathscr{E} \exp\left(t\frac{S_n}{s_n}\right) \\ &\geq \tfrac{1}{2} \exp\left\{\frac{t^2}{2}\left(1 - \frac{\beta^2}{4}\right)\right\} \end{aligned}$$

by (2.4.5). But

$$tJ_3 = t\int_{t(1-\beta)}^{t(1+\beta)} \exp(tx)q(x)\,dx \leq 2t^2\beta \exp\{t^2(1 + \beta)\}q(\epsilon),$$

so that

$$q(\epsilon) > \frac{1}{4t^2\beta}\exp\left\{-t^2(1+\beta) + \frac{t^2}{2}\left(1 - \frac{\beta^2}{4}\right)\right\}$$
$$= \frac{1}{4t^2\beta}\exp\left(\frac{t^2\beta^2}{8}\right)\exp\left\{-\frac{t^2}{2}\left(1 + 2\beta + \frac{\beta^2}{2}\right)\right\}.$$

Since $(1/4t^2\beta)\exp(t^2\beta^2/8) \to \infty$ as $t \to \infty$, we can choose $\epsilon(\gamma)$ sufficiently large and replace t by $\epsilon(1-\beta)$ to get

$$q(\epsilon) > \exp\left\{-\frac{\epsilon^2[1 + 2\beta + (\beta^2/2)]}{2(1-\beta)^2}\right\}.$$

For given $\gamma > 0$ we can choose β, $0 < \beta < 1$, such that

$$\frac{1 + 2\beta + (\beta^2/2)}{(1-\beta)^2} \leq 1 + \gamma.$$

Then for $\epsilon \geq \epsilon(\gamma)$, ϵc sufficiently small, $\epsilon c \leq \pi(\gamma)$ say, we get

$$q(\epsilon) > \exp\left\{-\frac{\epsilon^2}{2}(1+\gamma)\right\},$$

which is (2.4.3). This completes the proof. ■

We also need the following two lemmas.

Lemma 2.4.1 (Lévy Inequalities). Let $X_1, X_2, \ldots, X_n$ be independent random variables, and write $S_k = \sum_{j=1}^{k} X_j$, $k = 1, 2, \ldots, n$. Then for every $\epsilon > 0$

$$P\left\{\max_{1 \leq k \leq n}[S_k - \text{med}(S_k - S_n)] \geq \epsilon\right\} \leq 2P\{S_n \geq \epsilon\} \tag{2.4.13}$$

and

$$P\left\{\max_{1 \leq k \leq n}|S_k - \text{med}(S_k - S_n)| \geq \epsilon\right\} \leq 2P\{|S_n| \geq \epsilon\}. \tag{2.4.14}$$

Proof. Set $S_0 = 0$ and $M_k = \max_{1 \leq j \leq k}[S_j - \text{med}(S_j - S_n)]$ for $k = 1, 2, \ldots, n$. Write

$$A_k = \{M_{k-1} < \epsilon, S_k - \text{med}(S_k - S_n) \geq \epsilon\}$$

and

$$B_k = \{S_n - S_k - \text{med}(S_n - S_k) \leq 0\}$$

for $k = 1, 2, \ldots, n$, where $\text{med}(S_k - S_n) = -\text{med}(S_n - S_k)$. Clearly the A_k are disjoint with $\bigcup_{k=1}^n A_k = \{M_n \geq \epsilon\}$. Also A_k, B_k are independent, and $P(B_k) \geq \frac{1}{2}$ for $k = 1, 2, \ldots, n$. It follows that

$$P\{S_n \geq \epsilon\} \geq \sum_{k=1}^n P(A_k)P(B_k) \geq \frac{1}{2} \sum_{k=1}^n P(A_k)$$
$$= \tfrac{1}{2}P\{M_n \geq \epsilon\},$$

which is (2.4.13). To prove (2.4.14) we replace X_i by $-X_i$, $1 \leq i \leq n$, in (2.4.13) to get

$$P\{S_n \leq -\epsilon\} \geq \tfrac{1}{2}P\left\{\min_{1 \leq k \leq n} [S_k - \text{med}(S_k - S_n)] \leq -\epsilon\right\}.$$

Combining this last inequality with (2.4.13), we obtain

$$P\{|S_n| \geq \epsilon\} = P\{S_n \geq \epsilon\} + P\{S_n \leq -\epsilon\}$$
$$\geq \tfrac{1}{2}P\left\{\max_{1 \leq k \leq n} |S_k - \text{med}(S_k - S_n)| \geq \epsilon\right\}$$

as asserted. ■

Corollary 1. If the X_i are independent and symmetric random variables, then

$$P\{S_n \geq \epsilon\} \geq \tfrac{1}{2}P\left\{\max_{1 \leq k \leq n} S_k \geq \epsilon\right\}$$

and

$$P\{|S_n| \geq \epsilon\} \geq \tfrac{1}{2}P\left\{\max_{1 \leq k \leq n} |S_k| \geq \epsilon\right\}.$$

Corollary 2. Let $X_1, X_2, \ldots, X_n$ be independent random variables with $\mathscr{E}X_i = 0$ and $\mathscr{E}X_i^2 < \infty$ for all $1 \leq i \leq n$. Then

(2.4.15) $$P\left\{\max_{1 \leq k \leq n} S_k \geq \epsilon\right\} \leq 2P\left\{S_n \geq \epsilon - \sqrt{2 \sum_{k=1}^n \mathscr{E}X_k^2}\right\}.$$

Proof. For any random variable X with finite variance σ^2

$$P\{|X - \mathscr{E}X| \geq \sqrt{2\sigma^2}\} \leq \tfrac{1}{2},$$

so that if follows from the definition of a median that

(2.4.16) $$\mathscr{E}X - \sqrt{2\sigma^2} \leq \text{med}(X) \leq \mathscr{E}X + \sqrt{2\sigma^2}.$$

Applying (2.4.16) to $(S_k - S_n)$ and noting that $\mathscr{E}(S_k - S_n) = 0$, we get

$$|\text{med}(S_k - S_n)| \le \sqrt{2 \sum_{j=k+1}^{n} \mathscr{E}X_k^2} \le \sqrt{2 \sum_{k=1}^{n} \mathscr{E}X_k^2},$$

so that

$$-\text{med}(S_k - S_n) \ge -\sqrt{2 \sum_{k=1}^{n} \mathscr{E}X_k^2}.$$

Therefore from (2.4.13)

$$P\left\{\max_{1 \le k \le n} \left[S_k - \sqrt{2 \sum_{k=1}^{n} \mathscr{E}X_k^2}\right] \ge \epsilon\right\}$$

$$\le P\left\{\max_{1 \le k \le n} [S_k - \text{med}(S_k - S_n)] \ge \epsilon\right\} \le 2P\{S_n \ge \epsilon\}.$$

Replacing ϵ by $\epsilon - \sqrt{2 \sum_{k=1}^{n} \mathscr{E}X_k^2}$, we get (2.4.15).

Lemma 2.4.2. Let $\{B_n\}$ be a nondecreasing sequence of positive real numbers satisfying

$$B_n \to \infty \qquad \text{and} \qquad \frac{B_n}{B_{n+1}} \to 1 \qquad \text{as } n \to \infty.$$

Then for every $\tau > 0$ there exists an eventually increasing sequence of positive integers $\{n_k\}$ such that $B_{n_k} \sim (1 + \tau)^k$ as $k \to \infty$.

Proof. Since $B_{n+1}/B_n \to 1$ as $n \to \infty$, there exists an integer k_0 such that for $k > k_0$ every interval of the form $[(1 + \tau)^k, (1 + \tau)^{k+1})$ contains at least one B_n, for otherwise $\limsup_n (B_{n+1}/B_n) \ge (1 + \tau) > 1$. Take $n_1 = n_2 = \cdots = n_{k_0}$, and for $k > k_0$ let

$$n_k = \text{smallest integer such that } B_{n_k} \ge (1 + \tau)^k.$$

Then $n_k < n_{k+1}$ for all $k > k_0$, and

$$\frac{B_{n_k - 1}}{B_{n_k}} < \frac{(1 + \tau)^k}{B_{n_k}} \le 1.$$

Since $(B_{n_k - 1}/B_{n_k}) \to 1$, we must have $B_{n_k} \sim (1 + \tau)^k$ as asserted. ∎

2.4.3 The Law of the Iterated Logarithm

We are now ready to prove Kolmogorov's law of the iterated logarithm.

Theorem 2.4.1 (Kolmogorov). Let $\{X_n\}$ be a sequence of independent random variables with $\mathscr{E}X_n = 0$ and $\mathscr{E}X_n^2 < \infty$ for all $n \ge 1$. Set $S_n =$

$\sum_{k=1}^{n} X_k$ and $s_n^2 = \sum_{k=1}^{n} \mathscr{E} X_k^2$ for $n = 1, 2, \ldots$. Let $\{k_n\}$ be a sequence of positive constants such that $k_n \to 0$ as $n \to \infty$. If the following conditions hold:

(i) $s_n^2 \to \infty$,
(ii) $|X_n| \le k_n s_n (\ln \ln s_n^2)^{-1/2}$ a.s. for all $n \ge 1$,

then

$$P\left\{\limsup_{n\to\infty} \frac{S_n}{(2s_n^2 \ln \ln s_n^2)^{1/2}} = 1\right\} = 1. \tag{2.4.17}$$

Proof. Set $t_n = (2 \ln \ln s_n^2)^{1/2}$. Then it suffices to show that for every $\epsilon > 0$

$$P\{S_n \ge (1+\epsilon) t_n s_n \text{ i.o.}\} = 0 \tag{2.4.18}$$

and

$$P\{S_n \ge (1-\epsilon) t_n s_n \text{ i.o.}\} = 1. \tag{2.4.19}$$

We first note that since $s_n^2 \to \infty$ it follows from (ii) that

$$1 \le \frac{s_{n+1}^2}{s_n^2} = 1 + \frac{\mathscr{E} X_{n+1}^2}{s_n^2} \le 1 + \frac{s_{n+1}^2}{s_n^2} \frac{k_{n+1}^2}{\ln \ln s_{n-1}^2},$$

so that

$$\frac{s_{n+1}^2}{s_n^2}\left(1 - \frac{k_{n+1}^2}{\ln \ln s_{n+1}^2}\right) \le 1.$$

Since $k_n \to 0$ and $\ln \ln s_{n+1}^2 \to \infty$ as $n \to \infty$, it follows that $s_{n+1}^2/s_n^2 \to 1$ as $n \to \infty$. From Lemma 2.4.2 it follows that for every $\tau > 0$ there exists a nondecreasing sequence $\{n_k\}$, $n_k = n_k(\tau) > 0$, of integers such that as $k \to \infty$

$$s_{n_k}^2 \sim (1+\tau)^k \tag{2.4.20}$$

and

$$s_{n_k}^2 - s_{n_{k-1}}^2 = s_{n_k}^2\left(1 - \frac{s_{n_{k-1}}^2}{s_{n_k}^2}\right) \sim \frac{\tau}{1+\tau} s_{n_k}^2. \tag{2.4.21}$$

We first prove that (2.4.18) holds. Setting $M_{n_k} = \max_{1 \le n \le n_k} S_n$, we show that for every $\delta > 0$

$$\sum_{k=1}^{\infty} P\{M_{n_k} \ge (1+\delta) t_{n_k}\} < \infty. \tag{2.4.22}$$

From (2.4.15) we have

$$P\{M_{n_k} \ge (1+\delta) t_{n_k} s_{n_k}\} \le 2P\{S_{n_k} \ge (1+\delta) t_{n_k} s_{n_k} - \sqrt{2 s_{n_k}^2}\}.$$

Since $t_{n_k} \to \infty$,

$$(1+\delta)t_{n_k}s_{n_k} - \sqrt{2s_{n_k}^2} = t_{n_k}s_{n_k}\left(1+\delta - \frac{\sqrt{2}}{t_{n_k}}\right) > t_{n_k}s_{n_k}(1+\delta_1)$$

for every $\delta_1 < \delta$ and sufficiently large k. Therefore, for $\delta_1 < \delta$ and k sufficiently large,

(2.4.23) $$P\{M_{n_k} \geq (1+\delta)t_{n_k}s_{n_k}\} \leq 2P\{S_{n_k} \geq (1+\delta_1)t_{n_k}s_{n_k}\}.$$

From (ii), for each $i \leq n_k$,

$$\frac{|X_i|}{s_{n_k}} \leq \frac{2k_i s_i}{t_i s_{n_k}} \text{ a.s.}$$

Set

$$c_k = \max_{1 \leq i \leq n_k} \frac{|X_i|}{s_{n_k}}.$$

Then

$$c_k \leq \max_{1 \leq i \leq n_k} \frac{2k_i s_i}{t_i s_{n_k}} \text{ a.s.}$$

Now

$$c_k(1+\delta_1)t_{n_k} \leq \max_{1 \leq i \leq n_k} 2(1+\delta_1)\frac{k_i s_i}{t_i}\frac{t_{n_k}}{s_{n_k}} \to 0$$

as $k \to \infty$. Using (2.4.1) in the right-hand side of (2.4.23), we see that, for any γ, $0 < \gamma < 1$, and k sufficiently large,

$$\begin{aligned} P\{M_{n_k} \geq (1+\delta)t_{n_k}s_{n_k}\} &\leq \exp\{-(1+\delta_1)^2 \ln\ln s_{n_k}^2(1-\gamma)\} \\ &= (\ln s_{n_k}^2)^{-(1-\gamma)(1+\delta_1)^2}. \end{aligned}$$

We choose γ, $0 < \gamma < 1$, such that

$$(1+\delta_1)^2(1-\gamma) > 1.$$

Then, since $s_{n_k}^2 \sim (1+\tau)^k$, $\tau > 0$, we have

$$\sum_{k=1}^{\infty} (\ln s_{n_k}^2)^{-(1-\gamma)(1+\delta_1)^2} < \infty,$$

so that (2.4.22) holds.

For every $\epsilon > 0$ we have

$$\begin{aligned} P\{S_n \geq (1+\epsilon)t_n s_n \text{ i.o.}\} &\leq P\left\{\max_{n_{k-1} \leq n \leq n_k} S_n \geq (1+\epsilon)t_{n_{k-1}}s_{n_{k-1}} \text{ i.o.}\right\} \\ &\leq P\{M_{n_k} \geq (1+\epsilon)t_{n_{k-1}}s_{n_{k-1}} \text{ i.o.}\}. \end{aligned}$$

In view of (2.4.20)

$$t_{n_k} s_{n_k} < (1 + 2\tau)^{1/2} t_{n_{k-1}} s_{n_{k-1}}$$

for k sufficiently large. Thus

$$P\{S_n \geq (1 + \epsilon) t_n s_n \text{ i.o.}\} \leq P\left\{M_{n_k} \geq \frac{1 + \epsilon}{\sqrt{1 + 2\tau}} t_{n_k} s_{n_k} \text{ i.o.}\right\}.$$

Let $\epsilon > 0$ be fixed. Choose δ, $0 < \delta < \epsilon$, and suppose that τ satisfies

$$\frac{1 + \epsilon}{\sqrt{1 + 2\tau}} > 1 + \delta.$$

Then

$$P\{S_n \geq (1 + \epsilon) t_n s_n \text{ i.o.}\} \leq P\{M_{n_k} \leq (1 + \delta) t_{n_k} s_{n_k} \text{ i.o.}\}.$$

In view of (2.4.22) and the Borel–Cantelli lemma it follows that (2.4.18) holds. Thus

$$\limsup_n \frac{S_n}{t_n s_n} \leq 1 \qquad \text{a.s.}$$

We next show that

$$\limsup_n \frac{S_n}{t_n s_n} \geq 1 \qquad \text{a.s.,}$$

that is, that (2.4.19) holds. It suffices to show that

$$P\{S_{n_k} \geq (1 - \epsilon) t_{n_k} s_{n_k} \text{ i.o.}\} = 1$$

holds for every $0 < \epsilon < 1$ and some sequence $\{n_k\}$. Let $\{n_k\}$ and τ be as selected above. Let

$$u_k^2 = s_{n_k}^2 - s_{n_{k-1}}^2 \sim s_{n_k}^2 \left(\frac{\tau}{1 + \tau}\right) \qquad \text{[from (2.4.21)]}$$

and

$$v_k^2 = (2 \ln \ln u_k^2) \sim 2 \ln\left(\ln s_{n_k}^2 + \ln \frac{\tau}{1 + \tau}\right)$$

$$\sim 2 \ln \ln s_{n_k}^2 = t_{n_k}^2.$$

Set $A_k = \{S_{n_k} - S_{n_{k-1}} \geq (1 - \epsilon) u_k v_k\}$. We show that $P\{A_k \text{ i.o.}\} = 1$. Since $S_{n_k} - S_{n_{k-1}}$ are independent random variables, it suffices to show that $\sum_{k=1}^{\infty} P(A_k) = \infty$ [see Proposition 2.2.1(b)]. Note that

$$(1 - \epsilon) v_k \sim (1 - \epsilon) t_{n_k} \to \infty \qquad \text{as } k \to \infty$$

and

$$\max_{n_{k-1}<n\leq n_k} \frac{|X_n|}{u_k} \leq \max_{n_{k-1}<n\leq n_k} \frac{k_n s_n}{(\ln \ln s_n^2)^{1/2}} \sqrt{\frac{1+\tau}{\tau}} s_{n_k}^{-1}$$
$$\to 0 \quad \text{as} \quad k \to \infty.$$

We can therefore apply (2.4.3) with γ chosen such that $(1 + \gamma)(1 - \epsilon)^2 < 1$. We have

$$\begin{aligned}
P(A_k) &> \exp\{-\tfrac{1}{2}(1+\gamma)(1-\epsilon)^2 v_k^2\} \\
&= \exp\{-(1+\gamma)(1-\epsilon)^2 \ln \ln(s_{n_k}^2 - s_{n_{k-1}}^2)\} \\
&= [\ln(s_{n_k}^2 - s_{n_{k-1}}^2)]^{-(1+\gamma)(1-\epsilon)^2} \\
&\sim \left\{\ln\left[(1+\tau)^k \cdot \frac{\tau}{1+\tau}\right]\right\}^{-(1+\gamma)(1-\epsilon)^2} \\
&\sim ck^{-(1+\gamma)(1-\epsilon)^2},
\end{aligned}$$

where $c > 0$ is some constant independent of k. Since $(1 + \gamma)(1 - \epsilon)^2 < 1$, it follows that

$$\sum_{k=1}^{\infty} P(A_k) > c \sum_{k=1}^{\infty} k^{-(1+\gamma)(1-\epsilon)^2} = \infty,$$

so that

$$P(A_k \text{ i.o.}) = 1.$$

It remains to show that $S_{n_{k-1}}$ appearing in A_k can be dropped. Let $B_k = \{|S_{n_{k-1}}| < 2s_{n_{k-1}}t_{n_{k-1}}\}$. Note that the proof of (2.4.18) given above remains valid if we replace each X_k by $-X_k$. It follows that the proof remains valid for $|S_{n_{k-1}}|$, so with $\epsilon = 1$ we have

$$P\{|S_{n_{k-1}}| \geq 2t_{n_{k-1}}s_{n_{k-1}} \text{ i.o.}\} = 0,$$

that is, $P(B_k^c \text{ i.o.}) = 0$. Therefore the probability that all but a finite number of B_k occur is 1. Hence $|S_{n_{k-1}}(\omega)| < 2t_{n_{k-1}}s_{n_{k-1}}$ for $n > n_0(\omega)$ and for all $\omega \in \Omega$ except for a null set. If follows that $P(A_k B_k \text{ i.o.}) = 1$. We show that this implies (2.4.19). We have

$$\begin{aligned}
A_k \cap B_k &= [S_{n_k} - S_{n_{k-1}} \geq (1-\epsilon)u_k v_k] \cap [|S_{n_{k-1}}| < 2t_{n_{k-1}}s_{n_{k-1}}] \\
&\subset [S_{n_k} \geq (1-\epsilon)u_k v_k + S_{n_{k-1}}] \cap [S_{n_{k-1}} > -2t_{n_{k-1}}s_{n_{k-1}}] \\
&\subset [S_{n_k} \geq (1-\epsilon)u_k v_k - 2t_{n_{k-1}}s_{n_{k-1}}],
\end{aligned}$$

where

$$(1-\epsilon)u_k v_k - 2t_{n_{k-1}}s_{n_{k-1}} = t_{n_k}s_{n_k}\left[(1-\epsilon)\frac{u_k v_k}{t_{n_k}s_{n_k}} - 2\frac{t_{n_{k-1}}s_{n_{k-1}}}{t_{n_k}s_{n_k}}\right]$$

$$\sim t_{n_k}s_{n_k}\left[(1-\epsilon)\left(\frac{\tau}{1+\tau}\right)^{1/2} - \frac{2}{(1+\tau)^{1/2}}\right].$$

If we choose $\epsilon' > \epsilon$ and then choose τ sufficiently large so that

$$(1-\epsilon)\left(\frac{\tau}{1+\tau}\right)^{1/2} - \frac{2}{(1+\tau)^{1/2}} > 1 - \epsilon',$$

then

$$(A_k \cap B_k \text{ i.o.}) \subset [S_{n_k} \geq (1-\epsilon')t_{n_k}s_{n_k} \text{ i.o.}],$$

and it follows that for every $\epsilon' > 0$

$$P\{S_{n_k} \geq (1-\epsilon')t_{n_k}s_{n_k} \text{ i.o.}\} = 1.$$

This completes the proof of Theorem 2.4.1. ∎

Remark 2.4.1. Under the assumptions of Theorem 2.4.1, if we replace every X_n by $-X_n$, we obtain

$$P\left\{\liminf_{n\to\infty} \frac{S_n}{s_n t_n} = -1\right\} = 1.$$

Therefore the conclusion holds also for the sequence $\{|S_n|\}$.

Remark 2.4.2. The result obtained in Theorem 2.4.1 is sharp in the sense that we cannot replace condition (ii) by

(2.4.24) $$|X_n| \leq Ns_n(\ln\ln s_n^2)^{-1/2} \qquad \text{a.s.}$$

for all $n \geq 1$, where $N > 0$ is a constant. Marcinkiewicz and Zygmund [61] have given an example of a sequence of random variables satisfying (2.4.24) but having

$$\limsup_{n\to\infty} \frac{S_n}{(2s_n^2 \ln\ln s_n^2)^{1/2}} < 1 \qquad \text{a.s.}$$

Remark 2.4.3. In the case where the X_n are unbounded but are (independent and) identically distributed with a common finite second moment, Hartman and Wintner [32] have show that (2.4.17) holds. Their proof depends on a truncation argument and a refinement of the above methods. We will not carry out the proof here but will content ourselves by proving the following corollary to Theorem 2.4.1.

Corollary 1 to Theorem 2.4.1. Let $\{X_n\}$ be a sequence of independent, identically distributed random variables with $0 < \mathscr{E}|X_1|^{2+\delta} < \infty$ for some $\delta > 0$ and $\mathscr{E}X_1 = 0$. Then

$$P\left\{\limsup_{n\to\infty} \frac{S_n}{[2n\mathscr{E}X_1^2 \ln \ln(n\mathscr{E}X_1^2)]^{1/2}} = 1\right\} = 1.$$

Proof. Without loss of generality we assume that $\mathscr{E}X_1^2 = 1$, so that $s_n^2 = \sum_{k=1}^n \mathscr{E}X_k^2 = n$. Let

$$X_k' = \begin{cases} X_k & \text{if } |X_k| < k^{1/2-\eta} \\ 0 & \text{otherwise} \end{cases}$$

$(k = 1, 2, \ldots)$, where η, $0 < \eta < \frac{1}{2}$, is chosen such that

$$(\tfrac{1}{2} - \eta)(2 + \delta) > 1.$$

Let $Y_k = X_k' - \mathscr{E}X_k'$, $k = 1, 2, \ldots$. Clearly

$$|Y_k| \le 2k^{1/2-\eta} \qquad \text{a.s.,}$$

$\mathscr{E}Y_k^2 \to \mathscr{E}X_1^2 = 1$ as $k \to \infty$, and $\sum_{k=1}^n \mathscr{E}Y_k^2/n \to \mathscr{E}X_1^2 = 1$ as $n \to \infty$, so that $\sum_{k=1}^\infty \mathscr{E}Y_k^2 = \infty$. Let

$$(s_n')^2 = \sum_{k=1}^n \operatorname{var}(Y_k) \qquad \text{for } n = 1, 2, \ldots.$$

Then for the sequence $\{Y_n\}$ of independent random variables with $\mathscr{E}Y_n = 0$ and $\mathscr{E}Y_n^2 < \infty$ we have

$$(s_n')^2 \to \infty$$

and a.s.

$$\begin{aligned} |Y_n| &\le 2n^{(1/2)-\eta} = 2s_n^{1-2\eta} \\ &= \left\{\frac{2s_n[\ln \ln(s_n')^2]^{1/2}}{s_n'} s_n^{-2\eta}\right\} s_n'[\ln \ln(s_n')^2]^{-1/2} \end{aligned}$$

where the quantity in the braces $\to 0$ as $n \to \infty$. It follows immediately from Theorem 2.4.1 that

$$P\left\{\limsup_{n\to\infty} \frac{\sum_{k=1}^n Y_k}{[2(s_n')^2 \ln \ln(s_n')^2]^{1/2}} = 1\right\} = 1.$$

Since $n^{-1}(s_n')^2 = n^{-1}\sum_{k=1}^n \mathscr{E}Y_k^2 \to 1$, it follows that

$$P\left\{\limsup_{n\to\infty} \frac{\sum_{k=1}^n Y_k}{(2n \ln \ln n)^{1/2}} = 1\right\} = 1.$$

Note that

$$\sum_{k=1}^{\infty} P\{X'_k \neq X_k\} = \sum_{k=1}^{\infty} P\{|X_k| \geq k^{1/2-\eta}\}$$

$$= \sum_{k=1}^{\infty} P\{|X_1|^{2+\delta} \geq k^{(2+\delta)(1/2-\eta)}\}$$

$$\leq \sum_{k=1}^{\infty} P\{|X_1|^{2+\delta} \geq k\} < \infty$$

by the moments lemma and in view of the assumption $\mathscr{E}|X_1|^{2+\delta} < \infty$ and the choice of η. Since

$$\sum_{k=1}^{n} Y_k = \sum_{k=1}^{n} X'_k - \sum_{k=1}^{n} \mathscr{E}X'_k$$

and $\{X_k\}$ and $\{X'_k\}$ are convergence equivalent, we need only show that $\sum_{k=1}^{n} \mathscr{E}X'_k/(2n \ln \ln n)^{1/2} \to 0$ as $n \to \infty$. This is done as follows. By a simple rearrangement of the series and use of the fact that $\mathscr{E}|X_1|^{2+\delta} < \infty$, it is easy to show that

$$\sum_{k=1}^{\infty} (2k \ln \ln k)^{-1/2} \int_{|X_1| \geq k^{1/2-\eta}} |X_1| \, dP < \infty.$$

From Kronecker's lemma therefore

$$(2n \ln \ln n)^{-1/2} \sum_{k=1}^{n} \int_{|X_1| \geq k^{1/2-\eta}} |X_1| \, dP \to 0 \qquad \text{as } n \to \infty.$$

But

$$\sum_{k=1}^{n} \mathscr{E}X'_k = \sum_{k=1}^{n} \int_{|X_k| < k^{1/2-\eta}} X_k \, dP$$

$$= \sum_{k=1}^{n} \left(\int X_k \, dP - \int_{|X_k| \geq k^{1/2-\eta}} X_k \, dP \right)$$

$$= - \sum_{k=1}^{n} \int_{|X_1| \geq k^{1/2-\eta}} X_1 \, dP.$$

since $\mathscr{E}X_k = 0$. It follows that

$$(2n \ln \ln n)^{-1/2} \sum_{k=1}^{n} \mathscr{E}X'_k \to 0 \qquad \text{as } n \to \infty.$$

This completes the proof.

Corollary 2 to Theorem 2.4.1. Under the hypothesis of Theorem 2.4.1, we have

$$P\left\{\limsup_{n\to\infty} \frac{T_n}{(2s_n^2 \ln \ln s_n^2)^{1/2}} = 1\right\} = 1$$

when T_n is taken to be any one of the random variables S_n, $|S_n|$, $\max_{1<k\le n} S_k$, $\max_{1\le k\le n} |S_k|$.

Proof. We need only prove the result for $T_n = \max_{1\le k\le n} S_k$ and $T_n = \max_{1\le k\le n} |S_k|$. Consider, for example, the second case. Let $\varphi(x) \uparrow \infty$ as $x \uparrow \infty$, and write

$$A_1 = \{\omega: |S_n(\omega)| \ge \varphi(s_n) \text{ i.o.}\},$$

$$A_2 = \left\{\omega: \max_{1\le k\le n} |S_k(\omega)| \ge \varphi(s_n) \text{ i.o.}\right\}.$$

We show that $A_1 = A_2$. Clearly $A_1 \subset A_2$. Let $\omega \in A_2$ be fixed. We show that $\omega \in A_1$. There exists a sequence $\{k_n\}$ with $1 \le k_n \le n$ such that

$$S_{k_n}(\omega) = \max_{1\le k\le n} S_k(\omega) \ge \varphi(s_n) \ge \varphi(s_{k_n}).$$

Since $\varphi(s_n) \to \infty$, $k_n \to \infty$ and we can extract a subsequence $\{k_n'\}$ of $\{k_n\}$ such that $k_n' \uparrow \infty$ and $S_{k_n'}(\omega) \ge \varphi(s_{k_n'})$. It follows that $\omega \in A_1$ as asserted.

The final result of this section is the following converse to the Hartman–Wintner version of the law of the iterated logarithm. The proof given here is due to Feller.

Theorem 2.4.2. Let $\{X_n\}$ be a sequence of independent, identically distributed random variables. If $\mathscr{E}X_1^2 = \infty$, then

$$P\left\{\limsup_{n\to\infty} \frac{|S_n|}{(2n \ln \ln n)^{1/2}} = \infty\right\} = 1. \tag{2.4.25}$$

Proof. Let us suppose that the X_i are symmetric. Let $N > 0$ be fixed, and let Y_i be X_i truncated at $c = c(N)$, where $c > 0$ is chosen such that $\mathscr{E}Y_i^2 \ge c$. Let

$$S_n = \sum_{i=1}^n X_i \qquad \text{and} \qquad T_n = \sum_{i=1}^n Y_i,\ n \ge 1.$$

Since the X_i are symmetric random variables for each $i \ge 1$, X_i and $Y_i - X_i\chi_{\{|X_i|\ge c\}}$ are identically distributed. It follows that the events

$$A_n = \{T_n \ge (Nn \ln \ln n)^{1/2}, S_n - T_n \ge 0 \text{ i.o.}\}$$

and

$$B_n = \{T_n \geq (Nn \ln \ln n)^{1/2}, S_n - T_n \leq 0 \text{ i.o.}\}$$

have the same probability. Since both the events are exchangeable in the sense of Definition 2.2.1, if follows from the Hewitt–Savage zero-one law (Theorem 2.2.2) that both events have probability either 0 or 1. But by Theorem 2.4.1

$$P\{T_n \geq (Nn \ln \ln n)^{1/2} \text{ i.o.}\} = 1,$$

and since

$$\{T_n \geq (Nn \ln \ln n)^{1/2} \text{ i.o.}\} = A_n \cup B_n,$$

it follows that

$$P(A_n \text{ i.o.}) = 1.$$

Therefore for all $N > 0$

$$P\{S_n \geq (Nn \ln \ln n)^{1/2} \text{ i.o.}\} \geq P\{A_n \text{ i.o.}\} = 1,$$

that is,

$$P\left\{\limsup_{n\to\infty} \frac{S_n}{(2n \ln \ln n)^{1/2}} = \infty\right\} = 1.$$

Since the result also holds if we replace each X_n by $-X_n$, it follows that (2.4.25) holds for symmetric summands.

In the general case we use the symmetrization procedure to obtain the same result. ∎

Remark 2.4.4. Let $\{X_n\}$ be a sequence of independent, identically distributed random variables. If

$$P\left\{\limsup_{n\to\infty} \frac{|S_n|}{(2n \ln \ln n)^{1/2}} < \infty\right\} > 0,$$

then from Kolmogorov's zero-one law

$$P\left\{\limsup_{n\to\infty} \frac{|S_n|}{(2n \ln \ln n)^{1/2}} < \infty\right\} = 1.$$

It follows therefore that $n^{-1}S_n \xrightarrow{\text{a.s.}} 0$, and hence that $\mathscr{E}X_1 = 0$. Also $\mathscr{E}X_1^2 < \infty$ in view of Theorem 2.4.2.

2.5 APPLICATIONS

In this section we consider some applications of the two laws of large numbers proved in Sections 2.1 and 2.3.

(a) An Application in Analysis. As an application of the weak law of large numbers (Theorem 2.1.1) we prove the Weierstrass approximation theorem.

Proposition 2.5.1. Let f be a continuous function defined on $[a, b]$. Then there exists a sequence of polynomials $\{B_n\}$ for which $\lim_{n\to\infty} B_n(x) = f(x)$ uniformly in $a \le x \le b$.

Proof. Since the function $f(a + (b - a)x)$ is continuous on $[0, 1]$, it is sufficient to prove the result for $a = 0$ and $b = 1$.

Let S_n have a binomial distribution with parameters n and x, $0 \le x \le 1$. Set $\bar{X}_n = n^{-1}S_n$, $n = 1, 2, \ldots$, and define B_n by

$$B_n(x) = \mathscr{E}\{f(\bar{X}_n)\}, \qquad 0 \le x \le 1.$$

Then

$$\mathscr{E}\{f(\bar{X}_n)\} = \sum_{k=0}^{n} f\left(\frac{k}{n}\right)\binom{n}{k}x^k(1 - x)^{n-k}.$$

The right-hand side is a polynomial of degree n and is known as the Bernstein polynomial of degree n associated with the given function f.

Since f is continuous, f is bounded and uniformly continuous on $[a, b]$. Therefore there exists a constant c such that $|f(x)| \le c$, $0 \le x \le 1$, and for any $\epsilon > 0$ there exists a constant δ such that

$$|f(x) - f(y)| < \epsilon \qquad \text{whenever } |x - y| < \delta, \qquad 0 \le x, y \le 1.$$

For any $x \in [0, 1]$ we have

$$\begin{aligned}
|f(x) - B_n(x)| &= \left|\sum_{k=0}^{n}\left[f(x) - f\left(\frac{k}{n}\right)\right]\binom{n}{k}x^k(1 - x)^{n-k}\right| \\
&\le \left\{\sum_{|k/n - x| < \delta} + \sum_{|k/n - x| \ge \delta}\right\}\left|f(x) - f\left(\frac{k}{n}\right)\right|\binom{n}{k}x^k(1 - x)^{n-k} \\
&\le \epsilon + 2c\sum_{|k/n - x| \ge \delta}\binom{n}{k}x^k(1 - x)^{n-k} \\
&= \epsilon + 2cP\{|\bar{X}_n - x| \ge \delta\} \\
&\le \epsilon + 2c\frac{x(1 - x)}{n\delta^2} \\
&\le \epsilon + 2c\frac{1}{4n\delta^2}
\end{aligned}$$

by the weak law of large numbers (or Chebyshev's inequality). Choosing n so large that $n \geq c/2\delta^2\epsilon$, we see that $|f(x) - B_n(x)| \leq 2\epsilon$ uniformly in $x \in [0, 1]$. Thus

$$\lim_{n\to\infty} \sup_{0 \leq x \leq 1} |f(x) - B_n(x)| = 0$$

as asserted. ■

We note that the Weierstrass theorem only asserts the possibility of uniform approximation to any desired degree of accuracy by some polynomial. The present result, due to Bernstein, is somewhat stronger in that it exhibits the approximating polynomials.

Under some additional conditions on f we can get a fairly good (in fact, the best possible) estimate of $|f(x) - B_n(x)|$. For example, let f satisfy the Hölder condition with exponent α, $0 \leq \alpha \leq 1$, that is, for every $x, y \in [0, 1]$ let

$$|f(x) - f(y)| < M|x - y|^\alpha,$$

where M is a constant independent of x, y. (For $\alpha = 1$ this is the Lipschitz condition.) For such an f we have

$$\begin{aligned} |f(x) - B_n(x)| &\leq \mathscr{E}|f(x) - f(\overline{X}_n)| \\ &\leq M\mathscr{E}|x - \overline{X}_n|^\alpha \\ &= M\mathscr{E}\left|\frac{\sum_{i=1}^n (X_i - x)}{n}\right|^\alpha. \end{aligned}$$

From Hölder's inequality (1.2.14) with

$$p = \frac{2}{\alpha}, \qquad q = \frac{2}{2-\alpha}, \qquad X = \left|\frac{\sum_{i=1}^n (X_i - x)}{n}\right|^\alpha, \qquad \text{and} \qquad Y \equiv 1,$$

we get

$$\begin{aligned} |f(x) - B_n(x)| &\leq M\left[\mathscr{E}\left(\frac{\sum_{i=1}^n (X_i - x)}{n}\right)^2\right]^{\alpha/2} \\ &= M\left[\frac{x(1-x)}{n}\right]^{\alpha/2} \\ &\leq M2^{-\alpha}n^{-\alpha/2}. \end{aligned}$$

(b) An Application in Number Theory. A number $x \in [0, 1]$ is said to be *normal with respect to the base b* if each digit in the expansion of x to the base b has the same limiting relative frequency (namely, $1/b$). We say that x is *normal* if it is normal to the base b for every $b > 1$. A well-known result in number theory is that except for a set of Lebesgue measure zero all real numbers are normal. We derive this result as a simple application of the strong law (Theorem 2.3.1).

Proposition 2.5.2 (Borel). Let $\Omega = [0, 1]$, and P be the Lebesgue measure on Ω. Almost every (with respect to P) real number in Ω is normal.

Proof. Since there are only a countable number of bases, it suffices to show that almost every real number in Ω is normal to the base 10, say. Let $\omega \in \Omega$, and consider the decimal expansion $\omega = 0 \cdot x_1 x_2, \ldots, x_i \in \{0, 1, 2, \ldots, 9\}$, with any convention adopted to make the expansion unique. (The set of ω's which have more than one expansion has P-measure zero.) Let $N_n^{(k)}(\omega)$ denote the number of times k appears in $x_1, x_2, \ldots, x_n$, $k = 0, 1, \ldots, n$. We wish to show that $N_n^{(k)}/n \to \frac{1}{10}$ as $n \to \infty$ for $k = 0, 1, 2, \ldots, n$.

For each $n \geq 1$, let $X_n(\omega)$ denote the nth number in the expansion of ω for each $\omega \in \Omega$. Then the X_n are independent, identically distributed random variables with common discrete uniform distribution on $\{0, 1, 2, \ldots, 9\}$, that is, $P\{X_n = k\} = \frac{1}{10}$ for each $k = 0, 1, \ldots, 9$ and $n \geq 1$. Let $k \in \{0, 1, \ldots, 9\}$ be fixed, and define random variable Y_i by

$$Y_i(\omega) = \begin{cases} 1 & \text{if } X_i(\omega) = k, \\ 0 & \text{otherwise} \end{cases}$$

($\omega \in \Omega$). Then $\{Y_i\}$ is a sequence of independent, identically distributed Bernoulli random variables with $\mathscr{E} Y_i = P\{X_i = k\} = \frac{1}{10}$. It follows from the strong law of large numbers that

$$n^{-1} \sum_{i=1}^{n} Y_i \xrightarrow{\text{a.s.}} \mathscr{E} Y_1 = \frac{1}{10} \qquad \text{as } n \to \infty.$$

Since $n^{-1} \sum_{i=1}^{n} Y_i(\omega) = n^{-1} N_n^{(k)}(\omega)$, the proof is complete. ■

Note that rational numbers are not normal, although they may be with respect to a particular base. Also, although nearly every real number is normal, it is quite difficult to exhibit a normal number. A simple example in decimal notation is

$$0.123456789101112131415161718192021\ldots,$$

where we have written all positive integers in succession after the decimal point.

Finally, according to Proposition 2.5.2 and in the notation used there,

$$N_n^{(k)}(\omega) = O(n) \qquad \text{as } n \to \infty.$$

A much better estimate is provided by the law of the iterated logarithm. According to the law of the iterated logarithm, the best possible estimate is given by

$$\limsup_{n\to\infty} \frac{N_n^{(k)}(\omega) - 0.10n}{(n \ln \ln n)^{1/2}} = 0.3\sqrt{3}$$

for almost all $\omega \in [0, 1]$.

(c) Monte Carlo Simulation. Let $f: [0, 1] \to [0, 1]$ be a continuous function, and suppose that the integral $\int_0^1 f(x)\,dx$ is to be evaluated numerically. Suppose we use a random number generator to generate independent random numbers in $[0, 1]$. Let $X_1, Y_1, X_2, Y_2, \ldots$ be independent, indentically distributed random variables with common probability density function

$$g(x) = 1 \quad \text{for } 0 \le x \le 1 \qquad \text{and} \qquad g(x) = 0 \quad \text{otherwise.}$$

For $i = 1, 2, \ldots$, define

$$Z_i = \begin{cases} 1 & \text{if } f(X_i) > Y_i, \\ 0 & \text{otherwise.} \end{cases}$$

The numbers Z_i can be obtained with the help of some computing device which for each pair (X_i, Y_i) tells us whether $f(X_i) > Y_i$ or $f(X_i) \le Y_i$. Clearly the Z_i are independent, identically distributed random variables with

$$\mathscr{E}Z_1 = P\{f(X_1) > Y_1\} = \int_0^1 f(x)\,dx.$$

By the strong law of large numbers it follows that

$$n^{-1}\sum_{i=1}^{n} Z_i \xrightarrow{\text{a.s.}} \int_0^1 f(x)\,dx \qquad \text{as } n \to \infty.$$

Therefore, by using a random number generator to generate (X_i, Y_i) and a computing device to compute $n^{-1}\sum_{i=1}^{n} Z_i$, we can evaluate the integral numerically.

(d) Applications in Statistics. Let $(\Omega, \mathscr{S}, P)$ be a probability space, and let X be a random variable defined on it with distribution function F. We say that $X_1, X_2, \ldots, X_n$ constitute a *random sample of size n* from a population with distribution F, if $X_1, X_2, \ldots, X_n$ are independently and identically distributed random variables with common distribution function F. An important problem in statistics is to estimate F from the given sample $X_1, X_2, \ldots, X_n$. For this purpose we first introduce the concept of an *empirical* (or a *sample*) *distribution function.*

Definition.2.5.1. Let $X_1, X_2, \ldots, X_n$ be a random sample of size n from a distribution function F. Let $\hat{F}_n$ be the step function defined by

$$\hat{F}_n(x) = \frac{\text{number of } X_i \text{ that are} \leq x}{n}, \qquad x \in \mathbb{R}. \tag{2.5.1}$$

Then $\hat{F}_n$ is called the empirical (or sample) distribution function based on the sample $X_1, X_2, \ldots, X_n$.

Remark 2.5.1. Clearly we can rewrite $\hat{F}_n$ as

$$\hat{F}_n(x) = \frac{1}{n} \sum_{k=1}^{n} \chi[X_k \leq x].$$

It then follows at once that $n\hat{F}_n(x)$, for fixed x, is the sum of n independently and identically distributed Bernoulli random variables with parameter $F(x)$. Thus $n\hat{F}_n(x)$ has a binomial distribution. From Theorem 2.3.1 it then follows immediately that, for every fixed x, $\hat{F}_n(x) \xrightarrow{\text{a.s.}} F(x)$ as $n \to \infty$. The question that naturally arises is whether this convergence is uniform in x a.s. We now prove a theorem due to Glivenko–Cantelli which provides an affirmative answer to this question.

Theorem 2.5.1 (Glivenko–Cantelli). Let $\{X_n\}$ be a sequence of independent, identically distributed random variables with common distribution function F. Let $\hat{F}_n$ be the empirical distribution function based on a sample of size n. Then

$$P\left\{\lim_{n \to \infty} \sup_{-\infty < x < \infty} |\hat{F}_n(x) - F(x)| = 0\right\} = 1. \tag{2.5.2}$$

Proof. Let r run through the set $\mathbb{Q}$ of all rational numbers. Since the supremum of a countable set of random variables is again a random variable and since

$$\sup_{r \in \mathbb{Q}} |\hat{F}_n(r) - F(r)| = \sup_{x \in \mathbb{R}} |\hat{F}_n(x) - F(x)|,$$

it follows that $\sup_{x \in \mathbb{R}} |\hat{F}_n(x) - F(x)|$ is a random variable.

Let $v \geq 2$ be a positive integer. Then for $k = 1, 2, \ldots, v - 1$ we set

$$x_{v,k} = \min\left\{x \in \mathbb{R}: \frac{k}{v} \leq F(x)\right\},$$

$$x_{v,0} = -\infty, \qquad \text{and} \qquad x_{v,v} = +\infty.$$

Without loss of generality we assume that $[x_{\nu,k}, x_{\nu,k+1}) \neq \emptyset$. Let $x \in [x_{\nu,k}, x_{\nu,k+1})$. Then

$$\begin{aligned}
\hat{F}_n(x) - F(x) &\leq \hat{F}_n(x_{\nu,k+1} - 0) - F(x_{\nu,k}) \\
&= [\hat{F}_n(x_{\nu,k+1} - 0) - F(x_{\nu,k+1} - 0)] \\
&\quad + [F(x_{\nu,k+1} - 0) - F(x_{\nu,k})] \\
&\leq \hat{F}_n(x_{\nu,k+1} - 0) - F(x_{\nu,k+1} - 0) + \frac{1}{\nu}.
\end{aligned}$$

In a similar manner we show that

$$\hat{F}_n(x) - F(x) \geq \hat{F}_n(x_{\nu,k}) - F(x_{\nu,k}) - \frac{1}{\nu}$$

for $x \in [x_{\nu,k}, x_{\nu,k+1})$. It follows therefore that for every $x \in \mathbb{R}$

$$\begin{aligned}
|\hat{F}_n(x) - F(x)| \leq \max_{\substack{1 \leq k \leq \nu - 1 \\ 1 \leq j \leq \nu - 1}} &\{|\hat{F}_n(x_{\nu,k}) - F(x_{\nu,k})|, \\
&|\hat{F}_n(x_{\nu,j} - 0) - F(x_{\nu,j} - 0)|\} + \frac{1}{\nu}.
\end{aligned}$$

We now take the supremum of both sides over all $x \in \mathbb{R}$ and then take the limit of both sides as $n \to \infty$. It follows that

$$\lim_{n \to \infty} \sup_{x \in \mathbb{R}} |\hat{F}_n(x) - F(x)| \leq \frac{1}{\nu} \qquad \text{a.s.}$$

Since ν is arbitrary, the conclusion follows immediately. ∎

Consider a population whose distribution is known to have a probability density p, but p is otherwise unknown. Let $x_1, x_2, \ldots, x_N$ be N observations from this population. In elementary statistics courses one estimates the probability density by drawing a histogram based on $x_1, x_2, \ldots, x_N$. Suppose we divide the real line into a sequence of disjoint subintervals $[a_k, a_{k+1})$, $k = 0, 1, 2, \ldots$. Let f_k be the number of x_i which lie in the interval $I_k = [a_k, a_{k+1})$, $k = 0, 1, 2, \ldots$. Then f_k is called the *class frequency* of the interval I_k, and $p_k = f_k/N$ is called the *relative frequency*. Clearly $\sum_{k=0}^{\infty} p_k = 1$. It is usually asserted that the histogram based on $x_1, x_2, \ldots, x_N$ approximates the probability density function p in the sense that, if N is large enough and the subdivision of $\mathbb{R}$ is fine enough, the histogram is "close" to the curve p.

We shall show that $p_k \to \int_{I_k} p(x)\, dx$ as $N \to \infty$. To do so we consider a sequence $\{X_n\}$ of independent random variables with common probability density function p. We choose and fix a decomposition of $\mathbb{R}$ into subintervals

$I_k = [a_k, a_{k+1}), k = 0, 2, \ldots,$ and let Y_k be the number of X_j $(j = 1, 2, \ldots, N)$ that lie in I_k, $k \geq 0$. Clearly Y_k is a random variable. We show that if $\mathscr{E}|X_1| < \infty$ then

$$\lim_{N\to\infty} \frac{Y_k}{N} = \int_{I_k} p(x)\, dx, \qquad k \geq 0,$$

with probability 1.

To see this let us define for each $j = 1, 2, \ldots$ and fixed $k \geq 0$ the random variable Z_j by

$$Z_j(\omega) = \begin{cases} 1 & \text{if } X_j(\omega) \in I_k, \\ 0 & \text{otherwise.} \end{cases}$$

Then $P\{Z_j = 1\} = \int_{I_k} p(x)\, dx$ and $P\{Z_j = 0\} = 1 - P\{Z_j = 1\}$. Moreover, since the X_j are independent and identically distributed, so are the Z_j. Clearly

$$\mathscr{E}Z_j = P\{Z_j = 1\} = \int_{I_k} p(x)\, dx.$$

It follows immediately from the strong law of large numbers (Theorem 2.3.1) that

$$\frac{1}{N}\sum_{j=1}^{N} Z_j = \frac{Y_k}{N} \to \int_{I_k} p(x)\, dx \qquad \text{as } N \to \infty$$

on some event E_k with probability 1. Hence

$$\frac{Y_k}{N} \to \int_{I_k} p(x)\, dx \qquad \text{as } N \to \infty$$

for all k on the event $E = \bigcap_{k=1}^{\infty} E_k$, which also has probability 1.

Another frequent application is in statistical estimation theory. Let F_θ be a distribution function on $\mathbb{R}$ where $\theta \in \Theta \subseteq \mathbb{R}$ is an unknown parameter to be estimated from observations $X_1, X_2, \ldots, X_n$ on F_θ. ($X_1, X_2, \ldots, X_n$ are independent and identically distributed with common distribution function F_θ.) A *statistic* $T_n = T(X_1, \ldots, X_n)$ is a Borel-measurable map of $\mathbb{R}_n \to \mathbb{R}$. We say that $\{T_n\}$ is *consistent* for θ if $T_n \xrightarrow{P} \theta$ as $n \to \infty$ for all $\theta \in \Theta$. We say that $\{T_n\}$ is *strongly consistent* for θ if $T_n \xrightarrow{\text{a.s.}} \theta$. The two laws of large numbers can easily be used to derive consistent sequences of estimators (statistics). For example, if θ is the mean of F_θ, then $\overline{X}_n = \sum_{i=1}^{n} X_i/n$ is clearly strongly consistent for θ. If θ is the variance of F_θ, then $\sum_{i=1}^{n} (X_i - \overline{X}_n)^2/n$ is strongly consistent for θ. Sometimes one has an *asymptotically unbiased estimator* T_n

for θ, that is, $\mathscr{E}_\theta T_n \to \theta$ as $n \to \infty$. Then $\{T_n\}$ is consistent for θ provided that $\text{var}_\theta(T_n) \to 0$ for each $\theta \in \Theta$. This follows trivially from Chebyshev's inequality. In fact,

$$\begin{aligned} P_\theta\{|T_n - \theta| \geq \epsilon\} &\leq \epsilon^{-2}\mathscr{E}_\theta(T_n - \theta)^2 \\ &= \epsilon^{-2}\{\text{var}_\theta(T_n) + (\mathscr{E}_\theta T_n - \theta)^2\} \to 0 \end{aligned}$$

as $n \to \infty$ for each $\theta \in \Theta$. (Here $\mathscr{E}_\theta$ means that expectation is computed with respect to F_θ.)

(e) Renewal Theory. Let $\{X_n\}$ be a sequence of independent, identically distributed random variables, and suppose that $X_n > 0$ a.s. for all $n \geq 1$. Let $S_0 = 0$ and, for $n \geq 1$, $S_n = \sum_{k=1}^n X_k$. Then $\{S_n\}$ is an increasing sequence of a.s. positive random variables. In many applications where a sequence of events occurs in time in an irregular manner, S_n can be regarded as the (waiting) time until the occurrence of the nth event after time $t = 0$. Thus, if at time $t = 0$ a piece of equipment is placed into service, and remains in use until, after a time X_1, it fails and is then replaced by a second identical piece of the same equipment, which fails after a time X_2 and is itself replaced, and so on, then S_n describes the instant at which the nth piece of equipment fails and is replaced. Often $\{S_n\}$ is called a *renewal process*.

Suppose that $\mathscr{E}X_n = \mu$, $0 < \mu < \infty$, for all $n \geq 1$. By the strong law of large numbers $n^{-1}S_n \xrightarrow{\text{a.s.}} \mu$ as $n \to \infty$, and it follows that $S_n \xrightarrow{\text{a.s.}} \infty$ as $n \to \infty$. Therefore the number of points S_n in any finite interval is a.s. finite.

For $t > 0$, let $N(t)$ denote the number of points S_n in the interval $(0, t]$. Formally, let

(2.5.3) $$N(t) = \max_{n \geq 0} \{n: 0 < S_n \leq t\}.$$

For any nonnegative integer n

(2.5.4) $$\{\omega: N(t)(\omega) = n\} = \{\omega: S_n(\omega) \leq t < S_{n+1}(\omega)\}.$$

It follows that $N(t)$ is a nonnegative integer-valued random variable defined on Ω. For every $k \geq 1$ we have

(2.5.5) $$\{N(t) < k\} = \bigcup_{n=0}^{k-1} \{N(t) = n\} = \{S_k > t\}.$$

Lemma 2.5.1. We have

$$N(t) \xrightarrow{\text{a.s.}} \infty \qquad \text{as } t \to \infty.$$

Proof. For every fixed $\omega \in \Omega$, $N(t)(\omega)$ is nondecreasing, so that $\lim_{t \to \infty} N(t)(\omega)$ is either $+\infty$ or finite. Suppose that

$$P\left\{\omega : \lim_{t \to \infty} N(t) < \infty\right\} > 0.$$

Then there exists an integer $M > 0$ such that

$$P\left\{\sup_{t \in [0, \infty)} N(t) < M\right\} > 0.$$

In view of (2.5.5) this immediately implies that

$$P\left\{\sup_{t \in [0, \infty)} S_M > t\right\} > 0,$$

so that $P\{S_M = +\infty\} > 0$, which is a contradiction since S_M is a.s. finite. Hence $N(t) \xrightarrow{\text{a.s.}} \infty$ as $t \to \infty$. ∎

Theorem 2.5.2. Let $\{X_n\}$ be a sequence of independent, identically distributed random variables which are a.s. positive. Suppose that $\mathscr{E}X_1 = \mu, 0 < \mu < \infty$. Let S_n, $N(t)$ be as defined above. Then as $t \to \infty$

$$t^{-1}N(t) \xrightarrow{\text{a.s.}} \mu^{-1}. \tag{2.5.6}$$

Proof. From (2.5.4) it follows that for every $\omega \in \Omega$

$$S_{N(t)(\omega)}(\omega) \leq t < S_{N(t)(\omega)+1}(\omega).$$

Since $N(t) \xrightarrow{\text{a.s.}} \infty$, we can choose t sufficiently large so that $N(t)(\omega) > 0$. Then

$$\frac{S_{N(t)(\omega)}(\omega)}{N(t)(\omega)} \leq \frac{t}{N(t)(\omega)} < \frac{S_{N(t)(\omega)+1}(\omega)}{N(t)(\omega) + 1} \cdot \frac{N(t)(\omega) + 1}{N(t)(\omega)}.$$

It follows immediately from the strong law of large numbers that as $t \to \infty$

$$\mu \leq \liminf_{t \to \infty} \frac{t}{N(t)} \leq \limsup_{t \to \infty} \frac{t}{N(t)} \leq \mu,$$

which yields (2.5.6). ∎

The expected number of renewals during $(0, t]$: $\mathscr{E}N(t) = M(t)$ is called the *renewal function.* We will show that, for every $t > 0$, $\mathscr{E}N(t) < \infty$. Let F_n be the distribution function of S_n. If F is the distribution function of $X_i (i = 1, 2, \ldots)$, then

$$F_n(x) = P\{S_n \leq x\} = \int_0^\infty F_{n-1}(x - y)\, dF(y) = \int_0^x F_{n-1}(x)\, dF(y)$$

for $n = 2, 3, \ldots$, where $F_1 = F$. Note that

(2.5.7) $$M(t) = \mathscr{E}N(t) = \sum_{n=1}^{\infty} P\{N(t) \geq n\} = \sum_{n=1}^{\infty} F_n(t).$$

Now, for $1 < m \leq n - 1$ and $t > 0$,

$$F_n(t) = P\{S_{n-m} + S_m \leq t\} = \int_0^t F_{n-m}(t - y)\, dF_m(y)$$
$$\leq F_{n-m}(t)F_m(t),$$

so, in particular,

$$F_{lm+n}(t) \leq F_{(l-1)m+n}(t)F_m(t)$$

for any integers l, m, n. By iteration it follows that

(2.5.8) $$F_{lm+n}(t) \leq [F_m(t)]^l F_n(t), \qquad 0 \leq n < m - 1.$$

Since the X_i are a.s. positive random variables, $F(0) = 0$; hence there exists some number x_0 such that $F(x_0) < 1$. Similarly for each $t > 0$ we see easily (by induction, for example) that there must exist an integer m such that $F_m(t) < 1$. It follows immediately from (2.5.7) and (2.5.8) that $M(t) < \infty$ for all t.

In view of Theorem 2.5.2 it is natural to ask whether (2.5.6) holds if we replace N by its expectation M.

Theorem 2.5.3 (Elementary Renewal Theorem). Let $\{X_n\}$ be a sequence of independent, identically distributed random variables which are a.s. positive. Suppose that $\mathscr{E}X_1 = \mu$, $0 < \mu < \infty$. Let S_n, $N(t)$, and $M(t)$ be as defined above. Then

(2.5.9) $$\lim_{t \to \infty} \{t^{-1}M(t)\} = \mu^{-1}.$$

Proof. Clearly there exists a $\lambda > 0$ such that

$$P\{X_n \geq \lambda\} > 0 \qquad \text{for all } n.$$

Consider the sequence $\{X_n'\}$ of truncated random variables, defined by

$$X_n' \doteq \begin{cases} \lambda, & X_n \geq \lambda, \\ 0, & \text{otherwise} \end{cases}$$

($n = 1, 2, \ldots$). Then the X_n' are independent, identically distributed random variables such that $X_n' \leq X_n$ a.s. for $n = 1, 2, \ldots$. Let $S_n' = \sum_{k=1}^n X_k'$, and $N'(t)$ be the number of $S_1', S_2', \ldots$ that $<t$. Then $S_n' \leq S_n$ a.s. and $N'(t) \geq N(t)$ for all t. Let $P\{X_n \geq \lambda\} = p$. Then each X_n'/λ is a Bernoulli random variable

with parameter p. Note that, if t is an integer, $N(t+1) - N(t)$ is the number of times the partial sums take on the value t. Also

$$N(t) = \sum_{j=0}^{[t]-1} \{N(j+1) - N(j)\},$$

where $[t]$ is the largest integer $\le t$. A simple calculation now shows that

$$\mathscr{E}\left(\frac{N(t)}{t}\right)^2 \le \mathscr{E}\left(\frac{N'(t)}{t}\right)^2 = O(1) \qquad \text{as } t \to \infty,$$

so that

$$\limsup_{t\to\infty} \mathscr{E}\left(\frac{N(t)}{t}\right)^2 \le O(1).$$

Since $N(t)/t \xrightarrow{P} 1/\mu$, it follows that (2.5.9) holds. ■

2.6 PROBLEMS

SECTION 2.2

1. Let $\{X_k\}$ be a sequence of random variables which are independent and identically distributed. Suppose that $\mathscr{E}X_k = 0$ and $0 < \mathscr{E}X_k^2 < \infty$. Determine values of α for which $\sum_{k=1}^{n} X_k/n^\alpha$ converges in probability.

2. Let $\{X_n\}$ be a sequence of independent random variables with common distribution having the Cauchy probability density function

$$f(x) = \frac{1}{\pi}\frac{1}{1+x^2}, \qquad x \in \mathbb{R}.$$

Show that the weak law of large numbers does not hold for the sequence $\{X_n\}$.

3. Consider a sequence of independent and identically distributed random variables $\{X_k\}$ with common distribution as follows:

(a) $P\{X_k = \pm 2n\} = 1/(n \ln n)$, $P\{X_k = 0\} = 1 - 2/(n \ln n)$, $n \ge 2$.
(b) $P\{X_k = n\} = c/(n \ln n)^2$, $n \ge 2$, where $c = \{\sum_{n=2}^{\infty} 1/(n \ln n)^2\}^{-1}$, $n \ge 2$.
(c) $P\{X_k = 2^{n-2\ln n - 2\ln\ln n}\} = 1/2^{n-1}$, $n \ge 2$.

Does the weak law of large numbers hold for the corresponding sequence $\{X_k\}$?

4. Show that the weak law of large numbers holds for the sequence of independent random variables $\{X_n\}$ for which

$$P\{X_n = \pm n^\alpha\} = \tfrac{1}{2}$$

if and only if $\alpha < \frac{1}{2}$.

5. Show that the weak law of large numbers holds for the sequence of independent random variables $\{X_n\}$ for which

$$\max_{1 \le k \le n} \int_{|x| \ge a} |x| \, dF_k(x) \to 0 \qquad \text{as } a \to \infty.$$

Here F_k is the distribution function of X_k, $k \ge 1$.

6. Let $\{X_k\}$ be independent with common distribution function F. In order that there exist constants $\{A_n\}$ such that $\{X_k\}$ obeys the weak law of large numbers show that it is necessary and sufficient that

$$nP\{|X_1| > n\} \to 0 \qquad \text{as } n \to \infty.$$

In this case we may choose $A_n = \int_{-n}^{n} x \, dF(x)$.

7. Let $\{X_n\}$ be a sequence of independent, identically distributed a.s. positive random variables. Let F be the common distribution function. In order that there exist constants B_n such that $P\{|B_n^{-1} \sum_{k=1}^{n} X_k - 1| \ge \epsilon\} \to 0$ for $\epsilon > 0$ show that it is necessary and sufficient that

$$\int_0^t \frac{x \, dF(x)}{nP\{X_1 \ge n\}} \to 0 \qquad \text{as } n \to \infty.$$

In this case show that there exist constants t_n such that $n \int_0^{t_n} x \, dF(x) = t_n$ and $B_n^{-1} \sum_{k=1}^{n} X_k \xrightarrow{P} 1$ with $B_n = n \int_0^{t_n} x \, dF(x)$.

8. Let $\{X_n\}$ be a sequence of independent and identically distributed random variables with $\mathscr{E} X_1 = 0$. Show that $\mathscr{E}|S_n|/n \to 0$ as $n \to \infty$.

SECTION 2.2

9. Let $\{A_n\}$ be a sequence of independent events with $P(A_n) < 1$ for all $n \ge 1$. Show that $P(A_n \text{ i.o.}) = 1 \Leftrightarrow P(\bigcup_{n=1}^{\infty} A_n) = 1$.

10. For any sequence of events $\{A_n\}$ satisfying $P(A_n) \to 0$ as $n \to \infty$ and

$$\sum_{n=1}^{\infty} P(A_n \bigcap A_{n+1}^c) < \infty,$$

show that $P(A_n \text{ i.o.}) = 0$.

11. (a) Let $(\Omega, \mathscr{S}, P)$ be a probability space. If, for every $A \in \mathscr{S}$ with $P(A) > 0$, $\sum_{k=1}^{\infty} P(A \cap A_k) = \infty$, where $A_k \in \mathscr{S}$ for all k, show that $P(\limsup_k A_k) = 1$.

(b) If there is an $A \in \mathscr{S}$ such that $\sum_{k=1}^{\infty} P(A \cap A_k) < \infty$, show that $P(\limsup_k A_k) \le 1 - P(A)$.

(c) In order that with probability 1 only finitely many of the A_k occur, show that it is necessary and sufficient that for every $\epsilon > 0$ there exists an event $D = D(\epsilon)$ having $P(D) > 1 - \epsilon$ and satisfying $\sum_{k=1}^{\infty} P(A_k \cap D) < \infty$.

12. Let $\{A_n\}$ and $\{B_n\}$ be sequences of events on $(\Omega, \mathscr{S}, P)$, and suppose that A_n is independent of $\{B_n, B_{n+1}, \ldots\}$ for each sufficiently large n. Suppose further than $P(B_n \text{ i.o.}) = 1$ and $\liminf_{n \to \infty} P(A_n) \ge \delta > 0$. Show that $P(A_n \cap B_n \text{ i.o.}) \ge \delta$.

13. Let $\{X_n\}$ be independent, identically distributed random variables, and suppose that $\mathscr{E}|X_1| = +\infty$. Show that, for every $\alpha > 0$, $P\{|X_n| > n\alpha \text{ i.o.}\} = 1$. Also show that $P\{|S_n| > n\alpha \text{ i.o.}\} = 1$.

14. In a sequence of independent tossings of a coin with probability p of heads, let A_n be the event that a run of n consecutive heads occurs between the 2^nth and 2^{n+1}st trial. Show that A_n occurs i.o. if and only if $p \geq \frac{1}{2}$.

15. (a) Let $\{X_n\}$ be a sequence of random variables. Show that

$$\liminf_n X_n, \qquad \limsup_n X_n, \qquad \liminf_n \frac{1}{n} \sum_{k=1}^{n} X_k, \qquad \text{and} \qquad \limsup_n \frac{1}{n} \sum_{k=1}^{n} X_k$$

are tail functions.

(b) Show that the events

$$\left\{\sum_{n=1}^{\infty} X_n \text{ converges}\right\}, \qquad \left\{\lim_{n\to\infty} X_n \text{ exists}\right\}, \qquad \text{and} \qquad \{X_n < c \text{ i.o.}\}$$

are tail events.

16. Construct an example to show that the Hewitt–Savage zero-one law may not hold if the assumption of identical distribution is dropped.

17. Let $\{X_n\}$ be a sequence of independent random variables with common uniform distribution on (0, 1). Let $\{\lambda_n\}$ be a sequence of positive constants. Show that the infinite product $\prod_{n=1}^{\infty} X_n^{\lambda_n}$ converges a.s. if and only if $\sum_{n=1}^{\infty} \lambda_n < \infty$.

SECTION 2.3

18. Let $\{X_n\}$ be a sequence of random variables, and write $S_n = \sum_{k=1}^{n} X_k$, $n \geq 1$. Let $\{Y_n\}$ be a sequence of independent random variables with zero means and finite variances. Set $Z_k = X_k - Y_k$ for $k \geq 1$. Let $\{c_n\}$ be a nonincreasing sequence of positive real numbers. Then, for any $\epsilon > 0$, $0 < \eta < 1$, and any positive integers n and m, $m > n$, show that

$$P\left\{\max_{n\leq k\leq m} c_k|S_k| \geq \epsilon\right\} \leq (1-\eta)^{-2}\epsilon^{-2}\left(c_n^2 \sum_{k=1}^{n} \mathscr{E}Y_k^2 + \sum_{k=n+1}^{m} c_k^2 \mathscr{E}Y_k^2\right)$$
$$+ \sum_{k=n+1}^{m} P\{X_k \neq Y_k\} + P\left\{c_n \sum_{k=1}^{n} |Z_k| \geq \eta\epsilon\right\}.$$

In particular, if $\mathscr{E}X_k = 0$, $\mathscr{E}X_k^2 < \infty$ for all k, take $X_k = Y_k$ to get the Hájek–Rényi inequality

$$P\left\{\max_{n\leq k\leq m} c_k|S_k| \geq \epsilon\right\} \leq \epsilon^{-2}\left\{c_n^{-2} \sum_{k=1}^{n} \mathscr{E}Y_k^2 + \sum_{k=n+1}^{\infty} c_k^2 \mathscr{E}Y_k^2\right\},$$

which is a generalization of Kolmogorov's inequality (2.3.1).

19. Does the strong law of large numbers hold in each of the three cases in Problem 3?

20. Let $\{X_n\}$ be a sequence of a.s. nonnegative and independent random variables. Let $Y_n = \min(1, X_n)$. Show that $\sum_{n=1}^{\infty} X_n$ converges a.s.if and only if $\sum_{n=1}^{\infty} \mathscr{E}Y_n$ converges.

Moreover, show that $\sum_{n=1}^{\infty} X_n$ converges a.s. implies $\sum_{n=1}^{\infty} \mathscr{E} Y_n^p < \infty$ and $\sum_{n=1}^{\infty} (\mathscr{E} Y_n)^p < \infty$ for all $p \geq 1$.

21. Let $\{X_n\}$ be a sequence of independent random variables. Show that $\sum_{n=1}^{\infty} |X_n|$ converges a.s. if and only if for some $c > 0$ the series $\sum_{n=1}^{\infty} P\{|X_n| > c\}$ and $\sum_{n=1}^{\infty} \mathscr{E}|X_n|^c$ both converge. (Here $|X_n|^c$ is $|X_n|$ truncated at c.)

22. Show that the Kolmogorov sufficient condition for a.s. convergence in Proposition 2.3.7 is the best possible in the following sense. Let $\{\sigma_k\}$ be a sequence of positive constants such that $\sum_{k=1}^{\infty} \sigma_k^2/k^2 = \infty$. Then there exists a sequence of random variables $\{X_n\}$ which does not obey the strong law and for which $\mathscr{E}X_n = 0$ and $\mathscr{E}X_n^2 = \sigma_n^2$.

23. Show that the condition $\sum_{k=1}^{\infty} \sigma_k^2/k^2 < \infty$ is not a necessary condition for the strong law to hold.

24. Let $\{X_n\}$ be a sequence of independent random variables with $\mathscr{E}X_n = \mu$ and $\operatorname{var}(X_n) = \sigma_n^2$, $n = 1, 2, \ldots$. Suppose that $\sum_{k=1}^{\infty} 1/\sigma_k^2 = \infty$. Show that

$$\left\{\sum_{k=1}^{n} X_k/\sigma_k^2\right\}\left\{\sum_{k=1}^{n} 1/\sigma_k^2\right\}^{-1} \xrightarrow{P} \mu \qquad \text{as } n \to \infty.$$

(*Hint*: Use Proposition 2.3.3.)

25. Let $\{a_k\} = \{a_k(\omega)\}$ $(k = 0, 1, 2, \ldots)$ be an infinite sequence of complex-valued random variables defined on a probability space $(\Omega, \mathscr{S}, P)$. The series

$$f(z, \omega) = \sum_{k=0}^{\infty} a_k(\omega) z^k \qquad \text{for } \omega \in \Omega,\ z \in \mathbb{C},$$

is known as a random power series with coefficients $\{a_k\}$. For fixed $\omega \in \Omega$, $f(z, \omega)$ is a power series with radius of convergence $r(\omega) = \{\limsup_{n\to\infty} \sqrt{|a_n(\omega)|}\}^{-1}$, and for fixed z, the power series $f(z, \omega)$ is an infinite sum of random variables. We say that the series $\sum_{k=0}^{\infty} a_k(\omega)z^k$ converges (a.s., or in probability, or in $\mathscr{L}_p$, or in distribution) at the point $z \in \mathbb{C}$ if the sequence of partial sums $\{\sum_{k=0}^{n} a_k(\omega)z^k\}$ converges (a.s., or in probability, or in $\mathscr{L}_p$, or in distribution).

(a) Given an arbitrary random power series, show that there exist circles of convergence for the convergence a.s., the absolute convergence in probability, and the absolute convergence in distribution. The radii of convergence are all equal to

$$r_0 = \sup\{x : P\{r(\omega) \geq x\} = 1\}.$$

[A random power series is said to converge absolutely (a.s., ...) at $z \in \mathbb{C}$ if the sequence of random variables $\{\sum_{k=0}^{n} |a_k(\omega)z^k|\}$ converges (a.s., ...) at $z \in \mathbb{C}$.]

(b) Let $a_n \in \mathscr{L}_p$ for all $n \geq 1$, $p > 0$. Show that the circles of convergence exist for the absolute and ordinary convergence in $\mathscr{L}_p$. Their radii are equal, and equal to

$$r_p = \left\{\limsup_{n\to\infty} (\mathscr{E}|a_n|^p)^{1/n}\right\}^{-1/p}.$$

Moreover, if $0 < q \leq p$, then show that

$$r_p \leq r_q \leq r_0.$$

(c) For convergence in distribution or in probability show that no circle of convergence exists, in general.

26. Let $\sum_{n=0}^{\infty} a_n(\omega)z^n$ be a random power series with $r(\omega)$ as its radius of convergence. Let $G(x) = P\{r(\omega) \leq x\}$, $x \in [0, \infty]$. Show that corresponding to any distribution function G on $[0, \infty]$ there exists a random power series with radius of convergence $r(\omega)$ that has the prescribed distribution function G.

27. Consider a random power series $\sum_{k=0}^{\infty} a_k(\omega)z^k$, where $|a_k|$ are identically distributed and $P\{a_k = 0\} < 1$. Show that, if $|a_k|$ are, in addition, independent, $r_0 \leqq 1$. Moreover if $a_k \in \mathscr{L}_p$, show that $r_p = r_0 = 1$.

28. Let $\sum_{k=0}^{\infty} a_k(\omega)z^k$ be a random power series such that $|a_k|$ are independent and identically distributed with distribution function F. Show that

$$r_0 = \begin{cases} 1 \\ 0 \end{cases} \Leftrightarrow \int_1^{\infty} \ln x \, dF(x) \begin{cases} < \infty \\ = \infty \end{cases}.$$

29. Let $\{X_n\}$ be a sequence of independent, identically distributed random variables such that $\mathscr{E}|X_1| = \infty$. Show that for any sequence of constants $\{a_n\}$ we have

$$\limsup_{n \to \infty} |n^{-1}S_n - a_n| = \infty \qquad \text{a.s.}$$

30. (Kolmogorov–Marcinkiewicz Law of Large Numbers) Let $\{X_n\}$ be a sequence of independent, identically distributed random variables, and let $S_n = \sum_{k=1}^{n} X_k$, $n \geq 1$.

(a) If $\mathscr{E}|X_1|^r < \infty$ for some $0 < r < 2$, show that

$$\frac{S_n - nc_r}{n^{1/r}} \xrightarrow{\text{a.s.}} 0,$$

where $c_r = 0$ if $0 < r < 1$ and $c_r = \mathscr{E}X_1$ if $1 \leq r < 2$.

(b) Conversely, if $(S_n - b_n)/n^{1/r} \xrightarrow{\text{a.s.}} 0$ for some $0 < r < 2$ and some constants $\{b_n\}$, show that $\mathscr{E}|X_1|^r < \infty$.

(c) Let $0 < r < 2$. Show that

$$\frac{S_n - b_n}{n^{1/r}} \xrightarrow{\text{a.s.}} 0 \Leftrightarrow \mathscr{E}|X_1|^r < \infty \qquad \text{and} \qquad \frac{b_n - nc_r}{n^{1/r}} \to 0,$$

where c_r is as defined above.

31. Let $\{X_n\}$ be a sequence of independent random variables, and let $S_n = \sum_{k=1}^{n} X_k$, $n \geq 1$.

(a) If $\sum_{i=1}^{\infty} X_i^2 < \infty$ a.s., show that S_n converges a.s. if and only if

$$\sum_{k=1}^{\infty} \int_{|x| \leq 1} x \, dF_k(x) < \infty,$$

where F_k is the distribution function of X_k, $k \geq 1$.

(b) Suppose S_n converges a.s. Show that the following statements are equivalent:

(1) $\sum_{i=1}^{\infty} X_i^2 < \infty$ a.s.,
(2) $\sum_{k=1}^{\infty} \{\int_{|x|\le 1} |x|\, dF_k(x)\}^2 < \infty$,
(3) $\sum_{k=1}^{\infty} \{\int_{|x|\le 1} x\, dF_k(x)\}^2 < \infty$,

where F_k is the distribution function of X_k. (*Hint*: Use the three-series criterion.)

32. Let $\{X_n\}$ be a sequence of independent random variables. Show that

$$\sum_{i=1}^{\infty} X_i^2 < \infty \qquad \text{a.s. if and only if } \sum_{i=1}^{\infty} \mathscr{E}\, \frac{X_i^2}{1 + X_i^2} < \infty.$$

33. (Symmetrization Inequalities) Let $\{Y_n\}$ be a sequence of random variables, and let $\{b_n\}$ be a sequence of constants. Show that for any $\epsilon > 0$ and any $m \ge 1$ the inequalities

$$P\left\{\sup_{n\ge m} (Y_n - \text{med}(Y_n)) \ge \epsilon\right\} \le 2P\left\{\sup_{n\ge m} Y_n^s \ge \epsilon\right\}$$

and

$$P\left\{\sup_{n\ge m} |Y_n - \text{med}(Y_n)| \ge \epsilon\right\} \le 2P\left\{\sup_{n\ge m} |Y_n^s| \ge \epsilon\right\}$$
$$\le 4P\left\{\sup_{n\ge m} |Y_n - b_n| \ge \frac{\epsilon}{2}\right\}$$

hold. Here $\{Y_n^s\}$ is the symmetrized sequence $\{Y_n\}$. (*Hint*: The procedure is similar to the proof of Lévy inequalities, Lemma 2.4.1.)

34. Let $\{Y_n\}$ be a sequence of random variables, and $\{b_n\}$ a sequence of real numbers. Show that $Y_n - b_n \xrightarrow{\text{a.s.}} 0$ if and only if $Y_n^s \xrightarrow{\text{a.s.}} 0$ and $b_n - \text{med}(Y_n) \to 0$. (*Hint*: Use the symmetrization inequalities of Problem 33.)

35. Let $\{X_n\}$ be a sequence of independent random variables, and let $S_n = \sum_{k=1}^{n} X_k$, $n \ge 1$. Suppose that $a_n \uparrow \infty$ and that there exist a subsequence $\{a_{n_k}\}$ and constants c_1, c_2 such that $1 < c_1 \le a_{n_{k+1}}/a_{n_k} \le c_2 < \infty$ for all sufficiently large k. Write

$$S_{n_0} = 0 \qquad \text{and} \qquad T_k = a_{n_k}^{-1}(S_{n_k} - S_{n_{k-1}}).$$

Show that $a_n^{-1}(S_n - \text{med}(S_n)) \xrightarrow{\text{a.s.}} 0$ if and only if either condition (i) or (ii) holds:

(i) $T_k - \text{med}(T_k) \xrightarrow{\text{a.s.}} 0$.
(ii) $\sum_{k=1}^{\infty} P\{|T_k - \text{med}(T_k)| \ge \epsilon\} < \infty$ for every $\epsilon > 0$.

SECTION 2.4

36. Let $\{X_n\}$ be independent and identically distributed with common (nondegenerate) distribution. Let $S_n = \sum_{k=1}^{n} X_k$, $n = 1, 2, \ldots$.

(a) If $\mathscr{E}|X_n|^\alpha < \infty$ for $\alpha \ge 0$, show that

$$|S_n| = O(n^{1/2+\epsilon}) \qquad \text{a.s. for any } \epsilon > 0.$$

(b) If $|X_n| \le c$ a.s., where $c > 0$ is a constant, show that

$$\mathscr{E}\exp(xS_n) \le \exp\left\{\frac{nx^2\sigma^2}{2}(1 + xc)\right\}$$

for any $0 \le x \le 2c^{-1}$, where $\sigma^2 = \text{var}(X_n)$. Hence show that for $0 \le \epsilon \le 2\sigma^2 nc^{-1}$ we have

$$P\{S_n \ge \epsilon\} \le \exp\left\{-\frac{\epsilon^2}{2n\sigma^2}\left(1 - \frac{c\epsilon}{n\sigma^2}\right)\right\}.$$

Finally, show that if $|X_i| \le c$ a.s. then

$$|S_n| = O(\sqrt{n \ln n}) \qquad \text{a.s.}$$

(c) (Khintchine) Let $\{X_n\}$ be independent and identically distributed such that $|X_n| \le c$ a.s. Show that

$$P\left\{\limsup_{n\to\infty} \frac{|S_n|}{\sqrt{n \ln \ln n}} \le \sigma\sqrt{2}\right\} = 1.$$

37. (a) Let $\{X_n\}$ be a sequence of independent, identically distributed random variables with finite mean $\mu = 0$. Suppose that, for some $0 < \delta < 1$, $\mathscr{E}|X|^{1+\delta} = \infty$. Show that $P\{|S_n| > a_n \text{ i.o.}\} = 1$ if $a_n = n^{1/(1+\delta)}$ and $=0$ if $a_n = n$. For any sequence $\{a_n\}$ of constants for which there exists an ϵ, $0 \le \epsilon < 1$, such that

$$a_n n^{-1/(1+\epsilon)} \uparrow \qquad \text{and} \qquad a_n n^{-1} \downarrow,$$

show that

$$P\{|S_n| > a_n \text{ i.o.}\} = \begin{cases} 1 & \text{if } \sum_{n=1}^{\infty} P\{|X_n| \ge a_n\} = \infty, \\ 0 & \text{if } \sum_{n=1}^{\infty} P\{|X_n| \ge a_n\} < \infty. \end{cases}$$

(b) If $\mathscr{E}|X| = \infty$, and $a_n = n$, then $P\{|S_n| > a_n \text{ i.o.}\} = 1$. For any sequence $\{a_n\}$ with $a_n n^{-1} \uparrow$, show that

$$P\{|S_n| > a_n \text{ i.o.}\} = \begin{cases} 1 & \text{if } \sum_{n=1}^{\infty} P\{|X_n| \ge a_n\} = \infty, \\ 0 & \text{if } \sum_{n=1}^{\infty} P\{|X_n| \ge a_n\} < \infty. \end{cases}$$

(Feller [22])

38. Let $\{X_n\}$ be a sequence of independent random variables with $\mathscr{E}X_n = 0$ and $\mathscr{E}X_n^2 < \infty$ for all $n \ge 1$. Set $S_n = \sum_{k=1}^n X_k$ and $s_n^2 = \sum_{k=1}^n \mathscr{E}X_k^2$ for $n = 1, 2, \ldots$. Let $\{k_n\}$ be a sequence of positive constants such that $k_n \to 0$ as $n \to \infty$, and suppose that $s_n^2 \to \infty$ as $n \to \infty$. Show that, if

$$\sum_{n=1}^{\infty} (k_n s_n)^{-2} \ln \ln s_n^2 \int_{|x| > s_n k_n/\ln \ln s_n^2} x^2 \, dF_n(x) < \infty,$$

then

$$\limsup_{n\to\infty} \frac{S_n}{(2s_n^2 \ln \ln s_n^2)^{1/2}} = 1 \qquad \text{a.s.}$$

(Here F_n is the distribution function of X_n.)

Hence or otherwise show that, if $\liminf_{n\to\infty} s_n^2/n > 0$ and $\mathscr{E}|X_n|^{2+\delta} \le c < \infty$ for $n \ge 1$ and some $\delta > 0$, then

$$\limsup_{n\to\infty} \frac{S_n}{(2s_n^2 \ln \ln s_n^2)^{1/2}} = 1 \qquad \text{a.s.}$$

(Petrov [65])

39. Let $\{X_n\}$ be a sequence of independent, identically distributed random variables, and write $S_n = \sum_{k=1}^n X_k$. Suppose that $\mathscr{E}X_1 = 0$, $\mathscr{E}X_1^2 = 1$, and $\mathscr{E}\exp(tX_1) < \infty$ in some neighborhood of the origin. Show that there exists a sequence of numbers $a_n \sim (2n \ln \ln n)^{1/2}$ such that

$$P\{|S_n| \ge a_n \quad \text{for some } n \ge m\} = O((\ln \ln m)^{-1}) \qquad \text{as } m \to \infty.$$

(Darling and Robbins [15])

40. Show that there exists a sequence of independent random variables $\{X_n\}$ with a common symmetric distribution and a sequence of constants $\{a_n\}$ such that $\mathscr{E}X_1^2 = \infty$, $a_n \uparrow \infty$, and $\limsup_{n\to\infty} S_n/a_n = 1$ a.s.

(Freedman; see Strassen [78])

41. Let $\{a_n\}$ be a sequence of constants, $a_n \uparrow \infty$, and such that $\limsup_{n\to\infty} |S_n|/a_n < \infty$ a.s., where $S_n = \sum_{k=1}^n X_k$ and X_k are independent and identically distributed. Show that at least one of the following two conditions is satisfied:

(i) $a_n^{-1}(S_n - \text{med}(S_n)) \xrightarrow{\text{a.s.}} 0$.

(ii) $\liminf_{x\to\infty} x^2 P\{|X_1| \ge x\}/\int_{|y|\le x} y^2\, dP\{X_1 \le y\} = 0$.

(Heyde [35])

[Condition (ii) is a necessary and sufficient condition for $\{X_n\}$ to belong to the domain of partial attraction of the normal distribution. See Lévy [54], page 113, and also Section 5.6.]

42. Let $\{X_n\}$ be a sequence of independent, identically distributed symmetric random variables for which condition (ii) of Problem 41 does not hold. Let $\{a_n\}$ be a sequence of positive constants with $a_n \to \infty$ as $n \to \infty$. Show that, for any $\epsilon > 0$, $P\{|S_n| > \epsilon a_n \text{ i.o.}\} = 0$ or $=1$ according as $\sum_{n=1}^\infty P\{|X_1| > a_n\}$ converges or diverges. Equivalently, $a_n^{-1}S_n \xrightarrow{\text{a.s.}} 0$ if and only if $\sum_{n=1}^\infty P\{|X_1| > a_n\} < \infty$.

(Heyde [35])

SECTION 2.5

43. Let f and g be measurable real-valued functions defined on $[0, 1]$ such that $0 \le f \le cg < \infty$ a.e. (Lebesgue) for some constant $c > 0$. Suppose that g is integrable. Then show that

$$\lim_{n\to\infty} \int_0^1 \int_0^1 \cdots \int_0^1 \frac{\sum_{i=1}^n f(x_i)}{\sum_{i=1}^n g(x_i)}\, dx_1\, dx_2 \ldots dx_n = \frac{\int_0^1 f(x)\, dx}{\int_0^1 g(x)\, dx}.$$

44. (Hille) Let f be a bounded continuous function on $(0, \infty)$. Show that

$$\lim_{n\to\infty} \sum_{k=1}^{\infty} f\left(x + \frac{k}{n}\right) e^{-n\epsilon} \frac{(n\epsilon)^k}{k!} = f(x + \epsilon)$$

holds for every $\epsilon > 0$ and $x > 0$.

45. (Widder's Inversion Formula) Let f be continuous and bounded in the interval $[0, \infty)$, and let g be the Laplace transform of f:

$$g(t) = \int_0^\infty e^{-tx} f(x)\, dx.$$

Show that

$$f(x) = \lim_{n\to\infty} \frac{(-1)^{n-1} n^n g^{(n-1)}(n/x)}{x^n (n-1)!}, \qquad x > 0,$$

where $g^{(k)}$ is the kth derivative of g, uniformly in every finite interval. (*Hint*: Let $X_1, X_2, \ldots, X_n, \ldots$ be independent, identically distributed random variables with common distribution

$$P\{X_1 \le x\} = 1 - e^{-\lambda x},$$

where $\lambda > 0$ is a constant. Then $\mathscr{E}\{f(S_n)\} = (\lambda g(\lambda))^n$. Choose $\lambda = n/x$.)

46. Let $(\Omega, \mathscr{S}, P)$ be a probability space, and suppose that $A_1, A_2, \ldots, A_n \in \mathscr{S}$ such that $A_i \cap A_j = \phi$, $i \ne j$, and $\bigcup_{i=1}^n A_i = \Omega$. Let $p_k = P(A_k) > 0$, $k = 1, 2, \ldots, n$. In N independent trials of the experiment $(\Omega, \mathscr{S}, P)$, let X_k be the number of times A_k happens, $k = 1, 2, \ldots, n$. Then

$$\pi_N = P\{X_1 = x_1, \ldots, X_n = x_n\} = \prod_{k=1}^n p_k^{x_k}$$

and

$$\mathscr{E}\{N^{-1} \log_2 \pi_N^{-1}\} = -\sum_{k=1}^n p_k \log_2 p_k.$$

Define $H(\mathfrak{A}) = -\sum_{k=1}^n p_k \log_2 p_k$. Then $H(\mathfrak{A})$ is called the entropy of the system of events $\mathfrak{A} = \{A_1, A_2, \ldots, A_n\}$. Show that

$$N^{-1} \log_2 \pi_N^{-1} \xrightarrow{\text{a.s.}} H(\mathfrak{A}) \qquad \text{as } N \to \infty.$$

(*Hint*: By the strong law $X_k/N \xrightarrow{\text{a.s.}} p_k$, $k = 1, 2, \ldots, n$.)

NOTES AND COMMENTS

For a more comprehensive treatment of the laws of large numbers we refer the reader to Révész [69]. In Chapter 7 we shall extend the weak and the strong laws to the case of random variables taking values in a Banach space. There has been considerable interest in the study of rates of convergence in the laws of large numbers. For this topic and related ones we refer to Stout [77] and Petrov [66].

The law of the iterated logarithm proved in Theorem 2.4.1 is due to Kolmogorov [47]. Hartman and Wintner [32] proved the law of the iterated logarithm for independent, identically distributed random variables, and the converse was first obtained by Strassen [78]. Simpler proofs of the converse are due to Heyde [34] and Feller [23].

CHAPTER 3

Distribution and Characteristic Functions

In this chapter we consider some tools from analysis which will be used in the study of limit theorems of probability theory. We shall see that many results in probability theory and mathematical statistics can be derived by using the methods of characteristic functions as discussed in this chapter. In particular, we shall see in Chapter 5 that the solution of the celebrated central limit problem depends crucially on these tools of characteristic functions.

In Section 3.1 we introduce the concept of weak convergence of a sequence of uniformly bounded, nondecreasing, and right-continuous functions on $\mathbb{R}$. Weak convergence of a sequence of finite measures on $(\mathbb{R}, \mathscr{B})$ is considered in Section 3.5, whereas that of probability measures on complete separable metric spaces is discussed in Section 3.9. In Sections 3.2 to 3.6 we consider some properties of characteristic functions. In Section 3.3, for example, we prove the uniqueness and inversion theorems, and in Section 3.5 we prove the continuity theorem. These three results are basic to the study of limit theorems and have a variety of applications which are considered in Section 3.7. Section 3.8 deals with the theory of convolution semigroups. In Section 3.9 we consider the recent work of Prokhorov on weak convergence and tightness on a metric space. It will be shown that the concept of tightness simplifies many results, which is the reason why it is used as an important tool in modern probability theory.

3.1 WEAK CONVERGENCE

3.1.1 Convergence of Distribution Functions

Let $\{F_n\}$ be a sequence of distribution functions. We recall that F_n is a nondecreasing, right-continuous function which satisfies $F_n(-\infty) = 0$ and $F_n(+\infty) = 1$. To investigate in detail the convergence properties of the sequence $\{F_n\}$, it is necessary to introduce the concept of weak convergence in the general framework of bounded, nondecreasing, right-continuous functions. We note that such functions can have at most a countable number of discontinuity points (Proposition 1.1.2).

Definition. 3.1.1. Let $\{F_n\}$ be a sequence of uniformly bounded, nondecreasing, right-continuous functions defined on $\mathbb{R}$. We say that F_n converges weakly to a bounded, nondecreasing, right-continuous function F on $\mathbb{R}$ if

$$F_n(x) \to F(x) \qquad \text{as } n \to \infty$$

at all continuity points x of F. In this case we write $F_n \xrightarrow{w} F$.

Remark 3.1.1. We note that the weak limit of the sequence $\{F_n\}$, if it exists, is unique. In fact, if $F_n \xrightarrow{w} F$ and $F_n \xrightarrow{w} F'$, then $F = F'$ on the set $C_F \cap C_{F'}$, where C_F and $C_{F'}$ are, respectively, the sets of continuity points of F and F'. From Proposition 1.1.3 it follows that $F = F'$.

Definition. 3.1.2. Let $\{F_n\}$ be as in Definition 3.1.1. Then $\{F_n\}$ is said to converge completely to F if (i) $F_n \xrightarrow{w} F$, and (ii) $F_n(\mp\infty) \to F(\mp\infty)$ as $n \to \infty$. In this case we write $F_n \xrightarrow{c} F$.

Remark 3.1.2. Let $\{F_n\}$ and $\{F\}$ be as in Definition 3.1.1, and suppose that $F_n \xrightarrow{w} F$. Then $F_n \xrightarrow{c} F$ if and only if

$$\sup_{n\geq 1} \{F_n(\infty) - F_n(-\infty) - [F_n(a) - F_n(-a)]\} \to 0 \qquad \text{as } a \to \infty$$

or

$$F_n(\infty) - F_n(-\infty) \to F(\infty) - F(-\infty) \qquad \text{as } n \to \infty.$$

Example 3.1.1. Let G be a distribution function. Then $G(-\infty) = 0$, and $G(+\infty) = 1$. For each $n \geq 1$ define

$$F_n(x) = G(x + n), \qquad x \in \mathbb{R}.$$

Then $\{F_n\}$ is a sequence of uniformly bounded, nondecreasing, right-continuous functions on $\mathbb{R}$. Clearly

$$F_n(x) \to G(+\infty) = 1 \qquad \text{as } n \to \infty \quad \text{for all } x \in \mathbb{R},$$

so that $F_n \to F \equiv 1$ but F_n does not converge completely.

Example 3.1.2. Let F_n be the distribution function defined by

$$F_n(x) = \frac{n}{\sqrt{2\pi}} \int_{-\infty}^{x} e^{-n^2y^2/2}\, dy, \qquad n = 1, 2, \ldots.$$

Then

$$\lim_{n\to\infty} F_n(x) = \begin{cases} 0 & \text{if } x < 0, \\ \frac{1}{2} & \text{if } x = 0, \\ 1 & \text{if } x > 0. \end{cases}$$

Let $F(x) = 0$ if $x < 0$, and $=1$ if $x \geq 0$. Then $F_n \xrightarrow{w} F$. Note that F is a distribution function so that $F_n \xrightarrow{c} F$.

In cases where $\{F_n\}$ and F are distribution functions such that $F_n \xrightarrow{c} F$, let X_n and X be random variables corresponding to F_n and F, respectively. Then we say that X_n *converges in law* to X, and write $X_n \xrightarrow{L} X$.

Theorem 3.1.1 (Kolmogorov). Let $\{X_n\}$ be a sequence of random variables such that $X_n \xrightarrow{P} X$. Then $X_n \xrightarrow{L} X$.

Proof. Let $x', x \in \mathbb{R}$. Then

$$P\{X \leq x'\} \leq P\{X_n \leq x\} + P\{X \leq x', X_n > x\}.$$

Suppose that $x' < x$. Then

$$P\{X \leq x'\} \leq P\{X_n \leq x\} + P\{|X_n - X| \geq x - x'\},$$

and since $X_n \xrightarrow{P} X$ it follows that

$$F(x') \leq \liminf_{n\to\infty} F_n(x),$$

where F_n and F are the distribution functions of X_n and X, respectively. Letting $x' \uparrow x$, we obtain

$$F(x - 0) \leq \liminf_{n\to\infty} F_n(x).$$

By considering $x'' > x$ and interchanging X_n and X, we obtain

$$F_n(x) \leq F(x'') + P\{|X_n - X| \geq x'' - x\}.$$

Letting $n \to \infty$ and then $x'' \downarrow x$, we obtain

$$\limsup_{n\to\infty} F_n(x) \le F(x).$$

Therefore

$$F(x-0) \le \liminf_{n\to\infty} F_n(x) \le \limsup_{n\to\infty} F_n(x) \le F(x).$$

Clearly $\lim_{n\to\infty} F_n(x) = F(x)$ at all continuity points x of F, and since F is a distribution function $F_n \xrightarrow{c} F$, that is, $X_n \xrightarrow{L} X$. ∎

Corollary. Let $\{X_n\}$ be a sequence of random variables, and c be a constant. Then

$$X_n \xrightarrow{L} c \Leftrightarrow X_n \xrightarrow{P} c.$$

Proof. The proof of this corollary is straightforward.

Next we prove the weak compactness criterion due to Helly.

Theorem 3.1.2 (Helly's Theorem). Let $\{F_n\}$ be a sequence of uniformly bounded, nondecreasing, right-continuous functions. Then $\{F_n\}$ contains a subsequence $\{F_{n_k}\}$ which converges weakly to a bounded, nondecreasing, right-continuous function.

For the proof of Theorem 3.1.2 we need the following two lemmas.

Lemma 3.1.1. Let $\{F_n\}$ be as above. Suppose, in addition, that F_n converges to a function G pointwise on a set D which is everywhere dense in $\mathbb{R}$. Then $\{F_n\}$ converges to G at all continuity points of G.

Proof. Let $x, x', x'' \in \mathbb{R}$ be such that $x' < x < x''$. Then for every $n \ge 1$

$$F_n(x') \le F_n(x) \le F_n(x'').$$

Let $x', x'' \in D$. Letting $n \to \infty$, we obtain

$$G(x') \le \liminf_{n\to\infty} F_n(x) \le \limsup_{n\to\infty} F_n(x) \le G(x'').$$

Now letting $x' \uparrow x$ and $x'' \downarrow x$, we obtain the required result. ∎

Lemma 3.1.2. Let $\{F_n\}$ be as above, and let $D = \{x_\nu, \nu \ge 1\}$ be a countable subset of $\mathbb{R}$. Then $\{F_n\}$ contains a subsequence $\{F_{n_k}\}$ which converges at every point of D.

Proof. Consider the set of real numbers $\{F_n(x_1), n \geq 1\}$. Clearly this is a bounded set, and hence, according to the Bolzano–Weierstrass theorem, $\{F_n\}$ contains a subsequence $\{F_n^{(1)}\}$ such that $\{F_n^{(1)}(x_1)\}$ converges to some real number, say a_1.

Next consider the sequence of real numbers $\{F_n^{(1)}(x_2), n \geq 1\}$. If we proceed as above, it follows that there exists a subsequence $\{F_n^{(2)}\} \subset \{F_n^{(1)}\}$ such that $\{F_n^{(2)}(x_2)\}$ converges to a real number, say a_2. Clearly $F_n^{(2)}(x_1) \to a_1$ as $n \to \infty$. We continue this process indefinitely and thus obtain a countable set of convergent subsequences $\{F_n^{(k)}\}$ such that

$$\{F_n\} \supset \{F_n^{(1)}\} \supset \{F_n^{(2)}\} \supset \cdots \supset \{F_n^{(k)}\} \supset \cdots,$$

and, moreover, the relation

$$\lim_{n\to\infty} F_n^{(k)}(x_k) = a_k$$

holds for all $k \geq 1$. Now consider the diagonal sequence $\{F_n^{(n)}, n \geq 1\}$. We show that, for every $x \in D$, $\{F_n^{(n)}(x), n \geq 1\}$ converges. Let $k \geq 1$ be fixed, and consider the point $x_k \in D$. Then $\{F_n^{(n)}(x_k), n \geq k\}$ clearly is a subsequence of $\{F_n^{(k)}(x_k), n \geq 1\}$, so that $F_n^{(n)}(x_k) \to a_k$ as $n \to \infty$. This completes the proof of Lemma 3.1.2. ■

Proof of Theorem 3.1.2. Let $D = \{x_\nu, \nu \geq 1\}$ be a countable set which is everywhere dense in $\mathbb{R}$. Then from Lemma 3.1.2 $\{F_n\} \supset \{F_{n_k}\}$ such that $\{F_{n_k}\}$ converges at every point of D. Define G on $\mathbb{R}$ by setting

$$G(x_\nu) = \lim_{k\to\infty} F_{n_k}(x_\nu) \qquad \text{for } x_\nu \in D$$

and

$$G(x) = \sup_{\substack{x_\nu \leq x \\ x_\nu \in D}} G(x_\nu) \qquad \text{for } x \in \mathbb{R} - D.$$

Clearly G is bounded. We show that $G \not\downarrow$.

CASE 1. Let $x \in D$ and $y \in D$ such that $x < y$. Clearly $G(x) \leq G(y)$ from the definition of G.

CASE 2. Let $x \notin D$ and $y \in D$ be such that $x < y$. Let $x_\nu \in D$ be such that $x_\nu < x$. Then $x_\nu < y$ so that $G(x_\nu) \leq G(y)$ from Case 1. Hence

$$G(x) = \sup_{\substack{x_\nu < x \\ x_\nu \in D}} G(x_\nu) \leq G(y).$$

CASE 3. Let $x \in D$ and $y \notin D$ such that $x < y$. Clearly

$$G(y) = \sup_{\substack{x_\nu < y \\ x_\nu \in D}} G(x_\nu) \geq G(x).$$

CASE 4. Let $x \notin D$ and $y \notin D$ with $x < y$. Then

$$\{x_\nu \in D\colon x_\nu < x\} \subset \{x_\nu \in D\colon x_\nu < y\}$$

so that $G(x) \leq G(y)$.

We note that $\{F_{n_k}\}$ converges to G on D. In view of Lemma 3.1.1, $\{F_{n_k}\}$ converges to G at all continuity points of G.

Next we define F by setting

$$F(x) = G(x) \qquad \text{if } x \in C_G,$$
$$F(x) = G(x+0) \qquad \text{if } x \notin C_G.$$

Clearly F is a bounded, nondecreasing, right-continuous function, and, moreover, $F_{n_k} \xrightarrow{w} F$. This completes the proof of Theorem 3.1.2. ■

3.1.2 Convergence of Sequences of Integrals

We now prove two results on convergence of sequences of integrals.

Theorem 3.1.3 (Helly–Bray Theorem). Let g be a real-valued, continuous function defined on a closed bounded interval $[a, b]$, and let $\{F_n\}$ be a sequence of uniformly bounded, nondecreasing, right-continuous functions which converges weakly to some function F on $[a, b]$, where $a, b \in C_F$. Then

$$\lim_{n\to\infty} \int_a^b g\, dF_n = \int_a^b g\, dF.$$

Proof. It follows from Theorem 3.1.2 that F is bounded and nondecreasing. Let μ_n and μ denote the finite measures corresponding to F_n and F, respectively. Let $a = x_{0,N} < x_{1,N} < \cdots < x_{\nu_N, N} = b$ be a sequence of subdivisions of $[a, b]$ such that $\Delta_N = \max_{1 \leq \nu \leq \nu_N} (x_{\nu,N} - x_{\nu-1,N}) \to 0$ as $N \to \infty$.

Set

$$g_N(x) = \sum_{\nu=1}^{\nu_N} g(x_{\nu,N}) \chi_{(x_{\nu-1,N},\, x_{\nu,N}]}(x), \qquad x \in [a, b].$$

Then g can be uniformly approximated on $[a, b]$ by the sequence $\{g_N\}$, that is,

$$\sup_{a \leq x \leq b} |g_N(x) - g(x)| \to 0 \qquad \text{as } N \to \infty.$$

Since $\{g_N\}$ is uniformly bounded, we have from the Lebesgue dominated convergence theorem

$$\int_a^b g\,dF_n = \int_{(a,b]} g\,d\mu_n = \lim_{N\to\infty}\int_{(a,b]} g_N\,d\mu_n$$

for every $n \geq 1$. Similarly

$$\int_a^b g\,dF = \int_{(a,b]} g\,d\mu = \lim_{N\to\infty}\int_{(a,b]} g_N\,d\mu.$$

We first show that for every $N \geq 1$

$$\lim_{n\to\infty}\int_{(a,b]} g_N\,d\mu_n = \int_{(a,b]} g_N\,d\mu.$$

For this purpose select the points of subdivision $x_{\nu,N}(\nu = 1, 2, \ldots, \nu_N; N = 1, 2, \ldots)$ to be the continuity points of F. Since $F_n \xrightarrow{w} F$, we have

$$\begin{aligned}\lim_{n\to\infty}\mu_n(x_{\nu-1,N}, x_{\nu,N}] &= \lim_{n\to\infty}[F_n(x_{\nu,N}) - F_n(x_{\nu-1,N})]\\ &= F(x_{\nu,N}) - F(x_{\nu-1,N})\\ &= \mu(x_{\nu-1,N}, x_{\nu,N}].\end{aligned}$$

Hence

$$\begin{aligned}\lim_{n\to\infty}\int_{(a,b]} g_N\,d\mu_n &= \lim_{n\to\infty}\sum_{\nu=1}^{\nu_N} g(x_{\nu,N})\mu_n(x_{\nu-1,N}, x_{\nu,N}]\\ &= \sum_{\nu=1}^{\nu_N} g(x_{\nu,N})\mu(x_{\nu-1,N}, x_{\nu,N}]\\ &= \int_{(a,b]} g_N\,d\mu.\end{aligned}$$

Writing $M_N = \sup_{a\leq x\leq b}|g_N(x) - g(x)|$, we see that

$$\begin{aligned}\left|\int_a^b g\,dF_n - \int_a^b g\,dF\right| &\leq \int_{(a,b]}|g - g_N|\,d\mu_n + \left|\int_{(a,b]} g_N\,d\mu_n - \int_{(a,b]} g_N\,d\mu\right| + \int_{(a,b]}|g - g_N|\,d\mu\\ &\leq M_N\mu_n(a,b] + \left|\int_{(a,b]} g_N\,d\mu_n - \int_{(a,b]} g_N\,d\mu\right|\\ &\quad + M_N\mu(a,b].\end{aligned}$$

Since $M_N \to 0$ as $N \to \infty$, by letting $n \to \infty$ and then $N \to \infty$ we obtain the result. ■

Theorem 3.1.4 (Extended Helly–Bray Theorem). Let g be a bounded, real-valued, continuous function on $\mathbb{R}$. Let $\{F_n\}$ be a sequence of uniformly bounded, nondecreasing, right-continuous functions which converges completely to some function F on $\mathbb{R}$. Then

$$\lim_{n\to\infty}\int_{-\infty}^{\infty} g\,dF_n = \int_{-\infty}^{\infty} g\,dF.$$

Proof. Clearly F is bounded, nondecreasing, and right-continuous. Let $a, b \in C_F$ such that $a < b$, and let

$$M = \sup_{x\in\mathbb{R}} |g(x)|.$$

Then

$$\begin{aligned}
\left|\int_{-\infty}^{\infty} g\,dF_n - \int_{-\infty}^{\infty} g\,dF\right|
&\le \left|\int_{-\infty}^{\infty} g\,dF_n - \int_a^b g\,dF_n\right| \\
&\quad + \left|\int_a^b g\,dF_n - \int_a^b g\,dF\right| + \left|\int_a^b g\,dF - \int_{-\infty}^{\infty} g\,dF\right| \\
&\le M[F_n(a) - F_n(-\infty) + F_n(+\infty) - F_n(b)] + \left|\int_a^b g\,dF_n - \int_a^b g\,dF\right| \\
&\quad + M[F(a) - F(-\infty) + F(+\infty) - F(b)].
\end{aligned}$$

Since $\lim_{n\to\infty} F_n(\pm\infty) = F(\pm\infty)$, $\lim_{n\to\infty} F_n(a) = F(a)$, and $\lim_{n\to\infty} F_n(b) = F(b)$, the first and the third term on the right-hand side of the last inequality $\to 0$ as $n \to \infty$, $a \to -\infty$, and $b \to +\infty$, while the second term $\to 0$ as $n \to \infty$ in view of Theorem 3.1.3. Hence we have the proof. ∎

Remark 3.1.3. Theorems 3.1.3 and 3.1.4 remain valid when g is a complex-valued function on $\mathbb{R}$. The proof follows by applying the above results to the real and imaginary parts of g separately.

Example 3.1.3. Let G and F_n be as defined in Example 3.1.1. Then $F_n \xrightarrow{w} F \equiv 1$, but not completely. Let $g(x) \equiv 1$, $x \in \mathbb{R}$. Then g is a bounded, real-valued, continuous function on $\mathbb{R}$, and

$$\int_{-\infty}^{\infty} g\,dF_n = \int_{-\infty}^{\infty} dG(x+n) = G(+\infty) - G(-\infty) = 1, \qquad n \ge 1.$$

It follows that $\int_{-\infty}^{\infty} g\, dF_n \to 1$ as $n \to \infty$. However,

$$\int_{-\infty}^{\infty} g\, dF = \int_{-\infty}^{\infty} dF = 0,$$

so that $\int_{-\infty}^{\infty} g\, dF_n \not\to \int_{-\infty}^{\infty} g\, dF$.

3.1.3 Convergence of Moments

We now study the relationship between the convergence of a sequence of distribution functions and the convergence of the corresponding sequence of moments. For this purpose we first introduce the concept of *uniform integrability* with respect to a sequence of uniformly bounded, nondecreasing, right-continuous functions.

Definition 3.1.3. Let $\{F_n\}$ be a sequence of uniformly bounded, nondecreasing, right-continuous functions, and let g be a continuous, real-valued function on $\mathbb{R}$ such that $\int_{-\infty}^{\infty} |g|\, dF_n < \infty, n = 1, 2, \ldots$. Then we say that g is uniformly integrable with respect to F_n if

$$\sup_{n \geq 1} \int_{|x| \geq a_m} |g|\, dF_n \to 0 \qquad \text{as } a_m \to \infty.$$

Proposition 3.1.1. Let $F_n \xrightarrow{w} F$. Then:

(a) g is uniformly integrable with respect to $F_n \Rightarrow \int g\, dF_n \to \int g\, dF$.
(b) $\int |g|\, dF_n \to \int |g|\, dF < \infty \Leftrightarrow g$ is uniformly integrable with respect to F_n

Proof. (a) Let $\epsilon > 0$. By the uniform integrability of g there exists an $a_\epsilon > 0$ such that

$$\int_{|x| \geq a} |g|\, dF_n < \epsilon \qquad \text{for all } a \geq a_\epsilon.$$

Clearly

$$\left| \int g\, dF_n - \int g\, dF \right| \leq \int_{|x| \geq a} |g|\, dF_n + \left| \int_{-a}^{a} g\, dF_n - \int_{-a}^{a} g\, dF \right| + \int_{|x| \geq a} |g|\, dF.$$

The first term on the right-hand side can be made arbitrarily small by choosing $\epsilon > 0$ small, the second term tends to zero as $n \to \infty$ by choosing $\pm a \in C_F$, and the last term can be made arbitrarily small by choosing a sufficiently large.

(b) Let $\int |g|\, dF_n \to \int |g|\, dF < \infty$. Then for any $a > 0$

$$\int_{|x| \geq a} |g|\, dF_n$$

$$\leq \left| \int |g|\, dF_n - \int |g|\, dF \right|$$

$$+ \int_{|x| \geq a} |g|\, dF + \left| \int_{-a}^{a} |g|\, dF_n - \int_{-a}^{a} |g|\, dF \right|.$$

Let $\epsilon > 0$. Choose $a_0 = a(\epsilon)$ so large that for $a \geq a_0$

$$\int_{|x| \geq a} |g|\, dF < \frac{\epsilon}{3},$$

and then choose $n \geq n_0$ large enough so that

$$\left| \int_{|x| < a} |g|\, dF_n - \int_{|x| < a} |g|\, dF \right| < \frac{\epsilon}{3} \qquad \text{for } n \geq n_0$$

in view of Helly's theorem, and such that

$$\left| \int |g|\, dF_n - \int |g|\, dF \right| < \frac{\epsilon}{3} \qquad \text{for } n \geq n_0$$

by hypothesis. Then, for all $a \geq a_0$ and $n \geq n_0$, we have

$$\int_{|x| \geq a} |g|\, dF_n < \epsilon.$$

Now choose $a_\epsilon = \max(a_0, a_1, \ldots, a_{n_0 - 1})$, where the a_k, $k = 1, 2, \ldots, n_0 - 1$, are such that $\int_{|x| \geq a_k} |g|\, dF_k < \epsilon$. This is possible, since $\int |g|\, dF_n$ is a convergent sequence of real numbers and hence is bounded. Clearly for $a \geq a_\epsilon$ (and independently of n)

$$\int_{|x| \geq a} |g|\, dF_n < \epsilon.$$

Conversely, let $|g|$ be uniformly integrable with respect to F_n. Then from (a)

$$\int |g|\, dF_n \to \int |g|\, dF < \infty.$$

This completes the proof of Proposition 3.1.1. ∎

Theorem 3.1.5 (Moment Convergence Theorem). Let $r_0 > 0$ be fixed, and suppose that $|x|^{r_0}$ is uniformly integrable with respect to a sequence of distribution functions $\{F_n\}$. Then:

(a) The sequence $\{F_n\}$ is completely compact.†
(b) For every subsequence $F_{n'} \xrightarrow{c} F$ and all $0 \le k, r \le r_0$ (k integral),

$$\alpha_k^{(n')} \to \alpha_k \text{ (finite)} \qquad \text{and} \qquad \beta_r^{(n')} \to \beta_r < \infty.$$

Here $\alpha_k^{(n')} = \int x^k \, dF_{n'}(x)$ and $\beta_r^{(n')} = \int |x|^r \, dF_{n'}(x)$.

Proof. By Helly's weak compactness criterion (Theorem 3.1.2) F_n contains a subsequence $\{F_{n'}\}$ such that $F_{n'} \xrightarrow{w} F$, where F is some bounded, nondecreasing, right-continuous function on $\mathbb{R}$. From the uniform integrability of $|x|^{r_0}$ it follows that

$$\sup_{n \ge 1} \int_{|x| \ge a} |x|^{r_0} \, dF_n(x) \to 0 \qquad \text{as } a \to \infty.$$

Thus for $r \le r_0$

$$\int_{|x| \ge a} |x|^r \, dF_{n'}(x) \le a^{r - r_0} \int_{|x| \ge a} |x|^{r_0} \, dF_{n'}(x) \to 0$$

as $a \to \infty$ independently of n', and it follows that $|x|^r$ is uniformly integrable with respect to $F_{n'}$. Then part (b) follows immediately from Proposition 3.1.1.

For $r = 0$ we get $F_{n'}(+\infty) - F_{n'}(-\infty) \to F(+\infty) - F(-\infty)$, and (a) follows from Remark 3.1.2. ∎

Corollary 1. If the sequence $\beta_{r_0+\delta}^{(n)}$ is bounded for all n and some $\delta > 0$, Theorem 3.1.5 holds.

Proof. We have

$$\int_{|x| \ge a} |x|^{r_0} \, dF_n(x) \le a^{-\delta} \int_{|x| \ge a} |x|^{r_0 + \delta} \, dF_n$$
$$< A a^{-\delta} \to 0 \qquad \text{as } a \to \infty.$$

Here $\beta_{r_0+\delta}^{(n)} \le A < \infty$ for all $n \ge 1$.

Corollary 2. Let k_0 be a fixed but otherwise arbitrary positive integer. Let

$$\lim_{n \to \infty} \alpha_k^{(n)} = \alpha_k \quad \text{(finite)}.$$

† A sequence $\{F_n\}$ of distribution functions is said to be completely compact if it contains a subsequence $\{F_{n_k}\}$ which converges completely to some distribution function.

Then there exists at least one distribution function F with moments α_k; moreover, one can extract a subsequence $\{F_{n'}\} \subset \{F_n\}$ such that $F_{n'} \xrightarrow{c} F$.

Proof. Let $k < k_0$, and choose an even integer $m \geq k_0$. Then for sufficiently large n

$$\int x^m \, dF_n(x) = \int |x|^m \, dF_n(x) < \alpha_m + 1$$

by hypothesis. It follows that

$$\beta_m^{(n)} = \beta_{k+(m-k)}^{(n)}$$

is bounded for all $n \geq 1$. The proof is now completed with the help of Corollary 1.

3.2 SOME ELEMENTARY PROPERTIES OF CHARACTERISTIC FUNCTIONS

Let F be a distribution function, and let φ be its characteristic function. Recall that

$$\varphi(t) = \int_{-\infty}^{\infty} e^{itx} \, dF(x), \qquad t \in \mathbb{R}.$$

In this section we study some elementary properties of φ.

Proposition 3.2.1. The characteristic function φ is uniformly continuous on $\mathbb{R}$ and satisfies three conditions:

(i) $\varphi(0) = 1$,
(ii) $|\varphi(t)| \leq 1$,
(iii) $\varphi(-t) = \overline{\varphi}(t)$,

where $\overline{\varphi}$ is the complex conjugate of φ.

Proof. Clearly (i), (ii), and (iii) follow from the definition. We show that φ is uniformly continuous on $\mathbb{R}$. Let $t \in \mathbb{R}$ and $h \in \mathbb{R}$. Then

$$|\varphi(t+h) - \varphi(t)| \leq \int_{-\infty}^{\infty} |e^{ihx} - 1| \, dF(x)$$

$$= 2 \int_{-\infty}^{\infty} \left| \sin \frac{hx}{2} \right| dF(x),$$

where the right-hand side is independent of t. Using the Lebesgue dominated convergence theorem, we see that

$$\sup_{t \in \mathbb{R}} |\varphi(t + h) - \varphi(t)| \to 0 \qquad \text{as } h \to 0. \qquad \blacksquare$$

Next we prove an important result which relates the characteristic function of a distribution to its moments.

Theorem 3.2.1. Let F be a distribution function with finite moments up to order n. Then the characteristic function φ of F has continuous derivatives up to order n, and the relation

$$\varphi^{(k)}(0) = i^k \alpha_k \tag{3.2.1}$$

holds for $k = 1, 2, \ldots$. Moreover, φ admits the expansion

$$\varphi(t) = 1 + \sum_{k=1}^{n} \alpha_k \frac{(it)^k}{k!} + o(t^n), \qquad t \to 0. \tag{3.2.2}$$

Conversely, suppose that the characteristic function φ of a distribution function F has an expansion of the form

$$\varphi(t) = 1 + \sum_{k=1}^{n} a_k \frac{(it)^k}{k!} + o(t^n) \qquad \text{as } t \to 0. \tag{3.2.3}$$

Then F has finite moments up to order n if n is even, but up to order $n - 1$ if n is odd. Moreover, in this case $a_k = \alpha_k$ for all k.

Proof. Suppose that $|\alpha_k| < \infty$ for $k = 1, 2, \ldots, n$. Clearly $\int_{-\infty}^{\infty} x^k e^{itx}\, dF(x)$ converges uniformly and absolutely for all $t \in \mathbb{R}$ and $k = 1, 2, \ldots, n$. First note that

$$\frac{\varphi(t + h) - \varphi(t)}{h} = \int_{-\infty}^{\infty} e^{itx} \frac{e^{ihx} - 1}{h}\, dF(x),$$

and $|e^{ihx} - 1| \le |hx|$, so that

$$\left| \frac{\varphi(t + h) - \varphi(t)}{h} \right| \le \int_{-\infty}^{\infty} |x|\, dF(x) < \infty.$$

Using the Lebesgue dominated convergence theorem, we obtain

$$\frac{d\varphi(t)}{dt} = \lim_{h \to 0} \frac{\varphi(t + h) - \varphi(t)}{h} = i \int_{-\infty}^{\infty} e^{itx} x\, dF(x).$$

Thus the first derivative $\varphi^{(1)}$ of φ exists, and (3.2.1) holds. Proceeding inductively, we conclude that φ has all derivatives up to order n and relation

(3.2.1) holds. It also follows immediately that $\varphi^{(k)}$ is continuous on $\mathbb{R}$ for $1 \le k \le n$.

In the neighborhood of $t = 0$ we develop φ into a Maclaurin's expansion, given by

$$\varphi(t) = 1 + \sum_{k=1}^{n} \varphi^{(k)}(0) \frac{t^k}{k!} + R_n(t),$$

where

$$R_n(t) = \frac{t^n}{n!} [\varphi^{(n)}(\theta t) - \varphi^{(n)}(0)], \qquad 0 < \theta < 1.$$

Clearly

$$|R_n(t)| \le \frac{|t|^n}{n!} \int_{-\infty}^{\infty} |x|^n |e^{i\theta tx} - 1| \, dF(x),$$

so that from the Lebesgue dominated convergence theorem we conclude that

$$R_n(t) = o(t^n), \qquad t \to 0.$$

This completes the proof of the first part.

Conversely, suppose that φ has an expansion of the form (3.2.3), where n is even, say $n = 2m$. Then φ has a finite derivative of order $2m$ at $t = 0$, which is given by

$$\begin{aligned} \varphi^{(2m)}(0) &= \lim_{h \to 0} \int_{-\infty}^{\infty} \left(\frac{e^{ihx} - e^{-ihx}}{2h} \right)^{2m} dF(x) \\ &= (-1)^m \lim_{h \to 0} \int_{-\infty}^{\infty} \left(\frac{\sin hx}{h} \right)^{2m} dF(x). \end{aligned}$$

By Fatou's lemma we conclude that

$$\begin{aligned} \int_{-\infty}^{\infty} x^{2m} \, dF(x) &\le \lim_{h \to 0} \int_{-\infty}^{\infty} \left(\frac{\sin hx}{h} \right)^{2m} dF(x) \\ &< \infty, \end{aligned}$$

which proves the existence of α_{2m}. If n is odd, we repeat the same procedure with $n - 1$ to conclude that $\alpha_{n-1} < \infty$. Clearly $a_k = \alpha_k$ for all k, and the proof is complete. ■

Corollary 1. In the notation of Theorem 3.2.1 φ has continuous derivatives of all orders if and only if F has finite moments of all orders.

Corollary 2. Let φ be of the form $\varphi(t) = 1 + o(t^{2+\delta})$ for $\delta > 0$ as $t \to 0$. Then φ is the characteristic function of the distribution degenerate at zero.

Proof. Clearly $\varphi(t) = 1 + o(t^2)$ as $t \to 0$, and it follows from Theorem 3.2.1 that $\alpha_1 = \alpha_2 = 0$. Hence we have the result.

Corollary 3. Let $a(t) = o(t)$ as $t \to 0$, and suppose that $a(-t) = -a(t)$. Then the only characteristic function of the form $\varphi(t) = 1 + a(t) + o(t)$, as $t \to 0$, is the function $\varphi(t) \equiv 1$.

Proof. We have

$$\begin{aligned} |\varphi(t)|^2 &= [1 + a(t) + o(t)][1 - a(t) + o(t)] \\ &= 1 + o(t^2) \qquad \text{as } t \to 0, \end{aligned}$$

and since $|\varphi(t)|^2$ is a characteristic function, it follows that $|\varphi(t)|^2 = \varphi(t)\varphi(-t) \equiv 1$, which implies that $\varphi(t) = e^{i\alpha t}$, where $\alpha \in \mathbb{R}$. Therefore

$$\varphi(t) = 1 + i\alpha t - \tfrac{1}{2}\alpha^2 t^2 + o(t^2),$$

so that φ is of the form $1 + a(t) + o(t)$ only if $\alpha = 0$. Hence $\varphi(t) \equiv 1$.

Example 3.2.1. The functions e^{-t^4}, $e^{-|t|^r}$ for $r > 2$, and $(1 + t^4)^{-1}$ are not characteristic functions.

3.3 THE INVERSION AND UNIQUENESS THEOREMS

In this section we prove two results which show the importance of characteristic functions in probability theory. The first of these is the inversion theorem, which enables us to compute the distribution function from its characteristic function. The second result is the uniqueness theorem, which allows us to work with characteristic functions instead of distribution functions in the study of limit theorems.

Theorem 3.3.1 (Inversion Theorem). Let F be a distribution function, and let φ be its characteristic function. Then the relation

$$F(a + h) - F(a - h) = \lim_{T \to \infty} \frac{1}{\pi} \int_{-T}^{T} \frac{\sin ht}{t} e^{-ita}\varphi(t)\, dt \tag{3.3.1}$$

holds for $a \in \mathbb{R}$ and $h > 0$ whenever the points $a \mp h \in C_F$.

Proof. For the proof we first note that the integral $\int_0^x (\sin t/t)\, dt$ exists, is bounded for all $x > 0$, and tends to the limit $\pi/2$ as $x \to +\infty$. We set

$$\theta(h, T) = \frac{2}{\pi} \int_0^T \frac{\sin ht}{t}\, dt.$$

Clearly $\theta(h, T)$ is uniformly bounded for all $h \in \mathbb{R}$ and all $T > 0$ such that

$$\theta(-h, T) = -\theta(h, T).$$

Moreover, the relation

$$\lim_{T\to\infty} \theta(h, T) = \begin{cases} 1 & \text{if } h > 0, \\ 0 & \text{if } h = 0, \\ -1 & \text{if } h < 0, \end{cases}$$

holds. For $T > 0$ we set

$$\begin{aligned} I_T &= \frac{1}{\pi}\int_{-T}^{T} \frac{\sin ht}{t} e^{-ita}\varphi(t)\,dt \\ &= \frac{1}{\pi}\int_{-T}^{T} \frac{\sin ht}{t} e^{-ita}\left[\int_{-\infty}^{\infty} e^{itx}\,dF(x)\right] dt, \end{aligned}$$

and, noting that for $h > 0$

$$\left|\frac{\sin ht}{t} e^{it(x-a)}\right| \le h,$$

we conclude that I_T exists and is finite. Using Fubini's theorem to change the order of integration, we obtain

$$\begin{aligned} I_T &= \frac{1}{\pi}\int_{-\infty}^{\infty} dF(x) \int_{-T}^{T} \frac{\sin ht}{t} e^{it(x-a)}\,dt \\ &= \frac{2}{\pi}\int_{-\infty}^{\infty} dF(x) \int_{0}^{T} \frac{\sin ht}{t} \cos[(x-a)t]\,dt \\ &= \int_{-\infty}^{\infty} g(x, T)\,dF(x), \end{aligned}$$

where

$$\begin{aligned} g(x, T) &= \frac{2}{\pi}\int_{0}^{T} \frac{\sin ht}{t} \cos[(x-a)t]\,dt \\ &= \frac{1}{\pi}\int_{0}^{T} \frac{\sin[(x-a+h)t]}{t}\,dt - \frac{1}{\pi}\int_{0}^{T} \frac{\sin[(x-a-h)t]}{t}\,dt \\ &= \tfrac{1}{2}\theta(x-a+h, T) - \tfrac{1}{2}\theta(x-a-h, T). \end{aligned}$$

Clearly $g(x, T)$ is uniformly bounded for all $x \in \mathbb{R}$ and all $T > 0$. Moreover,

$$\lim_{T\to\infty} g(x, T) = \begin{cases} 0 & \text{if } x < a - h, \\ \frac{1}{2} & \text{if } x = a - h, \\ 1 & \text{if } a - h < x < a + h, \\ \frac{1}{2} & \text{if } x = a + h, \\ 0 & \text{if } x > a + h. \end{cases}$$

Using the Lebesgue dominated convergence theorem, we can take the limit as $T \to \infty$ under the integral sign and thus obtain

$$\lim_{T\to\infty} I_T = \int_{-\infty}^{\infty} \left[\lim_{T\to\infty} g(x, T)\right] dF(x)$$

$$= \int_{a-h}^{a+h} dF(x) = F(a + h) - F(a - h),$$

since $a \mp h \in C_F$. This completes the proof. ■

Corollary 1. Let $a, b \in C_F$ be such that $a < b$. Then

(3.3.2) $$F(b) - F(a) = \lim_{T\to\infty} \frac{1}{2\pi} \int_{-T}^{T} \frac{e^{-ita} - e^{-itb}}{it} \varphi(t)\, dt.$$

Corollary 2 (Uniqueness Theorem). Let F_1 and F_2 be two distribution functions with characteristic functions φ_1 and φ_2, respectively. Suppose that $\varphi_1(t) = \varphi_2(t)$ for every $t \in \mathbb{R}$. Then $F_1 \equiv F_2$.

Proof. Let $a, b \in C_{F_1} \cap C_{F_2}$ be such that $a < b$. Then (3.3.2) yields

$$F_1(b) - F_1(a) = F_2(b) - F_2(a).$$

Letting $a \to -\infty$, we obtain $F_1(b) = F_2(b)$, so that $F_1 = F_2$ on $C_{F_1} \cap C_{F_2}$. Hence $F_1 \equiv F_2$.

Corollary 3. A distribution function F is symmetric if and only if its characteristic function is real and even.

Proof. Let φ be the characteristic function corresponding to F. If φ is real, $\varphi(t) = \varphi(-t) = \bar{\varphi}(t)$, and from uniqueness we see that φ and $\bar{\varphi}$ have the same distribution function, that is, F is symmetric.

Conversely, if F is symmetric, $\varphi(t) = \bar{\varphi}(t)$, so that $\varphi(t) = [\varphi(t) + \bar{\varphi}(t)]/2$. Thus

$$\varphi(t) = \int_{-\infty}^{\infty} \cos tx\, dF(x),$$

and φ is real and even.

Remark 3.3.1. We note that $\varphi_1(t) = \varphi_2(t)$ for all t in some interval on $\mathbb{R}$ does not imply that $F_1 = F_2$ on $\mathbb{R}$. See, for example, Problem 3.10.20.

Example 3.3.1. Let X_1 and X_2 be independent random variables such that X_i has a normal distribution with mean μ_i and variance σ_i^2, $i = 1, 2$. Let $X = X_1 + X_2$. Then

$$\begin{aligned}\varphi_X(t) &= \mathscr{E}e^{itX} = (\mathscr{E}e^{itX_1})(\mathscr{E}e^{itX_2})\\ &= \exp\left\{i(\mu_1 + \mu_2)t - (\sigma_1^2 + \sigma_2^2)\frac{t^2}{2}\right\}.\end{aligned}$$

It follows from the uniqueness theorem that X has a normal distribution with mean $\mu_1 + \mu_2$ and variance $\sigma_1^2 + \sigma_2^2$.

We next consider an important special case of Theorem 3.3.1 where φ is absolutely integrable on $\mathbb{R}$.

Theorem 3.3.2 (Fourier Inversion Theorem). Suppose that φ is absolutely integrable on $\mathbb{R}$. Then the corresponding distribution function F is absolutely continuous. Moreover, the probability density function $f = F'$ exists, is bounded and uniformly continuous on $\mathbb{R}$, and is given by

$$f(x) = \frac{1}{2\pi}\int_{-\infty}^{\infty} e^{-itx}\varphi(t)\,dt, \qquad x \in \mathbb{R}. \tag{3.3.3}$$

Proof. Since $\int_{-\infty}^{\infty} |\varphi(t)|\,dt < \infty$, the integrand on the right-hand side of (3.3.1) is dominated by an absolutely integrable function. Consequently we can write

$$F(x + h) - F(x - h) = \frac{1}{\pi}\int_{-\infty}^{\infty} \frac{\sin ht}{t} e^{-itx}\varphi(t)\,dt \tag{3.3.4}$$

whenever $x \mp h \in C_F$. Taking the limits of both sides as $h \downarrow 0$, we obtain

$$F(x) - F(x - 0) = 0.$$

It follows that F is continuous on $\mathbb{R}$. Using a similar argument on

$$\frac{F(x + h) - F(x - h)}{2h} = \frac{1}{2\pi}\int_{-\infty}^{\infty} \frac{\sin ht}{ht} e^{-itx}\varphi(t)\,dt,$$

we conclude that F is differentiable everywhere on $\mathbb{R}$ and for $x \in \mathbb{R}$ we have

$$\begin{aligned}F'(x) &= \frac{1}{2\pi}\int_{-\infty}^{\infty} \lim_{h\downarrow 0}\frac{\sin ht}{ht} e^{-itx}\varphi(t)\,dt\\ &= \frac{1}{2\pi}\int_{-\infty}^{\infty} e^{-itx}\varphi(t)\,dt.\end{aligned}$$

Hence F is absolutely continuous, and $f = F'$ exists and is bounded on $\mathbb{R}$. Finally we note that

$$|f(x+h) - f(x)| \le \frac{1}{2\pi}\int_{-\infty}^{\infty} \left|\sin\frac{ht}{2}\right| |\varphi(t)|\,dt,$$

so that the uniform continuity of f follows from the Lebesgue dominated convergence theorem. ∎

Remark 3.2.2. There exist absolutely continuous distribution functions whose characteristic functions are not absolutely integrable on $\mathbb{R}$. See Example 3.6.2.

Example 3.3.2. Consider the characteristic function $\varphi(t) = (\cos ht)^{-1}$ for $t \in \mathbb{R}$. Since φ is real and even, it corresponds to a symmetric distribution function. Now

$$\int_{-\infty}^{\infty} |\varphi(t)|\,dt = \int_{-\infty}^{\infty} \operatorname{sech} t\,dt = \tan^{-1}(\sinh t)|_{-\infty}^{\infty} < \infty,$$

so by the Fourier inversion theorem the probability density function corresponding to φ is given by

$$f(x) = \frac{1}{2\pi}\int_{-\infty}^{\infty} e^{-ixt}\varphi(t)\,dt.$$

We leave it to the reader to show that $f(x) = 1/2\cosh(\pi x/2)$.

Finally we give a generalization of the inversion theorem in $\mathbb{R}_n$, and consider an application of this generalization.

Theorem 3.3.3 (Inversion Theorem in $\mathbb{R}_n$). Let F be a distribution function on $\mathbb{R}_n$, and let φ be its characteristic function. Let $\mathbf{a} = (a_1, a_2, \ldots, a_n)$ and $\mathbf{b} = (b_1, b_2, \ldots, b_n)$ be two points in $\mathbb{R}_n$ such that $a_j < b_j$ and a_j, b_j are continuity points of the marginal distribution function

$$F_j(x_j) = \lim_{\substack{x_i \to \infty \\ i=1,2,\ldots,n,\, i \ne j}} F(x_1, x_2, \ldots, x_j, \ldots, x_n), \qquad j = 1, 2, \ldots, n.$$

Then

$$F(\mathbf{b}) - F(\mathbf{a}) = \frac{1}{\pi^n} \lim_{\substack{T_j \to \infty \\ j=1,2,\ldots,n}} \int_{-T_n}^{T_n} \cdots \int_{-T_1}^{T_1} \prod_{j=1}^{n} \frac{e^{-it_j a_j} - e^{-it_j b_j}}{it_j} \varphi(t_1, t_2, \ldots, t_n)\,dt_1\,dt_2 \ldots dt_n. \tag{3.3.5}$$

Corollary 1. The distribution function F is determined uniquely by its characteristic function φ.

Corollary 2 (Fourier Inversion Theorem in $\mathbb{R}_n$). Suppose that φ is absolutely integrable on $\mathbb{R}_n$. Then the corresponding distribution function F is absolutely continuous on $\mathbb{R}_n$. Moreover, the probability density function $f = \partial^n F/\partial x_1\, \partial x_2 \cdots \partial x_n$ exists, is bounded and uniformly continuous on $\mathbb{R}_n$, and is given by

$$f(x_1, x_2, \ldots, x_n) = \frac{1}{(2\pi)^n} \int_{\mathbb{R}_n} \exp(-i\mathbf{t}\mathbf{x}')\varphi(\mathbf{t})\, d\mathbf{t}, \tag{3.3.6}$$

where $\mathbf{t} = (t_1, t_2, \ldots, t_n)$ and $\mathbf{x} = (x_1, x_2, \ldots, x_n)$.

Proof. The proofs of Theorem 3.3.3 and its corollaries are straightforward.

We consider now the following important application of the above result.

Theorem 3.3.4. Let $\mathbf{X} = (X_1, X_2, \ldots, X_n)$ be a random vector with distribution function F and characteristic function φ. Let F_j and φ_j be the distribution function and the characteristic function, respectively, of X_j, $1 \le j \le n$. Then the following three assertions are equivalent:

(a) $X_1, X_2, \ldots, X_n$ are independent.
(b) $\varphi(t_1, t_2, \ldots, t_n) = \prod_{j=1}^n \varphi_j(t_j)$ for all $t_1, t_2, \ldots, t_n \in \mathbb{R}$.
(c) $F(x_1, x_2, \ldots, x_n) = \prod_{j=1}^n F_j(x_j)$ for all $x_1, x_2, \ldots, x_n \in \mathbb{R}$.

Proof. That (a) $\Rightarrow$ (b) follows trivially from the definition of independence.

We show that (b) $\Rightarrow$ (c). For this purpose we use (3.3.5) and Fubini's theorem to conclude that

$$F(\mathbf{b}) - F(\mathbf{a}) = \prod_{j=1}^{n} [F_j(b_j) - F_j(a_j)],$$

where $\mathbf{a}, \mathbf{b}$ are defined as in the statement of Theorem 3.3.3. Taking the limits on both sides as $a_j \to -\infty$ for $1 \le j \le n$, we obtain

$$F(\mathbf{b}) = \prod_{j=1}^{n} F_j(b_j),$$

which holds for all $b_j \in C_{F_j}$, $1 \le j \le n$. Since (c) $\Rightarrow$ (a) from the definition of independence, the proof is complete. ∎

3.4 CONVOLUTION OF DISTRIBUTION FUNCTIONS

Let F_1 and F_2 be two distribution functions. Set

$$F(x) = \int_{-\infty}^{\infty} F_1(x - y)\, dF_2(y), \qquad x \in \mathbb{R}, \tag{3.4.1}$$

where the integral on the right-hand side is an improper Riemann-Stieltjes integral. Note that F_1 is bounded and has a countable set of discontinuity points, so that (3.4.1) is defined. It can be easily verified that F as defined in (3.4.1) is a distribution function.

Definition 3.4.1. The distribution function F defined in (3.4.1) is said to be the convolution of F_1 and F_2 and is denoted by $F = F_1 * F_2$.

Remark 3.4.1. If at least one of F_1 and F_2 is continuous (absolutely continuous), then F is continuous (absolutely continuous). The converse is not true.

Remark 3.4.2. Suppose F_1 and F_2 are both absolutely continuous with corresponding probability density functions f_1 and f_2. Then the probability density function f of F is given by

$$f(x) = \int_{-\infty}^{\infty} f_1(x - y) f_2(y)\, dy, \qquad x \in \mathbb{R}. \tag{3.4.2}$$

In this case we write $f = f_1 * f_2$.

Remark 3.4.3. If F_1 and F_2 are discrete distribution functions, so also is $F = F_1 * F_2$. If F is discrete, both F_1 and F_2 must be discrete.

We next prove the following basic result.

Theorem 3.4.1 (Convolution Theorem). Let F, F_1, F_2 be three distribution functions with characteristic functions φ, φ_1, φ_2, respectively. Then

$$F = F_1 * F_2 \Leftrightarrow \varphi = \varphi_1 \varphi_2.$$

Proof. Suppose that $F = F_1 * F_2$. Let $[a, b]$ be an arbitrary closed bounded interval in $\mathbb{R}$, and let

$$a = x_{0,N} < x_{1,N} < \cdots < x_{\nu_N, N} = b$$

be a sequence of subdivisions of $[a, b]$ such that

$$\Delta_N = \max_{1 \le \nu \le \nu_N} (x_{\nu, N} - x_{\nu-1, N}) \to 0 \qquad \text{as } N \to \infty.$$

Using the definition of the Riemann–Stieltjes integral, we obtain for $t \in \mathbb{R}$

(3.4.3)

$$\int_a^b e^{itx}\, dF(x) = \lim_{N\to\infty} \sum_{\nu=1}^{\nu_N} e^{itx_{\nu,N}}[F(x_{\nu,N}) - F(x_{\nu-1,N})]$$

$$= \lim_{N\to\infty} \int_{-\infty}^{\infty} \sum_{\nu=1}^{\nu_N} e^{it(x_{\nu,N}-y)}[F_1(x_{\nu,N}-y) - F_1(x_{\nu-1,N}-y)]e^{ity}\, dF_2(y).$$

We note that

$$\left| \sum_{\nu=1}^{\nu_N} e^{it(x_{\nu,N}-y)}[F_1(x_{\nu,N}-y) - F_1(x_{\nu-1,N}-y)] \right|$$

$$\leq \sum_{\nu=1}^{\nu_N} [F_1(x_{\nu,N}-y) - F_1(x_{\nu-1,N}-y)]$$

$$= F_1(b) - F_1(a) \leq 1,$$

so that the integrand on the right-hand side of (3.4.3) is uniformly bounded. Using the Lebesgue dominated convergence theorem, we see that for $t \in \mathbb{R}$

$$\int_a^b e^{itx}\, dF(x) = \int_{-\infty}^{\infty} \left[\int_{a-y}^{b-y} e^{itx}\, dF_1(x)\right] e^{ity}\, dF_2(y).$$

Next we take the limits on both sides as $a \to -\infty$ and $b \to +\infty$ to obtain $\varphi = \varphi_1\varphi_2$.

Conversely, suppose that $\varphi = \varphi_1\varphi_2$. Let $G = F_1 * F_2$, and let θ be the characteristic function of G. Then from what has been shown above it follows that $\theta = \varphi_1\varphi_2 = \varphi$. It follows at once from the uniqueness theorem that $G = F$. ■

Remark 3.4.4. The operation $*$ is commutative and associative. It follows that the set $\mathscr{F}$ of all distribution functions on $\mathbb{R}$ forms a commutative semi-group with $*$ as the operation of multiplication and with the degenerate distribution $\epsilon(x)$ as its identity element.

Remark 3.4.5. Let X_1 and X_2 be two independent random variables with respective distribution functions F_1 and F_2, and characteristic functions φ_1 and φ_2. Then $X = X_1 + X_2$ has distribution function $F = F_1 * F_2$ and characteristic function $\varphi = \varphi_1\varphi_2$.

Example 3.4.1. Let (X_1, X_2) be jointly distributed with probability density function

$$f(x, y) = \begin{cases} \frac{1}{4}[1 + xy(x^2 - y^2)] & \text{if } |x| \le 1 \quad \text{and} \quad |y| \le 1, \\ 0 & \text{otherwise.} \end{cases}$$

Then X_1, X_2 are not independent. Let $X = X_1 + X_2$. It is easy to show that X has the characteristic function

$$\varphi(t) = \left(\frac{\sin t}{t}\right)^2, \qquad t \in \mathbb{R}.$$

Clearly X_1 and X_2 are identically distributed, and the common characteristic function is given by

$$\varphi_1(t) = \varphi_2(t) = \frac{\sin t}{t}, \qquad t \in \mathbb{R}.$$

It follows that $\varphi(t) = \varphi_1(t)\varphi_2(t)$ for $t \in \mathbb{R}$, but X_1 and X_2 are not independent.

Theorem 3.4.2. Let F, F_1, F_2 be three distribution functions, and let $\mathscr{C}_0(\mathbb{R}) = \mathscr{C}_0$ be the set of all real-valued, bounded, continuous functions on $\mathbb{R}$. Then $F = F_1 * F_2$ if and only if the relation

$$\int_{-\infty}^{\infty} g\, dF = \int_{-\infty}^{\infty} \int_{-\infty}^{\infty} g(x + y)\, dF_1(x)\, dF_2(y) \tag{3.4.4}$$

holds for every $g \in \mathscr{C}_0$.

Proof. The sufficiency of (3.4.4) follows at once from Theorem 3.4.1 by taking $g(x) = \cos tx(\sin tx)$. The proof of the necessity part can be carried out exactly as in Theorem 3.4.1. ■

Next we consider the convolution of probability measures. Let P_1 and P_2 be two probability measures on $(\mathbb{R}, \mathscr{B})$. For every $E \subset \mathbb{R}$ and $a \in \mathbb{R}$ we write $E + a = \{x + a: x \in E\}$. Set

$$P(E) = \int_{\mathbb{R}} P_1(E - y)\, dP_2(y), \qquad E \in \mathscr{B}. \tag{3.4.5}$$

Clearly P is a probability measure on $\mathscr{B}$.

Definition 3.4.2. The probability measure P defined in (3.4.5) is called the convolution of P_1 and P_2, and we write $P = P_1 * P_2$.

We recall that, if P is a probability measure on $\mathscr{B}$, its Fourier-transform φ is defined by

$$\varphi(t) = \int_{\mathbb{R}} e^{itx}\, dP(x), \qquad t \in \mathbb{R}. \tag{3.4.6}$$

Theorem 3.4.3. Let P, P_1, P_2 be probability measures on $\mathscr{B}$, and let $\varphi, \varphi_1, \varphi_2$, respectively, be their Fourier transforms. Then the following assertions are equivalent:

(a) $P = P_1 * P_2$.
(b) $\varphi = \varphi_1 \varphi_2$.
(c) For every $g \in \mathscr{C}_0$, the relation

$$\int_{\mathbb{R}} g\, dP = \int_{\mathbb{R}\times\mathbb{R}} g(x+y)\, d(P_1 \times P_2)(x, y) \tag{3.4.7}$$

holds, where $P_1 \times P_2$ is the product measure of P_1 and P_2.

Proof. The proof of Theorem 3.4.3 is analogous to the proofs of Theorems 3.4.1 and 3.4.2. ■

Remark 3.4.6. Note that the convolution $\mu_1 * \mu_2$ of two finite measures μ_1 and μ_2 on $(\mathbb{R}, \mathscr{B})$ can be defined exactly as in (3.4.5). The Fourier transform of a finite measure μ can also be defined as in (3.4.6). In that case Theorem 3.4.3 holds for finite measures on $\mathscr{B}$.

Remark 3.4.7. Let $\mathscr{P}$ be the class of all probability measures on $\mathscr{B}$. Then $\mathscr{P}$ is a commutative semigroup with $*$ as the operation of multiplication and with P_0 [the probability measure associated with the degenerate distribution $\epsilon(x)$] as its identity element.

3.5 CONTINUITY THEOREM

We now give a necessary and sufficient condition for the complete convergence of a sequence of distribution functions to a distribution function. The following result is due to P. Lèvy.

Theorem 3.5.1 (Continuity Theorem). Let $\{F_n\}$ be a sequence of distribution functions, and let $\{\varphi_n\}$ be the sequence of corresponding characteristic functions. Then F_n converges completely to a distribution function F if and only if $\varphi_n \to \varphi$ as $n \to \infty$ on $\mathbb{R}$, where φ is continuous at $t = 0$. In this case the limit function φ is the characteristic function of the limit distribution function F.

Proof. First suppose that $F_n \xrightarrow{c} F$, where F is a distribution function. Then it follows from Theorem 3.1.4 that for every $t \in \mathbb{R}$

$$\int_{\mathbb{R}} \cos tx \, dF_n(x) \to \int_{\mathbb{R}} \cos tx \, dF(x)$$

and

$$\int_{\mathbb{R}} \sin tx \, dF_n(x) \to \int_{\mathbb{R}} \sin tx \, dF(x),$$

so that

$$\varphi_n(t) \to \varphi(t) \qquad \text{for every } t \in \mathbb{R} \quad \text{as } n \to \infty,$$

where φ is the characteristic function of F.

Conversely, suppose that $\varphi_n \to \varphi$ on $\mathbb{R}$, where φ is continuous at $t = 0$. We first show that, if F is a distribution function with characteristic function φ, then for every $h > 0$

$$\text{(3.5.1)} \qquad \int_0^h F(y)\, dy - \int_{-h}^0 F(y)\, dy = \frac{1}{\pi} \int_{-\infty}^{\infty} \frac{1 - \cos ht}{t^2} \varphi(t)\, dt.$$

Let $a > 0$, and let G be the uniform distribution on $[-a, a]$ with characteristic function $\theta(t) = (\sin at)/at$. Consider the distribution function $H = F * G$. Clearly H is continuous on $\mathbb{R}$ and is given by

$$\text{(3.5.2)} \qquad H(x) = \frac{1}{2a} \int_{x-a}^{x+a} F(y)\, dy.$$

Let ψ be the characteristic function of H. Then it follows from Theorem 3.4.1 that

$$\psi(t) = \varphi(t)\theta(t) = \varphi(t) \frac{\sin at}{at}.$$

Applying the inversion theorem to H and ψ, we obtain for all $x \in \mathbb{R}$

$$H(x + a) - H(x - a) = \lim_{T\to\infty} \frac{1}{\pi} \int_{-T}^{T} \frac{\sin^2 at}{at^2} e^{-itx}\varphi(t)\,dt$$

$$= \frac{1}{2\pi a} \int_{-\infty}^{\infty} \frac{1 - \cos 2at}{t^2} e^{-itx}\varphi(t)\,dt.$$

In particular, for $x = 0$ we have

$$H(a) - H(-a) = \frac{1}{2\pi a} \int_{-\infty}^{\infty} \frac{1 - \cos 2at}{t^2} \varphi(t)\,dt. \tag{3.5.3}$$

The proof of (3.5.1) then follows at once on substituting (3.5.2) in (3.5.3) and writing $h = 2a$.

We now return to the proof of the theorem. Since $\{F_n\}$ is a sequence of distribution functions, Helly's theorem (Theorem 3.1.2) implies that there exists a subsequence $\{F_{n_k}\}$ of $\{F_n\}$ which converges weakly to a bounded, nondecreasing, and right-continuous function F. We show that F is a distribution function. From (3.5.1) we obtain

$$\int_0^h F_{n_k}(y)\,dy - \int_{-h}^0 F_{n_k}(y)\,dy = \frac{1}{\pi} \int_{-\infty}^{\infty} \frac{1 - \cos ht}{t^2} \varphi_{n_k}(t)\,dt$$

for every $h > 0$. We now take the limit of both sides as $k \to \infty$. First note that the integrals on the left-hand side are taken over finite intervals, and, moreover, the F_{n_k} are uniformly bounded and converge to F a.e. on $\mathbb{R}$. In view of the Lebesgue dominated convergence theorem we can take the limit under the integral sign as $k \to \infty$. On the other hand, the integrand on the right-hand side is dominated by the absolutely integrable function $(1 - \cos ht)/t^2$, so that we can again take the limit under the integral sign. Thus we obtain

$$\int_0^h F(y)\,dy - \int_{-h}^0 F(y)\,dy = \frac{1}{\pi} \int_{-\infty}^{\infty} \frac{1 - \cos ht}{t^2} \varphi(t)\,dt.$$

Dividing both sides by h, we have

$$\frac{1}{h}\left[\int_0^h F(y)\,dy - \int_{-h}^0 F(y)\,dy\right] = \frac{1}{\pi} \int_{-\infty}^{\infty} \frac{1 - \cos t}{t^2} \varphi\left(\frac{t}{h}\right) dt. \tag{3.5.4}$$

Since φ is continuous at $t = 0$, we have

$$\lim_{h\to\infty} \varphi\left(\frac{t}{h}\right) = \varphi(0) = \lim_{n\to\infty} \varphi_n(0) = 1.$$

Letting $h \to \infty$ in (3.5.4), we obtain

$$F(+\infty) - F(-\infty) = \frac{1}{\pi} \int_{-\infty}^{\infty} \frac{1 - \cos t}{t^2}\, dt = 1.$$

Since $0 \le F \le 1$, it follows that $F(-\infty) = 0$ and $F(+\infty) = 1$, so that F is a distribution function. We then conclude from the first part that φ is the characteristic function corresponding to F.

Now suppose that $\{F_n\}$ contains another subsequence which converges to a limit, say F^*. Then, proceeding as above, we see that F^* is a distribution function and φ is the corresponding characteristic function. By the uniqueness theorem it follows that $F = F^*$, which implies that every weakly convergent subsequence of $\{F_n\}$ has the same limit F. This shows that $F_n \xrightarrow{c} F$ and that φ is the characteristic function of F. ■

We now give an alternative version of Theorem 3.5.1.

Theorem 3.5.2. Let $\{F_n\}$ be a sequence of distribution functions, and let $\{\varphi_n\}$ be the sequence of corresponding characteristic functions. Then F_n converges completely to a distribution function F if and only if $\varphi_n \to \varphi$ uniformly in every finite t-interval. In this case φ is the characteristic function of F.

Proof. Suppose first that $F_n \xrightarrow{c} F$. Then, in view of the continuity theorem, φ is the characteristic function of F. We now show that this convergence is uniform in every finite interval. In fact, we have

$$|\varphi_n(t) - \varphi(t)| \le \left| \int_a^b e^{itx}\, dF_n(x) - \int_a^b e^{itx}\, dF(x) \right| + [1 - F_n(b) + F_n(a)] + [1 - F(b) + F(a)],$$

where $[a, b]$ is a closed, bounded interval. Let $\epsilon > 0$. Choose $a, b \in C_F$ such that $1 - F(b) + F(a) < \epsilon/3$. Since $F_n \xrightarrow{c} F$, it follows that for sufficiently large n

$$1 - F_n(b) + F_n(a) < 1 - F(b) + F(a) + \frac{\epsilon}{6} < \frac{\epsilon}{3},$$

and consequently for sufficiently large n

$$|\varphi_n(t) - \varphi(t)| \le \left| \int_a^b e^{itx}\, dF_n(x) - \int_a^b e^{itx}\, dF(x) \right| + \frac{\epsilon}{2}.$$

Let $T > 0$. We now show that for sufficiently large n and for all $t \in [-T, T]$

(3.5.5) $$\left| \int_a^b e^{itx}\, dF_n(x) - \int_a^b e^{itx}\, dF(x) \right| < \frac{\epsilon}{2}.$$

For this purpose consider a subdivision

$$a = x_0 < x_1 < \cdots < x_N = b$$

of $[a, b]$ such that (a) $x_\nu \in C_F$ for $\nu = 0, 1, \ldots, N$, and (b) $\lambda = \max_{1 \le \nu \le N} (x_\nu - x_{\nu-1}) < \epsilon/8T$. Then

$$\begin{aligned}
&\left| \int_a^b e^{itx}\, dF_n(x) - \int_a^b e^{itx}\, dF(x) \right| \\
&\quad \le \left| \sum_{\nu=1}^{N} \left[\int_{x_{\nu-1}}^{x_\nu} e^{itx_\nu}\, dF_n(x) - \int_{x_{\nu-1}}^{x_\nu} e^{itx_\nu}\, dF(x) \right] \right| \\
&\qquad + \left| \sum_{\nu=1}^{N} \left[\int_{x_{\nu-1}}^{x_\nu} (e^{itx} - e^{itx_\nu})\, dF_n(x) - \int_{x_{\nu-1}}^{x_\nu} (e^{itx} - e^{itx_\nu})\, dF(x) \right] \right| \\
&\quad \le \sum_{\nu=1}^{N} \left| \int_{x_{\nu-1}}^{x_\nu} dF_n(x) - \int_{x_{\nu-1}}^{x_\nu} dF(x) \right| \\
&\qquad + \sum_{\nu=1}^{N} \int_{x_{\nu-1}}^{x_\nu} |e^{itx} - e^{itx_\nu}|\, dF_n(x) \\
&\qquad + \sum_{\nu=1}^{N} \int_{x_{\nu-1}}^{x_\nu} |e^{itx} - e^{itx_\nu}|\, dF(x) \\
&\quad \le \sum_{\nu=1}^{N} |[F_n(x_\nu) - F(x_\nu)] - [F_n(x_{\nu-1}) - F(x_{\nu-1})]| \\
&\qquad + \lambda T \sum_{\nu=1}^{N} \int_{x_{\nu-1}}^{x_\nu} dF_n(x) + \lambda T \sum_{\nu=1}^{N} \int_{x_{\nu-1}}^{x_\nu} dF(x) \\
&\quad < \sum_{\nu=1}^{N} |[F_n(x_\nu) - F(x_\nu)] - [F_n(x_{\nu-1}) - F(x_{\nu-1})]| \\
&\qquad + \frac{\epsilon}{4}.
\end{aligned}$$

Since $F_n(x_\nu) \to F(x_\nu)$ for $\nu = 0, 1, 2, \ldots, N$, it follows that for sufficiently large n

$$\sum_{\nu=1}^{N} |[F_n(x_\nu) - F(x_\nu)] - [F_n(x_{\nu-1}) - F(x_\nu)]| < \frac{\epsilon}{4}.$$

This proves (3.5.5), and hence for sufficiently large n and $t \in [-T, T]$

$$|\varphi_n(t) - \varphi(t)| < \epsilon$$

as asserted.

Conversely, if $\varphi_n \to \varphi$ uniformly in every finite t-interval, φ is continuous on $\mathbb{R}$ and, in particular, at $t = 0$. The proof is an immediate consequence of Theorem 3.5.1. ■

Corollary. Let $\mathscr{C}_0(\mathbb{R}) = \mathscr{C}_0$ be the space of all bounded, real-valued, continuous functions on $\mathbb{R}$. Let $\{F_n\}$ be a sequence of distribution functions, and let F be a distribution function. Then

$$F_n \xrightarrow{c} F \Leftrightarrow \int_{-\infty}^{\infty} g\,dF_n \to \int_{-\infty}^{\infty} g\,dF \qquad \text{for every } g \in \mathscr{C}_0.$$

Proof. The proof is an immediate consequence of the Helly–Bray and continuity theorems.

Motivated by the above corollary, we now introduce the concept of weak convergence of a sequence of probability measures defined on $(\mathbb{R}, \mathscr{B})$. Let $\{P_n\}$ be a sequence of probability measures, and P be a probability measure on $(\mathbb{R}, \mathscr{B})$. We say that $\{P_n\}$ *converges weakly* to P if

$$\int_{\mathbb{R}} g\,dP_n \to \int_{\mathbb{R}} g\,dP \qquad \text{for every } g \in \mathscr{C}_0,$$

and in this case we write $P_n \Rightarrow P$.

Theorem 3.5.3. Let $\{P_n\}$ be a sequence of probability measures, and let P be a probability measure on $(\mathbb{R}, \mathscr{B})$. Let F_n, F be the corresponding distribution functions. Then the following statements are equivalent:

(a) $P_n \Rightarrow P$.
(b) $F_n \xrightarrow{c} F$.
(c) $\int_{\mathbb{R}} g\,dP_n \to \int_{\mathbb{R}} g\,dP$, $g \in \mathscr{C}_0$.
(d) $\int_{-\infty}^{\infty} g\,dF_n \to \int_{-\infty}^{\infty} g\,dF$, $g \in \mathscr{C}_0$.
(e) $P_n(B) \to P(B)$ for every Borel set $B \in \mathscr{B}$ for which the boundary has P-measure zero.

Proof. Clearly (a) $\Leftrightarrow$ (c) $\Leftrightarrow$ (d) $\Leftrightarrow$ (b). Next we show that (c) $\Rightarrow$ (e). Let $B \in \mathscr{B}$ be such that $P(\bar{B} - B^0) = 0$, where $\bar{B}$ is the closure of B and B^0 is the interior of B. Since $B^0 \subset B \subset \bar{B}$, we conclude that

$$P(\bar{B}) = P(B) = P(B^0).$$

Let $\epsilon > 0$. Since $\mathscr{B}$ coincides with the σ-field generated by the class of all open sets in $\mathbb{R}$, there exists an open set $O \supset \bar{B}$ such that $P(O) < P(\bar{B}) + \epsilon$. Then $\bar{B}$ and $\mathbb{R} - O$ are disjoint closed subsets of $\mathbb{R}$. It follows from the Urysohn lemma (Kelley [46], p. 115) that there exists a $g \in \mathscr{C}_0$ such that $0 \leq g \leq 1$ on $\mathbb{R}$, $g = 1$ on $\bar{B}$ and $=0$ on $\mathbb{R} - O$. Clearly

$$\int_{\mathbb{R}} g\,dP_n \geq P_n(\bar{B})$$

and

$$\int_{\mathbb{R}} g\,dP = \int_{O} g\,dP \leq P(O) < P(\bar{B}) + \epsilon.$$

It follows from (c) that

$$P_n(B) \leq P_n(\bar{B}) \leq \int_{\mathbb{R}} g\,dP_n \to \int_{\mathbb{R}} g\,dP < P(\bar{B}) + \epsilon.$$

Taking limits as $n \to \infty$, we get

$$\limsup_{n\to\infty} P_n(B) \leq P(\bar{B}) + \epsilon = P(B) + \epsilon. \tag{3.5.6}$$

On the other hand, for given $\epsilon > 0$ there exists a closed set $F \subset B^0$ such that $P(B^0) < P(F) + \epsilon$. Clearly F and $\mathbb{R} - B^0$ are disjoint closed subsets of $\mathbb{R}$, so that there exists an $h \in \mathscr{C}_0$ such that $0 \leq h \leq 1$ on $\mathbb{R}$, $h = 1$ on F and $=0$ on $\mathbb{R} - B^0$. Proceeding as above, we obtain

$$P_n(B) \geq P_n(B^0) \geq \int_{\mathbb{R}} h\,dP_n \to \int_{\mathbb{R}} h\,dP \geq P(F) > P(B^0) - \epsilon,$$

which yields

$$\liminf_{n\to\infty} P_n(B) \geq P(B^0) - \epsilon = P(B) - \epsilon. \tag{3.5.7}$$

Combining (3.5.6) and (3.5.7), we obtain

$$P(B) - \epsilon \leq \liminf_{n\to\infty} P_n(B) \leq \limsup_{n\to\infty} P_n(B) \leq P(B) + \epsilon,$$

so that

$$\lim_{n\to\infty} P_n(B) = P(B).$$

Finally we prove (e) $\Rightarrow$ (c). It suffices to show that (e) $\Rightarrow$ (b). Let $x, a \in \mathbb{R}$ be such that $x > a$ and $x, a \in C_F$. Let $B = (a, x]$. Then $P(\bar{B} - B^0) = 0$, and it follows that

$$P_n(B) \to P(B),$$

that is, $F_n(x) - F_n(a) \to F(x) - F(a)$.

Let $\epsilon > 0$, and choose $a \in C_F$ sufficiently small so that $F(a) < \epsilon$. Then

$$F_n(x) \geq F_n(x) - F_n(a) \to F(x) - F(a) > F(x) - \epsilon,$$

so that

$$\liminf_{n\to\infty} F_n(x) \geq F(x) - \epsilon.$$

Next, for given $\epsilon > 0$, we choose $b \in C_F$, $b > x$, sufficiently large so that $F(b) > 1 - \epsilon$. Taking $B = (x, b]$, we obtain

$$1 - F_n(x) \geq F_n(b) - F_n(x) \to F(b) - F(x) > 1 - \epsilon - F(x),$$

and hence

$$\limsup_{n\to\infty} F_n(x) \leq F(x) + \epsilon.$$

Thus

$$F(x) - \epsilon \leq \liminf_{n\to\infty} F_n(x) \leq \limsup_{n\to\infty} F_n(x) \leq F(x) + \epsilon,$$

which yields

$$\lim_{n\to\infty} F_n(x) = F(x), \qquad x \in C_F.$$

Clearly $F_n(\pm\infty) \to F(\pm\infty)$, so that $F_n \xrightarrow{c} F$. This completes the proof of the theorem. ■

Finally we extend the concept of weak convergence to a sequence of finite signed measures on $\mathscr{B}$.

Definition 3.5.1. Let $\{\nu_n\}$ be a sequence of finite signed measures, and let ν be a finite signed measure on $(\mathbb{R}, \mathscr{B})$. We say that ν_n converges weakly to ν, and we write $\nu_n \Rightarrow \nu$ if

$$\int_{\mathbb{R}} g\, d\nu_n \to \int_{\mathbb{R}} g\, d\nu \qquad \text{for every } g \in \mathscr{C}_0. \tag{3.5.8}$$

We show that the weak convergence of finite signed measures (or finite measures) on $(\mathbb{R}, \mathscr{B})$ coincides with the usual concept of convergence induced by the weak topology. Let $\mathscr{S} = \mathscr{S}(\mathbb{R})$ be the set of all finite signed measures defined on $(\mathbb{R}, \mathscr{B})$. Then $\mathscr{S}$ is a real Banach space with respect to the norm $\|\nu\| = |\nu|(\mathbb{R})$, $\nu \in \mathscr{S}$, where $|\nu|$ is the total variation of the signed measure ν (Halmos [31], p. 123). In this case we note that the space $\mathscr{C}_0 = \mathscr{C}_0(\mathbb{R})$, which is a real Banach space with respect to the norm $\|g\| = \sup_{x\in\mathbb{R}} |g(x)|$, is the dual of $\mathscr{S}$. Here the duality relation is defined by

$$\langle g, \nu\rangle = \int_{\mathbb{R}} g\, d\nu \qquad \text{for } g \in \mathscr{C}_0, \nu \in \mathscr{S}. \tag{3.5.9}$$

Let $v_0 \in \mathscr{S}$. Then a *weak neighborhood* of v_0 is defined by the relation

$$W(v_0; g_1, g_2, \ldots, g_n; \epsilon) = \{v \in \mathscr{S}: |\langle g_i, v\rangle - \langle g_i, v_0\rangle| < \epsilon, 1 \le i \le n\}, \tag{3.5.10}$$

where $g_i \in \mathscr{C}_0$ for $1 \le i \le n$, and $\epsilon > 0$. It is easy to verify that the family of sets $W(v_0; g_1, g_2, \ldots, g_n; \epsilon)$ obtained by varying $g_1, g_2, \ldots, g_n \in \mathscr{C}_0$ and $\epsilon > 0$ forms a fundamental system of neighborhoods for $v_0 \in \mathscr{S}$. The topology thus defined is known as the *weak topology* in the space $\mathscr{S}$. Clearly, a sequence $v_n \in \mathscr{S}$ converges in the weak topology to $v \in \mathscr{S}$ if and only if (3.5.8) holds. In particular, let $\mathscr{P} = \mathscr{P}(\mathbb{R}) \subset \mathscr{S}$ be the set of all probability measures on $(\mathbb{R}, \mathscr{B})$. We consider $\mathscr{P}$ as a topological subspace of $\mathscr{S}$ with respect to the weak topology on $\mathscr{S}$ as defined above. Then a sequence $P_n \in \mathscr{P}$ converges in the weak topology to $P \in \mathscr{P}$ if and only if

$$\int_{\mathbb{R}} g\, dP_n \to \int_{\mathbb{R}} g\, dP, \qquad g \in \mathscr{C}_0.$$

Thus we have shown that the weak convergence of probability measures defined earlier coincides with the convergence with respect to the weak topology on $\mathscr{P}$.

3.6 SOME CRITERIA FOR CHARACTERISTIC FUNCTIONS

In this section we derive some important necessary and sufficient conditions for a complex-valued function on $\mathbb{R}$ to be a characteristic function. For this purpose we first introduce the concept of a *positive definite function* on $\mathbb{R}$, which is due to Bochner.

Definition 3.6.1. Let φ be a complex-valued function defined on $\mathbb{R}$. Then φ is said to be positive definite on $\mathbb{R}$ if for every positive integer N, for every $t_1, t_2, \ldots, t_N \in \mathbb{R}$, and for every $\omega_1, \omega_2, \ldots, \omega_N \in \mathbb{C}$ the inequality

$$\sum_{j=1}^{N} \sum_{k=1}^{N} \omega_j \bar{\omega}_k \varphi(t_j - t_k) \ge 0 \tag{3.6.1}$$

holds.

We note the following elementary property of a positive definite function.

Proposition 3.6.1. Let φ be positive definite on $\mathbb{R}$. Then:

(a) $\varphi(0) \ge 0$.
(b) $\varphi(-t) = \bar{\varphi}(t)$.
(c) $|\varphi(t)| \le \varphi(0)$.

Proof. Part (a) follows by setting $N = 1, t_1 = 0, \omega_1 = 1$.

For the proof of (b), set $N = 2$, $t_1 = 0$, and $t_2 = t$, and take $\omega_1 = \omega \in \mathbb{C}$ and $\omega_2 = 1$. From (3.6.1) it then follows that

$$(1 + |\omega|^2)\varphi(0) + \omega\varphi(-t) + \bar{\omega}\varphi(t) \geq 0,$$

so that, using (a), we see that $\omega\varphi(-t) + \bar{\omega}\varphi(t)$ is real. Now set $\omega = 1$ and $\omega = i = \sqrt{-1}$ successively to get

$$\varphi(-t) + \varphi(t) = a \qquad (a \text{ real}),$$
$$\varphi(t) - \varphi(-t) = ib \qquad (b \text{ real}).$$

Part (b) now follows easily.

As for (c), consider first the case where $\varphi(0) = 0$. Then, setting $N = 2$, $t_1 = 0$, $t_2 = t$, $\omega_1 = \varphi(t)$, and $\omega_2 = -1$, we get, using (b), $-2|\varphi(t)|^2 \geq 0$, so that $\varphi(t) = 0$. Finally, when $\varphi(0) > 0$, we set $N = 2$, $t_1 = 0$, $t_2 = t$, $\omega_1 = \varphi(t)/\varphi(0)$, and $\omega_2 = -1$ to get

$$\varphi(0) - \frac{|\varphi(t)|^2}{\varphi(0)} \geq 0,$$

which yields (c). ■

Remark 3.6.1. If a positive definite function φ satisfies $\varphi(0) = 0$, then $\varphi(t) \equiv 0$. We say that a positive definite function $\varphi \not\equiv 0$ is *normalized* if $\varphi(0) = 1$.

Theorem 3.6.1 (Bochner). Let φ be a complex-valued function defined on $\mathbb{R}$. Then φ is a continuous, normalized, positive definite function on $\mathbb{R}$ if and only if φ is a characteristic function.

Proof. First suppose that φ is the characteristic function of a distribution function F. Clearly φ is continuous on $\mathbb{R}$, and $\varphi(0) = 1$. Let $N \geq 1$, $t_1, t_2, \ldots, t_N \in \mathbb{R}$, and $\omega_1, \omega_2, \ldots, \omega_N \in \mathbb{C}$. Then

$$\sum_{j=1}^{N} \sum_{k=1}^{N} \omega_j \bar{\omega}_k \varphi(t_j - t_k)$$
$$= \sum_{j=1}^{N} \sum_{k=1}^{N} \omega_j \bar{\omega}_k \int_{-\infty}^{\infty} e^{i(t_j - t_k)x}\, dF(x)$$
$$= \int_{-\infty}^{\infty} \left| \sum_{j=1}^{N} \omega_j e^{it_j x} \right|^2 dF(x) \geq 0,$$

which proves (3.6.1).

Conversely, let φ be a continuous, normalized, positive definite function on $\mathbb{R}$. We show that φ is a characteristic function. For this purpose we need the following lemma, due to Herglotz.

Lemma 3.6.1. Let $\{\theta(s),\ s = 0,\ \pm 1,\ \pm 2, \ldots\}$ be a sequence of complex numbers satisfying the following conditions:

(i) $\theta(0) = 1$.
(ii) For every positive integer $N \geq 1$ and for every $\omega_0, \omega_1, \ldots, \omega_N \in \mathbb{C}$ the inequality

$$\sum_{j=0}^{N} \sum_{k=0}^{N} \omega_j \bar{\omega}_k \theta(j - k) \geq 0 \tag{3.6.2}$$

holds.

Then there exists a distribution function G concentrated on $[-\pi, \pi]$ such that the integral representation

$$\theta(s) = \int_{-\pi}^{\pi} e^{isx}\, dG(x) \tag{3.6.3}$$

holds for $s = 0, \pm 1, 2, \ldots$.

Proof of Lemma 3.6.1. We set $\omega_j = e^{-ijx}$, $j = 0, 1, 2, \ldots, N - 1$, $x \in \mathbb{R}$, in (3.6.2) and obtain

$$g_N(x) = \frac{1}{N} \sum_{j=0}^{N-1} \sum_{k=0}^{N-1} e^{-i(j-k)x}\theta(j - k) \geq 0 \tag{3.6.4}$$

for all $N \geq 1$, $x \in \mathbb{R}$. We set $j - k = r$. Then r is an integer which occurs in $N - |r|$ terms of the sum in (3.6.4). Moreover, $-N + 1 \leq r \leq N - 1$. Hence we can rewrite (3.6.4) as

$$g_N(x) = \sum_{r=-N}^{N} \left(1 - \frac{|r|}{N}\right) e^{-irx}\theta(r) \geq 0. \tag{3.6.5}$$

Let s be an integer, $-N \leq s \leq N$. Multiplying both sides of (3.6.5) by e^{isx} and integrating with respect to the Lebesgue measure on $[-\pi, \pi]$, we obtain

$$\int_{-\pi}^{\pi} e^{isx} g_N(x)\, dx = 2\pi\left(1 - \frac{|s|}{N}\right)\theta(s).$$

We introduce the function G_N, defined by

$$G_N(x) = \begin{cases} 0, & x < -\pi, \\ \dfrac{1}{2\pi} \displaystyle\int_{-\pi}^{x} g_N(y)\, dy, & -\pi \leq x < \pi, \\ 1, & x \geq \pi. \end{cases}$$

Clearly

$$\left(1 - \frac{|s|}{N}\right)\theta(s) = \int_{-\pi}^{\pi} e^{isx}\, dG_N(x), \qquad -N \le s \le N, \tag{3.6.6}$$

so that

$$\int_{-\pi}^{\pi} dG_N(x) = \theta(0) = 1$$

by (*i*). Hence G_N is a distribution function concentrated on $[-\pi, \pi]$. Since $\{G_N\}$ is a sequence of uniformly bounded, nondecreasing, right-continuous functions, it follows from Helly's theorem that $\{G_N\}$ contains a subsequence $\{G_{N_k}\}$ ($N_k \to \infty$ as $k \to \infty$) which converges weakly to a bounded, nondecreasing, right-continuous function, say G on $\mathbb{R}$. In this case we note that, for every $N \ge 1$ and for every $\epsilon > 0$, $G_N(-\pi - \epsilon) = 0$ and $G_N(\pi + \epsilon) = 1$. Hence $G(-\pi - \epsilon) = 0$ and $G(\pi + \epsilon) = 1$, so that G is a distribution function concentrated on $[-\pi, \pi]$. From (3.6.6) we obtain

$$\left(1 - \frac{|s|}{N_k}\right)\theta(s) = \int_{-\pi}^{\pi} e^{isx}\, dG_{N_k}(x).$$

Taking limits on both sides as $k \to \infty$ and using the Helly–Bray theorem, we get

$$\theta(s) = \int_{-\pi}^{\pi} e^{isx}\, dG(x), \qquad s = 0, \pm 1, \ldots .$$

This completes the proof of Lemma 3.6.1. ■

Returning now to the proof of Theorem 3.6.1, we note that for every fixed positive integer $n \ge 1$ the sequence $\varphi(s/n)$, $s = 0, \pm 1, \pm 2, \ldots$, satisfies the conditions of Lemma 3.6.1. It follows that there exists a distribution function G_n concentrated on $[-\pi, \pi]$ such that

$$\varphi\left(\frac{s}{n}\right) = \int_{-\pi}^{\pi} e^{isx}\, dG_n(x), \qquad s = 0, \pm 1, \pm 2, \ldots . \tag{3.6.7}$$

Set $F_n(x) = G_n(x/n)$, $x \in \mathbb{R}$. Then F_n is a distribution function concentrated on $[-n\pi, n\pi]$. Let φ_n be the characteristic function of F_n. Then

$$\varphi_n(t) = \int_{-n\pi}^{n\pi} e^{itx}\, dF_n(x) = \int_{-\pi}^{\pi} e^{itny}\, dG_n(y). \tag{3.6.8}$$

It follows from (3.6.7) and (3.6.8) that for every $n \ge 1$

$$\varphi\left(\frac{s}{n}\right) = \varphi_n\left(\frac{s}{n}\right), \qquad s = 0, \pm 1, \pm 2, \ldots . \tag{3.6.9}$$

Let $t \in \mathbb{R}$ and $n \geq 1$ be fixed. Then there exists an integer $k = k(t, n)$ such that $0 \leq t - (k/n) < 1/n$. Setting $\theta = t - (k/n)$, we have $0 \leq \theta < 1/n$. Also,

$$\begin{aligned}\left|\varphi_n(t) - \varphi_n\left(\frac{k}{n}\right)\right| &= \left|\varphi_n\left(\theta + \frac{k}{n}\right) - \varphi_n\left(\frac{k}{n}\right)\right| \\ &\leq \int_{-n\pi}^{n\pi} |e^{i\theta x} - 1| \, dF_n(x) \\ &\leq \left[\int_{-n\pi}^{n\pi} |e^{i\theta x} - 1|^2 \, dF_n(x)\right]^{1/2} \\ &\qquad\qquad \text{(Cauchy–Schwartz inequality)} \\ &= \left[2 \int_{-n\pi}^{n\pi} (1 - \cos \theta x) \, dF_n(x)\right]^{1/2}.\end{aligned}$$

Note that

$$1 - \cos \theta x \leq 1 - \cos \frac{x}{n} \qquad \text{for } 0 \leq \theta < \frac{1}{n} \quad \text{and} \quad -n\pi \leq x \leq n\pi.$$

Thus

$$\begin{aligned}\left|\varphi_n(t) - \varphi_n\left(\frac{k}{n}\right)\right| &\leq \left[2 \int_{-n\pi}^{n\pi} \left(1 - \cos \frac{x}{n}\right) dF_n(x)\right]^{1/2} \\ &= \left\{2\left[1 - \operatorname{Re} \varphi_n\left(\frac{1}{n}\right)\right]\right\}^{1/2} \\ &= \left[2\left(1 - \operatorname{Re} \varphi\left(\frac{1}{n}\right)\right)\right]^{1/2} \qquad \text{(from (3.6.9))}.\end{aligned}$$

Since φ is continuous on $\mathbb{R}$ and $\varphi(0) = 1$, we conclude that

$$\lim_{n \to \infty} \left|\varphi_n(t) - \varphi_n\left(\frac{k}{n}\right)\right| = 0.$$

Now we note that

$$\begin{aligned}\varphi_n(t) &= \left[\varphi_n(t) - \varphi_n\left(\frac{k}{n}\right)\right] + \varphi_n\left(\frac{k}{n}\right) \\ &= \left[\varphi_n(t) - \varphi_n\left(\frac{k}{n}\right)\right] + \varphi\left(\frac{k}{n}\right) \\ &\to \varphi(t) \qquad \text{as } n \to \infty,\end{aligned}$$

so that the continuity theorem implies that φ is a characteristic function. This completes the proof of Theorem 3.6.1. ■

Remark 3.6.2. We note that a positive definite function φ need not be continuous. Consider the function defined by $\varphi(t) = 1$ if $t = 0$, and $=0$ if $t \neq 0$. Note, however, that if a positive definite function φ is continuous at 0, it is uniformly continuous (Proposition 3.2.1).

Example 3.6.1.

(a) Let φ_i be a characteristic function for every $i = 1, 2, \ldots,$ and let $\alpha_1, \alpha_2, \ldots$ be nonnegative real numbers such that $\sum_{j=1}^{\infty} \alpha_j = 1$. Then it follows from Bochner's theorem that $\sum_{j=1}^{\infty} \alpha_j \varphi_j$ is also a characteristic function. In particular, if φ is a characteristic function, then so is $e^{\varphi - 1}$; more generally, $e^{\lambda(\varphi - 1)}$ is a characteristic function for each $\lambda \geq 0$. Some other examples are $(1 - \alpha)\varphi(t)/[1 - \alpha\varphi(t)]$, $0 \leq \alpha < 1$, and $\sinh \varphi(t)/\sinh 1$.

(b) Let φ be a characteristic function. It follows immediately from Bochner's theorem that $\bar{\varphi}$, $|\varphi|^2$, and $\operatorname{Re} \varphi(t)$ are also characteristic functions.

We next consider another criterion for a complex-valued function to be a characteristic function, which is due to Cramér.

Theorem 3.6.2. Let φ be a bounded, continuous, complex-valued function on $\mathbb{R}$. Then φ is a characteristic function if and only if the following conditions hold:

(i) $\varphi(0) = 1$.
(ii) For every $T > 0$ and $x \in \mathbb{R}$

$$\psi(x, T) = \int_0^T \int_0^T \varphi(t - u) e^{ix(t-u)}\, dt\, du \geq 0.$$

Proof. Let φ be a characteristic function with distribution function F. Then clearly (i) holds, and

$$\varphi(t - u) = \int_{-\infty}^{\infty} e^{i(t-u)y}\, dF(y),$$

so that for $T > 0$ and $x \in \mathbb{R}$

$$\psi(x, T) = \int_0^T \int_0^T \left[\int_{-\infty}^{\infty} e^{i(t-u)y}\, dF(y)\right] e^{i(t-u)x}\, dt\, du. \tag{3.6.10}$$

Interchanging the order of integration in (3.6.10), which is permissible because of Fubini's theorem, we have

$$\psi(x, T) = \int_{-\infty}^{\infty} \left[\int_0^T e^{it(x+y)}\, dt\right]\left[\int_0^T e^{-iu(x+y)}\, du\right] dF(y)$$
$$= 2\int_{-\infty}^{\infty} \frac{1 - \cos T(x + y)}{(x + y)^2}\, dF(y)$$
$$\geq 0,$$

so (ii) holds.

Conversely, let φ be a bounded, continuous, complex-valued function on $\mathbb{R}$ satisfying (i) and (ii). Then, for every $T > 0$ and $x \in \mathbb{R}$,

$$f(x, T) = \frac{1}{2\pi T}\psi(x, T) = \frac{1}{2\pi T}\int_0^T \int_0^T \varphi(u - v)e^{i(u-v)x}\, du\, dv$$
$$\geq 0.$$

Write $t = u - v$ and $w = v$. After some simple computation we obtain

$$f(x, T) = \frac{1}{2\pi T}\int_{-T}^{0} \varphi(t)e^{itx}\left(\int_{-t}^{T} dw\right) dt + \frac{1}{2\pi T}\int_0^T \varphi(t)e^{itx}\left(\int_0^{T-t} dw\right) dt.$$
$$= \frac{1}{2\pi}\int_{-\infty}^{\infty} \varphi_T(t)e^{itx}\, dt,$$

where

$$\varphi_T(t) = \begin{cases} \left(1 - \dfrac{|t|}{T}\right)\varphi(t) & \text{if } |t| \leq T, \\ 0 & \text{otherwise.} \end{cases}$$

For every $W > 0$, set

$$J(u; T, W) = \int_{-W}^{W} \left(1 - \frac{|x|}{W}\right) f(x, T)e^{iux}\, dx.$$

Then

$$J(u; T, W) = \frac{1}{2\pi}\int_{-W}^{W} \left(1 - \frac{|x|}{W}\right)\left[\int_{-\infty}^{\infty} \varphi_T(t)e^{i(t+u)x}\, dt\right] dx.$$

Interchanging the order of integration, we obtain, after some simplification,

$$J(u; T, W) = \frac{1}{\pi}\int_{-\infty}^{\infty} \frac{1 - \cos[(u + t)W]}{(u + t)^2 W}\, \varphi_T(t)\, dt,$$

and writing $v = W(t + u)$, we obtain

$$J(u; T, W) = \frac{1}{\pi} \int_{-\infty}^{\infty} \frac{1 - \cos v}{v^2} \varphi_T\left(\frac{v}{W} - u\right) dv.$$

Using the Lebesgue dominated convergence theorem, we have

$$\begin{aligned}\lim_{W \to \infty} J(u; T, W) &= \frac{1}{\pi} \int_{-\infty}^{\infty} \frac{1 - \cos v}{v^2} \lim_{W \to \infty} \varphi_T\left(\frac{v}{W} - u\right) dv \\ &= \frac{2}{\pi} \int_{0}^{\infty} \frac{1 - \cos v}{v^2} \varphi_T(-u)\, dv \\ &= \varphi_T(-u).\end{aligned}$$

Since $f(x, T) \geq 0$, we easily verify that the function $J(u; T, W)/J(0; T, W)$ is a characteristic function for every $W > 0$. From what has been shown above and the continuity theorem, we conclude that $\varphi_T(-u)$ is a characteristic function, since $\varphi_T(0) = 1$. Clearly

$$\varphi(t) = \lim_{T \to \infty} \varphi_T(t),$$

so that φ is also a characteristic function. ∎

Remark 3.6.3. Theorem 3.6.1 can be restated in terms of a probability measure on $(\mathbb{R}, \mathscr{B})$ as follows. Let φ be a complex-valued function defined on $\mathbb{R}$. Then φ is a continuous, normalized, positive definite function on $\mathbb{R}$ if and only if there exists a probability measure P on $(\mathbb{R}, \mathscr{B})$, determined uniquely by φ, such that φ is the Fourier transform of P.

Finally we give a set of useful sufficient conditions for a function to be a characteristic function.

Theorem 3.6.3 (Pólya). Let φ be a real-valued continuous function on $\mathbb{R}$ satisfying the following conditions:

(i) $\varphi(0) = 1$.
(ii) $\varphi(-t) = \varphi(t)$.
(iii) φ is convex on $(0, \infty)$.
(iv) $\lim_{t \to \infty} \varphi(t) = 0$.

Then φ is the characteristic function of an absolutely continuous distribution function.

Proof. Since φ is convex on $(0, \infty)$, it has a right-hand derivative everywhere which is nondecreasing on $(0, \infty)$. (See, for example, Krasnoselsky and Rutitsky [51], p. 4.) We denote this by ψ. We now show the following:

(a) $\psi(t) \leq 0$ for $t > 0$.
(b) $\lim_{t\to\infty} \psi(t) = 0$.

Suppose that (a) does not hold. Then there exists a $t_0 > 0$ such that $\psi(t_0) > 0$, which implies that $\psi(t) > 0$ for all $t \geq t_0$. Consequently φ is strictly increasing for $t \geq t_0$. Let $t_1, t_2 \in \mathbb{R}$ and $t_1 \geq t_0, t_2 \geq t_0$. By the convexity of φ we have

$$\varphi\left(\frac{t_1 + t_2}{2}\right) \leq \tfrac{1}{2}[\varphi(t_1) + \varphi(t_2)].$$

Letting $t_2 \to \infty$ and using (iv) of Theorem 3.6.3, we see that $\varphi(t_1) \geq 0$ for all $t_1 \geq t_0$. Since φ is strictly increasing for $t \geq t_0$, this contradicts (iv). This proves (a). We note that (a) implies that φ is nonincreasing on $(0, \infty)$, so it has a derivative a.e. on $(0, \infty)$. Let us denote it by φ'. Clearly φ' is nondecreasing and ≤ 0 a.e. on $(0, \infty)$. It is sufficient to show that $\varphi'(t) \to 0$ as $t \to \infty$. Suppose this is not the case. Then there exists an $\alpha < 0$ such that $\lim_{t\to\infty} \varphi'(t) = \alpha$. Hence $\varphi'(t) \leq \alpha$ for all $t > 0$. Then for $t > 0$

$$\int_0^t \varphi'(s)\, ds \leq \alpha t,$$

which gives $\varphi(t) \leq \alpha t + 1$. Letting $t \to \infty$, we see that this last inequality contradicts (iv). This proves (b).

Next we note that the integral

$$\int_{-\infty}^{\infty} e^{-itx}\varphi(t)\, dt = 2\int_0^{\infty} \cos tx\, \varphi(t)\, dt$$

exists for all $x \neq 0$. Indeed, let $T > 0$. Integrating by parts, we see that

$$\int_0^T \cos tx\, \varphi(t)\, dt = \varphi(T)\frac{\sin Tx}{x} - \frac{1}{x}\int_0^T \sin tx\, \varphi'(t)\, dt.$$

Using (iv) and the fact that $\varphi'(t) \leq 0$ a.e., we conclude that the limit of the right-hand side of this last equation as $T \to \infty$ exists.

We now set

$$\begin{aligned} f(x) &= \frac{1}{2\pi}\int_{-\infty}^{\infty} e^{-itx}\varphi(t)\, dt \\ &= \frac{1}{\pi}\int_0^{\infty} \cos tx\, \varphi(t)\, dt. \end{aligned}$$

We use the following version of the Fourier inversion theorem, due to Pringsheim. (See Titchmarsh [80], p. 16.)

Pringsheim's Lemma. Let φ be nonincreasing on $(0, \infty)$ and integrable over every finite interval $(0, a)$ $(a > 0)$. Suppose further that $\lim_{t\to\infty} \varphi(t) = 0$. Then for every $t > 0$ the inversion formula

$$\tfrac{1}{2}[\varphi(t+0) + \varphi(t-0)] = \frac{2}{\pi}\int_0^\infty \cos tu\left[\int_0^\infty \varphi(y)\cos uy\, dy\right] du$$

holds.

Thus we obtain

$$\begin{aligned}\varphi(t) &= \tfrac{1}{2}[\varphi(t+0) + \varphi(t-0)]\\ &= \frac{2}{\pi}\int_0^\infty \cos tx\left[\int_0^\infty \varphi(y)\cos xy\, dy\right] dx\\ &= 2\int_0^\infty \cos tx f(x)\, dx = \int_{-\infty}^\infty e^{itx} f(x)\, dx,\end{aligned}$$

so that φ is the Fourier transform of f. Finally we show that f is a probability density function. Clearly, in view of (i),

$$\int_{-\infty}^\infty f(x)\, dx = \varphi(0) = 1.$$

It suffices to show that $f(x) \geq 0$ for $x \in \mathbb{R}$ $(x \neq 0)$. For this purpose we integrate by parts and obtain

$$f(x) = \frac{1}{\pi x}\int_0^\infty \sin tx[-\varphi'(t)]\, dt.$$

Note that $-\varphi'$ is nonincreasing, is nonnegative a.e. on $(0, \infty)$, and tends to zero as $t \to \infty$. Then for $x > 0$

$$\begin{aligned} f(x) &= \frac{1}{\pi x}\sum_{j=0}^\infty \int_{j\pi/x}^{(j+1)\pi/x} \sin tx[-\varphi'(t)]\, dt\\ &= \frac{1}{\pi x}\int_0^{\pi/x} \sin tx\left\{\sum_{j=0}^\infty (-1)^j\left[-\varphi'\left(t + \frac{j\pi}{x}\right)\right]\right\} dt. \qquad (3.6.11)\end{aligned}$$

For $x > 0$, the series $\sum_{j=0}^\infty (-1)^j\{-\varphi'[t + (j\pi/x)]\}$ is an alternating series whose terms are nonincreasing in absolute value; moreover, the first term ≥ 0. It follows that

$$\sum_{j=0}^\infty (-1)^j\left[-\varphi'\left(t + \frac{j\pi}{x}\right)\right] \geq 0 \qquad \text{for } x > 0.$$

On the other hand, $\sin tx \geq 0$ for $t \in [0, \pi/x]$. Therefore we conclude that the integrand on the right-hand side of (3.6.11) is nonnegative. Hence $f(x) \geq 0$ for $x > 0$. Since f is even, it follows that $f(x) \geq 0$ for all $x \neq 0$. Thus we have shown that f is a probability density function whose characteristic is φ. ■

Example 3.6.2. Some examples of functions which satisfy the conditions of Theorem 3.6.3 are as follows:

(a) $\varphi(t) = e^{-t^2/2}$.
(b) $\varphi(t) = e^{-|t|}$.
(c) $\varphi(t) = (1 + |t|)^{-1}$.
(d) $\varphi(t) = \begin{cases} 1 - |t| & \text{for } 0 < |t| \leq \frac{1}{2}, \\ 1/4|t| & \text{for } |t| > \frac{1}{2}. \end{cases}$
(e) $\varphi(t) = \begin{cases} 1 - |t| & \text{for } |t| \leq 1, \\ 0 & \text{for } |t| > 1. \end{cases}$

Here the characteristic functions in (a), (b), and (e) are absolutely integrable, so that we can apply the Fourier inversion formula (3.3.3) to get the corresponding probability density functions. But the characteristic functions in (c) and (d) are not absolutely integrable, and yet they are characteristic functions of absolutely continuous distribution functions. (See Remark 3.3.2.) It is clear that the probability density function of a characteristic function which satisfies the conditions of Pólya's theorem can always be obtained by means of formula (3.3.3), even though φ may not be absolutely integrable.

Note that the two distinct characteristic functions in (d) and (e) coincide over a finite interval. (See Remark 3.3.1.) It follows that there exist characteristic functions φ_1, φ_2, and φ_3 such that $\varphi_1\varphi_3 = \varphi_2\varphi_3$ with $\varphi_1 \neq \varphi_2$.

Example 3.6.3. Let $0 < \alpha \leq 1$. Then

$$\varphi(t) = (1 + |t|^\alpha)^{-1}$$

is a characteristic function. This follows since $\varphi''(t) > 0$, so that φ is convex and the conditions of Pólya's theorem are easily seen to be satisfied. For any $N > 0$ the function $\varphi(t/N)$ is also a characteristic function. Choose N such that $n = N^\alpha$ is an integer. Then

$$\left[\varphi\left(\frac{t}{N}\right)\right]^n = \left[1 + \frac{|t|^\alpha}{n}\right]^{-n}$$

is also a characteristic function. Letting $n \to \infty$ and using the continuity theorem, we see that

$$\lim_{n\to\infty} \left[\varphi\left(\frac{t}{N}\right)\right]^n = e^{-|t|^\alpha}$$

is also a characteristic function. This is the well-known characteristic function of a symmetric stable law with exponent $\alpha(0 < \alpha \leq 1)$. That $e^{-|t|^\alpha}$ is a characteristic function also follows from Pólya's theorem. We note that $\varphi(t) = 1/(1 + |t|^\alpha)$ is also a characteristic function for $1 < \alpha \leq 2$.

3.7 APPLICATIONS

We now consider some applications of characteristic functions. We will show how many results in probability and statistics can be established with ease by the use of characteristic functions. As we shall see, most applications of characteristic functions involve one or more of the three basic properties, namely, the inversion, uniqueness, and continuity theorems.

(a) Applications in Statistics. In statistics one of the fundamental problems is to determine the distribution of certain test statistics or estimators. This can frequently be done by using properties of characteristic functions. For example, let $X_1, X_2, \ldots, X_n$ be a sample from some population distribution function F. Let μ be the mean of F. One of the most commonly used estimators of μ is $\overline{X} = \sum_{i=1}^n X_i/n$, the *sample mean.* It has many nice properties. It also appears frequently as the test statistic for testing hypothesis concerning μ. It is therefore of interest to know the distribution of $\overline{X}$. The characteristic function of $\overline{X}$ is given by

$$\begin{aligned}\varphi_{\overline{X}}(t) &= \mathscr{E} \exp\left(\frac{it \sum_{k=1}^n X_k}{n}\right) = \prod_{k=1}^n \mathscr{E} \exp\left(\frac{itX_k}{n}\right)\\ &= \left[\varphi\left(\frac{t}{n}\right)\right]^n,\end{aligned}$$

where φ is the common characteristic function of X_k, $k = 1, 2, \ldots, n$. Therefore one can invert $\varphi_{\overline{X}}$ to get the distribution of $\overline{X}$. Frequently this is not necessary, however, since one can recognize the characteristic function of $\overline{X}$ as one of the standard known forms. For example, if $X_1, X_2, \ldots, X_n$ are normal with common mean μ and common variance σ^2, then

$$\begin{aligned}\varphi_{\overline{X}}(t) &= \left[\exp\left(i\mu\frac{t}{n} - \frac{\sigma^2}{2}\frac{t^2}{n^2}\right)\right]^n\\ &= \exp\left(i\mu t - \frac{\sigma^2}{2n}t^2\right),\end{aligned}$$

and it follows that $\overline{X}$ is also normal with mean μ and variance σ^2/n.

If $X_1, X_2, \ldots, X_n$ are Bernoulli random variables with parameter p, $0 \le p \le 1$, the characteristic function of $S_n = \sum_{k=1}^{n} X_k$ is

$$\varphi_{S_n}(t) = [(1 - p + p\exp(it))]^n,$$

and it follows that S_n has a binomial distribution with parameter p. Thus

$$P\left\{\bar{X} = \frac{k}{n}\right\} = \binom{n}{k} p^k (1-p)^{n-k}, \qquad k = 0, 1, 2, \ldots, n.$$

Sometimes it is of interest to determine the distribution of the geometric mean of a set of a.s. nonnegative, independent random variables. Let $X_1, \ldots, X_n$ be a sample from a population with distribution F such that $F(0) = 0$. Let $Z = (X_1 X_2 \ldots X_n)^{1/n}$ be the geometric mean of the X_i. Then

$$\ln Z = \frac{1}{n} \sum_{k=1}^{n} \ln X_k.$$

Let $Y_k = \ln X_k$, $k = 1, 2, \ldots, n$. Then the characteristic function of Y_k is given by

$$\varphi(t) = \int_0^\infty \exp(it \ln x)\, dF(x) = \int_0^\infty x^{it}\, dF(x),$$

and the characteristic function of $\bar{Y}$ is given by

$$\varphi_{\bar{Y}}(t) = \left[\varphi\left(\frac{t}{n}\right)\right]^n.$$

The corresponding distribution function $F_{\bar{Y}}$ of $\bar{Y}$ can be obtained from the inversion theorem. Since $Z = \exp(\bar{Y})$, the distribution function of Z is given by

$$F_Z(t) = F_{\bar{Y}}(\exp(t)).$$

As an example, let $X_1, \ldots, X_n$ be a sample from the uniform distribution on (0, 1). Then

$$\varphi(t) = \int_0^1 x^{it}\, dx = \frac{1}{1 + it},$$

so that

$$\varphi_{\bar{Y}}(t) = \left(1 + i\frac{t}{n}\right)^{-n} \quad \text{and} \quad \varphi_{-\bar{Y}}(t) = \left(1 - i\frac{t}{n}\right)^{-n}.$$

From Example 1.2.6 it follows that $-\bar{Y}$ has a gamma distribution with parameters n and $1/n$. In particular, $-2\sum_{k=1}^{n} \ln Y_k$ is a chi-square random variable with $2n$ degrees of freedom. The probability density function of Z is easily computed.

We next consider the important special case of sampling from a normal population. Let $X_1, X_2, \ldots, X_n$ be a random sample on a random variable X with distribution function F. Let $\bar{X}$ be as defined above, and let

$$S^2 = \frac{\sum_{i=1}^n (X_i - \bar{X})^2}{n-1},$$

where S^2 is known as the *sample variance*. If $\mathscr{E}|X| < \infty$, then $\mathscr{E}\bar{X} = \mathscr{E}X$. If $\mathscr{E}|X|^2 < \infty$, it is easy to check that $\mathscr{E}S^2 = \text{var}(X)$, and $S^2 \xrightarrow{\text{a.s.}} \text{var}(X)$. Since S^2 has many other nice properties, it is frequently used as an estimator for $\text{var}(X)$. In particular, if F is the normal distribution with mean μ and variance σ^2, then $\bar{X}$ and S^2 are independent. This result plays a key role in distribution theory. We now show how characteristic functions can be used to prove this result and obtain various associated distributions.

Without loss of generality we assume that $\mu = 0$. For real numbers $t_1, t_2, \ldots, t_n$ we have

$$\sum_{k=1}^n t_k(X_k - \bar{X}) = \sum_{k=1}^n \left(t_k - \frac{t_1 + \cdots + t_n}{n}\right)X_k,$$

so that for $s \in \mathbb{R}$

$$\sum_{k=1}^n t_k(X_k - \bar{X}) + s\bar{X} = \sum_{k=1}^n \left[\frac{s}{n} + (t_k - \bar{t})\right]X_k,$$

where $\bar{t} = (1/n)\sum_{k=1}^n t_k$. Therefore the joint characteristic function of $(X_1 - \bar{X}, \ldots, X_n - \bar{X})$ and $\bar{X}$ is given by

$$\begin{aligned}\mathscr{E}\exp\left\{i\sum_{k=1}^n t_k(X_k - \bar{X}) + s\bar{X}\right\} &= \prod_{k=1}^n \mathscr{E}\exp\left\{i\left[\frac{s}{n} + (t_k - \bar{t})\right]X_k\right\}\\ &= \exp\left\{-\frac{\sigma^2}{2n}s^2 - \frac{\sigma^2}{2}\sum_{k=1}^n t_k - \bar{t})^2\right\}\\ &= \mathscr{E}\exp(is\bar{X})\mathscr{E}\left[i\sum_{k=1}^n t_k(X_k - \bar{X})\right],\end{aligned}$$

which holds for all $s, t_1, \ldots, t_n \in \mathbb{R}$. It follows from Theorem 3.4.1 that $\bar{X}$ and $(X_1 - \bar{X}, \ldots, X_n - \bar{X})$ are independent, and hence $\bar{X}$ and S^2 are independent. To derive the distribution of S^2 note that X_k^2/σ^2 is a gamma variate with parameters $\frac{1}{2}$ and 2 (see Example 1.2.6). Hence

$$\mathscr{E}\exp(itX_k^2) = (1 - 2i\sigma^2 t)^{-1/2}, \qquad k = 1, 2, \ldots, n.$$

Also

$$\sum_{k=1}^n (X_k - \bar{X})^2 + n\bar{X}^2 = \sum_{k=1}^n X_k^2.$$

Since X_k^2 are independent and identically distributed,

$$\mathscr{E} \exp\left(it \sum_{k=1}^{n} X_k^2\right) = \{1 - 2i\sigma^2 t\}^{-n/2}.$$

The independence of $\overline{X}^2$ and S^2 implies that

$$\mathscr{E} \exp\left\{it \sum_{k=1}^{n} (X_k - \overline{X})^2\right\} \cdot \mathscr{E} \exp(itn\overline{X}^2) = \{1 - 2i\sigma^2 t\}^{-n/2}.$$

Since $\overline{X}$ is normal with mean 0 and variance σ^2/n, $n\overline{X}^2/\sigma^2$ is gamma with parameters $\frac{1}{2}$ and 2, so that

$$\mathscr{E} \exp(itn\overline{X}^2) = \{1 - 2i\sigma^2 t\}^{-1/2}.$$

Hence

$$\mathscr{E} \exp\left\{it \sum_{k=1}^{n} (X_k - \overline{X})^2\right\} = \{1 - 2it\sigma^2\}^{-(n-1)/2},$$

so that

$$\mathscr{E} \exp\left\{\frac{it(n-1)S^2}{\sigma^2}\right\} = \{1 - 2it\}^{-(n-1)/2}.$$

By the uniqueness theorem it follows that $(n - 1)S^2/\sigma^2$ is distributed as a gamma random variable with parameters $(n - 1)/2$ and 2, that is, $(n - 1)S^2/\sigma^2$ has a *chi-square* distribution with $n - 1$ degrees of freedom.

If X is normal with mean μ and variance σ^2, Y/σ^2 is a chi-square random variable with n degrees of freedom, and X and Y are independent, the ratio $(X - \mu)/\sqrt{Y/n}$ is known as *Student's t-statistic* with n degrees of freedom. In view of what has been shown above, $[(\overline{X} - \mu)/S]\sqrt{n}$ is a t-statistic with $n - 1$ degrees of freedom. This result is very useful in mathematical statistics.

Actually the converse of what we have shown is also true, namely, if the sample mean and the sample variance are independent, X must be normal. There is a proof based on characteristic functions, due to Lukacs [58], but we will not present it here. (See Problem 3.10.47.) It should be noted that the method used above can also be applied to exhibit the independence of the sample mean vector and the sample variance-covariance matrix if we are sampling from a multivariate normal distribution.

Yet another application of characteristic functions is in characterization problems. We have already mentioned the use of characteristic functions to show that the normal distribution can be characterized by the independence of $\overline{X}$ and S^2. Let us consider the following example. Let $X_1, X_2, \ldots, X_n$ be independent random variables with common (Cauchy) probability density function

$$f(x) = \frac{\mu}{\pi} \frac{1}{\mu^2 + (x - \theta)^2}, \qquad x \in \mathbb{R},$$

where $\mu > 0$ and $\theta \in \mathbb{R}$. Then the characteristic function of f is given by (see Example 1.2.5)

$$\varphi(t) = \exp(it\theta - \mu|t|), \qquad t \in \mathbb{R}.$$

It follows that the characteristic function of $\overline{X}$ is given by

$$\begin{aligned}\varphi_{\overline{X}}(t) &= \left[\exp\left(i\frac{t}{n}\theta - \mu\frac{|t|}{n}\right)\right]^n \\ &= \exp(it\theta - \mu|t|),\end{aligned}$$

so that $\overline{X}$ itself has the same Cauchy distribution. This reproductive property of random variables will be studied in detail in Section 5.4. Note that $\overline{X}$ does not obey the law of large numbers. Neither does $\overline{X}$ converge to the normal law. The interesting fact is that this property characterizes the Cauchy distribution. To see this, let F be a nondegenerate distribution function with the property that $\overline{X}$, the sample mean of any size n, has the same distribution. If φ is the characteristic function of F, we wish to solve the functional equation

$$\varphi(t) = \left[\varphi\left(\frac{t}{n}\right)\right]^n.$$

Let m, n be positive integers. Then

$$\varphi\left(\frac{m}{n}\right) = [\varphi(m)]^{1/n} = [\varphi(1)]^{m/n},$$

so that for every rational $r > 0$

$$\varphi(r) = [\varphi(1)]^r.$$

Since φ is continuous, it follows that for any real $t \geq 0$

$$\varphi(t) = [\varphi(1)]^t = [\rho e^{i\theta}]^t.$$

Also $\varphi(-t) = \overline{\varphi(t)}$, so $\varphi(t) = \rho^{|t|}e^{it\theta}$ for $t \in \mathbb{R}$. Since φ is bounded, we can write $\rho = e^{-\mu}$, $\mu \geq 0$, and it follows that

$$\varphi(t) = \exp(-\mu|t| + it\theta).$$

If $\mu = 0$, then φ is degenerate, so that $\mu > 0$ and F must be a Cauchy distribution.

(b) Some Applications of the Continuity Theorem. Some applications of the continuity theorem have already appeared (for example, in the proofs of Theorems 3.6.1 and 3.6.2 and in Example 3.6.3). There will be many more in later chapters, specially Chapter 5.

Let us first give a very short proof of Khintchine's weak law of large numbers (Theorem 2.1.3). Let $X_1, X_2, \ldots$ be independent, identically distributed random variables with common mean μ. Let $\overline{X}_n = n^{-1} \sum_{k=1}^{n} X_k$. If φ is the characteristic function of X_k, then

$$\mathscr{E} \exp(it\overline{X}_n) = \left[\varphi\left(\frac{t}{n}\right)\right]^n = \left[1 + i\frac{t}{n}\varphi'(0) + o\left(\frac{t}{n}\right)\right]^n$$

from Theorem 3.2.1. Since $\varphi'(0) = \mu$, it follows that

$$\lim_{n\to\infty} \mathscr{E} \exp(it\overline{X}_n) = \exp(it\mu),$$

and the continuity theorem implies that $\overline{X}_n \xrightarrow{L} \mu$ and hence $\overline{X}_n \xrightarrow{P} \mu$.

Consider now the following probability density function:

$$\begin{aligned} f(x) &= e(\ln|x| + 1)[2x^2 \ln^2|x|]^{-1} && \text{if } |x| > e \\ &= 0 && \text{otherwise.} \end{aligned}$$

Let $X_1, X_2, \ldots, X_n$ be independent and identically distributed with probability density function f. Then $\mathscr{E}X_k$ does not exist, and Khintchine's law of large numbers cannot be applied. However, if φ is the corresponding characteristic function, then, as $t \to 0$,

$$\frac{1 - \varphi(t)}{t} = \frac{e}{t}\int_e^\infty \frac{(1 - \cos xt)(1 + \ln x)}{x^2 \ln^2 x}\, dx = o(1).$$

Hence $\varphi'(0) = 0$. It follows, as above, that

$$\mathscr{E} \exp(it\overline{X}_n) = \left[\varphi\left(\frac{t}{n}\right)\right]^n = \left[1 + i\varphi'(0)\frac{t}{n} + o\left(\frac{t}{n}\right)\right]^n,$$

and the continuity theorem again implies that $\overline{X}_n \xrightarrow{P} 0$. Actually, if $\overline{X}_n \xrightarrow{P} c$ for some constant c, then $\varphi'(0)$ must exist and equal ic. This follows since

$$\begin{aligned} \left[\varphi\left(\frac{t}{n}\right)\right]^n \to \exp(itc) &\Rightarrow \lim_{n\to\infty} \ln\left\{\frac{\varphi(t/n)}{(t/n)}\right\} = ic \\ &\Rightarrow (\ln \varphi(t))' = \frac{\varphi'(t)}{\varphi(t)} \end{aligned}$$

exists at $t = 0$, and hence so does $\varphi'(0)$.

As another example of how the methods of characteristic functions sometimes lead to simple and quick proofs, we consider the celebrated Lévy central limit theorem. (This will appear as Corollary 1 to Theorem 5.1.1.) Let $X_1, X_2, \ldots$ be independent, identically distributed random variables with mean 0 and variance 1. Let $S_n = \sum_{k=1}^n X_k$, $n = 1, 2, \ldots$. Let φ be the characteristic function of X_k, $k = 1, 2, \ldots$. Then φ can be developed into a Taylor series expansion as

$$\varphi(x) = \varphi(0) + x\varphi'(0) + \tfrac{1}{2}x^2\varphi''(0) + o(x^2) \qquad \text{as } x \to 0.$$

Let $t \in \mathbb{R}$, and set $x = t/\sqrt{n}$. Since $\varphi'(0) = \mathscr{E}X_k = 0$ and $\varphi''(0) = \text{var}(X_k) = 1$, it follows that

$$\left[\varphi\left(\frac{t}{\sqrt{n}}\right)\right]^n = \left[1 - \frac{t^2}{2n} + o\left(\frac{1}{n}\right)\right]^n \to \exp\left(\frac{-t^2}{2}\right) \qquad \text{as } n \to \infty.$$

By virtue of the continuity theorem we conclude that $n^{-1/2}S_n \xrightarrow{L} Z$, where Z is standard normal.

In Example 3.6.3 we saw how the continuity theorem can sometimes be used to prove that a given function is a characteristic function. Let us consider another example of this nature. Consider the function

$$\varphi(t) = (\cosh t)^{-1}, \qquad t \in \mathbb{R}.$$

Note that $\cosh t$ is an entire function (see Section 4.2.3) with zeros at $e^t + e^{-t} = 0$, or $e^{2t} = -1$, so that $t = i\pi(n - \frac{1}{2})$, $n = 0, \pm 1, \pm 2, \ldots$. By the Weierstrass theorem on the factorization of entire functions we have

$$\varphi(t) = \prod_{k=1}^{\infty} \left[1 + \frac{4t^2}{(2k-1)^2\pi^2}\right]^{-1} = \prod_{k=1}^{\infty} \varphi_k(t),$$

say. Now

$$\varphi_k(t) = \theta\left(\frac{2}{(2k-1)\pi} t\right),$$

where $\theta(t) = 1/(1 + t^2)$ is the characteristic function of the Laplace distribution (Example 3.6.3). Thus φ_k is a characteristic function. Hence so also is $\psi_n = \prod_{k=1}^n \varphi_k$. Since $\psi_n(t) \to \varphi(t)$ as $n \to \infty$ and φ is continuous at 0, the continuity theorem implies that φ is a characteristic function.

Finally we consider an important application of the continuity theorem (in $\mathbb{R}_n$) to mathematical statistics. A commonly used test in statistics is the chi-square test of goodness of fit. It is based on the limiting distribution of the Pearson statistic, which we define below. Let X be a random variable with distribution function F. Let $A_1, A_2, \ldots, A_k$ be a partition of $\mathbb{R}$ where the A_j are Borel sets. Let $p_j = P\{X \in A_j\}$, $j = 1, 2, \ldots, k$. Then $p_j \geq 0$, $\sum_{j=1}^k p_j = 1$. Let $X_1, X_2, \ldots, X_n$ be a random sample from F, and suppose that n_j is the number of X_i that belong to A_j. Then $n_1, n_2, \ldots, n_k$ are integer-valued random variables with $n_1 + n_2 + \cdots + n_k = n$. For a given j, n_j has a binomial distribution. Consider the statistic

$$V = \sum_{j=1}^{k} \frac{(n_j - np_j)^2}{np_j};$$

V is called *Pearson's* χ^2*-statistic*. We show how the multidimensional analogue of the continuity theorem implies the convergence in law of V to a chi-square random variable with $k - 1$ degrees of freedom.

First note that the joint distribution of $(n_1, n_2, \ldots, n_k)$ is given by

$$\frac{n!}{\nu_1!\,\nu_2!\ldots\nu_k!}\,p_1^{\nu_1}p_2^{\nu_2}\ldots p_k^{\nu_k},$$

where $\nu_1, \nu_2, \ldots, \nu_k$ is any set of nonnegative integers with $\sum_{j=1}^{k}\nu_j = n$. The joint characteristic function of $n_1, \ldots, n_k$ is therefore given by

$$\varphi(t_1, t_2, \ldots, t_k) = \left[\sum_{j=1}^{k} p_j \exp(it_j)\right]^n.$$

Let us write

$$Y_j = \frac{n_j - np_j}{\sqrt{np_j}}, j = 1, 2, \ldots, k.$$

Then $V = \sum_{j=1}^{k} Y_j^2$ and $\sum_{j=1}^{k} Y_j\sqrt{p_j} = 0$. Thus the Y_j are linearly dependent with characteristic function

$$\varphi_1(t_1, \ldots, t_k) = \exp\left(-\sum_{j=1}^{k} it_j\sqrt{np_j}\right)\left[\sum_{j=1}^{k} p_j \exp\left(i\frac{t_j}{\sqrt{np_j}}\right)\right]^n.$$

It follows that

$$\begin{aligned}
\ln \varphi_1(t_1, \ldots, t_k) &= -i\sqrt{n}\sum_{j=1}^{k} t_j\sqrt{p_j} \\
&\quad + n \ln\left[1 + \frac{i}{\sqrt{n}}\sum_{j=1}^{k} t_j\sqrt{p_j} - \frac{1}{2n}\sum_{j=1}^{k} t_j^2 + o\left(\frac{1}{n}\right)\right] \\
&= -i\sqrt{n}\sum_{j=1}^{k} t_j\sqrt{p_j} + \ln(1 + z),
\end{aligned}$$

where

$$z = \frac{i}{\sqrt{n}}\sum_{j=1}^{k} t_j\sqrt{p_j} - \frac{1}{2n}\sum_{j=1}^{k} t_j^2 + o\left(\frac{1}{n}\right) \qquad \text{as } n \to \infty.$$

Since $z \to 0$ as $n \to \infty$, we can, for sufficiently large n, expand $\ln(1 + z)$ in a Maclaurin series, and we have

$$n \ln(1 + z) - i\sqrt{n} \sum_{j=1}^{k} t_j \sqrt{p_j} \to -\frac{1}{2}\left[\sum_{j=1}^{k} t_j^2 - \left(\sum_{j=1}^{k} t_j \sqrt{p_j}\right)^2\right].$$

It follows that

$$\lim_{n \to \infty} \ln \varphi_1(t_1, t_2, \ldots, t_k) = \exp\left\{-\frac{1}{2}\left[\sum_{j=1}^{k} t_j^2 - \left(\sum_{j=1}^{k} t_j \sqrt{p_j}\right)^2\right]\right\}.$$

The quadratic form

$$Q(t_1, t_2, \ldots, t_k) = \sum_{j=1}^{k} t_j^2 - \left(\sum_{j=1}^{k} t_j \sqrt{p_j}\right)^2$$

has the matrix $\mathbf{A} = \mathbf{I} - \mathbf{p}\mathbf{p}'$, where $\mathbf{I}$ is the unit matrix and $\mathbf{p}$ denotes the column vector $\mathbf{p} = (\sqrt{p_1}, \ldots, \sqrt{p_k})$. Changing $t_1, t_2, \ldots, t_k$ into $u_1, u_2, \ldots, u_k$ by means of an orthogonal transformation

$$u_j = \sum_{p=1}^{k} a_{jp} t_p, \qquad j = 1, 2, \ldots, k,$$

with

$$a_{kj} = \sqrt{\pi_j}, \qquad j = 1, 2, \ldots, k,$$

we see that

$$Q(t_1, t_2, \ldots, t_k) = \sum_{j=1}^{k-1} u_j^2.$$

Such an orthogonal transformation exists. It follows that the characteristic function of the random vector $(Y_1, \ldots, Y_k)$ converges as $n \to \infty$ to the characteristic function of a random vector with $k - 1$ independent components, each of which is standard normal. By the multivariate version of the continuity theorem it follows that $(Y_1, \ldots, Y_k)$ has a limiting normal distribution. Since V is the sum of squares of the Y_j for $j = 1, \ldots, k$, the result follows immediately.

3.8 CONVOLUTION SEMIGROUPS

In this section we develop the basic theory of convolution semigroups. In Chapter 4 we will indicate the connection between convolution semigroups of probability distributions and the so-called infinitely divisible laws which appear as limit distributions of certain sums of independent random variables. Although the concept of a semigroup of probability distributions is very fruitful in dealing with infinitely divisible distributions, we emphasize that the methods of characteristic functions, whenever applicable, lead to sharper results.

3.8.1 Preliminary Results

Let $\mathscr{C}_0 = \mathscr{C}_0(\mathbb{R})$ be the space of all real-valued, bounded, continuous functions on $\mathbb{R}$. Let $\overline{\mathbb{R}} = \mathbb{R} \cup \{-\infty, \infty\}$ be the extended real line. We note that $\overline{\mathbb{R}}$ is the one-point compactification of $\mathbb{R}$. Let $\overline{\mathscr{C}}_0 \subset \mathscr{C}_0$ be the subspace of all functions $g \in \mathscr{C}_0$ such that $g(\pm\infty)$ exist and are finite.

Let F be a distribution function on $\mathbb{R}$. For every $g \in \mathscr{C}_0$ we define the convolution $g * F$ of g and F by the formula

$$(g * F)(x) = \int_{-\infty}^{\infty} g(x - y)\, dF(y), \qquad x \in \mathbb{R}. \tag{3.8.1}$$

We see easily that $g * F \in \mathscr{C}_0$. Moreover, if $g \in \overline{\mathscr{C}}_0$, then $g * F \in \overline{\mathscr{C}}_0$. We now consider $\mathscr{C}_0$ as a Banach space with norm $\|\cdot\|$, defined by

$$\|g\| = \sup_{x \in \mathbb{R}} |g(x)|. \tag{3.8.2}$$

Clearly

$$\|g * F\| \leq \|g\|, \qquad g \in \mathscr{C}_0. \tag{3.8.3}$$

We say that a sequence $g_n \in \mathscr{C}_0$ *converges* to an element $g \in \mathscr{C}_0$ if $\|g_n - g\| \to 0$ as $n \to \infty$.

We now prove the following result.

Theorem 3.8.1. Let $\{F_n, n \geq 1\}$ be a sequence of distribution functions. Then the sequence $\{F_n\}$ converges completely to a distribution function F if and only if for every $g \in \overline{\mathscr{C}}_0$ the sequence $\{g * F_n\}$ converges to an element $h \in \overline{\mathscr{C}}_0$. Moreover, in this case $h = g * F$.

Proof. First suppose that F_n converges completely to a distribution function F. We show that $\|g * F_n - g * F\| \to 0$ as $n \to \infty$ for every $g \in \overline{\mathscr{C}}_0$. For convenience we set $h_n = g * F_n$, $n \geq 1$. Since g is bounded and uniformly continuous on $\overline{\mathbb{R}}$, we see easily that $\{h_n\} \in \overline{\mathscr{C}}_0$ is uniformly bounded and equicontinuous. In view of the Arzela–Ascoli theorem (Yosida [91], p. 85) we conclude that $\{h_n\}$ is relatively compact, that is, $\{h_n\}$ contains an infinite subsequence $\{h_{n_k}\}$ such that h_{n_k} converges to a limit, h_0 say, on $\mathscr{C}_0$. On the other hand, since $F_n \xrightarrow{c} F$, it follows that $F_{n_k} \xrightarrow{c} F$. Using the extended Helly–Bray theorem, we conclude that

$$h_{n_k} = g * F_{n_k} \to g * F = h \text{ (pointwise)}.$$

In view of the uniqueness of the limit we must have $h_0 = h = g * F$, so that $\|h_{n_k} - h\| \to 0$ as $k \to \infty$. Suppose $\{h_n\}$ contains another infinite subsequence $\{h_{n'_k}\}$, which converges to some limit h'_0. Proceeding as above, we conclude that $h'_0 = h$ and, moreover, $\|h_{n'_k} - h\| \to 0$ as $k \to \infty$. Therefore it follows that $\|h_n - h\| \to 0$ as $n \to \infty$. Clearly $h \in \overline{\mathscr{C}}_0$.

Conversely, suppose that $\{g * F_n\}$ converges to some element $h \in \mathscr{C}_0$ for every $g \in \overline{\mathscr{C}}_0$. We show that there exists a distribution function F such that $F_n \xrightarrow{c} F$. In view of Helly's weak compactness theorem (Theorem 3.1.2) $\{F_n\}$ contains an infinite subsequence $\{F_{n_k}\}$ which converges weakly to a bounded, nondecreasing, right-continuous function F. We set $h_n = g * F_n$. Then from the extended Helly–Bray theorem we conclude that $h_{n_k} = g * F_{n_k} \to g * F$ as $k \to \infty$ for every $g \in \overline{\mathscr{C}}_0$. Clearly $h = g * F$, and, moreover, $\|h_{n_k} - h\| \to 0$ as $k \to \infty$. Next we show that F is a distribution function, that is, $F(-\infty) = 0$ and $F(\infty) = 1$. Let $\epsilon > 0$. Since $\|h_{n_k} - h\| \to 0$ as $k \to \infty$, there exists a $k_0 = k_0(\epsilon)$ such that for all $k \geq k_0$ and all $x \in \overline{\mathbb{R}}$

$$|h_{n_k}(x) - h_{n_{k_0}}(x)| < \epsilon.$$

We now select a $g \in \overline{\mathscr{C}}_0$ such that g is nondecreasing on $\mathbb{R}$ and $g(-\infty) = 0$, $g(\infty) = 1$. Then, for every $k \geq 1$, h_{n_k} is also nondecreasing on $\overline{\mathbb{R}}$ and $h_{n_k}(-\infty) = 0$, $h_{n_k}(\infty) = 1$. We can choose an $a > 0$ sufficiently large that $0 \leq h_{n_{k_0}}(-a) < \epsilon$, so that for all $k \geq k_0$

$$0 \leq h_{n_k}(-a) < 2\epsilon.$$

On the other hand, we have

$$h_{n_k}(x) = \int_{-\infty}^{\infty} g(x - y)\, dF_{n_k}(y).$$

Using the conditions on g, we have

$$h_{n_k}(-a) \geq \int_{-\infty}^{-2a} g(-a - y)\, dF_{n_k}(y) \geq g(a)F_{n_k}(-2a).$$

Hence, for all $k \geq k_0$ and $a > 0$ sufficiently large, we have

$$g(a)F_{n_k}(-2a) \leq h_{n_k}(-a) < 2\epsilon.$$

Since $g(x) \uparrow 1$ as $x \uparrow \infty$, we conclude from the above that for all $k \geq k_0$ and sufficiently large $a > 0$ we have

$$F_{n_k}(-2a) < 4\epsilon.$$

Since $F_{n_k} \xrightarrow{w} F$, it follows that $F(-\infty) = 0$. Repeating the argument with $F_{n_k}(-a)$ replaced by $1 - F_{n_k}(a)$, we show that $F(\infty) = 1$, so that F is a distribution function. Moreover, $F_{n_k} \xrightarrow{c} F$. Proceeding exactly as in the first part of the proof, we show that $F_n \xrightarrow{c} F$ and, moreover, $h = g * F$. ∎

As an application of Theorem 3.8.1 we prove the following useful result.

Proposition 3.8.1. Let $\overline{\mathscr{C}}_\infty \subset \overline{\mathscr{C}}_0$ be the subspace of all infinitely differentiable functions. Then $\overline{\mathscr{C}}_\infty$ is everywhere dense in $\overline{\mathscr{C}}_0$, that is, for a given $\epsilon > 0$ and $g \in \overline{\mathscr{C}}_0$ there exists an $h = h_\varepsilon \in \overline{\mathscr{C}}_\infty$ such that $\|g - h\| < \epsilon$.

Proof. Let F be a distribution function, and let $b > 0$. Define the distribution function F_b by setting

$$F_b(x) = F\left(\frac{x}{b}\right), \qquad x \in \mathbb{R}.$$

Consider the collection of distribution functions $\{F_b, b > 0\}$. Clearly, as $b \to 0$, $F_b(x) \xrightarrow{c} \epsilon(x)$ for $x \in \mathbb{R}$. Let $g \in \overline{\mathscr{C}}_0$. In view of Theorem 3.8.1 we conclude that $\|g * F_b - g\| \to 0$ as $b \to 0$, since

$$(g * \epsilon)(x) = \int_{-\infty}^{\infty} g(x - y)\, d\epsilon(y) = g(x) \qquad \text{for all } x \in \mathbb{R}.$$

Suppose F is absolutely continuous with probability density function f. Then for $x \in \mathbb{R}$

$$\begin{aligned}(g * F_b)(x) &= \int_{-\infty}^{\infty} g(x - y) f\left(\frac{y}{b}\right) \frac{dy}{b} \\ &= \int_{-\infty}^{\infty} g(y) f\left(\frac{x - y}{b}\right) \frac{dy}{b}.\end{aligned}$$

Assume, in addition, that f has a bounded derivative everywhere on $\mathbb{R}$. Then, using the Lebesgue dominated convergence theorem, we can differentiate the last expression under the integral sign and conclude that $g * F_b$ is differentiable everywhere on $\mathbb{R}$. In particular, choosing F to be the standard normal distribution function, we conclude that $g * F_b \in \overline{\mathscr{C}}_\infty$ for every $b > 0$. Consequently, for a given $\epsilon > 0$ we can select a $b = b(\epsilon)$ sufficiently small so that $h = g * F_b$ satisfies the condition

$$\|g - h\| < \epsilon.$$

This completes the proof. ∎

3.8.2 Probability Operators

Let F be a distribution function. We now define an operator on $\mathscr{C}_0$ by setting

$$A_F g = g * F, \qquad g \in \mathscr{C}_0. \tag{3.8.4}$$

Here A_F is called the *probability operator* associated with F. We note that A_F is a linear operator. Moreover,

$$\|A_F g\| \leq \|g\|, \qquad g \in \mathscr{C}_0, \tag{3.8.5}$$

so that A_F is a bounded operator and

$$\|A_F\| \equiv \sup_{\substack{g \in \mathscr{C}_0 \\ g \neq 0}} \frac{\|A_F g\|}{\|g\|} \leq 1.$$

Hence A_F is a *contraction* operator. It also follows from the definition of A_F that, if $g \in \mathscr{C}_0$ with $g \geq 0$, then $A_F g \geq 0$. Moreover, for every $g \in \overline{\mathscr{C}}_0$, $A_F g \in \overline{\mathscr{C}}_0$, so that A_F leaves $\overline{\mathscr{C}}_0 \subset \mathscr{C}_0$ invariant.

Let F_1 and F_2 be two distribution functions, and let $F = F_1 * F_2$. Let A_F, A_{F_1}, and A_{F_2} be the associated probability operators. Then it is easy to see that

$$A_F = A_{F_1} A_{F_2} = A_{F_2} A_{F_1}, \tag{3.8.6}$$

so that A_{F_1} and A_{F_2} commute.

Definition 3.8.1. Let $\{A_n, n \geq 1\}$ be a sequence of bounded linear operators on $\mathscr{C}_0$, and let A be a bounded linear operator also on $\mathscr{C}_0$. We say that A_n converges (strongly) to A if for every $g \in \mathscr{C}_0$

$$\|A_n g - Ag\| \to 0 \qquad \text{as } n \to \infty.$$

In this case we write $A_n \xrightarrow{s} A$.

Theorem 3.8.2. Let $\{F_n\}$ be a sequence of distribution functions with associated probability operators $\{A_{F_n}\}$. Then $\{F_n\}$ converges completely to a distribution function F if and only if for every $g \in \overline{\mathscr{C}}_0$ the sequence $\{A_{F_n} g\}$ converges to an element $h \in \overline{\mathscr{C}}_0$. Moreover, in this case $h = A_F g$, where A_F is the probability operator associated with F.

Proof. The proof is an immediate consequence of Theorem 3.8.1. ∎

The following inequality will be used later.

Proposition 3.8.2. Let $A_1, A_2, \ldots, A_n$ and $B_1, B_2, \ldots, B_n$ be two finite sets of probability operators on $\mathscr{C}_0$. Set $A_0 = A_1 A_2 \ldots A_n$ and $B_0 = B_1 B_2 \ldots B_n$. Then, for every $g \in \mathscr{C}_0$, we have

$$\|A_0 g - B_0 g\| \leq \sum_{k=1}^{n} \|A_k g - B_k g\|. \tag{3.8.7}$$

In particular, for every integer $n \geq 1$ and every pair of probability operators A and B on $\mathscr{C}_0$, the inequality

$$\|A^n g - B^n g\| \leq n\|Ag - Bg\| \tag{3.8.8}$$

holds for every $g \in \mathscr{C}_0$.

Proof. It is sufficient to prove (3.8.7) for the case $n = 2$, since the proof in the general case can be carried out by induction. Let $A_0 = A_1 A_2$ and $B_0 = B_1 B_2$. Then in view of (3.8.6) A_0 and B_0 are also probability operators. For $g \in \mathscr{C}_0$ we have

$$\begin{aligned}
\|A_0 g - B_0 g\| &= \|(A_1 - B_1)A_2 g + (A_2 - B_2)B_1 g\| \\
&\leq \|A_2\| \, \|A_1 g - B_1 g\| + \|B_1\| \, \|A_2 g - B_2 g\| \\
&\leq \|A_1 g - B_1 g\| + \|A_2 g - B_2 g\|,
\end{aligned}$$

since A_2 and B_1 are contraction operators. ■

As an application of (3.8.7) we obtain the following result.

Proposition 3.8.3. Let $\{F_n\}$ and $\{G_n\}$ be two sequences of distribution functions such that $F_n \xrightarrow{c} F$ and $G_n \xrightarrow{c} G$. Then $F_n * G_n \xrightarrow{c} F * G$.

Remark 3.8.1. Note that Proposition 3.8.3 is also an immediate consequence of the continuity theorem.

For further investigation we restrict ourselves to the subspace $\overline{\mathscr{C}}_\infty \subset \overline{\mathscr{C}}_0$ consisting of all functions g which are infinitely differentiable on $\mathbb{R}$ and have derivatives of all orders belonging to $\overline{\mathscr{C}}_\infty$. In view of Proposition 3.8.1 we note that $\overline{\mathscr{C}}_\infty$ is everywhere dense in $\overline{\mathscr{C}}_0$. Moreover, for every $g \in \overline{\mathscr{C}}_\infty$ and every distribution function F we can show that $A_F g \in \overline{\mathscr{C}}_\infty$, so that A_F leaves $\overline{\mathscr{C}}_\infty$ invariant. We now prove the following important result.

Theorem 3.8.3. Let $\{F_n\}$ be a sequence of distribution functions, and let $\{A_{F_n}\}$ be the corresponding sequence of probability operators. Suppose that for every $h \in \overline{\mathscr{C}}_\infty$ the sequence $\{A_{F_n} h\}$ converges to an element in $\overline{\mathscr{C}}_0$. Then there exists a distribution function F such that $F_n \xrightarrow{c} F$.

Proof. Let $g \in \overline{\mathscr{C}}_0$, and let $\epsilon > 0$. Then in view of Proposition 3.8.1 there exists an $h \in \overline{\mathscr{C}}_\infty$ such that $\|g - h\| < \epsilon/3$. Let $m, n \geq 1$. Then

$$\begin{aligned}\|A_{F_n}g - A_{F_m}g\| &\leq \|A_{F_n}g - A_{F_n}h\| + \|A_{F_n}h - A_{F_m}h\| + \|A_{F_m}h - A_{F_m}g\| \\ &\leq \|g - h\| + \|A_{F_n}h - A_{F_m}h\| + \|g - h\| \\ &< \frac{2\epsilon}{3} + \|A_{F_n}h - A_{F_m}h\|.\end{aligned}$$

In view of our hypothesis, for given $\epsilon > 0$ there exists an $n_0 = n_0(\epsilon)$ such that for all $n, m \geq n_0$

$$\|A_{F_n}h - A_{F_m}h\| < \frac{\epsilon}{3},$$

so that for $n, m \geq n_0$

$$\|A_{F_n}g - A_{F_m}g\| < \epsilon.$$

Hence for every $g \in \overline{\mathscr{C}}_0$ the sequence $\{A_{F_n}g\}$ converges to an element in $\overline{\mathscr{C}}_0$. The assertion is now an immediate consequence of Theorem 3.8.2. ■

Example 3.8.1. Let $X_1, X_2, \ldots$ be a sequence of independent, identically distributed random variables having finite expectations. Let $\mathscr{E}X_i = \mu$ for all $i \geq 1$. Set $S_n = \sum_{i=1}^n X_i, n \geq 1$. Then $S_n/n \xrightarrow{P} \mu$ as $n \to \infty$. This is Khintchine's weak law of large numbers (Theorem 2.1.3). We give an alternative proof using the method of probability operators. Without loss of generality we assume that $\mu = 0$. We show that $S_n/n \xrightarrow{P} 0$. Let F be the common distribution function, and for every $n \geq 1$ set $F_n(x) = F(nx)$, $x \in \mathbb{R}$. Then for $x \in \mathbb{R}$

$$\begin{aligned}P\left(\frac{S_n}{n} \leq x\right) &= P(S_n \leq nx) = \underbrace{(F * F * \cdots * F)}_{n\text{-fold}}(nx) \\ &= \underbrace{(F_n * \cdots * F_n)}_{n\text{-fold}}(x).\end{aligned}$$

Let A_{F_n} be the probability operator associated with F_n. Then for $g \in \overline{\mathscr{C}}_\infty$ we have

$$\begin{aligned}\|A_{F_n}^n g - g\| &= \|A_{F_n}^n g - I^n g\| \\ &\leq n\|A_{F_n}g - g\|.\end{aligned}$$

Here I is the identity operator associated with the distribution function $\epsilon(x)$, and $A_{F_n}^n$ is the probability operator associated with $P(S_n/n \leq x)$. Clearly

$$\|A_{F_n}^n g - g\| \leq n \sup_{x \in \mathbb{R}} \left| \int_{-\infty}^{\infty} [g(x - y) - g(x)]\, dF_n(y) \right|.$$

For fixed $x \in \mathbb{R}$ we have

$$n \int_{-\infty}^{\infty} [g(x-y) - g(x)]\, dF_n(y) = n \int_{-\infty}^{\infty} \left[g\left(x - \frac{y}{n}\right) - g(x) \right] dF(y)$$

$$= \int_{|y| \le n} n \left[g\left(x - \frac{y}{n}\right) - g(x) \right] dF(y)$$

$$+ \int_{|y| > n} n \left[g\left(x - \frac{y}{n}\right) - g(x) \right] dF(y).$$

Since g is bounded and $\mathscr{E}|X_1| < \infty$, we see that

$$\left| \int_{|y|>n} n \left[g\left(x - \frac{y}{n}\right) - g(x) \right] dF(y) \right| \le 2\|g\| nP\{|X_1| > n\} \to 0$$

as $n \to \infty$. Note that $\int_{-\infty}^{\infty} y\, dF(y) = 0$ so that, using Taylor's expansion, we have

$$\left| \int_{|y| \le n} n \left[g\left(x - \frac{y}{n}\right) - g(x) \right] dF(y) \right|$$

(3.8.9) $$\le \|g'\| \left| \int_{|y|>n} y\, dF(y) \right| + \|g''\| \frac{1}{n} \int_{|y| \le n} y^2\, dF(y).$$

The first term on the right hand side of (3.8.9) tends to zero as $n \to \infty$, since $\mathscr{E}X_1 = 0$. That the second term also $\to 0$ as $n \to \infty$ is easily seen on integration by parts. Theorem 3.8.3 immediately implies that $S_n/n \xrightarrow{P} 0$.

3.8.3 Convolution Semigroups

We now study an important class of probability measures on $(\mathbb{R}, \mathscr{B})$. Let $\mathbb{R}_+^\times$ be the multiplicative group of positive real numbers. Then a class $\mathscr{P} = \{P_t, t \in \mathbb{R}_+^\times\}$ of probability measures on $\mathscr{B}$ is said to be a *convolution semigroup* if the relation

(3.8.10) $$P_s * P_t = P_{s+t}$$

holds for all $s, t \in \mathbb{R}_+^\times$. Here $*$ denotes the operation of convolution as defined in (3.4.5).

Let $\mathscr{F} = \{F_t : t \in \mathbb{R}_+^\times\}$ be a class of distribution functions on $\mathbb{R}$. Then $\mathscr{F}$ is said to be a *convolution semigroup* if the relation

(3.8.11) $$F_s * F_t = F_{s+t}$$

holds for all $s, t \in \mathbb{R}_+^\times$.

Example 3.8.2 (Translation Semigroup). For each $t \in \mathbb{R}_+^\times$ let ϵ_t denote the distribution on $\mathbb{R}$ with unit mass at t. Then $\epsilon_t(x) = \epsilon(x - t)$, $x \in \mathbb{R}$. For $t, s \in \mathbb{R}_+^\times$ we have

$$\begin{aligned}(\epsilon_t * \epsilon_s)(x) &= \int_{-\infty}^{\infty} \epsilon_t(x - y)\, d\epsilon_s(y) \\ &= \int_{-\infty}^{\infty} \epsilon(x - u - t - s)\, d\epsilon(u) \\ &= \epsilon_{s+t}(x), \qquad x \in \mathbb{R},\end{aligned}$$

so that $\{\epsilon_t, t \in \mathbb{R}_+^\times\}$ is a convolution semigroup. Here $\{\epsilon_t\}$ is known as a *translation semigroup.*

Example 3.8.3. Consider the family $\mathscr{F} = \{F_t, t \in \mathbb{R}_+^\times\}$, where F_t is the distribution function of a normal random variable with variance σt, $\sigma > 0$, $t > 0$. For convenience we assume that all $F \in \mathscr{F}$ have zero mean. Then, for $s, t \in \mathbb{R}_+^\times$ and $x \in \mathbb{R}$,

$$\begin{aligned}(F_s * F_t)(x) &= \int_{-\infty}^{\infty} F_s(x - y)\, dF_t(y) \\ &= (2\pi\sigma\sqrt{st})^{-1} \int_{-\infty}^{\infty} \left[\int_{-\infty}^{x-y} \exp\left(-\frac{u^2}{2\sigma s}\right) du\right] \exp\left(\frac{-y^2}{2\sigma t}\right) dy \\ &= (2\pi\sigma\sqrt{st})^{-1} \int_{-\infty}^{x} \exp\left\{-\frac{u^2}{2\sigma(s + t)}\right\} G(u)\, du,\end{aligned}$$

where

$$\begin{aligned}G(u) &= \int_{-\infty}^{\infty} \exp\left\{-\frac{s + t}{2\sigma st}\left(y - \frac{tu}{s + t}\right)^2\right\} dy \\ &= \left(2\pi \frac{\sigma st}{s + t}\right)^{1/2}.\end{aligned}$$

Hence

$$\begin{aligned}(F_s * F_t)(x) &= \frac{1}{\sqrt{2\pi\sigma(s + t)}} \int_{-\infty}^{x} \exp\left\{-\frac{u^2}{2\sigma(s + t)}\right\} du \\ &= F_{s+t}(x),\end{aligned}$$

and it follows that $\mathscr{F}$ is a convolution semigroup.

Other examples include the Poisson distribution with parameter $t > 0$, the Cauchy distribution centered at the origin with scale parameter $t > 0$, the one-sided stable distribution with probability density function

$$f_t(x) = \frac{t}{\sqrt{2\pi}} \frac{1}{\sqrt{x^3}} e^{-(1/2)t^2/x}, \qquad x > 0, \quad t > 0,$$

and the gamma distribution with density function

$$f(x) = \frac{1}{\Gamma(t)} \beta^t x^{t-1} e^{-\beta x}, \qquad x > 0, \quad t > 0.$$

Let $\mathscr{F} = \{F_t, t \in \mathbb{R}_t^\times\}$ be a convolution semigroup of distribution functions. For every $t \in \mathbb{R}_+^\times$ we define the operator T_t on $\mathscr{C}_0$ by the relation

(3.8.12) $$(T_t g)(x) = \int_{-\infty}^{\infty} g(x - y)\, dF_t(y)$$

for $g \in \mathscr{C}_0$, $x \in \mathbb{R}$, and $F_t \in \mathscr{F}$. We see easily that

(3.8.13) $$T_s T_t = T_{s+t}$$

holds for all $s, t \in \mathbb{R}_+^\times$. We note that T_t is the probability operator associated with F_t as defined in (3.8.4).

Definition 3.8.2. The class of probability operators $\{T_t, t \in \mathbb{R}_+^\times\}$ defined on $\mathscr{C}_0$ by (3.8.12) is said to be a convolution semigroup of probability operators associated with the convolution semigroup $\mathscr{F}$.

Remark 3.8.2. If $g \in \mathscr{C}_0$ such that $0 \le g \le 1$, then $0 \le T_t g \le 1$. Moreover, $T_t 1 = 1$ for all $t \in \mathbb{R}_+^\times$.

We consider $\mathscr{C}_0$ as a Banach space with norm $\|\cdot\|$ as defined above. Clearly $\{T_t, t \in \mathbb{R}_+^\times\}$ is a *contraction semigroup* of bounded linear operators. Moreover, for any $t, h \in \mathbb{R}_+^\times$ the relation

(3.8.14) $$\|T_{t+h} g - T_t g\| \le \|T_h g - g\|$$

holds for every $g \in \mathscr{C}_0$.

Let $\overline{\mathscr{C}}_\infty \subset \overline{\mathscr{C}}_0 \subset \mathscr{C}_0$ be subspaces as defined above. Then we note that for every $t \in \mathbb{R}_+^\times$ the operator T_t leaves each of the subspaces $\overline{\mathscr{C}}_\infty$ and $\overline{\mathscr{C}}_0$ invariant.

Throughout the following investigation, unless stated otherwise, we take $\overline{\mathscr{C}}_0$ as the domain of definition of operators $\{T_t, t \in \mathbb{R}_+^\times\}$.

Definition 3.8.3. The semigroup $\{T_t, t \in \mathbb{R}_+^\times\}$ associated with the convolution semigroup $\mathscr{F} = \{F_t, t \in \mathbb{R}_+^\times\}$ is said to be continuous if for every $g \in \overline{\mathscr{C}}_0$

$$\lim_{t \to 0+} \|T_t g - g\| = 0.$$

Remark 3.8.3. It follows immediately from (3.8.14) that, if $\{T_t\}$ is continuous, then

$$\lim_{h \to 0+} \|T_{t+h} g - T_t g\| = 0, \qquad g \in \overline{\mathscr{C}}_0,$$

uniformly in t.

Remark 3.8.4. In view of Theorem 3.8.2 it follows that $\{T_t, t \in \mathbb{R}_+^\times\}$ is continuous if and only if $F_t(x) \to \epsilon(x)$ at all continuity points of $\epsilon(x)$ as $t \to 0+$.

In the following, $\{T_t, t \in \mathbb{R}_+^\times\}$ denotes a continuous semigroup of convolution operators acting on $\overline{\mathscr{C}}_0$. Let $\mathscr{D}$ be the set of all $g \in \overline{\mathscr{C}}_0$ such that the limit

$$\lim_{t \to 0+} \left\| \frac{T_t g - g}{t} \right\|$$

exists and is finite. We define an operator A on $\mathscr{D}$ by the relation

$$Ag = \lim_{t \to 0+} \left\| \frac{T_t g - g}{t} \right\|, \qquad g \in \mathscr{D}, \tag{3.8.15}$$

or, equivalently, by

$$(Ag)(x) = \lim_{t \to 0+} t^{-1} \int_{-\infty}^{\infty} [g(x - y) - g(x)]\, dF_t(y), \qquad g \in \mathscr{D}. \tag{3.8.15'}$$

Definition 3.8.4. The operator A defined in (3.8.15) or (3.8.15′) is called the infinitesimal generator associated with the semigroup $\{T_t\}$ with the domain of definition $\mathscr{D} \subset \overline{\mathscr{C}}_0$.

Clearly $\mathscr{D}$ is a linear subspace of $\overline{\mathscr{C}}_0$, and the generator A on $\mathscr{D}$ is a linear operator. It can be shown using the analytic theory of semigroups (see, for example, Dynkin [20], pp. 23 and 30) that $\mathscr{D}$ is everywhere dense in $\overline{\mathscr{C}}_0$ and, moreover, that A is a closed operator, that is,

$$g_n \in \mathscr{D},\ g, h \in \overline{\mathscr{C}}_0 \quad \text{such that } \|g_n - g\| \to 0,\ \|Ag_n - h\| \to 0$$
$$\Rightarrow g \in \mathscr{D} \quad \text{and} \quad Ag = h.$$

Let I be the identity operator on $\overline{\mathscr{C}}_0$. We can rewrite (3.8.15) in the form

$$A = \lim_{t \to 0+} \frac{T_t - I}{t}. \tag{3.8.16}$$

Clearly, as $h \to 0+$,

$$\frac{T_{t+h} - T_t}{h} = \frac{T_h - I}{h} T_t = T_t \frac{T_h - I}{h} \to AT_t = T_t A, \tag{3.8.17}$$

so that A commutes with T_t for all $t \in \mathbb{R}_+^\times$.

Example 3.8.4. For $t \in \mathbb{R}_+^\times$ let τ_t be the probability operator associated with the distribution ϵ_t defined in Example 3.8.2. Then for every $g \in \overline{\mathscr{C}}_0$ and $t \in \mathbb{R}_+^\times$

$$(\tau_t g)(x) = (g * \epsilon_t)(x) = \int_{-\infty}^{\infty} g(x - y)\, d\epsilon_t(y) = g(x - t).$$

Clearly

$$\tau_t \tau_s = \tau_{t+s} \qquad \text{for all } s, t \in \mathbb{R}_+^\times.$$

A simple computation shows that the infinitesimal generator of $\{\tau_t\}$ is $-D$, where D is the differential operator d/dx with its domain of definition $\mathscr{D} \supset \overline{\mathscr{C}}_\infty$. In view of Remark 3.8.4 it is clear that $\{\tau_t, t \in \mathbb{R}_+^\times\}$ is continuous.

Example 3.8.5. Let $\mathscr{F} = \{F_t, t \in \mathbb{R}_+^\times\}$ be the convolution semigroup of normal distributions considered in Example 3.8.3. For each $t \in \mathbb{R}_+^\times$ let T_t be the probability operator associated with F_t. Then for every $g \in \overline{\mathscr{C}}_\infty$ and $x \in \mathbb{R}$

$$\begin{aligned}(T_t g)(x) &= \frac{1}{\sqrt{2\pi\sigma t}} \int_{-\infty}^{\infty} g(x - y) \exp\left(-\frac{y^2}{2\sigma t}\right) dy \\ &= \frac{1}{\sqrt{2\pi\sigma t}} \int_{-\infty}^{\infty} g(u) \exp\left\{-\frac{(x - u)^2}{2\sigma t}\right\} du.\end{aligned}$$

Clearly $T_t T_s = T_{t+s}$ for all $s, t \in \mathbb{R}_+^\times$. First note that $\{T_t\}$ is continuous. In fact, for every $g \in \overline{\mathscr{C}}_0$

$$(T_t g)(x) - g(x) = \frac{1}{\sqrt{2\pi\sigma t}} \int_{-\infty}^{\infty} \exp\left\{-\frac{(x - u)^2}{2\sigma t}\right\} [g(u) - g(x)]\, du.$$

Changing the variable to $v = (x - u)/\sqrt{\sigma t}$, we have

$$(T_t g)(x) - g(x) = \frac{1}{\sqrt{2\pi}} \int_{-\infty}^{\infty} e^{-v^2/2} [g(x - v\sqrt{\sigma t}) - g(x)]\, dv.$$

Note that all moments of F_t exist and that the mean is assumed to be zero. By developing $[g(x - v\sqrt{\sigma t}) - g(x)]$ into a Taylor series and letting $t \to 0+$, we see easily that

$$\lim_{t\to 0+} \sup_{x\in\mathbb{R}} |T_t g(x) - g(x)| = 0,$$

so that $\{T_t, t \in \mathbb{R}_+^\times\}$ is continuous.

To find the infinitesimal generator associated with $\{T_t, t \in \mathbb{R}_+^\times\}$ we use Taylor's expansion. For every $g \in \overline{\mathscr{C}}_\infty$ and $x, y \in \mathbb{R}$ we have

$$g(x - y) - g(x) = -yg'(x) + \tfrac{1}{2}y^2 g''(x) - \frac{1}{3!}\, y^3 g'''(x - \zeta y),$$

where $\zeta = \zeta(x, y)$ and $0 < \zeta < 1$. It follows that

$$t^{-1}[(T_t g)(x) - g(x)] = t^{-1} \int_{-\infty}^{\infty} [g(x - y) - g(x)]\, dF_t(y)$$

$$= t^{-1} \int_{-\infty}^{\infty} [-yg'(x) + \tfrac{1}{2}y^2 g''(x) - \tfrac{1}{6}y^3 g'''(x - \zeta y)]\, dF_t(y)$$

and

$$\lim_{t\to 0+} t^{-1}[(T_t g)(x) - g(x)] = Ag = \tfrac{1}{2}\sigma g''(x).$$

Therefore the differential operator $(\sigma/2)(d^2/dx^2)$ is the generator of the semigroup associated with $\{F_t, t \in \mathbb{R}_\times^+\}$.

Example 3.8.6. For each $t \in \mathbb{R}_+^\times$ let F_t be the distribution function of a Poisson random variable with parameter λt, where $\lambda > 0$. Let T_t be the probability operator associated with F_t. Then for every $g \in \mathscr{C}_0$, $x \in \mathbb{R}$, and $t \in \mathbb{R}_+^\times$ we have

$$(T_t g)(x) = \sum_{k=0}^{\infty} e^{-\lambda t} \frac{(\lambda t)^k}{k!}\, g(x - k).$$

It is easy to check that $\{T_t, t \in \mathbb{R}_+^\times\}$ is a continuous convolution semigroup of probability operators. To find the infinitesimal generator we write

$$(T_t g)(x) - g(x) = \sum_{k=0}^{\infty} e^{-\lambda t} \frac{(\lambda t)^k}{k!} [g(x - k) - g(x)],$$

so that

$$\frac{(T_t g)(x) - g(x)}{t} = \lambda[g(x - 1) - g(x)] + \sum_{k=2}^{\infty} e^{-\lambda t} \frac{\lambda^k}{k!}\, t^{k-1}[g(x - k) - g(x)].$$

Letting $t \to 0+$, we see that the infinitesimal generator of $\{T_t\}$ is the *difference operator* A, given by

$$(Ag)(x) = \lambda[g(x - 1) - g(x)].$$

If, instead, we take F_t to be the distribution function of a Poisson random variable taking values $0, \mu, 2\mu, \ldots$ ($\mu > 0$) and having parameter λt ($\lambda > 0$), the infinitesimal generator of the corresponding convolution semigroup $\{T_t, t \in \mathbb{R}_+^\times\}$ is the difference operator A, given by

$$(Ag)(x) = \lambda[g(x - \mu) - g(x)].$$

We next study some properties of the infinitesimal generator associated with a continuous convolution semigroup of probability operators.

Proposition 3.8.4. Let $\{T_t, t \in \mathbb{R}_+^\times\}$ and $\{T'_t, t \in \mathbb{R}_+^\times\}$ be two continuous convolution semigroups of probability operators on $\overline{\mathscr{C}}_0$. Let A and A', respectively, be their infinitesimal generators. Suppose that $A = A'$. Then $T_t = T'_t$ for all $t \in \mathbb{R}_+^\times$.

Proof. Let $\mathscr{D}$ and $\mathscr{D}'$ be the domains of definitions of A and A', respectively. First we show that for every $g \in \mathscr{D} \cap \mathscr{D}'$ and for every $t \in \mathbb{R}_+^\times$ the inequality

$$\|T_t g - T'_t g\| \le t\|Ag - A'g\| \tag{3.8.18}$$

holds. We note that T_t and T'_t commute. Hence for every integer $n \ge 1$ we conclude from (3.8.8) that for every $g \in \overline{\mathscr{C}}_0$

$$\begin{aligned} \|T_t g - T'_t g\| &= \|T^n_{t/n} g - T'^n_{t/n} g\| \\ &\le n\|T_{t/n} g - T'_{t/n} g\|. \end{aligned} \tag{3.8.19}$$

Therefore

$$\|T_t g - T'_t g\| \le t\left\|\frac{T_{t/n} - I}{t/n} g - \frac{T'_{t/n} - I}{t/n} g\right\|.$$

Taking the limit as $n \to \infty$ and using (3.8.16), we obtain (3.8.18).

Suppose $A = A'$. This implies that $\mathscr{D} = \mathscr{D}'$ and, moreover, $Ag = A'g$ for all $g \in \mathscr{D}$. Hence it follows immediately from (3.8.18) that

$$T_t g = T'_t g$$

for all $t \in \mathbb{R}_+^\times$ and all $g \in \mathscr{D}$. Let $g \in \overline{\mathscr{C}}_0$. Since $\mathscr{D}$ is everywhere dense in $\overline{\mathscr{C}}_0$, there exists a sequence $g_n \in \mathscr{D}$ such that

$$\|g_n - g\| \to 0 \qquad \text{as } n \to \infty.$$

Clearly

$$\|T_t g_n - T_t g\| \leq \|g_n - g\| \to 0$$

and also

$$\|T'_t g_n - T'_t g\| \leq \|g_n - g\| \to 0.$$

But $T_t g_n = T'_t g_n$ for all $n \geq 1$, so that $T_t g = T'_t g$ for all $g \in \overline{\mathscr{C}}_0$. ■

Proposition 3.8.5. Let $\{T'_t\}$ and $\{T''_t\}$ be two continuous convolution semigroups of probability operators on $\overline{\mathscr{C}}_0$, and let A' and A'', respectively, be their infinitesimal generators. Then the operator $A = A' + A''$ is the infinitesimal generator of the convolution semigroup $T_t = T'_t T''_t$. (Here $\{T_t\}$ is the semigroup associated with the convolutions of the distribution functions associated with $\{T'_t\}$ and $\{T''_t\}$.)

Proof. For any $t \in \mathbb{R}_+^\times$ we have

$$\frac{T'_t T''_t - I}{t} = \frac{T'_t - I}{t} + T'_t \frac{T''_t - I}{t}.$$

Letting $t \to 0+$, we get the result. ■

Corollary. Let $\{T_t, t \in \mathbb{R}_+^\times\}$ be the convolution semigroup associated with the semigroup of distribution functions $\{F_t, t \in \mathbb{R}_+^\times\}$. Let A be the corresponding infinitesimal generator. Let $b > 0$. Then $A - bD(D = d/dx)$ is the infinitesimal generator of the semigroup $\{T_t^0, t \in \mathbb{R}_+^\times\}$ associated with the semigroup of distribution functions $\{F_t^0, t \in \mathbb{R}_+^\times\}$, where $F_t^0(x) = F_t(x - bt)$.

Proof. The proof is an immediate consequence of Proposition 3.8.6 and Example 3.8.4.

Theorem 3.8.4. For every positive integer n let $\{T_t^{(n)}, t \in \mathbb{R}_+^\times\}$ be a continuous convolution semigroup, and let A_n be the associated infinitesimal generator. Suppose that the domain of definition $\mathscr{D}_n$ of A_n contains $\overline{\mathscr{C}}_\infty$ for every $n \geq 1$. Let $\mathscr{D} = \bigcap_{n=1}^\infty \mathscr{D}_n$. Assume further that, for every $g \in \mathscr{D}$, $A_n g$ converges to an element in $\overline{\mathscr{C}}_0$. Define the operator A on $\mathscr{D}$ by the formula

$$Ag = \lim_{n \to \infty} A_n g, \qquad g \in \mathscr{D}.$$

Then there exists a continuous convolution semigroup $\{T_t, t \in \mathbb{R}_+^\times\}$ with A as its infinitesimal generator. Moerover, $T_t^{(n)} \xrightarrow{s} T_t$ as $n \to \infty$ for every $t \in \mathbb{R}_+^\times$.

Proof. Clearly $\mathscr{D} \supset \overline{\mathscr{C}}_\infty$, so that $\mathscr{D}$ is everywhere dense in $\overline{\mathscr{C}}_0$. According to our assumption for $g \in \mathscr{D}$, the sequence $\{A_n g\}$ is a Cauchy sequence of elements in $\mathscr{D}$, that is, $\|A_n g - A_m g\| \to 0$ as $m, n \to \infty$. From (3.8.18) for every $g \in \mathscr{D}$ and every $t \in \mathbb{R}_+^\times$ we have the inequality

$$\|T_t^{(n)}g - T_t^{(m)}g\| \le t\|A_n g - A_m g\|, \tag{3.8.20}$$

which tends to zero as $m, n \to \infty$. Hence $\{T_t^{(n)}g\}$ converges to an element in $\overline{\mathscr{C}}_0$. It follows from Theorem 3.8.3 that there exists a distribution function F_t for every $t \in \mathbb{R}_+^\times$ such that $F_t^{(n)} \xrightarrow{c} F_t$ as $n \to \infty$. Here $F_t^{(n)}$ is the distribution function associated with the probability operator $T_t^{(n)}$, $n \ge 1$. Proposition 3.8.3 now implies that $\{F_t, t \in \mathbb{R}_+^\times\}$ is a convolution semigroup. Let T_t be the probability operator associated with F_t, for every $t \in \mathbb{R}_+^\times$. It follows from Theorem 3.8.1 that for every $g \in \overline{\mathscr{C}}_0$ and for every $t \in \mathbb{R}_+^\times$ the sequence $\{T_t^{(n)}g\}$ converges to $T_t g$ in $\overline{\mathscr{C}}_0$. We see easily that $\{T_t, t \in \mathbb{R}_+^\times\}$ is continuous.

Finally we show that A is the infinitesimal generator of $\{T_t, t \in \mathbb{R}_+^\times\}$. Let $g \in \mathscr{D}$. We write

$$\left\|\frac{T_t - I}{t}g - Ag\right\| \le \left\|\frac{T_t g - T_t^{(n)}g}{t}\right\| + \left\|\frac{T_t^{(n)} - I}{t}g - A_n g\right\| + \|A_n g - Ag\|.$$

We take the limit of (3.8.20) as $m \to \infty$ and obtain

$$\|T_t^{(n)}g - T_t g\| \le t\|A_n g - Ag\|$$

for every $g \in \mathscr{D}$, $t \in \mathbb{R}_+^\times$, and $n \ge 1$. Hence

$$\left\|\frac{T_t - I}{t}g - Ag\right\| \le 2\|A_n g - Ag\| + \left\|\frac{T_t^{(n)} - I}{t}g - A_n g\right\|.$$

We note that the first term on the right-hand side $\to 0$ as $n \to \infty$, while for every fixed n the second term $\to 0$ at $t \to 0+$ in view of the definition of A_n. This completes the proof of Theorem 3.8.4. ■

We finally prove the following main result.

Theorem 3.8.5. Let $\{T_t, t \in \mathbb{R}_+^\times\}$ be a continuous convolution semigroup. Let $\{t_k, k \ge 1\}$ be a sequence of positive real numbers such that $t_k \to 0$ as $k \to \infty$. Suppose there exists an operator A such that

$$\frac{T_{t_k} - I}{t_k} \xrightarrow{s} A \qquad \text{as } k \to \infty.$$

Then A is the infinitesimal generator of $\{T_t\}$.

Proof. For every $k \ge 1$ we set

$$A_k = \frac{T_{t_k} - I}{t_k}.$$

We first show that A_k is the infinitesimal generator of a convolution semigroup. For this purpose we proceed as follows. Let $\lambda > 0$ and F be a distribution function. For every $t \in \mathbb{R}_+^\times$ and $x \in \mathbb{R}$ set

$$G_t(x) = e^{-\lambda t} \sum_{n=0}^{\infty} \frac{(\lambda t)^n}{n!} \underbrace{(F * F * \cdots * F)}_{n\text{-fold}}(x). \tag{3.8.21}$$

Then G_t is a distribution function satisfying

$$G_t * G_s = G_{t+s}.$$

Let W_t be the probability operator associated with G_t for each $t \in \mathbb{R}_+^\times$, and let T be the probability operator associated with F. Then for every $g \in \overline{\mathscr{C}}_0$ we have

$$\begin{aligned} W_t g - g &= g * G_t - g \\ &= e^{-\lambda t} \sum_{n=0}^{\infty} \frac{(\lambda t)^n}{n!} g * (F * F * \cdots * F) - g \\ &= e^{-\lambda t} \sum_{n=0}^{\infty} \frac{(\lambda t)^n}{n!} T^n g - g, \end{aligned}$$

where $T^0 = I$. Hence

$$W_t g - g = (e^{-\lambda t} - 1)g + e^{-\lambda t} \sum_{n=1}^{\infty} \frac{(\lambda t)^n}{n!} T^n g. \tag{3.8.22}$$

Dividing both sides of (3.8.22) by t and letting $t \to 0+$, we obtain

$$\lim_{t \to 0+} \frac{W_t g - g}{t} = -\lambda g + \lambda T g = \lambda(T - I)g,$$

so that $\lambda(T - I)$ is the infinitesimal generator of $\{W_t, t \in \mathbb{R}_+^\times\}$. The domain of definition $\mathscr{D}$ of $\lambda(T - I)$ contains $\overline{\mathscr{C}}_\infty$.

For every $k \geq 1$ we set $\lambda = 1/t_k$ and $T = T_{t_k}$. Then A_k is the infinitesimal generator of the convolution semigroup $\{W_t, t \in \mathbb{R}_+^\times\}$ associated with the distribution functions $\{G_t\}$ defined by (3.8.21) with λ replaced by $1/t_k$ and F by the distribution function associated with T_{t_k}. Moreover, the domain of definition of each A_k contains $\overline{\mathscr{C}}_\infty$. According to our hypothesis, $A_k \xrightarrow{s} A$ as $k \to \infty$. In view of Theorem 3.8.4 we conclude that there exists a convolution semigroup $\{T'_t, t \in \mathbb{R}_+^\times\}$ with A as its infinitesimal generator. It follows at once from Proposition 3.8.4 that $T_t = T'_t$ for all $t \in \mathbb{R}_+^\times$. ■

Example 3.8.7. Let $\epsilon > 0$. Consider a distribution function F concentrated on $\{x \in \mathbb{R}, |x| > \epsilon\}$ having mean μ_1 and variance μ_2. For every $t \in \mathbb{R}_+^\times$ and

$x \in \mathbb{R}$ define G_t as in (3.8.21). As we have shown in the proof of Theorem 3.8.5, $\{G_t, t \in \mathbb{R}_+^\times\}$ is a convolution semigroup of distribution functions with probability operators $\{W_t\}$. The semigroup $\{W_t, t \in \mathbb{R}_+^\times\}$ is generated by $\lambda(T - I)$, where T is the operator corresponding to F. Clearly the distribution function G_t has mean $\lambda\mu_1 t$ and variance $\lambda\mu_2 t$.

For each $t \in \mathbb{R}_+^\times$ define F_t by

$$F_t(x) = G_t(x + \lambda\mu_1 t), \qquad x \in \mathbb{R}.$$

The F_t have zero means. It follows immediately from the corollary to Proposition 3.8.5 that the semigroup of probability operators associated with $\{F_t\}$ is generated by $\lambda(T - I + \mu_1 D)$, where D is the differential operator d/dx. Clearly

$$\begin{aligned}[\lambda(T - I + \mu_1 D)g](x) &= \lambda \int_{|y|>\epsilon} [g(x-y) - g(x) + yg'(x)]\, dF(y) \\ &= \alpha \int_{|y|>\epsilon} \frac{g(x-y) - g(x) + yg'(x)}{y^2}\, dH(y),\end{aligned}$$

where $H(x) = \int_{-\infty}^{x} y^2\, dF(y)/\mu_2$ and $\alpha = \lambda\mu_2$.

3.8.4 Representation of the Infinitesimal Generator

We now derive the representation of the infinitesimal generator of a continuous convolution semigroup of distribution functions. We restrict ourselves to the finite variance case.

Let $\{F_t, t \in \mathbb{R}_+^\times\}$ be a convolution semigroup of distribution functions, and suppose that F_t has mean zero and finite variance σ_t^2, $t \in \mathbb{R}_+^\times$. The semigroup property implies that the relation

$$\sigma_{s+t}^2 = \sigma_t^2 + \sigma_s^2 \tag{3.8.23}$$

holds for all $s, t \in \mathbb{R}_+^\times$. It is well known (see, for example, Aczel [1], p. 32) that the only positive solution of the functional equation (3.8.23) is of the form $\sigma_t^2 = \lambda t$, where $\lambda > 0$ is a constant.

For each $t \in \mathbb{R}_+^\times$ and $x \in \mathbb{R}$ set

$$G_t(x) = (\lambda t)^{-1} \int_{-\infty}^{x} y^2\, dF_t(y). \tag{3.8.24}$$

Then G_t is clearly a distribution function. A simple application of Helly's theorem (Theorem 3.1.2) shows that there exists a sequence $\{t_n\}$, $t_n \in \mathbb{R}_+^\times$ with $t_n \to 0+$, such that $G_{t_n} \xrightarrow{w} G$, where G is a bounded, nondecreasing, and right-continuous function.

We need the following result.

Proposition 3.8.6. The limit function G defined above is independent of the choice of the sequence $\{t_n\}$, $t_n \to 0+$, so that $G_t \xrightarrow{w} G$ as $t \to 0+$.

Proof. Let $\{T_t, t \in \mathbb{R}_+^\times\}$ be the convolution semigroup of probability operators associated with $\{F_t, t \in \mathbb{R}_+^\times\}$. If follows from (3.8.24) that for every $g \in \overline{\mathscr{C}}_0$

$$\begin{aligned}(T_t g)(x) &= \int_{-\infty}^{\infty} g(x-y)\, dF_t(y)\\ &= \lambda t \int_{-\infty}^{\infty} y^{-2} g(x-y)\, dG_t(y).\end{aligned}$$

Since $\int_{-\infty}^{\infty} y\, dF_t(y) = 0$, we can write for every $g \in \overline{\mathscr{C}}_\infty$

$$\left(\frac{T_t - I}{t} g\right)(x) = \lambda \int_{-\infty}^{\infty} \frac{g(x-y) - g(x) + y g'(x)}{y^2}\, dG_t(y). \tag{3.8.25}$$

We note that for fixed $x \in \mathbb{R}$ the integrand on the right-hand side of (3.8.25) is continuous in y and assumes the value $\frac{1}{2}g''(x)$ at $y = 0$. Moreover, the integrand, as well as its derivative, vanishes at $y = \pm\infty$. We see easily from Theorem 3.1.4 that for every fixed $x \in \mathbb{R}$ and $g \in \overline{\mathscr{C}}_\infty$

$$\begin{aligned}\lim_{n\to\infty} &\int_{-\infty}^{\infty} \frac{g(x-y) - g(x) + y g'(x)}{y^2}\, dG_{t_n}(y)\\ &= \int_{-\infty}^{\infty} \frac{g(x-y) - g(x) + y g'(x)}{y^2}\, dG(y).\end{aligned}$$

We now define the operator A on $\overline{\mathscr{C}}_\infty$ by setting

$$(Ag)(x) = \lambda \int_{-\infty}^{\infty} \frac{g(x-y) - g(x) + y g'(x)}{y^2}\, dG(y), \qquad g \in \overline{\mathscr{C}}_\infty. \tag{3.8.26}$$

In view of Theorem 3.8.5 it follows immediately that A defined by (3.8.26) is the infinitesimal generator of $\{T_t, t \in \mathbb{R}_+^\times\}$.

We now show that the representation of A is unique, that is, A determines G uniquely. For this purpose let $f \in \overline{\mathscr{C}}_\infty$ be arbitrary, and set

$$g_f(x) = 1 + \frac{x^2}{1+x^2} f(-x), \qquad x \in \mathbb{R}.$$

Clearly $g_f \in \overline{\mathscr{C}}_\infty$; moreover, for all $f \in \overline{\mathscr{C}}_\infty$

$$(Ag_f)(0) = \lambda \int_{-\infty}^{\infty} \frac{f(y)}{1+y^2}\, dG(y). \tag{3.8.27}$$

Consequently A determines uniquely the function G_1 defined on $\mathbb{R}$ by

$$G_1(x) = \int_{-\infty}^{x} (1 + y^2)^{-1} \, dG(y).$$

Hence the function G occurring in representation (3.8.26) is determined uniquely by A. In view of this uniqueness we conclude that the limit function G is independent of the choice of the sequence $\{t_n\}, t_n \to 0+$. Therefore $G_t \xrightarrow{w} G$ as $t \to 0+$. This completes the proof. ∎

We now prove the following basic result.

Theorem 3.8.6. Let $\{F_t, t \in \mathbb{R}_+^\times\}$ be a continuous convolution semigroup of distribution functions with means zero and variances λt, $\lambda > 0$, $t \in \mathbb{R}_+^\times$. Let $\{T_t\}$ be the associated semigroup of probability operators. Then the infinitesimal generator A of $\{T_t\}$ has a representation of the form (3.8.26), where G is a distribution function determined uniquely by A and hence $\{T_t\}$.

Conversely, every operator A of the form (3.8.26), where G is a distribution function, is the infinitesimal generator of a continuous convolution semigroup $\{T_t, t \in \mathbb{R}_+^\times\}$ associated with distribution functions having means zero and variances λt, $\lambda > 0$.

Proof. First suppose that $\{F_t\}$ is a continuous convolution semigroup of distribution functions with means zero and variances λt, $\lambda > 0$. From the proof of Proposition 3.8.6 we conclude that the infinitesimal generator A of the semigroup $\{T_t\}$ associated with $\{F_t\}$ has a representation of the form (3.8.26). Moreover, the function G occurring in representation (3.8.26) is a bounded, nondecreasing, and right-continuous function with $G(-\infty) = 0$ and $G(\infty) \leq 1$, which is determined uniquely by A. It remains to show that $G(\infty) = 1$. Suppose that $G(\infty) = \alpha < 1$. We show that in this case the operator A is the infinitesimal generator of a semigroup of probability operators associated with distribution functions that have means zero and variances $\leq \lambda \alpha t < \lambda t$. This contradiction will prove that $G(\infty) = 1$.

Let $\epsilon > 0$ be fixed. Define the operator A_ϵ by

$$(3.8.28) \quad (A_\epsilon g)(x) = \lambda \int_{\mathbb{R} - \{0 < |y| \leq \epsilon\}} \frac{g(x-y) - g(x) + yg'(x)}{y^2} \, dG(y)$$

for $x \in \mathbb{R}$ and $g \in \overline{\mathscr{C}}_\infty$. Clearly $A_\epsilon \to A$ as $\epsilon \to 0$. Suppose that G assigns mass $\beta \geq 0$ to the origin, and mass $\alpha_\epsilon \geq 0$ to the set $\{|y| > \epsilon\}$. Clearly $0 \leq \beta + \alpha_\epsilon \leq \alpha$ and $\beta + \alpha_\epsilon \to \alpha$ as $\epsilon \to 0$.

Consider the measure degenerate at zero with mass β. Let A_1 be the corresponding operator, defined by (3.8.26). Clearly, for $g \in \overline{\mathscr{C}}_\infty$ and $x \in \mathbb{R}$,

$$(A_1 g)(x) = \tfrac{1}{2}\lambda \beta g''(x).$$

It follows from Example 3.8.5 that A_1 is the infinitesimal generator of the semigroup of normal distributions with means zero and variances $\lambda\beta t$, $t > 0$.

Next consider the function G_ϵ defined on $\mathbb{R}$ by

$$G_\epsilon(x) = \begin{cases} G(x) & \text{if } |x| > \epsilon, \\ 0 & \text{otherwise.} \end{cases}$$

Let A_2 be the corresponding operator defined by (3.8.26). It is clear that $A_\epsilon = A_1 + A_2$. Moreover, it follows from Example 3.8.7 that A_2 is the infinitesimal generator of a semigroup with variance $\lambda\alpha_\epsilon t$. From Proposition 3.8.5 we see that A_ϵ is the infinitesimal generator of a convolution semigroup with variances

$$\lambda\beta t + \lambda\alpha_\epsilon t = \lambda t(\beta + \alpha_\epsilon) \leq \lambda\alpha t < \lambda t.$$

Since $A_\epsilon \to A$ as $\epsilon \to 0$, we conclude that A itself is the infinitesimal generator of a convolution semigroup with variances $\leq \lambda t\alpha < \lambda t$. It follows therefore that $\alpha = G(\infty) = 1$, so that G is indeed a distribution function.

Conversely suppose that an operator A has a representation of the form (3.8.26), where G is a distribution function. We have shown that, if G occurring in (3.8.26) is such that $G(-\infty) = 0$ and $G(\infty) = \alpha \leq 1$, then A is the infinitesimal generator of a semigroup with means zero and variances $\lambda\alpha t$. Consequently, if $G(\infty) = 1$, then A is the generator of a continuous convolution semigroup associated with means zero and variances λt. ■

3.9 WEAK CONVERGENCE AND TIGHTNESS ON A METRIC SPACE

Ever since Prokhorov published his fundamental paper in 1956 [67], there has been a great deal of interest in the theory of weak convergence of probability measures on metric spaces. Indeed the general theory of stochastic processes may now be regarded as the theory of probability measures on complete separable metric spaces. In this section we describe the main features of this development. Our objective is to prove the basic theorem due to Prokhorov (Theorem 3.9.2) and indicate some of its applications.

3.9.1 Preliminaries

Let $(\mathfrak{X}, \rho)$ be a metric space, and let $\mathscr{B}$ be the σ-field of subsets of $\mathfrak{X}$ generated by the class of all open sets. Then every $\mathscr{B}$-measurable set is said to be a *Borel set* in $\mathfrak{X}$. A real-valued set function μ defined on $\mathscr{B}$ is said to be a *measure* if it satisfies the following conditions:

(i) $\mu(E) \geq 0, E \in \mathscr{B}$.
(ii) μ is countably additive, that is, for every disjoint sequence $\{E_n\}$ of Borel sets we have

$$\mu\left(\bigcup_{n=1}^{\infty} E_n\right) = \sum_{n=1}^{\infty} \mu(E_n).$$

(iii) $\mu(\varnothing) = 0$.

A measure μ on $\mathscr{B}$ is said to be finite if $\mu(\mathfrak{X}) < \infty$. In particular, μ is said to be a *probability measure* if $\mu(\mathfrak{X}) = 1$.

Definition 3.9.1. Let μ be a finite measure on $\mathscr{B}$. A set $E \in \mathscr{B}$ is said to be μ-regular if

$$\begin{aligned}\mu(E) &= \sup\{\mu(C): C \subset E, C \text{ closed}\}\\ &= \inf\{\mu(O): E \subset O, O \text{ open}\}.\end{aligned}$$

We say that μ is a regular measure if every $E \in \mathscr{B}$ is μ-regular.

Let $\mathfrak{X}$ be a metric space, and μ a finite measure on $\mathscr{B}$. It follows immediately from the definition that $E \in \mathscr{B}$ is μ-regular if and only if for each $\epsilon > 0$ there exist an open set G and a closed set F such that $F \subset E \subset G$ and $\mu(G - F) < \epsilon$.

Proposition 3.9.1. Let $\mathfrak{X}$ be a metric space. Then every finite measure μ on $\mathscr{B}$ is regular.

Proof. Let μ be a finite measure on $\mathscr{B}$. Let $\mathscr{R}$ be the class of all μ-regular sets in $\mathscr{B}$. We show that $\mathscr{R} = \mathscr{B}$. Clearly $\mathscr{R} \subset \mathscr{B}$, and since $\varnothing$ and $\mathfrak{X}$ are both open as well as closed sets, $\varnothing$ and $\mathfrak{X} \in \mathscr{R}$. We next show that $\mathscr{R}$ is a σ-field. In fact, let $E \in \mathscr{R}$ and let $\epsilon > 0$. Then it follows from the definition of E that there exists an open set $O_\epsilon \supset E$ and a closed set $C_\epsilon \subset E$ such that

$$\mu(O_\epsilon) - \frac{\epsilon}{2} < \mu(E) < \mu(C_\epsilon) + \frac{\epsilon}{2}.$$

Then $O_\epsilon \supset C_\epsilon$; moreover,

$$\mu(O_\epsilon - C_\epsilon) = \mu(O_\epsilon) - \mu(C_\epsilon) < \epsilon.$$

Taking complements, we have $O_\epsilon^c \subset E^c \subset C_\epsilon^c$ and $C_\epsilon^c - O_\epsilon^c = O_\epsilon - C_\epsilon$, where O_ϵ^c is closed and C_ϵ^c is open. It follows, then, that $E^c \in \mathscr{R}$.

Next let $E_n \in \mathscr{R}, n = 1, 2, \ldots$, and let $E = \bigcup_{n=1}^{\infty} E_n$. We show that $E \in \mathscr{R}$. Let $\epsilon > 0$. Since $E_n \in \mathscr{R}$, there exist an open set $O_{n,\epsilon} \supset E_n$ and a closed set $C_{n,\epsilon} \subset E_n$ such that

$$\mu(O_{n,\epsilon} - C_{n,\epsilon}) < \frac{\epsilon}{3^n} \qquad \text{for every } n \geq 1.$$

Set $O_\epsilon = \bigcup_{n=1}^{\infty} O_{n,\epsilon}$ and $C = \bigcup_{n=1}^{\infty} C_{n,\epsilon}$. Since μ is finite, we see easily that there exists an $n_0 = n_0(\epsilon)$ such that

$$\mu\left(C - \bigcup_{n=1}^{n_0} C_{n,\epsilon}\right) < \frac{\epsilon}{2}.$$

Set $C_\epsilon = \bigcup_{n=1}^{n_0} C_{n,\epsilon}$. Clearly O_ϵ is open, C_ϵ is closed, and, moreover,

$$C_\epsilon \subset E \subset O_\epsilon$$

and

$$\begin{aligned}\mu(O_\epsilon - C_\epsilon) &\le \mu(O_\epsilon - C) + \mu(C - C_\epsilon)\\ &\le \sum_{n=1}^{\infty} \mu(O_{n,\epsilon} - C_{n,\epsilon}) + \mu(C - C_\epsilon)\\ &< \sum_{n=1}^{\infty} \frac{\epsilon}{3^n} + \frac{\epsilon}{2} = \epsilon.\end{aligned}$$

This proves that $E \in \mathscr{R}$, that is, $\mathscr{R}$ is closed under countable unions. Thus we have shown that $\mathscr{R}$ is a subσ-field of $\mathscr{B}$.

The proof will be completed if we can show that $\mathscr{R}$ contains all open subsets of $\mathfrak{X}$ or, equivalently, that $\mathscr{R}$ contains all closed subsets of $\mathfrak{X}$. Let C be a closed subset of $\mathfrak{X}$, and let $\epsilon > 0$. We note that C is a G_δ set, that is, C can be written as a countable intersection of open sets in $\mathfrak{X}$. In fact, setting $\rho(x, C) = \inf_{y \in C} \rho(x, y)$ for $x \in \mathfrak{X}$, we see easily that

$$C = \bigcap_{n=1}^{\infty} \left\{x : \rho(x, C) < \frac{1}{n}\right\},$$

and since $x \to \rho(x, C)$ is continuous on $\mathfrak{X}$, it follows that C is a G_δ set. Hence there exists a nonincreasing sequence of open sets $\{O_n\}$ such that $C = \lim_{n\to\infty} O_n = \bigcap_{n=1}^{\infty} O_n$. Since μ is finite, $\mu(C) = \lim_{n\to\infty} \mu(O_n)$. Let $\epsilon > 0$. Then it follows that there exists an $N_0 = N_0(\epsilon)$ such that $\mu(O_{N_0} - C) < \epsilon$. Set $C_\epsilon = C$ and $O_\epsilon = O_{N_0}$. We see that $\mu(O_\epsilon - C_\epsilon) < \epsilon$, so that $C \in \mathscr{R}$. This completes the proof of Proposition 3.9.1. ■

According to Proposition 3.9.1, a finite measure μ on $\mathscr{B}$ is determined by the values of $\mu(F)$ for closed sets F.

We now extend the concept of weak convergence of a sequence of finite measures to a metric space $\mathfrak{X}$.

Definition 3.9.2. Let $\mathscr{C}_0 = \mathscr{C}_0(\mathfrak{X})$ be the space of all bounded, real-valued, continuous functions on $\mathfrak{X}$. Let $\{\mu_n\}$ be a sequence of finite measures, and μ a finite measure defined on $(\mathfrak{X}, \mathscr{B})$. We say that μ_n converges weakly to μ if

$$\int_{\mathfrak{X}} g\, d\mu_n \to \int_{\mathfrak{X}} g\, d\mu \qquad \text{for every } g \in \mathscr{C}_0 \quad \text{as } n \to \infty. \tag{3.9.1}$$

In this case we write $\mu_n \Rightarrow \mu$.

Definition 3.9.3. A set $E \in \mathscr{B}$ is said to be a continuity set of measure μ if the boundary of E has μ-measure 0, that is,

$$\mu(\bar{E} - E^O) = 0,$$

where $\bar{E}$ is the closure of E, and E^O the interior of E.

Remark 3.9.1. Although we will concern ourselves only with the weak convergence of finite measures in this section, it should be noted that the notion of convergence in law can be defined for random variables (elements) taking values in a metric space (see also Chapter 7). Let $\mathfrak{X}$ be a metric space with Borel σ-field $\mathscr{B}$, and let $(\Omega, \mathscr{S}, P)$ be a probability space. An $\mathscr{S}$-measurable function $X: \Omega \to \mathfrak{X}$ is called a random variable, that is, $X: \Omega \to \mathfrak{X}$ is a random variable if $\{\omega: X(\omega) \in B\} \in \mathscr{S}$ for all $B \in \mathscr{B}$. The random variable X induces a probability measure P_X on $\mathscr{B}$ by the correspondence

$$P_X(B) = PX^{-1}(B), \qquad B \in \mathscr{B}$$

The probability measure P_X is called the probability distribution of X. Suppose there is a sequence of $\mathfrak{X}$-valued random variables $\{X_n\}$ defined on $(\Omega, \mathscr{S}, P)$ with probability distribution $P_n = PX_n^{-1}$. If $P_n \Rightarrow P$, we say that X_n converges in distribution (or in law) to X, and we write $X_n \Rightarrow X$. It is clear that every result about weak convergence of probability measures has an analogue about convergence in law, and conversely.

The following result characterizes weak convergence as defined in (3.9.1), in terms of the convergence of measures of certain Borel sets. This is an extension of Theorem 3.5.3 to a metric space.

Theorem 3.9.1. Let $\{\mu_n\}$ be a sequence of finite measures on $\mathscr{B}$, and let μ be a finite measure on $\mathscr{B}$. Then the following statements are equivalent:

(a) $\mu_n \Rightarrow \mu$.
(b) $\mu(C) \geq \limsup_{n \to \infty} \mu_n(C)$ for every closed set $C \subset \mathfrak{X}$.
(c) $\mu(O) \leq \liminf_{n \to \infty} \mu_n(O)$ for every open set $O \subset \mathfrak{X}$.
(d) $\mu_n(E) \to \mu(E)$ as $n \to \infty$ for every continuity set E of μ.

Proof. We first show that (a) $\Rightarrow$ (b). Let $C \subset \mathfrak{X}$ be an arbitrary closed set. For every $N \geq 1$ set $S_N = \{x \in \mathfrak{X}: \rho(x, C) < 1/N\}$, where $\rho(x, C) = \inf_{y \in C}$

$\rho(x, y)$ for $x \in \mathfrak{X}$. Then C and S_N^c are disjoint closed sets in $\mathfrak{X}$. Hence by Urysohn's lemma there exists a function $g_N \in \mathscr{C}_0$ such that $0 \leq g_N \leq 1$, $g_N(x) = 1$ for $x \in C$, and $g_N(x) = 0$ for $x \in S_N^c$. We note also that $S_N \downarrow C$. It follows that for every $N \geq 1$ we have

$$\begin{aligned}
\limsup_{n\to\infty} \mu_n(C) &\leq \limsup_{n\to\infty} \int_{\mathfrak{X}} g_N \, d\mu_n \\
&= \int_{\mathfrak{X}} g_N \, d\mu \qquad (\text{since } \mu_n \Rightarrow \mu) \\
&= \int_{S_N} g_N \, d\mu \\
&\leq \mu(S_N).
\end{aligned}$$

Since $S_N \downarrow C$ and μ is finite, it follows that $\mu(S_N) \to \mu(C)$. Hence

$$\mu(C) \geq \limsup_{n\to\infty} \mu_n(C).$$

Clearly (b) $\Leftrightarrow$ (c), since open sets are the complements of closed sets, and conversely.

Next we show that (c) $\Rightarrow$ (d). Let E be a continuity set of μ. Let $\bar{E}$ be the closure, and E^O the interior, of E. Then $\mu(\bar{E} - E^O) = 0$. Since $\bar{E}$ is closed and E^O open, we have

$$\mu(\bar{E}) \geq \limsup_{n\to\infty} \mu_n(\bar{E}) \geq \limsup_{n\to\infty} \mu_n(E)$$

and

$$\mu(E^O) \leq \liminf_{n\to\infty} \mu_n(E^O) \leq \liminf_{n\to\infty} \mu_n(E)$$

in view of (b) and (c), respectively. In view of our hypothesis $\mu(\bar{E}) = \mu(E^O) = \mu(E)$, so that (d) follows.

Finally we show that (d) $\Rightarrow$ (a). Suppose that $\mu_n(E) \to \mu(E)$ for every continuity set E of μ. We show that (3.9.1) holds. Let $g \in \mathscr{C}_0$, and μ^g be the set function defined on the Borel σ-field of the real line by

$$\mu^g(E) = \mu\{x \in \mathfrak{X} : g(x) \in E\}$$

for every Borel set E of $\mathbb{R}$. Clearly μ^g is a finite measure, and since g is bounded there exists a closed bounded interval $[a, b]$ such that $a < g(x) \leq b$ and, moreover, $\mu^g \equiv 0$ outside $[a, b]$. We also note that μ^g can have at most a countable number of mass points. Let $\epsilon > 0$. Choose a subdivision $a = t_0 < t_1 < \cdots < t_N = b$ such that

$$t_j - t_{j-1} < \epsilon \qquad \text{for } j = 1, 2, \ldots, N$$

and

$$\mu\{x \in \mathfrak{X}: g(x) = t_j\} = 0 \qquad \text{for } j = 1, 2, \ldots, N.$$

Set $E_j = \{x \in \mathfrak{X}: t_{j-1} < g(x) \le t_j\}$, $j = 1, 2, \ldots, N$. Then $E_1, E_2, \ldots, E_N$ are disjoint Borel sets in $\mathfrak{X}$ such that

$$\mathfrak{X} = \bigcup_{j=1}^{N} E_j.$$

Moreover, $\bar{E}_j - E_j^O \subset \{x: g(x) = t_{j-1}\} \cup \{x: g(x) = t_j\}$, so that

$$\mu(\bar{E}_j - E_j^O) = 0 \qquad \text{for } j = 1, 2, \ldots, N.$$

Consequently

$$\lim_{n\to\infty} \mu_n(E_j) = \mu(E_j), \qquad 1 \le j \le N.$$

Let h be the simple function on $\mathfrak{X}$ defined by

$$h = \sum_{j=1}^{N} t_j \chi_{E_j}.$$

Then

$$\sup_{x\in\mathfrak{X}} |g(x) - h(x)| < \epsilon.$$

For $n \ge 1$ we have

$$\begin{aligned}\left|\int_{\mathfrak{X}} g\, d\mu_n - \int_{\mathfrak{X}} g\, d\mu\right| &\le \left\{\int_{\mathfrak{X}} |g - h|\, d\mu_n \right.\\ &\quad \left. + \left|\int_{\mathfrak{X}} h\, d\mu_n - \int_{\mathfrak{X}} h\, d\mu\right| + \int_{\mathfrak{X}} |g - h|\, d\mu\right\}\\ &\le \epsilon\mu_n(\mathfrak{X}) + \sum_{j=1}^{N} |t_j|\,|\mu_n(E_j) - \mu(E_j)| + \epsilon\mu(\mathfrak{X}).\end{aligned}$$

Since $\mathfrak{X}$ is a continuity set of μ, we see easily that

$$\limsup_{n\to\infty} \left|\int_{\mathfrak{X}} g\, d\mu_n - \int_{\mathfrak{X}} g\, d\mu\right| \le \gamma\epsilon$$

for some constant $0 < \gamma < \infty$. Since $\epsilon > 0$ is arbitrary, (a) follows. ■

Corollary 1. Suppose that μ and ν are two finite measures on $(\mathfrak{X}, \mathscr{B})$ satisfying

$$\int_{\mathfrak{X}} g\, d\mu = \int_{\mathfrak{X}} g\, d\nu \qquad \text{for all } g \in \mathscr{C}_0.$$

Then $\mu = \nu$.

Proof. Set $\mu_n \equiv \nu$ for all $n \geq 1$. Then $\mu_n \Rightarrow \mu$, and Theorem 3.9.1 implies that $\nu(C) \leq \mu(C)$ for all closed sets C. When ν and μ are interchanged, it follows that μ and ν coincide for all closed sets in $\mathfrak{X}$. Since μ and ν are regular from Proposition 3.9.1, it follows immediately that $\mu = \nu$.

We note that the proof of Corollary 1 is also an immediate consequence of the fact that the space $\mathscr{C}_0 = \mathscr{C}_0(\mathfrak{X})$ is the dual of the Banach space $\mathscr{S} = \mathscr{S}(\mathfrak{X}, \mathscr{B})$ of all finite signed measures on $(\mathfrak{X}, \mathscr{B})$. (See also discussion following Definition 3.5.1.)

Corollary 2. If $\mu_n \Rightarrow \mu$ and $\mu_n \Rightarrow \nu$, then $\mu = \nu$.

Proof. Clearly

$$\lim_{n\to\infty} \int_{\mathfrak{X}} g \, d\mu_n = \int_{\mathfrak{X}} g \, d\mu$$

$$= \int_{\mathfrak{X}} g \, d\nu,$$

so that $\mu = \nu$.

Definition 3.9.4. Let $\mathscr{M} = \{\mu_t, t \in T\}$ be a class of finite measures on $(\mathfrak{X}, \mathscr{B})$. We say that $\mathscr{M}$ is relatively (weakly) compact if every infinite sequence of elements in $\mathscr{M}$ has an infinite subsequence which converges weakly to a finite measure μ (not necessarily in $\mathscr{M}$) on $\mathscr{B}$. We say that the class $\mathscr{M}$ is (uniformly) tight if for a given $\epsilon > 0$ there exists a compact set $K_\epsilon \subset \mathfrak{X}$ such that

$$\sup_{t\in T} \mu_t(\mathfrak{X} - K_\epsilon) < \epsilon.$$

In particular, if $\mathscr{M}$ consists only of probability measures, we say that $\mathscr{M}$ is relatively compact if every infinite sequence of elements in $\mathscr{M}$ has an infinite subsequence which converges weakly to a probability measure on $\mathscr{B}$.

As an important special case let $\mathfrak{X} = \mathbb{R}$ and $\mathscr{F} = \{F_t, t \in T\}$ be a class of bounded, nondecreasing, right-continuous functions on $\mathbb{R}$. Let $\mathscr{M} = \{\mu_t, t \in T\}$ be the corresponding class of finite measures on $\mathscr{B}$. Then $\mathscr{F}$ is said to be tight (relatively compact) if $\mathscr{M}$ is tight (relatively compact). A sequence of real-valued random variables is said to be *stochastically bounded* if the corresponding sequence of induced probability measures on $\mathscr{B}$ is tight. Since each closed and bounded interval on $\mathbb{R}$ is compact, a sequence $\{X_n\}$ of random variables is stochastically bounded if for every $\epsilon > 0$ there exists an $a = a(\epsilon) > 0$ such that

$$P\{-a \leq X_n \leq a\} \geq 1 - \epsilon \qquad \text{for all } n \geq 1.$$

Example 3.9.1. Let $\mathfrak{X} = \mathbb{R}$ and $\mathscr{M} = \{P_n, n \geq 1\}$, where P_n is the probability measure corresponding to the normal distribution with mean μ_n and variance 1. Suppose that $\mu_n \to -\infty$ as $n \to \infty$. Then $\mathscr{M}$ is not tight.

Example 3.9.2. Let $\mathscr{M} = \{P_n, n \geq 1\}$, where P_n is the Poisson measure

$$P_n\{x\} = e^{-\lambda_n} \frac{\lambda_n^x}{x!}, \qquad x = 0, 1, 2, \ldots, \quad \lambda_n > 0.$$

Suppose that $\{\lambda_n, n \geq 1\}$ are bounded away from zero. Then $\mathscr{M}$ is tight as well as relatively compact.

Remark 3.9.2. Let $\mathfrak{X} = \mathbb{R}$, and $\mathscr{M}$ be a family of probability measures on $(\mathbb{R}, \mathscr{B})$. Let $P_n \in \mathscr{M}$ be a sequence of probability measures with distribution function F_n. From Helly's theorem (Theorem 3.1.2) there exists a subsequence $\{F_{n_k}\} \subset \{F_n\}$ and a bounded, nondecreasing, right-continuous function F such that $F_{n_k} \xrightarrow{w} F$ as $k \to \infty$. Let μ be the finite measure on $(\mathbb{R}, \mathscr{B})$ determined by the correspondence

$$\mu(a, b] = F(b) - F(a).$$

Since F is not necessarily a distribution function, $\mu(\mathbb{R})$ may be <1. Thus $\mathscr{M}$ is not necessarily relatively compact. Suppose, however, that μ is a probability measure, that is, $\mu(\mathbb{R}) = 1$. Then $F_{n_k} \xrightarrow{c} F$, and hence $\mu_{n_k} \Rightarrow \mu$ (Theorem 3.5.3).

If each of these limiting measures μ is a probability measure, $\mathscr{M}$ will be relatively compact. This will be the case if, for every sequence $\{P_n\} \subset \mathscr{M}$ and every $\epsilon > 0$,

$$P_n[a, b] > 1 - \epsilon \qquad \text{for all } n \geq 1.$$

It follows from Theorem 3.9.1 (b) that $\mu[a, b] > 1 - \epsilon$, that is, each limit μ is a probability measure. Thus $\mathscr{M}$ is relatively compact if for every $\epsilon > 0$ there exist $a, b \in \mathbb{R}$ such that $P[a, b] > 1 - \epsilon$ for all $P \in \mathscr{M}$. Thus we see that tightness of $\mathscr{M}$ is sufficient for relative compactness. Actually this condition is also necessary, for if the condition fails, then for some $\epsilon > 0$ and for every $n \geq 1$ there exists a $P_n \in \mathscr{M}$ such that $P_n[-n, n] \leq 1 - \epsilon$. Suppose that $\{P_n\}$ contains a subsequence $\{P_{n_k}\}$ which converges weakly to a probability measure Q. In that case for $x \in \mathbb{R}$ we have

$$\begin{aligned} Q(-x, x) \leq \liminf_{k \to \infty} P_{n_k}(-x, x) &\leq \liminf_{k \to \infty} P_{n_k}[-n_k, n_k] \\ &\leq 1 - \epsilon, \end{aligned}$$

so that $Q(\mathfrak{X}) \leq 1 - \epsilon < 1$, which is a contradiction. Thus $\mathscr{M}$ is relatively compact if and only if it is tight. This result holds more generally on a complete, separable metric space, as is shown in Theorem 3.9.2 below.

3.9.2 Prokhorov's Theorem

The following result of Prokhorov is basic in the study of weak convergence of probability measures on metric spaces. It shows the equivalence of tightness and relative compactness for a family $\mathcal{M}$ of probability measures defined on a complete separable metric space.

Theorem 3.9.2 (Prokhorov). Let $(\mathfrak{X}, \rho)$ be a complete separable metric space, and let $\mathcal{B}$ be the σ-field of Borel sets on $\mathfrak{X}$. Let $\mathcal{M}$ be a class of probability measures defined on $\mathcal{B}$. Then $\mathcal{M}$ is tight if and only if it is relatively compact.

Proof. Let $\mathcal{M}$ be relatively compact. To show that $\mathcal{M}$ is tight we need to establish, for every $\epsilon > 0$, the existence of a compact set $K_\epsilon \subset \mathfrak{X}$ such that

$$\mu(\mathfrak{X} - K_\epsilon) < \epsilon \qquad \text{for all } \mu \in \mathcal{M},$$

or, equivalently,

$$\mu(K_\epsilon) > 1 - \epsilon \qquad \text{for all } \mu \in \mathcal{M}.$$

Let $\delta > 0$. Since $\mathfrak{X}$ is separable, there exists a sequence $\{x_n\}$ of elements of $\mathfrak{X}$ which is everywhere dense in $\mathfrak{X}$. For every $n \geq 1$ let $S_n \subset \mathfrak{X}$ be the open sphere with center x_n and radius δ. We see easily that $\mathfrak{X} = \bigcup_{n=1}^{\infty} S_n$. For every $n \geq 1$ set $U_n = \bigcup_{j=1}^{n} S_j$. Let $\epsilon > 0$. We now show that there exists an $n_0 = n_0(\epsilon)$ such that $\mu(U_{n_0}) > 1 - \epsilon$ for all $\mu \in \mathcal{M}$. Suppose this is not the case. Then, for some $\epsilon_0 > 0$ and for every positive integer n, there exists a $\mu_n \in \mathcal{M}$ such that $\mu_n(U_n) \leq 1 - \epsilon_0$. Since $\mathcal{M}$ is relatively compact, the sequence $\{\mu_n\}$ contains an infinite subsequence $\{\mu_{n_k}\}$ which converges weakly to a probability measure μ_0. In view of Theorem 3.9.1 we have for every $n \geq 1$

$$\mu_0(U_n) \leq \liminf_{k\to\infty} \mu_{n_k}(U_n) \leq \liminf_{k\to\infty} \mu_{n_k}(U_{n_k})$$
$$\leq 1 - \epsilon_0.$$

Since, however, $U_n \uparrow \mathfrak{X}$, this is not possible. This contradiction proves the result.

Thus, for given $\epsilon > 0$ and $\delta > 0$, there exist finitely many open spheres $S_1, S_2, \ldots, S_{n_0}$, each of radius δ, such that

$$\mu\left(\bigcup_{j=1}^{n_0} S_j\right) > 1 - \epsilon \qquad \text{for all } \mu \in \mathcal{M}.$$

Let $k \geq 1$ be a positive integer. Choose open spheres $S_{k,1}, S_{k,2}, \ldots, S_{k,n_k}$, each of radius $1/k$, such that

$$\mu\left(\bigcup_{j=1}^{n_k} S_{k,j}\right) > 1 - \frac{\epsilon}{2^k}.$$

For every $k \geq 1$, $\bigcup_{j=1}^{n_k} S_{k,j}$ is totally bounded, so that $\bigcap_{k=1}^{\infty} \bigcup_{j=1}^{n_k} S_{k,j}$ is totally bounded. Let K_ϵ be the closure of $\bigcap_{k=1}^{\infty} \bigcup_{j=1}^{n_k} S_{k,j}$ in $\mathfrak{X}$. Then K_ϵ is complete, totally bounded, and hence compact. Taking complements, we see easily that

$$\mu(K_\epsilon) > 1 - \epsilon \qquad \text{for all } \mu \in \mathscr{M}.$$

This completes the proof of the sufficiency part of the theorem.

Next suppose that $\mathscr{M}$ is tight. We show that $\mathscr{M}$ is relatively compact. For each $n \geq 1$ there exists a compact set $K_n \subset \mathfrak{X}$ such that $K_n \subset K_{n+1}$, and $\mu(K_n) > 1 - 1/n$ for every $\mu \in \mathscr{M}$. Since $\mathfrak{X}$ is separable, there exists a countable sequence $\{S_n\}$ of open spheres which forms a basis for the topology of $\mathfrak{X}$. Write $\mathfrak{A} = \{S_n, n \geq 1\}$. Let $\mathscr{F}$ be the class of subsets of $\mathfrak{X}$ which consists of the finite unions of sets of the form $\bar{S} \cap K_n$ with $S \in \mathfrak{A}$, $\bar{S}$ the closure of S, and $n \geq 1$. We note that $\mathscr{F}$ is countable, and it is closed under the formation of finite unions. Moreover, each set in $\mathscr{F}$ is compact.

Let $\mu_n \in \mathscr{M}$, $n > 1$, be an infinite sequence. Since $\mathscr{F}$ is countable, we can, by proceeding exactly as in the proof of Theorem 3.1.2, select an infinite subsequence $\{\mu_{n_k}\} \subset \{\mu_n\}$ such that for every $F \in \mathscr{F}$ the limit

$$\lim_{k \to \infty} \mu_{n_k}(F) = \nu(F) \tag{3.9.2}$$

exists and is finite. Clearly the set function ν defined on $\mathscr{F}$ by (3.9.2) is nonnegative and has the properties

$$\nu(F_1) \leq \nu(F_2) \qquad \text{for } F_1 \subset F_2,\ F_1, F_2 \in \mathscr{F}, \tag{3.9.3}$$

$$\nu(F_1 \cup F_2) \leq \nu(F_1) + \nu(F_2) \qquad \text{for } F_1, F_2 \in \mathscr{F}. \tag{3.9.4}$$

$$\nu(F_1 \cup F_2) = \nu(F_1) + \nu(F_2) \qquad \text{for } F_1 \cap F_2 = \varnothing,\ F_1, F_2 \in \mathscr{F}. \tag{3.9.5}$$

Suppose there exists a probability measure μ defined on $\mathscr{B}$ such that for every open set $O \subset \mathfrak{X}$

$$\mu(O) = \sup_{\substack{F \subset O \\ F \in \mathscr{F}}} \nu(F). \tag{3.9.6}$$

Then we show that $\mu_{n_k} \Rightarrow \mu$ as $k \to \infty$. In fact, let O be an arbitrary open subset of $\mathfrak{X}$, and let $F \in \mathscr{F}$ be such that $F \subset O$. Then from (3.9.2) we have

$$\begin{aligned} \nu(F) &= \lim_{k \to \infty} \mu_{n_k}(F) \\ &\leq \liminf_{k \to \infty} \mu_{n_k}(O) \end{aligned}$$

so that (3.9.6) gives

$$\mu(O) \leq \liminf_{k \to \infty} \mu_{n_k}(O)$$

for every open set $O \subset \mathfrak{X}$. Hence it follows from Theorem 3.9.1 that $\mu_{n_{n_k}} \Rightarrow \mu$. Therefore it is sufficient to construct a probability measure μ on $\mathscr{B}$ satisfying (3.9.6).

Let $\mathscr{O}$ be the class of all open subsets of $\mathfrak{X}$. We define a set function λ on $\mathscr{O}$ by setting

$$\lambda(O) = \sup_{\{F \subset O,\, F \in \mathscr{F}\}} \nu(F). \tag{3.9.7}$$

Let $\mathscr{S} = \mathscr{S}(\mathfrak{X})$ be the class of all subsets of $\mathfrak{X}$. Define a set function λ^* on $\mathscr{S}$ by setting

$$\lambda^*(E) = \inf_{\{E \subset O,\, O \in \mathscr{O}\}} \lambda(O), \qquad E \in \mathscr{S}. \tag{3.9.8}$$

Clearly $\lambda^*(O) = \lambda(O)$ for every $O \in \mathscr{O}$.

We first show that λ^* is an outer measure on $\mathscr{S}$. Clearly λ^* is nonnegative, $\lambda^*(\varnothing) = 0$, and $\lambda^*(E_1) \leq \lambda^*(E_2)$ for $E_1 \subset E_2$, $E_1, E_2 \in \mathscr{S}$. We need only prove that λ^* is countably subadditive on $\mathscr{S}$. Let $E_n \in \mathscr{S}$, and let $E = \bigcup_{n=1}^{\infty} E_n$. We show that

$$\lambda^*(E) \leq \sum_{n=1}^{\infty} \lambda^*(E_n). \tag{3.9.9}$$

Let $\epsilon > 0$. Then for every $n \geq 1$ we select an open set $O_n \in \mathscr{O}$ such that $E_n \subset O_n$ and

$$\lambda(O_n) < \lambda^*(E_n) + \frac{\epsilon}{2^n}.$$

Then

$$\sum_{n=1}^{\infty} \lambda(O_n) < \sum_{n=1}^{\infty} \lambda^*(E_n) + \epsilon. \tag{3.9.10}$$

On the other hand, $E \subset \bigcup_{n=1}^{\infty} O_n$, where $\bigcup_{n=1}^{\infty} O_n \in \mathscr{O}$, so that

$$\lambda^*(E) \leq \lambda\left(\bigcup_{n=1}^{\infty} O_n\right). \tag{3.9.11}$$

We now show that the set function λ is countably subadditive on $\mathscr{O}$. For this purpose let $O_n \in \mathscr{O}$, and let $O = \bigcup_{n=1}^{\infty} O_n$. Let $F \in \mathscr{F}$ be such that $F \subset O$. Since F is compact, it follows from the Heine–Borel theorem that there exist a finite number of open sets, say $\{O_1, O_2, \ldots, O_{n_0}\} \subset \{O_n, n \geq 1\}$, such that

$$F \subset \bigcup_{n=1}^{n_0} O_n.$$

Note that λ is finitely subadditive on $\mathscr{O}$. In fact, let $O_1, O_2 \in \mathscr{O}$, and let $F \in \mathscr{F}$ be such that $F \subset O_1 \cup O_2$. Define

$$C_1 = \{x \in F: \rho(x, O_1^c) \geq \rho(x, O_2^c)\}$$

and

$$C_2 = \{x \in F: \rho(x, O_2^c) \geq \rho(x, O_1^c)\}.$$

Clearly C_1 and C_2 are closed subsets of the compact set F and hence are both compact. Moreover, $C_1 \subset O_1$, for otherwise there exists an $x \in C_1 \subset F$ such that $x \notin O_1$. Hence $x \in O_2$. Then $\rho(x, O_1^c) = 0$. But since O_2^c is closed, $\rho(x, O_2^c) > 0$ so that $\rho(x, O_1^c) < \rho(x, O_2^c)$, which contradicts the assumption that $x \in C_1$. Similarly we see that $C_2 \subset O_2$.

We next show that there exist sets $F_1, F_2 \in \mathscr{F}$ such that $C_1 \subset F_1 \subset O_1$ and $C_2 \subset F_2 \subset O_2$. Let $x \in C_1$. Since O_1 is open, there exists an open sphere $S(x; \epsilon)$ with center x and radius $\epsilon > 0$ such that $x \in S(x; \epsilon) \subset O_1$. Clearly there exists a positive integer n_x such that $x \in S_{n_x} \subset S(x; \epsilon/2)$. Then

$$x \in S_{n_x} \subset \bar{S}_{n_x} \subset S(x; \epsilon) \subset O_1,$$

so that

$$C_1 \subset \bigcup_{x \in C_1} S_{n_x} \subset \bigcup_{x \in C_1} \bar{S}_{n_x} \subset O_1.$$

Since C_1 is compact, there exist a finite number of elements $x_1, x_2, \ldots, x_N \in C_1$ such that

$$C_1 \subset \bigcup_{j=1}^{N} S_{n_{x_j}} \subset \bigcup_{j=1}^{N} \bar{S}_{n_{x_j}} \subset O_1.$$

Since $F \in \mathscr{F}$, there exists a positive integer n_0 such that $F \subset K_{n_0}$. Hence

$$C_1 \subset \bigcup_{j=1}^{N} \bar{S}_{n_{x_j}} \cap K_{n_0} \subset O_1.$$

Set $F_1 = \bigcup_{j=1}^{N} \bar{S}_{n_{x_j}} \cap K_{n_0}$. Clearly $C_1 \subset F_1 \subset O_1$, where $F_1 \in \mathscr{F}$. In a similar manner we show that there exists an $F_2 \in \mathscr{F}$ such that $C_2 \subset F_2 \subset O_2$.

Now

$$F = C_1 \cup C_2 \subset F_1 \cup F_2,$$

so that

$$\nu(F) \leq \nu(F_1) + \nu(F_2) \leq \lambda(O_1) + \lambda(O_2).$$

Since $F \subset O_1 \cup O_2$ is arbitrary, we conclude that

$$\lambda(O_1 \cup O_2) \leq \lambda(O_1) + \lambda(O_2).$$

This shows that λ is finitely subadditive on $\mathcal{O}$. It now follows from (3.9.7) that

$$\nu(F) \leq \lambda\left(\bigcup_{n=1}^{n_0} O_n\right) \leq \sum_{n=1}^{n_0} \lambda(O_n) \leq \sum_{n=1}^{\infty} \lambda(O_n).$$

Taking the supremum over $F \subset O$, $F \in \mathcal{F}$, we get

$$\lambda(O) \leq \sum_{n=1}^{\infty} \lambda(O_n). \tag{3.9.12}$$

Using (3.9.10), (3.9.11), and (3.9.12), we obtain

$$\lambda^*(E) \leq \lambda\left(\bigcup_{n=1}^{\infty} O_n\right) \leq \sum_{n=1}^{\infty} \lambda(O_n) < \sum_{n=1}^{\infty} \lambda^*(E_n) + \epsilon.$$

Since $\epsilon > 0$ is arbitrary, this proves (3.9.9). Hence λ^* is an outer measure on $\mathcal{S}$.

Next we show that every closed set in $\mathfrak{X}$ is λ^*-measurable, that is, for every closed set $C \in \mathcal{S}$ and every $E \in \mathcal{S}$ the inequality

$$\lambda^*(E) \geq \lambda^*(E \cap C) + \lambda^*(E \cap C^c) \tag{3.9.13}$$

holds. It is sufficient to prove that

$$\lambda(O) \geq \lambda^*(O \cap C) + \lambda^*(O \cap C^c) \tag{3.9.14}$$

holds for every $O \in \mathcal{O}$. This follows from the fact that $O \supset E$, and $O \in \mathcal{O}$, together with the montone property of λ^*, implies

$$\lambda(O) \geq \lambda^*(E \cap C) + \lambda^*(E \cap C^c).$$

Taking the infimum over $O \in \mathcal{O}$ such that $O \supset E$, we get (3.9.13).

To prove (3.9.14) let $\epsilon > 0$, and let $F_0 \in \mathcal{F}$ be such that $F_0 \subset O \cap C^c$ and $\nu(F_0) > \lambda(O \cap C^c) - \epsilon/2$. Next let $F_1 \in \mathcal{F}$ be such that $F_1 \subset O \cap F_0^c$ and $\nu(F_1) > \lambda(O \cap F_0^c) - \epsilon/2$. Clearly $F_0 \cap F_1 = \varnothing$ and $F_0 \cup F_1 \subset O$. Hence

$$\begin{aligned} \lambda(O) &\geq \nu(F_0 \cup F_1) \\ &= \nu(F_0) + \nu(F_1) \qquad \text{[by (3.9.5)]} \\ &\geq \lambda(O \cap C^c) + \lambda(O \cap F_0^c) - \epsilon \\ &\geq \lambda^*(O \cap C^c) + \lambda^*(O \cap C) - \epsilon. \end{aligned}$$

Since $\epsilon > 0$ is arbitrary, this proves (3.9.14) and hence (3.9.13).

Let $\mathscr{S}^*$ be the class of all λ^*-measurable subsets of $\mathfrak{X}$. Proceeding exactly as in the Caratheodary extension theorem (Halmos [31], p. 54), we see that $\mathscr{S}^*$ is a σ-field of subsets of $\mathfrak{X}$, and the restriction of λ^* to $\mathscr{S}^*$ is a measure. Moreover, $\mathscr{B} \subset \mathscr{S}^*$, so that we define μ on $\mathscr{B}$ as the restriction of λ^* to $\mathscr{B}$. Then for every $O \in \mathscr{O}$ we have

$$\mu(O) = \lambda^*(O) = \lambda(O),$$

so that (3.9.6) holds. Finally we note that μ is a probability measure on $\mathscr{B}$. This follows from the fact that

$$\begin{aligned} 1 \geq \mu(\mathfrak{X}) = \lambda(\mathfrak{X}) &\geq \sup_{n \geq 1} \nu(K_n) \\ &\geq \sup_{n \geq 1} \left(1 - \frac{1}{n}\right). \end{aligned}$$

This completes the proof of Theorem 3.9.2. ∎

Remark 3.9.3. Theorem 3.9.2 remains valid if $\mathscr{M}$ is a class of uniformly bounded finite measures on $\mathscr{B}$, that is, there exists a positive number γ such that

$$\mu(\mathfrak{X}) \leq \gamma < \infty \qquad \text{for all } \mu \in \mathscr{M}.$$

Remark 3.9.4. Let $(\mathfrak{X}, \mathscr{B})$ be a complete, separable metric space. Then every finite measure μ on $\mathscr{B}$ is tight. This fact was used in the proof of Kolmogorov's consistency theorem (Theorem 1.1.7).

3.9.3 Applications of Tightness

We now consider some applications of tightness. For this purpose we will restrict ourselves to the case of $\mathfrak{X} = \mathbb{R}$. Our object is to give simple alternative proofs of some of the results proved earlier, using methods of this section. First we prove a useful inequality.

Lemma 3.9.1 (Truncation Inequality). Let F be a distribution function with characteristic function φ. Let $T > 0$. Then there exists a positive constant C such that

(3.9.15) $$\int_{|x| \geq T^{-1}} dF(x) \leq CT^{-1} \int_0^T [1 - \operatorname{Re} \varphi(t)]\, dt.$$

Proof. We have

$$T^{-1}\int_0^T [1 - \operatorname{Re}\varphi(t)]\,dt = T^{-1}\int_0^T \left[\int_{-\infty}^{\infty} (1 - \cos tx)\,dF(x)\right] dt$$

$$= \int_{-\infty}^{\infty} \left[T^{-1}\int_0^T (1 - \cos tx)\,dt\right] dF(x)$$

(Fubini's theorem)

$$= \int_{-\infty}^{\infty} \left(1 - \frac{\sin Tx}{Tx}\right) dF(x)$$

$$\geq \inf_{|\theta|\geq 1}\left(1 - \frac{\sin\theta}{\theta}\right)\int_{|Tx|\geq 1} dF(x)$$

$$\geq C^{-1}\int_{|x|\geq T^{-1}} dF(x),$$

since

$$\inf_{|\theta|\geq 1}\left(1 - \frac{\sin\theta}{\theta}\right) = 1 - \sin 1 \geq \tfrac{1}{7}.$$

We may therefore take $C = 7$. ■

We next introduce the concept of a *separating class of functions*, which is sometimes useful in checking whether or not a given sequence of tight distribution functions converges completely to a distribution function.

Definition 3.9.5. Let $\mathscr{F}$ be the class of all distribution functions on $\mathbb{R}$, and let $\mathscr{E} \subset \mathscr{C}_0$ be a set of bounded, real-valued, continuous functions on $\mathbb{R}$. We say that the class $\mathscr{E}$ is $\mathscr{F}$-separating if for any pair $F_1, F_2 \in \mathscr{F}$ the relation

$$\int_{-\infty}^{\infty} g\,dF_1 = \int_{-\infty}^{\infty} g\,dF_2 \qquad \text{for all } g \in \mathscr{E}$$

implies $F_1 = F_2$ on $\mathbb{R}$.

It is convenient to extend the concept of separating classes to the case of bounded, complex-valued, continuous functions on $\mathbb{R}$. Consider functions g of the form $g = g_1 + ig_2$, where g_1 and g_2 are bounded, real-valued, continuous functions on $\mathbb{R}$. For any $F \in \mathscr{F}$ we write

$$\int_{-\infty}^{\infty} g\,dF = \int_{-\infty}^{\infty} g_1\,dF + i\int_{-\infty}^{\infty} g_2\,dF.$$

Example 3.9.3. For every $t \in \mathbb{R}$ let χ_t be the complex-valued function defined on $\mathbb{R}$ by

$$\chi_t(x) = e^{itx}, \qquad x \in \mathbb{R}.$$

Then χ_t is continuous and bounded. It follows immediately from the uniqueness of characteristic functions (Corollary 2 to Theorem 3.3.1) that the class $\mathscr{E} = \{\chi_t, t \in \mathbb{R}\}$ is $\mathscr{F}$-separating. Note that for all $x, y \in \mathbb{R}$ and $t \in \mathbb{R}$

$$\chi_t(x + y) = \chi_t(x) \cdot \chi_t(y)$$

and

$$|\chi_t| = 1.$$

Clearly χ_t, for $t \in \mathbb{R}$, is a continuous homomorphism of the additive group of $\mathbb{R}$ into the circle group defined by $\{z \in \mathbb{C}^\times : z\bar{z} = 1\}$. In harmonic analysis such a function is called a *character* of the additive group of $\mathbb{R}$.

Example 3.9.4 (Moment Problem). For every nonnegative integer k let $g_k(x) = x^k$, $x \in \mathbb{R}$. Then the class of functions $\{g_k, k \geq 0\}$ is not an $\mathscr{F}$-separating class. Consider, for example, the log-normal distribution with probability density given by

$$\begin{aligned} f(x) &= (x\sqrt{2\pi})^{-1} e^{-(1/2)(\ln x)^2}, & x > 0, \\ &= 0, & x \leq 0. \end{aligned}$$

For $|\epsilon| \leq 1$ set

$$f_\epsilon(x) = f(x)[1 + \epsilon \sin(2\pi \ln x)], \qquad x \in \mathbb{R}.$$

Note that $f_\epsilon \geq 0$ for all ϵ, $|\epsilon| \leq 1$, and $\int_{-\infty}^{\infty} f_\epsilon(x)\,dx = 1$, so f_ϵ is a probability density function. Since, however,

$$\begin{aligned} \int_0^\infty x^k f(x) \sin(2\pi \ln x)\,dx &= \frac{1}{\sqrt{2\pi}} \int_{-\infty}^{\infty} e^{-(t^2/2)+kt} \sin(2\pi t)\,dt \\ &= \frac{1}{\sqrt{2\pi}} e^{k^2/2} \int_{-\infty}^{\infty} e^{-y^2/2} \sin(2\pi y)\,dy \\ &= 0, \end{aligned}$$

it follows that

$$\int_0^\infty x^k f_\epsilon(x)\,dx = \int_0^\infty x^k f(x)\,dx$$

for all ϵ, $|\epsilon| \le 1$, and $k = 0, 1, 2, \ldots$. Thus

$$\int_{-\infty}^{\infty} x^k \, dF_\epsilon(x) = \int_{-\infty}^{\infty} x^k \, dF(x)$$

for all $k \ge 0 \not\Rightarrow F = F_\epsilon$. Here F and F_ϵ are the distribution functions corresponding to f and f_ϵ, respectively. This example is due to C. C. Heyde.

Proposition 3.9.2. Let $\mathscr{E} \subset \mathscr{C}_0$ be an $\mathscr{F}$-separating class. Let $\{F_n \in \mathscr{F}, n \ge 1\}$ be a tight sequence of distribution functions. Then there exists an $F \in \mathscr{F}$ such that $F_n \xrightarrow{c} F$ if and only if $\lim_{n\to\infty} \int_{-\infty}^{\infty} g \, dF_n$ exists and is finite for all $g \in \mathscr{E}$. Moreover, in this case

$$\lim_{n\to\infty} \int_{-\infty}^{\infty} g \, dF_n = \int_{-\infty}^{\infty} g \, dF.$$

Proof. If $F_n \xrightarrow{c} F$, it follows immediately from Theorem 3.1.4 that

$$\int_{-\infty}^{\infty} g \, dF_n \to \int_{-\infty}^{\infty} g \, dF \qquad \text{for all } g \in \mathscr{E}.$$

Conversely, suppose that $\lim_{n\to\infty} \int_{-\infty}^{\infty} g \, dF_n$ exists and is finite for all $g \in \mathscr{E}$. Since $\{F_n\}$ is tight, it is relatively compact, and it follows that there exist a subsequence $\{F_{n_k}\} \subset \{F_n\}$ and a distribution function F such that $F_{n_k} \xrightarrow{c} F$ as $k \to \infty$. Let $\{F_{n'_k}\} \subset \{F_n\}$ be some other convergent subsequence, and suppose that $F_{n'_k} \xrightarrow{c} F^*$, $F^* \in \mathscr{F}$. By Theorem 3.1.4 we have for all $g \in \mathscr{E}$

$$\lim_{k\to\infty} \int_{-\infty}^{\infty} g \, dF_{n_k} = \int_{-\infty}^{\infty} g \, dF$$

and

$$\lim_{k\to\infty} \int_{-\infty}^{\infty} g \, dF_{n'_k} = \int_{-\infty}^{\infty} g \, dF^*.$$

Since $\lim_{n\to\infty} \int_{-\infty}^{\infty} g \, dF_n$ exists and is finite, we must have

$$\int_{-\infty}^{\infty} g \, dF = \int_{-\infty}^{\infty} g \, dF^* \qquad \text{for all } g \in \mathscr{E}.$$

Since, however, $\mathscr{E}$ is $\mathscr{F}$-separating, $F = F^*$, and it follows that all convergent subsequences of $\{F_n\}$ converge to the same limit F so that $F_n \xrightarrow{c} F$. ∎

Corollary. Let $\mathscr{E}$ and $\{F_n\}$ be as above. Let $F \in \mathscr{F}$ be such that

$$\int_{-\infty}^{\infty} g \, dF_n \to \int_{-\infty}^{\infty} g \, dF \qquad \text{for all } g \in \mathscr{E}.$$

Then $F_n \xrightarrow{c} F$.

We now give a simple alternative proof of Lévy's continuity theorem.

Lévy's Continuity Theorem (Theorem 3.5.1). Let $\{F_n\}$ be a sequence of distribution functions, and let $\{\varphi_n\}$ be the corresponding sequence of characteristic functions. Suppose that $\varphi_n \to \varphi$ on $\mathbb{R}$ (pointwise), where φ is continuous at $t = 0$. Then φ is the characteristic function of a distribution function F. Moreover, $F_n \xrightarrow{c} F$.

Proof. From Example 3.9.3, we see that the class $\mathscr{E} = \{\chi_t : x \to e^{itx}, t \in \mathbb{R}\}$ is $\mathscr{F}$-separating. Since $\lim_{n\to\infty} \int_{-\infty}^{\infty} e^{itx}\, dF_n(x)$ exists for all $t \in \mathbb{R}$, we need only show that $\{F_n\}$ is tight (Proposition 3.9.2). Applying Lemma 3.9.1 to F_n and φ_n, we have for all $T > 0$

$$\int_{|x| \geq T^{-1}} dF_n(x) \leq CT^{-1} \int_0^T [1 - \operatorname{Re} \varphi_n(t)]\, dt$$

$$\to CT^{-1} \int_0^T [1 - \operatorname{Re} \varphi(t)]\, dt$$

as $n \to \infty$ by the Lebesgue dominated convergence theorem. Thus

$$\limsup_{n\to\infty} \int_{|x| \geq T^{-1}} dF_n(x) \leq CT^{-1} \int_0^T [1 - \operatorname{Re} \varphi(t)]\, dt.$$

Now φ is continuous at $t = 0$, so that $\operatorname{Re} \varphi(t) \to 1$ as $t \to 0$, and it follows that

$$\lim_{T\to 0} \limsup_{n\to\infty} \int_{|x| \geq T^{-1}} dF_n(x) = 0.$$

Let $\epsilon > 0$. Then there exists a $T_0 = T_0(\epsilon)$ small enough so that

$$\int_{|x| \geq T_0^{-1}} dF_n(x) < \epsilon \qquad \text{for all } n \geq 1.$$

Hence $\{F_n\}$ is tight as asserted. ∎

Similarly we can construct a proof of Theorem 3.5.2 using methods of this section.

Theorem 3.5.2. Let $\{F_n\}$ be a sequence of distribution functions, and let $\{\varphi_n\}$ be the corresponding sequence of characteristic functions. Then $F_n \xrightarrow{c} F$, where F is some distribution function, if and only if $\varphi_n \to \varphi$ uniformly in every finite t-interval. In this case φ is the characteristic function of F.

Proof. We need only prove that $F_n \xrightarrow{c} F$ implies $\varphi_n \xrightarrow{c} \varphi$ uniformly in every finite t-interval. Let $I = [-T, T]$, where $T > 0$ is finite. Then for any n, t, and δ

$$\begin{aligned}|\varphi_n(t+\delta) - \varphi_n(t)| &= \left|\int_{-\infty}^{\infty} (e^{i(t+\delta)x} - e^{itx})\, dF_n(x)\right| \\ &\le \int_I |e^{i\delta x} - 1|\, dF_n(x) + 2\int_{|x|>T} dF_n(x) \\ &\le \sup_{x\in I} |e^{i\delta x} - 1| + 2\int_{|x|>T} dF_n(x).\end{aligned}$$

Since $F_n \xrightarrow{c} F$, $\{F_n\}$ is relatively compact and hence is tight by Theorem 3.9.2. Let $\epsilon > 0$. We can choose $T = T(\epsilon) > 0$, and hence I, such that $2\int_{|x|>T} dF_n < \epsilon/2$ for all n. Hence

$$\sup_{n\ge 1}\sup_{t\in\mathbb{R}} |\varphi_n(t+\delta) - \varphi_n(t)| < \epsilon$$

for sufficiently small $\delta = \delta(\epsilon)$, that is, $\{\varphi_n\}$ is equicontinuous on $\mathbb{R}$.

Now φ is uniformly continuous on $\mathbb{R}$. Since $\{\varphi_n\}$ is equicontinuous on $\mathbb{R}$, it follows that there is an $\eta > 0$ such that

$$|\varphi(t+\delta) - \varphi(t)| < \epsilon \quad \text{and} \quad |\varphi_n(t+\delta) - \varphi_n(t)| < \epsilon$$

for $|\delta| < \eta$, for all $t \in \mathbb{R}$ and all $n \ge 1$.

Let $-T = t_1 < t_2 < \cdots < t_m = T$ be a subdivision of I such that $0 < t_k - t_{k-1} < \eta$ for $k = 2, 3, \ldots, m$. Since $\varphi_n \to \varphi$ pointwise, there exists an N such that

$$|\varphi_n(t_k) - \varphi(t_k)| < \epsilon \qquad \text{for all } n > N \quad \text{and all } k.$$

Let $t \in I$. Then there is a k such that $t_{k-1} \le t \le t_k$, and for any $n > N$

$$|\varphi_n(t) - \varphi(t)| \le |\varphi_n(t) - \varphi_n(t_k)| + |\varphi(t) - \varphi(t_k)| + |\varphi_n(t_k) - \varphi(t_k)| < 3\epsilon$$

uniformly in $t \in I$. This completes the proof. ∎

Let $\mathscr{F}^+$ be the subclass of all distribution functions F of $\mathscr{F}$ such that $F(x) = 0$ for $x < 0$. Let $\mathscr{E}_0$ be the subclass of $\mathscr{C}_0$ defined by functions $x \to e^{-tx}$ for $x \ge 0$ and 0 for $x < 0$, $t \ge 0$. Then $\mathscr{E}_0$ is an $\mathscr{F}^+$-separating class. This follows from the uniqueness of Laplace–Stieltjes transforms of distribution functions (see, for example, Widder [88], p. 63).

Proposition 3.9.3. Let $F_n \in \mathscr{F}^+$, $n \ge 1$, and suppose that

$$\lim_{n\to\infty} \int_0^{\infty} e^{-tx}\, dF_n(x)$$

exists and is finite for all $t \geq 0$. Then there exists a unique distribution function F such that $F_n \xrightarrow{c} F$.

Proof. Set

$$\lim_{n\to\infty} \int_0^\infty e^{-tx}\, dF_n(x) = \theta(t).$$

Clearly θ is a continuous, real-valued function on $\mathbb{R}$. Let $T > 0$. Then

$$\frac{1}{T}\int_0^T \theta(t)\, dt = \frac{1}{T}\int_0^T \left[\lim_{n\to\infty} \int_0^\infty e^{-tx}\, dF_n(x)\right] dt,$$

and, using the Lebesgue dominated convergence theorem and Fubini's theorem, we see that

$$\frac{1}{T}\int_0^T \theta(t)\, dt = \lim_{n\to\infty} \int_0^\infty \frac{1 - e^{-Tx}}{Tx}\, dF_n(x)$$

holds for all $T > 0$. Letting $T \downarrow 0$, we see easily that

$$\lim_{t\downarrow 0} \theta(t) = 1. \tag{3.9.16}$$

To complete the proof we need to show that (3.9.16) implies that $\{F_n, n \geq 1\}$ is tight. This can be done exactly as in the proof of the continuity theorem above. ■

As a final application we prove the following criteria for convergence in probability in terms of convergence of characteristic functions. As a corollary we get the important result that for a sequence $\{X_n\}$ of independent random variables convergence in law of $\sum_{n=1}^\infty X_n$ is equivalent to convergence a.s. of $\sum_{n=1}^\infty X_n$.

Proposition 3.9.4. Let $\{X_n\}$ be a sequence of random variables with characteristic functions $\{\varphi_n\}$ and distribution functions $\{F_n\}$. Then $X_n \xrightarrow{L} 0$ if and only if $\varphi_n(t) \to 1$ for t in some neighborhood of the origin.

Proof. Clearly $X_n \xrightarrow{L} 0$ implies $\varphi_n(t) \to 1$ for all t (continuity theorem).

Conversely, let $\varphi_n(t) \to 1$ for $t \in [-\delta, \delta]$, $\delta > 0$. From Lemma 3.9.1 we have

$$\int_{|x|>1/\delta} dF_n(x) \leq C\delta^{-1} \int_0^\delta [1 - \operatorname{Re} \varphi_n(t)]\, dt,$$

where C is a constant. Letting $n \to \infty$, we see that $\{F_n\}$ is tight and hence relatively compact. It follows that every convergent subsequence of $\{F_n\}$

converges completely to a distribution function. If $\{F_{n'}\} \subset \{F_n\}$ is such that $F_{n'} \xrightarrow{c} F$, then $\varphi_{n'} \equiv 1$ in $[-\delta, \delta]$ and F is degenerate at zero. It follows that $F_n(x) \xrightarrow{w} \epsilon(x)$ as $n \to \infty$, where $\epsilon(x)$ is the distribution function of the random variable degenerate at zero, that is, $X_n \xrightarrow{L} 0$. ■

Corollary 1. Let $\{X_n\}$ be a sequence of independent random variables, and let φ_n be the characteristic function of X_n for each $n \geq 1$. Then $S_n = \sum_{k=1}^n X_k$ converges a.s. if and only if $\prod_{k=1}^\infty \varphi_k(t)$ converges to some characteristic function φ in some neighborhood of the origin, and $|\varphi(t)| > 0$ in this neighborhood.

Proof. Clearly S_n converges a.s. implies S_n converges in law, so that

$$\prod_{k=1}^{\infty} \varphi_k(t)$$

converges everywhere to a characteristic function by the continuity theorem.

Conversely, consider the sequence $\{S_n - S_m, n > m \geq 1\}$ with characteristic function $\prod_{k=m}^n \varphi_k$. Since $\prod_{k=1}^n \varphi_k(t) \to \varphi(t) \neq 0$ in some neighborhood of the origin, it follows that $\prod_{k=m}^n \varphi_k \to 1$ in this neighborhood. It follows from Proposition 3.9.4 that $S_n - S_m \xrightarrow{L} 0$ as $m, n \to \infty$, that is, $\{S_n\}$ converges in probability to some random variable.

Finally we show that $S_n \xrightarrow{P} S$ implies $S_n \xrightarrow{\text{a.s.}} S$. There exists a subsequence $\{S_{n_k}\} \subset \{S_n\}$ such that $S_{n_k} \xrightarrow{\text{a.s.}} S$. Moreover, we can choose $n_k < n \leq n_{k+1}$ such that

$$\sum_{k=1}^{\infty} P\left\{|S_{n_{k+1}} - S_{n_k}| \geq \frac{1}{2^k}\right\} < \infty.$$

Set $T_k = \max_{n \geq 1} |S_n - S_{n_k} - \text{med}(S_n - S_{n_{k+1}})|$. By Lévy's inequality (Lemma 2.4.1) we have

$$\sum_{k=1}^{\infty} P\left\{T_k \geq \frac{1}{2^k}\right\} \leq 2 \sum_{k=1}^{\infty} P\left\{|S_{n_{k+1}} - S_{n_k}| \geq \frac{1}{2^k}\right\} < \infty,$$

so that $T_k \xrightarrow{\text{a.s.}} 0$ as $k \to \infty$. Hence

$$\begin{aligned} |S_n - S - \text{med}(S_n - S_{n_{k+1}})| &\leq |S_n - S_{n_k} - \text{med}(S_n - S_{n_{k+1}})| + |S_{n_k} - S| \\ &\leq T_k + |S_{n_k} - S| \xrightarrow{\text{a.s.}} 0 \qquad \text{as } k \to \infty, \end{aligned}$$

that is, $S_n - \text{med}(S_n - S_{n_{k+1}}) \xrightarrow{\text{a.s.}} S$. It follows that $S_n - \text{med}(S_n - S_{n_{k+1}}) \xrightarrow{P} S$. Since, however, $S_n \xrightarrow{P} S$, we have $\text{med}(S_n - S_{n_{k+1}}) \to 0$ and hence $S_n \xrightarrow{\text{a.s.}} S$.

Corollary 2. For a sequence $\{X_n\}$ of independent random variables the series $\sum_{n=1}^\infty X_n$ converges in law if and only if it converges a.s.

3.10 PROBLEMS

SECTION 3.1

1. Construct an example to show that a sequence of uniformly bounded, nondecreasing, right-continuous functions may have some subsequences that converge weakly and some others that converge completely.

2. Let $\{F_n\}$ be a sequence of uniformly bounded, nondecreasing, and right-continuous functions defined on $\mathbb{R}$. Let $F_n \xrightarrow{w} F$, where F is also bounded, nondecreasing, and right-continuous on $\mathbb{R}$. Show that $F_n \xrightarrow{c} F$ if and only if

$$\sup_{n \geq 1}\{F_n(\infty) - F_n(-\infty) - [F_n(a) - F_n(-a)]\} \to 0 \qquad \text{as } a \to \infty$$

or

$$F_n(\infty) - F_n(-\infty) \to F(\infty) - F(-\infty) \qquad \text{as } n \to \infty.$$

3. (Slutsky) Let $\{X_n\}$ be a sequence of random variables, and let X be a random variable such that $X_n \xrightarrow{L} X$. Let F be the distribution function of X. Let $\{Y_n\}$ be a sequence of random variables such that $Y_n \xrightarrow{P} c$, where c is a constant. Show that:

(a) $X_n + Y_n \xrightarrow{L} X + c$.
(b) $X_n Y_n \xrightarrow{L} cX$ if $c \neq 0$, and $X_n Y_n \xrightarrow{P} 0$ if $c = 0$.
(c) $X_n/Y_n \xrightarrow{L} X/c$ if $c \neq 0$.

4. Let $\{F_n\}$ be a sequence of distribution function, and let F be a continuous distribution function such that $F_n \xrightarrow{c} F$. Show that the convergence is uniform in $x \in \mathbb{R}$.

5. Suppose that the random variables X_n have probability density functions f_n and $f_n(x) \to f(x)$ as $n \to \infty$, $x \in \mathbb{R}$. Show that, if f is a probability density function,

$$\int_{-\infty}^{\infty} |f_n(x) - f(x)|\, dx \to 0 \qquad \text{as } n \to \infty.$$

6. Let $X_1, X_2, \ldots$ be independent, identically distributed random variables with common distribution function F. Let F be the uniform distribution on $(0, \theta)$. Set $Y_n = \min(X_1, X_2, \ldots, X_n)$ and $Z_n = nY_n$. Does Z_n converge in law to some random variable Z? Find the distribution function of Z.

7. Let $X_1, X_2, \ldots, X_{2n}$ be independent, identically distributed random variables with common normal distribution having mean 0 and variance 1. Set

$$U_n = \sum_{i=0}^{n-1} \frac{X_{2i+1}}{X_{2i+2}}, \qquad V_n = \sum_{i=1}^{n} X_i^2, \qquad n \geq 1.$$

Find the limiting distribution of U_n/V_n.

8. We say that $\{F_n\}$ converges weakly to F if $F_n(x_1, x_2, \ldots, x_k) \to F(x_1, x_2, \ldots, x_k)$ for all continuity points $(x_1, x_2, \ldots, x_k)$ of F. In this case we write $F_n \xrightarrow{w} F$. If F is a distribution function on $\mathbb{R}_k$, we say that $\{F_n\}$ converges completely to F and write $F_n \xrightarrow{c} F$.

(a) State and prove the corresponding version of Helly's weak compactness theorem.
(b) State and prove the corresponding version of the Helly–Bray and the extended Helly–Bray theorems.

9. Let $\{F_n\}$ be a sequence of distribution functions on $\mathbb{R}$, and let $\{Q_n\}$ be the sequence of corresponding concentration functions. (See Problem 1.5.47.) For every $n \geq 1$ and every $l \geq 0$, let $x_l^{(n)} \in \mathbb{R}$ be such that

$$Q_n(l) = Q(X_n, l) = F_n(x_l^{(n)} + l) - F_n(x_l^{(n)} - 0).$$

Show that, if every weakly convergent subsequence of $\{F_n\}$ converges completely, the sequence $\{x_l^{(n)}\}$ is bounded for every fixed $l > 0$.

10. Let $\{X_n\}$ be a sequence of random variables such that $X_n \xrightarrow{L} X$. Let med(X) be any limit point of the sequence $\{\text{med}(X_n)\}$. Show that med(X) is a median of X. If, in particular, med(X) is the unique median of X, then $\text{med}(X_n) \to \text{med}(X)$ as $n \to \infty$.

SECTION 3.2

11. A purely discrete distribution is known as a lattice distribution if its discontinuity points are of the form $a + kd$, where a,d are constants ($d > 0$) and k is an integer. Show that a characteristic function φ is the characteristic function of a lattice distribution if and only if there exists a $t_0 \in \mathbb{R}$, $t_0 \neq 0$, such that $|\varphi(t_0)| = 1$. If, in particular, $|\varphi(t)| = 1$ for all t, then φ is the characteristic function of a degenerate distribution.

12. Let φ be the characteristic function of a random variable X. Show that

$$\frac{\lim_{t \to 0} [2 - \varphi(t) - \varphi(-t)]}{t^2} = \mathscr{E}X^2$$

whether $\mathscr{E}X^2 < \infty$ or $= \infty$. Deduce that $\mathscr{E}X^2 < \infty$ if and only if φ is twice differentiable.

13. Do there exist independent, identically distributed random variables such that $X - Y$ has a uniform distribution on $[-1, 1]$?

14. Are the following complex-valued functions defined on $\mathbb{R}$ characteristic functions?
(a) $\varphi(t) = \cos^2 t/(1 + 13t^2 - 12it^3)$.
(b) $\varphi(t) = \exp\{\sum_{r=3}^{\infty} [(it)^r/r!]\}$.
(c) $\varphi(t) = 1 + \sin t$.
(d) $\varphi(t) = [1 + \exp(-\alpha|t|)]/2$.

15. (Convergence of Types) Let X and Y be two random variables. We say that X and Y are of the *same positive type* if and only if there exist $a > 0, b \in \mathbb{R}$ such that $a^{-1}(Y - b)$ and X have the same distribution. We say that X and Y are of the *same type* if and only if X and $a^{-1}(Y - b)$ have the same distribution for some $a,b \in \mathbb{R}$, $a \neq 0$. The notion of type is preserved under convergence in law.

(a) Let $\{a_n\}$ and $\{b_n\}$ be sequences of constants, and suppose that $a_n > 0$ for all $n \geq 1$. Let $\{F_n\}$ be a sequence of distribution functions such that $F_n \xrightarrow{c} F$, where F is a nondegenerate distribution function. Show that if $F_n(a_n x + b_n) \to G(x)$, where G is a nondegenerate distribution function, $G(x) = F(ax + b)$, $a_n \to a$ and $b_n \to b$. In particular, if $F_n(a_n x + b_n) \to F(x)$, then $a_n \to 1$ and $b_n \to 0$. Show also that, if $a_n \to a$, $b_n \to b$, then $F_n(a_n x + b_n) \to F(ax + b)$.
(b) Let $\{F_n\}$ be a sequence of distribution functions. Then show that

$$F_n(b_n x + a_n) \xrightarrow{c} F(x)$$

and

$$F_n(\beta_n x + \alpha_n) \xrightarrow{c} F(x),$$

as $n \to \infty$, where $b_n > 0$, $\beta_n > 0$, $a_n \in \mathbb{R}$, $\alpha_n \in \mathbb{R}$, and F is a nondegenerate distribution function, hold simultaneously if and only if

$$\frac{\beta_n}{b_n} \to 1, \frac{a_n - \alpha_n}{b_n} \to 0 \qquad \text{as } n \to \infty.$$

16. Let φ be the characteristic function of a distribution function F defined on $\mathbb{R}_k$. Show that φ is uniformly continuous on $\mathbb{R}_k$.

17. Let X be a random variable with characteristic function φ. Show that, if $\varphi''(0) = 0$, then X must be degenerate. Use this result to show that $\varphi(t) = e^{-|t|^\alpha}$ is not a characteristic function for $\alpha > 2$.

18. Let X be a random variable with characteristic function φ, and let $T > 0$, $A > 0$, and $TA < 1$. Show that

$$P\{|X| \le A\} \ge \frac{|1/2T \int_{-T}^{T} \varphi(t)\, dt| - 1/TA}{1 - 1/TA}.$$

19. If the moment α_k exists and is finite, then

$$\ln \varphi(t) = \sum_{k=1}^{n} \frac{\kappa_k}{k!}(it)^k + o(t^k),$$

where κ_k are known as semi-invariants (or cumulants), and we have

$$\sum_{n=1}^{\infty} \frac{\kappa_n}{n!} z^n = \ln \sum_{n=0}^{\infty} \frac{\alpha_n}{n!} z^n, \qquad z \in \mathbb{C}.$$

Deduce the expressions for $\kappa_1, \kappa_2, \kappa_3$, and κ_4 in terms of moments, and conversely. Show that

$$|\kappa_k| \le k^k \beta_k.$$

(*Hint*: The series $\ln[1 + \sum_{k=1}^{n} (\alpha_k/k!)z^k]$ is majorized by the series

$$\sum_{k=1}^{\infty} k^{-1}(e^{\beta_k^{1/k} z} - 1)^k.)$$

SECTION 3.3

20. Construct an example to show that, if φ_1 and φ_2 are two characteristic functions such that $\varphi_1(t) = \varphi_2(t)$ for all t in some subinterval $[-a, +a]$ for $a > 0$ on $\mathbb{R}$, the corresponding distribution functions F_1 and F_2 need not be identical on $\mathbb{R}$.

21. Find the probability density function corresponding to the following characteristic functions:

(a) $\varphi(t) = (\cosh t)^{-1}$.
(b) $\varphi(t) = \cos at$.
(c) $\varphi(t) = t(\sinh t)^{-1}$.
(d) $\varphi(t) = 1 - |t|$ if $|t| \le 1$, and $=0$ if $|t| > 1$.
(e) $\varphi(t) = (\cosh t)^{-2}$.

22. Show that the characteristic function

$$\varphi(t) = \exp\left\{-c|t|^{1/2}\left[1 + i\beta\frac{t}{|t|}\right]\right\}, \qquad c > 0, \quad |\beta| \le 1,$$

corresponds to an absolutely continuous distribution function.

23. Let H_n be the Hermite polynomial of degree n, defined by the formula

$$H_n(x) = e^{x^2/2}\frac{d^n}{dx^n}e^{-x^2/2}.$$

Show that

$$|H_n(x)| \le 2^{n/2}\pi^{-1/2}\Gamma\left(\frac{n+1}{2}\right)e^{x^2/2},$$

and hence that the function $(-1)^n e^{-x^2/2}H_{2n}(x)$ is the characteristic function of an absolutely continuous distribution function. Determine the corresponding probability density function.

24. Let X be a random variable with characteristic function φ, given by $\varphi(t) = \max\{0, (1 - |t|)\}, t \in \mathbb{R}$. Show that X has a continuous and bounded probability density function. Find this probability density function.

25. Let φ be a bounded, measurable function which is Lebesgue integrable in $[-T, T]$, $T > 0$. Show that, if

$$f(x) = \int_{-T}^{T} e^{itx}\varphi(t)\,dt \ge 0, \qquad x \in \mathbb{R},$$

then f is integrable over $\mathbb{R}$.

26. Let F be a distribution function with characteristic function φ. Show that for every $x \in \mathbb{R}$ the limit

$$\lim_{T\to\infty}(2T)^{-1}\int_{-T}^{T} e^{-itx}\varphi(t)\,dt$$

exists and equals $F(x) - F(x - 0)$.

27. Let φ be an absolutely integrable characteristic function on $\mathbb{R}$. Show that

$$\lim_{t\to\pm\infty}\varphi(t) = 0.$$

28. Let φ be the characteristic function of the distribution function F. If $\varphi \ge 0$, show that φ is absolutely integrable over $\mathbb{R}$ if and only if F has a bounded probability density function.

29. Let φ_n $(n \ge 1)$ and φ be absolutely integrable characteristic functions such that $\int_{-\infty}^{\infty}|\varphi_n(t) - \varphi(t)|\,dt \to 0$ as $n \to \infty$. Show that, if f_n and f are the corresponding probability density functions, $f_n \to f$ as $n \to \infty$.

30. Let F be a distribution function with probability density function f and characteristic function φ. Then show that $|\varphi|^2$ is integrable on $\mathbb{R}$ if and only if f^2 is, and in this case

$$\int_{-\infty}^{\infty} f^2(x)\,dx = (2\pi)^{-1}\int_{-\infty}^{\infty} |\varphi(t)|^2\,dt.$$

SECTION 3.4

31. Let $F = F_1 * F_2$, where F_1, F_2 are distribution functions. If at least one of F_1 and F_2 is continuous (absolutely continuous), show that F is also continuous (absolutely continuous). In particular, if φ_1 is the characteristic function of a continuous (absolutely continuous) distribution function, show that $|\varphi|^2$ is also the characteristic function of a continuous (absolutely continuous) distribution function.

32. If F_1 and F_2 are discrete distribution functions, show that so also is $F = F_1 * F_2$, and conversely.

33. Let F be a distribution function, and for $t \neq 0$ define

$$\Phi(x) = t^{-1}\int_{x}^{x+t} F(u)\,du, \qquad \Psi(x) = (2t)^{-1}\int_{x-t}^{x+t} F(u)\,du.$$

Show that Φ and Ψ are also distribution functions.

34. Let $K(t, x)$ be a complex-valued function defined on $\mathbb{R}_2$ which is bounded and continuous in x. Define g_F on $\mathbb{R}$ by

$$g_F(t) = \int K(t, x)\,dF(x)$$

for every distribution function F on $\mathbb{R}$. Suppose that the uniqueness and composition properties hold, that is,

$$g_F \equiv g_G \quad \text{if and only if } F \equiv G, \qquad \text{and} \qquad F = F_1 * F_2 \Rightarrow g_F = g_{F_1} g_{F_2}.$$

Show that $K(t, x) = e^{ixh(t)}$, where h is a real-valued function. Moreover, if φ is the characteristic function corresponding to F, show that $\varphi(h(t)) = g_F(t)$, $t \in \mathbb{R}$.

SECTION 3.5

35. Let $\{F_n\}$ be a sequence of joint distribution functions defined on $\mathbb{R}_k$. Let φ_n be the characteristic function of F_n, $n \geq 1$. Show that a necessary and sufficient condition for the sequence $\{F_n\}$ to converge to a distribution function F on $\mathbb{R}_k$ is that, for every $(t_1, t_2, \ldots, t_k) \in \mathbb{R}_k$, the sequence $\{\varphi_n(t_1, t_2, \ldots, t_k)\}$ converges to a limit $\varphi(t_1, t_2, \ldots, t_k)$, which is continuous at the point $(0, 0, \ldots, 0)$. In this case the limit φ is the characteristic function of F.

36. Let $\psi\colon \mathbb{R}_k \to \mathbb{R}_m$ be a continuous mapping, and let $\mathbf{X}, \mathbf{X}_1, \mathbf{X}_2, \ldots$ be a sequence of k-dimensional random variables. Let F_n be the distribution function of $\mathbf{X}_n$, $n \geq 1$, and F that of $\mathbf{X}$. Suppose that $F_n \xrightarrow{c} F$. Show that $\psi(\mathbf{X}_n) \xrightarrow{L} \psi(\mathbf{X})$, that is, show that the sequence of distribution functions of $\psi(\mathbf{X}_n)$ converges completely to the distribution function of $\psi(\mathbf{X})$.

37. Let $X_n \xrightarrow{P} X$ as $n \to \infty$. Show that either X is degenerate or X_n and X are dependent (for all but a finite number of values of n).

38. (Bochner's Theorem on $\mathbb{R}_n$) A complex-valued function φ defined on $\mathbb{R}_n$ is said to be positive definite if, for every positive integer N, for every $\mathbf{t}_1, \mathbf{t}_2, \ldots, \mathbf{t}_N \in \mathbb{R}_n$, and for any complex number $\omega_1, \omega_2, \ldots, \omega_N$, the inequality

$$\sum_{j=1}^{N} \sum_{k=1}^{N} \omega_j \bar{\omega}_k \varphi(\mathbf{t}_j - \mathbf{t}_k) \geq 0$$

holds. Let φ be a complex-valued function on $\mathbb{R}_n$. Show that φ is a continuous, normalized, positive definite function on $\mathbb{R}_n$ if and only if φ is a characteristic function.

SECTION 3.6

39. Let φ be a characteristic function. Let φ_1 be a complex-valued function on $\mathbb{R}$ which is periodic with period $2a$ ($a > 0$) and coincides with φ on $[-a, a]$. Prove that φ_1 is the characteristic function of a lattice distribution.

40. Let P be a probability measure on $(\mathbb{R}, \mathscr{B})$, and let $\varphi = \varphi_P$ be the Fourier transform of P. Show that the mapping $P \to \varphi_P$ is an isomorphism of the convolution semigroup of all probability measures P on $\mathscr{B}$ onto the multiplicative semigroup of all continuous, positive definite functions φ on $\mathbb{R}$ such that $\varphi(0) = 1$.

41. Let $\{\zeta_t, t \in \mathbb{R}\}$ be a family of complex-valued random variables on $(\Omega, \mathscr{S}, P)$ such that:

(a) $\int_\Omega \zeta_t(\omega)\, dP(\omega) = 0$ for all $t \in \mathbb{R}$.
(b) $\int_\Omega \zeta_t(\omega)\overline{\zeta_s(\omega)}\, dP(\omega) = \theta(t - s)$, $t, s \in \mathbb{R}$.

Suppose that $\theta(0) = 1$ and, moreover, that θ is continuous on $\mathbb{R}$. Show that there exists a unique probability measure P on $\mathscr{B}$ such that θ is the Fourier transform of P.

42. Show that a continuous function f on $\mathbb{R}$ is positive definite if and only if, for any measurable bounded function ξ which vanishes outside some finite interval,

$$\int_{-\infty}^{\infty} \int_{-\infty}^{\infty} f(t - s)\xi(t)\overline{\xi(s)}\, dt\, ds \geq 0.$$

43. Show that the function

$$\varphi(t) = \begin{cases} 1 - \dfrac{t}{2e}, & 0 < t \leq e, \\[2ex] \dfrac{1}{2 \ln t}, & t > e, \end{cases}$$

with $\varphi(t) = \varphi(-t)$, is the characteristic function of an absolutely continuous distribution function.

44. Let φ be a real-valued, continuous function on $\mathbb{R}$ such that $\varphi(0) = 1$, $\varphi(-t) = \varphi(t)$, φ is convex in $(0, r]$, φ is periodic with period $2r$, and $\varphi(r) = 0$. Then show that φ is the characteristic function of a lattice distribution.

SECTION 3.7

45. Let φ be a characteristic function. Show that $\exp\{(\varphi(t) - 1)\}$ is also a characteristic function for any $\alpha > 0$.

46. (a) Let X and Y be independent random variables with the same distribution having a finite variance. Suppose that $X + Y$ and $X - Y$ are independent. Show that both X and Y must be normal. (The condition of finite variance is not necessary and is made for convenience.)

(b) Prove the result in (a) if the condition of finite variance and identical distribution is replaced by the condition that X and Y are symmetric and independent.

47. Let $X_1, X_2, \ldots, X_n$ be independent and identically distributed with finite variance. Suppose that $\sum_{i=1}^{n} X_i/n$ and $\sum_1^n (X_i - \bar{X})^2$ are independent. Then show that $X_1, X_2, \ldots, X_n$ have a common normal distribution. (Again, the condition of finite variance is not necessary and may be dropped.)

(a) Let X_1, X_2 be independent and identically distributed with $\mathscr{E}X_1 = 0$ and $\mathscr{E}X_1^2 < \infty$. Suppose that $(X_1 + X_2)/\sqrt{2}$ has the same distribution as X_1. Show that X_1 and X_2 have a normal distribution.

(b) Let F be a distribution function with mean 0 and variance 1. If the family of distributions

$$\left\{F\left(\frac{x-\mu}{\sigma}\right), \sigma > 0, \mu \in \mathbb{R}\right\}$$

is closed under the operation of convolution, show that F must be the standard normal distribution function.

48. Let $(X_1, X_2, \ldots, X_{k-1})$ have a multinomial distribution with parameters n, p_1, $p_2, \ldots, p_{k-1}$. Set $V = \sum_{j=1}^{k} [(X_j - np_j)^2/np_j]$, where $p_k = 1 - p_1 - \cdots - p_{k-1}$, and $X_k = n - X_1 - X_2 - \cdots - X_{k-1}$. Find $\mathscr{E}V$ and $\text{var}(V)$.

49. Let $X_1, X_2, \ldots$ be independent random variables with common distribution $P\{X_n = \pm 1\} = \frac{1}{2}$, $n \geq 1$. Let $Z_n = \sum_{j=1}^{n} X_j/2^j$. Show that $Z_n \xrightarrow{L} V$, where V is the uniform distribution on $[-1, 1]$.

50. Let $X_1, X_2, \ldots$ be a sequence of independent, identically distributed random variables with common absolutely continuous distribution function F. Let $M_n = \max(X_1, X_2, \ldots, X_n)$, and set $Y_n = n[1 - F(M_n)]$, $n \geq 1$. Find the limiting distribution of Y_n.

51. Let $\{X_n\}$ be a sequence of independent random variables, and let $\{F_n\}$ be the corresponding sequence of distribution functions. Suppose that, as $n \to \infty$,

$$\sum_{k=1}^{n} \int_{|x|>n} dF_k(x) \to 0 \qquad \text{and} \qquad n^{-2} \sum_{k=1}^{n} \int_{|x| \leq n} x^2\, dF_k(x) \to 0.$$

Show that $\sum_{k=1}^{n} (X_k - \mu_{k,n})/n$ converges in probability to zero where

$$\mu_{k,n} = \int_{|x| \leq n} x\, dF_k(x), \qquad 1 \leq k \leq n, \quad n \geq 1.$$

52. In Remark 2.1.2 we saw that a sufficient condition for a sequence $\{X_n\}$ of random variables to obey the weak law of large numbers is that $n^{-2}\operatorname{var}(\sum_{k=1}^{n} X_k) \to 0$ as $n \to \infty$. This condition is not necessary. When the X_n are independent, we can use the methods of characteristic functions (continuity theorem) to see whether the sequence $\{X_n\}$ obeys the weak law. For the following sequences of independent random variables does the weak law of large numbers hold?

(a) $P\{X_k = \pm 2^k\} = \frac{1}{2}$.
(b) $P\{X_k = \pm k\} = 1/(2\sqrt{k})$, $P\{X_k = 0\} = 1 - 1/\sqrt{k}$.
(c) $P\{X_k = \pm\sqrt{k}\} = \frac{1}{2}$.
(d) X_n has probability density function

$$f_n(x) = (\sigma_n\sqrt{2})^{-1}\exp(-|x|\sqrt{2}/\sigma_n), \quad x \in \mathbb{R},$$

where $\sigma_n^2 = n^{2+\delta}$, $\delta \geq 0$.

SECTION 3.8

53. Let $\{X_n\}$ be a sequence of independent, identically distributed random variables with $\mathscr{E}X_n = 0$, $\mathscr{E}X_n^2 = 1$. Show, by the methods of this section, that $\sum_{k=1}^{n} X_k/\sqrt{n}$ converges in law to the standard normal random variable.

54. Let $\{\mathbf{X}_n\}$ be a sequence of independent, k-dimensional random variables. Suppose that $\mathscr{E}\mathbf{X}_n = \mathbf{0}$, and the variance-covariance matrix is given by

$$\mathbf{\Sigma} = \begin{pmatrix} \sigma_1^2 & \rho\sigma_1\sigma_2 & \cdots & \rho\sigma_1\sigma_n \\ \rho\sigma_1\sigma_2 & \sigma_2^2 & \cdots & \rho\sigma_2\sigma_n \\ \cdot & \cdot & \cdots & \cdot \\ \cdot & \cdot & \cdots & \cdot \\ \cdot & \cdot & \cdots & \cdot \\ \rho\sigma_n\sigma_1 & \rho\sigma_n\sigma_2 & \cdots & \sigma_n^2 \end{pmatrix}.$$

Show, by the methods of this section, that $\sum_{k=1}^{n} \mathbf{X}_k/\sqrt{n}$ converges in law to a k-variate normal random vector with mean vector zero and variance-covariance matrix $\mathbf{\Sigma}$.

55. Let $\{X_n\}$ be a sequence of independent random variables with means $\mathscr{E}X_n = 0$ and variances σ_n^2, $n \geq 1$. Let F_n be the distribution function of X_n, $n \geq 1$, and set $s_n^2 = \sum_{k=1}^{n} \sigma_k^2$. If for $\epsilon > 0$

$$s_n^{-2} \sum_{k=1}^{n} \int_{|y| > \epsilon s_n} x^2\, dF_k(x) \to 0 \qquad \text{as } n \to \infty,$$

show that $\sum_{k=1}^{n} X_k/s_n$ converges in law to the standard normal random variable.

56. Let $\{N_n\}$ be a sequence of positive integer-valued random variables such that $n^{-1}N_n \xrightarrow{P} 1$ as $n \to \infty$. Let $\{X_n\}$ be a sequence of independent, indentically distributed random variables with $\mathscr{E}X_n = 0$, $\mathscr{E}X_n^2 = 1$ and such that $\{X_k\}$ and $\{N_k\}$ are independent. Show that $\sum_{j=1}^{N_n} X_j/\sqrt{n}$ converges in law to the standard normal random variable.

57. Consider the family of one-sided stable distributions with probability density function

$$f_t(x) = t(2\pi x^3)^{-1/2} e^{-t^2/2x}, \qquad x > 0, \quad t > 0.$$

Show that this family forms a continuous convolution semigroup with generator A given by

$$(Ag)(x) = \frac{1}{\sqrt{2\pi}} \int_0^\infty \frac{g(x-y) - g(x)}{y^{3/2}}\, dy.$$

58. Consider the family of probability densities defined by

$$f_{\alpha,\beta}(x) = [\Gamma(\alpha)]^{-1}\beta^\alpha x^{\alpha-1}e^{-\beta x}, \qquad x > 0, \quad \alpha > 0, \quad \beta > 0.$$

Show that the family of probability densities $\{f_{\alpha,\beta}(x)\}$ enjoys the convolution property

$$f_{\mu,\beta} * f_{\nu,\beta} = f_{\mu+\nu,\beta}, \qquad \mu > 0, \quad \nu > 0,$$

which implies that the corresponding family of distributions forms a continuous convolution semigroup with infinitesimal generator A given by

$$(Ag)(x) = \int_0^\infty y^{-1}[g(x-y) - g(x)]e^{-y}\, dy.$$

SECTION 3.9

59. Let $\{\lambda_n\}$, $\{\mu_n\}$, and $\{\nu_n\}$ be three sequences of probability measures on $(\mathbb{R}, \mathscr{B})$ such that $\lambda_n = \mu_n * \nu_n$ for each $n \geq 1$. If $\{\lambda_n\}$ and $\{\mu_n\}$ are relatively compact, show that so also is $\{\nu_n\}$.

60. Let $\{P_n\}$ be a sequence of probability measures on $\mathbb{R}_k$ such that $\{P_n^{*k}\}$ is relatively compact. (Here $P_n^{*k} = \underbrace{P_n * P_n * \cdots * P_n}_{\kappa\text{-fold}}$.) Show that $\{P_n\}$ is relatively compact.

61. Let $\{P_n\}$ be a relatively compact sequence of probability measures on $(\mathbb{R}, \mathscr{B})$. Let $\{Q_n\}$ be the corresponding sequence of concentration functions.

$$\left[Q_n(l) = \sup_{x \in \mathbb{R}} P_n x, x + l, l \geq 0.\right]$$

Show that $\lim_{l\to\infty} \inf_n Q_n(l) = 1$.

62. Let $(\mathfrak{X}, \rho)$ be a metric space, and let $\mathscr{B}$ be the σ-field of Borel sets in $\mathfrak{X}$. Let $\{\mu_n\}$ be a sequence of finite measures on $\mathscr{B}$, and let μ be a finite measure on $\mathscr{B}$. Show that $\mu_n \Rightarrow \mu$ if and only if $\lim_{n\to\infty} \int_{\mathfrak{X}} f\, d\mu_n = \int_{\mathfrak{X}} f\, d\mu$ for all bounded, real-valued, uniformly continuous functions on $\mathfrak{X}$. Also show that in Corollary 1 to Theorem 3.9.1 it is sufficient to require that $\int_{\mathfrak{X}} g\, d\mu = \int_{\mathfrak{X}} g\, d\nu$ for all bounded, real-valued, uniformly continuous functions on $\mathfrak{X}$.

63. Let $(\Omega, \mathscr{S}, P)$ be a probability space, and let $\{X_n\}$ be a sequence of independent, identically distributed $\mathfrak{X}$-valued random variables defined on Ω. Here $\mathfrak{X}$ is a separable metric space. Let $\mu = PX_n^{-1}$ be the common induced measure. For each $\omega \in \Omega$ let μ_n^ω denote the measure which assigns mass $1/n$ to each of the n points $X_1(\omega), \ldots, X_n(\omega)$. Then μ_n^ω is known as the sample distribution of μ based on $X_1, X_2, \ldots, X_n$ at ω. Show that $P\{\omega: \mu_n^\omega \Rightarrow \mu\} = 1$. (This result generalizes the Glivenko-Cantelli lemma of Section 2.5.)

64. Let $\{\mathbf{X}_n\}$ be a sequence of k-dimensional random vectors, and let Q_n be the probability distribution of $\mathbf{X}_n$, $n \geq 1$. Let $\{\mathbf{Y}_n\}$ be another sequence of l-dimensional random vectors, and suppose that P_n is the probability distribution of $\mathbf{Y}_n$, $n \geq 1$. Suppose that $Q_n \Rightarrow Q$, where Q is the probability distribution of some k-dimensional random vector, and $P_n \Rightarrow P_{\mathbf{c}}$, where $\mathbf{c}$ is a vector of constants. (This means that $\mathbf{Y}_n \xrightarrow{P} \mathbf{c}$.) Show that $Q_{\mathbf{X}_n, \mathbf{Y}_n} \Rightarrow Q_{\mathbf{X}, \mathbf{c}}$, where $Q_{\mathbf{X}_n, \mathbf{Y}_n}$ is the joint distribution of $(\mathbf{X}_n, \mathbf{Y}_n)$, and $Q_{\mathbf{X}, \mathbf{c}}$ that of $(\mathbf{X}, \mathbf{c})$. (See Problem 3 above.)

NOTES AND COMMENTS

For a more comprehensive treatment of the theory of characteristic functions we refer to Lukacs [60], and for applications of characteristic functions we cite Lukacs and Laha [59]. We also refer the reader to the more recent monographs of Kawata [44] and Cuppens [14]. The analytic theory of semigroups of operators on a Banach space is treated in Hille and Phillips [36] and also in Dynkin [20]. The theory of convolution semigroups of probability operators was first presented by Feller [25]. In this chapter we have closely followed his development. It should be noted that Trotter [82] first used the idea of probability operators to give a proof of the Lindeberg central limit theorem.

Probability measures on metric spaces were first studied systematically by Prokhorov [67]. This work plays a fundamental role in the study of limit theorems on abstract spaces (in particular, linear topological spaces and locally compact topological groups). For a detailed treatment of some aspects of this subject we refer to the monographs of Parthasarathy [64] and Billingsley [8]. The proof of Prokhorov's theorem (Theorem 3.9.2) presented here is from Billingsley [9].

CHAPTER 4

Some Further Results on Characteristic Functions

In this chapter we consider some further aspects of the theory of characteristic functions. In particular, we study a special class of characteristic functions known as infinitely divisible characteristic functions (Section 4.1). This class plays an important role in the study of limit theorems, as well as in the study of the decomposition of probability distributions. In Section 4.2 we discuss some important properties of analytic characteristic functions, while in Section 4.3 we consider the decomposition of characteristic functions. In particular, we consider the decomposition of the normal and Poisson distributions.

4.1 INFINITELY DIVISIBLE DISTRIBUTIONS

4.1.1 Basic Definitions

We note that the product of two characteristic functions is a characteristic function. It is therefore natural to consider the problem of decomposing a characteristic function as a product of two or more characteristic functions. Clearly every characteristic function φ can be decomposed as $\varphi = \varphi_1\varphi_2$, where $\varphi_1(t) = e^{iat}$, $a \in \mathbb{R}$, and $\varphi_2(t) = \varphi(t) \cdot e^{-iat}$. To avoid such trivial decompositions we introduce the concept of a decomposable characteristic function.

Definition 4.1.1. A characteristic function φ is said to be decomposable if it can be written as $\varphi = \varphi_1\varphi_2$, where φ_1, φ_2 are characteristic functions of nondegenerate distributions. In this case φ_1, φ_2 are called the factors of φ. A characteristic function φ which admits only trivial decompositions is called indecomposable.

Example 4.1.1. Let φ be the characteristic function of a purely discrete distribution function with only two discontinuity points. Then φ is indecomposable.

Remark 4.1.1. The theory of decomposition of characteristic functions is often known as the *arithmetic of distribution functions*, since the decomposition of a characteristic function into indecomposable factors is analogous to the decomposition of positive integers into prime factors. In this connection we note that the decomposition of a characteristic function into indecomposable factors is not unique.

Example 4.1.2. Let $\varphi(t) = \frac{1}{6}\sum_{j=0}^{5} e^{itj}$, and let

$$\varphi_1(t) = \tfrac{1}{3}\sum_{k=0}^{2} e^{2itk}, \qquad \varphi_2(t) = \tfrac{1}{2}(1 + e^{it}),$$

$$\varphi_3(t) = \tfrac{1}{3}\sum_{k=0}^{2} e^{itk}, \qquad \varphi_4(t) = \tfrac{1}{2}(1 + e^{3it}).$$

Then clearly $\varphi_1, \varphi_2, \varphi_3, \varphi_4$ are all distinct indecomposable characteristic functions, and it is easily verified that $\varphi = \varphi_1\varphi_2 = \varphi_3\varphi_4$.

We now study an important class of distribution functions which plays a fundamental role in the study of the arithmetic of distribution functions as well as in the study of limit theorems.

Definition 4.1.2. A distribution function F is infinitely divisible (i.d.) if for every positive integer n there exists a distribution function F_n such that

$$F = \underbrace{F_n * F_n * \cdots * F_n}_{n\text{ times}}.$$

Equivalently, a characteristic function φ is said to be i.d. if for every positive integer n there exists a characteristic function φ_n such that $\varphi = [\varphi_n]^n$.

Example 4.1.3.

(a) Let φ be the characteristic function of a normal random variable with mean μ and variance σ^2. Then

$$\begin{aligned}\varphi(t) &= \exp\left(it\mu - \frac{\sigma^2 t^2}{2}\right)\\ &= \left[\exp\left(it\frac{\mu}{n} - \frac{\sigma^2}{n}\cdot\frac{t^2}{2}\right)\right]^n,\end{aligned}$$

and we see that φ is i.d.

(b) The Poisson distribution is i.d. For

$$\begin{aligned}\varphi(t) &= \exp\{\lambda(e^{it} - 1)\}\\ &= \left[\exp\left\{\frac{\lambda}{n}(e^{it} - 1)\right\}\right]^n.\end{aligned}$$

(c) Let φ be the characteristic function of the gamma distribution. Then

$$\begin{aligned}\varphi(t) &= \left(1 - \frac{it}{\beta}\right)^{-\alpha}\\ &= \left[\left(1 - \frac{it}{\beta}\right)^{-\alpha/n}\right]^n,\end{aligned}$$

and φ is i.d.

4.1.2 Some Elementary Properties

We next study some elementary properties of i.d. characteristic functions.

Proposition 4.1.1. Let φ be an i.d. characteristic function. Then φ has no real zeros.

Proof. We first note that a real-valued characteristic function θ satisfies the following inequality:

$$1 - \theta(2t) \leq 4[1 - \theta(t)]. \tag{4.1.1}$$

In fact, let G be the distribution function corresponding to θ. Then, since θ is real-valued, we have

$$\theta(t) = \int_{-\infty}^{\infty} \cos tx \, dG(x),$$

so that

$$\begin{aligned}1 - \theta(2t) &= \int_{-\infty}^{\infty} [1 - \cos 2tx] \, dG(x)\\ &= 2\int_{-\infty}^{\infty} (1 - \cos tx)(1 + \cos tx) \, dG(x)\\ &\leq 4\int_{-\infty}^{\infty} (1 - \cos tx) \, dG(x)\\ &= 4[1 - \theta(t)],\end{aligned}$$

as asserted.

Next let θ be an arbitrary characteristic function. Then we show that the inequality

$$1 - |\theta(2t)| \leq 8[1 - |\theta(t)|] \tag{4.1.2}$$

holds. For this purpose we apply (4.1.1) to the real-valued characteristic function $|\theta(t)|^2 = \theta(t)\theta(-t)$, and we obtain

$$1 - |\theta(2t)|^2 \leq 4[1 - |\theta(t)|^2].$$

Since $0 \leq |\theta(t)| \leq 1$, we conclude that

$$\begin{aligned} 1 - |\theta(2t)| \leq 1 - |\theta(2t)|^2 &\leq 4[1 - |\theta(t)|][1 + |\theta(t)|] \\ &\leq 8[1 - |\theta(t)|]. \end{aligned}$$

This proves (4.1.2).

Finally we return to the proof of the proposition. Since φ is continuous and $\varphi(0) = 1$, there exists a $\delta > 0$ such that $\varphi(t) \neq 0$ for $|t| < \delta$. On the other hand, since φ is i.d. for every positive integer n, there exists a φ_n such that $\varphi = [\varphi_n]^n$. Hence, for $|t| < \delta$, $\varphi_n(t) \neq 0$.

Moreover, the relation

$$\begin{aligned} 1 - |\varphi_n(t)| &= 1 - \exp\left\{\frac{1}{n} \ln |\varphi(t)|\right\} \\ &= -\frac{1}{n} \ln |\varphi(t)| + O\left(\frac{1}{n^2}\right) \end{aligned} \tag{4.1.3}$$

holds for every $n \geq 1$ and for $|t| < \delta$. Let $0 < \epsilon < 1$. Then it follows from (4.1.3) that for sufficiently large n and for $|t| < \delta$

$$1 - |\varphi_n(t)| < \frac{\epsilon}{8}.$$

Using (4.1.2), we see that

$$1 - |\varphi_n(2t)| < \epsilon,$$

so that for sufficiently large n and $|t| < \delta$

$$\varphi_n(2t) \neq 0.$$

This implies that $\varphi_n(t) \neq 0$ and hence $\varphi(t) \neq 0$ for $|t| < 2\delta$. Proceeding in this manner, we can show that $\varphi(t) \neq 0$ for $|t| < 2^N\delta$, where N is an arbitrary positive integer. This completes the proof. ■

Example 4.1.4. Let X have the uniform distribution on $[-1, 1]$. Then $\varphi(t) = (\sin t)/t$, $t \in \mathbb{R}$. Since φ vanishes for some t, it cannot be i.d.

The converse of Proposition 4.1.1 does not hold.

Remark 4.1.2. We recall from the elementary theory of complex variables that, if θ is a complex-valued, continuous function defined on $\mathbb{R}$ such that $\theta(t) \neq 0$ for all $t \in \mathbb{R}$ and $\theta(0) = 1$, we can write

$$\theta(t) = |\theta(t)| e^{i\omega(t)}, \qquad t \in \mathbb{R},$$

where ω is a real-valued, continuous function on $\mathbb{R}$ determined uniquely by θ and satisfying $\omega(0) = 0$. In this case we define $\ln \theta(t)$ by the relation

$$\ln \theta(t) = \ln |\theta(t)| + i\omega(t), \qquad t \in \mathbb{R}. \tag{4.1.4}$$

In particular, let φ be an i.d. characteristic function. Then φ is a complex-valued, continuous function on $\mathbb{R}$ that has no real zeros (Proposition 4.1.1). We note that $\varphi(0) = 1$. It follows therefore that $\ln \varphi$, defined uniquely by (4.1.4), exists and is finite. Moreover, for every $n \geq 1$ the relation

$$\varphi_n = [\varphi]^{1/n} = \exp\left(\frac{1}{n} \ln \varphi\right)$$

holds, where φ_n is a characteristic function.

Proposition 4.1.2. The convolution of two i.d. distribution functions is i.d.

Proof. Let F and G be two i.d. distribution functions, and let $H = F * G$. Let φ, θ, and ψ be the characteristic functions corresponding to F, G, and H, respectively. Then $\psi = \varphi\theta$. It follows from the definition of F and G that for every $n \geq 1$ there exist characteristic functions φ_n and θ_n such that

$$\varphi = [\varphi_n]^n \qquad \text{and} \qquad \theta = [\theta_n]^n \quad \text{so that } \psi = [\varphi_n \theta_n]^n = [\psi_n]^n,$$

where $\psi_n = \varphi_n \theta_n$ is a characteristic function. It follows that H is i.d. ∎

Corollary 1. The product of a finite number of i.d. characteristic functions is i.d.

The converse of Corollary 1 is not true.

Corollary 2. If φ is i.d., then so is $|\varphi|$.

Proof. For the proof of Corollary 2 first note that φ i.d. $\Rightarrow \varphi(-t) = \overline{\varphi(t)}$ is i.d. Thus $|\varphi(t)|^2$ is i.d., and hence so is $|\varphi(t)|$.

Note that the result in Corollary 2 cannot be improved, since if φ is a characteristic function $|\varphi|$ may not be a characteristic function.

Example 4.1.5. Let φ be the characteristic function of the Laplace distribution, that is, let $\varphi(t) = 1/(1 + t^2)$. Then

$$\begin{aligned}\varphi(t) &= \left(\frac{1}{1+it}\right)\left(\frac{1}{1-it}\right)\\ &= \left[\frac{1}{(1+it)^{1/n}}\frac{1}{(1-it)^{1/n}}\right]^n,\end{aligned}$$

and it follows from Example 4.1.3(c) and Corollary 1 that φ is i.d.

Proposition 4.1.3. The weak limit of a sequence of i.d. distribution functions (if it exists) is i.d.

Proof. Let $\{F^{(k)}\}$ be a sequence of i.d. distribution functions which converges weakly (= completely) to a distribution function F. Let $\varphi^{(k)}$ and φ be the characteristic functions corresponding to $F^{(k)}$ and F, respectively. By the continuity theorem we have

$$\lim_{k\to\infty} \varphi^{(k)}(t) = \varphi(t).$$

Since $\varphi^{(k)}$ is i.d., for every $n \geq 1$ we can write

$$\varphi^{(k)}(t) = [\varphi_n^{(k)}(t)]^n,$$

where $\varphi_n^{(k)}$ is a characteristic function. First we show that φ has no real zeros. Suppose this is not the case. Then there exists a $t_0 \in \mathbb{R}$ such that $\varphi(t_0) = 0$. Let $T > 0$ be such that $t_0 \in [-T, T]$. Let $\epsilon > 0$ (to be chosen later). Since the convergence of $\{\varphi^{(k)}\}$ to φ is uniform in $[-T, T]$, for given $\epsilon > 0$, there exists a $k_0 = k_0(\epsilon) \geq 1$ (independent of $t \in [-T, T]$) such that, for all $k \geq k_0$, $|\varphi^{(k)}(t) - \varphi(t)| < \epsilon$ for all $t \in [-T, T]$. In particular, $|\varphi^{(k_0)}(t_0) - \varphi(t_0)| < \epsilon$, that is, $|\varphi^{(k_0)}(t_0)| < \epsilon$. Note that $\varphi^{(k_0)}$ is i.d., so that it has no real zeros. Since $\varphi^{(k_0)}$ is continuous on $\mathbb{R}$, $|\varphi^{(k_0)}|$ is also continuous on $\mathbb{R}$, and hence there exists a t', say, $t' \in [-T, T]$, such that

$$|\varphi^{(k_0)}(t')| = \min_{t\in[-T,T]} |\varphi^{(k_0)}(t)| > 0.$$

Choosing $\epsilon = |\varphi^{(k_0)}(t')|$, we obtain the desired contradiction. This proves that φ has no real zeros. Hence for every $n \geq 1$ we have

$$\begin{aligned}\lim_{k\to\infty} \varphi_n^{(k)}(t) &= \lim_{k\to\infty} [\varphi^{(k)}(t)]^{1/n}\\ &= [\varphi(t)]^{1/n} = \varphi_n(t), \text{ say.}\end{aligned}$$

Here φ_n is continuous at $t = 0$, so it follows from the continuity theorem that φ_n is a characteristic function. Clearly

$$\varphi(t) = [\varphi_n(t)]^n$$

for every $n \geq 1$, and hence φ is i.d. ■

We next give a characterization of i.d. distributions which will be useful in our subsequent investigation. For this purpose we need the following definition.

Definition 4.1.3. A distribution function F is said to be of the Poisson type if its characteristic function φ is of the form

$$\varphi(t) = \exp\{i\alpha t + \lambda(e^{i\beta t} - 1)\}, \tag{4.1.5}$$

where $\lambda \geq 0$ and $\alpha, \beta \in \mathbb{R}$.

Clearly a characteristic function of type (4.1.5) is i.d.

Proposition 4.1.4. A distribution function F is i.d. if and only if it is the weak limit of a sequence of finite convolutions of Poisson-type distributions.

Proof. Let F be the weak limit of a sequence of finite convolutions of Poisson-type distributions, and let φ be its characteristic function. Then φ is the limit of a sequence of characteristic functions φ_n, where

$$\varphi_n(t) = \prod_{k=1}^{k_n} \exp\{i\alpha_{n,k}t + \lambda_{n,k}(e^{i\beta_{n,k}t} - 1)\},$$

$\alpha_{n,k}$, $\beta_{n,k} \in \mathbb{R}$, and $\lambda_{n,k} \geq 0$ for $k = 1, 2, \ldots, k_n$ and $n \geq 1$. It follows immediately from Propositions 4.1.2 and 4.1.3 that φ is i.d.

Conversely, let φ be i.d. Clearly $\ln \varphi$ exists and is finite on $\mathbb{R}$; moreover, the relation

$$\varphi^{1/n} = \exp\left(\frac{1}{n} \ln \varphi\right) = 1 + \frac{1}{n} \ln \varphi + O\left(\frac{1}{n^2}\right)$$

holds, so that

$$\ln \varphi = \lim_{n\to\infty} n(\varphi^{1/n} - 1) = \lim_{n\to\infty} n(\varphi_n - 1), \tag{4.1.6}$$

where $\varphi_n = \varphi^{1/n}$ is a characteristic function. Let F_n be the distribution function corresponding to φ_n. Then

$$\ln \varphi = \lim_{n\to\infty} n \int_{-\infty}^{\infty} (e^{itx} - 1)\, dF_n(x)$$

$$= \lim_{n\to\infty} \lim_{\substack{A\to-\infty \\ B\to\infty}} n \int_A^B (e^{itx} - 1)\, dF_n(x).$$

Let

$$A = x_{0,N} < x_{1,N} < \cdots < x_{\nu_N - 1, N} < x_{\nu_N, N} = B$$

be a sequence of subdivisions of [A, B] such that $\max_{1\le\nu\le\nu_N}(x_{\nu,N} - x_{\nu-1,N}) \to 0$ as $N \to \infty$. Then it follows from the definition of the Riemann–Stieltjes integral that

$$\int_A^B (e^{itx} - 1)\, dF_n(x) = \lim_{N\to\infty} \sum_{\nu=1}^{\nu_N} (e^{itx_{\nu,N}} - 1)[F_n(x_{\nu,N}) - F_n(x_{\nu-1,N})],$$

so that

$$n \int_A^B (e^{itx} - 1)\, dF_n(x) = \lim_{N\to\infty} \sum_{\nu=1}^{\nu_N} \lambda_{n,\nu,N}(e^{it\beta_{n,\nu,N}} - 1),$$

where $\lambda_{n,\nu,N} = n[F_n(x_{\nu,N}) - F_n(x_{\nu-1,N})] \ge 0$ and $\beta_{n,\nu,N} = x_{\nu,N}$. Then for every $n \ge 1$

$$\theta_n(t; A, B) = \exp\left\{n \int_A^B (e^{itx} - 1)\, dF_n(x)\right\}$$

$$= \lim_{N\to\infty} \prod_{\nu=1}^{\nu_N} \exp\{\lambda_{n,\nu,N}(e^{it\beta_{n,\nu,N}} - 1)\}$$

is the limit of a sequence of finite products of Poisson-type distributions. Clearly

$$\varphi(t) = \lim_{n\to\infty} \lim_{\substack{A\to-\infty \\ B\to+\infty}} \exp\{\theta_n(t; A, B)\},$$

so that the proof is immediate. ■

Remark 4.1.3. The class of all i.d. distribution functions coincides with the class of limit distributions (in the sense of weak convergence) of sequences of finite sums of independent random variables having Poisson-type distributions.

Example 4.1.6. Let $p > 1$, and consider the characteristic function

$$\varphi(t) = (p-1)(p-e^{it})^{-1}, \qquad t \in \mathbb{R}.$$

We show that φ is i.d. We note that

$$\begin{aligned} \ln \varphi(t) &= \ln\left(1 - \frac{1}{p}\right) - \ln\left(1 - \frac{e^{it}}{p}\right) \\ &= \sum_{k=1}^{\infty} \frac{1}{kp^k}(e^{itk} - 1), \end{aligned}$$

so that

$$\varphi(t) = \prod_{k=1}^{\infty} \exp\left\{\frac{1}{kp^k}(e^{itk} - 1)\right\}.$$

From Proposition 4.1.4 it is immediate that φ is i.d.

4.1.3 Canonical Representations of Infinitely Divisible Characteristic Functions

We now derive a fundamental result which gives a representation of an i.d. characteristic function.

Theorem 4.1.1 (Lévy–Khintchine Representation). A complex-valued function φ defined on $\mathbb{R}$ is an i.d. characteristic function if and only if $\ln \varphi$ admits the representation

$$\ln \varphi(t) = i\alpha t + \int_{-\infty}^{\infty} \left(e^{itx} - 1 - \frac{itx}{1+x^2}\right) \frac{1+x^2}{x^2}\, dG(x), \tag{4.1.7}$$

where $\alpha \in \mathbb{R}$ and G is a bounded, nondecreasing, right-continuous function on $\mathbb{R}$ such that $G(-\infty) = 0$ and $G(+\infty) < \infty$. Here the value of the integrand at $x = 0$ is defined by continuity as

$$\left[\left(e^{itx} - 1 - \frac{itx}{1+x^2}\right)\frac{1+x^2}{x^2}\right]_{x=0} = -\frac{t^2}{2}.$$

Moreover, α and G are determined uniquely by φ. Here G is known as the Lévy spectrum of φ.

Proof. Let φ be a complex-valued function on $\mathbb{R}$ such that $\ln \varphi$ has the representation given in (4.1.6). Let $\epsilon (0 < \epsilon < 1)$ be fixed. Consider the integral

$$I_\epsilon(t) = \int_\epsilon^{1/\epsilon} \left(e^{itx} - 1 - \frac{itx}{1+x^2}\right)\frac{1+x^2}{x^2}\, dG(x).$$

Note that the integrand is bounded in x and is continuous in t (respectively, x) for every fixed x (respectively, t). Let

$$\epsilon = x_{0,N} < x_{1,N} < \cdots < x_{\nu_N,N} = \frac{1}{\epsilon}$$

be a sequence of subdivisions of $[\epsilon, 1/\epsilon]$ such that

$$\delta_N = \max_{1 \le \nu \le \nu_N} (x_{\nu,N} - x_{\nu-1,N}) \to 0 \qquad \text{as } N \to \infty.$$

Then

$$I_\epsilon(t) = \lim_{N\to\infty} \sum_{\nu=1}^{\nu_N} \left(e^{itx_{\nu,N}} - 1 - \frac{itx_{\nu,N}}{1 + x_{\nu,N}^2}\right) \frac{1 + x_{\nu,N}^2}{x_{\nu,N}^2} [G(x_{\nu,N}) - G(x_{\nu-1,N})]$$

$$= \lim_{N\to\infty} \sum_{\nu=1}^{\nu_N} [it\alpha_{\nu,N} + \lambda_{\nu,N}(e^{it\beta_{\nu,N}} - 1)],$$

where

$$\lambda_{\nu,N} = \frac{1 + x_{\nu,N}^2}{x_{\nu,N}^2} [G(x_{\nu,N}) - G(x_{\nu-1,N})] \ge 0,$$

$$\alpha_{\nu,N} = -\lambda_{\nu,N} \frac{x_{\nu,N}}{1 + x_{\nu,N}^2},$$

and

$$\beta_{\nu,N} = x_{\nu,N}.$$

We note that I_ϵ is continuous at $t = 0$, and hence it follows from the continuity theorem that e^{I_ϵ} is the limit of a sequence of finite products of characteristic functions of Poisson-type distributions. We conclude from Proposition 4.1.4 that e^{I_ϵ} is an i.d. characteristic function. Set

$$I_+(t) = \lim_{\epsilon\to 0+} I_\epsilon(t) = \int_{x>0} \left(e^{itx} - 1 - \frac{itx}{1 + x^2}\right) \frac{1 + x^2}{x^2}\, dG(x).$$

Clearly I_+ is continuous at $t = 0$, so that it follows from Proposition 4.1.3 that e^{I_+} is an i.d. characteristic function. Next set

$$I_-(t) = \int_{x<0} \left(e^{itx} - 1 - \frac{itx}{1 + x^2}\right) \frac{1 + x^2}{x^2}\, dG(x),$$

and by proceeding as above we see that e^{I_-} is an i.d. characteristic function. Finally note that

$$\ln \varphi(t) = (i\alpha t - \tfrac{1}{2}\beta^2 t^2) + I_+(t) + I_-(t),$$

where $\beta^2 \geq 0$ is the jump of G at $x = 0$. The term $i\alpha t - \frac{1}{2}\beta^2 t^2$ is the logarithm of the characteristic function of a normal distribution which is i.d. From Proposition 4.1.2 it follows at once that φ is an i.d. characteristic function.

Next we show that $\alpha \in \mathbb{R}$ and G in (4.1.6) are determined uniquely by φ. We set $\psi = \ln \varphi$, and note that for $h > 0$

$$\psi(t) - \tfrac{1}{2}[\psi(t+h) + \psi(t-h)] = \int_{-\infty}^{\infty} e^{itx}(1 - \cos hx)\frac{1+x^2}{x^2}\, dG(x).$$

Define

$$\theta(t) = \int_0^1 \{\psi(t) - \tfrac{1}{2}[\psi(t+h) + \psi(t-h)]\}\, dh.$$

Then

$$\theta(t) = \int_0^1 \left[\int_{-\infty}^{\infty} e^{itx}(1 - \cos hx)\frac{1+x^2}{x^2}\, dG(x)\right] dh.$$

Since the integrand is bounded, we can interchange the order of integration by using Fubini's theorem and thus obtain

$$\theta(t) = \int_{-\infty}^{\infty} e^{itx}\left(1 - \frac{\sin x}{x}\right)\frac{1+x^2}{x^2}\, dG(x).$$

First note that there exist two positive real numbers a_1, a_2 such that

$$0 < a_1 \leq \left(1 - \frac{\sin x}{x}\right)\frac{1+x^2}{x^2} \leq a_2 < \infty, \qquad x \in \mathbb{R}.$$

Set

$$H(x) = \int_{-\infty}^{x} \left(1 - \frac{\sin y}{y}\right)\frac{1+y^2}{y^2}\, dG(y), \qquad x \in \mathbb{R}.$$

Clearly H is a bounded, nondecreasing function on $\mathbb{R}$ such that $H(-\infty) = 0$ and

$$0 < a_1 G(+\infty) \leq H(+\infty) \leq a_2 G(+\infty) < \infty.$$

Moreover, G can be expressed in terms of H by the relation

$$G(x) = \int_{-\infty}^{x} \left(1 - \frac{\sin y}{y}\right)^{-1} \frac{y^2}{1+y^2}\, dH(y),$$

and

$$\theta(t) = \int_{-\infty}^{\infty} e^{itx}\, dH(x)$$

is the Fourier transform of H. It follows from the uniqueness property of Fourier transforms that H is determined uniquely by θ. Consequently the function G is determined uniquely by φ, and hence also is α.

Conversely, suppose that φ is an i.d. characteristic function. In view of Proposition 4.1.1 $\psi = \ln \varphi$ exists, and the relation [see (4.1.6)]

$$\psi = \lim_{n\to\infty} n[(\varphi)^{1/n} - 1] = \lim_{n\to\infty} n(\varphi_n - 1)$$

holds, where φ_n is some characteristic function. Let F_n be the distribution function corresponding to φ_n. Then we have for $n \geq 1$

$$\begin{aligned} n[\varphi_n(t) - 1] &= n \int_{-\infty}^{\infty} (e^{itx} - 1)\, dF_n(x) \\ &= itn \int_{-\infty}^{\infty} \frac{x}{1 + x^2}\, dF_n(x) + n \int_{-\infty}^{\infty} \left(e^{itx} - 1 - \frac{itx}{1 + x^2}\right) dF_n(x). \end{aligned}$$

We now set for every $n \geq 1$

$$\psi_n = n(\varphi_n - 1),$$

$$\alpha_n = n \int_{-\infty}^{\infty} \frac{x}{1 + x^2}\, dF_n(x),$$

and

$$G_n(x) = n \int_{-\infty}^{x} \frac{y^2}{1 + y^2}\, dF_n(y), \qquad x \in \mathbb{R}.$$

We see easily that for every $n \geq 1$ G_n is a bounded, nondecreasing, right-continuous function on $\mathbb{R}$ such that $G_n(-\infty) = 0$. Clearly

$$\psi(t) = \lim_{n\to\infty} \psi_n(t), \tag{4.1.8}$$

where

$$\psi_n(t) = it\alpha_n + \int_{-\infty}^{\infty} \left(e^{itx} - 1 - \frac{itx}{1 + x^2}\right) \frac{1 + x^2}{x^2}\, dG_n(x). \tag{4.1.9}$$

Next we show that there exists a real number α and a bounded, nondecreasing, right-continuous function G on $\mathbb{R}$ with $G(-\infty) = 0$ such that $\alpha_n \to \alpha$ and $G_n \xrightarrow{w} G$ as $n \to \infty$. We note from (4.1.9) and the first part of the proof of the theorem that ψ_n is the logarithm of an i.d. characteristic function. Proceeding exactly as in the first part of the proof of this theorem, we set for $n \geq 1$

$$\theta_n(t) = \int_0^1 \{\psi_n(t) - \tfrac{1}{2}[\psi_n(t + h) + \psi_n(t - h)]\}\, dh$$

and thus obtain

$$\theta_n(t) = \int_{-\infty}^{\infty} e^{itx}\left(1 - \frac{\sin x}{x}\right)\frac{1 + x^2}{x^2}\, dG_n(x).$$

Setting

$$H_n(x) = \int_{-\infty}^{x} \left(1 - \frac{\sin y}{y}\right)\frac{1 + y^2}{y^2}\, dG_n(y), \qquad x \in \mathbb{R},$$

we have

$$\theta_n(t) = \int_{-\infty}^{\infty} e^{itx}\, dH_n(x).$$

Since $\{\psi_n\}$ converges pointwise to the continuous function ψ on $\mathbb{R}$, we see easily that the sequence $\{\theta_n\}$ also converges pointwise to the continuous function θ on $\mathbb{R}$, given by

$$\theta(t) = \int_0^1 \{\psi(t) - \tfrac{1}{2}[\psi(t + h) + \psi(t - h)]\}\, dh.$$

Note that $\{H_n\}$ is a sequence of nondecreasing, right-continuous functions on $\mathbb{R}$ such that $H_n(-\infty) = 0$. Moreover, $\{H_n\}$ is uniformly bounded on $\mathbb{R}$, since

$$H_n(+\infty) = \theta_n(0) \to \theta(0) = -\int_0^1 \frac{\psi(h) + \psi(-h)}{2}\, dh < \infty.$$

By using an argument similar to the one used in the proof of the continuity theorem applied to the sequence $\{\theta_n\}$, which is the Fourier transform of H_n, we conclude that there exists a bounded, nondecreasing, right-continuous function H on $\mathbb{R}$ such that $H_n \xrightarrow{w} H$.

Clearly

$$G_n(x) = \int_{-\infty}^{x} \left(1 - \frac{\sin y}{y}\right)^{-1} \frac{y^2}{1 + y^2}\, dH_n(y), \qquad x \in \mathbb{R}.$$

Now we set

$$G(x) = \int_{-\infty}^{x} \left(1 - \frac{\sin y}{y}\right)^{-1} \frac{y^2}{1 + y^2}\, dH(y), \qquad x \in \mathbb{R}.$$

Clearly G is a bounded, nondecreasing, right-continuous function on $\mathbb{R}$ such that $G(-\infty) = 0$. We show that $G_n \xrightarrow{w} G$. In fact, it follows from the extended Helly–Bray theorem that $G_n(x) \to G(x)$ for every $x \in C_H$ as $n \to \infty$. On the other hand, since the integrand in the definition of G is positive, it is easily verified that $C_H = C_G$. Hence we conclude that $G_n \xrightarrow{w} G$.

Next we set

$$I_n(t) = \int_{-\infty}^{\infty} \left(e^{itx} - 1 - \frac{itx}{1 + x^2}\right) \frac{1 + x^2}{x^2}\, dG_n(x), \qquad n \geq 1$$

and

$$I(t) = \int_{-\infty}^{\infty} \left(e^{itx} - 1 - \frac{itx}{1 + x^2}\right) \frac{1 + x^2}{x^2}\, dG(x).$$

Then from the extended Helly–Bray theorem

$$\lim_{n\to\infty} I_n(t) = I(t), \qquad t \in \mathbb{R}.$$

Finally, since $\psi_n \to \psi$ and $I_n \to I$, we conclude that the sequence of real numbers $\{\alpha_n\}$ must also converge to a real number α as $n \to \infty$. Thus we obtain

$$\begin{aligned}
\ln \varphi(t) &= \psi(t) \\
&= \lim_{n\to\infty} \psi_n(t) \\
&= \lim_{n\to\infty} it\alpha_n + \lim_{n\to\infty} I_n(t) \\
&= it\alpha + I(t),
\end{aligned}$$

so that $\ln \varphi$ has representation (4.1.7). This completes the proof of Theorem 4.1.1. ■

Example 4.1.7.

(a) Let X have the characteristic function given by

$$\varphi(t) = \exp\left(i\mu t - \sigma^2 \frac{t^2}{2}\right).$$

Then φ is of the form (4.1.7) with

$$\alpha = \mu \qquad \text{and} \qquad G(x) = 0 \quad \text{if } x < 0, \text{and } = \sigma^2 \quad \text{if } x \geq 0.$$

(b) Let φ be the characteristic function of the Poisson distribution. Then

$$\begin{aligned}
\ln \varphi(t) &= \lambda(e^{it} - 1) \\
&= i\frac{\lambda}{2}t + \lambda\left(e^{it} - 1 - \frac{it}{2}\right),
\end{aligned}$$

so that by taking

$$\alpha = \frac{\lambda}{2} \qquad \text{and} \qquad G(x) = 0 \quad \text{if } x < 1, \text{and } = \frac{\lambda}{2} \quad \text{if } x \geq 1$$

we see that φ has representation (4.1.7). Note that $\alpha \neq \mathscr{E}X = \lambda$ in this case.

Example 4.1.8. Let X_1 and X_2 be independent and i.d. such that $X = X_1 + X_2$ is a normal random variable. Let (α_1, G_1) and (α_2, G_2) correspond to X_1 and X_2, respectively. Then $(\alpha_1 + \alpha_2, G_1 + G_2)$ corresponds to X, and since X is normal, $G_1 + G_2$ must have a jump of $\sigma^2 = \text{var}(X)$ at $x = 0$. It follows that both G_1 and G_2 increase only at $x = 0$ and hence must be normally distributed.

Example 4.1.9. Let $0 < a \le b < 1$, and consider the complex-valued function

$$\varphi(t) = \frac{1-b}{1+a}\frac{1+ae^{-it}}{1-be^{it}}.$$

Since φ is continuous at $t = 0$ and

$$\varphi(t) = \frac{1-b}{1+a}\left[ae^{-it} + (1+ab)\sum_{k=0}^{\infty} b^k e^{ikt}\right],$$

it is clear that φ is the characteristic function of the random variable X with distribution

$$P\{X = -1\} = \frac{1-b}{1+a}a; \qquad P\{X = k\} = \frac{1-b}{1+a}(1+ab)b^k, \quad k = 0, 1, 2, \dots.$$

Now

$$\ln \varphi(t) = \sum_{k=1}^{\infty}\left[(-1)^{k-1}\frac{a^k}{k}(e^{-ikt} - 1) + \frac{b^k}{k}(e^{ikt} - 1)\right],$$

so that by taking

$$\alpha = \sum_{k=1}^{\infty} \frac{b^k + (-1)^k a^k}{k^2 + 1}$$

and G to be a function of bounded variation with jumps of magnitudes

$$\frac{kb^k}{k^2+1} \qquad \text{at } x = k$$

and

$$(-1)^{k-1}\frac{ka^k}{k^2+1} \qquad \text{at } x = -k, \quad k = 1, 2, 3, \dots,$$

we can write $\ln \varphi$ in the form (4.1.7). But G is clearly not monotone, and it follows that φ cannot be i.d. Note, however, that $\bar{\varphi}$ is also not i.d., but $|\varphi|$ is i.d. For

$$\bar{\varphi}(t) = \frac{1-b}{1+a}\frac{1+ae^{it}}{1-be^{-it}},$$

so that

$$\ln \bar{\varphi}(t) = \sum_{k=1}^{\infty} \left[\frac{b^k}{k}(e^{-ikt} - 1) + (-1)^{k-1}\frac{a^k}{k}(e^{ikt} - 1)\right].$$

Let

$$\theta(t) = |\varphi(t)|^2 = \varphi(t)\bar{\varphi}(t),$$

so that

$$\ln \theta(t) = \sum_{k=1}^{\infty} \frac{1}{k}[b^k + (-1)^{k-1}a^k](e^{-ikt} - 1)$$
$$+ \sum_{k=1}^{\infty} \frac{1}{k}[b^k + (-1)^{k-1}a^k](e^{ikt} - 1).$$

Taking $\alpha = 0$ and G to be a nondecreasing function with jumps of magnitude

$$\frac{k}{k^2 + 1}[b^k + (-1)^{k-1}a^k] \qquad \text{at } \pm k, \quad k = 1, 2, \ldots,$$

we see that θ has the unique representation (4.1.7). Thus θ is i.d.

We next derive the canonical representation due to Kolmogorov when the i.d. distribution has a finite variance.

Theorem 4.1.2 (Kolmogorov Representation). The function φ is the characteristic function of an i.d. distribution with finite variance if and only if $\ln \varphi$ has the representation

$$\ln \varphi(t) = it\gamma + \int_{-\infty}^{\infty} (e^{itx} - 1 - itx)\frac{1}{x^2}\,dK(x), \tag{4.1.10}$$

where $\gamma \in \mathbb{R}$ and K is a bounded, nondecreasing, right-continuous function on $\mathbb{R}$ such that $K(-\infty) = 0$ and $K(+\infty) < \infty$. Here the value of the integrand at $x = 0$ is defined by continuity as

$$\left.(e^{itx} - 1 - itx)\frac{1}{x^2}\right]_{x=0} = \frac{-t^2}{2}.$$

Moreover, γ and K are determined uniquely by φ.

Proof. First let φ be the characteristic function of an i.d. distribution with finite variance. Then φ and hence $\psi = \ln \varphi$ can be differentiated twice. Also, it follows from Theorem 4.1.1 that ψ has the representation

$$\psi(t) = it\alpha + \int_{-\infty}^{\infty} \left(e^{itx} - 1 - \frac{itx}{1 + x^2}\right)\frac{1 + x^2}{x^2}\,dG(x).$$

Clearly

$$\psi''(0) = \lim_{h\to 0} \frac{\psi(2h) - 2\psi(0) + \psi(-2h)}{(2h)^2}$$

exists and is bounded. After some elementary computation we conclude that

$$0 < -\psi''(0) = \lim_{h\to 0} \int_{-\infty}^{\infty} \frac{\sin^2 hx}{h^2} \frac{1 + x^2}{x^2} \, dG(x) < \infty.$$

By Fatou's lemma it follows immediately that

$$\int_{-\infty}^{\infty} (1 + x^2) \, dG(x) < \infty,$$

and consequently

$$\int_{-\infty}^{\infty} |x| \, dG(x) < \infty.$$

Now set

$$K(x) = \int_{-\infty}^{x} (1 + y^2) \, dG(y), \qquad x \in \mathbb{R},$$

and

$$\gamma = \alpha - \int_{-\infty}^{\infty} y \, dG(y)$$

to get representation (4.1.10).

Conversely, suppose that $\psi = \ln \varphi$ has representation (4.1.10). Set

$$G(x) = \int_{-\infty}^{x} \frac{1}{1 + y^2} \, dK(y), \qquad x \in \mathbb{R},$$

and

$$\alpha = \gamma + \int_{-\infty}^{\infty} \frac{y}{1 + y^2} \, dK(y).$$

Then G is a bounded, nondecreasing, right-continuous function on $\mathbb{R}$ such that $G(-\infty) = 0$, $G(+\infty) < \infty$. Clearly $\psi = \ln \varphi$ admits representation (4.1.7), so that φ is an i.d. characteristic function.

Finally we see easily that the expression on the right-hand side of (4.1.10) can be differentiated twice with respect to t so that the distribution function corresponding to φ has finite variance. The uniqueness of γ and K is immediate. ∎

Example 4.1.10.

(a) Let X be a normally distributed random variable with mean μ and variance σ^2. Let φ be the characteristic function of X. Then

$$\ln \varphi = i\mu t - \frac{\sigma^2 t^2}{2}.$$

Taking $\gamma = \mu$ and $K(x) = \sigma^2 \epsilon(x)$, $x \in \mathbb{R}$, we see that $\ln \varphi$ has the Kolmogorov representation (4.1.10).

(b) Let

$$\varphi(t) = \exp\{\lambda(e^{it} - 1)\}$$

be the characteristic function of a Poisson random variable. Then

$$\begin{aligned} \ln \varphi(t) &= \lambda(e^{it} - 1) \\ &= it\lambda + \lambda(e^{it} - it - 1). \end{aligned}$$

Here $\gamma = \lambda$, and $K(x) = \lambda\epsilon(x - 1)$, $x \in \mathbb{R}$.

4.1.4 Infinitely Divisible Distributions and Convolution Semigroups

In this section we show the connection between i.d. distributions and continuous convolution semigroups. In Chapter 5 we will show that the class of i.d. distributions coincides with the class of limit distributions of the row sums of certain triangular arrays.

Theorem 4.1.3. The class of all i.d. distributions on $\mathbb{R}$ coincides with the class of all distributions associated with continuous convolution semigroups.

Proof. First let $\{F_t, t \in \mathbb{R}_+^\times\}$ be a (continuous) convolution semigroup of distribution functions. We show that, for every $t \in \mathbb{R}_+^\times$, F_t is i.d. In fact, for every $n \geq 1$ we can write

$$F_t = \underbrace{F_{t/n} * F_{t/n} * \cdots * F_{t/n}}_{n\text{-fold}},$$

so that F_t is i.d.

Conversely, let F be an i.d. distribution function on $\mathbb{R}$. We show that there exists a continuous convolution semigroup $\{F_t, t \in \mathbb{R}_+^\times\}$ of distribution functions, determined uniquely by F, such that $F_1 = F$. Let φ be the characteristic function of F. We note that $\varphi \neq 0$ on $\mathbb{R}$. From Remark 4.1.2 there exists a continuous, real-valued function ω on $\mathbb{R}$, determined uniquely by φ, such that $\omega(0) = 0$ and

$$\varphi(u) = |\varphi(u)| e^{i\omega(u)}, \qquad u \in \mathbb{R}.$$

Clearly, for every $n \geq 1$ there exists a unique distribution function $F_{1/n}$ with characteristic function $\varphi_{1/n}$, given by

$$\varphi_{1/n}(u) = [\varphi(u)]^{1/n} = |\varphi(u)|^{1/n} e^{i\omega(u)/n}, \qquad u \in \mathbb{R}.$$

Let $r = m/n$, where $m, n \geq 1$ are integers. Then

$$F_r = \underbrace{F_{1/n} * F_{1/n} * \cdots * F_{1/n}}_{m\text{-fold}}$$

is the unique m-fold distribution function with characteristic function φ_r, given by

$$\varphi_r(u) = [\varphi_{1/n}(u)]^m = |\varphi(u)|^r e^{ir\omega(u)}, \qquad u \in \mathbb{R}.$$

Let $t \in \mathbb{R}_+^\times$. Then there exists a sequence $\{r_n\}$ of positive rational numbers such that $r_n \to t$ as $n \to \infty$. Consequently the sequence of characteristic functions $\{\varphi_{r_n}\}$ converges to the function φ_t, given by

$$\varphi_t(u) = |\varphi(u)|^t e^{it\omega(u)}, \qquad u \in \mathbb{R}.$$

Clearly φ_t is continuous on $\mathbb{R}$, and, in view of the continuity theorem, φ_t is a characteristic function. Let F_t be the corresponding distribution function. We see easily that

$$\varphi_s \varphi_t = \varphi_{s+t} \qquad \text{for all } s, t \in \mathbb{R}_+^\times,$$

so that

$$F_s * F_t = F_{s+t} \text{ for all } s, t \in \mathbb{R}_+^\times.$$

Hence $\{F_t, t \in \mathbb{R}_+^\times\}$ is a convolution semigroup of distribution functions with $F_1 = F$. Clearly $\varphi_t \to 1$ as $t \to 0+$, so that $F_t(x) \to \epsilon(x)$ as $t \to 0+$. Hence $\{F_t, t \in \mathbb{R}_+^\times\}$ is a continuous convolution semigroup.

To complete the proof, suppose $\{G_t,\ t \in \mathbb{R}_+^\times\}$ is another continuous convolution semigroup of distribution functions with $G_1 = F$. Then clearly

$$F = G_1 = \underbrace{G_{1/n} * G_{1/n} * \cdots * G_{1/n}}_{n\text{-fold}},$$

so that

$$\varphi(u) = [\theta_{1/n}(u)]^n, \qquad u \in \mathbb{R},$$

where θ_t is the characteristic function of G_t. This implies that

$$\theta_{1/n}(u) = |\varphi(u)|^{1/n} e^{i\omega(u)/n}, \qquad u \in \mathbb{R}.$$

Proceeding exactly as above and using the fact that $\{G_t\}$ is a continuous convolution semigroup, we see that for any $t \in \mathbb{R}_+^\times$

$$\begin{aligned}\theta_t(u) &= |\varphi(u)|^t e^{it\omega(u)} \\ &= \varphi_t(u), \qquad u \in \mathbb{R}.\end{aligned}$$

It follows from the uniqueness theorem that $F_t = G_t$ for all $t \in \mathbb{R}_+^\times$. ∎

Corollary. Let F be an i.d. distribution with zero mean and finite variance, and let $\{F_t, t \in \mathbb{R}_+^\times\}$ be the corresponding continuous convolution semigroup of distribution functions such that $F_1 = F$. Let $\{T_t, t \in \mathbb{R}_+^\times\}$ be the associated semigroup of probability operators. Then the infinitesimal generator A of $\{T_t, t \in \mathbb{R}_+^\times\}$ is of the form (3.8.26), where G is a distribution function which is determined uniquely by A.

Conversely, every operator A of the form (3.8.26), where G is a distribution function, is the infinitesimal generator of a convolution semigroup $\{T_t, t \in \mathbb{R}_+^\times\}$ associated with i.d. distribution functions with zero means and finite (nonzero) variances.

Proof. The proof immediately follows from Theorems 3.8.6 and 4.1.3.

4.2 ANALYTIC CHARACTERISTIC FUNCTIONS

In this section we investigate some properties of a special class of characteristic functions which plays an important role in probability theory and mathematical statistics. In the following, t and v denote real variables, and $z = t + iv (i = \sqrt{-1})$ denotes a complex variable.

4.2.1 Analytic Characteristic Functions

Definition 4.2.1. A characteristic function φ is said to be an analytic characteristic function if there exist a complex-valued function, θ, of the complex variable z which is holomorphic (or regular) in a circle $|z| < \rho$ $(\rho > 0)$ and a positive real number δ such that $\varphi(t) = \theta(t)$ for $|t| < \delta$. In other words, an analytic characteristic function is a characteristic function which coincides with a holomorphic function in some neighborhood of zero.

As some well-known examples of distributions with analytic characteristic functions we mention the binomial, Poisson, normal, and gamma distributions. The Cauchy distribution is an example of a distribution which does not have an analytic characteristic function. In fact, the characteristic function $\varphi(t) = e^{-|t|^\alpha}$ is not analytic for any α in (0, 2).

Theorem 4.2.1. Let φ be an analytic characteristic function, and let F be the corresponding distribution function. Then φ can be continued analytically as a function which is holomorphic in a horizontal strip $-\alpha < \operatorname{Im} z < \beta$ $(\alpha, \beta > 0)$ containing the real axis. Moreover, φ admits the Fourier integral representation

$$\varphi(z) = \int_{-\infty}^{\infty} e^{izx}\, dF(x) \tag{4.2.1}$$

in the strip $-\alpha < \operatorname{Im} z < \beta$. This strip is either the whole complex plane, in which case φ is an entire characteristic function, or it has one or two horizontal boundary lines. The purely imaginary points $-i\alpha$ and $i\beta$ on the boundary of this strip are the singularities of φ nearest to the origin.

Proof. Let φ be an analytic characteristic function. Clearly φ has continuous derivatives of all orders at zero. It follows immediately from Theorem 3.2.1 that moments of all orders of F exist, and, moreover, φ admits the expansion

$$\varphi(z) = \sum_{k=0}^{\infty} \frac{\alpha_k}{k!} (iz)^k, \qquad |z| < \rho, \tag{4.2.2}$$

where $\rho > 0$ is the radius of convergence of this series. Set

$$\varphi_0(z) = \tfrac{1}{2}[\varphi(z) + \varphi(-z)]$$

and

$$\varphi_1(z) = \tfrac{1}{2}[\varphi(z) - \varphi(-z)].$$

Clearly the series

$$\varphi_0(z) = \sum_{k=0}^{\infty} \frac{(-1)^k \alpha_{2k}}{(2k)!} z^{2k} \quad \text{and} \quad \varphi_1(z) = \sum_{k=1}^{\infty} \frac{i^{2k-1} \alpha_{2k-1}}{(2k-1)!} z^{2k-1} \tag{4.2.3}$$

also converge in some circles around the origin of radii at least ρ. Let ρ_0 and ρ_1, respectively, be the radii of convergence of these series. Since, for $k \geq 1$, $|x|^{2k-1} \leq (x^{2k} + x^{2k-2})/2$, it follows that

$$\frac{\alpha_{2k-1}}{(2k-1)!} \leq \frac{\beta_{2k-1}}{(2k-1)!} \leq \frac{1}{2}\left[\frac{\alpha_{2k}}{(2k)!}(2k) + \frac{\alpha_{2k-2}}{(2k-2)!}\right], \tag{4.2.4}$$

where, as usual, β_k denotes the absolute moment of F of order $k > 0$. Using (4.2.2) through (4.2.4), we conclude that $\rho_1 \geq \rho_0 \geq \rho$. Let ρ^* be the radius of convergence of the series $\sum_{k=0}^{\infty} \beta_k (z^k/k!)$. Then it follows from (4.2.4) that $\rho^* \geq \rho_0 \geq \rho$.

Next we show that φ is holomorphic at least in the strip $|\text{Im } z| < \rho$. For this purpose let a be a real number, and denote the radius of convergence of the series of φ_0 (respectively, φ_1) about the point a by $\rho_0(a)$ [respectively, $\rho_1(a)$]. In view of the inequalities

$$|\varphi^{(2k)}(a)| \leq \beta_{2k} \qquad \text{and} \qquad |\varphi^{(2k-1)}(a)| \leq \beta_{2k-1}$$

it follows at once that

$$\rho_0(a) \geq \rho^* \geq \rho_0 \geq \rho \qquad \text{and} \qquad \rho_1(a) \geq \rho^* \geq \rho_0 \geq \rho.$$

Consequently the series expansions of φ_0 and φ_1 about the point a converge in circles of radii at least equal to ρ. The same statement is also true for the series expansion of φ about the point a, so that φ is holomorphic at least in the strip $|\text{Im } z| < \rho$.

We now show that the integral $\int_{-\infty}^{\infty} e^{izx}\, dF(x)$ converges at least in the strip $|\text{Im } z| < \rho$ and is holomorphic inside this strip. First note that the series $\sum_{k=0}^{\infty} (\beta_k/k!)|v|^k$ converges for $|v| < \rho$. On the other hand, for any $A > 0$, $B > 0$ we have the estimate

$$\begin{aligned} \int_{-A}^{B} e^{|vx|}\, dF(x) &= \sum_{k=0}^{\infty} \frac{|v|^k}{k!} \int_{-A}^{B} |x|^k\, dF(x) \\ &\leq \sum_{k=0}^{\infty} \frac{|v|^k}{k!} \beta_k, \end{aligned}$$

and therefore we conclude that the integral $\int_{-\infty}^{\infty} e^{|vx|}\, dF(x)$ exists and is finite for $|v| < \rho$. Moreover, for $z = t + iv$ we have

$$|e^{izx}| \leq e^{|vx|},$$

so that the integral $\int_{-\infty}^{\infty} e^{izx}\, dF(x)$ converges for all t and for $|v| < \rho$. Note that this integral is holomorphic in its strip of convergence $|v| < \rho$ and coincides with φ for all $z = t$ real. Consequently, it must coincide with φ for all complex $z = t + iv$ for which $|v| < \rho$. Thus φ admits the Fourier integral representation in a strip $|\text{Im } z| < \rho$. Suppose that φ is holomorphic in the strip $-\alpha < \text{Im } z < \beta$, where $\alpha \geq \rho$, $\beta \geq \rho$. Proceeding exactly as above, we can show that φ admits the Fourier integral representation (4.2.1) in the strip $-\alpha < \text{Im } z < \beta$.

Finally, we write

$$\begin{aligned} \varphi(z) &= \int_{-\infty}^{0} e^{izx}\, dF(x) + \int_{0}^{\infty} e^{izx}\, dF(x) \\ &= \varphi^-(z) + \varphi^+(z). \end{aligned}$$

Note that the functions φ^+ and φ^- converge in the half-planes $\operatorname{Im} z > -\alpha$ and $\operatorname{Im} z < \beta$, respectively. Set $w = iz$ so that $w = it - v$ and

$$\varphi^+(z) = \int_0^\infty e^{izx}\, dF(x) = \int_0^\infty e^{wx}\, dF(x) = \Phi(w),$$

say. Then $\Phi(w)$ converges for $\operatorname{Re} w = -v < \alpha$. Clearly Φ has the power series representation with nonnegative coefficients. Hence it follows from a theorem of Pringsheim (Hille [37], p. 133) that the singularity of Φ nearest to the origin $w = 0$ must be on the positive half of the real axis. Therefore it must be at the point $w = \alpha$. Consequently the point $-i\alpha$ is the singularity of φ^+ and hence of φ nearest to the origin $z = 0$. Arguing in a similar manner with φ^-, we show that $i\beta$ is the other singularity of φ nearest to the origin. This completes the proof of Theorem 4.2.1. ■

Remark 4.2.1. Let φ be an analytic characteristic function which is holomorphic in the strip $-\alpha < \operatorname{Im} z < \beta$. Then the inequality

$$\sup_{t \in \mathbb{R}} |\varphi(t + iv)| \le \varphi(iv) \tag{4.2.5}$$

holds for all v in $-\alpha < v < \beta$. This follows, since

$$|\varphi(z)| = \left| \int_{-\infty}^\infty e^{i(t+iv)x}\, dF(x) \right| \le \int_{-\infty}^\infty e^{-vx}\, dF(x) = \varphi(iv).$$

It follows from (4.2.5) that φ cannot have zeros for $z = iv$, $-\alpha < v < \beta$, that is, $\varphi(iv) \ne 0$ within the strip of regularity.

4.2.2 Analytic Characteristic Functions and Their Distribution Functions

We now investigate the relationship between an analytic characteristic function and the corresponding distribution function.

Theorem 4.2.2. Let φ be a characteristic function, and let F be the corresponding distribution function. Then φ is analytic if and only if the following conditions hold:

(i) F has moments α_k of all orders k.
(ii) There exists a positive number γ such that

$$|\alpha_k| \le k!\,\gamma^k \qquad \text{for all } k \ge 1.$$

Proof. Clearly (i) and (ii) are equivalent to the statement that the series

$$\varphi(z) = \sum_{k=0}^\infty \frac{\alpha_k}{k!} (iz)^k$$

converges and represents a function which is holomorphic at least in the circle $|z| < \gamma^{-1}$ and which coincides with the characteristic function of F for all real z. ■

Remark 4.2.2. We see easily that a distribution function F has an analytic characteristic function if and only if its moment-generating function exists. Consequently conditions (i) and (ii) of Theorem 4.2.2 are a set of necessary and sufficient conditions for the existence of a moment-generating function.

Next we give another set of necessary and sufficient conditions for a characteristic function to be analytic in terms of the corresponding distribution function.

Theorem 4.2.3. Let φ be a characteristic function with corresponding distribution function F. Then φ is analytic if and only if there exists a positive real number R such that the relations

$$\begin{aligned} &1 - F(x) = O(e^{-rx}) \qquad \text{as} \quad x \to \infty \\ \text{and} \\ &F(-x) = O(e^{-rx}) \qquad \text{as} \quad x \to \infty \end{aligned} \tag{4.2.6}$$

hold for every $0 < r < R$. Moreover, in this case φ is holomorphic at least in the strip $|\mathrm{Im}\, z| < R$.

[In particular, if $R = +\infty$, (4.2.6) holds for all $r > 0$ and φ is an entire characteristic function.]

Proof. We first prove the sufficiency of conditions (4.2.6) for φ to be analytic. Let $v \neq 0$ be a real number such that $|v| < R$, and let $r > 0$ be such that $|v| < r < R$. Let $k \geq 1$ be an integer. Then

$$\int_{k-1}^{k} e^{|v|x}\, dF(x) \leq e^{|v|k}[1 - F(k-1)].$$

But $1 - F(k-1) = O(e^{-r(k-1)})$ as $k \to \infty$, so that there exists a constant $c_1 > 0$ such that $1 - F(k-1) \leq c_1 e^{-rk}$ for sufficiently large k. Hence there exists an integer $K > 0$ such that for $k > K$

$$\int_{k-1}^{k} e^{|v|x}\, dF(x) \leq c_1 e^{-k(r-|v|)}.$$

Consequently for real numbers $a \geq K$ and $b > 0$ we have

$$\begin{aligned}\int_a^{a+b} e^{|v|x}\, dF(x) &\leq \sum_{k=K}^{\infty} \int_{k-1}^{k} e^{|v|x}\, dF(x) \\ &\leq c_1 \sum_{k=K}^{\infty} e^{-k(r-|v|)} \\ &= c_1 \frac{e^{-(r-|v|)K}}{1-e^{-(r-|v|)}} < \infty.\end{aligned}$$

It now easily follows that the integral $\int_0^\infty e^{|vx|}\, dF(x)$ exists and is finite for all $|v| < r < R$ and hence for $|v| < R$.

Proceeding in a similar manner, we show that the integral $\int_{-\infty}^0 e^{|vx|}\, dF(x)$ exists and is finite for all $|v| < R$. It follows therefore that the integral $\int_{-\infty}^\infty e^{|vx|}\, dF(x)$ exists and is finite for all $|v| < R$. Let $z = t + iv$. Clearly the integral $\int_{-\infty}^\infty e^{izx}\, dF(x)$ converges in the strip $|\text{Im } z| < R$ for any t and represents a holomorphic function. Moreover, it coincides with the characteristic function φ for all real t. This completes the proof of the sufficiency part.

We next show that if φ is analytic relations (4.2.6) hold. In view of Theorem 4.2.1 the function $\varphi(z) = \int_{-\infty}^\infty e^{izx}\, dF(x)$ is holomorphic in a strip $-\alpha < \text{Im } z < \beta$, say, where $\alpha > 0$, $\beta > 0$. Let $R = \min(\alpha, \beta)$ so that $R > 0$. Let $x > 0$. Clearly the integrals

$$\int_x^\infty e^{vy}\, dF(y) \qquad \text{and} \qquad \int_{-\infty}^{-x} e^{vy}\, dF(y)$$

exist and are finite for all $|v| < R$. Let $0 < r < R$, and choose r_1 such that $0 < r < r_1 < R$. Then there exists a constant $C > 0$ (independent of x) such that

$$\int_x^\infty e^{r_1 y}\, dF(y) < C \qquad \text{for all } x > 0,$$

so that

$$C > \int_x^\infty e^{r_1 y}\, dF(y) \geq e^{r_1 x}[1 - F(x)] \geq 0 \qquad (x > 0).$$

Hence for $x > 0$

$$0 \leq [1 - F(x)]e^{rx} \leq Ce^{-(r_1 - r)x}.$$

Since $r_1 > r$, the last expression tends to 0 as $x \to \infty$. It follows that

$$1 - F(x) = o(e^{-rx}) \qquad \text{as } x \to \infty$$

for all $0 < r < R$. Proceeding in a similar manner, we show that

$$F(-x) = o(e^{-rx}) \qquad \text{as } x \to \infty$$

for all $0 < r < R$. Clearly conditions (4.2.6) hold. This completes the proof. ∎

Remark 4.2.3. Let F be a distribution function which is concentrated in a finite interval $[a, b]$; that is, let $F(x) = 0$ for all $x \leq a$, and $F(x) = 1$ for all $x \geq b$. Clearly $1 - F(x) + F(-x) = 0$ for all $x > \max(|a|, |b|)$. Therefore conditions (4.2.6) are trivially satisfied for all $r > 0$, so that the corresponding characteristic function is an entire characteristic function.

4.2.3 Entire Characteristic Functions

We now consider some properties of entire characteristic functions. First we give the definition of the order of an entire function. Let f be an entire function. Set

$$M(r, f) = \sup_{|z| \leq r} |f(z)| \qquad \text{for } r > 0. \tag{4.2.7}$$

It is well known that the function f attains the value $M(r, f)$ on the circle $|z| = r$.

Definition 4.2.2. The function f is said to be an entire function of finite order if there exists a positive number $A > 0$ such that

$$M(r, f) = O(e^{r^A}) \qquad \text{as } r \to \infty. \tag{4.2.8}$$

In this case the order ρ of the function f is defined by the relation

$$\rho = \limsup_{r \to \infty} \cdot \frac{\ln \ln M(r, f)}{\ln r}. \tag{4.2.9}$$

Clearly $0 \leq \rho \leq \infty$.

Remark 4.2.4. Let φ be an entire characteristic function, and let $M(r, \varphi) = \sup_{|z| \leq r} |\varphi(z)|$ for $r > 0$. Then it follows immediately from (4.2.5) that $M(r, \varphi) = \max\{\varphi(ir), \varphi(-ir)\}$.

Example 4.2.1. Let $\varphi(t) = \exp(i\mu t - \frac{1}{2}\sigma^2 t^2)$ be the characteristic function of the normal distribution with mean μ and variance σ^2. Clearly φ is an entire characteristic function of order $\rho = 2$. Similarly, if φ is the characteristic function of a Poisson distribution, its order is $\rho = +\infty$. In this case $\ln \varphi$ is an entire function of order $\rho = 1$.

Theorem 4.2.4. Let φ be an entire characteristic function. Then either $\varphi \equiv 1$, or the order of φ must be ≥ 1.

Proof. Let F be the distribution function that corresponds to φ. Then

$$\varphi(z) = \int_{-\infty}^{\infty} e^{izx}\, dF(x),$$

where the integral on the right-hand side converges for all complex z. If $F(x) = \epsilon(x)$, clearly $\varphi \equiv 1$. Otherwise, F has at least another point of increase, say $x_0 \neq 0$. Setting $z = \pm iv \operatorname{sgn} x_0$, we see that

$$\begin{aligned}\varphi(\pm iv \operatorname{sgn} x_0) &= \int_{-\infty}^{\infty} \exp(\mp vx \operatorname{sgn} x_0)\, dF(x)\\ &\geq \int_{x_0-\epsilon}^{x_0+\epsilon} \exp(\mp vx \operatorname{sgn} x_0)\, dF(x).\end{aligned}$$

Hence

$$\begin{aligned}\max\{\varphi(-iv \operatorname{sgn} x_0), \varphi(iv \operatorname{sgn} x_0)\} &\geq \int_{x_0-\epsilon}^{x_0+\epsilon} \exp(|v| x \operatorname{sgn} x_0)\, dF(x)\\ &= \int_{x_0-\epsilon}^{x_0+\epsilon} \exp(|vx|)\, dF(x)\end{aligned}$$

holds if we choose $\epsilon > 0$ sufficiently small. Now

$$\int_{x_0-\epsilon}^{x_0+\epsilon} dF(x) = F(x_0 + \epsilon) - F(x_0 - \epsilon) = \epsilon_1 > 0,$$

since x_0 is a point of increase of F. Therefore we have for sufficiently small $\epsilon > 0$

$$\max\{\varphi(iv \operatorname{sgn} x_0), \varphi(-iv \operatorname{sgn} x_0)\} \geq \epsilon_1 \exp\{|v|(|x_0| - \epsilon)\}.$$

Then for $r > 0$ we see easily (see Remark 4.2.4) that

$$M(r, \varphi) = \sup_{|z| \leq r} |\varphi(z)| \geq \epsilon_1 e^{c_0 r},$$

where $c_0 > 0$ is a constant. It follows immediately that the order ρ of φ must be ≥ 1. This completes the proof of Theorem 4.2.4. ■

Remark 4.2.5. Let F be a distribution function which is concentrated in a finite interval. Then the corresponding characteristic function φ is an entire characteristic function of order 1. To see this we note that, if ρ is of the order of φ, then $\rho \leq 1$, so that from Theorem 4.2.4 we have $\rho = 1$.

Finally we prove a result due to Marcinkiewicz on entire characteristic functions which will be used later. The following simplified proof is due to Dugué.

Theorem 4.2.5. Let P_n be a polynomial of degree $n \geq 1$ with complex coefficients. Let $\varphi(t) = e^{P_n(t)}$, $t \in \mathbb{R}$. Suppose that φ is a characteristic function. Then n must be ≤ 2.

Proof. First we note that φ is an entire characteristic function. Let F be the distribution function corresponding to φ. Then the integral representation

$$\varphi(z) = e^{P_n(z)} = \int_{-\infty}^{\infty} e^{izx}\, dF(x)$$

holds for all $z \in \mathbb{C}$. Set

$$\theta(z) = \varphi(-iz) \qquad \text{for } z \in \mathbb{C}.$$

Then it can be shown by using the Hermitian property of φ that θ is of the form

$$\theta(z) = \exp\left(\sum_{k=1}^{n} a_k z^k\right),$$

where every a_k is a real number. In view of the inequality

$$|\theta(t + iv)| \leq \theta(t), \qquad t \in \mathbb{R}, \quad v \in \mathbb{R},$$

we conclude that

$$\operatorname{Re} \sum_{k=1}^{n} a_k(t + iv)^k \leq \sum_{k=1}^{n} a_k t^k$$

holds for all $t \in \mathbb{R}$ and all $v \in \mathbb{R}$. We now set $t = r \cos \theta$ and $v = r \sin \theta$. Then we see easily that for sufficiently large $r > 0$ the inequality

$$a_n \cos n\theta \leq a_n(\cos \theta)^n$$

holds for all θ. On the other hand, it can be verified that for $n > 3$ neither of the following inequalities:

(a) $\cos n\theta - (\cos \theta)^n \leq 0$,
(b) $\cos n\theta - (\cos \theta)^n \geq 0$,

holds for all θ. Hence for $n > 3$ we must have $a_n = a_{n-1} = \cdots = a_4 = 0$.

Next we consider the case $n = 3$. First we note that $\ln \varphi(iv)$ is a convex function of the real variable v. This implies that $-a_1 v + a_2 v^2 - a_3 v^3$ is convex. Hence it follows that $a_3 = 0$, so that the degree of P_n must be ≤ 2. This completes the proof. ■

Remark 4.2.6. If φ is a characteristic function of the form $\varphi(t) = \exp\{P(t)\}$, where P is a polynomial in t, then φ is the characteristic function of a normal (possibly degenerate) distribution.

4.3 DECOMPOSITION OF CHARACTERISTIC FUNCTIONS

4.3.1 General Decomposition Theorems

In Section 4.1 we introduced the notion of a decomposable characteristic function. We now prove some general results concerning the decomposition of characteristic functions. The following result is due to Khintchine.

Theorem 4.3.1. Let φ be the characteristic function of a nondegenerate distribution. Then φ admits a decomposition

$$\varphi(t) = \theta(t)\psi(t), \qquad t \in \mathbb{R},$$

where ψ is a characteristic function which does not have any indecomposable factor, while θ can be written as the convergent product of at most a countable sequence of indecomposable characteristic functions.

For the proof of Theorem 4.3.1 we need some lemmas which are of independent interest.

Let φ be a characteristic function. Then clearly there exists a positive real number a such that $|\varphi(t)| > 0$ for $0 \leq t \leq a$. For fixed $a > 0$ with this property we set

$$N_a(\varphi) = -\int_0^a \ln |\varphi(t)|\, dt. \qquad (4.3.1)$$

Lemma 4.3.1. The function $N_a(\varphi)$ defined in (4.3.1) has the following properties:

(a) $N_a(\varphi) \geq 0$.
(b) $N_a(e^{itb}) = 0$ for $b \in \mathbb{R}$.
(c) $\varphi = \varphi_1\varphi_2 \Rightarrow N_a(\varphi) = N_a(\varphi_1) + N_a(\varphi_2)$.
(d) $N_a(\varphi) \geq \int_0^a [1 - |\varphi(t)|]\, dt$.
(e) $N_a(\varphi) = 0$ if and only if φ is the characteristic function of a degenerate distribution.

Proof. The proof of Lemma 4.3.1 is straightforward. ■

Lemma 4.3.2. Let $\{F_n : n \geq 1\}$ be a sequence of distribution functions, and let $\{\varphi_n\}$ be the corresponding sequence of characteristic functions. Suppose that the point $x = 0$ is a median of F_n for every n and that there exists a real number $a > 0$ such that $\lim_{n\to\infty} N_a(\varphi_n) = 0$. Then F_n converges completely to the distribution that is degenerate at zero.

Proof. Clearly, for sufficiently large n and $0 \leq t \leq a$, $\varphi_n(t) \neq 0$. Then, using (d) of Lemma 4.3.1, we see that

$$\int_0^a [1 - |\varphi_n(t)|^2]\, dt \leq 2 \int_0^a [1 - |\varphi_n(t)|]\, dt$$

$$\leq 2N_a(\varphi_n)$$

for n sufficiently large. On the other hand, using inequality (4.1.1), we conclude that

$$\int_0^{2a} [1 - |\varphi_n(t)|^2]\, dt = 2 \int_0^a [1 - |\varphi_n(2t)|^2]\, dt \leq 8 \int_0^a [1 - |\varphi_n(t)|^2]\, dt.$$

Since $N_a(\varphi_n) \to 0$ as $n \to \infty$, we conclude that

$$\lim_{n\to\infty} \int_0^{2a} [1 - |\varphi_n(t)|^2]\, dt = 0.$$

Clearly, then, for sufficiently large n, $\varphi_n(t) \neq 0$ for $0 \leq t \leq 2a$. Repeating the above argument, we conclude that for every $T > 0$

$$\lim_{n\to\infty} \int_0^T [1 - |\varphi_n(t)|^2]\, dt = 0. \tag{4.3.2}$$

For every n set $\tilde{F}_n(x) = 1 - F_n(-x - 0)$, and write

$$F_n^s = F_n * \tilde{F}_n.$$

Then F_n^s is a symmetric distribution function with characteristic function $|\varphi_n|^2$. Let Φ be the distribution function of the standard normal distribution, and set

$$G_n = F_n^s * \Phi \qquad \text{for } n \geq 1.$$

Then G_n has characteristic function θ_n, given by

$$\theta_n(t) = e^{-t^2/2} |\varphi_n(t)|^2.$$

Note that G_n is absolutely continuous and $\theta_n(t)$ is absolutely integrable on $\mathbb{R}$ so that, using inversion formula (3.3.1), we have

$$G_n(x) - G_n(-x) = \frac{1}{\pi} \int_{-\infty}^{\infty} \frac{\sin tx}{t} e^{-t^2/2} |\varphi_n(t)|^2\, dt$$

for every $x \in \mathbb{R}$. Since G_n is symmetric and the integrand on the right-hand side is an even function in t, we obtain

$$G_n(x) - \frac{1}{2} = \frac{1}{\pi} \int_0^\infty \frac{\sin tx}{t} e^{-t^2/2} |\varphi_n(t)|^2 \, dt$$

$$= \frac{1}{\pi} \int_0^\infty \frac{\sin tx}{t} e^{-t^2/2} \, dt$$

$$+ \frac{1}{\pi} \int_0^\infty \frac{\sin tx}{t} e^{-t^2/2} [|\varphi_n(t)|^2 - 1] \, dt.$$

Setting

$$J_n(x) = \frac{1}{\pi} \int_0^\infty \frac{\sin tx}{t} e^{-t^2/2} [|\varphi_n(t)|^2 - 1] \, dt$$

and using the Fourier inversion formula, we obtain

$$G_n(x) = \Phi(x) + J_n(x).$$

Now, for any $T > 0$,

$$\left| \frac{1}{\pi} \int_0^T \frac{\sin tx}{t} e^{-t^2/2} [|\varphi_n(t)|^2 - 1] \, dt \right| \le \frac{|x|}{\pi} \int_0^T [1 - |\varphi_n(t)|^2] \, dt,$$

so that it follows from (4.3.2) that

$$\lim_{n \to \infty} J_n(x) = 0 \qquad \text{for } x \in \mathbb{R}.$$

Consequently, $G_n \xrightarrow{w} \Phi$ as $n \to \infty$. By the continuity theorem we conclude that

$$\lim_{n \to \infty} \theta_n(t) = e^{-t^2/2} \lim_{n \to \infty} |\varphi_n(t)|^2 = e^{-t^2/2},$$

so that

$$\lim_{n \to \infty} |\varphi_n(t)|^2 = 1 \qquad \text{for } t \in \mathbb{R}.$$

Thus F_n^s converges completely to the distribution degenerate at zero.

Writing

$$F_n^s(x) = \int_{-\infty}^\infty F_n(x - y) \, d\tilde{F}_n(y),$$

we obtain

$$F_n^s(x) \ge \int_{-\infty}^\epsilon F_n(x - y) \, d\tilde{F}_n(y)$$

$$\ge F_n(x - \epsilon)[1 - F_n(-\epsilon - 0)]$$

for every $\epsilon > 0$. Since $x = 0$ is a median of F_n, we also have

$$F_n^s(x) \geq \tfrac{1}{2}F_n(x - \epsilon), \qquad \epsilon > 0, \quad x \in \mathbb{R},$$

so that for $x < 0$, since $F_n^s(x) \to 0$, we have

$$\lim_{n\to\infty} F_n(x) = 0.$$

Proceeding in a similar manner, we show that

$$\lim_{n\to\infty} F_n(x) = 1 \qquad \text{for } x > 0.$$

This completes the proof of Lemma 4.3.2. ∎

Proof of Theorem 4.3.1. Clearly there exists an $a > 0$ such that $\varphi(t) \neq 0$ for $|t| \leq a$. Fix a, and set $N_a(\varphi) = \alpha$. Since φ is the characteristic function of a nondegenerate distribution, we see that $\alpha > 0$. If φ has no indecomposable factors, there is nothing to prove. Otherwise, suppose that φ has indecomposable factors.

Suppose φ has an indecomposable factor p_1 such that $N_a(p_1) > \alpha/2$. Then φ admits a factorization of the form $\varphi(t) = p_1(t)\varphi_1(t)$, where $N_a(\varphi_1) < \alpha/2$.

We repeat the same argument with φ_1, using $\alpha/4$ in place of $\alpha/2$. If φ_1 has an indecomposable factor p with $N_a(p) > \alpha/4$, we have the decomposition

$$\varphi(t) = p_1(t)p_2(t)\varphi_2(t),$$

where $N_a(p_j) > \alpha/4$, $j = 1, 2$, while every indecomposable factor p of φ_2 satisfies $N_a(p) < \alpha/4$.

On the other hand, if φ has no indecomposable factor p with $N_a(p) > \alpha/2$, we search for indecomposable factors p with $N_a(p) > \alpha/4$. In this case we easily obtain a decomposition into at most four factors:

$$\varphi(t) = p_1(t) \cdots p_{n_2}(t)\varphi_2(t),$$

where $1 \leq n_2 \leq 3$ and the p_j are indecomposable factors with $N_a(p_j) > \alpha/4$, while every indecomposable factor p of φ_2 satisfies $N_a(p) < \alpha/4$.

We continue this process and see easily that φ admits the decomposition

$$\varphi(t) = p_1(t)p_2(t) \cdots p_{n_k}(t)\varphi_k(t), \tag{4.3.3}$$

where $1 \leq n_k \leq 2^k - 1$ and p_j is indecomposable with $N_a(p_j) > \alpha/2^k$, $j = 1, 2, \ldots, n_k$, while every indecomposable factor p of φ_k satisfies $N_a(p) < \alpha/2^k$.

If for some $k \geq 1$ the characteristic function φ_k has no indecomposable factors, the above process terminates with a finite number of indecomposable factors, and the theorem holds trivially. Otherwise, decomposition (4.3.3) can be continued indefinitely, and the factors $\{p_j\}$ then form a countably infinite sequence. Since $N_a(\varphi) > \sum_{j=1}^{k} N_a(p_j)$ for $k \geq 1$, the series $\sum_{j=1}^{\infty} N_a(p_j)$ converges so that $\sum_{j=n+1}^{n+m} N_a(p_j) \to 0$ as $n \to \infty$ uniformly in m (≥ 1). From Lemma 4.3.2 we conclude that there exist real numbers $\alpha_{n,n'}$ ($n' > n \geq 1$) such that

$$\lim_{n\to\infty} e^{it\alpha_{n,n'}} \prod_{j=n}^{n'} p_j(t) = 1 \tag{4.3.4}$$

uniformly in every finite t-interval $|t| \leq T$.

We now use the polar transformation

$$p_j(t) = \rho_j(t)e^{i\omega_j(t)}$$

in (4.3.4), where ω_j and ρ_j are real-valued with $\omega_j(0) = 0$. On taking logarithms and then equating real and imaginary parts, we see that, as $n \to \infty$,

$$t\alpha_{n,n'} + \sum_{j=n}^{n'} \omega_j(t) = 2\pi\beta_{n,n'}(t) + o(1) \tag{4.3.5}$$

uniformly in $|t| \leq T$, where $\beta_{n,n'}(t)$ are some integers. Clearly the left-hand side of (4.3.5) is continuous in t, and, moreover, $\beta_{n,n'}(0) = 0$, so that it follows that $\beta_{n,n'}(t) = 0$ for sufficiently large n and $|t| \leq T$. Therefore we have

$$t\alpha_{n,n'} + \sum_{j=n}^{n'} \omega_j(t) = o(1) \qquad \text{as } n \to \infty \quad (n' > n)$$

uniformly in $|t| \leq T$. Without loss of generality we assume that $\omega_j(1) = 0$, since otherwise we can multiply each $p_j(t)$ by $e^{-it\omega_j(1)}$. It follows immediately that $\alpha_{n,n'} = o(1)$ as $n \to \infty$, so that $\sum_{j=n}^{n'} \omega_j(t) = o(1)$ as $n \to \infty$ uniformly in $|t| \leq T$. Thus from (4.3.4) we conclude that

$$\lim_{n\to\infty} \prod_{j=n}^{n'} p_j(t) = 1 \qquad (n' > n) \tag{4.3.6}$$

uniformly in $|t| \leq T$. This implies that the infinite product $\prod_{j=1}^{\infty} p_j$ converges to a characteristic function, say θ.

On the other hand, it follows easily from (4.3.3) that

$$\varphi_k(t) = \varphi_{k+m}(t) \prod_{j=k+1}^{k+m} p_j(t),$$

so that

$$|\varphi_k(t) - \varphi_{k+m}(t)| \leq \left| \prod_{j=k+1}^{k+m} p_j(t) - 1 \right|.$$

In view of (4.3.6)

$$\lim_{k \to \infty} |\varphi_k(t) - \varphi_{k+m}(t)| = 0$$

uniformly in $|t| \leq T$ and $m \geq 1$. This implies that $\{\varphi_k\}$ also converges to a characteristic function, say ψ. It follows from (4.3.3) that

$$\varphi(t) = \theta(t)\psi(t).$$

Finally we show that ψ has no indecomposable factors. Suppose that ψ has an indecomposable factor p such that $N_a(p) = \beta > 0$. Then it follows from (4.3.3) that p is an indecomposable factor of φ_k for every $k \geq 1$, so that $\beta = N_a(p) < \alpha/2^k$ for every k, which is a contradiction. This completes the proof of Theorem 4.3.1. ■

Remark 4.3.1. The factorization of a characteristic function φ in Theorem 4.3.1 is not unique, as was demonstrated in Example 4.1.2.

We now show that the factor ψ in Theorem 4.3.1 which has no indecomposable factors is i.d.

Theorem 4.3.2 (Khintchine). A characteristic function that has no indecomposable factors is i.d.

Proof. Let φ be a characteristic function which has no indecomposable factors. Choose and fix $a > 0$ such that $\varphi(t) \neq 0$ for $|t| \leq a$. Denote by D a decomposition of φ of the form

$$\varphi(t) = \varphi_1(t)\varphi_2(t) \cdots \varphi_n(t), \tag{4.3.7}$$

and for this decomposition set

$$\nu(D) = \max_{1 \leq j \leq n} N_a(\varphi_j),$$

where N_a is as defined in (4.3.1). Next set

$$\nu = \inf_D \nu(D),$$

where the infimum is taken over all possible decompositions D of φ of the form (4.3.7). We first show that $\nu = 0$.

In fact, from the definition of ν it follows that for every positive integer n there exists a decomposition D_n, say

$$\varphi(t) = \prod_{j=1}^{k_n} \varphi_{n,j}(t), \tag{4.3.8}$$

such that

$$\nu \le \nu(D_n) < \nu + \frac{1}{n} \qquad (n \ge 1).$$

We now investigate the decomposition D_n of (4.3.8). Let $\varphi_1^{(n)}$ be a factor of φ in (4.3.8) for which $\nu(D_n) = N_a(\varphi_1^{(n)})$. For convenience we denote by $\varphi_2^{(n)}$ the product of all the remaining factors of φ. Then it follows that for every $n \ge 1$

$$\varphi(t) = \varphi_1^{(n)}(t)\varphi_2^{(n)}(t)$$

(4.3.9) and

$$\nu \le N_a(\varphi_1^{(n)}) < \nu + \frac{1}{n}.$$

Let F, $F_1^{(n)}$, and $F_2^{(n)}$ be the distribution functions corresponding to φ, $\varphi_1^{(n)}$, and $\varphi_2^{(n)}$, respectively. Since a translation does not affect $N_a(\varphi)$, we may assume without loss of generality that $x = 0$ is a median of $F_2^{(n)}$ for every $n \ge 1$.

In view of Theorem 3.1.2 the sequence $\{F_1^{(n)}\}$ contains a subsequence which converges weakly to a bounded, nondecreasing, right-continuous function, say F_1. For simplicity in notation we may denote this subsequence by $\{F_1^{(n)}\}$ itself. We show that F_1 is a distribution function. It suffices to show that, for every $\epsilon > 0$ and for sufficiently large $b > 0$ and sufficiently large n, we have

$$F_1^{(n)}(-b) < \epsilon \qquad \text{and} \qquad F_1^{(n)}(b) > 1 - \epsilon. \tag{4.3.10}$$

Suppose that one of these inequalities, say the second, is not satisfied. Then, for some $\epsilon_0 > 0$, sufficiently large $b > 0$, and sufficiently large n, we have the estimate

$$\begin{aligned} 1 - F(b) &= \int_{-\infty}^{\infty} [1 - F_1^{(n)}(b - y)]\, dF_2^{(n)}(y) \\ &\ge \int_{-1}^{\infty} [1 - F_1^{(n)}(b - y)]\, dF_2^{(n)}(y) \\ &\ge [1 - F_1^{(n)}(b + 1)][1 - F_2^{(n)}(-1)] \\ &\ge \tfrac{1}{2}\epsilon_0, \end{aligned}$$

which contradicts the fact that F is a distribution function. Therefore the second inequality in (4.3.10) holds. Similarly we can show that the first inequality in (4.3.10) also holds for sufficiently large n and $b > 0$ and for every $\epsilon > 0$. Consequently $F_1^{(n)} \xrightarrow{c} F_1$. Let φ_1 be the characteristic function of F_1.

We next consider the sequence $\{F_2^{(n)}\}$. Proceeding in a similar manner, we can show that $\{F_2^{(n)}\}$ contains a subsequence which converges completely to a distribution function F_2. For ease in notation we identify this subsequence with $\{F_2^{(n)}\}$ itself. Let φ_2 be the characteristic function of F_2. Then we have

$$\varphi_1(t) = \lim_{n\to\infty} \varphi_1^{(n)}(t), \qquad \varphi_2(t) = \lim_{n\to\infty} \varphi_2^{(n)}(t),$$

so that it follows from (4.3.9) that

$$\varphi(t) = \varphi_1(t)\varphi_2(t) \qquad \text{and} \qquad N_a(\varphi_1) = v. \tag{4.3.11}$$

Our next step is to show that $v < \frac{1}{2}N_a(\varphi)$. Suppose $v \geq \frac{1}{2}N_a(\varphi)$. From Lemma 4.3.1 and (4.3.11) we see that $N_a(\varphi_2) \leq v$. If φ_1 and φ_2 are decomposed once more, we obtain a decomposition, say D^*:

$$\varphi(t) = \theta_1(t)\theta_2(t)\theta_3(t)\theta_4(t).$$

Clearly $v(D^*) < v$, which contradicts the definition of v. Therefore we have $v < \frac{1}{2}N_a(\varphi)$. It now follows from the definition of v that there exists a decomposition D of φ such that

$$v < v(D) < \tfrac{1}{2}N_a(\varphi).$$

Since each factor of φ in decomposition D has no indecomposable factors, we can apply the above result to one such factor of φ in D and thus conclude that

$$v < \tfrac{1}{2}v(D) < \tfrac{1}{4}N_a(\varphi).$$

We repeat this procedure and thus obtain $v < (1/2^n)N_a(\varphi)$ for every $n \geq 1$. It therefore follows that $v = 0$.

We now return to the proof of Theorem 4.3.2. Let $\epsilon_n > 0$, $\epsilon_n \downarrow 0$ as $n \to \infty$. It follows that for every $n \geq 1$ there exists a decomposition D_n of the form (4.3.8) such that

$$N_a(\varphi_{n,j}) < \epsilon_n$$

for $j = 1, 2, \ldots, k_n$, $n \geq 1$, and, also, $k_n \to \infty$ as $n \to \infty$.

From Lemma 4.3.2 there exist constants $\alpha_{n,j}$ ($j = 1, 2, \ldots, k_n$; $n \geq 1$) such that

$$\lim_{n\to\infty} \{\varphi_{n,j}(t)e^{it\alpha_{n,j}}\} = 1 \tag{4.3.12}$$

uniformly in every finite interval $|t| \leq T$ and uniformly for $1 \leq j \leq k_n$. We now use the polar transformation

$$\begin{cases} \varphi_{n,j}(t) = \rho_{n,j}(t)e^{i\omega_{n,j}(t)}, \\ \varphi(t) = \rho(t)e^{i\omega(t)}, \end{cases} \tag{4.3.13}$$

where ω and ρ are real-valued and $\omega_{n,j}(0) = \omega(0) = 0$. Using (4.3.13) in (4.3.12), we obtain easily

$$\lim_{n\to\infty} \rho_{n,j}(t) = 1 \qquad \text{and} \qquad \lim_{n\to\infty} [\omega_{n,j}(t) + t\alpha_{n,j}] = 0$$

uniformly in $|t| \le T$ and for $1 \le j \le k_n$. In particular,

$$\lim_{n\to\infty} [\omega_{n,j}(1) + \alpha_{n,j}] = 0,$$

so that

$$\lim_{n\to\infty} [\omega_{n,j}(t) - t\omega_{n,j}(1)] = 0.$$

Set

$$\theta_{n,j}(t) = \varphi_{n,j}(t) \exp\left\{it\left[\frac{\omega(1)}{k_n} - \omega_{n,j}(1)\right]\right\}.$$

Then in view of (4.3.13) we get

$$\theta_{n,j}(t) = \rho_{n,j}(t) \exp\left\{i\left[\omega_{n,j}(t) - t\omega_{n,j}(1) + \frac{t\omega(1)}{k_n}\right]\right\}. \tag{4.3.14}$$

It follows that

$$\lim_{n\to\infty} \theta_{n,j}(t) = 1 \tag{4.3.15}$$

uniformly in every finite interval $|t| \le T$ and for $1 \le j \le k_n$.

Now

$$\varphi(t) = \prod_{j=1}^{k_n} \varphi_{n,j}(t) = \prod_{j=1}^{k_n} [\rho_{n,j}(t) e^{i\omega_{n,j}(t)}],$$

and in view of (4.3.14) we see easily that

$$\varphi(t) = \prod_{j=1}^{k_n} \theta_{n,j}(t).$$

In view of (4.3.15) it now follows from Theorem 5.2.1 that φ is i.d. The proof of Theorem 4.3.2 is now complete. ∎

Example 4.3.1. We construct an example of an i.d. characteristic function which is the product of an indecomposable characteristic function and an i.d. characteristic function. Let $0 < \alpha < 1$. Then the function

$$\varphi(t) = \frac{1-\alpha}{1-\alpha e^{it}} = \exp\{\ln(1-\alpha) - \ln(1-\alpha e^{it})\}$$

$$= \exp\left\{\sum_{j=0}^{\infty} \frac{\alpha^j}{j} (e^{ijt} - 1)\right\}$$

is the characteristic function of an i.d. distribution, as shown in Example 4.1.6. Actually φ is the characteristic function of the geometric distribution given by

$$P\{X = j\} = (1 - \alpha)\alpha^j, \qquad j = 0, 1, 2, \ldots, \quad 0 < \alpha < 1.$$

On the other hand, for $|x| < 1$ we can write $(1 - x)^{-1}$ as

$$(1 - x)^{-1} = \prod_{j=0}^{\infty}(1 + x^{2^j}),$$

so that we can rewrite

$$\varphi(t) = \prod_{j=0}^{\infty}\left(\frac{1 + \alpha^{2^j}e^{2^j it}}{1 + \alpha^{2^j}}\right).$$

Here φ admits a representation as a product of a countable sequence of characteristic functions

$$\varphi_j(t) = \frac{1 + \alpha^{2^j}e^{2^j it}}{1 + \alpha^{2^j}}, \qquad j = 0, 1, 2, \ldots$$

Note that φ_j is the characteristic function of the random variable which takes values 0 and 2^j with probabilities $1/(1 + \alpha^{2^j})$ and $\alpha^{2^j}/(1 + \alpha^{2^j})$, respectively, and hence φ_j is indecomposable for each $j = 0, 1, 2, \ldots$.

4.3.2 Decomposition of Analytic Characteristic Functions

We now consider some useful results on the decomposition of analytic characteristic functions. The following result is due to Raikov.

Theorem 4.3.3. Let φ be an analytic characteristic function which is holomorphic in the strip $-\alpha < \text{Im } z < \beta (\alpha, \beta > 0)$. Suppose φ admits the decomposition $\varphi = \varphi_1\varphi_2$. Then both φ_1 and φ_2 are analytic characteristic functions which are holomorphic at least in the strip $-\alpha < \text{Im } z < \beta$.

Proof. Let F, F_1, and F_2 be the distribution functions corresponding to φ, φ_1, and φ_2, respectively. Then for $x \in \mathbb{R}$

$$\text{(4.3.16)} \quad F(x) = \int_{-\infty}^{\infty} F_1(x - y)\, dF_2(y) = \int_{-\infty}^{\infty} F_2(x - y)\, dF_1(y).$$

Clearly

$$\text{(4.3.17)} \quad F(a_2) - F(a_1) \geq \int_{-A}^{B} [F_1(a_2 - y) - F_1(a_1 - y)]\, dF_2(y)$$

for $A > 0$, $B > 0$, and $a_1, a_2 \in \mathbb{R}$ such that $a_2 > a_1$. Let $v \in \mathbb{R}$ be fixed such that $-\alpha < v < \beta$. Since φ is holomorphic in $-\alpha < \operatorname{Im} z < \beta$, it follows that $\int_{-\infty}^{\infty} e^{vx}\, dF(x)$ exists and is finite and that the inequality

$$\int_{-\infty}^{\infty} e^{vx}\, dF(x) \geq \int_{a}^{b} e^{vx}\, dF(x)$$

holds for $a, b \in \mathbb{R}$ with $a < b$. Consider the sequence of subdivisions of $[a, b]$ defined by

$$x_{v,n} = a + (v - 1)\frac{b-a}{2^n}, \qquad v = 1, 2, \ldots, 2^n + 1, \quad n = 1, 2, \ldots .$$

Clearly

(4.3.18) $$x_{2v-1,n+1} = x_{v,n}, \qquad v = 1, 2, \ldots, 2^n + 1.$$

In view of the definition of a Riemann–Stieltjes integral we have

$$\int_a^b e^{vx}\, dF(x) = \lim_{n\to\infty} \sum_{v=1}^{2^n} e^{vx_{v,n}}[F(x_{v+1,n}) - F(x_{v,n})]$$

(4.3.19) $$= \lim_{n\to\infty} \sum_{v=1}^{2^n} e^{vx_{v+1,n}}[F(x_{v+1,n}) - F(x_{v,n})].$$

Set

$$h_{v,n}(y, v) = \begin{cases} e^{vx_{v,n}}[F_1(x_{v+1,n} - y) - F_1(x_{v,n} - y)] & \text{if } v > 0, \\ e^{vx_{v+1,n}}[F_1(x_{v+1,n} - y) - F_1(x_{v,n} - y)] & \text{if } v < 0, \end{cases}$$

for $v = 1, 2, \ldots, 2^n$ and $y \in \mathbb{R}$. Let

$$g_n(y, v) = \sum_{v=1}^{2^n} h_{v,n}(y, v).$$

Using inequality (4.3.17) in (4.3.19), we conclude that

(4.3.20) $$\int_a^b e^{vx}\, dF(x) \geq \lim_{n\to\infty} \int_{-A}^{B} g_n(y, v)\, dF_2(y).$$

On the other hand, we note that

$$x_{2v-1,n+1} < x_{2v,n+1} < x_{2v+1,n+1},$$

so that in view of (4.3.18) we obtain, after some elementary computation,

$$h_{v,n}(y, v) \leq h_{2v,n+1}(y, v) + h_{2v-1,n+1}(y, v).$$

Summing over v, we obtain

$$0 \leq g_n(y, v) \leq g_{n+1}(y, v) \qquad \text{for } n \geq 1, \quad y \in \mathbb{R}.$$

Now, from the definition of a Riemann–Stieltjes integral, it follows that

$$\lim_{n\to\infty} g_n(y, v) = \int_{a-y}^{b-y} e^{v(y+u)}\, dF_1(u) < \infty.$$

In view of the monotone convergence theorem, (4.3.20) gives

$$\int_a^b e^{vx}\, dF(x) \geq \int_{-A}^{B} \lim_{n\to\infty} g_n(y, v)\, dF_2(y)$$

$$(4.3.21) \qquad = \int_{-A}^{B} \left[\int_{a-y}^{b-y} e^{v(y+u)}\, dF_1(u)\right] dF_2(y).$$

Note that $-A \leq y \leq B$, so that

$$\int_{a-y}^{b-y} e^{vu}\, dF_1(u) \geq \int_{a+A}^{b-B} e^{vu}\, dF_1(u).$$

Then it follows from (4.3.21) and Fubini's theorem that

$$\infty > \int_{-\infty}^{\infty} e^{vx}\, dF(x) \geq \int_a^b e^{vx}\, dF(x)$$

$$\geq \left[\int_{-A}^{B} e^{vy}\, dF_2(y)\right]\left[\int_{a+A}^{b-B} e^{vu}\, dF_1(u)\right]$$

holds for all $A > 0$, $B > 0$ and all a, b $(a < b)$ real. It follows immediately that

$$\int_{-\infty}^{\infty} e^{vx}\, dF_1(x) \qquad \text{and} \qquad \int_{-\infty}^{\infty} e^{vy}\, dF_2(y)$$

both exist and are finite; moreover, the inequality

$$(4.3.22) \qquad \int_{-\infty}^{\infty} e^{vx}\, dF(x) \geq \left(\int_{-\infty}^{\infty} e^{vx}\, dF_1(x)\right)\left(\int_{-\infty}^{\infty} e^{vy}\, dF_2(y)\right)$$

holds for $-\alpha < v < \beta$. Clearly, then, the integrals

$$\varphi_1(z) = \int_{-\infty}^{\infty} e^{izx}\, dF_1(x) \qquad \text{and} \qquad \varphi_2(z) = \int_{-\infty}^{\infty} e^{izy}\, dF_2(y)$$

both exist and are finite for all complex z such that $-\alpha < \text{Im } z < \beta$. In other words, φ_1 and φ_2 are both analytic characteristic functions which are holomorphic at least in the strip $-\alpha < \text{Im } z < \beta$. This completes the proof. ■

Remark 4.3.2. By analytic continuation we note that the relation $\varphi(z) = \varphi_1(z)\varphi_2(z)$ holds for all z such that $-\alpha < \text{Im } z < \beta$. This implies that

$$\int_{-\infty}^{\infty} e^{izx}\, dF(x) = \int_{-\infty}^{\infty} e^{izx}\, dF_1(x) \int_{-\infty}^{\infty} e^{izx}\, dF_2(x) \tag{4.3.23}$$

holds for all z with $-\alpha < \text{Im } z < \beta$.

Corollary 1 to Theorem 4.3.3. With the same notations and assumptions as in Theorem 4.3.3, there exist $C_1 > 0$, $C_2 > 0$, and a_1, $a_2 > 0$ such that the inequalities

$$\varphi_1(-iv) \le C_1 e^{a_1|v|}\varphi(-iv) \quad \text{and} \quad \varphi_2(-iv) \le C_2 e^{a_2|v|}\varphi(-iv) \tag{4.3.24}$$

hold for all v in $-\alpha < v < \beta$.

Proof. Let c_1, c_2 be two real numbers such that

$$F_2(c_1) > 0 \qquad \text{while } F_2(c_2) < 1.$$

Then it follows that for $v > 0$

$$\int_{-\infty}^{\infty} e^{vx}\, dF_2(x) \ge \int_{c_2}^{\infty} e^{vx}\, dF_2(x) \ge e^{c_2 v}[1 - F_2(c_2)].$$

Similarly, for $v < 0$

$$\int_{-\infty}^{\infty} e^{vx}\, dF_2(x) \ge e^{c_1 v} F_2(c_1).$$

Set $C_1^{-1} = \min\{F_2(c_1),\ 1 - F_2(c_2)\}$ and $a_1 = \max\{|c_1|, |c_2|\}$. Clearly $C_1 > 0$ and $a_1 > 0$. Using inequality (4.3.22), we obtain

$$\int_{-\infty}^{\infty} e^{vx}\, dF_1(x) \le C_1 e^{a_1|v|} \int_{-\infty}^{\infty} e^{vx}\, dF(x),$$

which proves the first inequality in (4.3.24). The second inequality can be proved in a similar manner.

In the case where φ is an entire characteristic function the following result holds.

Corollary 2 to Theorem 4.3.3. Let φ be an entire characteristic function which admits a decomposition $\varphi = \varphi_1\varphi_2$. Then both φ_1 and φ_2 are entire characteristic functions whose orders cannot exceed the order of φ.

Proof. That φ_1 and φ_2 are entire characteristic functions follows immediately from Theorem 4.3.3. For $r > 0$ let $M(r, \varphi)$ and $M(r, \varphi_1)$ be as defined in (4.2.6). It follows from Corollary 1 that

$$M(r, \varphi_1) \leq C_1 e^{a_1 r} M(r, \varphi), \qquad r > 0.$$

The proof is now easily completed by using the definition of the order given by (4.2.8).

Remark 4.3.3. Let φ be an entire characteristic function without zeros, so that φ can be written in the form $\varphi(z) = e^{\psi(z)}$, where ψ is an entire function. Then every factor φ_1 of φ is also an entire characteristic function without zeros and hence can be written as $\varphi_1(z) = e^{\psi_1(z)}$, where ψ_1 is an entire characteristic function.

Remark 4.3.4. We know (Proposition 4.1.1) that an infinitely divisible characteristic function does not vanish for real values of its argument. Let φ be an analytic characteristic function which is infinitely divisible. Then $\varphi(z)$, z complex, cannot have zeros in the strip in which it is holomorphic. Indeed, since φ is i.d., $[\varphi(t)]^{1/n}$ is a characteristic function for every $n \geq 1$ and is a factor of φ. In view of Theorem 4.3.3 $[\varphi(z)]^{1/n}$ is an analytic characteristic function which is holomorphic at least in the strip in which $\varphi(z)$ is. If $\varphi(z)$ had a zero inside this strip, say at z_0, then $[\varphi(z)]^{1/n}$ would have a singularity at z_0 for sufficiently large n. But this is impossible. It also follows immediately that, if φ is an entire characteristic function which is i.d., φ has no zeros.

We conclude our discussion with the following remark on the factorization of i.d. characteristic functions.

Remark 4.3.5. It is easy to construct examples of characteristic functions (even i.d. characteristic functions) which admit several decompositions. But the cancellation law does not hold unless all the characteristic functions are i.d. Let $\varphi_1(t) = 1 - |t|$ if $|t| \leq 1$ and $= 0$ if $|t| > 1$, and let φ_2 be a real-valued periodic function with period 2 defined by the Fourier series

$$\varphi_2(t) = \frac{1}{2} + \sum_{n=0}^{\infty} \frac{4\cos(2n+1)\pi t}{(2n+1)^2\pi^2}.$$

Then φ_2 agrees with φ_1 for $|t| \leq 1$, and for all $t \in \mathbb{R}$ we have

$$\varphi_1(t)\varphi_2(t) = \varphi_1(t)\varphi_1(t).$$

But $\varphi_1 \neq \varphi_2$. If, on the other hand, φ is an i.d. characteristic function that can be decomposed into two i.d. factors φ_1 and φ_2, then φ and φ_1 determine φ_2

uniquely, so that the cancellation law holds. This result is an immediate consequence of the uniqueness of the canonical representation of an i.d. characteristic function. Note, however, that these are not necessarily the only possible decompositions.

Example 4.3.2. Let $\alpha, \beta > 0$ with $\beta > 2\sqrt{2}\alpha$, and write $z = \alpha + i\beta$. Define φ by

$$\varphi(t) = \left\{1 + \frac{it}{z}\right\}\left\{1 + \frac{it}{\bar{z}}\right\}\left\{1 - \frac{it}{\alpha}\right\}^{-1}\left\{1 - \frac{it}{z}\right\}^{-1}\left\{1 - \frac{it}{\bar{z}}\right\}^{-1}.$$

By expanding φ into partial fractions and using Theorem 3.3.2, integrating term by term, we see that φ is a characteristic function. Clearly $\varphi(-t)$ is also a characteristic function, and so also is

$$|\varphi(t)|^2 = \varphi(t)\varphi(-t) = \frac{\alpha^2}{\alpha^2 + t^2}.$$

In Example 4.1.5 we showed that $|\varphi(t)|^2$ is infinitely divisible, and hence so is $|\varphi(t)|$. It is easy to show that φ is analytic with zeros in the strip in which it is holomorphic, so that, in view of Remark 4.3.4, $\varphi(t)$ and hence also $\varphi(-t)$ are not i.d. Thus the i.d. characteristic function $|\varphi(t)|^2$ admits two decompositions:

$$|\varphi(t)|^2 = |\varphi(t)||\varphi(t)| = \varphi(t)\varphi(-t),$$

where the first decomposition has two i.d. factors, while the second has two characteristic functions neither of which is i.d.

4.3.3 Decomposition of Normal and Poisson Distributions

We now prove a result concerning the decomposition of a normal distribution.

Theorem 4.3.4 (Cramér). Let $\varphi(t) = \exp(i\mu t - \frac{1}{2}\sigma^2 t^2)$ be the characteristic function of a normal distribution with mean μ and variance σ^2. Suppose that $\varphi = \varphi_1\varphi_2$ is a decomposition of φ. Then both φ_1 and φ_2 are characteristic functions of normal distributions.

Proof. We note that φ is an entire function of order 2 without zeros. It follows from Corollary 2 to Theorem 4.3.3 and Remark 4.3.3 that φ_1 and φ_2 are entire characteristic functions without zeros with orders not exceeding 2. Moreover, we can write $\varphi_j(z) = e^{\psi_j(z)}$, $j = 1, 2$, where ψ_j is an entire function. Since the order of φ_j cannot exceed 2, it follows from the Hadamard factorization theorem (Titchmarsh [81], p. 250) that $\psi_j(z) =$

$P_j(z)$, where P_j is a polynomial of degree not exceeding 2. We see easily that P_j must be of the form

$$P_j(t) = i\mu_j t - \tfrac{1}{2}\sigma_j^2 t^2, \qquad j = 1, 2,$$

where $\mu_j \in \mathbb{R}$ and $\sigma_j^2 > 0$. This completes the proof. ■

The next result is due to Raikov and concerns the decomposition of a Poisson distribution.

Theorem 4.3.5 Let $\varphi(t) = \exp\{\lambda(e^{it} - 1)\}$ be the characteristic function of a Poisson distribution with parameter $\lambda > 0$. Let $\varphi = \varphi_1\varphi_2$ be a decomposition of φ. Then both φ_1 and φ_2 are characteristic functions of Poisson distributions.

Proof. First note that both φ_1 and φ_2 must be characteristic functions of discrete distributions. Moreover, since φ has its jump points concentrated only on the set $\{0, 1, 2, \ldots\}$, we may assume without loss of generality that the jump points of φ_1 and φ_2 are also concentrated on the same set. On the other hand, since φ is an entire characteristic function without zeros, the same is true for φ_1 and φ_2, and thus we can write

$$\varphi_1(z) = \sum_{n=0}^{\infty} a_n e^{inz}, \qquad \varphi_2(z) = \sum_{n=0}^{\infty} b_n e^{inz},$$

where $a_n, b_n \geq 0$ and $\sum_{n=0}^{\infty} a_n = 1 = \sum_{n=0}^{\infty} b_n$, and the two series converge for all complex z. Let us write $w = e^{iz}$, and set

$$\psi_1(w) = \sum_{n=0}^{\infty} a_n w^n, \qquad \psi_2(w) = \sum_{n=0}^{\infty} b_n w^n, \qquad \text{and} \qquad \psi(w) = e^{-\lambda} \sum_{n=0}^{\infty} \frac{\lambda^n}{n!} w^n.$$

Then, using the relation $\psi(w) = \psi_1(w)\psi_2(w)$ and comparing the coefficients of w^n on both sides, we obtain

$$(4.3.25) \qquad e^{-\lambda}\frac{\lambda^n}{n!} = a_0 b_n + a_1 b_{n-1} + \cdots + a_{n-1}b_1 + a_n b_0 \qquad \text{for all } n \geq 0.$$

Since $a_n \geq 0$, $b_n \geq 0$, and $a_0 b_0 = e^{-\lambda} \neq 0$, it follows from (4.3.25) that

$$(4.3.26) \qquad a_n \leq b_0^{-1} e^{-\lambda} \frac{\lambda^n}{n!} \qquad n = 1, 2, \cdots.$$

Clearly ψ_1 is also an entire function, and for every $r > 0$ we have

$$M(r, \psi_1) \leq b_0^{-1} M(r, \psi),$$

so that the order of ψ_1 cannot exceed the order of ψ. But $\psi(w) = \exp\{\lambda(w-1)\}$ is an entire function of order 1 without zeros. Therefore ψ_1 is also an entire function of order not exceeding 1 without zeros. It now follows from the Hadamard factorization theorem that ψ_1 must be exactly of order 1. Since $\psi_1(1) = 1$, we see that ψ_1 must be of the form

$$\psi_1(w) = \exp\{\lambda_1(w-1)\}, \qquad \lambda_1 > 0.$$

Clearly $\varphi_1(t) = \exp\{\lambda_1(e^{it}-1)\}$ is the characteristic function of a Poisson distribution. A similar argument can be carried out to show that φ_2 is also the characteristic function of a Poisson distribution. ∎

4.4 PROBLEMS

SECTION 4.1

1. (a) Let φ be a characteristic function, and let $b > 0$ and $0 < c < 1$ be constants. If $|\varphi(t)| \le c$ for $|t| \ge b$, show that

$$|\varphi(t)| \le 1 - \frac{1-c^2}{8b^2}t^2 \qquad \text{for } |t| < b.$$

(b) Let φ be the characteristic function of a nondegenerate distribution. Show that there exist constants $\delta > 0$, $\epsilon > 0$ such that $|\varphi(t)| < 1 - \epsilon t^2$ for $|t| \le \delta$.

2. Construct an example to show that, if φ is an i.d. characteristic function which can be written as a finite product of characteristic functions, the characteristic functions in this product need not be i.d.

3. Are the following characteristic functions i.d.?

(a) $\varphi(t) = \exp\{i\mu t - \theta|t|\}$, $\theta > 0$.
(b) $\varphi(t) = (1 + t^2/\beta^2)^{-\alpha}$, $\alpha > 0$, $\beta > 0$.
(c) $\varphi(t) = \{p[1-(1-p)e^{it}]^{-1}\}^r$, $0 < p < 1$, $r \ge 1$ is an integer.
(d) $\varphi(t) = \lambda[\lambda + 1 - \varphi(t)]^{-1}$, $\lambda > 0$.

4. Let φ be an i.d. characteristic function, and let $\alpha > 0$ be a real number. Show that $[\varphi(t)]^\alpha$ is also an i.d. characteristic function. The converse also holds.

5. Find the Lévy–Khintchine and Kolmogorov canonical representations for the following i.d. characteristic functions:

(a) Gamma: $\varphi(t) = (1 - it/\beta)^{-\alpha}$, $\alpha > 0$, $\beta > 0$.
(b) Cauchy: $\varphi(t) = e^{-\theta|t|}$, $\theta > 0$.
(c) Negative binomial: $\varphi(t) = \{p[1-(1-p)e^{it}]^{-1}\}^r$, $0 < p < 1$, $r \ge 1$ is an integer.
(d) Laplace: $\varphi(t) = (1 + t^2)^{-1}$.

6. Suppose that X_1 and X_2 are independent random variables with i.d. characteristic functions such that $X_1 + X_2$ has a Poisson distribution. Show that X_1 and X_2 both must have the Poisson distribution.

7. (Lévy Canonical Representation) Show that a complex-valued function φ on $\mathbb{R}$ is an i.d. characteristic function if and only if it can be represented in the form

$$\varphi(t) = \exp\left\{i\gamma t - \tfrac{1}{2}\sigma^2 t^2 + \left(\int_{-\infty}^{0-} + \int_{0+}^{\infty}\right)\left(e^{itx} - 1 - \frac{itx}{1+x^2}\right) dM(x)\right\}.$$

Here γ, σ^2 are constants, $\gamma \in \mathbb{R}$ and $\sigma^2 > 0$, and the function M defined on $\mathbb{R} - \{0\}$ is nondecreasing on $(-\infty, 0)$ and on $(0, \infty)$ and satisfies these conditions:

(i) $M(-\infty) = M(\infty) = 0$,
(ii) $\int_{-1}^{0} u^2\, dM(u) + \int_0^1 u^2\, dM(u) < \infty$.

Also prove that this representation is unique.

8. Find the Lévy representations of the normal and the Poisson distributions.

9. Let $\mathscr{L}$ be a class of i.d. characteristic functions such that $\varphi_1 \in \mathscr{L}$, $\varphi_2 \in \mathscr{L} \Rightarrow \alpha\varphi_1 + (1-\alpha)\varphi_2 \in \mathscr{L}, 0 \leq \alpha \leq 1$, and $\varphi_1\varphi_2 \in \mathscr{L}$. Show that $\varphi \in \mathscr{L} \Rightarrow \varphi$ is real and positive (and hence even).

10. Let φ be a real-valued function which is even, continuous, and convex on $(0, \infty)$, and satisfies $\varphi(0) = 1$ and $\varphi(t) \to p$, $0 \leq p \leq 1$, as $t \to \infty$. Show that φ is a characteristic function. Also show that, if a characteristic function φ is even, positive, and log-convex on $(0, \infty)$, it is i.d.

(Keilson and Steutel [45]; the first part generalizes Polya's theorem of Section 3.6.)

11. Let ζ be the Riemann zeta function, defined for $t > 1$ by

$$\zeta(t + iv) = \sum_n n^{-t-iv} = \prod_p (1 - p^{-t-iv})^{-1},$$

where p varies over all rational primes. Show that

$$\varphi_t(v) = \frac{\zeta(t+iv)}{\zeta(t)}$$

is an i.d. characteristic function.

SECTION 4.2

12. Show that the following characteristic functions are analytic, and determine in each case the strip of analyticity.

(a) $\varphi(t) = (1 - p + pe^{it})^n$, $0 < p < 1$.
(b) $\varphi(t) = \exp\{\lambda(e^{it} - 1)\}$, $\lambda > 0$.
(c) $\varphi(t) = \exp\{i\mu t - (t^2/2)\sigma^2\}$, $\sigma > 0$, $\mu \in \mathbb{R}$.
(d) $\varphi(t) = (1 - \beta i t)^{-\alpha}$, $\alpha > 0$, $\beta > 0$.
(e) $\varphi(t) = [(1 - (it/a)(1 - (it/a + ib)(1 - (it/a - ib))]^{-1}$, $a \geq b > 0$.

13. Show that the following characteristic functions are not analytic.
(a) $\varphi(t) = e^{-|t|^\alpha}$, $0 < \alpha < 2$.
(b) $\varphi(t) = \begin{cases} 1 - |t|, & |t| \leq 1, \\ 0, & |t| > 1. \end{cases}$

14. Which of the functions in Problem 12 are entire characteristic functions? In each case compute the order.

15. (a) Let F be a distribution function, and for $t \in \mathbb{R}$ set $M(t) = \int_{-\infty}^{\infty} e^{tx}\,dF(x)$. Show that:

(1) M is defined for all t in $(-\alpha, \beta)$, where

$$\alpha = \liminf_{x\to\infty} \left\{\frac{-\ln[1 - F(x)]}{x}\right\}$$

and

$$\beta = \liminf_{x\to\infty} \left\{\frac{-\ln F(-x)}{x}\right\},$$

and where we take α (respectively, β) to be ∞ if $F(-x) = 0$ [respectively, if $1 - F(x) = 0$] for some $x > 0$.

(2) M does not exist for $t < -\alpha$ (if α is finite) or for $t > \beta$ (if β is finite).

(b) Show that a necessary and sufficient condition for F to have an analytic characteristic function is that $\alpha > 0$, $\beta > 0$.

(c) Show that F is a distribution function of an entire characteristic function if and only if $\alpha = \beta = \infty$.

16. Let φ be a characteristic function which is analytic in $-\alpha < \text{Im } z < \beta$. Let $t \in \mathbb{R}$ and $v \in (-\alpha, \beta)$. Show that $\ln \varphi(iv)$ is a convex function.

17. Let $\mathscr{L}$ be the class of i.d. characteristic functions defined in Problem 9. If $\varphi \in \mathscr{L}$ $(\varphi \not\equiv 1)$, show that φ cannot have a finite variance. Moreover, φ cannot be analytic in any neighborhood of $t = 0$. Show that $\varphi \in \mathscr{L}$ cannot have a strip of convergence.

(Keilson and Steutel [45])

18. (The Problem of Moments) The problem of characterizing the distribution function from its moments is known as the problem of moments. According to Remark 4.2.2, conditions (i) and (ii) of Theorem 4.2.2 give a set of necessary and sufficient conditions under which the moments determine the distribution uniquely.

(a) Let X be a random variable with probability density function

$$f(x) = ce^{-|x|^\alpha}, \qquad 0 < \alpha < 1, \quad x \in \mathbb{R},$$

where $c = (\int_{-\infty}^{\infty} e^{-|x|^\alpha}\,dx)^{-1}$. Show that $\mathscr{E}|X|^n < \infty$ for every $n \geq 0$ but the moment-generating function of X does not exist.

(In Example 3.9.4 we saw that, when the moment-generating function does not exist but moments of all orders do, they may not determine the distribution function.)

(b) For the random variable X with probability density function

$$f(x) = \tfrac{1}{2}\exp(-|\sqrt{x}|), \qquad x > 0,$$
$$= 0 \qquad \text{otherwise},$$

show that condition (i) of Theorem 4.2.2 holds, but condition (ii) is violated. Hence the moment-generating function does not exist.

(In this case, however, $\sum_{n=1}^{\infty} (\mathscr{E}X^n)^{-1/(2n)} = \infty$ holds, which is the Carleman [11] sufficient condition for the unique determination of a distribution function from its moments.)

(c) (Stieltjes) Let (a, b) be a finite interval, and let F_1, F_2 be two distribution functions that assign all the mass to (a, b). Show that, if

$$\int_a^b x^k \, dF_1(x) = \int_a^b x^k \, dF_2(x)$$

holds for all $k = 0, 1, 2, \ldots$, then $F_1 = F_2$ on (a, b).

(d) (Hausdorff) Show that there exists a unique distribution function F, with mass on $[0, 1]$, such that

$$\int_0^1 x^n \, dF(x) = c_n \qquad (n = 1, 2, \ldots, c_0 = 1)$$

if and only if

$$(-1)^n \Delta^n c_k \geq 0 \qquad (n \geq 0).$$

Here

$$\Delta c_k = c_{k+1} - c_k, \Delta^2 c_k = c_{k+2} - 2c_{k+1} + c_k, \ldots, \Delta^r c_k = \sum_{j=0}^{r} \binom{r}{j} (-1)^{r-j} c_{k+j}.$$

19. (a) Let f be a probability density function which has a normal component. Show that $f(0) > 0$.

(b) Show that the function φ defined on $\mathbb{R}$ by

$$\varphi(t) = (1 - t^2)e^{-t^2/2}$$

is a characteristic function of an absolutely continuous distribution function.

(c) Show that the characteristic function φ defined in (b) has no normal component and hence is indecomposable.

20. Show that the converse of Theorem 4.3.2 is not true. In other words, show that there do exist i.d. laws which have indecomposable factors.

21. Consider the indecomposable characteristic function $\varphi(t) = (1 - \frac{1}{2}t^2)e^{-t^2/4}$, and define $\psi(t) = [\varphi(t)]^2$, $t \in \mathbb{R}$. Show that ψ has a normal component.

22. (a) Show that the characteristic function

$$\varphi_1(t) = \left(\frac{\lambda^2 - 1}{\lambda^2 - e^{2it}} \right)^{1/2}, \qquad t \in \mathbb{R}, \quad \lambda > 1,$$

has no Poisson factor.

(b) Let

$$\varphi_2(t) = \frac{\lambda + e^{it}}{\lambda + 1}, \qquad t \in \mathbb{R}, \quad \lambda > 1,$$

and let φ_1 be as defined in (a). Let $\varphi = \varphi_1 \varphi_2$. Show that the characteristic function φ has the Poisson factor $\exp\{\lambda^{-1}(e^{it} - 1)\}$.

23. We say that two characteristic functions φ_1 and φ_2 are equivalent, and write $\varphi_1 \sim \varphi_2$ if $\varphi_1(t) = \varphi_2(t)e^{iat}$ for some $a \in \mathbb{R}$. Two characteristic functions are said to be of the same type if there exists a $c > 0$ such that $\varphi_1(\mathrm{t}) \sim \varphi_2(t/c)$.

Let φ be a decomposable characteristic function, and suppose that all factors of φ are of the same type (as φ). Show that φ is an i.d. characteristic function.

24. Let $\{X_n\}$ be a sequence of independent, identically distributed random variables. Let $0 < c < 1$, and write $S_n = \sum_{k=1}^{n} c^k X_k$.

(a) Show that a law is a limit law for sums of the above type if and only if its characteristic function φ satisfies the relation

$$\varphi(t) = \varphi(ct)\varphi_c(t),$$

where φ_c is a characteristic function.

(b) If, in particular, $\{X_n\}$ are independent and identically distributed with common distribution $P\{X_1 = \pm 1\} = \frac{1}{2}$ and $c = \frac{1}{2}$, show that the limit law is uniform on $(-1, 1)$.

NOTES AND COMMENTS

The theory of i.d. characteristic functions is considered in detail in Gnedenko and Kolmogorov [28]. Its applications to limit theorems are also discussed in this monograph. For a detailed treatment of the theory of analytic characteristic functions we refer the reader to Lukacs [60]. The general theory of decomposition of characteristic functions is treated in the monographs of Linnik [56], Lukacs [60], and Linnik and Ostrovskii [57]. For an overview of the moment problem we refer to Shohat and Tamarkin [75] and Akhiezer [2].

CHAPTER 5

The Central Limit Problem

In many physical applications the problem can be reduced, in mathematical terms, to the study of the limiting behavior of sums of independent random variables. For example, in statistical inference it is frequently of interest to know the asymptotic distribution of the partial sums $S_n, n \geq 1$, of independent, identically distributed random variables.

Let $\{X_n\}$ be a sequence of independent random variables, and write $S_n = \sum_{k=1}^{n} X_k$, $n \geq 1$. In this chapter we investigate, in detail, the limiting behavior of the sequence $\{S_n\}$. The type of problems considered here may be classified into the following broad categories:

(a) What are the conditions under which there exist sequences of constants $\{A_n\}$ and $\{B_n\}$ such that the sequence $\{(S_n - A_n)/B_n\}$ of centered and normed partial sums converges in law to a nondegenerate random variable?

(b) If there exist sequences $\{A_n\}$ and $\{B_n\}$ such that the sequence

$$\left\{\frac{S_n - A_n}{B_n}\right\}$$

converges in law to a nondegenerate random variable Z with distribution V, say, what are these limit laws?

(c) What are the sufficient conditions under which convergence to a particular one of these possible limit laws takes place?

We begin with the bounded variance case in Section 5.1. The case of most importance is that of the convergence in law of centered and normed partial sums to a normal random variable. This is explored in Section 5.1. In Section 5.2 we study the general case where the condition of bounded variances is dropped. It is shown that under certain mild conditions the class of limit laws coincides with the class of i.d. laws. In Section 5.3 we obtain the criterion for convergence to the normal, degenerate, and Poisson distributions. In Section 5.4 we restrict ourselves to the case of independent, identically distributed summands. It is shown that in this special case the class of limit laws coincides with the class of stable laws. Section 5.5 is devoted to some applications of the central limit theory.

5.1 THE BOUNDED VARIANCE CASE

Let S_n be the number of successes in n independent trials with probability p of success in each trial, $0 < p < 1$. Then S_n is a binomial random variable with $\mathscr{E}S_n = np$ and $\text{var}(S_n) = np(1 - p)$. DeMoivre and Laplace were the first to show that the sequence $\{(S_n - \mathscr{E}S_n)/\sqrt{\text{var}(S_n)}\}$ converges in law to the standard normal random variable as the number of trials n increases indefinitely. This result can clearly be reformulated as follows. Let $\{X_n\}$ be a sequence of independent, identically distributed random variables with $P\{X_1 = 1\} = p$, $P\{X_1 = 0\} = 1 - p$, $0 < p < 1$. Let $S_n = \sum_{k=1}^{n} X_k$, $n \geq 1$. Then

$$\frac{S_n - np}{\sqrt{np(1 - p)}} \xrightarrow{L} X \qquad \text{as } n \to \infty,$$

where X has the standard normal distribution. Motivated by this result, Lévy investigated the case of independent, identically distributed random variables with finite variance and obtained the most useful version of the celebrated central limit theorem. Later investigation centered around dropping the condition of identical distribution. In this direction Lindeberg obtained a set of sufficient conditions (which were later shown also to be necessary by Feller) for the convergence of suitably centered and normed S_n to the normal distribution. This leads us naturally to the general bounded variance case, which may be stated as follows. Let $\{X_{nk}, k = 1, 2, \ldots, k_n; n \geq 1\}$ be a sequence of independent random variables such that $k_n \to \infty$ as $n \to \infty$ and $\text{var}(X_{nk}) < \infty$. Set $S_n = \sum_{k=1}^{k_n} X_{nk}$. Find the conditions under which the sequence $\{S_n\}$ when suitably centered converges in law to some more general random variable. This is precisely the subject of our investigation in this section.

5.1.1 Lindeberg–Feller Central Limit Theorem

We first prove the classical result of DeMoivre and Laplace.

Proposition 5.1.1. Let $X_1, X_2, \ldots$ be a sequence of independent, identically distributed Bernoulli random variables with common distribution given by

$P\{X_j = 1\} = p, 0 < p < 1$, and $P\{X_j = 0\} = 1 - p = q$. Let $S_n = \sum_{j=1}^n X_j$, $n = 1, 2, \ldots$. Then for every $x \in \mathbb{R}$

$$\lim_{n\to\infty} P\left\{\frac{S_n - np}{\sqrt{npq}} \le x\right\} = \frac{1}{\sqrt{2\pi}} \int_{-\infty}^{x} e^{-u^2/2}\,du.$$

Proof. We give a simple proof, using the continuity theorem. Let φ_n be the characteristic function of $(S_n - np)/\sqrt{npq}$. Then

$$\varphi_n(t) = \left[q \exp\left(-it\sqrt{\frac{p}{nq}}\right) + p \exp\left(it\sqrt{\frac{q}{np}}\right)\right]^n.$$

By using the expansion for e^z we have

$$q \exp\left(-it\sqrt{\frac{p}{nq}}\right) + p \exp\left(it\sqrt{\frac{q}{np}}\right) = 1 - \frac{t^2}{2n}[1 + R_n(t)],$$

where

$$R_n(t) = 2\sum_{j=3}^{\infty} \frac{1}{j!}\left(\frac{it}{\sqrt{n}}\right)^{j-2} \frac{pq^j + q(-p)^j}{(pq)^{j/2}}.$$

Clearly $R_n(t) \to 0$ as $n \to \infty$ uniformly in every finite t-interval, so that

$$\varphi_n(t) = \left\{1 - \frac{t^2}{2n}[1 + R_n(t)]\right\}^n \to e^{-t^2/2}$$

uniformly in every finite t-interval. The result is an immediate consequence of the continuity theorem. ■

We now consider the most general result in this direction for the case of independent but not necessarily identically distributed random variables.

Theorem 5.1.1 (Lindeberg–Feller Central Limit Theorem). Let $\{X_n\}$ be a sequence of independent but not necessarily identically distributed random variables with $\operatorname{var}(X_n) = \sigma_n^2 < \infty$, $n = 1, 2, \ldots$. Let $\mathscr{E}X_n = \alpha_n$, and $S_n = \sum_{j=1}^n X_j$. Set $\operatorname{var}(S_n) = B_n^2$. Let F_n be the distribution function of X_n. Then the following two conditions:

(i) $\lim_{n\to\infty} \max_{1\le k\le n} (\sigma_k^2/B_n^2) = 0$,
(ii) $\lim_{n\to\infty} P\{[S_n - \mathscr{E}(S_n)]/B_n \le x\} = 1/\sqrt{2\pi} \int_{-\infty}^{x} e^{-u^2/2}\,du$,

for every $x \in \mathbb{R}$, hold if and only if for every $\epsilon > 0$ the condition

$$\lim_{n\to\infty} \frac{1}{B_n^2} \sum_{k=1}^{n} \int_{|x-\alpha_k|\ge \epsilon B_n} (x - \alpha_k)^2\,dF_k(x) = 0 \tag{5.1.1}$$

is satisfied.

[Condition (5.1.1) is known as the Lindeberg condition. It should be noted that the convergence in (ii) is uniform in x.]

Proof SUFFICIENCY OF (5.1.1). For every $n \geq 1$ we set

$$X_{nk} = \frac{X_k - \alpha_k}{B_n}, \qquad 1 \leq k \leq n.$$

Let F_{nk} be the distribution function of X_{nk}. Then

$$\mathscr{E} X_{nk} = \int_{-\infty}^{\infty} x \, dF_{nk}(x) = 0$$

and

$$\operatorname{var}(X_{nk}) = \int_{-\infty}^{\infty} x^2 \, dF_{nk}(x) = \frac{\sigma_k^2}{B_n^2},$$

so that

$$\sum_{k=1}^{n} \operatorname{var}(X_{nk}) = 1.$$

We see easily that (5.1.1) is equivalent to

$$\text{(5.1.2)} \qquad \lim_{n\to\infty} \sum_{k=1}^{n} \int_{|x|\geq\epsilon} x^2 \, dF_{nk}(x) = 0 \qquad \text{for } \epsilon > 0.$$

Now, for $n \geq 1$ and $1 \leq k \leq n$, we can write

$$\begin{aligned}\frac{\sigma_k^2}{B_n^2} &= \int_{|x|<\epsilon} x^2 \, dF_{nk}(x) + \int_{|x|\geq\epsilon} x^2 \, dF_{nk}(x) \\ &\leq \epsilon^2 + \int_{|x|\geq\epsilon} x^2 \, dF_{nk}(x),\end{aligned}$$

so that

$$\max_{1\leq k\leq n} \frac{\sigma_k^2}{B_n^2} \leq \epsilon^2 + \sum_{k=1}^{n} \int_{|x|\geq\epsilon} x^2 \, dF_{nk}(x).$$

In view of (5.1.2) we see that condition (i) holds.

For the proof of (ii), let φ_{nk} be the characteristic function of X_{nk}, and let φ_n be the characteristic function of $\sum_{k=1}^{n} X_{nk} = [S_n - E(S_n)]/B_n$. Clearly

$$\varphi_n(t) = \prod_{k=1}^{n} \varphi_{nk}(t), \qquad t \in \mathbb{R}.$$

We first prove that

$$\lim_{n\to\infty} \max_{1\le k\le n} |\varphi_{nk}(t) - 1| = 0 \tag{5.1.3}$$

uniformly in every finite t-interval.

Clearly

$$\varphi_{nk}(t) - 1 = \int_{-\infty}^{\infty} (e^{itx} - 1 - itx)\, dF_{nk}(x),$$

so that

$$\begin{aligned} |\varphi_{nk}(t) - 1| &\le \int_{-\infty}^{\infty} |e^{itx} - 1 - itx|\, dF_{nk}(x) \\ &\le \frac{t^2}{2}\int_{-\infty}^{\infty} x^2\, dF_{nk}(x) \\ &= \frac{t^2}{2}\frac{\sigma_k^2}{B_n^2}, \end{aligned}$$

and hence for $T > 0$

$$\max_{|t|\le T}\max_{1\le k\le n} |\varphi_{nk}(t) - 1| \le \frac{T^2}{2}\max_{1\le k\le n}\frac{\sigma_k^2}{B_n^2}.$$

In view of (i) result (5.1.3) holds.

Next we show that

$$\lim_{n\to\infty}\left\{\ln \varphi_n(t) - \sum_{k=1}^{n}[\varphi_{nk}(t) - 1]\right\} = 0 \tag{5.1.4}$$

holds uniformly in every finite t-interval. To this end let $T > 0$ be fixed. From (5.1.3) we can choose n sufficiently large so that

$$|\varphi_{nk}(t) - 1| < \tfrac{1}{2}$$

for all $1 \le k \le n$ and $|t| \le T$. Then we have the expansion

$$\ln \varphi_n(t) = \sum_{k=1}^{n} \ln \varphi_{nk}(t) = \sum_{k=1}^{n} [\varphi_{nk}(t) - 1] + R_n(t),$$

where

$$R_n(t) = \sum_{k=1}^{n}\sum_{j=2}^{\infty} \frac{(-1)^{j-1}}{j} [\varphi_{nk}(t) - 1]^j,$$

so that

$$
\begin{aligned}
|R_n(t)| &\le \sum_{k=1}^{n} \sum_{j=2}^{\infty} \frac{|\varphi_{nk}(t) - 1|^j}{j} \\
&\le \frac{1}{2} \sum_{k=1}^{n} \frac{|\varphi_{nk}(t) - 1|^2}{1 - |\varphi_{nk}(t) - 1|} \\
&< \sum_{k=1}^{n} |\varphi_{nk}(t) - 1|^2 \qquad \text{(for } n \text{ sufficiently large)} \\
&\le \max_{1 \le k \le n} |\varphi_{nk}(t) - 1| \sum_{k=1}^{n} |\varphi_{nk}(t) - 1|.
\end{aligned}
$$

Now

$$\sum_{k=1}^{n} |\varphi_{nk}(t) - 1| \le \frac{t^2}{2} \sum_{k=1}^{n} \int_{-\infty}^{\infty} x^2 \, dF_{nk}(x) = \frac{t^2}{2},$$

so that for $T > 0$

$$\max_{|t| \le T} |R_n(t)| \le \frac{T^2}{2} \max_{|t| \le T} \max_{1 \le k \le n} |\varphi_{nk}(t) - 1|.$$

Then (5.1.4) follows easily.

We now return to the proof of condition (ii) and write

$$\sum_{k=1}^{n} [\varphi_{nk}(t) - 1] = -\frac{t^2}{2} + \rho_n(t),$$

where

$$\rho_n(t) = \frac{t^2}{2} + \sum_{k=1}^{n} \int_{-\infty}^{\infty} (e^{itx} - 1 - itx) \, dF_{nk}(x).$$

Let $\epsilon > 0$. Since

$$\sum_{k=1}^{n} \int_{-\infty}^{\infty} x^2 \, dF_{nk}(x) = 1,$$

we can rewrite $\rho_n(t)$ as

$$
\begin{aligned}
\rho_n(t) = &\sum_{k=1}^{n} \int_{|x| < \epsilon} \left[e^{itx} - 1 - itx - \frac{(itx)^2}{2} \right] dF_{nk}(x) \\
&+ \sum_{k=1}^{n} \int_{|x| \ge \epsilon} \left(e^{itx} - 1 - itx + \frac{t^2 x^2}{2} \right) dF_{nk}(x),
\end{aligned}
$$

so that

$$|\rho_n(t)| \le \frac{|t|^3}{6} \sum_{k=1}^{n} \int_{|x|<\epsilon} |x|^3 \, dF_{nk}(x) + t^2 \sum_{k=1}^{n} \int_{|x|\ge\epsilon} x^2 \, dF_{nk}(x).$$

Hence for $T > 0$

$$\begin{aligned}\max_{|t|\le T} |\rho_n(t)| &\le \frac{T^3}{6} \epsilon \sum_{k=1}^{n} \int_{|x|<\epsilon} x^2 \, dF_{nk}(x) + T^2 \sum_{k=1}^{n} \int_{|x|\ge\epsilon} x^2 \, dF_{nk}(x) \\ &= \frac{T^3}{6}\epsilon + T^2\left(1 - \frac{T\epsilon}{6}\right) \sum_{k=1}^{n} \int_{|x|\ge\epsilon} x^2 \, dF_{nk}(x).\end{aligned}$$

Using (5.1.3), we conclude that $\rho_n(t) \to 0$ as $n \to \infty$ uniformly in every finite t-interval. Finally, using (5.1.4), we see that $\varphi_n(t) \to e^{-t^2/2}$ as $n \to \infty$ uniformly in every finite t-interval. This completes the proof of (ii) in view of the continuity theorem.

NECESSITY OF (5.1.1). We use the same notation as in the first part of the proof. Since condition (i) holds, (5.1.3) holds and consequently (5.1.4) holds. On the other hand, using (ii) and the continuity theorem, we conclude that $\ln \varphi_n(t) \to -t^2/2$ as $n \to \infty$ uniformly in every finite t-interval. Combining the two results, we obtain

$$\sum_{k=1}^{n} [\varphi_{nk}(t) - 1] = -\frac{t^2}{2} + o(1) \qquad \text{as } n \to \infty$$

uniformly in every finite t-interval. Taking the real part on both sides, we obtain

$$\frac{t^2}{2} - \sum_{k=1}^{n} \int_{-\infty}^{\infty} (1 - \cos tx) \, dF_{nk}(x) = o(1) \qquad \text{as } n \to \infty.$$

Let $\epsilon > 0$. Then we rewrite the last relation as

$$\frac{t^2}{2} - \sum_{k=1}^{n} \int_{|x|<\epsilon} (1 - \cos tx) \, dF_{nk}(x) = \sum_{k=1}^{n} \int_{|x|\ge\epsilon} (1 - \cos tx) \, dF_{nk}(x) + o(1) \qquad \text{as } n \to \infty. \tag{5.1.5}$$

Clearly

$$\begin{aligned}\sum_{k=1}^{n}\int_{|x|<\epsilon}(1-\cos tx)\,dF_{nk}(x) &\le \frac{t^2}{2}\sum_{k=1}^{n}\int_{|x|<\epsilon}x^2\,dF_{nk}(x)\\ &= \frac{t^2}{2}\left[1-\sum_{k=1}^{n}\int_{|x|\ge\epsilon}x^2\,dF_{nk}(x)\right],\end{aligned}$$

so that

$$\frac{t^2}{2}-\sum_{k=1}^{n}\int_{|x|<\epsilon}(1-\cos tx)\,dF_{nk}(x) \ge \frac{t^2}{2}\sum_{k=1}^{n}\int_{|x|\ge\epsilon}x^2\,dF_{nk}(x). \tag{5.1.6}$$

On the other hand,

$$\begin{aligned}\sum_{k=1}^{n}\int_{|x|\ge\epsilon}(1-\cos tx)\,dF_{nk}(x) &\le 2\sum_{k=1}^{n}\int_{|x|\ge\epsilon}dF_{nk}(x)\\ &\le \frac{2}{\epsilon^2}\sum_{k=1}^{n}\int_{|x|\ge\epsilon}x^2\,dF_{nk}(x)\\ &\le \frac{2}{\epsilon^2}. \qquad (5.1.7)\end{aligned}$$

Combining (5.1.5), (5.1.6), and (5.1.7), we get

$$0 \le \sum_{k=1}^{n}\int_{|x|\ge\epsilon}x^2\,dF_{nk}(x) \le \frac{2}{t^2}\left(\frac{2}{\epsilon^2}+o(1)\right) \qquad \text{as } n\to\infty.$$

Taking the limits on both sides, first as $n\to\infty$ and then as $|t|\to\infty$, we see that (5.1.1) is satisfied. This completes the proof of Theorem 5.1.1. ∎

Corollary 1 (Lévy Central Limit Theorem). Let $\{X_n\}$ be a sequence of independent, identically distributed random variables with $0<\text{var}(X_n)=\sigma^2<\infty$. Let $S_n=\sum_{k=1}^{n}X_k$. Then for every $x\in\mathbb{R}$

$$\lim_{n\to\infty}P\left\{\frac{S_n-\mathscr{E}(S_n)}{\sigma\sqrt{n}}\le x\right\}=\frac{1}{\sqrt{2\pi}}\int_{-\infty}^{x}e^{-u^2/2}\,du.$$

Proof. It suffices to check that the Lindeberg condition holds. Writing $\mathscr{E}X_n=\alpha$ and letting F be the common distribution function of the X_n we have for $\epsilon>0$

$$\frac{1}{\text{var}(S_n)}\sum_{k=1}^{n}\int_{|x-\alpha|\ge\epsilon\sqrt{\text{var}(S_n)}}(x-\alpha)^2\,dF(x)=\frac{1}{\sigma^2}\int_{|x-\alpha|\ge\epsilon\sigma\sqrt{n}}(x-\alpha)^2\,dF(x).$$

Since $0 < \sigma^2 < \infty$, the integral on the right-hand side $\to 0$ as $n \to \infty$ for every $\epsilon > 0$.

Corollary 2 (Lyapounov). Let $\{X_n\}$ be a sequence of independent random variables such that $E|X_k|^{2+\delta} < \infty$ for some $\delta > 0$ and $k = 1, 2, \ldots$. Let

$$\sum_{k=1}^{n} E|X_k|^{2+\delta} = o(B_n^{2+\delta}) \qquad \text{as } n \to \infty,$$

where B_n is defined in Theorem 5.1.1. Then conditions (i) and (ii) of Theorem 5.1.1 hold.

Proof. We have

$$\frac{1}{B_n^2} \sum_{k=1}^{n} \int_{|x-\alpha_k| \geq \epsilon B_n} (x - \alpha_k)^2 \, dF_k(x)$$
$$\leq \frac{1}{\epsilon^\delta B_n^{2+\delta}} \sum_{k=1}^{n} \int_{-\infty}^{\infty} |x|^{2+\delta} \, dF_k(x) \to 0 \qquad \text{as } n \to \infty,$$

so that (5.1.1) holds.

Example 5.1.1. Let $X_1, X_2, \ldots$ be a sequence of independent, identically distributed random variables with common distribution given by

$$P\{X_j = 1\} = p, \quad 0 < p < 1, \qquad \text{and} \qquad P\{X_j = 0\} = 1 - p = q, \qquad j = 1, 2, \cdots.$$

Let $S_n = \sum_{k=1}^{n} X_k$. Then $0 < \operatorname{var}(X_j) = pq < \infty$, and it follows from Corollary 1 to Theorem 5.1.1 that

$$\frac{S_n - np}{\sqrt{npq}} \xrightarrow{L} Z \qquad \text{as } n \to \infty,$$

where Z is a standard normal random variable. This is Proposition 5.1.1.

Example 5.1.2. Let $\{X_n\}$ be a sequence of uniformly bounded, independent random variables; that is, there exists a $\gamma > 0$ such that $|X_k| \leq \gamma$ a.s. for all k. Suppose that $B_n^2 = \operatorname{var}(S_n) \to \infty$ as $n \to \infty$. Since $|X_k - \alpha_k| \leq 2\gamma$ a.s., it follows that

$$\int_{|x-\alpha_k| \geq \epsilon B_n} (x - \alpha_k)^2 \, dF_k(x) \leq (2\gamma)^2 P\{|X_k - \alpha_k| \geq \epsilon B_n\}$$
$$\leq (2\gamma)^2 \frac{\operatorname{var}(X_k)}{\epsilon^2 B_n^2}$$

by Chebyshev's inequality. Hence

$$\frac{1}{B_n^2}\sum_{k=1}^{n}\int_{|x-\alpha_k|\geq \epsilon B_n}(x-\alpha_k)^2\,dF_k(x)\leq \frac{4\gamma^2}{\epsilon^2 B_n^2}\to 0 \qquad \text{as } n\to 0,$$

and the Lindeberg condition is satisfied. It follows that conditions (i) and (ii) of Theorem 5.1.1 hold.

5.1.2 The General Bounded Variance Case

In the Lindeberg–Feller central limit theorem we have shown that the sequence of normed and centered sums $\{(S_n - \mathscr{E}S_n)/B_n\}$ of independent random variables converges in law to the standard normal random variable, provided that condition (5.1.1) is satisfied. Let us write

$$\frac{S_n - \mathscr{E}S_n}{B_n} = \sum_{k=1}^{n}\frac{X_k - \mathscr{E}X_k}{B_n}.$$

Then we can write

$$\frac{S_n - \mathscr{E}S_n}{B_n} = \sum_{k=1}^{n} X_{nk},$$

where $X_{nk} = (X_k - \mathscr{E}X_k)/B_n$, $k = 1, 2, \ldots, n$, $n \geq 1$, and see that we are really dealing with a special case of the following triangular array (a double sequence). For each $n \geq 1$ let $X_{n1}, X_{n2}, \ldots, X_{nk_n}$ be k_n random variables, $k_n \to \infty$ as $n \to \infty$.

$$\begin{array}{l} X_{11}, X_{12}, \ldots, X_{1k_1}, \\ X_{21}, X_{22}, \ldots, X_{2k_2}, \\ \vdots \quad\ \vdots \\ X_{n1}, X_{n2}, \ldots, X_{nk_n}. \end{array}$$

The random variables with n as the first subscript will be referred to as those in the nth row. In the following assume that the random variables in each row are independent. Assume further that $\sigma_{nk}^2 = \operatorname{var}(X_{nk}) < \infty$ and, moreover,

$$(5.1.8)\qquad \begin{cases} \max\limits_{1\leq k\leq k_n} \sigma_{nk}^2 \to 0 & \text{as } n \to \infty \\ \text{and} & \\ \sum\limits_{k=1}^{k_n} \sigma_{nk}^2 \leq C < \infty & \text{for all } n \geq 1, \end{cases}$$

where $C > 0$ is a constant (independent of n). Set $S_n = \sum_{k=1}^{k_n} X_{nk}$, $n \geq 1$. Then it follows from the proof of Theorem 5.1.1 that the sequence $\{S_n\}$, when suitably centered and normalized, converges in law to the standard normal random variable provided that the Lindeberg condition (5.1.1) is satisfied. We now investigate the conditions under which the sequence $\{S_n\}$ when suitably centered converges in law to some more general random variable. For convenience we first consider the case when $\mathscr{E}X_{nk} = 0$ for all k and n.

Let F_{nk} be the distribution function of X_{nk} with corresponding characteristic function φ_{nk}, and let F_n be the distribution function of S_n with characteristic function φ_n. Then clearly

$$\varphi_n = \prod_{k=1}^{k_n} \varphi_{nk}, \qquad n \geq 1.$$

We first prove some lemmas which will be needed for the proof of the main result.

Lemma 5.1.1. The relation

$$\lim_{n\to\infty} \max_{1\leq k\leq k_n} |\varphi_{nk}(t) - 1| = 0 \tag{5.1.9}$$

holds uniformly in every finite t-interval.

Proof. Since $\mathscr{E}X_{nk} = 0$, we can write

$$|\varphi_{nk}(t) - 1| \leq \int_{-\infty}^{\infty} |e^{itx} - 1 - itx| \, dF_{nk}(x)$$
$$< \frac{t^2}{2} \sigma_{nk}^2.$$

Let $T > 0$. Then we have

$$\max_{|t|\leq T} \max_{1\leq k\leq k_n} |\varphi_{nk}(t) - 1| \leq \frac{T^2}{2} \max_{1\leq k\leq k_n} \sigma_{nk}^2 \to 0$$

as $n \to \infty$ by (5.1.8). ∎

Lemma 5.1.2. Let

$$\theta_n(t) = \sum_{k=1}^{k_n} [\varphi_{nk}(t) - 1].$$

Then, for sufficiently large n, $\psi_n = \ln \varphi_n$ exists; moreover, the relation

$$\lim_{n\to\infty} [\psi_n(t) - \theta_n(t)] = 0 \tag{5.1.10}$$

holds uniformly in every finite t-interval.

Proof. Let $T > 0$. Then in view of (5.1.9) we can choose n sufficiently large so that $|\varphi_{nk}(t) - 1| < \frac{1}{2}$ for all $k = 1, 2, \ldots, k_n$ and all $|t| \leq T$. Then $\ln \varphi_{nk}$ exists and is finite, and we have the expansion

$$\begin{aligned}\psi_n(t) &= \sum_{k=1}^{k_n} \ln \varphi_{nk}(t) \\ &= \sum_{k=1}^{k_n} [\varphi_{nk}(t) - 1] + R_n(t),\end{aligned}$$

where

$$R_n(t) = \sum_{k=1}^{k_n} \sum_{j=2}^{\infty} \frac{(-1)^{j-1}}{j} [\varphi_{nk}(t) - 1]^j.$$

We need only to show that $R_n(t) \to 0$ as $n \to \infty$ uniformly in every finite t-interval. Clearly

$$|R_n(t)| \leq \frac{1}{2} \sum_{k=1}^{k_n} \frac{|\varphi_{nk}(t) - 1|^2}{1 - |\varphi_{nk}(t) - 1|},$$

so that for sufficiently large n

$$\begin{aligned}|R_n(t)| &\leq \sum_{k=1}^{k_n} |\varphi_{nk}(t) - 1|^2 \\ &\leq \max_{1 \leq k \leq k_n} |\varphi_{nk}(t) - 1| \sum_{k=1}^{k_n} |\varphi_{nk}(t) - 1|.\end{aligned}$$

On the other hand, in view of (5.1.8)

$$\sum_{k=1}^{k_n} |\varphi_{nk}(t) - 1| \leq \frac{t^2}{2} \sum_{k=1}^{k_n} \sigma_{nk}^2 \leq \frac{t^2}{2} C,$$

and hence for sufficiently large n we have

$$\max_{|t| \leq T} |R_n(t)| \leq \frac{T^2}{2} C \max_{|t| \leq T} \max_{1 \leq k \leq k_n} |\varphi_{nk}(t) - 1|.$$

It follows from Lemma 5.1.1 that $R_n(t) \to 0$ as $n \to \infty$ uniformly in every finite t-interval. ■

Lemma 5.1.3. The function θ_n as defined in Lemma 5.1.2 has the representation

(5.1.11) $$\theta_n(t) = \int_{-\infty}^{\infty} (e^{itx} - 1 - itx) \frac{dK_n(x)}{x^2},$$

where K_n is a nondecreasing, right-continuous function defined on $\mathbb{R}$ such that $K_n(-\infty) = 0$ and $K_n(+\infty) \leq C < \infty$, where C is as defined in (5.1.8). The value of the integrand in (5.1.11) at $x = 0$ is defined by continuity to be $-t^2/2$.

Proof. Note that

$$\theta_n(t) = \sum_{k=1}^{k_n} \int_{-\infty}^{\infty} (e^{itx} - 1 - itx)\, dF_{nk}(x). \tag{5.1.12}$$

Define the function K_n on $\mathbb{R}$ by the formula

$$K_n(x) = \sum_{k=1}^{k_n} \int_{-\infty}^{x} y^2\, dF_{nk}(y), \qquad x \in \mathbb{R}. \tag{5.1.13}$$

Then we see easily that K_n is a nondecreasing, right-continuous function on $\mathbb{R}$ such that $K_n(-\infty) = 0$ and $K_n(+\infty) \leq C < \infty$. Combining (5.1.12) and (5.1.13), we obtain the result in (5.1.11). ∎

Remark 5.1.1. It follows from the Kolmogorov representation of an i.d. characteristic function that θ_n is the logarithm of the characteristic function of an i.d. random variable with mean zero and variance $\sigma_n^2 = K_n(+\infty) \leq C < \infty$.

Let us denote by $\mathscr{D}_0$ the class of all i.d. distribution functions with zero mean and finite variance. Clearly the characteristic function of every distribution function in $\mathscr{D}_0$ is of the form e^{ψ}, where

$$\psi(t) = \int_{-\infty}^{\infty} (e^{itx} - 1 - itx) \frac{dK(x)}{x^2}. \tag{5.1.14}$$

Here K is a nondecreasing, right-continuous function on $\mathbb{R}$ such that $K(-\infty) = 0$ and $K(+\infty) < \infty$. Moreover, K is uniquely determined by ψ.

We now prove the main result of this section.

Theorem 5.1.2. Let $\{X_{nk}\}$ be a sequence of independent random variables with $\mathscr{E}X_{nk} = 0$ and $\operatorname{var}(X_{nk}) = \sigma_{nk}^2 < \infty$, and suppose that the σ_{nk} satisfy (5.1.8). Let $S_n = \sum_{k=1}^{k_n} X_{nk}$. Then the class of limit distributions of $\{S_n\}$ coincides with the class $\mathscr{D}_0$.

Moreover, the sequence $\{S_n\}$ converges in law to a random variable with distribution function belonging to $\mathscr{D}_0$ and with characteristic function of the form e^{ψ} if and only if $K_n \xrightarrow{w} K$, where K_n and K are defined by (5.1.13) and (5.1.14), respectively.

Proof. First suppose that $S_n \xrightarrow{L} X$, where X is a random variable. We show that the distribution function of X is in $\mathscr{D}_0$. Let φ_n and φ be the characteristic functions of S_n and X, respectively. Then $\varphi_n \to \varphi$ by the continuity theorem. From Lemma 5.1.2 it follows that $\theta_n \to \ln \varphi$. Since $\{K_n\}$ is a sequence of uniformly bounded, nondecreasing, right-continuous functions, we conclude from Helly's theorem that there exists a subsequence $\{K_{n'}\} \subset \{K_n\}$ such that $K_{n'} \xrightarrow{w} K$, where K is a bounded, nondecreasing, right-continuous function on $\mathbb{R}$. Clearly $K(-\infty) = 0$ and $K(+\infty) \leq C < \infty$. By the Helly–Bray theorem and Lemma 5.1.3 we conclude that $\theta_{n'} \to \psi$, where ψ is given by (5.1.14). It follows that $\psi = \ln \varphi$, so that $\varphi = e^{\psi}$. This completes the proof of the first part.

We next show that $K_n \xrightarrow{w} K$. Suppose that there exists a subsequence $\{K_{n''}\} \subset \{K_n\}$ such that $K_{n''} \xrightarrow{w} K^*$, say where K^* is a bounded, nondecreasing, right-continuous function on $\mathbb{R}$ with $K^*(-\infty) = 0$ and $K^*(+\infty) \leq C < \infty$. By the same argument as was used above, $\theta_{n''} \to \psi^*$, say where ψ^* is of the form (5.1.14) with K replaced by K^*. It follows that $\psi = \psi^* = \ln \varphi$, so that $K = K^*$ by uniqueness. Consequently $K_n \xrightarrow{w} K$.

Conversely, suppose that $K_n \xrightarrow{w} K$. Again by using the Helly–Bray theorem we conclude that $\theta_n \to \psi$, and hence from Lemma 5.1.2 $\ln \varphi_n = \psi_n \to \psi$, so that $\varphi_n \to e^{\psi} = \varphi$ as $n \to \infty$. The proof is completed with the help of the continuity theorem. ∎

Finally we consider the general case where the X_{nk} are not necessarily centered at $\mathscr{E}X_{nk}$. In this case the following result holds.

Theorem 5.1.3. Let $\{X_{nk}\}$ be a sequence of independent random variables with $\mathscr{E}X_{nk} = \gamma_{nk}$ and $\sigma_{nk}^2 = \operatorname{var}(X_{nk}) < \infty$, where $\{\sigma_{nk}^2\}$ satisfies (5.1.8). Let $S_n = \sum_{k=1}^{k_n} X_{nk}$. Then the class of limit distributions of the sequence $\{S_n\}$ coincides with the class $\mathscr{D}$ of i.d. distribution functions with characteristic functions of the form e^{ψ}, where

$$\psi(t) = i\gamma t + \int_{-\infty}^{\infty} (e^{itx} - 1 - itx)\frac{dK(x)}{x^2}. \tag{5.1.15}$$

Here $\gamma \in \mathbb{R}$, and K is a nondecreasing, right-continuous function on $\mathbb{R}$ such that $K(-\infty) = 0$ and $K(+\infty) \leq C < \infty$. In this case γ and K are determined uniquely by ψ.

Moreover, the sequence $\{S_n\}$ converges in law to a random variable whose distribution function is in $\mathscr{D}$ and which has a characteristic function of the form e^{ψ} if and only if

$$\gamma_n = \sum_{k=1}^{k_n} \gamma_{nk} \to \gamma$$

and

$$K_n \xrightarrow{w} K,$$

where

$$K_n(x) = \sum_{k=1}^{k_n} \int_{-\infty}^{x} y^2 \, dF_{nk}(y + \gamma_{nk}), \qquad x \in \mathbb{R},$$

and γ and K are defined by (5.1.15).

Proof. The proof follows easily from Theorem 5.1.2. ■

We next show that the Lindeberg–Feller central limit theorem follows as an immediate consequence of Theorem 5.1.2. For this purpose we use the same notations and assumptions as in Theorem 5.1.1. We set

$$X_{nk} = \frac{X_k - \mathscr{E}X_k}{B_n} \qquad \text{for } 1 \leq k \leq n, \quad n \geq 1.$$

Then $\mathscr{E}X_{nk} = 0$ and $\sigma_{nk}^2 = \text{var}(X_{nk}) = \sigma_k^2/B_n^2 < \infty$. Suppose that condition (5.1.1) or, equivalently, (5.1.2) holds. Then clearly condition (i) of Theorem 5.1.1 also holds; moreover,

$$\sum_{k=1}^{n} \sigma_{nk}^2 = \sum_{k=1}^{n} \frac{\sigma_k^2}{B_n^2} = 1 < \infty \qquad \text{for all } n \geq 1,$$

so that (5.1.8) holds. Then from (5.1.2) and (5.1.13) we conclude that the function K_n on $\mathbb{R}$ defined by

$$\text{(5.1.16)} \qquad K_n(x) = \sum_{k=1}^{n} \int_{-\infty}^{x} y^2 \, dF_{nk}(y), \qquad x \in \mathbb{R}, \quad n \geq 1,$$

satisfies

$$K_n(-\epsilon) + 1 - K_n(\epsilon - 0) \to 0 \qquad \text{as } n \to \infty$$

for every $\epsilon > 0$. Consequently $K_n \xrightarrow{w} K$ as $n \to \infty$, where

$$K(-\epsilon) + 1 - K(\epsilon - 0) = 0 \qquad \text{for every } \epsilon > 0.$$

This implies that

$$K(x) = \epsilon(x), \qquad x \in \mathbb{R},$$

and hence

$$\psi(t) = -\frac{t^2}{2}.$$

It follows therefore from Theorem 5.1.2 that the sequence $S_n = \sum_{k=1}^{n} X_{nk}$ converges in law to the standard normal random variable.

Conversely, if conditions (i) and (ii) of Theorem 5.1.1 hold, then clearly (5.1.8) holds. We conclude from Theorem 5.1.2 and (ii) of Theorem 5.1.1 that

$$K_n(x) \xrightarrow{w} \epsilon(x) \qquad \text{as } n \to \infty, \quad x \in \mathbb{R},$$

where K_n is defined in (5.1.16). Hence for every $\epsilon > 0$

$$K_n(-\epsilon) + 1 - K_n(\epsilon - 0) \to 0 \qquad \text{as } n \to \infty,$$

which is (5.1.2) and hence equivalent to (5.1.1). This completes our demonstration.

5.2 THE GENERAL CENTRAL LIMIT PROBLEM

Let

$$\begin{matrix} X_{11}, X_{12}, \ldots, X_{1k_1}, \\ X_{21}, X_{22}, \ldots, X_{2k_2}, \\ \vdots \quad \vdots \quad \vdots \\ X_{n1}, X_{n2}, \ldots, X_{nk_n}, \end{matrix} \tag{5.2.1}$$

be a sequence of random variables satisfying the following condition:

$$\lim_{n\to\infty} \max_{1 \le k \le k_n} P\{|X_{nk}| \ge \epsilon\} = 0 \qquad \text{for every } \epsilon > 0. \tag{5.2.2}$$

A sequence of random variables satisfying (5.2.2) is known as a *uniformly asymptotically negligible* (u.a.n.) or *infinitesimal* sequence, and condition (5.2.2) is called the *u.a.n.* condition. The u.a.n. condition ensures that the contribution of any one random variable to $S_n = \sum_{k=1}^{k_n} X_{nk}$ is negligible.

We first prove the following important theorem.

Theorem 5.2.1. Let $\{X_{nk}\}$ be a sequence of random variables as in (5.2.1) satisfying (5.2.2). Assume further that the random variables in each row are independent, that is, assume that $X_{n1}, X_{n2}, \ldots, X_{nk_n}$ are independent for each $n \ge 1$. Let $S_n = \sum_{k=1}^{k_n} X_{nk}$, $n \ge 1$. Then the class of limit laws of the sequence $\{S_n\}$ coincides with the class of i.d. laws, that is, the class of laws with characteristic function of the form e^{ψ}, where

$$\psi(t) = i\alpha t + \int_{-\infty}^{\infty} \left(e^{itx} - 1 - \frac{itx}{1 + x^2}\right) \frac{1 + x^2}{x^2}\, dG(x). \tag{5.2.3}$$

Here $\alpha \in \mathbb{R}$, and G is a bounded, nondecreasing, right-continuous function on $\mathbb{R}$ with $G(-\infty) = 0$.

Moreover, the sequence $\{S_n\}$ converges in law to a random variable with characteristic function of the form e^{ψ} if and only if

$$G_n \xrightarrow{c} G \quad \text{and} \quad \alpha_n \to \alpha \quad \text{as } n \to \infty, \tag{5.2.4}$$

where

$$G_n(x) = \sum_{k=1}^{k_n} \int_{-\infty}^{x} \frac{y^2}{1 + y^2} \, dF_{nk}(y + \alpha_{nk}), \qquad x \in \mathbb{R} \tag{5.2.5}$$

$$\alpha_n = \sum_{k=1}^{k_n} \left[\alpha_{nk} + \int_{-\infty}^{\infty} \frac{x}{1 + x^2} \, dF_{nk}(x + \alpha_{nk}) \right], \tag{5.2.6}$$

and

$$\alpha_{nk} = \int_{|x| < \gamma} x \, dF_{nk}(x). \tag{5.2.7}$$

Here $\gamma > 0$ is finite, and F_{nk} is the distribution function of the X_{nk}.

For the proof of Theorem 5.2.1 we need some lemmas. Lemmas 5.2.1 and 5.2.2 explore the ramifications of the u.a.n. condition.

Lemma 5.2.1. The u.a.n. condition (5.2.2) is equivalent to each of the following two conditions:

$$\lim_{n \to \infty} \max_{1 \le k \le k_n} |\varphi_{nk}(t) - 1| = 0 \tag{5.2.8}$$

uniformly in every finite t-interval, where φ_{nk} is the characteristic function of the X_{nk}, and

$$\lim_{n \to \infty} \max_{1 \le k \le k_n} \int_{-\infty}^{\infty} \frac{x^2}{1 + x^2} \, dF_{nk}(x) = 0. \tag{5.2.9}$$

Proof. We first note that the equivalence of (5.2.2) and (5.2.9) follows immediately from the inequalities

$$\max_{1 \le k \le k_n} \int_{-\infty}^{\infty} \frac{x^2}{1 + x^2} \, dF_{nk}(x) \le \epsilon^2 + \max_{1 \le k \le k_n} P\{|X_{nk}| \ge \epsilon\}$$

and

$$\max_{1 \le k \le k_n} P\{|X_{nk}| \ge \epsilon\} \le \frac{1 + \epsilon^2}{\epsilon^2} \max_{1 \le k \le k_n} \int_{-\infty}^{\infty} \frac{x^2}{1 + x^2} \, dF_{nk}(x).$$

Next we show that (5.2.2) $\Rightarrow$ (5.2.8). Let $T > 0$ be fixed. For $|t| \le T$ and $\epsilon > 0$ we have

$$\max_{1 \le k \le k_n} |\varphi_{nk}(t) - 1| \le \max_{1 \le k \le k_n} \int_{|x| < \epsilon} |e^{itx} - 1| \, dF_{nk}(x)$$

$$+ 2 \max_{1 \le k \le k_n} \int_{|x| \ge \epsilon} dF_{nk}(x)$$

$$\le T\epsilon + 2 \max_{1 \le k \le k_n} P\{|X_{nk}| \ge \epsilon\},$$

so that (5.2.8) follows on taking limits as $n \to \infty$ and then $\epsilon \to 0$.

Next we show that (5.2.8) $\Rightarrow$ (5.2.2). Suppose that (5.2.8) holds. Clearly, as $n \to \infty$, $|\varphi_{nk}(t) - 1| \to 0$ uniformly in k, $1 \le k \le k_n$, and uniformly in every finite t-interval. This implies that, as $n \to \infty$, $X_{nk} \xrightarrow{P} 0$ uniformly in k, $1 \le k \le k_n$. Let $\epsilon > 0$. Then, for given n and ϵ, let $k_0 = k_0(n, \epsilon)$ be the integer defined by

$$P\{|X_{nk_0}| \ge \epsilon\} = \max_{1 \le k \le k_n} P\{|X_{nk}| \ge \epsilon\}.$$

Clearly $1 \le k_0 \le k_n$. Since $X_{nk_0} \xrightarrow{P} 0$ as $n \to \infty$, the u.a.n. condition (5.2.2) holds. This completes the proof of the lemma. ■

Lemma 5.2.2. Let m_{nk} be a median of X_{nk}, and let α_{nk} be as defined in (5.2.7). Suppose that the random variables in the sequence $\{X_{nk}\}$ satisfy the u.a.n. condition. Then the following relations hold:

(a) $\lim_{n\to\infty} \max_{1 \le k \le k_n} |m_{nk}| = 0$.
(b) $\lim_{n\to\infty} \max_{1 \le k \le k_n} |\alpha_{nk}| = 0$.

Proof. In view of the u.a.n. condition, we can, given $\epsilon > 0$, choose an $n_0 = n_0(\epsilon)$ such that $P\{|X_{nk}| < \epsilon\} > \frac{3}{4}$ for $n \ge n_0$. It follows from the definition of m_{nk} that $|m_{nk}| < \epsilon$ for $n \ge n_0$, which yields (a) immediately.

As for (b), let $0 < \epsilon < \gamma$. Then we have

$$|\alpha_{nk}| \le \left| \int_{|x| < \epsilon} x \, dF_{nk}(x) \right| + \left| \int_{\epsilon \le |x| < \gamma} x \, dF_{nk}(x) \right|$$

$$\le \epsilon + \gamma P\{|X_{nk}| \ge \epsilon\},$$

so that

$$\max_{1 \le k \le k_n} |\alpha_{nk}| \le \epsilon + \gamma \max_{1 \le k \le k_n} P\{|X_{nk}| \ge \epsilon\}.$$

Result (b) now follows on taking limits as $n \to \infty$ and then $\epsilon \to 0$. ■

Throughout the present investigation we fix $\gamma > 0$.

Corollary. Under the u.a.n. condition

$$\lim_{n\to\infty} \max_{1\le k\le k_n} |\tilde{\varphi}_{nk}(t) - 1| = 0 \tag{5.2.10}$$

uniformly in every finite t-interval where $\tilde{\varphi}_{nk}(t) = \int_{-\infty}^{\infty} e^{itx}\, dF_{nk}(x + \alpha_{nk})$ and α_{nk} is defined in (5.2.7).

Proof. In view of part (b) of Lemma 5.2.2, the sequence $\{X_{nk} - \alpha_{nk}\}$ satisfies the u.a.n. condition. The proof now follows immediately from (5.2.8).

Remark 5.2.1. The random variables in the sequence $\{X_{nk}\}$ are said to be *asymptotically constant* if there exist constants c_{nk} such that

$$\lim_{n\to\infty} \max_{1\le k\le k_n} P\{|X_{nk} - c_{nk}| \ge \epsilon\} = 0 \qquad \text{for every } \epsilon > 0.$$

According to Lemma 5.2.2, if the X_{nk} satisfy the u.a.n. condition, they are asymptotically constant with $c_{nk} = \alpha_{nk}$ or m_{nk}.

Lemma 5.2.3. Let X and X' be two independently, identically distributed random variables. Let $X^s = X - X'$ be the symmetrized X. Let g on $\mathbb{R}$ be defined by

$$g(x) = \begin{cases} x^2 & \text{for } |x| \le 1, \\ 1 & \text{for } |x| > 1. \end{cases}$$

Then for any median m of X the inequality

$$\mathscr{E}g(X^s) \ge \tfrac{1}{2}\mathscr{E}(g(X) - m) \tag{5.2.11}$$

holds.

Proof. Without loss of generality we assume that $m = 0$. Let F be the distribution function of X. Then

$$\begin{aligned}
\mathscr{E}g(X^s) &= \int_{-\infty}^{\infty}\int_{-\infty}^{\infty} g(x-y)\, dF(x)\, dF(y) \\
&\ge \iint_{\{x\ge 0,\, y\le 0\}} g(x-y)\, dF(x)\, dF(y) \\
&\quad + \iint_{\{x< 0,\, y> 0\}} g(x-y)\, dF(x)\, dF(y) \\
&\ge \int_{y\le 0} dF(y) \int_{x\ge 0} g(x)\, dF(x) + \int_{y>0} dF(y)\int_{x<0} g(x)\, dF(x) \\
&\ge \tfrac{1}{2}\mathscr{E}g(X).
\end{aligned}$$

■

Lemma 5.2.4. Let $\{X_{nk}\}$ be a sequence of random variables satisfying the u.a.n. condition. Suppose further that $\prod_{k=1}^{k_n} |\varphi_{nk}| \to |\varphi|$ as $n \to \infty$, where φ is continuous on $\mathbb{R}$. Then there exists a constant $C = C(\gamma) > 0$ independent of n such that

$$\sup_{n \geq 1} \sum_{k=1}^{k_n} \int_{-\infty}^{\infty} \frac{x^2}{1 + x^2} \, dF_{nk}(x + \alpha_{nk}) < C. \tag{5.2.12}$$

Proof. Let G_{nk} be the distribution function of the random variable Y_{nk} which is a symmetrization of X_{nk}, and let θ_{nk} be the corresponding characteristic function. Then

$$0 < \theta_{nk}(t) = |\varphi_{nk}(t)|^2 \leq 1.$$

It follows that $\prod_{k=1}^{k_n} \theta_{nk}(t) \to |\varphi(t)|^2$ as $n \to \infty$. We first show that there exists a constant C_1 independent of n such that

$$\sum_{k=1}^{k_n} \int_{-\infty}^{\infty} \frac{x^2}{1 + x^2} \, dG_{nk}(x) < C_1, \qquad n \geq 1. \tag{5.2.13}$$

In fact, writing $\theta_{nk}(t) = \int_{-\infty}^{\infty} \cos tx \, dG_{nk}(x)$, we see that for $T > 0$ (fixed)

$$\begin{aligned} 0 \leq \int_0^T (1 - \theta_{nk}(t)) \, dt &= \int_0^T \int_{-\infty}^{\infty} (1 - \cos tx) \, dG_{nk}(x) \, dt \\ &= T \int_{-\infty}^{\infty} \left(1 - \frac{\sin Tx}{Tx}\right) dG_{nk}(x) \end{aligned}$$

in view of Fubini's theorem. Since there exists a $C(T) > 0$ such that

$$\left(1 - \frac{\sin Tx}{Tx}\right) \frac{1 + x^2}{x^2} \geq C(T) > 0$$

for all $x \in \mathbb{R}$, we obtain

$$\int_0^T (1 - \theta_{nk}(t)) \, dt \geq TC(T) \int_{-\infty}^{\infty} \frac{x^2}{1 + x^2} \, dG_{nk}(x),$$

so that

$$\int_{-\infty}^{\infty} \frac{x^2}{1 + x^2} \, dG_{nk}(x) \leq \frac{1}{TC(T)} \int_0^T (1 - \theta_{nk}(t)) \, dt.$$

On the other hand, in view of (5.2.8) we note that $\ln \varphi_{nk}(t)$ exists and is finite for sufficiently large n and for all $1 \leq k \leq k_n$. Consequently, for sufficiently large n and $1 \leq k \leq k_n$, $\ln \theta_{nk}(t)$ exists and is finite. Moreover, $\theta_{nk}(t) = |\varphi_{nk}(t)|^2$, so that we have the inequality

$$0 \leq 1 - \theta_{nk}(t) \leq -\ln \theta_{nk}(t)$$

for all $t \in \mathbb{R}$ and sufficiently large n. Therefore

$$\int_{-\infty}^{\infty} \frac{x^2}{1 + x^2} \, dG_{nk}(x) \leq -\frac{1}{TC(T)} \int_0^T \ln \theta_{nk}(t) \, dt,$$

so that for sufficiently large n

$$\sum_{k=1}^{k_n} \int_{-\infty}^{\infty} \frac{x^2}{1+x^2}\, dG_{nk}(x) \le -\frac{1}{TC(T)} \sum_{k=1}^{k_n} \int_0^T \ln \theta_{nk}(t)\, dt. \tag{5.2.14}$$

In view of our hypothesis $\sum_{k=1}^{k_n} \ln \theta_{nk}(t) \to 2 \ln |\varphi(t)|$, so that

$$-\sum_{k=1}^{k_n} \int_0^T \ln \theta_{nk}(t)\, dt \to -2 \int_0^T \ln|\varphi(t)|\, dt < \infty.$$

Hence for sufficiently large n there exists a constant $C'(T) > 0$ such that

$$0 < -\sum_{k=1}^{k_n} \int_0^T \ln \theta_{nk}(t)\, dt < C'(T). \tag{5.2.15}$$

The proof of (5.2.13) now follows easily from (5.2.14) and (5.2.15).

Next we show that

$$\sum_{k=1}^{k_n} \int_{-\infty}^{\infty} \frac{x^2}{1+x^2}\, dF_{nk}(x + m_{nk}) < C_2 \qquad \text{for } n \ge 1, \tag{5.2.16}$$

where $C_2 > 0$ is a constant independent of n. Let g on $\mathbb{R}$ be as defined in Lemma 5.2.3. Then, using the inequality

$$\frac{x^2}{1+x^2} \le g(x) \le 2\frac{x^2}{1+x^2}, \qquad x \in \mathbb{R},$$

and Lemma 5.2.3, we obtain easily, after some elementary computation,

$$\int_{-\infty}^{\infty} \frac{x^2}{1+x^2}\, dF_{nk}(x + m_{nk}) \le 4 \int_{-\infty}^{\infty} \frac{x^2}{1+x^2}\, dG_{nk}(x), \tag{5.2.17}$$

so that (5.2.16) follows at once from (5.2.13).

Finally we return to the proof of (5.2.12). Using the elementary inequality $(a + b)^2 \le 2(a^2 + b^2)$ and Lemma 5.2.2, we obtain

$$\begin{aligned} &\int_{-\infty}^{\infty} \frac{x^2}{1+x^2}\, dF_{nk}(x + \alpha_{nk}) \\ &= \int_{-\infty}^{\infty} \frac{(x + m_{nk} - \alpha_{nk})^2}{1 + (x + m_{nk} - \alpha_{nk})^2}\, dF_{nk}(x + m_{nk}) \\ &\le 2 \int_{-\infty}^{\infty} \frac{x^2}{1 + (x + m_{nk} - \alpha_{nk})^2}\, dF_{nk}(x + m_{nk}) + 2(m_{nk} - \alpha_{nk})^2 \\ &\le c \int_{-\infty}^{\infty} \frac{x^2}{1+x^2}\, dF_{nk}(x + m_{nk}) + 2(m_{nk} - \alpha_{nk})^2 \end{aligned} \tag{5.2.18}$$

for sufficiently large n and $1 \le k \le k_n$, where $c > 0$ is a constant independent of n and k. Moreover, it follows from the definition of α_{nk} and the fact that

$(a - b)^2 \leq 2(a^2 + b^2)$ that

$$(\alpha_{nk} - m_{nk})^2 = \left[\int_{|x|<\gamma} (x - m_{nk})\, dF_{nk}(x) - m_{nk} \int_{|x|\geq\gamma} dF_{nk}(x)\right]^2$$

$$\leq 2\left[\int_{|x+m_{nk}|<\gamma} |x|\, dF_{nk}(x + m_{nk})\right]^2 \tag{5.2.19}$$

$$+ 2m_{nk}^2\left[\int_{|x+m_{nk}|\geq\gamma} dF_{nk}(x + m_{nk})\right]^2.$$

We choose n sufficiently large so that $|m_{nk}| < \gamma/2$ for $1 \leq k \leq k_n$. Then from (5.2.19) we get

$$(\alpha_{nk} - m_{nk})^2 \leq 2\left[\int_{|x|<3\gamma/2} |x|\, dF_{nk}(x + m_{nk})\right]^2 + 2m_{nk}^2\left[\int_{|x|>\gamma/2} dF_{nk}(x + m_{nk})\right]^2.$$

By the Cauchy–Schwartz inequality applied to the first integral we obtain for sufficiently large n

$$\begin{aligned}(\alpha_{nk} - m_{nk})^2 &\leq 2\int_{|x|<3\gamma/2} x^2\, dF_{nk}(x + m_{nk}) \\ &\quad + 2m_{nk}^2 \int_{|x|>\gamma/2} dF_{nk}(x + m_{nk}) \\ &\leq 2\left(1 + \frac{9\gamma^2}{4}\right)\int_{|x|<3\gamma/2} \frac{x^2}{1 + x^2}\, dF_{nk}(x + m_{nk}) \\ &\quad + \frac{2m_{nk}^2(4 + \gamma^2)}{\gamma^2}\int_{|x|>\gamma/2} \frac{x^2}{1 + x^2}\, dF_{nk}(x + m_{nk}),\end{aligned}$$

so that

$$(\alpha_{nk} - m_{nk})^2 \leq c' \int_{-\infty}^{\infty} \frac{x^2}{1 + x^2}\, dF_{nk}(x + m_{nk}) \tag{5.2.20}$$

holds for sufficiently large n, where

$$c'(\gamma) = c' = \left[2\left(1 + \frac{9\gamma^2}{4}\right) + \left(\frac{4 + \gamma^2}{2}\right)\right].$$

Combining (5.2.18) and (5.2.20), we obtain

$$\int_{-\infty}^{\infty} \frac{x^2}{1 + x^2}\, dF_{nk}(x + \alpha_{nk}) \leq c'' \int_{-\infty}^{\infty} \frac{x^2}{1 + x^2}\, dF_{nk}(x + m_{nk})$$

for sufficiently large n and $1 \le k \le k_n$, where $c'' = c''(\gamma)$ is a constant independent of n and k. The proof of (5.2.12) now follows from (5.2.16). This completes the proof of Lemma 5.2.4. ■

In the following we set

$$\tilde{F}_{nk}(x) = F_{nk}(x + \alpha_{nk}) \qquad \text{and} \qquad \tilde{\varphi}_{nk}(t) = \int_{-\infty}^{\infty} e^{itx}\, d\tilde{F}_{nk}(x).$$

Lemma 5.2.5. Let $\{X_{nk}\}$ satisfy the conditions of Lemma 5.2.4. Let $T > 0$ be fixed. Then there exists a constant $C(T) > 0$ independent of n such that

$$\sum_{k=1}^{k_n} |\tilde{\varphi}_{nk}(t) - 1| \le C(T) \tag{5.2.21}$$

holds for sufficiently large n and all $|t| \le T$.

Proof. We first show that for sufficiently large n and $1 \le k \le k_n$ the inequality

$$\max_{|t| \le T} |\tilde{\varphi}_{nk}(t) - 1| \le C'(T) \int_{-\infty}^{\infty} \frac{x^2}{1 + x^2}\, d\tilde{F}_{nk}(x) \tag{5.2.22}$$

holds, where $C'(T) > 0$ is a constant independent of n and k. Indeed,

$$\begin{aligned} |\tilde{\varphi}_{nk}(t) - 1| &= \left| \int_{-\infty}^{\infty} (e^{itx} - 1)\, d\tilde{F}_{nk}(x) \right| \\ &= \left| \int_{|x| < \gamma} (e^{itx} - 1 - itx)\, d\tilde{F}_{nk}(x) + \int_{|x| \ge \gamma} (e^{itx} - 1)\, d\tilde{F}_{nk}(x) \right. \\ &\qquad \left. + it \int_{|x| < \gamma} x\, d\tilde{F}_{nk}(x) \right| \\ &\le \frac{t^2}{2} \int_{|x| < \gamma} x^2\, d\tilde{F}_{nk}(x) + 2 \int_{|x| \ge \gamma} d\tilde{F}_{nk}(x) \\ &\qquad + |t| \left| \int_{|x| < \gamma} x\, d\tilde{F}_{nk}(x) \right|, \end{aligned}$$

so that for $T > 0$

$$\begin{aligned} \max_{|t| \le T} |\tilde{\varphi}_{nk}(t) - 1| &\le \frac{T^2}{2} \int_{|x| < \gamma} x^2\, d\tilde{F}_{nk}(x) + 2 \int_{|x| \ge \gamma} d\tilde{F}_{nk}(x) \\ &\qquad + T \left| \int_{|x| < \gamma} x\, d\tilde{F}_{nk}(x) \right|. \end{aligned} \tag{5.2.23}$$

In view of Lemma 5.2.2 we can choose n so large that $|\alpha_{nk}| < \gamma/2$ for $k = 1, 2, \ldots, k_n$. Then

$$\left| \int_{|x|<\gamma} x\, d\tilde{F}_{nk}(x) - \int_{|x+\alpha_{nk}|<\gamma} x\, d\tilde{F}_{nk}(x) \right| \le \int_{\gamma/2<|x|<3\gamma/2} |x|\, d\tilde{F}_{nk}(x)$$

$$\le \frac{3\gamma}{2} \int_{|x|>\gamma/2} d\tilde{F}_{nk}(x). \tag{5.2.24}$$

On the other hand,

$$\int_{|x+\alpha_{nk}|<\gamma} x\, d\tilde{F}_{nk}(x) = \int_{|x|<\gamma} (x - \alpha_{nk})\, dF_{nk}(x)$$

$$= \alpha_{nk} \int_{|x|\ge\gamma} dF_{nk}(x),$$

so that for sufficiently large n and $1 \le k \le k_n$ we have

$$\left| \int_{|x+\alpha_{nk}|<\gamma} x\, d\tilde{F}_{nk}(x) \right| \le \frac{\gamma}{2} \int_{|x+\alpha_{nk}|\ge\gamma} d\tilde{F}_{nk}(x)$$

$$\le \frac{\gamma}{2} \int_{|x|>\gamma/2} d\tilde{F}_{nk}(x). \tag{5.2.25}$$

Combining (5.2.24) and (5.2.25), we see that the inequality

$$\left| \int_{|x|<\gamma} x\, d\tilde{F}_{nk}(x) \right| \le 2\gamma \int_{|x|>\gamma/2} d\tilde{F}_{nk}(x) \tag{5.2.26}$$

holds for sufficiently large n and $1 \le k \le k_n$. Using (5.2.26) and (5.2.23), we get for n sufficiently large

$$\max_{|t|\le T} |\tilde{\varphi}_{nk}(t) - 1| \le \frac{T^2}{2} \int_{|x|<\gamma} x^2\, d\tilde{F}_{nk}(x) + 2 \int_{|x|\ge\gamma} d\tilde{F}_{nk}(x)$$

$$+ 2T\gamma \int_{|x|>\gamma/2} d\tilde{F}_{nk}(x)$$

$$\le \frac{T^2}{2}(1 + \gamma^2) \int_{|x|<\gamma} \frac{x^2}{1+x^2}\, d\tilde{F}_{nk}(x)$$

$$+ \frac{2(1+\gamma^2)}{\gamma^2} \int_{|x|\ge\gamma} \frac{x^2}{1+x^2}\, d\tilde{F}_{nk}(x)$$

$$+ 2T\gamma \frac{4+\gamma^2}{\gamma^2} \int_{|x|>\gamma/2} \frac{x^2}{1+x^2}\, d\tilde{F}_{nk}(x)$$

$$\le C'(T) \int_{-\infty}^{\infty} \frac{x^2}{1+x^2}\, d\tilde{F}_{nk}(x),$$

where

$$C'(T) = \frac{T^2}{2}(1+\gamma^2) + \frac{2(1+\gamma^2)}{\gamma^2} + \frac{2T(4+\gamma^2)}{\gamma}$$

which is (5.2.22). The proof of the lemma is now easily completed with the help of Lemma 5.2.4. ■

Lemma 5.2.6. Let $\{X_{nk}\}$ satisfy the conditions of Lemma 5.2.4. Then

$$\lim_{n\to\infty}\left\{\sum_{k=1}^{k_n}\left[\ln \varphi_{nk}(t) - it\alpha_{nk} - \int_{-\infty}^{\infty}(e^{itx}-1)\,d\tilde{F}_{nk}(x)\right]\right\} = 0 \tag{5.2.27}$$

uniformly in every finite t-interval.

Proof. Let $T > 0$ be fixed. In view of (5.2.10) we can choose n sufficiently large so that

$$|\tilde{\varphi}_{nk}(t) - 1| < \tfrac{1}{2} \qquad \text{for } k = 1, 2, \ldots, k_n \quad \text{and} \quad |t| \le T.$$

Then

$$\begin{aligned}\sum_{k=1}^{k_n} \ln \varphi_{nk}(t) &= \sum_{k=1}^{k_n}\{it\alpha_{nk} + \ln[1 + (\tilde{\varphi}_{nk}(t) - 1)]\}\\ &= \sum_{k=1}^{k_n}\{it\alpha_{nk} + [\tilde{\varphi}_{nk}(t) - 1]\} + R_n(t),\end{aligned} \tag{5.2.28}$$

where

$$R_n(t) = \sum_{k=1}^{k_n}\sum_{j=2}^{\infty}(-1)^{j-1}\frac{[\tilde{\varphi}_{nk}(t) - 1]^j}{j}.$$

Hence

$$\begin{aligned}|R_n(t)| &\le \sum_{k=1}^{k_n}\sum_{j=2}^{\infty}\frac{|\tilde{\varphi}_{nk}(t) - 1|^j}{j}\\ &\le \sum_{k=1}^{k_n}|\tilde{\varphi}_{nk}(t) - 1|^2\\ &\le \max_{1\le k\le k_n}|\tilde{\varphi}_{nk}(t) - 1|\sum_{k=1}^{k_n}|\tilde{\varphi}_{nk}(t) - 1|.\end{aligned}$$

From Lemma 5.2.5 we conclude that for sufficiently large n

$$\max_{|t|\le T}|R_n(t)| \le C(T)\max_{|t|\le T}\max_{1\le k\le k_n}|\tilde{\varphi}_{nk}(t) - 1|.$$

In view of (5.2.10) we see that $R_n(t) \to 0$ as $n \to \infty$ uniformly in every finite t-interval. This completes the proof of (5.2.27). ■

We are now in a position to prove Theorem 5.2.1.

Proof of Theorem 5.2.1. Set $\varphi_n(t) = \prod_{k=1}^{k_n} \varphi_{nk}(t)$. Clearly φ_n is the characteristic function of S_n. Suppose that $\{S_n\}$ converges in law so that φ_n converges to some characteristic function φ. It follows from Lemma 5.2.6 that φ is the limit of a sequence of i.d. characteristic functions and hence itself is i.d.

Conversely, let φ be i.d., and for $n \geq 1$ let $X_{n1}, X_{n2}, \ldots, X_{nn}$ be independent random variables with common characteristic function $\varphi_n = [\varphi]^{1/n}$. Clearly

$$\varphi_n = \exp\left(\frac{1}{n} \ln \varphi\right) \to 1 \qquad \text{as } n \to \infty,$$

so that the X_{nk} satisfy the u.a.n. condition. Set $S_n = \sum_{k=1}^{n} X_{nk}$. Then φ is the characteristic function of S_n for every n.

Next suppose that $\{S_n\}$ converges in law to a random variable with characteristic function of the form e^{ψ}, where ψ is defined in (5.2.3). Then in view of Lemma 5.2.6

$$\sum_{k=1}^{k_n} \left[it\alpha_{nk} + \int_{-\infty}^{\infty} (e^{itx} - 1)\, d\tilde{F}_{nk}(x) \right] \to \psi(t)$$

as $n \to \infty$ uniformly in every finite t-interval. Let G and α_n be as defined in (5.2.5) and (5.2.6), respectively. Now, proceeding as in the latter part of the proof of Theorem 4.1.1, we conclude that $G_n \xrightarrow{c} G$ and $\alpha_n \to \alpha$ as $n \to \infty$.

Conversely, suppose that (5.2.4) holds. It follows from the extended Helly–Bray theorem that $\psi_n \to \psi$, where

$$\psi_n(t) = i\alpha_n t + \int_{-\infty}^{\infty} \left(e^{itx} - 1 - \frac{itx}{1 + x^2} \right) \frac{1 + x^2}{x^2}\, dG_n(x).$$

We see easily that

$$\psi_n(t) = it \sum_{k=1}^{k_n} \alpha_{nk} + \sum_{k=1}^{k_n} \int_{-\infty}^{\infty} (e^{itx} - 1)\, d\tilde{F}_{nk}(x).$$

Then the conditions of Lemma 5.2.6 are satisfied, and we conclude from (5.2.27) that

$$\prod_{k=1}^{k_n} \varphi_{nk} \to e^{\psi} \qquad \text{as } n \to \infty.$$

This completes the proof of Theorem 5.2.1. ■

Remark 5.2.2. It may happen that under the u.a.n. condition the sequence $\{S_n\}$ does not converge in law, but the sequence $\{S_n - c_n\}$ does for suitably chosen constants c_n. For instance, in the classical central limit theorem case with $X_{nk} = X_k/B_n$, $k = 1, 2, \ldots, n$, $n \geq 1$, where $B_n^2 = \sum_{k=1}^n \operatorname{var}(X_k)$, $S_n = \sum_{k=1}^n X_{nk}$ converges in law with $c_n = \mathscr{E}S_n/B_n$. In such a situation $\prod_{k=1}^{k_n} \varphi_{nk}(t)$ is to be replaced by $e^{-itc_n} \prod_{k=1}^{k_n} \varphi_{nk}(t)$. Clearly Lemma 5.2.4 can still be applied. Consequently Lemmas 5.2.5 and 5.2.6 also hold.

The following extended version of Theorem 5.2.1 is therefore immediate.

Theorem 5.2.2. Let $\{X_{nk}\}$ be a sequence of row-wise independent random variables satisfying the u.a.n. condition. Then the class of limit laws of $\{S_n - c_n\}$, where c_n is a sequence of constants, coincides with the class of i.d. laws, that is, the class of laws with characteristic function of the form e^{ψ}, where ψ is given by (5.2.3).

Moreover, there exist constants c_n such that $\{S_n - c_n\}$ converges in law to a random variable with characteristic function of the form e^{ψ} if and only if $G_n \xrightarrow{c} G$ for some G, where G_n is given by (5.2.5). In this case the c_n are of the form

$$c_n = \alpha_n - c + o(1) \qquad \text{as } n \to \infty,$$

where c is a (finite) constant and

$$\alpha_n = \sum_{k=1}^{k_n} \left[\alpha_{nk} + \int_{-\infty}^{\infty} \frac{x}{1 + x^2} \, dF_{nk}(x + \alpha_{nk}) \right].$$

Remark 5.2.3. Let $\{X_{nk}\}$ be a sequence of row-wise independent random variables satisfying condition (5.1.8). Then, for every $\epsilon > 0$,

$$\max_{1 \leq k \leq k_n} P\{|X_{nk} - \mathscr{E}X_{nk}| \geq \epsilon\} \leq \epsilon^{-2} \max_{1 \leq k \leq k_n} \sigma_{nk}^2 \to 0 \qquad \text{as } n \to \infty,$$

and the u.a.n. condition is satisfied. It follows that the general bounded variance model of Section 5.1.2 is a particular case of the general central limit problem.

5.3 NORMAL, DEGENERATE, AND POISSON CONVERGENCE

5.3.1 Convergence to the Normal Distribution

We now consider some important particular cases of Theorem 5.2.1. For this purpose it is convenient first to prove the following result.

Theorem 5.3.1 (Central Convergence Criterion). Let $\{X_{nk}\}$, $1 \leq k \leq k_n$, $n \geq 1$, be a row-wise independent sequence of random variables satisfying the u.a.n. condition. Let $S_n = \sum_{k=1}^{k_n} X_{nk}$, $n \geq 1$. Then the sequence $\{S_n\}$ converges in law to a random variable with characteristic function of the form e^{ψ}, with ψ given by (5.2.3), if and only if the following three conditions, given by (5.3.1), (5.3.2), and (5.3.3), hold.

At every continuity point $x(\neq 0)$ of G

$$(5.3.1)\quad \begin{cases} \lim_{n\to\infty} \sum_{k=1}^{k_n} F_{nk}(x) = \int_{-\infty}^{x} \frac{1+y^2}{y^2}\, dG(y) & \text{for } x < 0 \\ \text{and} & \\ \lim_{n\to\infty} \sum_{k=1}^{k_n} [1 - F_{nk}(x)] = \int_{x}^{\infty} \frac{1+y^2}{y^2}\, dG(y) & \text{for } x > 0, \end{cases}$$

where F_{nk} is the distribution function of the X_{nk}.

$$\lim_{\epsilon\to 0} \limsup_{n\to\infty} \sum_{k=1}^{k_n} \left\{ \int_{|x|<\epsilon} x^2\, dF_{nk}(x) - \left(\int_{|x|<\epsilon} x\, dF_{nk}(x) \right)^2 \right\}$$

$$= G(0) - G(0-)$$

$$(5.3.2)\quad = \lim_{\epsilon\to 0} \liminf_{n\to\infty} \sum_{k=1}^{k_n} \left\{ \int_{|x|<\epsilon} x^2\, dF_{nk}(x) - \left(\int_{|x|<\epsilon} x\, dF_{nk}(x) \right)^2 \right\}.$$

For every fixed $\gamma > 0$ such that $\pm\gamma \in C_G$

$$(5.3.3)\quad \lim_{n\to\infty} \sum_{k=1}^{k_n} \int_{|x|<\gamma} x\, dF_{nk}(x) = \alpha + \int_{|x|<\gamma} x\, dG(x) - \int_{|x|\geq\gamma} \frac{1}{x}\, dG(x).$$

Proof. In view of Theorem 5.2.1 it suffices to show that (5.3.1), (5.3.2), and (5.3.3) are equivalent to (5.2.4). Let G_n be as defined in (5.2.5). First we show that $G_n \xrightarrow{c} G$ if and only if (5.3.1) and (5.3.2) hold. Let $x \in C_G$. Then $G_n \xrightarrow{c} G$ is equivalent to

$$G_n(x) \to G(x) \qquad \text{for } x < 0,$$

$$G_n(+\infty) - G_n(x) \to G(+\infty) - G(x) \qquad \text{for } x > 0$$

and

$$\lim_{\epsilon\to 0} \limsup_{n\to\infty} [G_n(\epsilon - 0) - G_n(-\epsilon)] = G(0) - G(0-)$$

$$= \lim_{\epsilon\to 0} \liminf_{n\to\infty} [G_n(\epsilon - 0) - G_n(-\epsilon)].$$

Using the extended Helly-Bray theorem, we conclude that $G_n \xrightarrow{c} G$ if and only if

$$(5.3.4)\quad \begin{cases} \lim\limits_{n\to\infty} \sum\limits_{k=1}^{k_n} F_{nk}(x + \alpha_{nk}) = \displaystyle\int_{-\infty}^{x} \frac{1 + y^2}{y^2}\, dG(y) & \text{for } x < 0, \\ \lim\limits_{n\to\infty} \sum\limits_{k=1}^{k_n} [1 - F_{nk}(x + \alpha_{nk})] = \displaystyle\int_{x}^{\infty} \frac{1 + y^2}{y^2}\, dG(y) & \text{for } x > 0, \end{cases}$$

and

$$\begin{aligned} &\lim_{\epsilon\to 0} \limsup_{n\to\infty} \sum_{k=1}^{k_n} \int_{|x|<\epsilon} \frac{x^2}{1 + x^2}\, dF_{nk}(x + \alpha_{nk}) \\ &\quad = G(0) - G(0-) \\ (5.3.5)\qquad &\quad = \lim_{\epsilon\to 0} \liminf_{n\to\infty} \sum_{k=1}^{k_n} \int_{|x|<\epsilon} \frac{x^2}{1 + x^2}\, dF_{nk}(x + \alpha_{nk}). \end{aligned}$$

Set $\beta_n = \max_{1 \le k \le k_n} \int_{|x|<\gamma} |x|\, dF_{nk}(x)$. Clearly

$$|\alpha_{nk}| \le \beta_n \to 0 \qquad \text{as } n \to \infty.$$

Moreover, for every $x \in \mathbb{R}$,

$$\sum_{k=1}^{k_n} F_{nk}(x - \beta_n) \le \sum_{k=1}^{k_n} F_{nk}(x + \alpha_{nk}) \le \sum_{k=1}^{k_n} F_{nk}(x + \beta_n).$$

We note that every continuity point of G is also a continuity point of the integrals on the right-hand side in (5.3.1). It follows immediately that the first parts of (5.3.4) and (5.3.1) are equivalent, and similarly for the second parts. Hence the equivalence of (5.3.1) and (5.3.4) follows.

Next note that

$$\begin{aligned} \frac{1}{1 + \epsilon^2} \sum_{k=1}^{k_n} \int_{|x|<\epsilon} x^2\, dF_{nk}(x + \alpha_{nk}) &\le \sum_{k=1}^{k_n} \int_{|x|<\epsilon} \frac{x^2}{1 + x^2}\, dF_{nk}(x + \alpha_{nk}) \\ &\le \sum_{k=1}^{k_n} \int_{|x|<\epsilon} x^2\, dF_{nk}(x + \alpha_{nk}), \end{aligned}$$

so that (5.3.5) is equivalent to

$$\begin{aligned} &\lim_{\epsilon\to 0} \limsup_{n\to\infty} \sum_{k=1}^{k_n} \int_{|x|<\epsilon} x^2\, dF_{nk}(x + \alpha_{nk}) \\ &\quad = \lim_{\epsilon\to 0} \liminf_{n\to\infty} \sum_{k=1}^{k_n} \int_{|x|<\epsilon} x^2\, dF_{nk}(x + \alpha_{nk}) \\ (5.3.6)\qquad &\quad = G(0) - G(0-). \end{aligned}$$

On the other hand,

$$\left|\sum_{k=1}^{k_n}\int_{|x|<\epsilon} x^2\,dF_{nk}(x+\alpha_{nk}) - \sum_{k=1}^{k_n}\int_{|x|<\epsilon}(x-\alpha_{nk})^2\,dF_{nk}(x)\right|$$

$$\le \sum_{k=1}^{k_n}\int_{\substack{|x|\ge\epsilon\\ |x-\alpha_{nk}|<\epsilon}}(x-\alpha_{nk})^2\,dF_{nk}(x)$$

$$\le \epsilon^2\sum_{k=1}^{k_n}\int_{|x|\ge\epsilon} dF_{nk}(x).$$

Let $0<\epsilon<\gamma$. Then

$$\sum_{k=1}^{k_n}\int_{|x|<\epsilon}(x-\alpha_{nk})^2\,dF_{nk}(x) - \sum_{k=1}^{k_n}\left\{\int_{|x|<\epsilon} x^2\,dF_n(x) - \left[\int_{|x|<\epsilon} x\,dF_{nk}(x)\right]^2\right\}$$

$$= \sum_{k=1}^{k_n}\left[\int_{\epsilon\le|x|<\gamma} x\,dF_{nk}(x)\right]^2 - \sum_{k=1}^{k_n}\alpha_{nk}^2\int_{|x|\ge\epsilon} dF_{nk}(x)$$

$$\le (\gamma\beta_n+\beta_n^2)\sum_{k=1}^{k_n}\int_{|x|\ge\epsilon} dF_{nk}(x).$$

In view of (5.3.1)

$$\lim_{n\to\infty}\sum_{k=1}^{k_n}\int_{|x|\ge\epsilon} dF_{nk}(x) = \left(\int_{-\infty}^{-\epsilon} + \int_{\epsilon}^{\infty}\right)\frac{1+y^2}{y^2}\,dG(y),$$

so that, as $n\to\infty$ and then $\epsilon\to 0$, we conclude that under (5.3.1) conditions (5.3.2) and (5.3.6) or (5.3.5) are equivalent. Thus we have shown that $G_n \xrightarrow{c} G$ if and only if (5.3.1) and (5.3.2) hold.

Finally, we show that under $G_n \xrightarrow{c} G$ condition (5.3.3) is equivalent to $\alpha_n \to \alpha$ as $n\to\infty$. It suffices to show that under (5.3.1) and (5.3.2) $\alpha_n\to\alpha$ if and only if (5.3.3) holds. Let $\gamma>0$ be fixed such that $\pm\gamma\in C_G$.

First note that

$$\sum_{k=1}^{k_n}\int_{-\infty}^{\infty}\frac{x}{1+x^2}\,dF_{nk}(x+\alpha_{nk})$$

$$= \sum_{k=1}^{k_n}\int_{|x|<\gamma} x\,dF_{nk}(x+\alpha_{nk}) - \sum_{k=1}^{k_n}\int_{|x|<\gamma}\frac{x^3}{1+x^2}\,dF_{nk}(x+\alpha_{nk})$$

$$+ \sum_{k=1}^{k_n}\int_{|x|\ge\gamma}\frac{x}{1+x^2}\,dF_{nk}(x+\alpha_{nk}).$$

Moreover, by using the extended Helly-Bray theorem we have

$$\sum_{k=1}^{k_n} \int_{|x|<\gamma} \frac{x^3}{1+x^2}\, dF_{nk}(x+\alpha_{nk}) = \int_{|x|<\gamma} x\, dG_n(x) \to \int_{|x|<\gamma} x\, dG(x)$$

and

$$\sum_{k=1}^{k_n} \int_{|x|\geq\gamma} \frac{x}{1+x^2}\, dF_{nk}(x+\alpha_{nk}) = \int_{|x|\geq\gamma} \frac{1}{x}\, dG_n(x) \to \int_{|x|\geq\gamma} \frac{1}{x}\, dG(x).$$

Therefore

$$\lim_{n\to\infty} \left\{ \sum_{k=1}^{k_n} \int_{|x|<\gamma} x\, dF_{nk}(x+\alpha_{nk}) - \sum_{k=1}^{k_n} \int_{-\infty}^{\infty} \frac{x}{1+x^2}\, dF_{nk}(x+\alpha_{nk}) \right\}$$
$$= \int_{|x|<\gamma} x\, dG(x) - \int_{|x|\geq\gamma} \frac{1}{x}\, dG(x).$$

We need only prove that

$$\sum_{k=1}^{k_n} \int_{|x|<\gamma} x\, dF_{nk}(x+\alpha_{nk}) \to 0.$$

We have

$$\left| \sum_{k=1}^{k_n} \int_{|x|<\gamma} x\, dF_{nk}(x+\alpha_{nk}) \right|$$
$$\leq \left| \sum_{k=1}^{k_n} \int_{|x|<\gamma} (x+\alpha_{nk})\, dF_{nk}(x) \right| + \left| \sum_{k=1}^{k_n} \int_{|x-\alpha_{nk}|<\gamma} (x-\alpha_{nk})\, dF_{nk}(x) \right.$$
$$\left. - \sum_{k=1}^{k_n} \int_{|x|<\gamma} (x-\alpha_{nk})\, dF_{nk}(x) \right|$$
$$\leq \beta_n \sum_{k=1}^{k_n} \int_{|x|\geq\gamma} dF_{nk}(x) + (\gamma+\beta_n) \sum_{k=1}^{k_n} \int_{\gamma\leq|x|<\gamma+\beta_n} dF_{nk}(x)$$
$$\to 0$$

by (5.3.1) and the fact that $\beta_n \to 0$. This completes the proof of Theorem 5.3.1. ■

Remark 5.3.1. As noted in Remark 5.2.2, under the u.a.n. condition the sequence $\{S_n\}$ may not converge in law, but the sequence $\{S_n - c_n\}$ does for suitably chosen constants c_n. In this case one can derive from Theorem 5.2.2 the following analogue of Theorem 5.3.1. Let $\{X_{nk}\}$, $1 \le k \le k_n$, $n \ge 1$, be a row-wise independent sequence of random variables satisfying the u.a.n. condition. Let $S_n = \sum_{k=1}^{k_n} X_{nk}$, $n \ge 1$. Then for suitably chosen constants $\{c_n\}$ the sequence $\{S_n - c_n\}$ converges in law to a random variable with characteristic function of the form e^{ψ}, with ψ given by (5.2.3), if and only if (5.3.1) and (5.3.2) hold. In this case the constants c_n are of the form

$$c_n = \sum_{k=1}^{k_n} \int_{|x|<\gamma} x\, dF_{nk}(x) - \alpha - \int_{|x|<\gamma} x\, dG(x) + \int_{|x|\ge\gamma} x^{-1}\, dG(x) + o(1) \qquad \text{as } n \to \infty,$$

where α, G, and $\gamma > 0$ are as in Theorem 5.3.1.

We now specialize the central convergence criterion to yield conditions for convergence to the normal law. For this purpose let $\gamma > 0$, and set

$$\alpha_{nk}(\gamma) = \int_{|x|<\gamma} x\, dF_{nk}(x)$$

and

$$\sigma_{nk}^2(\gamma) = \int_{|x|<\gamma} x^2\, dF_{nk}(x) - \left[\int_{|x|<\gamma} x\, dF_{nk}(x)\right]^2.$$

Note that the normal distribution with mean μ and variance σ^2 corresponds to the characteristic function

$$\varphi(t) = i\mu t - \frac{\sigma^2}{2} t^2,$$

so that

$$\alpha = \mu \qquad \text{and} \qquad G(x) = 0 \quad \text{if } x < 0 \qquad \text{and} \quad = \sigma^2 \quad \text{if } x > 0.$$

Theorem 5.3.1 immediately implies the following assertion.

Proposition 5.3.1. Let $\{X_{nk}\}$ be a sequence of row-wise independent random variables satisfying the u.a.n. condition. Let F_{nk} be the distribution function of X_{nk}. Then the sequence of sums $\{S_n\}$ converges in law to a normal random variable with mean μ and variance σ^2 if and only if the following conditions are satisfied:

(i) $\sum_{k=1}^{k_n} P\{|X_{nk}| \geq \epsilon\} \to 0$ for every $\epsilon > 0$.
(ii) $\lim_{\epsilon\to 0} \limsup_{n\to\infty} \sum_{k=1}^{k_n} \sigma_{nk}^2(\epsilon) = \lim_{\epsilon\to 0} \liminf_{n\to\infty} \sum_{k=1}^{k_n} \sigma_{nk}^2(\epsilon) = \sigma^2$.
(iii) $\sum_{k=1}^{k_n} \alpha_{nk}(\epsilon) \to \mu$ for every $\epsilon > 0$.

This result yields the following proposition.

Proposition 5.3.2. Let $\{X_{nk}: k = 1, 2, \ldots, k_n, n \geq 1\}$ be a sequence of row-wise independent random variables, and let F_{nk} be the distribution function of the X_{nk}. Let $S_n = \sum_{k=1}^{k_n} X_{nk}$, $n \geq 1$. Then the sequence $\{S_n\}$ converges in law to a normal random variable with mean μ and variance σ^2, and the X_{nk} satisfy the u.a.n. condition if and only if for every fixed $\epsilon > 0$ the following conditions are satisfied:

(A) $\sum_{k=1}^{k_n} P\{|X_{nk}| \geq \epsilon\} \to 0$ for every $\epsilon > 0$.
(B) $\sum_{k=1}^{k_n} \sigma_{nk}^2(\epsilon) \to \sigma^2$.
(C) $\sum_{k=1}^{k_n} \alpha_{nk}(\epsilon) \to \mu$.

Proof. Clearly (A), (B), and (C) $\Rightarrow$ (i), (ii), and (iii) of Proposition 5.3.1. We need only show that the random variables satisfy the u.a.n. condition. We have

$$\max_{1\leq k\leq k_n} P\{|X_{nk}| \geq \epsilon\} \leq \sum_{k=1}^{k_n} P\{|X_{nk}| \geq \epsilon\} \to 0,$$

so that the sufficiency of (A), (B), and (C) has been established.

As for the necessity, it suffices to show that under (i) of Proposition 5.3.1 condition (ii) $\Rightarrow$ (B) for every $\epsilon > 0$. Let $\epsilon > 0$ be fixed. Then for every $\delta < \epsilon$ we have

$$\begin{aligned}\sum_{k=1}^{k_n} \sigma_{nk}^2(\epsilon) &= \sum_{k=1}^{k_n} \sigma_{nk}^2(\delta)\\ &\quad + \sum_{k=1}^{k_n} \left\{\int_{\delta\leq|x|<\epsilon} x^2\, dF_{nk}(x) - \left[\int_{\delta\leq|x|<\epsilon} x\, dF_{nk}(x)\right]^2\right.\\ &\quad \left. - 2\int_{|x|<\delta} x\, dF_{nk}(x) \int_{\delta\leq|x|<\epsilon} x\, dF_{nk}(x)\right\}.\end{aligned}$$

Now

$$\begin{aligned}0 &\leq \sum_{k=1}^{k_n} \left\{\int_{\delta\leq|x|<\epsilon} x^2\, dF_{nk}(x) - \left(\int_{\delta\leq|x|<\epsilon} x\, dF_{nk}(x)\right)^2\right\}\\ &\leq \epsilon^2 \sum_{k=1}^{k_n} P\{|X_{nk}| \geq \delta\} \to 0\end{aligned}$$

and

$$\begin{aligned}\sum_{k=1}^{k_n} \left| \int_{|x|<\delta} x\, dF_{nk}(x) \right| \cdot \left| \int_{\delta \le |x| < \epsilon} x\, dF_{nk}(x) \right| &\le \delta\epsilon \sum_{k=1}^{k_n} P\{|X_{nk}| \ge \delta\} \\ &\to 0\end{aligned}$$

in view of (i). Therefore

$$\limsup_{n\to\infty} \sum_{k=1}^{k_n} \sigma_{nk}^2(\epsilon) = \limsup_{n\to\infty} \sum_{k=1}^{k_n} \sigma_{nk}^2(\delta).$$

A similar relation holds for the limit inferior, so that these limits are independent of ϵ. In view of (ii) it follows that the limit $\lim_{n\to\infty} \sum_{k=1}^{k_n} \sigma_{nk}^2(\epsilon)$ exists and equals σ^2, so that (B) holds. ∎

The following formulation of Proposition 5.3.2 is useful.

Theorem 5.3.2. Let $\{X_{nk}\}$ be a sequence of row-wise independent random variables. Then the sequence of partial sums $\{S_n\}$ converges in law to a normal random variable with mean μ and variance σ^2, and $\{X_{nk}\}$ satisfies the u.a.n. condition if and only if for every (fixed) $\epsilon > 0$ and some $\gamma > 0$, as $n \to \infty$,

(5.3.7) $$\sum_{k=1}^{k_n} P\{|X_{nk}| \ge \epsilon\} \to 0,$$

(5.3.8) $$\sum_{k=1}^{k_n} \sigma_{nk}^2(\gamma) \to \sigma^2,$$

and

(5.3.9) $$\sum_{k=1}^{k_n} \alpha_{nk}(\gamma) \to \mu.$$

Proof. Clearly, it is sufficient to show that, if (5.3.7), (5.3.8), and (5.3.9) hold for every $\epsilon > 0$ and some $\gamma > 0$, then (B) and (C) of Proposition 5.3.2 will be satisfied for every $\epsilon > 0$. Let $0 < \epsilon < \gamma$. In view of (5.3.7)

$$\begin{aligned}\left| \sum_{k=1}^{k_n} \sigma_{nk}^2(\gamma) - \sum_{k=1}^{k_n} \sigma_{nk}^2(\epsilon) \right| &\le \sum_{k=1}^{k_n} \int_{\epsilon \le |x| < \gamma} x^2\, dF_{nk}(x) + 3\gamma \left| \sum_{k=1}^{k_n} \int_{\epsilon \le |x| < \gamma} x\, dF_{nk}(x) \right| \\ &\le 4\gamma^2 \sum_{k=1}^{k_n} \int_{|x|<\gamma} dF_{nk}(x) \to 0 \qquad \text{as } n \to \infty.\end{aligned}$$

Interchanging ϵ and γ, we see that the same result holds for $0 < \gamma < \epsilon$.

Finally

$$\left| \sum_{k=1}^{k_n} \alpha_{nk}(\gamma) - \sum_{k=1}^{k_n} \alpha_{nk}(\epsilon) \right|$$
$$\leq \sum_{k=1}^{k_n} \int_{\min(\gamma,\epsilon) \leq |x| < \max(\gamma,\epsilon)} |x| \, dF_{nk}(x) \to 0,$$

and the proof is complete. ■

Remark 5.3.2. In view of Remark 5.3.1 the following analogue of Theorem 5.3.2 can be easily derived. Let $\{X_{nk}\}$ be a sequence of row-wise independent random variables. Then, in order that the sequence $\{S_n - c_n\}$ converge in law to a normal random variable with mean μ and variance σ^2 for a suitably chosen sequence of constants $\{c_n\}$ and $\{X_{nk}\}$ satisfies the u.a.n. condition, it is necessary and sufficient that (5.3.7) and (5.3.8) hold for every $\epsilon > 0$ and some $\gamma > 0$. In this case the c_n are given by

$$c_n = \sum_{k=1}^{n} \alpha_{nk}(\gamma) - \mu + o(1) \qquad \text{as } n \to \infty.$$

Theorem 5.3.3. Let $\{X_{nk}\}$ be a sequence of row-wise independent random variables such that the sequence of sums $\{\sum_{k=1}^{k_n} X_{nk}\}$ converges in law to some nondegenerate random variable. Then the limit law is normal, and the u.a.n. condition is satisfied if and only if

$$\sum_{k=1}^{k_n} P\{|X_{nk}| \geq \epsilon\} \to 0$$

for every fixed $\epsilon > 0$.

Proof. We need only show that, if $\sum_{k=1}^{k_n} P\{|X_{nk}| \geq \epsilon\} \to 0$ and the sequence $\{\sum_{k=1}^{k_n} X_{nk}\}$ converges in law to a nondegenerate random variable, the limit law is normal. This follows from Theorem 5.3.1, which implies that the function G in (5.3.1) vanishes for every $x \neq 0$. Since the limit law is nondegenerate, it must be normal. ■

Remark 5.3.3. The condition $\sum_{k=1}^{k_n} P\{|X_{nk}| \geq \epsilon\} \to 0$ is equivalent to the condition

$$\max_{1 \leq k \leq k_n} |X_{nk}| \xrightarrow{P} 0.$$

To see this write $p_{nk} = P\{|X_{nk}| \geq \epsilon\}$. Then

$$P\left\{\max_{1 \leq k \leq k_n} |X_{nk}| \geq \epsilon\right\} = 1 - \sum_{k=1}^{k_n} (1 - p_{nk})$$

due to row independence of the X_{nk}. Clearly

$$1 - \exp\left(-\sum_{k=1}^{k_n} p_{nk}\right) \leq 1 - \prod_{k=1}^{k_n} (1 - p_{nk}) \leq \sum_{k=1}^{k_n} p_{nk},$$

and our assertion follows. It should be noted that $\{X_{nk}\}$ may satisfy the u.a.n. condition but not the condition $\max_{1 \leq k \leq k_n} |X_{nk}| \xrightarrow{P} 0$. For example, let F_{nk}, $1 \leq k \leq n$, $n \geq 1$, be the distribution function of random variable X_{nk} which assigns mass $1 - (1/n)$ to $x = 0$ and $1/n$ to $x = 1$. Then for every ϵ $(0 < \epsilon < 1)$

$$\max_{1 \leq k \leq n} P\{|X_{nk}| \geq \epsilon\} = \frac{1}{n} \to 0$$

as $n \to \infty$, so that $\{X_{nk}\}$ satisfies the u.a.n. condition. But for $0 < \epsilon < 1$

$$P\left\{\max_{1 \leq k \leq n} |X_{nk}| \geq \epsilon\right\} = 1 - \prod_{k=1}^{n} P\{|X_{nk}| < \epsilon\} = 1 - \left(1 - \frac{1}{n}\right)^n \to 1 - e^{-1}$$

as $n \to \infty$.

The following result follows immediately from Proposition 5.3.2 and Theorem 5.3.2.

Proposition 5.3.3. Let $\{X_n\}$ be a sequence of independent random variables, and let F_n be the distribution function of X_n. Let $\{b_n\}$ be a sequence of positive constants. In order that $\{X_k/b_n\}$ satisfy the u.a.n. condition and the sequence $\{\sum_{k=1}^{n} X_k/b_n\}$ converge in law to a standard normal random variable, it is necessary and sufficient that

(5.3.10) $$\sum_{k=1}^{n} P\{|X_k| \geq \epsilon b_n\} \to 0 \qquad \text{for every fixed } \epsilon > 0,$$

(5.3.11) $$b_n^{-2} \sum_{k=1}^{n} \left\{\int_{|x| < b_n} x^2 \, dF_k(x) - \left[\int_{|x| < b_n} x \, dF_k(x)\right]^2\right\} \to 1,$$

and

(5.3.12) $$b_n^{-1} \sum_{k=1}^{n} \int_{|x| < b_n} x \, dF_k(x) \to 0.$$

Clearly (5.3.11) and (5.3.12) may be replaced by

$$(5.3.11') \qquad b_n^{-2} \sum_{k=1}^{n} \left\{ \int_{|x|<\epsilon b_n} x^2 \, dF_k(x) - \left[\int_{|x|<\epsilon b_n} x \, dF_k(x) \right]^2 \right\} \to 1$$

and

$$(5.3.12') \qquad b_n^{-1} \sum_{k=1}^{n} \int_{|x|<\epsilon b_n} x \, dF_k(x) \to 0$$

for every fixed $\epsilon > 0$.

Corollary to Proposition 5.3.3. Let $\{X_n\}$ be a sequence of independent random variables, and let F_n be the distribution function of the X_n. Let $\{b_n\}$, $b_n > 0$, be constants. Then (5.3.10) and

$$b_n^{-2} \sum_{k=1}^{n} \int_{|x|<b_n} x^2 \, dF_k(x) \to 1$$

and

$$b_n^{-1} \sum_{k=1}^{n} \left| \int_{|x|<b_n} x \, dF_k(x) \right| \to 0$$

imply that $b_n^{-1} \sum_{k=1}^{n} X_k \xrightarrow{L} Z$, where Z is a standard normal random variable.

Remark 5.3.4. Let $\{X_n\}$ be a sequence of independent random variables. Let $S_n = \sum_{k=1}^{n} X_k$, $n \geq 1$, and $\{b_n\}$ be a sequence of positive constants. Then $\{X_k/b_n\}$ satisfies the u.a.n. condition, and $\{b_n^{-1}S_n - c_n\}$ converges in law to a standard normal random variable for suitably chosen $\{c_n\}$ if and only if (5.3.10) and (5.3.11) hold. In this case the c_n are of the form

$$c_n = b_n^{-1} \sum_{k=1}^{n} \int_{|x|<\epsilon b_n} x \, dF_k(x) + o(1) \qquad \text{as } n \to \infty$$

for every (fixed) $\epsilon > 0$. Here F_n is the distribution function of X_n, and we may take $\epsilon = 1$.

Lemma 5.3.1. For every fixed $\epsilon > 0$ let $x_n(\epsilon) \to x$ as $n \to \infty$. Here x may be $\pm\infty$. Then there exists a sequence of real numbers $\{\epsilon_n\}$ such that $\epsilon_n \to 0$ and $x_n(\epsilon_n) \to x$ as $n \to \infty$.

Proof. First we consider the case where $|x| < \infty$. Without loss of generality we may assume that $x = 0$ and $x_n(\epsilon) \geq 0$, for otherwise we can take $y_n(\epsilon) = |x_n(\epsilon) - x|$. Let $z_n(\epsilon) = \max_{k \geq n} x_k(\epsilon)$. Then $z_n(\epsilon) \downarrow 0$ as $n \to \infty$. Clearly there exists an increasing sequence of integers $\{n_m\}$ such that

$$z_n(2^{-m}) \leq 2^{-m} \qquad \text{for } n \geq n_m, \; m = 1, 2, \ldots.$$

Take $\epsilon_n = 2^{1-m}$ for $n_{m-1} \leq n < n_m$. Then $\epsilon_n \to 0$, and

$$x_n(\epsilon_n) \leq \max_{n_{m-1} \leq n < n_m} z_n(\epsilon_n) \leq z_{n_{m-1}}(\epsilon_n) \leq z_{n_{m-1}}(2^{1-m}) \leq 2^{1-m} \to 0$$

as $n \to \infty$. Finally, if $x = \pm\infty$, then $[x_n(\epsilon)]^{-1} \to 0$ as $n \to \infty$, and the result follows from above. ■

Theorem 5.3.4 (Bernstein and Feller). Let $\{X_n\}$ be a sequence of independent random variables. In order that there exist sequences of constants $\{a_n\}$ and $\{b_n\}$ with $b_n > 0$ such that the sequence $\{b_n^{-1} \sum_{k=1}^n X_k - a_n\}$ converges in law to a standard normal random variable and that $\{X_k/b_n\}$ satisfies the u.a.n. condition, it is necessary and sufficient that there exist positive constants $c_n \to \infty$ as $n \to \infty$ such that

$$\sum_{k=1}^{n} P\{|X_k| \geq c_n\} \to 0 \qquad \text{as } n \to \infty \tag{5.3.13}$$

and

$$c_n^{-2} \sum_{k=1}^{n} \left\{ \int_{|x|<c_n} x^2 \, dF_k(x) - \left[\int_{|x|<c_n} x \, dF_k(x) \right]^2 \right\} \to \infty. \tag{5.3.14}$$

Here F_k is the distribution function of X_k.

Proof. For the proof of sufficiency of (5.3.13) and (5.3.14) write $b_n = +\sqrt{b_n^2}$, where

$$b_n^2 = \sum_{k=1}^{n} \left\{ \int_{|x|<c_n} x^2 \, dF_k(x) - \left[\int_{|x|<c_n} x \, dF_k(x) \right]^2 \right\}.$$

From (5.3.14) it follows that $b_n^2/c_n^2 \to \infty$ or $c_n/b_n \to 0$ as $n \to \infty$. Let $\epsilon > 0$. Then, for sufficiently large n, $c_n < b_n\epsilon$, so that

$$\sum_{k=1}^{n} \int_{|x| \geq c_n} dF_k(x) \geq \sum_{k=1}^{n} \int_{|x| \geq b_n\epsilon} dF_k(x).$$

In view of (5.3.13) it follows that $\{X_k/b_n, 1 \leq k \leq n\}$ satisfies the u.a.n. condition and also condition (5.3.10) with $X_{nk} = X_k/b_n$, $1 \leq k \leq n$. For n sufficiently large so that $c_n < b_n\epsilon$, set

$$\begin{aligned} K_n &= b_n^{-2} \sum_{k=1}^{n} \left\{ \int_{|x|<\epsilon b_n} x^2 \, dF_k(x) - \left(\int_{|x|<\epsilon b_n} x \, dF_k(x) \right)^2 \right\} \\ &= L_n + M_n + N_n, \end{aligned}$$

where

$$L_n = b_n^{-2} \sum_{k=1}^{n} \left\{ \int_{|x|<c_n} x^2 \, dF_k(x) - \left(\int_{|x|<c_n} x \, dF_k(x) \right)^2 \right\},$$

$$M_n = b_n^{-2} \sum_{k=1}^{n} \left\{ \int_{c_n \le |x| < b_n\epsilon} x^2 \, dF_k(x) - \left(\int_{c_n \le |x| < b_n\epsilon} x \, dF_k(x) \right)^2 \right\},$$

and

$$N_n = -2b_n^{-2} \sum_{k=1}^{n} \left\{ \int_{|x|<c_n} x \, dF_k(x) \int_{c_n \le |x| < b_n\epsilon} x \, dF_k(x) \right\}.$$

Clearly $L_n \equiv 1$ by the definition of b_n, and

$$0 \le M_n \le b_n^{-2} \sum_{k=1}^{n} \int_{c_n \le |x| < b_n\epsilon} x^2 \, dF_k(x)$$

$$\le \epsilon^2 \sum_{k=1}^{n} \int_{|x| \ge c_n} dF_k(x) \to 0 \qquad \text{by (5.3.13).}$$

Next,

$$|N_n| \le \frac{2\epsilon c_n}{b_n} \sum_{k=1}^{n} \int_{|x| \ge c_n} dF_k(x) \to 0,$$

again by (5.3.13) and the fact that $c_n/b_n \to 0$. Therefore $K_n \to 1$ as $n \to \infty$. Thus (5.3.11′) holds. In view of Remark 5.3.4, $\{b_n^{-1} \sum_{k=1}^{n} X_k - a_n\}$ converges in law to a standard normal random variable and $\{X_k/b_n\}$ satisfies the u.a.n. condition. Here a_n can be chosen to be $a_n = b_n^{-1} \sum_{k=1}^{n} \int_{|x|<\epsilon b_n} x \, dF_k(x)$.

As for the proof of the necessity part, suppose that $\{b_n^{-1} \sum_{k=1}^{n} X_k - a_n\}$ converges in law to a standard normal random variable and that $\{X_k/b_n\}$ satisfies the u.a.n. condition. Then $\max_{1 \le k \le n} X_k/b_n \xrightarrow{P} 0$ as $n \to \infty$. Moreover, $b_n \to \infty$ (see Lemma 5.4.1), and it follows from Proposition 5.3.3 and Remark 5.3.3 that as $n \to \infty$

$$\sum_{k=1}^{n} P\{|X_k| \ge \epsilon b_n\} \to 0 \tag{5.3.15}$$

and

$$b_n^{-2} \sum_{k=1}^{n} \left\{ \int_{|x|<\epsilon b_n} x^2 \, dF_k(x) - \left[\int_{|x|<\epsilon b_n} x \, dF_k(x) \right]^2 \right\} \to 1 \tag{5.3.16}$$

for every $\epsilon > 0$. In view of Lemma 5.3.1 we can select a sequence of positive constants $\{\epsilon_n\}$ such that $\epsilon_n \to 0$ with $\epsilon_n b_n \to \infty$ and

$$\sum_{k=1}^{n} P\{|X_k| > \epsilon_n b_n\} \to 0, \tag{5.3.17}$$

$$b_n^{-2} \sum_{k=1}^{n} \left\{ \int_{|x|<\epsilon_n b_n} x^2\, dF_k(x) - \left(\int_{|x|<\epsilon_n b_n} x\, dF_k(x) \right)^2 \right\} \to 1. \tag{5.3.18}$$

Set $c_n = \epsilon_n b_n$. Then $c_n \to \infty$, and (5.3.17) and (5.3.18) imply (5.3.13) and (5.3.14), respectively. This completes the proof. ■

We note that the Lindeberg–Feller central limit theorem can also be derived from Proposition 5.3.3. We leave the details to the reader.

5.3.2 Degenerate and Poisson Convergence

By considering the distribution degenerate at zero as a degenerate normal law with mean 0 and variance 0 we obtain the following criterion for degenerate convergence from Theorem 5.3.2.

Proposition 5.3.4 (Degenerate Convergence Criterion). Let $\{X_{nk}\}$ be a rowwise independent sequence. Then $S_n \xrightarrow{P} 0$ as $n \to \infty$, and the u.a.n. condition is satisfied if and only if, for every $\epsilon > 0$ and a $\gamma > 0$, as $n \to \infty$

$$\sum_{k=1}^{k_n} P\{|X_{nk}| \geq \epsilon\} \to 0, \tag{5.3.19}$$

$$\sum_{k=1}^{k_n} \sigma_{nk}^2(\gamma) \to 0, \tag{5.3.20}$$

and

$$\sum_{k=1}^{k_n} \alpha_{nk}(\gamma) \to 0. \tag{5.3.21}$$

Corollary 1. Let $\{X_n\}$ be a sequence of independent random variables. Let $S_n = \sum_{k=1}^{n} X_k$. Then $n^{-1}S_n \xrightarrow{P} 0$ as $n \to \infty$ if and only if, for every $\epsilon > 0$, as $n \to \infty$

$$\sum_{k=1}^{k_n} P\{|X_k| \geq n\epsilon\} \to 0, \tag{5.3.22}$$

$$n^{-2} \sum_{k=1}^{k_n} \left\{ \int_{|x|<n} x^2\, dF_k(x) - \left(\int_{|x|<n} x\, dF_k(x) \right)^2 \right\} \to 0, \tag{5.3.23}$$

and

$$n^{-1} \sum_{k=1}^{k_n} \int_{|x|<n} x\, dF_k(x) \to 0, \tag{5.3.24}$$

where F_k is the distribution function of X_k.

Proof. Let $X_{nk} = X_k/n$, $k = 1, 2, \ldots, n$, and $n = 1, 2, \ldots$. Then $F_{nk}(x) = F_k(nx)$. Take $\gamma = 1$ in Proposition 5.3.4. It only remains to show that $n^{-1}S_n \xrightarrow{P} 0 \Rightarrow \max_{1 \le k \le n} P\{|X_k| \ge n\epsilon\} \to 0$. For $\delta > 0$, $\epsilon > 0$ there exists an $n_{\epsilon,\delta}$ such that for $n \ge n_{\epsilon,\delta}$

$$P\{|S_n| < n\epsilon\} > 1 - \delta.$$

Therefore for $n > n_{\epsilon,\delta}$

$$\begin{aligned} P\{|X_n| < 2n\epsilon\} &= P\left\{\left|\frac{S_n}{n} - \frac{n-1}{n}\frac{S_{n-1}}{n-1}\right| < 2\epsilon\right\} \\ &\ge P\left\{\left|\frac{S_n}{n}\right| < \epsilon, \left|\frac{S_{n-1}}{n-1}\right| < \epsilon\right\} \\ &\ge 1 - P\left\{\left|\frac{S_n}{n}\right| \ge \epsilon\right\} - P\left\{\left|\frac{S_{n-1}}{n-1}\right| \ge \epsilon\right\} \\ &\ge 1 - 2\delta, \end{aligned}$$

so that $n^{-1}X_n \xrightarrow{P} 0$. Now $n^{-1}S_n \xrightarrow{P} 0 \Rightarrow (n-1)^{-1}S_{n-1} \xrightarrow{P} 0$, so, repeating the argument used above with $n^{-1}S_{n-1}$, we conclude that $n^{-1}X_{n-1} \xrightarrow{P} 0$, and so on. Thus $\max_{1 \le k \le n} X_k/n \xrightarrow{P} 0$ and $\{X_k/n\}$ satisfy the u.a.n. condition.

Corollary 2. Let $\{X_n\}$ be a sequence of independent, identically distributed random variables with common distribution function F. Then $n^{-1}S_n \xrightarrow{P} 0$ if and only if, as $n \to \infty$,

(5.3.25)
$$nP\{|X_1| \ge n\} \to 0$$

and

(5.3.26)
$$\int_{|x|<n} x\, dF(x) \to 0.$$

Proof. It is sufficient to show that (5.3.25) and (5.3.26) together imply

$$n^{-1}\left\{\int_{|x|<n} x^2\, dF(x) - \left(\int_{|x|<n} x\, dF(x)\right)^2\right\} \to 0.$$

Clearly it is sufficient to prove that $n^{-1}\int_{|x|<n} x^2\, dF(x) \to 0$. This follows on integration by parts and only requires (5.3.25). In fact,

$$\begin{aligned} n^{-1}\int_{|x|<n} x^2\, dF(x) &= -n^{-1}\int_0^n x^2\, dP\{|X_1| > x\} \\ &\le 2n^{-1}\int_0^n xP\{|X_1| > x\}\, dx. \end{aligned}$$

Let $\epsilon > 0$. Choose $N(\epsilon)$ sufficiently large so that for $n > N(\epsilon)$

$$nP\{|X_1| > n\} < \epsilon.$$

Then for $n > N(\epsilon)$

$$\begin{aligned} n^{-1}\int_0^n xP\{|X_1| > x\}\,dx &= n^{-1}\left\{\int_0^{N(\epsilon)} xP\{|X_1| > x\}dx \right. \\ &\quad \left. + \int_{N(\epsilon)}^n xP\{|X_1| > x\}\,dx\right\} \\ &\le n^{-1}(c + n\epsilon), \end{aligned}$$

where c is a constant. Letting $n \to \infty$ and then $\epsilon \to 0$, we get the required result.

Note that, if $\mathscr{E}X_1$ exists and equals zero, both (5.3.25) and (5.3.26) hold, and we get Khintchine's weak law of large numbers (Theorem 2.1.3).

Finally we specialize the central convergence criterion to yield the conditions for convergence to the Poisson distribution. Recall that the Poisson distribution with parameter λ corresponds to characteristic function $\varphi(t) = \exp\{\lambda(e^{it} - 1)\}$, so that $\alpha = \lambda/2$ and $G(x) = 0$ for $x < 1$, and $= \lambda/2$ for $x > 1$. We immediately get the following criterion.

Proposition 5.3.5 (Poisson Convergence Criterion). Let $\{X_{nk}\}$ be a sequence or row-wise independent random variables which satisfy the u.a.n. condition. Then the sequence $\{S_n\}$ converges in law to the Poisson random variable with parameter λ if and only if, for every ϵ, $0 < \epsilon < 1$, and for a γ, $0 < \gamma < 1$, as $n \to \infty$

$$(5.3.27) \qquad \begin{cases} \displaystyle\sum_{k=1}^{k_n} \int_{[|x| \ge \epsilon,\, |x-1| \ge \epsilon]} dF_{nk}(x) \to 0 \\ \text{and} \\ \displaystyle\sum_{k=1}^{k_n} \int_{|x-1| < \epsilon} dF_{nk}(x) \to \lambda, \end{cases}$$

$$(5.3.28) \qquad \sum_{k=1}^{k_n} \sigma_{nk}^2(\gamma) \to 0,$$

and

$$(5.3.29) \qquad \sum_{k=1}^{k_n} \alpha_{nk}(\gamma) \to 0.$$

5.4 STABLE DISTRIBUTIONS

5.4.1 Distributions of Class L

We now study an important special subclass of the class of all i.d. distributions. Let $\{X_n, n \geq 1\}$ be a sequence of independent random variables, and let $\{b_n\}$ be a sequence of positive real numbers such that the following condition holds:

$$\lim_{n\to\infty} \max_{1\leq k\leq n} \{|X_k| \geq b_n\epsilon\} = 0 \qquad \text{for every } \epsilon > 0. \tag{5.4.1}$$

By writing $X_{nk} = X_k/b_n$, $1 \leq k \leq n$, $n \geq 1$, we see that the sequence $\{X_{nk}\}$ of row-independent random variables satisfies the u.a.n. condition. Set $S_n = \sum_{k=1}^n X_k$ for $n \geq 1$. Let L be the class of distributions which are the weak limits of the distributions of the sums

$$b_n^{-1}S_n - a_n, \qquad n \geq 1,$$

where a_n and $b_n > 0$ are suitably chosen constants. It follows from the results of Section 5.2 that the class L is a subclass of the class of all i.d. distributions.

We first prove some preliminary lemmas.

Lemma 5.4.1. Let $\{X_n\}$ be a sequence of independent random variables, and $\{b_n\}$ a sequence of positive constants such that (5.4.1) is satisfied. Suppose that the sequence $\{b_n^{-1}S_n - a_n\}$ converges in law to a nondegenerate random variable for some sequence of constants $\{a_n\}$. Then

$$b_n \to \infty \qquad \text{and} \qquad \frac{b_{n+1}}{b_n} \to 1 \qquad \text{as } n \to \infty. \tag{5.4.2}$$

Proof. Let θ_k be the characteristic function of X_k, $k \geq 1$, and let φ_n be the characteristic function of $b_n^{-1}S_n - a_n$. Clearly

$$\varphi_n(t) = e^{-ita_n} \prod_{k=1}^n \theta_k\left(\frac{t}{b_n}\right). \tag{5.4.3}$$

According to our hypothesis, $\varphi_n \to \varphi$ as $n \to \infty$, where φ is the characteristic function of a nondegenerate random variable.

We first show that $b_n \to \infty$ as $n \to \infty$. Suppose that $b_n \not\to \infty$. Then there exists a subsequence of $\{b_n\}$ which is bounded and hence, according to the Bolzano–Weierstrass theorem, contains a further subsequence $\{b_{n_r}\}$ such that $\{b_{n_r}\}$ converges to a finite limit b, say as $n \to \infty$. Let $t \in \mathbb{R}$ be fixed. Set $t_r = b_{n_r}t$ for $r \geq 1$. Clearly $t_r \to bt$ as $r \to \infty$. In view of (5.4.1) and Lemma 5.2.1 we conclude that

$$|\theta_k(t)| = \left|\theta_k\left(\frac{t_r}{b_{n_r}}\right)\right| \to 1$$

as $r \to \infty$ for every k. This in turn implies that $|\theta_k(t)| = 1$ for every $k \geq 1$, and hence $|\varphi(t)| = 1$ for $t \in \mathbb{R}$. Then $\varphi \equiv 1$, which contradicts the hypothesis. Hence we must have $b_n \to \infty$ as $n \to \infty$.

Next we show that $b_{n+1}/b_n \to 1$ as $n \to \infty$. Let F be the distribution function corresponding to φ. Clearly, in view of (5.4.1),

$$b_n^{-1}S_n - a_n \qquad \text{and} \qquad b_{n+1}^{-1}S_n - a_{n+1}$$

converge in law to the same random variable with distribution function F. Let F_n be the distribution function of $b_n^{-1}S_n - a_n$. Then for $x \in \mathbb{R}$

$$P\{b_{n+1}^{-1}S_n - a_{n+1} \leq x\} = F_n(\beta_n x + \alpha_n),$$

where

$$\beta_n = \frac{b_{n+1}}{b_n} \qquad \text{and} \qquad \alpha_n = \frac{b_{n+1}}{b_n}a_{n+1} - a_n.$$

We recall (Problem 3.10.15) that, if $F_n(u_n x + v_n) \xrightarrow{w} F$ and $F_n(\gamma_n x + \delta_n) \xrightarrow{w} F$, then $\gamma_n/u_n \to 1$ and $(v_n - \delta_n)/u_n \to 0$. It follows immediately that $\beta_n = b_{n+1}/b_n \to 1$ as $n \to \infty$. ■

Lemma 5.4.2. A distribution function F with characteristic function φ belongs to the class L if and only if, for every $0 < c < 1$, there exists a characteristic function φ_c such that

$$\varphi(t) = \varphi(ct)\varphi_c(t) \qquad \text{for } t \in \mathbb{R}. \tag{5.4.4}$$

Proof. We first prove the sufficiency of (5.4.4). We note that φ has no real zeros. In fact, let $t_0 > 0$ be the smallest number such that $\varphi(2t_0) = 0$. Then $\varphi(t) \neq 0$ for $0 \leq t < 2t_0$. In view of (5.4.4) we conclude that $\varphi_c(2t_0) = 0$. Then

$$1 = 1 - |\varphi_c(2t_0)|^2 \leq 4(1 - |\varphi_c(t_0)|^2) \tag{5.4.5}$$

in view of (4.1.1) for every $c \in (0, 1)$. On the other hand, in view of the continuity of φ

$$\varphi_c(t_0) = \frac{\varphi(t_0)}{\varphi(ct_0)} \to 1 \qquad \text{as } c \to 1,$$

so that (5.4.5) cannot hold for c sufficiently close to 1. Thus we have shown that φ has no real zeros.

Let $X_1, X_2, \ldots, X_n$ be independent random variables such that X_k has the characteristic function

$$\theta_k(t) = \varphi_{(k-1)/k}(kt) = \frac{\varphi(kt)}{\varphi((k-1)t)}, \qquad 1 \leq k \leq n.$$

Then the characteristic function of $(1/n)\sum_{k=1}^{n} X_k$ equals

$$\prod_{k=1}^{n} \theta_k\left(\frac{t}{n}\right) = \prod_{k=1}^{n} \left[\frac{\varphi\left(k\frac{t}{n}\right)}{\varphi\left(\frac{k-1}{n}t\right)}\right]$$

$$= \varphi(t).$$

Since φ is continuous and has no real zeros, it follows that

$$\max_{1 \le k \le n} \left|\theta_k\left(\frac{t}{n}\right) - 1\right| \to 0 \qquad \text{as } n \to \infty$$

in every finite t-interval. Hence the sequence $\{X_n\}$ satisfies condition (5.4.1) with $b_n = n$. Consequently F belongs to the class L.

Conversely, let $\{X_n\}$ be a sequence of independent random variables. Let $\{a_n\}$ and $\{b_n\}$, with $b_n > 0$, be sequences of constants satisfying (5.4.1) such that $b_n^{-1}S_n - a_n$ converges in law to a random variable with distribution function F and characteristic function φ. We show that φ satisfies (5.4.4). Clearly we may assume that F is nondegenerate, since otherwise relation (5.4.4) holds trivially with $|\varphi(t)| = 1$, $t \in \mathbb{R}$. Using the notation of Lemma 5.4.1, we have $\varphi_n \to \varphi$ as $n \to \infty$, where φ_n is given by (5.4.3). Clearly φ is i.d., and hence it has no real zeros. Also note that $b_n \to \infty$ and $b_{n+1}/b_n \to 1$ as $n \to \infty$ (by Lemma 5.4.1). Let $0 < c < 1$ be fixed. Then we choose a sequence of positive integers $m = m(n) < n$ such that $m \to \infty$ and $b_m/b_n \to c$ as $n \to \infty$. We can rewrite $\varphi_n(t)$ in the form

$$\varphi_n(t) = \varphi_n^{(1)}(t) \cdot \varphi_n^{(2)}(t),$$

where

$$\varphi_n^{(1)}(t) = e^{-ita_m c} \prod_{k=1}^{m} \theta_k\left(\frac{b_m}{b_n} \cdot \frac{t}{b_m}\right),$$

$$\varphi_n^{(2)}(t) = e^{-it(a_n - a_m c)} \prod_{k=n+1}^{n} \theta_k\left(\frac{t}{b_n}\right).$$

Clearly $\varphi_n^{(1)}(t) \to \varphi(ct)$ as $n \to \infty$. Consequently the characteristic function $\varphi_n^{(2)}(t)$ converges to the ratio $\varphi_c(t) = \varphi(t)/\varphi(ct)$, which is continuous for all t. It follows from the continuity theorem that φ_c is a characteristic function. This completes the proof of the lemma. ■

Remark 5.4.1. If (5.4.4) holds, φ_c is an i.d. characteristic function for every $c \in (0, 1)$.

Since the distributions of the class L are i.d., it is natural to seek a characterization of the characteristic functions of these distributions in terms of the corresponding Lévy spectral function.

Theorem 5.4.1. An i.d. distribution function F belongs to the class L if and only if the following conditions are satisfied:

(i) The spectrum G in the canonical representation of the characteristic function φ of F is continuous and has left and right derivatives on $\mathbb{R} - \{0\}$.
(ii) The function $[(1 + x^2)/x]G'(x)$ is nonincreasing on $\mathbb{R} - \{0\}$, where G' is either the left or the right derivative of G.

Proof. First suppose that $F \in L$. Let $\psi = \ln \varphi$ and $c \in (0, 1)$. Then

$$\begin{aligned}\psi(ct) &= i\alpha ct + \int_{-\infty}^{\infty} \left(e^{ictx} - 1 - \frac{ictx}{1 + x^2}\right) \frac{1 + x^2}{x^2}\, dG(x) \\ &= i\alpha ct + \int_{-\infty}^{\infty} \left(e^{itx} - 1 - \frac{itx}{1 + c^{-2}x^2}\right) \frac{1 + c^{-2}x^2}{c^{-2}x^2}\, dG(c^{-1}x).\end{aligned}$$

Set $\psi_c(t) = \psi(t) - \psi(ct)$. Then it is easy to check that

$$\psi_c(t) = i\alpha_c t + \int_{-\infty}^{\infty} \left(e^{itx} - 1 - \frac{itx}{1 + x^2}\right) \frac{1 + x^2}{x^2}\, dG_c(x), \tag{5.4.6}$$

where

$$\left\{\begin{aligned} &G_c(x) = G(x) - \int_{-\infty}^{x} \frac{1 + c^{-2}y^2}{c^{-2}(1 + y^2)}\, dG(c^{-1}y) \\ &\text{and} \\ &\alpha_c = \alpha(1 - c). \end{aligned}\right. \tag{5.4.7}$$

Clearly α_c is finite and G_c, being the difference between two bounded, nondecreasing, right-continuous functions, is itself a right-continuous function of bounded variation on $\mathbb{R}$ such that $G_c(-\infty) = 0$. Hence ψ_c is the logarithm of an i.d. characteristic function if and only if G_c is nondecreasing on $\mathbb{R}$. First we note that

$$G_c(0) - G_c(0-) = (1 - c^2)[G(0) - G(0-)] \geq 0,$$

and hence according to our hypothesis G_c is nondecreasing on $\mathbb{R} - \{0\}$ for every $c \in (0, 1)$. Using (5.4.7), we conclude that for every $c \in (0, 1)$ and for every $x', x'' \in \mathbb{R} - \{0\}$, such that $x' < x''$ and $x'x'' > 0$, we have

$$(5.4.8) \qquad \int_{x'}^{x''} \frac{1+y^2}{y^2}\, dG_c(y) = \int_{x'}^{x''} \frac{1+y^2}{y^2}\, dG(y) - \int_{x'}^{x''} \frac{1+c^{-2}y^2}{c^{-2}y^2}\, dG(c^{-1}y) \geq 0.$$

We define

$$I_+(x) = -\int_{e^x}^{\infty} \frac{1+y^2}{y^2}\, dG(y), \qquad x \in \mathbb{R}.$$

Clearly I_+ is nondecreasing on $\mathbb{R}$. By setting $x' = e^{x-h}$, $x'' = e^x$, and $c = e^{-h}$ for $h > 0$ in (5.4.8), we obtain

$$I_+(x) - I_+(x-h) \geq I_+(x+h) - I_+(x),$$

that is,

$$(5.4.9) \qquad I_+(x) \geq \tfrac{1}{2}[I_+(x+h) + I_+(x-h)], \qquad x \in \mathbb{R}, \quad h > 0.$$

Therefore I_+ is also convex on $\mathbb{R}$, and consequently I_+ is continuous on $\mathbb{R}$ and has left and right derivatives which we denote indifferently by $I'_+(x)$. Moreover, I'_+ is nonincreasing on $\mathbb{R}$. Now it follows from the definition of $I_+(x)$ that

$$I'_+(x) = \frac{1+e^{2x}}{e^x}\, G'(e^x), \qquad x \in \mathbb{R},$$

so that, setting $e^x = y$, we conclude that the left and the right derivatives $G'(y)$ exist and, moreover, the function $[(1+y^2)/y]G'(y)$ is nonincreasing on $(0, \infty)$. Similarly, by writing

$$I_-(x) = \int_{-\infty}^{-e^x} \frac{1+y^2}{y^2}\, dG_c(y)$$

and proceeding as above, we can verify that the same assertion holds on $(-\infty, 0)$. This proves that conditions (i) and (ii) are necessary.

Conversely, suppose that G satisfies (i) and (ii). We use the same notation as in the first part. Then, for every $c \in (0, 1)$ and every $x', x'' \in \mathbb{R} - \{0\}$, such that $x' < x''$ and $x'x'' > 0$, we have

$$\begin{aligned} \int_{x'}^{x''} \frac{1+y^2}{y^2}\, dG(y) &= \int_{x'}^{x''} \frac{1+y^2}{y}\, G'(y) \frac{dy}{y} \\ &\geq \int_{x'}^{x''} \frac{1+c^{-2}y^2}{c^{-1}y}\, G'(c^{-1}y) \frac{dy}{y} \\ &= \int_{x'}^{x''} \frac{1+c^{-2}y^2}{c^{-2}y^2}\, dG(c^{-1}y), \end{aligned}$$

so that it follows from (5.4.7) and (5.4.8) that

$$\int_{x'}^{x''} \frac{1 + y^2}{y^2} \, dG_c(y) \geq 0.$$

This implies, as above, that G_c is nondecreasing on $\mathbb{R}$, so that ψ_c is the logarithm of an i.d. characteristic function. Hence φ satisfies (5.4.4) and consequently $F \in L$. ∎

Remark 5.4.2. The Poisson distribution with parameter λ does not belong to the class L, since it corresponds to spectrum G, which is discontinuous at some $x \neq 0$. The degenerate, normal, and Cauchy laws belong to class L.

5.4.2 Stable Distributions

We now restrict ourselves to independent, identically distributed random variables X_n, $n = 1, 2, \ldots$, and investigate the limit distributions of the sums

$$b_n^{-1} S_n - a_n, \qquad n \geq 1,$$

where $S_n = \sum_{k=1}^{n} X_k$, and a_n and $b_n > 0$ are suitably chosen constants. Clearly these limit distributions are a subset of the set of all i.d. distributions. This subset coincides with the set of *stable distributions*.

Definition 5.4.1. Let F be a distribution function with characteristic function φ. We say that F (or, equivalently, φ) is stable if for every pair of positive real numbers b_1 and b_2 there exist finite constants a and b (>0) such that

(5.4.10) $$\varphi(b_1 t)\varphi(b_2 t) = \varphi(bt)e^{iat}.$$

It is easy to see that a distribution function F is stable if and only if, for every $b_1 > 0$, $b_2 > 0$, c_1 and c_2 real, there exist a $b > 0$ and a $c \in \mathbb{R}$ such that

(5.4.11) $$F\left(\frac{x - c_1}{b_1}\right) * F\left(\frac{x - c_2}{b_2}\right) = F\left(\frac{x - c}{b}\right).$$

Let F be a stable distribution function with characteristic function φ. Writing $bt = u$, $b_1 = \alpha b$, and $b_2 = \alpha_1 b$, we see that

$$\begin{aligned} \varphi(u) &= e^{-(iau/b)}\varphi(\alpha u)\varphi(\alpha_1 u) \\ &= \varphi(\alpha u)\varphi_\alpha(u), \end{aligned}$$

where

$$\varphi_\alpha(u) = e^{-(iau/b)}\varphi(\alpha_1 u)$$

is a characteristic function. It follows that $F \in L$.

Example 5.4.1. The normal and Cauchy distributions are stable. The Laplace distribution with probability density function

$$f(x) = \frac{1}{2\lambda} e^{-(|x-\mu|/\lambda)}, \qquad x \in \mathbb{R}, \lambda > 0,$$

and characteristic function

$$\varphi(t) = e^{it\mu} \cdot (1 + t^2\lambda^2)^{-1}, \qquad t \in \mathbb{R},$$

is not stable but belongs to the class L.

Clearly degenerate laws are stable. We exclude degenerate laws from consideration.

Let L_1 be the family of distributions which appear as limit distributions of sums $b_n^{-1}S_n - a_n$ as $b_n \to \infty$, with $n \to \infty$. Since $b_n \to \infty$,

$$P\{|X_k| \geq b_n\epsilon\} = P\{|X_1| \geq b_n\epsilon\} \to 0,$$

and it follows that $\{X_k/b_n: 1 \leq k \leq n, n \geq 1\}$ satisfies the u.a.n. condition. Hence $L_1 \subset L$.

Theorem 5.4.2. $F \in L_1$ if and only if F is stable.

Proof. Let φ be the characteristic function corresponding to F. Let F be stable. Consider a sequence of independent random variables $\{X_n\}$ with common distribution function F. Then the partial sum $S_n = \sum_{k=1}^n X_k$ has the characteristic function

$$[\varphi(t)]^n = e^{ita_n}\varphi(b_n t),$$

and the sum $b_n^{-1}S_n - a_n$ has the characteristic function $\varphi(t)$. It follows that $F \in L_1$.

If F is degenerate, it is clearly stable, so that we may assume that F is nondegenerate. Let φ_0 be the characteristic function of X_1. Since $F \in L_1$, there exist constants $b_n > 0$ and $a_n \in \mathbb{R}$ such that

$$e^{-ita_n}\varphi_0^n\left(\frac{t}{b_n}\right) \to \varphi(t) \qquad \text{for every } t \in \mathbb{R}.$$

From Lemma 5.4.1 we see that $b_n \to \infty$ and $b_{n+1}/b_n \to 1$ as $n \to \infty$. Let $0 < c_2 < c_1$ be arbitrary constants, and let d_1 and d_2 be real numbers. Then there exists a sequence of integers $\{m_n\}$ such that, as $n \to \infty$,

$$\frac{b_{m_n}}{b_n} \to \frac{c_2}{c_1}.$$

Consider the sum

$$(5.4.12)\qquad c_1\left(\frac{S_n}{b_n} - a_n - d_1\right) + \frac{b_{m_n}}{b_n}c_1\left(\frac{S_{n+m_n} - S_n}{b_{m_n}} - a_{m_n} - d_2\right)$$

$$= c_1\frac{S_{n+m_n}}{b_n} - A_n,$$

where

$$A_n = c_1 a_n + c_1 d_1 + \frac{b_{m_n}}{b_n}c_1 a_{m_n} + \frac{b_{m_n}}{b_n}c_1 d_2.$$

Since $F \in L_1$, the distributions of the first term on the left-hand side of (5.4.12) converge to $F(c_1^{-1}x + d_1)$, and those of the second term to $F(c_2^{-1}x + d_2)$. It follows that the distribution functions of the left-hand side converge to a proper law $F(c_1^{-1}x + d_1) * F(c_2^{-1}x + d_2)$. The right-hand side must also have the same limiting distribution, and by convergence of types (Problem 3.10.15) it follows that F satisfies

$$F(c_1^{-1}x + d_1) * F(c_2^{-1}x + d_2) = F(c^{-1}x + d),$$

so that F is stable. ■

Theorem 5.4.3. In order that a distribution function F with characteristic function φ be in L_1 it is necessary and sufficient that F be i.d., and $\ln\varphi$ has either of the following representations:

$$(5.4.13)\qquad \ln\varphi(t) = i\alpha t + c_1\int_{-\infty}^{0}\left(e^{itx} - 1 - \frac{itx}{1+x^2}\right)\frac{dx}{|x|^{1+\beta}}$$

$$+ c_2\int_0^{\infty}\left(e^{itx} - 1 - \frac{itx}{1+x^2}\right)\frac{dx}{|x|^{1+\beta}},$$

with $c_1 \geq 0$, $c_2 \geq 0$, $c_1 + c_2 > 0$, and $0 < \beta < 2$, or

$$(5.4.14)\qquad \ln\varphi(t) = i\alpha t - \tfrac{1}{2}\sigma^2 t^2.$$

Proof. Let $F \in L_1$. Then φ is i.d. and can be represented in terms of the Lévy–Khintchine representation. Let $\psi = \ln\varphi$. Since $F \in L_1$ for every $b_1 > 0$, $b_2 > 0$, there exist a and b $(b > 0)$ such that

$$\varphi(b_1 t)\varphi(b_2 t) = \varphi(bt)e^{iat},$$

so that

$$(5.4.14)\qquad iat + \psi(bt) = \psi(b_1 t) + \psi(b_2 t).$$

Replacing ψ by its representation, we see that

$$\psi(bt) = i\alpha tb + \int_{-\infty}^{\infty} \left(e^{ity} - 1 - \frac{ity}{1 + b^{-2}y^2}\right) \frac{1 + b^{-2}y^2}{b^{-2}y^2}\, dG(b^{-1}y),$$

and since the function $z/(1 + b^2z^2)$ is bounded, the integral

$$b \int_{-\infty}^{\infty} \frac{z}{1 + b^2z^2}\, dG(z) = \int_{-\infty}^{\infty} \frac{y}{1 + y^2}\, dG(b^{-1}y)$$

exists. Writing

$$\alpha_b = ab + (1 - b^2) \int_{-\infty}^{\infty} \frac{y}{1 + y^2}\, dG(b^{-1}y),$$

we see after elementary computation that

$$\psi(bt) = it\alpha_b + \int_{-\infty}^{\infty} \left(e^{ity} - 1 - \frac{ity}{1 + y^2}\right) \frac{1 + b^{-2}y^2}{b^{-2}}\, dG(b^{-1}y). \tag{5.4.15}$$

It follows from the uniqueness of the representation that requirement (5.4.14) reduces to

(5.4.16)

$$\frac{1 + b^{-2}x^2}{b^{-2}}\, dG(b^{-1}x) = \frac{1 + b_1^{-2}x^2}{b_1^{-2}}\, dG(b_1^{-1}x) + \frac{1 + b_2^{-2}x^2}{b_2^{-2}}\, dG(b_2^{-1}x).$$

Now define

$$I_+(x) = -\int_{e^x}^{\infty} \frac{1 + y^2}{y^2}\, dG(y), \qquad x \in \mathbb{R},$$

and

$$I_-(x) = \int_{-\infty}^{-e^x} \frac{1 + y^2}{y^2}\, dG(y), \qquad x \in \mathbb{R},$$

as in the proof of Theorem 5.4.1, and set $e^{-h_1} = b_1$, $e^{-h_2} = b_2$, $e^{-h} = b$. Requirement (5.4.16) becomes

$$[G(0) - G(0-)](b_1^2 + b_2^2 - b^2) = 0 \tag{5.4.17}$$

and

$$\begin{cases} I_+(x + h) = I_+(x + h_1) + I_+(x + h_2) \\ I_-(x + h) = I_-(x + h_1) + I_-(x + h_2) \end{cases} \tag{5.4.18}$$

($x \in \mathbb{R}$). Here $h_1, h_2 \in \mathbb{R}$, and h is a function of h_1 and h_2.

Let $G(+\infty) - G(0) > 0$, so that I_+ does not vanish. From the relation for I_+ in (5.4.18) we see that for any $h_1, h_2, \ldots, h_n$ there exists an h such that

$$I_+(x + h) = I_+(x + h_1) + I_+(x + h_2) + \cdots + I_+(x + h_n),$$

and, setting $h_1 = h_2 = \cdots = h_n = 0$, there exists an $h = h(n)$ such that

$$I_+(x + h) = nI_+(x). \tag{5.4.19}$$

Let $s = p/q > 0$ be a rational number in its lowest terms. Define

$$h(s) = h(p) - h(q).$$

Then (5.4.19) implies that

$$\begin{aligned} sI_+(x) &= p\left[\frac{1}{q} I_+(x)\right] = pI_+(x - h(q)) = I_+(x + h(p) - h(q)) \\ &= I_+(x + h(s)), \end{aligned}$$

and it follows that for any rational number $r > 0$

$$I_+(x + h(r)) = rI_+(x). \tag{5.4.20}$$

Clearly I_+ is nondecreasing with $I_+ \leq 0$ and $I_+(\infty) = 0$. It follows that the function $h(r)$ defined on the positive rationals is nonincreasing and consequently has right and left limits at all positive real numbers $s > 0$. From (5.4.20) these must be equal, and hence $h(s)$ is defined as a nonincreasing, continuous function on $(0, \infty)$ satisfying

$$I_+(x + h(s)) = sI_+(x), \qquad s > 0. \tag{5.4.21}$$

Moreover,

$$\lim_{s\to 0} h(s) = \infty, \qquad \text{and } \lim_{s\to\infty} h(s) = -\infty.$$

Since I_+ does not vanish, we can assume without loss of generality (by shifting the origin if necessary) that $I_+(0) \neq 0$. Write $I_0 = I_+/I_+(0)$. Let x_1, x_2 be arbitrary, and choose s_1, s_2 such that $h(s_1) = x_1$ and $h(s_2) = x_2$. Then $s_1 I_+(0) = I_+(x_1)$, $s_2 I_+(0) = I_+(x_2)$, and $s_2 I_+(x_1) = I_+(x_1 + x_2)$, so that

$$I_0(x_1 + x_2) = I_0(x_1)I_0(x_2), \qquad x_1, x_2 \in \mathbb{R}. \tag{5.4.22}$$

It is well known (see, for example, Aczel [1], p. 38) that the only nonvanishing continuous solution of the functional equation (5.4.22), with $I_0(\infty) = 0$, is of the form $I_0(x) = e^{-\beta x}$, where $\beta > 0$. Setting $y = e^x$ and going back to G, we find that the derivative $G'(y)$ exists for $y > 0$, and

$$\frac{1 + y^2}{y} G'(y) = c_2 y^{-\beta}, \qquad c_2 \geq 0, \tag{5.4.23}$$

where we have taken account of the vanishing case. Since G is of bounded variation on $(0, \infty)$, the integral $\int_0^\epsilon y^{1-\beta}\,dy$ must converge for $\epsilon > 0$, and we have $\beta < 2$. In view of (5.4.18) we have, for $0 < \beta < 2$,

$$b^\beta = b_1^\beta + b_2^\beta.$$

By proceeding in a similar manner with I_-, we see that for $y < 0$

$$\frac{1 + y^2}{y} G'(y) = -c_1 |y|^{-\beta'}, \qquad c_1 \geq 0, \tag{5.4.24}$$

with

$$b^{\beta'} = b_1^{\beta'} + b_2^{\beta'}.$$

Setting $b_1 = b_2 = 1$, we see that $\beta = \beta'$. Moreover, (5.4.17) becomes

$$[G(0) - G(0-)](2 - b^2) = 0,$$

which is incompatible with $b^\beta = 2$, $0 < \beta < 2$, unless $G(0) - G(0-) = 0$. It follows that either $b^2 = b_1^2 + b_2^2$, so that I_+ and I_- vanish and φ is the characteristic function of the normal distribution, or $G(0) - G(0-) = 0$ and, for $y \neq 0$, $G'(y)$ is given by (5.4.23) and (5.4.24). The proof of Theorem 5.4.3 is now complete. ■

Theorem 5.4.4. In order that a distribution function F be stable it is necessary and sufficient that its characteristic function φ be expressible as

$$\ln \varphi(t) = i\alpha t - c|t|^\beta \left[1 + i\gamma \frac{t}{|t|} \omega(t, \beta)\right], \tag{5.4.25}$$

where α, β, γ, c are constants with $c \geq 0$, $0 < \beta \leq 2$, $|\gamma| \leq 1$, and

$$\omega(t, \beta) = \begin{cases} \tan \dfrac{\pi\beta}{2} & \text{if } \beta \neq 1, \\ \dfrac{2}{\pi} \ln |t| & \text{if } \beta = 1. \end{cases}$$

Here β is called the characteristic exponent or index of F and has the same meaning as in Theorem 5.4.3.

Proof. CASE. $0 < \beta < 1$. In this case both

$$\int_{-\infty}^0 \frac{x}{1 + x^2} \frac{dx}{|x|^{1+\beta}} \quad \text{and} \quad \int_0^\infty \frac{x}{1 + x^2} \frac{dx}{x^{1+\beta}}$$

are finite, so that (5.4.13) becomes

$$\text{(5.4.26)}\quad \ln \varphi(t) = i\alpha' t + c_1 \int_{-\infty}^{0} (e^{itx} - 1)\frac{dx}{|x|^{1+\beta}} + c_2 \int_{0}^{\infty} (e^{itx} - 1)\frac{dx}{x^{1+\beta}}$$

for some α'. Let $t > 0$, and set $v = -xt$ in the first integral, and $v = xt$ in the second integral on the right-hand side of the last relation. Then

$$\text{(5.4.27)}\quad \ln \varphi(t) = i\alpha' t + t^{\beta}\left[c_1 \int_{0}^{\infty} (e^{-iv} - 1)\frac{dv}{v^{1+\beta}} + c_2 \int_{0}^{\infty} (e^{iv} - 1)\frac{dv}{v^{1+\beta}}\right].$$

By the usual method of contour integration it is easily seen that

$$\int_{0}^{\infty} (e^{iv} - 1)\frac{dv}{v^{1+\beta}} = e^{-i(\pi/2)\beta} K(\beta),$$

where

$$K(\beta) = \int_{0}^{\infty} (e^{-y} - 1)\frac{dy}{y^{1+\beta}} = -\frac{\Gamma(1-\beta)}{\beta} < 0.$$

Since the first integral in (5.4.27) is the complex conjugate of the second, we get for $t > 0$

$$\ln \varphi(t) = it\alpha' + t^{\beta} K(\beta)\left[(c_1 + c_2)\cos\frac{\pi\beta}{2} + i(c_1 - c_2)\sin\frac{\pi\beta}{2}\right].$$

Now, setting

$$c = -K(\beta)(c_1 + c_2)\cos\frac{\pi\beta}{2}$$

and

$$\gamma = \frac{c_1 - c_2}{c_1 + c_2},$$

so that $c \geq 0$ and $|\gamma| \leq 1$, we see that for $t > 0$.

$$\ln \varphi(t) = it\alpha' - ct^{\beta}\left(1 + i\gamma \tan\frac{\pi\beta}{2}\right).$$

On the other hand, for $t < 0$

$$\ln \varphi(t) = \ln \overline{\varphi(-t)} = it\alpha' - c|t|^\beta\left(1 + i\frac{t}{|t|}\gamma \tan\frac{\pi\beta}{2}\right),$$

so that (5.4.25) holds for all $t \in \mathbb{R}$.

CASE. $1 < \beta < 2$. In this case we can take out of the bracket in (5.4.13) the term $itx - itx/(1 + x^2)$ to get

$$\ln \varphi(t) = i\alpha_1 t + c_1 \int_{-\infty}^{0} (e^{itx} - 1 - itx)\frac{dx}{|x|^{1+\beta}} + c_2 \int_{0}^{\infty} (e^{itx} - 1 - itx)\frac{dx}{x^{1+\beta}}$$

instead of (5.4.26). Exactly as in the case $0 < \beta < 1$, it is easy to show that

$$\int_{-\infty}^{0} (e^{itx} - 1 - itx)\frac{dx}{|x|^{1+\beta}} = t^\beta e^{-\pi i\beta/2}\ \Lambda(\beta)$$

and

$$\int_{0}^{\infty} (e^{itx} - 1 - itx)\frac{dx}{x^{1+\beta}} = t^\beta e^{\pi i\beta/2}\ \Lambda(\beta),$$

where

$$\Lambda(\beta) = \int_{0}^{\infty} (e^{-y} - 1 + y)\frac{dy}{y^{1+\beta}} = \frac{\Gamma(2-\beta)}{\beta(\beta-1)} > 0.$$

Setting

$$c = -\Lambda(\beta)(c_1 + c_2)\cos\frac{\pi\beta}{2},$$

so that $c > 0$ [since $\cos(\pi\beta/2) < 0$ for $1 < \beta < 2$], and defining γ as before, we see that (5.4.25) holds for $t \in \mathbb{R}$.

CASE. $\beta = 1$. In this case, since

$$\int_{0}^{\infty} \frac{\cos tx - 1}{x^2}\,dx = -\frac{\pi}{2}t,$$

we have for $t > 0$

$$\int_{0+}^{\infty}\left(e^{itx} - 1 - \frac{itx}{1 + x^2}\right)\frac{dx}{x^2}$$

$$= \int_0^{\infty} \frac{\cos tx - 1}{x^2}\, dx + i \int_{0+}^{\infty}\left(\sin tx - \frac{tx}{1 + x^2}\right)\frac{dx}{x^2}$$

$$= -\frac{\pi t}{2} + i \lim_{\epsilon \downarrow 0}\left\{\int_{\epsilon}^{\infty} \frac{\sin tx}{x^2}\, dx - t\int_{\epsilon}^{\infty} \frac{dx}{x(1 + x^2)}\right\}$$

$$= -\frac{\pi t}{2} + i \lim_{\epsilon \downarrow 0}\left\{- t\int_{\epsilon}^{\epsilon t} \frac{\sin tx}{x^2}\, dx\right.$$

$$\left. + t\int_{\epsilon}^{\infty}\left[\frac{\sin x}{x^2} - \frac{1}{x(1 + x^2)}\right] dx\right\}$$

$$= -\frac{\pi t}{2} - it \lim_{\epsilon \downarrow 0} \int_{\epsilon}^{\epsilon t} \frac{dx}{x} + it\int_0^{\infty}\left[\frac{\sin x}{x^2} - \frac{1}{x(1 + x^2)}\right] dx$$

$$= -\frac{\pi t}{2} - it \ln t + it\rho,$$

say where ρ is finite. Setting

$$c = \frac{\pi}{2}(c_1 + c_2)$$

and

$$\gamma = \frac{c_1 - c_2}{c_1 + c_2},$$

we see once again that $\ln \varphi$ has the form given in (5.4.25) for $t > 0$. Applying an argument similar to the one used in the case of $0 < \beta < 1$, we see that $\ln \varphi$ also has the form (5.4.25) for $t < 0$. This, together with (5.4.14) for $\beta = 2$, completes the proof of Theorem 5.4.4. ■

Remark 5.4.3. The value $c = 0$ corresponds to the degenerate distribution, and the value $\beta = 2$ to the normal distribution. The case $\gamma = 0$, $\beta = 1$ corresponds to the Cauchy law. The case $\beta = \frac{1}{2}$, $\gamma = 1$, $\alpha = 0$, $c = 1$ corresponds to the probability density function

$$f(x) = \begin{cases} 0 & \text{for } x < 0, \\ (\sqrt{2\pi})^{-1} e^{-1/2x} x^{-3/2} & \text{for } x > 0. \end{cases}$$

Remark 5.4.4. In view of Theorem 5.4.4, if φ is the characteristic function of a nondegenerate stable distribution function F, then $|\varphi(t)| = e^{-c|t|^\beta}$, $0 < \beta \le 2$. It is easily seen that φ is absolutely integrable over $(-\infty, \infty)$, so that Theorem 3.3.2 applies, and it follows that F has an everywhere continuous derivative.

5.4.3 Domains of Attraction

Let $\{X_n\}$ be a sequence of independent, identically distributed random variables with common distribution function F. Suppose there exist sequences of constants $\{a_n\}$ and $\{b_n\}$, with $b_n > 0$, such that the sequence of sums $\{b_n^{-1} \sum_{k=1}^n X_k - a_n\}$ converges in law to some random variable with distribution function G. Then we say that F is *attracted* to G. The set of all distribution functions that are attracted to G is called the *domain of attraction* of the distribution G. According to Theorem 5.4.2, only stable distributions have domains of attraction.

We now consider the problem of determination of the domain of attraction of a stable law and give its complete solution.

Theorem 5.4.5. The distribution function F belongs to the domain of attraction of a proper normal distribution if and only if

$$\int_{|x| \ge z} dF(x) = o\left(z^{-2} \int_{|x| < z} x^2\, dF(x)\right) \qquad \text{as } z \to \infty. \tag{5.4.28}$$

Proof. If F has a finite variance σ^2, then

$$z^2 \int_{|x| \ge z} dF(x) \to 0 \qquad \text{and} \qquad \int_{|x| < z} x^2\, dF(x) \to \mathscr{E} X^2 < \infty \qquad \text{as } z \to \infty,$$

so that (5.4.28) holds. Moreover, F belongs to the domain of attraction of the normal distribution by the corollary to the Lindeberg–Feller central limit theorem with $B_n = \sqrt{n\sigma^2}$.

Let us therefore concentrate on the case where F has infinite variance, that is, where

$$\int_{|x| < z} x^2\, dF(x) \to \infty \qquad \text{as } z \to \infty. \tag{5.4.29}$$

We first show that (5.4.29) implies

$$\left[\int_{|x| \le z} x\, dF(x)\right]^2 = o\left(\int_{|x| \le z} x^2\, dF(x)\right) \qquad \text{as } z \to \infty. \tag{5.4.30}$$

In fact, if $t(x) > 0$ is an unbounded function (as $x \to \pm\infty$) such that

$$C = \int_{-\infty}^{\infty} t^2(x)\, dF(x) < \infty,$$

then

$$\left[\int_{|x| \le z} x\, dF(x)\right]^2 \le \left[\int_{|x| \le z} t^2(x)\, dF(x)\right]\left[\int_{|x| \le z} x^2 t^{-2}(x)\, dF(x)\right]$$
$$\le C \int_{|x| \le z} x^2 t^{-2}(x)\, dF(x).$$

Since $t^{-2}(x) \to 0$ as $x \to \pm\infty$, there exists a z_0 such that for all $z > z_0$

$$\left[\int_{|x| \le z} x\, dF(x)\right]^2 \le C \int_{|x| \le z_0} x^2 t^{-2}(x)\, dF(x) + C\epsilon \int_{z_0 < |x| \le z} x^2\, dF(x),$$

and (5.4.30) follows.

From Theorem 5.3.4 F belongs to the domain of attraction of the normal law if and only if there exists a sequence of constants $\{C_n\}$, with $C_n \to \infty$ as $n \to \infty$, such that the following conditions hold:

(i) $n \displaystyle\int_{|x| \ge C_n} dF(x) \to 0,$

(ii) $nC_n^{-2}\left\{\displaystyle\int_{|x| < C_n} x^2\, dF(x) - \left[\int_{|x| < C_n} x\, dF(x)\right]^2\right\} \to \infty,$

as $n \to \infty$. In view of what has just been proved for distributions with infinite variance we can replace condition (ii) by

(5.4.31) $$nC_n^{-2} \int_{|x| < C_n} x^2\, dF(x) \to \infty \qquad \text{as } n \to \infty.$$

We show that (i) and (5.4.31) together imply (5.4.28). Indeed, since $C_n \to \infty$ as $n \to \infty$, for every z sufficiently large we can find an n such that

$$C_n \le z < C_{n+1}.$$

Let

$$q(z) = \int_{|x| \ge z} dF(x)$$

and

$$H(z) = z^{-2} \int_{|x| < z} x^2\, dF(x).$$

Then for sufficiently large z and n

$$q(C_n) \geq q(z) \geq q(C_{n+1})$$

and

$$\begin{aligned}
H(z) &= z^{-2} \int_{|x|<z} x^2 \, dF(x) \\
&\leq \int_{|x|<z} dF(x) \\
&= \int_{|x|<C_n} dF(x) + \int_{C_n \leq |x| < z} dF(x) \\
&\leq C_n^{-2} \int_{|x|<C_n} x^2 \, dF(x) + q(C_n) \\
&= H(C_n) + q(C_n).
\end{aligned}$$

Similarly

$$H(z) \geq H(C_{n+1}) - q(C_n),$$

and we have

$$\frac{q(C_n)}{H(C_{n+1}) - q(C_n)} \geq \frac{q(z)}{H(z)} \geq \frac{q(C_{n+1})}{H(C_n) + q(C_n)},$$

that is,

$$\begin{aligned}
&\frac{n \int_{|x| \geq C_n} dF(x)}{\dfrac{n}{n+1} \dfrac{n+1}{C_n^2} \int_{|x|<C_{n+1}} x^2 \, dF(x) - n \int_{|x| \geq C_n} dF(x)} \\
&\geq \frac{\int_{|x| \leq z} dF(x)}{\dfrac{1}{z^2} \int_{|x|<z} dF(x)} \\
&\geq \frac{(n+1) \int_{|x| \geq C_{n+1}} dF(x)}{\dfrac{n+1}{n} \left[\dfrac{n}{C_n^2} \int_{|x|<C_n} x^2 \, dF(x) + n \int_{|x| \geq C_n} dF(x) \right]}.
\end{aligned}$$

It follows from (i) and (5.4.31) that the first and the last fractions in the last inequality tend to zero as $n \to \infty$, so that (5.4.28) holds.

To prove the sufficiency of (5.4.28) we need only find a sequence of constants $\{C_n\}$ with $C_n \to \infty$ as $n \to \infty$ such that conditions (i) and (ii) hold. Let $\delta > 0$, and define $C_n(\delta)$ as follows:

$$C_n(\delta) = \inf_z \left\{ z : n \int_{|x| \geq z} dF(x) \leq \delta \right\}.$$

Since $\int_{-\infty}^{\infty} x^2\, dF(x) = \infty$, it follows that $C_n(\delta) \to \infty$ as $n \to \infty$. Let $\epsilon > 0$. Then it follows from (5.4.28) that there exists a z_0 such that

$$q(z) \leq \epsilon H(z) \qquad \text{for } z \geq z_0.$$

Thus

$$q(\tfrac{1}{2}C_n(\delta)) \leq \epsilon H(\tfrac{1}{2}C_n(\delta))$$

for sufficiently large n, and since $nq(\frac{1}{2}C_n(\delta)) > \delta$ we have

$$\delta < nq(\tfrac{1}{2}C_n(\delta)) \leq n\epsilon H(\tfrac{1}{2}C_n(\delta))$$

or

$$nH(\tfrac{1}{2}C_n(\delta)) > \frac{\delta}{\epsilon}$$

for sufficiently large n. Now

$$H(\tfrac{1}{2}C_n(\delta)) \leq \frac{4}{C_n^2(\delta)} \int_{|x| < C_n(\delta)} x^2\, dF(x) = 4H(C_n(\delta)),$$

so that for large n

$$nH(C_n(\delta)) > \frac{\delta}{4\epsilon}.$$

Thus for every $\delta > 0$, as $n \to \infty$,

$$nH(C_n(\delta)) \to \infty,$$

and it follows that there exists a sequence $\delta_n \to 0$ such that

$$nH(C_n(\delta_n)) \to \infty \qquad \text{as } n \to \infty.$$

But $nq(C_n(\delta_n)) \leq \delta_n \to 0$ as $n \to \infty$ by the definition of $C_n(\delta)$, so that we have proved that (5.4.28) implies (i) and (ii). This completes the proof of Theorem 5.4.5. ■

Theorem 5.4.6. The distribution function F belongs to the domain of attraction of a stable law with index β, $0 < \beta < 2$, if and only if, as $x \to \infty$,

$$\frac{F(-x)}{1 - F(x)} \to \frac{c_1}{c_2} \tag{5.4.32}$$

and

$$\frac{1 - F(x) + F(-x)}{1 - F(kx) + F(-kx)} \to k^{\beta} \qquad \text{for every constant } k > 0. \tag{5.4.33}$$

Here c_1, c_2 are nonnegative constants with $c_1 + c_2 > 0$ which are related to the stable law by (5.4.13).

Proof. First note that, in view of Theorems 5.3.1 and 5.4.3, F belongs to the domain of attraction of a stable law with index β $(0 < \beta < 2)$ if and only if, for some choice of constants B_n,

$$nF(B_n x) \to c_1 |x|^{-\beta} \qquad (x < 0), \tag{5.4.34}$$

$$n[1 - F(B_n x)] \to c_2 x^{-\beta} \qquad (x > 0), \tag{5.4.35}$$

and

$$\lim_{\epsilon \to 0} \limsup_{n \to \infty} n \left\{ \int_{|x| < \epsilon} x^2 \, dF(B_n x) - \left[\int_{|x| < \epsilon} x \, dF(B_n x) \right]^2 \right\} = 0. \tag{5.4.36}$$

We first show that (5.4.34), (5.4.35), and (5.4.36) together imply (5.4.32) and (5.4.33). Let $x > 0$ be fixed. For large $y > 0$ choose n so that

$$B_n x \leq y \leq B_{n+1} x.$$

Clearly

$$F(-B_{n+1} x) \leq F(-y) \leq F(-B_n x)$$

and

$$1 - F(B_{n+1} x) \leq 1 - F(y) \leq 1 - F(B_n x),$$

so that

$$\frac{(n+1)F(-B_{n+1}x)}{n[1 - F(B_n x)]} \left(\frac{n}{n+1} \right) \leq \frac{F(-y)}{1 - F(y)} \leq \frac{nF(-B_n x)}{(n+1)[1 - F(B_{n+1} x)]} \left(\frac{n+1}{n} \right).$$

Similarly, for $k > 0$

$$\frac{(n+1)[1 - F(B_{n+1}x) + F(-B_{n+1}x)]}{n[1 - F(kB_n x) + F(-kB_n x)]}\left(\frac{n}{n+1}\right)$$
$$\leq \frac{1 - F(y) + F(-y)}{1 - F(ky) + F(-ky)}$$
$$\leq \frac{n[1 - F(B_n x) + F(-B_n x)]}{(n+1)[1 - F(kB_{n+1}x) + F(-kB_{n+1}x)]}\left(\frac{n+1}{n}\right).$$

As $y \to \infty$, $n \to \infty$, and (5.4.34) and (5.4.35) imply (5.4.32) and (5.4.33).

We next show that conditions (5.4.34) through (5.4.36) follow from (5.4.32) and (5.4.33). For $x > 0$ let us write

$$q(x) = 1 - F(x) + F(-x),$$

and note that conditions (5.4.32) and (5.4.33) have meaning only if $q(x) > 0$ for every x. Let B_n be defined by

$$B_n = \inf\left\{x : q(x+0) \leq \frac{c_1 + c_2}{n} \leq q(x-0)\right\}. \tag{5.4.37}$$

Clearly $\lim_{n\to\infty} B_n = \infty$; and, applying (5.4.33) with x replaced by $x[B_n \pm (1/n)]$ and k by $1/x$, we get for every $x > 0$

$$nq(B_n x) = nx^{-\beta}[1 - F(B_n) + F(-B_n - 0)][1 + o(1)],$$
$$nq(B_n x) = nx^{-\beta}[1 - F(B_n - 0) + F(-B_n)][1 + o(1)],$$

as $n \to \infty$. From the definition of B_n we conclude that

$$nq(B_n x) = \frac{c_1 + c_2}{x^\beta}[1 + o(1)] \qquad \text{as } n \to \infty. \tag{5.4.38}$$

On the other hand, (5.4.32) immediately yields

$$nc_1[1 - F(B_n x)] = nc_2 F(-B_n x)[1 + o(1)]. \tag{5.4.39}$$

It is easy to derive (5.4.34) and (5.4.35) from (5.4.38) and (5.4.39). It only remains to show that (5.4.36) follows from (5.4.32) and (5.4.33). Since the expression in (5.4.36) is nonnegative, it suffices to show that

$$\lim_{\epsilon\to 0}\limsup_{n\to\infty} n\int_{|x|<\epsilon} x^2\, dF(B_n x) = 0.$$

We first show that $\int_{-\infty}^{\infty} x^2\, dF(x) = +\infty$. Let $0 < \epsilon < 1$, choose $k > 1$ so that

$$k^{2-\beta} > (1-\epsilon)^{-1} \qquad \text{and} \qquad k^\beta > (1-\epsilon)^{-1}.$$

In view of (5.4.33) we can choose an $x_0 > 0$ large enough so that

$$\frac{q(x_0 k^n)}{q(x_0 k^{n+1})} = \frac{k^\beta}{1 + \epsilon_n}, \tag{5.4.40}$$

where $|\epsilon_n| \leq \epsilon$ for $n = 0, 1, 2, \ldots$. Now

$$\begin{aligned} q(x_0 k^{n-1}) - q(x_0 k^n) &\geq q(x_0)\left[k^{-\beta(n-1)} \prod_{r=1}^{n-1} (1 + \epsilon_r) - k^{-\beta n} \prod_{r=1}^{n} (1 + \epsilon_r)\right] \\ &\geq q(x_0)(1 - \epsilon)^n k^{-\beta n}\left(\frac{k^\beta}{1 + \epsilon} - 1\right) > 0, \end{aligned}$$

since $q(x_0) > 0$ for every $x_0 > 0$ and $k^\beta > 1 + \epsilon$ by the choice of k. Therefore

$$\begin{aligned} \int_{-\infty}^{\infty} x^2\, dF(x) &= \int_{|x| \leq x_0} x^2\, dF(x) + \sum_{n=1}^{\infty} \int_{x_0 k^{n-1} < |x| \leq x_0 k^n} x^2\, dF(x) \\ &\geq \int_{|x| \leq x_0} x^2\, dF(x) + x_0^2 \sum_{n=1}^{\infty} k^{2(n-1)}[q(x_0 k^{n-1}) - q(x_0 k^n)] \\ &\geq \int_{|x| \leq x_0} x^2\, dF(x) \\ &\quad + x_0^2 q(x_0) k^{-2}\left(\frac{k^\beta}{1 + \epsilon} - 1\right) \sum_{n=1}^{\infty} (1 - \epsilon)^n k^{n(2-\beta)} \\ &= \infty, \end{aligned}$$

since $0 < \beta < 2$ and $(1 - \epsilon)k^{2-\beta} > 1$. In view of what has just been shown, we can choose n large enough so that

$$\int_{|x| \leq x_0} x^2\, dF(x) \leq \int_{x_0 < |x| \leq \epsilon B_n} x^2\, dF(x).$$

Let us choose an integer $s > 0$ such that

$$k^s x_0 < \epsilon B_n \leq k^{s+1} x_0. \tag{5.4.41}$$

Then

$$\begin{aligned} \int_{|x| \leq \epsilon B_n} x^2\, dF(x) &\leq 2 \int_{x_0 < |x| \leq \epsilon B_n} x^2\, dF(x) \\ &\leq 2 \sum_{r=0}^{s} \int_{k^r x_0 < |x| \leq k^{r+1} x_0} x^2\, dF(x) \\ &\leq 2x_0^2 \sum_{r=0}^{s} k^{2(r+1)}[q(x_0 k^r) - q(x_0 k^{r+1})] \\ &< 2x_0^2 \sum_{r=0}^{s} k^{2(r+1)} q(x_0 k^r) \\ &\leq 2\epsilon^2 B_n^2 \sum_{r=0}^{s} k^{2(r-s+1)} q(x_0 k^r). \end{aligned}$$

But from (5.4.40)

$$\begin{aligned}q(x_0 k^r) &\le k^\beta(1+\epsilon)q(x_0 k^{r+1})\\ &\le [k^\beta(1+\epsilon)]^{s-r+1} q(x_0 k^{s+1})\\ &\le [k^\beta(1+\epsilon)]^{s-r+1} q(\epsilon B_n) \qquad (s \ge r),\end{aligned}$$

where the last inequality follows from (5.4.41), and the last but one is obtained by iterating (5.4.40). Since $k^{\beta-2} < (1-\epsilon)$, $k^{\beta-2} < (1+\epsilon)^{-1}$ and

$$\begin{aligned}\int_{|x|\le\epsilon B_n} x^2\, dF(x) &\le 2\epsilon^2 B_n^2 k^4 q(\epsilon B_n) \sum_{r=0}^{s} [(1+\epsilon)k^{\beta-2}]^{s-r+1}\\ &\le 2\epsilon^2 B_n^2 (1+\epsilon) k^{\beta+2} q(\epsilon B_n) \frac{1}{1-(1+\epsilon)k^{\beta-2}}.\end{aligned}$$

Thus from (5.4.38)

$$\limsup_{n\to\infty} \frac{n}{B_n^2} \int_{|x|\le\epsilon B_n} x^2\, dF(x) \le 2\epsilon^{2-\beta}(c_1+c_2)\frac{(1+\epsilon)k^{\beta+2}}{1-(1+\epsilon)k^{\beta-2}},$$

and it follows that

$$\lim_{\epsilon\to 0}\limsup_{n\to\infty} \frac{n}{B_n^2}\int_{|x|\le\epsilon B_n} x^2\, dF(x) = 0.$$

This completes the proof of Theorem 5.4.6. ■

Theorem 5.4.7. If F belongs to the domain of attraction of a stable law with index β, then for any δ $(0 \le \delta < \beta)$

$$\int_{-\infty}^{\infty} |x|^\delta\, dF(x) < \infty.$$

Proof. If the variance of F is finite, the result is obvious. Let us assume therefore that F does not have a finite variance and that $\beta < 2$. By (5.4.40), for $0 < \epsilon < 1$ and $k > 1$ there exists an x_0 such that

$$\frac{q(x_0 k^n)}{q(x_0 k^{n+1})} = (1+\epsilon_n)^{-1} k^\beta,$$

where $|\epsilon_n| \le \epsilon$ for $n = 0, 1, 2, \ldots$. Suppose that $k^{\delta-\beta}(1+\epsilon) < 1$. Then

$$\begin{aligned}\int_{-\infty}^{\infty} |x|^\delta\, dF(x) &= \int_{|x|\le x_0} |x|^\delta\, dF(x) + \sum_{n=1}^{\infty}\int_{k^{n-1}x_0<|x|\le k^n x_0} |x|^\delta\, dF(x)\\ &\le \int_{|x|\le x_0} |x|^\delta\, dF(x) + x_0^\delta \sum_{n=1}^{\infty} k^{n\delta} q(x_0 k^{n-1})\\ &\le O(1) + x_0^\delta q(x_0) k^\beta \sum_{n=1}^{\infty} (1+\epsilon)^n k^{-n(\beta-\delta)}\\ &= O(1)\end{aligned}$$

because of the choice of ϵ and k.

If $\beta = 2$ and F does not have a finite variance, we integrate by parts to get

$$\psi(z) = \int_{|x| \leq z} x^2 \, dF(x) = -z^2 q(z) + 2 \int_0^z x q(x) \, dx.$$

Clearly $\psi(z)$ is nondecreasing, and $\psi(z) \to \infty$ as $z \to \infty$. In view of Theorem 5.4.5 it follows that, as $z \to \infty$,

$$\psi(z) \leq 2 \int_0^z x q(x) \, dx = o\left(\int_1^z \frac{\psi(x)}{x} \, dx \right).$$

Let $\epsilon > 0$, and set

$$t(z) = \sup_{1 \leq x \leq z} \{x^{-\epsilon} \psi(x)\}.$$

Clearly

$$\int_1^z \frac{\psi(x)}{x} \, dx \leq t(z) \int_1^z x^{\epsilon - 1} \, dx < \frac{z^\epsilon t(z)}{\epsilon},$$

and it follows that for every $\epsilon > 0$

$$\psi(z) = o(z^\epsilon t(z)) \qquad \text{as } z \to \infty,$$

so that $\psi(z) = o(z^\epsilon)$ for every $\epsilon > 0$. Let $0 < \delta < 2$. Then for sufficiently large $z > 0$

$$\int_{z < |x| \leq 2z} |x|^\delta \, dF(x) = \int_{z < |x| \leq 2z} |x|^{\delta - 2} |x|^2 \, dF(x)$$
$$< z^{\delta - 2} \psi(2z) < z^{(\delta/2) - 1}$$

and

$$\int_{-\infty}^{\infty} |x|^\delta \, dF(x) = \int_{0 < |x| \leq z} |x|^\delta \, dF(x) + \sum_{n=0}^{\infty} \int_{z 2^n < |x| \leq z 2^{n+1}} |x|^\delta \, dF(x)$$
$$< O(1) + z^{(\delta/2) - 1} \sum_{n=0}^{\infty} 2^{n[(\delta/2) - 1]} = O(1).$$

The proof of Theorem 5.4.7 is now complete. ∎

5.5 APPLICATIONS OF THE CENTRAL LIMIT THEORY

(a) Normal Approximation. The central limit theorem (Theorem 5.1.1 and, especially, Corollary 1 along with its various extensions to the multivariate case and the dependent variable case) occupies a central place in statistics and its applications.

Let $X_1, X_2, \ldots, X_n$ be independent, identically distributed random variables, and let $\mathscr{E}X_1 = \mu$, $\text{var}(X_1) = \sigma^2$, both assumed finite. Lévy's version of the central limit theorem enables us to approximate by the normal distribution probabilities of the type $P\{a \leq S_n \leq b\}$, where $S_n = \sum_{k=1}^n X_k$ and $a, b \in \mathbb{R}$, $a \leq b$. We have for large n

$$P\{a \leq S_n \leq b\} = P\left\{\frac{a - n\mu}{\sigma\sqrt{n}} \leq \frac{S_n - n\mu}{\sigma\sqrt{n}} \leq \frac{b - n\mu}{\sigma\sqrt{n}}\right\}$$

$$\simeq \Phi\left(\frac{b - n\mu}{\sigma\sqrt{n}}\right) - \Phi\left(\frac{a - n\mu}{\sigma\sqrt{n}}\right) \tag{5.5.1}$$

where Φ is the standard normal distribution function. The result in (5.5.1) may alternatively be stated as follows: S_n has an asymptotically normal distribution with mean $n\mu$ and variance $n\sigma^2$. This approximation is of special importance in cases where the exact distribution of S_n cannot be computed by any of the known methods (namely, the inversion theorem or the transformation of variables technique). In practical applications one frequently encounters the statistic $\overline{X}_n = n^{-1}S_n$. Clearly

$$P\{a \leq \overline{X}_n \leq b\} \simeq \Phi\left(\frac{b - \mu}{\sigma}\sqrt{n}\right) - \Phi\left(\frac{a - \mu}{\sigma}\sqrt{n}\right), \tag{5.5.2}$$

that is, $\overline{X}_n$ has an asymptotic normal distribution with mean μ and variance σ^2/n. This result can be used to construct large sample tests of hypotheses concerning μ or to construct approximate confidence intervals for μ.

At this stage it should be emphasized that it is wise to use the normal approximation with caution. The speed of convergence depends on the distribution being sampled, so that a very large n may be required before the approximation is useful. The best result in this direction was obtained by Berry [5] and Esseen [21], who gave an estimate of the remainder term $F_n(x) - \Phi(x)$, where F_n is the distribution function of $S_n = \sum_{k=1}^n X_k$ and Φ is the standard normal distribution function.

In statistics one of the oldest methods of estimation is the method of moments. According to this method, we estimate an unknown moment of the distribution by the corresponding sample moment. Thus, if $\alpha_r = \mathscr{E}X^r$ exists, we estimate α_r by $m_r = \sum_{i=1}^n X_i^r/n$. By the law of large numbers $m_r \xrightarrow{a.s.} \alpha_r$ as $n \to \infty$, so that m_r is consistent for α_r. It is trivially unbiased for μ_r, and if $\mathscr{E}X^{2r} < \infty$ the central limit theorem shows that m_r is asymptotically normal with mean α_r and variance $(\alpha_{2r} - \alpha_r^2)/n$. A similar result holds for central moments $\mu_r = \mathscr{E}(X - \mu)^r$, where $\mu = \mathscr{E}X$ and $\mathscr{E}|X|^{2r}$ is assumed to be finite. In fact, in many instances we are interested in inferences concerning functions involving ratios between powers of certain moments, such as the regression coefficient, the correlation coefficient, the coefficient of variation, and the coefficient of skewness. It can be shown (see, for example, Cramér [13], p. 366) that under certain mild conditions any sample characteristic based on (sample) moments has approximately a normal distribution for large n.

Yet another application of the classical central limit theorem is to give a short proof of Stirling's formula, namely,

$$\lim_{n\to\infty} \frac{(2\pi n)^{1/2} n^n e^{-n}}{n!} = 1. \tag{5.5.3}$$

Let $X_1, X_2, \ldots, X_n$ be a sequence of independent random variables with common Poisson distribution given by

$$P\{X = x\} = \frac{e^{-1}}{x!}, \qquad x = 0, 1, 2, \ldots.$$

Then $S_n = \sum_{k=1}^{n} X_k$ has a Poisson distribution with mean n, and by the Lévy central limit theorem

$$Z_n = \frac{S_n - n}{\sqrt{n}} \xrightarrow{L} Z,$$

where Z is a normal (0, 1) random variable. For any real-valued function f we set

$$f^- = -f \quad \text{if } f < 0, \qquad \text{and} = 0 \quad \text{if } f \geq 0.$$

It follows (see Problem 3.10.35) that $Z_n^- \xrightarrow{L} Z^-$ and hence, from Corollary 1 to Theorem 3.1.5, that

$$\mathscr{E}\left(\frac{S_n - n}{\sqrt{n}}\right)^- \to \mathscr{E}Z^- \qquad \text{as } n \to \infty.$$

But

$$\mathscr{E}\left(\frac{S_n - n}{\sqrt{n}}\right)^- = \sum_{k=0}^{n} e^{-n}\left(\frac{n-k}{\sqrt{n}}\right)\frac{n^k}{k!} = \frac{n^{n+1/2}e^{-n}}{n!}$$

and

$$\mathscr{E}Z^- = \frac{1}{\sqrt{2\pi}}.$$

Hence

$$\lim_{n\to\infty} \frac{\sqrt{2\pi n}\, n^n e^{-n}}{n!} = 1$$

as asserted.

Finally, we mention a large sample property of maximum likelihood estimators. Let $\mathbf{X} = (X_1, X_2, \ldots, X_n)$ be a random vector with joint probability density (or mass) function $f(\mathbf{x}, \boldsymbol{\theta})$, $\boldsymbol{\theta} \in \Theta \subseteq \mathbb{R}_k$. The function $f(\mathbf{x}, \boldsymbol{\theta})$ as a function of $\boldsymbol{\theta}$, for fixed $\mathbf{x}$, is usually referred to as the *likelihood function.* The principle of *maximum likelihood* estimation consists of choosing the value $\hat{\boldsymbol{\theta}}(\mathbf{x})$ of the parameter which maximizes $f(\mathbf{x}, \boldsymbol{\theta})$. In other words, if $\mathbf{X} = \mathbf{x}$, we choose $\hat{\boldsymbol{\theta}}(\mathbf{x})$ to satisfy

(5.5.4) $$f(\mathbf{x}, \hat{\boldsymbol{\theta}}(\mathbf{x})) = \sup_{\theta \epsilon \Theta} f(\mathbf{x}, \boldsymbol{\theta}).$$

If such a statistic $\hat{\boldsymbol{\theta}}$ exists, we call it a *maximum likelihood estimator* of $\boldsymbol{\theta}$. In statistics maximum likelihood estimators are widely used, especially when the sample size n is large. It turns out that maximum likelihood estimators have an asymptotically normal distribution. We will give a broad outline of the proof of this result. In the following we assume that Θ is an open subset of $\mathbb{R}$ and that all the indicated derivatives exist.

Note that, since $\ln f$ is an increasing function of f, it suffices to maximize

$$L(\theta, \mathbf{x}) = \ln f(\mathbf{x}, \theta);$$

and if L is differentiable in θ for fixed $\mathbf{x}$, the maximum likelihood estimator, if it exists, must satisfy the *likelihood equation*

(5.5.5) $$\frac{\partial}{\partial \theta} L(\theta, \mathbf{x}) = 0.$$

Suppose that the X_i are independent and identically distributed with common probability density (or probability mass function) f. Then the likelihood equation becomes

(5.5.6) $$\sum_{i=1}^{n} \frac{\partial}{\partial \theta} \ln f(x_i, \theta) = 0.$$

Setting $\varphi(x, t) = \partial \ln f(x, \theta)/\partial\theta\,|_{\theta=t}$, and assuming that the maximum likelihood estimator $\hat{\theta}_n$ of θ exists, is unique and satisfies

(5.5.7) $$\sum_{i=1}^{n} \varphi(x_i, \hat{\theta}_n(\mathbf{x})) = 0 \qquad \text{for all } x_1, x_2, \ldots, x_n,$$

we deduce from the mean value theorem that for θ_n^* between θ and $\hat{\theta}_n$ we have

(5.5.8) $$\sum_{i=1}^{n} \varphi(X_i, \hat{\theta}_n(\mathbf{X})) - \sum_{i=1}^{n} \varphi(X_i, \theta) = \left\{\sum_{i=1}^{n} \frac{\partial^2 \ln f(X_i, \theta_n^*)}{\partial \theta^2}\right\}(\hat{\theta}_n(\mathbf{X}) - \theta),$$

where

$$\frac{\partial^2 \ln f(X_i, \theta_n^*)}{\partial \theta^2} = \left.\frac{\partial^2 \ln f(X_i, \theta)}{\partial \theta^2}\right|_{\theta=\theta_n^*} = \varphi'(X_i, \theta_n^*).$$

We define $R_n = n^{-1}\sum_{i=1}^n [\varphi'(X_i, \theta_n^*) - \mathscr{E}\varphi'(X_i, \theta)]/\mathscr{E}\varphi'(X_1, \theta)$. In view of (5.5.7) and (5.5.8) we have

$$n^{1/2}(\hat{\theta}_n(\mathbf{X}) - \theta) = -\frac{n^{-1/2}\sum_{i=1}^n \varphi(X_i, \theta)}{\mathscr{E}\varphi'(X_1, \theta)\{1 - R_n\}}. \tag{5.5.9}$$

Under some suitable conditions we can show that $\hat{\theta}_n$ is consistent for θ and that, moreover, in view of Khintchine's weak law of large numbers

$$n^{-1}\sum_{i=1}^n \varphi'(X_i, \theta_n^*) \xrightarrow{P} \mathscr{E}\varphi'(X_1, \theta). \tag{5.5.10}$$

Clearly $R_n \xrightarrow{P} 0$. Also

$$\mathscr{E}\varphi(X_1, \theta) = \mathscr{E}\frac{\partial}{\partial\theta}\ln f(X_1, \theta) = 0; \tag{5.5.11}$$

moreover,

$$\begin{aligned} \mathscr{E}\varphi'(X_1, \theta) &= \mathscr{E}\frac{\partial^2 \ln f(X_1, \theta)}{\partial\theta^2} \\ &= \mathscr{E}\left\{\frac{(\partial^2/\partial\theta^2) f(X_1, \theta)}{f(X_1, \theta)}\right\} - \mathscr{E}\left\{\frac{\partial}{\partial\theta}\ln f(X_1, \theta)\right\}^2 \\ &= -\mathscr{E}\left\{\frac{\partial}{\partial\theta}\ln f(X_1, \theta)\right\}^2 = -\operatorname{var}(\varphi(X_1, \theta)), \end{aligned} \tag{5.5.12}$$

provided that $\mathscr{E}\{(\partial/\partial\theta)\ln f(X_1, \theta)\}^2 < \infty$. [In (5.5.11) we have assumed that we can differentiate $\mathscr{E}1 = 1$ under the expectation sign with respect to θ, whereas in (5.5.12) we have assumed that $\mathscr{E}1 = 1$ can be differentiated twice with respect to θ under the expectation sign.] It follows that, if $0 < \mathscr{E}\{(\partial/\partial\theta)\ln f(X_1, \theta)\}^2 < \infty$, the central limit theorem implies that $-n^{-1/2}\sum_{i=1}^n \varphi(X_i, \theta)/\{\mathscr{E}\varphi'(X_1, \theta)\}^{1/2}$ converges in law to a standard normal random variable. Since $1 - R_n \xrightarrow{P} 1$, it follows from Slutsky's theorem (Problem 3.10.3) that $n^{1/2}(\hat{\theta}_n(\overline{X}) - \theta)$ converges in law to a normal random variable with mean zero and variance $\{\mathscr{E}[(\partial/\partial\theta)\ln f(X_1, \theta)]^2\}^{-1}$. The asymptotic variance of $\hat{\theta}_n(\overline{X})$, therefore, is $\{n\mathscr{E}[(\partial/\partial\theta)\ln f(X_1, \theta)]^2\}^{-1}$, which is the lower bound given by the Fréchet, Cramér, and Rao inequality [70]. Thus the maximum likelihood estimator is asymptotically efficient [73] in a wide variety of models. For a detailed account of the precise regularity conditions and a rigorous proof we refer to Cramér [13] (for the single-parameter case) and Schmetterer [73] (for the multiparameter case).

(b) The Gravitational Field of Stars (Holtzmark Distribution) and Stable Laws. Consider n points, each of mass $m > 0$, thrown at random in the interval $[-l, l]$. Let $X_1, X_2, \ldots, X_n$ be the abscissas of these points. Then the X_i are independently distributed with common uniform distribution on $[-l, l]$. If the gravitational constant is taken to be unity, the force exerted on a unit mass at the origin, assuming Newton's inverse square law, is given by

$$Y_n = \sum_{j=1}^{n} m \frac{\operatorname{sgn}(X_j)}{X_j^2}.$$

Clearly, for $t \in \mathbb{R}$

$$\begin{aligned}\mathscr{E} \exp(itY_n) &= \left\{\frac{1}{2l}\int_{-l}^{l} \exp\left(imt\,\frac{\operatorname{sgn}(x)}{x^2}\right) dx\right\}^n \\ &= \left[\frac{1}{l}\int_0^l \cos\left(\frac{mt}{x^2}\right) dx\right]^n.\end{aligned}$$

We shall be interested in the limiting case where $n \to \infty$, $l \to \infty$ in such a way that the number of points per unit length

$$\mu = \frac{n}{l}$$

remains fixed. We have

$$\begin{aligned}\frac{1}{l}\int_0^l \cos\left(\frac{mt}{x^2}\right) dx &= 1 - \frac{1}{l}\int_0^l \left(1 - \cos\frac{mt}{x^2}\right) dx \\ &= 1 - \frac{1}{l}\int_{1/l}^{\infty} \frac{1 - \cos mty^2}{y^2}\, dy \\ &= 1 - \frac{1}{l}\left[\left(\int_0^{\infty} - \int_0^{1/l}\right)\left(\frac{1 - \cos mty^2}{y^2}\right) dy\right].\end{aligned}$$

Now

$$\int_0^{\infty} \frac{1 - \cos mty^2}{y^2}\, dy = \sqrt{m}\sqrt{|t|}\int_0^{\infty} \frac{1 - \cos y^2}{y^2}\, dy,$$

and as $l \to \infty$

$$\int_0^{1/l} \frac{1 - \cos mty^2}{y^2}\, dy \to 0$$

uniformly in every finite t-interval. Replacing n by μl, we have

$$\lim_{\substack{n,l\to\infty \\ n/l=\mu}} \mathscr{E}\exp(itY_n) = \exp\left\{-(m|t|)^{1/2}\mu \int_0^\infty \frac{1-\cos y^2}{y^2}\,dy\right\}.$$

Letting

$$\lambda = \mu m^{1/2} \int_0^\infty \frac{1-\cos y^2}{y^2}\,dy$$

and noting that the right-hand side is continuous, it follows from the continuity theorem that there exists a distribution function F such that

$$\int_{-\infty}^{\infty} \exp(itx)\,dF(x) = \exp(-\lambda|t|^{1/2}), \qquad t \in \mathbb{R}.$$

If the inverse-square law of attraction is changed to the inverse αth power, that is,

$$Y_n = \sum_{j=1}^{n} m \frac{\operatorname{sgn}(X_j)}{|X_j|^\alpha},$$

where $0 < \alpha^{-1} < 2$, the limiting distribution of Y_n would have the characteristic function

$$\varphi(t) = \exp\left(-m^{1/\alpha}\mu|t|^{1/\alpha} \int_0^\infty \frac{1-\cos y^\alpha}{y^2}\,dy\right).$$

For $\alpha \le \frac{1}{2}$ the integral on the right-hand side diverges, thus explaining the restriction $\alpha > \frac{1}{2}$.

The astronomer Holtzmark was the first one to raise and solve this problem (in 1920) in connection with the gravitational field of stars.

5.6 PROBLEMS

SECTION 5.1

1. Does the Lindeberg condition hold for the following sequences of independent random variables?

(a) $P\{X_1 = \pm 1\} = 1/2c$ and, for $n > 1$,

$$P\{X_n = \pm n\} = \frac{1}{2n^2}\left(1 - \frac{1}{c}\right), \qquad P\{X_n = 0\} = 1 - \frac{1}{n^2} - \frac{1}{c}\left(1 - \frac{1}{n^2}\right),$$

where $c > 1$ is a fixed real number.

(b) $P\{X_n = \pm n^\alpha\} = \frac{1}{2}$.

(c) For each $n \geq 1$, X_n has probability density function

$$f_n(x) = \frac{1}{2a_n} \quad \text{if } x \in (-a_n, a_n), \qquad \text{and } = 0 \quad \text{otherwise,}$$

where $a_n \neq 0$ for all n. Suppose that $|a_n| < a$ for all n and $\sum_{k=1}^n a_k^2 \to \infty$ as $n \to \infty$.

(d) $P\{X_n = \pm 1/2^n\} = \frac{1}{2}$.

(e) $P\{X_n = \pm 2^{n+1}\} = 2^{-n-3}$, $P\{X_n = 0\} = 1 - 2^{-n-2}$.

(f) $P\{X_n = \pm 1\} = (1 - 2^{-n})/2$, $P\{X_n = \pm 2^{-n}\} = 2^{-n-1}$.

(g) $\{X_n\}$ is a sequence of independent Poisson random variables with parameter λ_n, $n \geq 1$, such that $\sum_{k=1}^n \lambda_k \to \infty$ as $n \to \infty$.

2. Let $\{X_{nk}\}$ be a triangular array of row-independent random variables such that $\max_{1 \leq k \leq k_n} \sigma_{nk}^2 \to 0$ as $n \to \infty$ and $\sum_{k=1}^{k_n} \sigma_{nk}^2 \leq C < \infty$, where $\sigma_{nk}^2 = \text{var}(X_{nk})$. Show that the limit law of $S_n = \sum_{k=1}^n X_{nk}$ necessarily has a characteristic function of the form e^{ψ} if and only if $K_n \xrightarrow{w} K$ and $\sum_{k=1}^{k_n} \gamma_{nk} \to \gamma$, where ψ, K_n, K, γ_{nk}, and γ are as in Theorem 5.1.3. If the condition $\sum_{k=1}^{k_n} \sigma_{nk}^2 \leq C < \infty$ is replaced by $\sum_{k=1}^{k_n} \sigma_{nk} \to \text{var}(X)$, where X is the random variable corresponding to the limit law, show that $K_n \xrightarrow{c} K$. Derive the following normal and Poisson convergence criteria.

(a) (Normal Convergence Criterion) Let $\{X_{nk}\}$ be centered at expectations, and suppose that $\sum_{k=1}^{k_n} \sigma_{nk}^2 = 1$ for all n. Then $\{S_n\}$ has for the limit law the standard normal distribution, and $\max_{1 \leq k \leq k_n} \sigma_{nk}^2 \to 0$ if and only if, for every $\varepsilon > 0$,

$$\sum_{k=1}^{k_n} \int_{|x| \geq \epsilon} x^2 \, dF_{nk}(x) \to 0.$$

Here F_{nk} is the distribution function of X_{nk}.

(b) (Poisson Convergence Criterion) Suppose that $\max_{1 \leq k \leq k_n} \sigma_{nk}^2 \to 0$ and

$$\sum_{k=1}^{k_n} \sigma_{nk}^2 \to \lambda \, (>0).$$

Then the limit law of S_n is the Poisson distribution with parameter λ if and only if $\sum_{k=1}^{k_n} \mathscr{E} X_{nk} \to \lambda$ and, for every $\epsilon > 0$,

$$\sum_{k=1}^{k_n} \int_{|x-1| \geq \epsilon} x^2 \, dF_{nk}(x + \mathscr{E} X_{nk}) \to 0.$$

3. Construct an example to show that there exist sequences of independent random variables for which condition (i) of Theorem 5.1.1 is not satisfied but the central limit theorem holds.

4. In the notation of Theorem 5.1.1 show that condition (i) is equivalent to the conditions $\lim_{n\to\infty} B_n = \infty$ and $\lim_{n\to\infty} \sigma_n/B_n = 0$.

5. Let $\{X_n\}$ be a sequence of independent random variables with $\text{var}(X_n) = \sigma_n^2 < \infty$, $n \geq 1$. Let $B_n^2 = \sum_{k=1}^n \sigma_k^2$, $n \geq 1$. Suppose that the central limit theorem holds. Show that

$$P\left\{\left|\frac{1}{n} \sum_{k=1}^n (X_k - \mathscr{E} X_k)\right| < \varepsilon\right\} - \frac{1}{\sqrt{2\pi}} \int_{-\epsilon(n/B_n)}^{\epsilon(n/B_n)} e^{-t^2/2} \, dt \to 0$$

uniformly in ϵ (>0), as $n \to \infty$. Conclude that the sequence $\{X_n\}$ (for which the central limit theorem holds) does not satisfy the weak law of large numbers if the sequence $\{n/B_n\}$ is bounded.

SECTIONS 5.2 AND 5.3

6. Let $\{X_{nk}\}$ be a triangular array of row-independent random variables satisfying the u.a.n. condition. Suppose that $\mathscr{E}X_{nk} = 0$ and $\sum_{k=1}^{k_n} \operatorname{var}(X_{nk}) = 1$ for all n. Then show that the limit law of $\sum_{k=1}^{k_n} X_{nk}$ is the standard normal distribution if and only if

$$\sum_{k=1}^{k_n} X_{nk}^2 \xrightarrow{P} 1.$$

7. Derive the Lindeberg–Feller central limit theorem from Proposition 5.3.3.

8. Construct an example of a sequence $\{X_n\}$ of random variables which does not satisfy the u.a.n. condition (hence the Lindeberg condition does not hold) but for which the sequence $\{B_n^{-1} \sum_{k=1}^{n} (X_k - \mathscr{E}X_k)\}$ converges in law to the standard normal distribution.

9. Prove the analogue of Theorem 5.3.1 as stated in Remark 5.3.1.

10. Obtain the analogue of Theorem 5.3.2 as stated in Remark 5.3.2.

11. Prove the result stated in Remark 5.3.4.

SECTION 5.4

12. A positive function L, defined on $[0, \infty)$, is said to be slowly varying if, for all $c > 0$, $\lim_{x\to\infty} L(cx)/L(x) = 1$. A positive function l on $[0, \infty)$ is said to be regularly varying with exponent α if, for all $c > 0$, $\lim_{x\to\infty} l(cx)/l(x) = c^\alpha$. Show that for a slowly varying function L the following results hold:

(a) $\lim_{x\to\infty} \int_0^x L(u)\,du = \infty$.
(b) $\lim_{x\to\infty} \int_0^1 [L(cx)/L(x)]\,dc = \int_0^1 \lim_{x\to\infty} [L(cx)/L(x)]\,dc = 1$.
(c) (Karamata) If L is integrable on any finite interval, it has the representation

$$L(x) = b(x) \exp\left\{\int_\beta^x \frac{a(u)}{u}\,du\right\},$$

where $\lim_{x\to\infty} b(x) = b \neq 0$, $\lim_{x\to\infty} a(x) = 1$, and $\beta > 0$.

13. Deduce from Karamata's result in Problem 12 the following results:

(a) $\lim_{x\to\infty} L(x + c)/L(x) = 1$ for $c \geq 0$.
(b) For all $\delta > 0$

$$x^\delta L(x) \to \infty \qquad \text{and } x^{-\delta}L(x) \to 0 \quad \text{as } x \to \infty.$$

(c) $\lim_{k\to\infty} \sup_{2^k \leq c \leq 2^{k+1}} L(c)/L(2^k) = 1$.
(d) The passage to the limit in $L(cx)/L(x) \to 1$ as $x \to \infty$ is uniform in finite intervals $0 < a < x < b$.

14. Let L_1 and L_2 be two slowly varying functions.

(a) Show that $L_1 L_2$ is slowly varying.

(b) If L_1 and L_2 are in addition measurable, show that L_1/L_2 and $L_1 + L_2$ are also slowly varying.

(c) If $\delta > 0$ and L_1 and L_2 are measurable, slowly varying functions, show that

$$x^{-\delta}L_1(x) + L_2(x) \sim L_2(x) \qquad \text{as } x \to \infty.$$

15. Let $X_1, X_2, \ldots$ be a sequence of independent, identically distributed random variables. Let

$$Z_n = \frac{\sum_{k=1}^{n} X_k}{B_n} - A_n, \qquad n \geq 1,$$

where A_n, B_n are constants such that $\{Z_n\}$ converges in law to a (necessarily) stable random variable with index α, $0 < \alpha < 2$. Show that B_n must be of the form $B_n = n^{1/\alpha}L(n)$, $0 < \alpha < 2$, where L is a slowly varying function.

16. In order that a distribution function F belong to the domain of attraction of a stable law with exponent α $(0 < \alpha < 2)$, show that it is necessary and sufficient that, as $|x| \to \infty$,

$$F(x) = \frac{c_1 + o(1)}{(-x)^\alpha} L(x), \qquad x < 0,$$

and

$$F(x) = 1 - \frac{c_2 + o(1)}{x^\alpha} L(x), \qquad x > 0,$$

where L is slowly varying and c_1 and c_2 are nonnegative constants, such that $c_1 + c_2 > 0$, which are related to the stable law by (5.4.13).

17. Show that the distribution function F belongs to the domain of attraction of a normal law if and only if the function $U(x) = \int_{-x}^{x} u^2\, dF(u)$ is slowly varying.

18. In order that a distribution function F belong to the domain of attraction of a stable law with index α, show that it is necessary and sufficient that

$$\frac{t^2[1 - F(t) + F(-t)]}{\int_0^t x^2\, dP\{|X| > x\}} \to \frac{2 - \alpha}{\alpha} \qquad \text{as } t \to \infty.$$

19. A distribution function F is said to belong to the domain of normal attraction of a stable law G with characteristic exponent α if it is in the domain of attraction of G and if the normalizing constants B_n are of the form $B_n = cn^{1/\alpha}$, where $c > 0$ is a constant.

(a) Show that the distribution function F belongs to the domain of normal attraction of the normal distribution if and only if F has finite variance σ^2, and then $B_n = \sigma n^{1/2}$.

(b) In order that the distribution function F belong to the domain of normal attraction of the stable law G with index α, $0 < \alpha < 2$, and given constants c_1 and c_2, with $B_n = cn^{1/2}$, show that it is necessary and sufficient that

$$F(x) = [c_1 c^\alpha + a_1(x)]|x|^{-\alpha}, \qquad x < 0,$$

$$F(x) = 1 - [c_2 c^\alpha + a_2(x)]x^{-\alpha}, \qquad x > 0,$$

where $a_i(x) \to 0$ as $|x| \to \infty$, $i = 1, 2$.

20. Let F be in the domain of attraction of a stable law with characteristic exponent α, $0 < \alpha < 2$. Show that $\int_{-\infty}^{\infty} |x|^{\alpha}\, dF(x) < \infty$ if and only if $\sum_{n=1}^{\infty} B_n^{\alpha}/n^2 < \infty$ for any sequence of normalizing constants $\{B_n\}$ for F. Moreover, $\int_{-\infty}^{\infty} |x|^{\alpha+\delta}\, dF(x) = \infty$ for all $\delta > 0$. (Tucker [83])

21. It can sometimes happen that the distribution functions of the normed and centered sums $B_n^{-1} \sum_{k=1}^{n} X_k - A_n$ of a sequence if independent, identically distributed random variables $\{X_n\}$ do not converge for any choice of the constants B_n and A_n, but for some subsequence $\{n_k\}$ convergence does take place necessarily to an i.d. law. We say that a distribution function F belongs to the domain of partial attraction of the nondegenerate law V if there exists a subsequence $\{n_k\}$ such that the distribution functions of

$$B_{n_k}^{-1} \sum_{j=1}^{n_k} X_j - A_{n_k}$$

for suitable constants B_n, A_n converge to V. Show that every i.d. law has a (nonempty) domain of partial attraction.

22. Construct examples to show that there exist distributions that belong to no domain of partial attraction and others that belong to the domain of partial attraction of every i.d. distribution.

23. Let F be a distribution function such that $F(x) < 1$ for all $x \in \mathbb{R}$. Let $X_1, X_2, \ldots, X_n$ be independent and identically distributed with common distribution function F, and set $M_n = \max\{X_1, \ldots, X_n\}$. Show that in order that, with appropriate constants a_n, the distribution functions G_n of $a_n^{-1} M_n$ converge completely to a nondegenerate distribution function G it is necessary and sufficient that $1 - F$ varies regularly with an exponent $\rho < 0$. In this case $G(x) = e^{-cx^{\rho}}$ for $x > 0$, and $G(x) = 0$ for $x < 0$, $c > 0$.

24. Let F_1 and F_2 be two distribution functions such that

$$1 - F_i(x) = x^{-\rho} L_i(x) \qquad \text{as } x \to \infty,$$

where L_i $(i = 1, 2)$ is slowly varying. Show that $G = F_1 * F_2$ has regularly varying tails of the form

$$1 - G(x) \sim x^{-\rho}[L_1(x) + L_2(x)].$$

In particular, if F is a distribution function such that $1 - F(x) \sim x^{-\rho} L(x)$, show that $1 - F_n(x) \sim n x^{-\rho} L(x)$, where F_n is the n-fold convolution of F and L is slowly varying.

25. Let X be a random variable with probability density function f. Suppose that $f(0) \neq 0$ and that f is continuous at $x = 0$. Then prove that for any $r > 0$ the random variable $1/|X|^r$ belongs to the domain of attraction of a stable law. In fact, if $0 < r \leq \frac{1}{2}$, then $1/|X|^r$ belongs to the domain of attraction of a normal law; and if $r \geq \frac{1}{2}$, then $1/|X|^r$ belongs to a stable law with index $1/r$. (Shapiro [74])

26. Let $r \geq 1$, and let X be a random variable that belongs to the domain of attraction of a stable law with characteristic exponent α, $0 < \alpha < 2$. Show that $|X|^r$ belongs to the domain of attraction of a stable law with characteristic exponent α/r.

27. Let $\{X_n\}$ be a sequence of a.s. positive, independent, identically distributed random variables with common distribution function F. Set $S_n = \sum_{k=1}^{n} X_k$, $n \geq 1$. Let $\{t_n\}$ be a sequence of positive real numbers such that $t_n \to \infty$ in such a way that $n[1 - F(t_n)] \to 0$ as $n \to \infty$. If $1 - F$ varies slowly, show that

$$P\{S_n > t_n\} \sim nP\{X_1 > t_n\} \sim P\left\{\max_{1 \le k \le n} X_k > t_n\right\}$$

as $n \to \infty$. [Here $f(n) \sim g(n)$ means $f(n)/g(n) \to 1$ as $n \to \infty$.]

By taking $F(x) = 2[1 - \Phi(1/\sqrt{x})]$, $x > 0$, where Φ is the standard normal distribution function and $n^2 = o(t_n)$ as $n \to \infty$, show that the converse of the above result does not hold. (Andersen [4])

SECTION 5.5

28. Show that $\lim_{n\to\infty} e^{-n} \sum_{j=0}^{n} (n^j/j!) = \frac{1}{2}$.

29. Consider a particle that moves at random on the real axis according to a Bernoulli scheme. At each stage the particle moves one step of length 1 with probability $\frac{1}{2}$ to the left and with probability $\frac{1}{2}$ to the right. Let $x > 0$ be an integer, and let $\tau(x)$ be the first time that the particle reaches x.

(a) Show that

$$P\{\tau(x) = 2n - x\} = \frac{x}{2n - x}\binom{2n - x}{n}\frac{1}{2^{2n-x}}, \qquad n = x, x + 1, \ldots.$$

(b) Show that for $t > 0$

$$P\left\{\frac{\tau(x)}{x^2} \le t\right\} \sim 1 - \sqrt{\frac{2}{\pi}}\int_0^{t^{-1/2}} e^{-u^2/2}\,du \qquad \text{as } x \to \infty.$$

[The limiting distribution in (b) is a stable law with index $\frac{1}{2}$ and probability density function $(2\pi)^{-1/2}t^{-3/2}e^{-1/2t}$.]

30. Let $\{N_k\}$ be a sequence of positive, integral-valued random variables such that $k^{-1}N_k \xrightarrow{P} c$ as $k \to \infty$, where $0 < c < \infty$. Let $\{X_n\}$ be a sequence of independent, identically distributed random variables with $\mathscr{E}X_j = 0$ and $\mathscr{E}X_j^2 = 1, j \ge 1$. Show that the sequence $\{\sum_{j=1}^{N_n} X_j/\sqrt{N_n}\}$ converges in law to a standard normal random variable.

31. Let $\{X_n\}$ be a sequence of random variables. We say that the sequence is m-dependent if there exists an integer $m \ge 0$ such that any subsequence $\{X_{n_j}, j \ge 1\} \subset \{X_n\}$, with $n_j + m < n_{j+1}$ for every $j \ge 1$, is a sequence of independent random variables. Let $\{X_n\}$ be an m-dependent sequence of random variables such that $|X_n| \le c$ a.s. for some constant $c > 0$. Let $S_n = \sum_{k=1}^{n} X_k, n \ge 1$, and suppose that $\text{var}(S_n)/n^{2/3} \to \infty$ as $n \to \infty$. Show that $\{(S_n - \mathscr{E}S_n)/\sqrt{\text{var}(S_n)}\}$ converges in law to a standard normal random variable. (*Hint*: Set $n_j = [nj/k]$, $0 \le j \le k$, where $[x]$ is the largest integer $\le x$. For sufficiently large n define

$$Y_j = \sum_{k=0}^{n_{j+1}-n_j-m} X_{n_j+k} \qquad \text{and} \qquad Z_j = \sum_{k=1}^{m-1} X_{n_{j+1}-k}.$$

Show that $\sum_{j=0}^{k-1} Z_j/\sqrt{\text{var}(S_n)} \xrightarrow{P} 0$, whereas $\{\sum_{j=1}^{k-1} Y_j/\sqrt{\text{var}(S_n)}\}$ converges in law to a standard normal random variable.)

32. Let F be a distribution function with mean 0 and variance 1. If the family of distributions

$$\left\{F\left(\frac{x-\mu}{\sigma}\right), \mu \in \mathbb{R}, \sigma > 0\right\}$$

is closed with respect to the operation of convolution, use the Lévy central limit theorem to show that F must be the distribution function of a standard normal random variable.

33. Let $0 \leq x \leq 1$, and let $\epsilon_n(x)$ denote the nth digit in the decimal expansion of x. Set $S_n(x) = \sum_{k=1}^{n} \epsilon_n(x)$, $n = 1, 2, \ldots$. Define

$$E_n(y) = \left\{x \in [0, 1]: \frac{2S_n(x) - 9n}{\sqrt{33n}} \leq y\right\}, \qquad y \in \mathbb{R},$$

and let $\lambda(E_n(y))$ be the Lebesgue measure of $E_n(y)$. Show that

$$\lim_{n\to\infty} \lambda(E_n(y)) = (2\pi)^{-1/2} \int_{-\infty}^{y} e^{-t^2/2}\, dt.$$

(*Hint*: Let X be a random variable with uniform distribution on (0, 1). Then $X_n = \epsilon_n(X)$ are independent and identically distributed with $\mathscr{E}X_n = 9/2$ and $\text{var}(X_n) = \sqrt{33}/2$.)

34. Let $X_1, X_2, \ldots$ be independent, identically distributed random variables with $\mathscr{E}X_1 = 0$ and $\mathscr{E}X_1^2 = 1$. Find the limiting distribution of the sequences of random variables

$$Y_n = n^{1/2} \frac{\sum_{k=1}^{n} X_k}{\sum_{k=1}^{n} X_k^2}, \qquad n \geq 1,$$

and

$$Z_n = \frac{\sum_{k=1}^{n} X_k}{(\sum_{k=1}^{n} X_k^2)^{1/2}}, \qquad n \geq 1.$$

35. Let $\{X_n\}$ and $\{Y_n\}$ be sequences of random variables such that

$$\left\{\sqrt{n}\left(\frac{X_n - a}{\sigma_1}\right)\right\} \quad \text{and} \quad \left\{\sqrt{n}\left(\frac{Y_n - b}{\sigma_2}\right)\right\}$$

have limiting standard normal distributions. Here $\sigma_1 > 0$, $\sigma_2 > 0$, and $b \neq 0$. Find the limiting distribution of $\{\sqrt{n}(X_n - a)/Y_n\}$.

NOTES AND COMMENTS

An exhaustive treatment of the general central limit problem on $\mathbb{R}$ is given in Gnedenko and Kolmogorov [28]. For the multivariate extension of some of these results we refer the reader to Cuppens [14]. The infinite-dimensional case is developed in detail in Parthasarathy [64]. Some of these results will be discussed in Chapter 7.

The question of convergence leads inevitably to the speed of convergence and the asymptotic behavior of tail probabilities. The best result on the speed of convergence to the normal law was obtained independently by Berry [5] and Esseen [21]. For a discussion of these and other, related extensions we refer to Bhattacharya and Ranga Rao [6]. The asymptotic behavior of (large deviation) tail probabilities has been studied by several authors, including Linnik [55] and Heyde. A detailed account of this work appears in Ibragimov and Linnik [39] and Petrov [66].

CHAPTER 6

Dependence

In the preceding chapters we restricted ourselves, mainly, to independent random variables (and independent events). In many applications of probability theory, however, we have to deal with random variables (and events) which are not independent. No course in basic probability theory is complete without a study of some notions of dependence. In Section 6.1 we study the abstract notions of conditional probability and conditional expectation, which play a key role in probability theory (including the theory of stochastic processes) and mathematical statistics. The rest of the chapter is devoted to a special type of dependence, namely, martingale dependence and its applications.

6.1 CONDITIONING

6.1.1 Conditional Expectation

Let $(\Omega, \mathscr{S}, P)$ be a probability space. Let $A \in \mathscr{S}$ and $B \in \mathscr{S}$. Then the *conditional probability of A, given B, is defined to be*

$$(6.1.1) \qquad P\{A \mid B\} = \frac{P(A \cap B)}{P(B)},$$

provided that $P(B) > 0$. Clearly

$$P(A \cap B) = P(B)P\{A \mid B\}.$$

In a similar manner the conditional probability of B, given A, namely, $P\{B \mid A\}$, can be defined, provided that $P(A) > 0$. We note that $A \in \mathscr{S}$ and $B \in \mathscr{S}$, with $P(A) > 0$, $P(B) > 0$, are independent if and only if either one of the following two equivalent conditions holds:

$$P\{A|B\} = P(A),$$
$$P\{B|A\} = P(B).$$

Let $B \in \mathscr{S}$ with $P(B) > 0$ be fixed. Then $P\{\cdot|B\}$ defined in (6.1.1) is a probability measure on $(\Omega, \mathscr{S})$, so that $(\Omega, \mathscr{S}, P\{\cdot|B\})$ is a probability space.

Let X be a random variable defined on $(\Omega, \mathscr{S}, P)$ such that $\mathscr{E}X$ exists. Clearly X is also integrable with respect to $P\{\cdot|B\}$ on Ω. We set

$$\text{(6.1.2)} \qquad \mathscr{E}\{X|B\} = \int_\Omega X\, dP\{\cdot|B\},$$

and in this case $\mathscr{E}\{X|B\}$ is known as the *conditional expectation of* X, *given* B. Since

$$P\{\cdot|B\} = 0 \qquad \text{on the class } \{A \cap B^c: A \in \mathscr{S}\}$$

and

$$P\{\cdot|B\} = \frac{1}{P(B)} P(\cdot) \qquad \text{on the class } \{A \cap B: A \in \mathscr{S}\},$$

we can rewrite (6.1.2) as

$$\text{(6.1.3)} \qquad \mathscr{E}\{X|B\} = \frac{1}{P(B)} \int_B X\, dP = \frac{1}{P(B)} \mathscr{E}(X\chi_B),$$

where χ_B is the indicator function of event B. In particular, if $X = \chi_A$ for $A \in \mathscr{S}$, then

$$\mathscr{E}\{\chi_A|B\} = \frac{P(A \cap B)}{P(B)} = P\{A|B\}.$$

Consider the σ-field $\mathfrak{A} = \{\Omega, B, B^c, \varnothing\} \subset \mathscr{S}$ generated by $B \in \mathscr{S}$. By setting

$$\mathscr{E}\{X|\mathfrak{A}\} = \begin{cases} \mathscr{E}\{X|B\} & \text{if } \omega \in B, \\ \mathscr{E}\{X|B^c\} & \text{if } \omega \in B^c, \end{cases}$$

we see that $\mathscr{E}\{X|\mathfrak{A}\}$ is a two-point random variable which satisfies

$$\begin{aligned} \int_\Omega \mathscr{E}\{X|\mathfrak{A}\}\, dP &= P(B)\mathscr{E}\{X|B\} + P(B^c)\mathscr{E}\{X|B^c\} \\ &= \mathscr{E}\{X\chi_B\} + \mathscr{E}\{X\chi_{B^c}\} \\ &= \int_\Omega X\, dP. \end{aligned}$$

The $\mathfrak{A}$-measurable function $\mathscr{E}\{X \mid \mathfrak{A}\}$ defined on Ω is known as the *conditional expectation of X, given (the σ-field)* $\mathfrak{A}$.

We next consider the σ-field $\mathfrak{A} \subset \mathscr{S}$, generated by a finite number of sets $B_1, B_2, \ldots, B_n \in \mathscr{S}$, $(n \geq 2)$ such that $B_i \cap B_j = \varnothing$ for $i \neq j$, $\bigcup_{i=1}^{n} B_i = \Omega$ and $P(B_i) > 0$ for $i = 1, 2, \ldots, n$. Let X be a random variable such that $\mathscr{E}X$ exists. Then we set

$$\mathscr{E}\{X \mid \mathfrak{A}\} = \mathscr{E}\{X \mid B_i\} \qquad \text{if } \omega \in B_i, \quad i = 1, 2, \ldots, n, \tag{6.1.4}$$

where $\mathscr{E}\{X \mid B_i\}$ is as defined in (6.1.2). Then $\mathscr{E}\{X \mid \mathfrak{A}\}$ is called the *conditional expectation of X, given (the σ-field)* $\mathfrak{A}$. Clearly $\mathscr{E}\{X \mid \mathfrak{A}\}$ is a random variable. Using (6.1.3), we see that

$$P(B_i)\mathscr{E}\{X \mid B_i\} = \mathscr{E}(X\chi_{B_i}), \qquad i = 1, 2, \ldots, n,$$

so that

$$\begin{aligned}\sum_{i=1}^{n} P(B_i)\mathscr{E}\{X \mid B_i\} &= \sum_{i=1}^{n} \mathscr{E}(X\chi_{B_i}) \\ &= \mathscr{E}X = \int_{\Omega} X \, dP.\end{aligned}$$

On the other hand, $\mathscr{E}\{X \mid \mathfrak{A}\}$ is a simple random variable taking values $\mathscr{E}\{X \mid B_i\}$ on B_i for $i = 1, 2, \ldots, n$, so that it follows from the definition of the integral of a simple function that

$$\int_{\Omega} \mathscr{E}\{X \mid \mathfrak{A}\} \, dP = \sum_{i=1}^{n} P(B_i)\mathscr{E}\{X \mid B_i\} = \int_{\Omega} X \, dP. \tag{6.1.5}$$

Clearly $\mathscr{E}\{X \mid \mathfrak{A}\}$ is measurable with respect to $\mathfrak{A}$. We see easily that, if X itself is $\mathfrak{A}$-measurable, $\mathscr{E}\{X \mid \mathfrak{A}\} = X$ a.s. with respect to the probability measure induced by P on $\mathfrak{A}$.

More generally, let $\{B_n : n \geq 1\} \subset \mathscr{S}$ be a countable partition of Ω with $P(B_n) > 0$ for $n \geq 1$, and let $\mathfrak{A}$ be the σ-field generated by the B_n. Let X be a random variable with finite expectation. Then

$$\mathscr{E}\{X \mid \mathfrak{A}\} = \sum_{n=1}^{\infty} \mathscr{E}\{X \mid B_n\}\chi_{B_n} \tag{6.1.6}$$

defines the *conditional expectation of X, given* $\mathfrak{A}$. As before, $\mathscr{E}\{X \mid \mathfrak{A}\}$ is a discrete random variable taking a countable set of values $\{\mathscr{E}\{X \mid B_n\}, n \geq 1\}$, and it follows that

$$\int_{\Omega} \mathscr{E}\{X \mid \mathfrak{A}\} \, dP = \int_{\Omega} X \, dP. \tag{6.1.7}$$

Clearly $\mathscr{E}\{X \mid \mathfrak{A}\}$ is measurable with respect to $\mathfrak{A}$; in particular, if X itself is $\mathfrak{A}$-measurable, $\mathscr{E}\{X \mid \mathfrak{A}\} = X$ a.s. with respect to the probability measure induced by P on $\mathfrak{A}$.

In practical applications we need the concept of conditional expectation of a random variable X, given a σ-field generated by a random variable Y or, more generally, by a fixed collection of random variables $\{Y_\lambda, \lambda \in \Lambda\}$. For this purpose it is necessary to extend the definition of conditional expectation of a random variable to the case where the given σ-field is arbitrary.

Let $(\Omega, \mathscr{S}, P)$ be a probability space, and let $\mathscr{D} \subset \mathscr{S}$ be a σ-field. Let $P_{\mathscr{D}}$ be the probability measure induced by P on $\mathscr{D}$, that is, $P_{\mathscr{D}}(E) = P(E)$ for $E \in \mathscr{D}$. Let X be a random variable defined on $(\Omega, \mathscr{S}, P)$ such that $\mathscr{E}X$ exists. Then for every $E \in \mathscr{D}$ we can define the indefinite integral

$$Q_X(E) = \int_E X\, dP = \int_\Omega X\chi_E\, dP.$$

Clearly Q_X is a finite signed measure on $\mathscr{D}$ such that $Q_X(E) = 0$ for every $E \in \mathscr{D}$ for which $P_{\mathscr{D}}(E) = 0$. Hence $Q_X \ll P_{\mathscr{D}}$, so that in view of the Radon–Nikodym theorem there exists a $\mathscr{D}$-measurable function defined on Ω, which we denote by $\mathscr{E}\{X \mid \mathscr{D}\}$, such that the relation

$$\int_E \mathscr{E}\{X \mid \mathscr{D}\}\, dP_{\mathscr{D}} = Q_X(E) = \int_E X\, dP \tag{6.1.8}$$

holds for every $E \in \mathscr{D}$. Here the function $\mathscr{E}\{X \mid \mathscr{D}\}$ is determined uniquely with respect to $P_{\mathscr{D}}$ in the sense that, if there exists another $\mathscr{D}$-measurable function g on Ω satisfying (6.1.8) for every $E \in \mathscr{D}$, then $g = \mathscr{E}\{X \mid \mathscr{D}\}$ a.s. with respect to $P_{\mathscr{D}}$.

Definition 6.1.1. The $\mathscr{D}$-measurable function $\mathscr{E}\{X \mid \mathscr{D}\}$ defined by (6.1.8) is said to be the conditional expectation of the random variable X with respect to the σ-field $\mathscr{D}$. Here $\mathscr{E}\{X \mid \mathscr{D}\}$ is defined uniquely except for $\mathscr{D}$-measurable sets of $P_{\mathscr{D}}$-measure zero.

We note that

$$\int_\Omega \mathscr{E}\{X \mid \mathscr{D}\}\, dP_{\mathscr{D}} = \int_\Omega X\, dP = \mathscr{E}X. \tag{6.1.9}$$

In particular, let Y be another random variable defined on Ω, and let $\mathscr{D} = \sigma(Y)$, where $\sigma(Y)$ is the σ-field generated by Y. Then we write $\mathscr{E}\{X \mid Y\} = \mathscr{E}\{X \mid \mathscr{D}\}$. More generally, let $\{Y_\lambda, \lambda \in \Lambda\}$ be a collection of random variables defined on Ω, and let $\mathscr{D} = \sigma(\{Y_\lambda, \lambda \in \Lambda\})$. Then we write $\mathscr{E}\{X \mid Y_\lambda, \lambda \in \Lambda\} = \mathscr{E}\{X \mid \mathscr{D}\}$.

Remark 6.1.1. The $\mathscr{D}$-measurable function $\mathscr{E}\{X \mid \mathscr{D}\}$ is determined uniquely up to a $\mathscr{D}$-measurable null set by condition (6.1.8). The class of all $\mathscr{D}$-measurable functions satisfying (6.1.8) for all $D \in \mathscr{D}$ is an equivalence class under the relation of a.s. equality. Any member of this class is known as a *version* of the conditional expectation of X, given (or with respect to) $\mathscr{D}$. Note that, although $\mathscr{E}\{X \mid \mathscr{D}\}$ is a.s. uniquely determined on Ω, the exceptional set may depend on the random variable X under consideration.

Example 6.1.1. If $\mathscr{D} = \{\varnothing, \Omega\}$, only constant functions are $\mathscr{D}$-measurable, and it follows that $\mathscr{E}\{X \mid \mathscr{D}\} = \mathscr{E}X$ for all $\omega \in \Omega$. If $\mathscr{D} = \mathscr{S}$, then X is $\mathscr{D}$-measurable and $\mathscr{E}\{X \mid \mathscr{D}\} = X$ a.s.

Example 6.1.2. Let $\Omega = (0, 1)$, $\mathscr{B} = \mathscr{B}_{(0,1)}$ be the Borel σ-field on Ω, and λ be the Lebesgue measure on $\mathscr{B}$. Let

$$Y(\omega) = \begin{cases} 2, & 0 < \omega < \frac{1}{3}, \\ 5, & \frac{1}{3} \le \omega < 1. \end{cases}$$

Let $\mathscr{D} = \sigma(Y)$. Then $\mathscr{D} = \{\Omega, \varnothing, (0, \frac{1}{3}), [\frac{1}{3}, 1)\}$. Only simple random variables with a (single) jump at $\frac{1}{3}$ are $\mathscr{D}$-measurable. Let $X(\omega) = \omega$ for all $\omega \in (0, 1)$. Then

$$\mathscr{E}\{X \mid \mathscr{D}\} = \mathscr{E}\{X \mid Y\} = \begin{cases} \frac{1}{6} & \text{if } \omega \in (0, \frac{1}{3}), \\ \frac{2}{3} & \text{if } \omega \in [\frac{1}{3}, 1). \end{cases}$$

6.1.2 Elementary Properties

We now study some properties of the function $\mathscr{E}\{X \mid \mathscr{D}\}$ defined above.

Proposition 6.1.1. Let X be a random variable defined on $(\Omega, \mathscr{S}, P)$ such that $\mathscr{E}X$ exists, and let $\mathscr{D} \subset \mathscr{S}$ be a σ-field.

(a) Let X be $\mathscr{D}$-measurable. Then $\mathscr{E}\{X \mid \mathscr{D}\} = X$ a.s. $(P_{\mathscr{D}})$.
(b) Let $X = c$ a.s. (P), where c is a constant. Then $\mathscr{E}\{X \mid \mathscr{D}\} = c$ a.s. $(P_{\mathscr{D}})$.
(c) Let $X \ge 0$ a.s. (P). Then $\mathscr{E}\{X \mid \mathscr{D}\} \ge 0$ a.s. $(P_{\mathscr{D}})$.
(d) Let Y be another random variable on $(\Omega, \mathscr{S}, P)$ such that $\mathscr{E}Y$ exists. Let $a, b \in \mathbb{R}$. Then

$$\mathscr{E}\{aX + bY \mid \mathscr{D}\} = a\mathscr{E}\{X \mid \mathscr{D}\} + b\mathscr{E}\{Y \mid \mathscr{D}\} \text{ a.s. } (P_{\mathscr{D}}).$$

Proof. The proof immediately follows from the definition. ■

Proposition 6.1.2 (Conditional Monotone Convergence Theorem). Let $\{X_n\}$ be a nondecreasing sequence of nonnegative random variables which converge a.s. on Ω to a random variable X. Suppose that $\mathscr{E}X$ exists. Then

$$0 \leq \lim_{n\to\infty} \mathscr{E}\{X_n|\mathscr{D}\} = \mathscr{E}\{X|\mathscr{D}\} \text{ a.s. } (P_{\mathscr{D}}).$$

Proof. Clearly $\mathscr{E}X_n$ exists for $n \geq 1$. It follows from Proposition 6.1.1 that the sequence $\{\mathscr{E}\{X_n|\mathscr{D}\}, n \geq 1\}$ is a nondecreasing sequence of nonnegative $\mathscr{D}$-measurable random variables such that

$$0 \leq \mathscr{E}\{X_n|\mathscr{D}\} \leq \mathscr{E}\{X|\mathscr{D}\} \text{ a.s.}$$

Therefore it converges a.s. to a $\mathscr{D}$-measurable random variable, say Y. In view of the monotone convergence theorem for every $E \in \mathscr{D}$

$$\lim_{n\to\infty} \int_E X_n \, dP = \int_E X \, dP = \int_E \mathscr{E}\{X|\mathscr{D}\} \, dP_{\mathscr{D}}.$$

Also

$$\lim_{n\to\infty} \int_E \mathscr{E}\{X_n|\mathscr{D}\} \, dP_{\mathscr{D}} = \int_E Y \, dP_{\mathscr{D}}.$$

Using (6.1.8) with X replaced by X_n, we see that for $E \in \mathscr{D}$

$$\int_E \mathscr{E}\{X|\mathscr{D}\} \, dP_{\mathscr{D}} = \int_E Y \, dP_{\mathscr{D}}.$$

It follows that $Y = \mathscr{E}\{X|\mathscr{D}\}$ a.s. $(P_{\mathscr{D}})$. ∎

Proposition 6.1.3 (Conditional Lebesgue Dominated Convergence Theorem). Let $\{X_n\}$ be a sequence of random variables which converge a.s. to a random variable X on Ω. Suppose that there exists a random variable $Y \geq 0$ a.s. (P) such that $\mathscr{E}Y$ exists and $|X_n| \leq Y$ a.s. (P) for all $n \geq 1$. Then

$$\lim_{n\to\infty} \mathscr{E}\{X_n|\mathscr{D}\} = \mathscr{E}\{X|\mathscr{D}\} \text{ a.s. } (P_{\mathscr{D}}). \tag{6.1.10}$$

Proposition 6.1.4 (Conditional Fatou's Lemma). Let $\{X_n\}$ be a sequence of nonnegative random variables such that $\mathscr{E}X_n$ exists for $n \geq 1$ and $\liminf_{n\to\infty} \mathscr{E}X_n < \infty$. Then

$$\mathscr{E}\left\{\liminf_{n\to\infty} X_n|\mathscr{D}\right\} \leq \liminf_{n\to\infty} \mathscr{E}\{X_n|\mathscr{D}\} \text{ a.s. } (P_{\mathscr{D}}). \tag{6.1.11}$$

Proof. The proofs of Propositions 6.1.3 and 6.1.4 can be completed along the lines of the proof of Proposition 6.1.2, using, respectively, the Lebesgue dominated convergence theorem and Fatou's lemma. ∎

The following result is quite useful.

Proposition 6.1.5. Let X and Y be two random variables such that $\mathscr{E}XY$ and $\mathscr{E}Y$ exist. Suppose that X is $\mathscr{D}$-measurable. Then

$$\mathscr{E}\{XY|\mathscr{D}\} = X\mathscr{E}\{Y|\mathscr{D}\} \text{ a.s. } (P_{\mathscr{D}}). \tag{6.1.12}$$

Proof. First suppose that $X = \chi_F$, where $F \in \mathscr{D}$. Then for every $E \in \mathscr{D}$

$$\begin{aligned}\int_E \mathscr{E}\{\chi_F Y|\mathscr{D}\}\, dP_{\mathscr{D}} &= \int_E \chi_F Y\, dP_{\mathscr{D}} = \int_{E\cap F} Y\, dP_{\mathscr{D}}\\ &= \int_{E\cap F} \mathscr{E}\{Y|\mathscr{D}\}\, dP_{\mathscr{D}}\\ &= \int_E \chi_F \mathscr{E}\{Y|\mathscr{D}\}\, dP_{\mathscr{D}},\end{aligned}$$

so that (6.1.12) holds. Since every simple $\mathscr{D}$-measurable function is a finite linear combination of indicator functions of $\mathscr{D}$-measurable sets, (6.1.12) holds when X is a $\mathscr{D}$-measurable simple function.

Next suppose that $X \geq 0$, $Y \geq 0$ a.s. Then there exists a nondecreasing sequence $\{X_n\}$ of nonnegative $\mathscr{D}$-measurable simple functions such that $0 \leq X_n \uparrow X$ a.s. (P). Clearly $0 \leq X_n Y \uparrow XY$ a.s. In this case result (6.1.12) follows by using Proposition 6.1.2 and from what has been proved above.

In the general case Proposition 6.1.5 is proved by setting $X = X^+ - X^-$ and $Y = Y^+ - Y^-$ and applying the above result separately to X^+Y^+, X^+Y^-, X^-Y^+, and X^-Y^-. ∎

Corollary 1. Let $\mathscr{D}_1, \mathscr{D}_2$ be two sub-σ-fields of $\mathscr{S}$ such that $\mathscr{D}_1 \subset \mathscr{D}_2$. Let X and Y be random variables such that $\mathscr{E}XY$ and $\mathscr{E}Y$ exist and X is $\mathscr{D}_2$-measurable. Then

$$\mathscr{E}\{XY|\mathscr{D}_1\} = \mathscr{E}\{[X\mathscr{E}\{Y|\mathscr{D}_2\}]|\mathscr{D}_1\} \text{ a.s. } (P_{\mathscr{D}_1}).$$

Proof. Let $P_{\mathscr{D}_1}$ and $P_{\mathscr{D}_2}$ be the restrictions of P to $\mathscr{D}_1$ and $\mathscr{D}_2$, respectively. Then for $E \in \mathscr{D}_1$

$$\begin{aligned}\int_E \mathscr{E}\{[X\mathscr{E}\{Y|\mathscr{D}_2\}]|\mathscr{D}_1\}\, dP_{\mathscr{D}_1} &= \int_E X\mathscr{E}\{Y|\mathscr{D}_2\}\, dP_{\mathscr{D}_2}\\ &= \int_E \mathscr{E}\{XY|\mathscr{D}_2\}\, dP_{\mathscr{D}_2}\\ &= \int_E XY\, dP\\ &= \int_E \mathscr{E}\{XY|\mathscr{D}_1\}\, dP_{\mathscr{D}_1},\end{aligned}$$

and the proof is complete.

Corollary 2. Let $\mathscr{D}_1 \subset \mathscr{D}_2 \subset \mathscr{S}$ be two sub-σ-fields. If $\mathscr{E}Y$ exists, then

$$\mathscr{E}\{Y \mid \mathscr{D}_1\} = \mathscr{E}\{\mathscr{E}\{Y \mid \mathscr{D}_1\} \mid \mathscr{D}_2\} = \mathscr{E}\{\mathscr{E}\{Y \mid \mathscr{D}_2\} \mid \mathscr{D}_1\} \text{ a.s. } (P_{\mathscr{D}_1}).$$

Proof. In Corollary 1 note that $\mathscr{E}\{XY \mid \mathscr{D}_1\}$ is $\mathscr{D}_1$-and, hence, $\mathscr{D}_2$-measurable, so that by part (a) of Proposition 6.1.1

$$\mathscr{E}\{\mathscr{E}\{XY \mid \mathscr{D}_1\} \mid \mathscr{D}_2\} = \mathscr{E}\{XY \mid \mathscr{D}_1\} = \mathscr{E}\{[X\mathscr{E}\{Y \mid \mathscr{D}_2\}] \mid \mathscr{D}_1\} \text{ a.s.}$$

Take $X = 1$ a.s. to get the result.

Proposition 6.1.6. Let $\mathscr{D} \subset \mathscr{S}$ be a σ-field, and let X be a random variable such that $\mathscr{E}X$ exists and $\sigma(X)$ and $\mathscr{D}$ are independent. Then

$$\mathscr{E}\{X \mid \mathscr{D}\} = \mathscr{E}X \text{ a.s. } (P_{\mathscr{D}}).$$

Proof. Since $\mathscr{D}$ and $\sigma(X)$ are independent, the random variables χ_E and X are independent for every $E \in \mathscr{D}$. Therefore

$$\int_E \mathscr{E}\{X \mid \mathscr{D}\}\, dP_{\mathscr{D}} = \int_E X\, dP = \int_\Omega \chi_E X\, dP = \mathscr{E}(\chi_E X) = \mathscr{E}\chi_E \mathscr{E}X$$

$$= P(E)\mathscr{E}X = \int_E \mathscr{E}X\, dP,$$

so that $\mathscr{E}\{X \mid \mathscr{D}\} = \mathscr{E}X$ a.s. ■

Remark 6.1.2. In the following we will drop reference to the probability measure with respect to which a relation holds a.s. and simply write "a.s."

From Proposition 6.1.6 it follows that, if X and Y are independent, $\mathscr{E}\{X \mid Y\} = \mathscr{E}X$ a.s. Let X and Z be independent random variables, and let X be integrable. Does it necessarily follow that

$$\mathscr{E}\{X \mid Y, Z\} = \mathscr{E}\{X \mid Y\} \text{ a.s.?}$$

That the answer is no in general (unless both X and Y are independent of Z) is demonstrated in the following example.

Example 6.1.3. Let $\Omega = [0, 1]$, $\mathscr{S} = \mathscr{B}_\Omega$, the Borel σ-field of subsets of Ω, and P be the Lebesgue measure on $[0, 1]$. Let $X(\omega) = 1$ if $\omega \in [0, \frac{1}{2}]$, and $=0$ if $\omega \in (\frac{1}{2}, 1]$; $Y(\omega) = 1$ if $\omega \in [0, \frac{3}{4})$, and $=0$ if $\omega \in [\frac{3}{4}, 1]$; and $Z(\omega) = 1$ if $\omega \in [\frac{1}{4}, \frac{3}{4}]$, and $=0$ if $\omega \notin [\frac{1}{4}, \frac{3}{4}]$. Then X and Z are independent. However,

$$\mathscr{E}\{X \mid Y\} = \begin{cases} \frac{2}{3} & \text{if } \omega \in [0, \frac{3}{4}), \\ 0 & \text{otherwise,} \end{cases}$$

and

$$\mathscr{E}\{X \mid Y, Z\} = \begin{cases} 0 & \text{if } \omega \in [\frac{3}{4}, 1], \\ \frac{1}{2} & \text{if } \omega \in [\frac{1}{4}, \frac{3}{4}) \\ 1 & \text{if } \omega \in [0, \frac{1}{4}), \end{cases}$$

so that $\mathscr{E}\{X \mid Y\} \neq \mathscr{E}\{X \mid Y, Z\}$.

Proposition 6.1.7. Let X be integrable, and $\mathscr{D}_1$, $\mathscr{D}_2$ be two sub-σ-fields of $\mathscr{S}$ such that $\sigma(X)$ and $\mathscr{D}_1$ are independent of $\mathscr{D}_2$. Then

$$\mathscr{E}\{X \mid \mathscr{D}_1 \vee \mathscr{D}_2\} = \mathscr{E}\{X \mid \mathscr{D}_1\} \text{ a.s.,}$$

where $\mathscr{D}_1 \vee \mathscr{D}_2$ is the σ-field generated by $\mathscr{D}_1 \cup \mathscr{D}_2$.

Proof. Let $\mathscr{D}$ be the class of all sets in $\mathscr{S}$ which are finite disjoint unions of sets of the form $D_1 \cap D_2$, where $D_1 \in \mathscr{D}_1$, $D_2 \in \mathscr{D}_2$. Then $\mathscr{D}$ is a σ-field which contains both $\mathscr{D}_1$ and $\mathscr{D}_2$ and hence also the σ-field $\mathscr{D}_1 \vee \mathscr{D}_2$ generated by $\mathscr{D}_1$ and $\mathscr{D}_2$. Conversely, any σ-field containing $\mathscr{D}_1 \vee \mathscr{D}_2$ clearly contains the class $\mathscr{D}$, so that $\mathscr{D} = \mathscr{D}_1 \vee \mathscr{D}_2$.

The random variable $\mathscr{E}\{X \mid \mathscr{D}_1\}$ is $\mathscr{D}_1$- and, hence, $\mathscr{D}$-measurable, so that it will be sufficient to show that it is a version of $\mathscr{E}\{X \mid \mathscr{D}\}$. In other words, we need only show that for all $D \in \mathscr{D}$

$$\int_D X \, dP = \int_D \mathscr{E}\{X \mid \mathscr{D}_1\} \, dP_{\mathscr{D}}.$$

It suffices to show that for all $D_1 \in \mathscr{D}_1$ and $D_2 \in \mathscr{D}_2$

$$\int_{D_1 \cap D_2} X \, dP = \int_{D_1 \cap D_2} \mathscr{E}\{X \mid \mathscr{D}_1\} \, dP_{\mathscr{D}}.$$

We have for $D_1 \in \mathscr{D}_1$, $D_2 \in \mathscr{D}_2$

$$\begin{aligned} \int_{D_1 \cap D_2} X \, dP &= \int_\Omega X \chi_{D_1 \cap D_2} \, dP \\ &= \int_\Omega X \chi_{D_1} \chi_{D_2} \, dP \\ &= \mathscr{E}(X \chi_{D_1}) \mathscr{E}(\chi_{D_2}) \end{aligned}$$

in view of the fact that $\sigma(X)$ and $\mathscr{D}_1$ are independent of $\mathscr{D}_2$. Therefore, in view of Proposition 6.1.5 and (6.1.9), we have

$$\begin{aligned}\int_{D_1 \cap D_2} X\, dP &= \mathscr{E}(\chi_{D_2}) \int_\Omega \mathscr{E}\{X\chi_{D_1} \mid \mathscr{D}_1\}\, dP_{\mathscr{D}_1} \\ &= \mathscr{E}(\chi_{D_2}) \int_\Omega \mathscr{E}\{X\chi_{D_1} \mid \mathscr{D}_1\}\, dP_{\mathscr{D}} \\ &= \mathscr{E}(\chi_{D_2}) \int_\Omega \chi_{D_1}\mathscr{E}\{X \mid \mathscr{D}_1\}\, dP_{\mathscr{D}}.\end{aligned}$$

Now note that $\chi_{D_1}\mathscr{E}\{X \mid \mathscr{D}_1\}$ is $\mathscr{D}_1$-measurable and χ_{D_2} is $\mathscr{D}_2$-measurable, so that the inequality

$$\begin{aligned}\int_{D_1 \cap D_2} X\, dP &= \int_\Omega \chi_{D_2}\chi_{D_1}\mathscr{E}\{X \mid \mathscr{D}_1\}\, dP_{\mathscr{D}} \\ &= \int_{D_1 \cap D_2} \mathscr{E}\{X \mid \mathscr{D}_1\}\, dP_{\mathscr{D}}.\end{aligned}$$

This completes the proof. ■

Finally we derive an important inequality. Let g on $\mathbb{R}$ be a (continuous) convex function. Then, for $0 < \alpha < 1$ and all $x \in \mathbb{R}$, $y \in \mathbb{R}$, we have

$$g(\alpha x + (1 - \alpha)y) \leq \alpha g(x) + (1 - \alpha)g(y).$$

First we consider the case where $x < y$. Set $\alpha x + (1 - \alpha)y = z$. Then $x < z < y$ and

$$g(z) \leq \frac{z - y}{x - y}\, g(x) - \frac{z - x}{x - y}\, g(y),$$

so that the inequality

$$\frac{g(z) - g(y)}{z - y} \geq \frac{g(x) - g(y)}{x - y}$$

holds. Let

$$h(x) = \frac{g(x) - g(y)}{x - y}.$$

Then, for fixed y, $h(x) \leq h(z)$ for $x < z$, and h is a nondecreasing function of x. It follows that the limit $k(y) = \lim_{x \uparrow y} h(x)$ exists and is finite. Moreover,

$$g(x) - g(y) \geq k(y)(x - y) \qquad \text{for all } x \in \mathbb{R},\ y \in \mathbb{R},\ x < y. \tag{6.1.13}$$

Clearly k is nondecreasing on $\mathbb{R}$. Proceeding in a similar manner, we see easily that (6.1.13) also holds when $x > y$, so that (6.1.13) holds for $x \in \mathbb{R}$, $y \in \mathbb{R}$, $x \neq y$.

Proposition 6.1.8 (Jensen's Inequality). Let X be a random variable, defined on $(\Omega, \mathscr{S}, P)$, such that $\mathscr{E}X$ exists. Let g be a (continuous) convex function on $\mathbb{R}$ such that $\mathscr{E}g(X)$ exists. Let $\mathscr{D} \subset \mathscr{S}$ be a σ-field. Then

$$g(\mathscr{E}\{X \mid \mathscr{D}\}) \leq \mathscr{E}\{g(X) \mid \mathscr{D}\} \text{ a.s.} \tag{6.1.14}$$

Proof. In (6.1.13) replace x by X and y by $\mathscr{E}\{X \mid \mathscr{D}\}$. We have a.s.

$$g(X) - g(\mathscr{E}\{X \mid \mathscr{D}\}) \geq k(\mathscr{E}\{X \mid \mathscr{D}\})(X - \mathscr{E}\{X \mid \mathscr{D}\}).$$

Since g is continuous and k is nondecreasing, both $g(\mathscr{E}\{X \mid \mathscr{D}\})$ and $k(\mathscr{E}\{X \mid \mathscr{D}\})$ are $\mathscr{D}$-measurable functions. It follows that

$$\begin{aligned}
\mathscr{E}\{g(X) \mid \mathscr{D}\} - g(\mathscr{E}\{X \mid \mathscr{D}\}) &= \mathscr{E}\{[g(X) - g(\mathscr{E}\{X \mid \mathscr{D}\})] \mid \mathscr{D}\} \\
&\geq \mathscr{E}\{k(\mathscr{E}\{X \mid \mathscr{D}\})(X - \mathscr{E}\{X \mid \mathscr{D}\}) \mid \mathscr{D}\} \\
&= k(\mathscr{E}\{X \mid \mathscr{D}\})\mathscr{E}\{(X - \mathscr{E}\{X \mid \mathscr{D}\}) \mid \mathscr{D}\} \\
&= 0 \text{ a.s.}
\end{aligned}$$

■

Corollary 1. If g is convex and $\mathscr{E}X$ and $\mathscr{E}g(X)$ both exist, then

$$g(\mathscr{E}X) \leq \mathscr{E}g(X).$$

Proof. Take $\mathscr{D} = \{\varnothing, \Omega\}$. Then X and $\mathscr{D}$ are independent, and $\mathscr{E}\{X \mid \mathscr{D}\} = \mathscr{E}X$ a.s., $\mathscr{E}\{g(X) \mid \mathscr{D}\} = \mathscr{E}g(X)$ a.s.

Corollary 2. Let $p \geq 1$, and suppose that $\mathscr{E}|X|^p < \infty$. Then

$$|\mathscr{E}\{X \mid \mathscr{D}\}|^p \leq \mathscr{E}\{|X|^p \mid \mathscr{D}\} \text{ a.s.}$$

Proof. Take $g(x) = |x|^p$.

Corollary 3. If $\mathscr{E}X$ exists, then

$$[\mathscr{E}\{X \mid \mathscr{D}\}]^+ \leq \mathscr{E}\{X^+ \mid \mathscr{D}\} \text{ a.s.}$$

and

$$[\mathscr{E}\{X \mid \mathscr{D}\}]^- \leq \mathscr{E}\{X^- \mid \mathscr{D}\} \text{ a.s.}$$

Example 6.1.4. Let $X_1, X_2, \ldots, X_n$ be independent, identically distributed random variables. Set $S_n = \sum_{k=1}^n X_k$. Let us compute $\mathscr{E}\{X_j|S_n\}$, $j = 1$,

$2, \ldots, n$. Let $B \in \mathscr{B}$, and F be the common distribution function of $X_1, \ldots, X_n$. Then

$$\int_{\{S_n \in B\}} X_j \, dP = \mathscr{E}(X_j \chi_{\{S_n \in B\}})$$

$$= \int_{-\infty}^{\infty} \cdots \int_{-\infty}^{\infty} x_j \chi_B(x_1 + \cdots + x_j + \cdots + x_n) \prod_{i=1}^{n} dF(x_i)$$

$$= \int_{-\infty}^{\infty} \cdots \int_{-\infty}^{\infty} x_k \chi_B(x_1 + \cdots + x_n) \prod_{i=1}^{n} dF(x_i) = \int_{\{S_n \in B\}} x_k \, dP$$

by Fubini's theorem, so that

$$\mathscr{E}\{X_j | S_n\} = \mathscr{E}\{X_k | S_n\} \text{ a.s.,} \qquad k = 1, 2, \ldots, n.$$

Hence

$$\mathscr{E}\{X_j | S_n\} = \frac{1}{n} \sum_{k=1}^{n} \mathscr{E}\{X_k | S_n\} = \mathscr{E}\left\{\frac{S_n}{n} \middle| S_n\right\}$$

$$= \frac{S_n}{n} \text{ a.s.,} \qquad j = 1, 2, \ldots, n.$$

Example 6.1.5. Let $X_1, X_2, \ldots$ be a sequence of independent, identically distributed random variables, and write $S_n = \sum_{k=1}^{n} X_k$, $n = 1, 2, \ldots$. First note that $\sigma(S_n, S_{n+1}, S_{n+2}, \ldots) = \sigma(S_n, X_{n+1}, X_{n+2}, \ldots)$. In fact, since $X_{n+k} = S_{n+k} - S_{n+k-1}, S_n, X_{n+1}, X_{n+2}, \ldots$ are each $\sigma(S_n, S_{n+1}, S_{n+2}, \ldots)$-measurable, so that $\sigma(S_n, X_{n+1}, X_{n+2}, \ldots) \subset \sigma(S_n, S_{n+1}, S_{n+2}, \ldots)$. On the other hand, $S_{n+k} = S_n + \sum_{j=1}^{k} X_{n+j}$, so that $S_n, S_{n+1}, \ldots$ are each $\sigma(S_n, X_{n+1}, X_{n+2}, \ldots)$-measurable. Hence

$$\sigma(S_n, S_{n+1}, S_{n+2}, \ldots) \subset \sigma(S_n, X_{n+1}, X_{n+2}, \ldots).$$

Now

$$\mathscr{E}\{S_{n-1} | S_n, S_{n+1}, S_{n+2}, \ldots\} = \mathscr{E}\{S_{n-1} | S_n, X_{n+1}, X_{n+2}, \ldots\}.$$

Since (S_{n-1}, S_n) is independent of $(X_{n+1}, X_{n+2}, \ldots)$, it follows from Proposition 6.1.7 that

$$\mathscr{E}\{S_{n-1} | S_n, X_{n+1}, X_{n+2}, \ldots\} = \mathscr{E}\{S_{n-1} | S_n\}$$

$$= \sum_{i=1}^{n-1} \mathscr{E}\{X_i | S_n\}$$

$$= \frac{n-1}{n} S_n \text{ a.s.} \qquad (n \geq 2).$$

6.1.3 Conditional Probability

Let $(\Omega, \mathscr{S}, P)$ be a probability space, and let $\mathscr{D} \subset \mathscr{S}$ be a sub-σ-field. Let $A \in \mathscr{S}$. Then the indicator function χ_A is an $\mathscr{S}$-measurable simple random variable and $\mathscr{E}\{\chi_A | \mathscr{D}\}$ is well defined.

Definition 6.1.2. The conditional probability $P\{A | \mathscr{D}\}$ of an event $A \in \mathscr{S}$, given $\mathscr{D}$, is defined by

$$P\{A | \mathscr{D}\} = \mathscr{E}\{\chi_A | \mathscr{D}\}.$$

Clearly, for every $E \in \mathscr{D}$

$$\int_E P\{A | \mathscr{D}\}\, dP_{\mathscr{D}} = \int_E \chi_A\, dP = P(A \cap E). \tag{6.1.15}$$

We emphasize again that $P\{A | \mathscr{D}\}$ is a $\mathscr{D}$-measurable function which is determined uniquely only up to $\mathscr{D}$-measurable null sets. The class of all $\mathscr{D}$-measurable functions satisfying (6.1.15) is an equivalence class under the relation of a.s. equality, and any member of this class is known as a version of the conditional probability of A, given $\mathscr{D}$.

It is easy to check that:

(a) $0 \le P\{A | \mathscr{D}\} \le 1$ a.s. and $P\{\Omega | \mathscr{D}\} = 1$ a.s.
(b) If $A_1, A_2, \ldots$ are disjoint sets in $\mathscr{S}$, then

$$P\left\{\bigcup_{n=1}^{\infty} A_n \Big| \mathscr{D}\right\} = \sum_{n=1}^{\infty} P\{A_n | \mathscr{D}\} \text{ a.s.}$$

The results in (a) and (b) do not imply, however, that the set function $A \to P\{A | \mathscr{D}\}$ for $A \in \mathscr{S}$ is a probability measure on $\mathscr{S}$. The difficulty is that the exceptional (null) set depends on A, so that in (b) the exceptional null set depends on the sequence $\{A_n\}$. The union of these generally uncountably many null sets is usually not a null set. It need not even be in $\mathscr{S}$. This observation leads us to the following definition.

Definition 6.1.3. A function $\hat{P}$ defined on $\Omega \times \mathscr{S}$ is called a regular conditional probability function, given $\mathscr{D}$, if it satisfies the following conditions:

(i) For every fixed $\omega \in \Omega$, the set function $\hat{P}(\omega, \cdot)$ defined on $\mathscr{S}$ is a probability measure on $\mathscr{S}$.
(ii) For every fixed $A \in \mathscr{S}$, the function $\hat{P}(\cdot, A)$ is a $\mathscr{D}$-measurable function on Ω.
(iii) For every $A \in \mathscr{S}$ and $E \in \mathscr{D}$ the relation

$$\int_E \hat{P}(\omega, A)\, dP_{\mathscr{D}}(\omega) = P(A \cap E) \tag{6.1.16}$$

holds.

Remark 6.1.3. In view of Definition 6.1.2, $\hat{P}(\cdot, A) = P\{A|\mathscr{D}\}$ a.s.

Example 6.1.6. We show that a regular conditional probability function, given a σ-field, does not always exist. Let $\Omega = [0, 1]$, λ be the Lebesgue measure on Ω, and $\mathscr{D}$ be the Borel σ-field on Ω. Let $A \subset \Omega$ be a nonmeasurable set of inner Lebesgue measure 0 and outer Lebesgue measure 1, that is,

$$\sup_{B \in \mathscr{D}} \{\lambda(B): B \subset A\} = 0$$

and

$$\inf_{B \in \mathscr{D}} \{\lambda(B): B \supset A\} = 1.$$

Such a set A can be easily constructed. Let $\mathfrak{A}$ be the σ-field generated by A and $\mathscr{D}$, that is, $\mathfrak{A}$ is the class of sets of the form $(B_1 \cap A) \cup (B_2 \cap A^c)$, where $B_1, B_2 \in \mathscr{D}$. Define P on $\mathfrak{A}$ by

$$P\{(B_1 \cap A) \cup (B_2 \cap A^c)\} = \tfrac{1}{2}[\lambda(B_1) + \lambda(B_2)].$$

It is easily checked (see Problem 6.6.14) that P is well defined and defines a probability measure on $\mathfrak{A}$. For every $B \in \mathscr{D}$

$$P\{(B \cap A) \cup (B \cap A^c)\} = P(B) = \lambda(B),$$

so that P coincides with λ on $\mathscr{D}$. Moreover,

$$P(A) = P\{(\Omega \cap A) \cup (\varnothing \cap A^c)\} = \tfrac{1}{2}.$$

We show that there does not exist any regular conditional probability function on $\Omega \times \mathfrak{A}$, given $\mathscr{D}$.

Suppose there exists such a function $\hat{P}$ on $\mathfrak{A}$. Then, for every $B \in \mathscr{D}$,

$$\int_B \hat{P}(\omega, A)\, dP_{\mathscr{D}}(\omega) = P(A \cap B)$$

$$= \tfrac{1}{2}\lambda(B) = \tfrac{1}{2}P(B) = \int_B \tfrac{1}{2}\, dP,$$

so that $\hat{P}(\omega, A) = P\{A|\mathscr{D}\} = \frac{1}{2}$ a.s. Similarly $\hat{P}(\omega, A^c) = \frac{1}{2}$ a.s.

Next suppose that $B \in \mathscr{D}$. Then for every $D \in \mathscr{D}$

$$\int_D \hat{P}(\omega, B)\, dP_{\mathscr{D}}(\omega) = P(B \cap D) = \int_D \chi_B(\omega)\, dP(\omega),$$

so that $P\{B|\mathscr{D}\} = \hat{P}(\omega, B) = \chi_B(\omega)$ a.s.

Let $\omega \in \Omega$ be fixed. Clearly $\{\omega\} \in \mathscr{D}$, so that

$$\hat{P}(\omega, \{\omega\}) = \chi_{\{\omega\}}(\omega) = 1.$$

But if $\omega \in A$, then

$$\hat{P}(\omega, \{\omega\}) \le \hat{P}(\omega, A) = \tfrac{1}{2};$$

and if $\omega \in A^c$, then

$$\hat{P}(\omega, \{\omega\}) \le \hat{P}(\omega, A^c) = \tfrac{1}{2}.$$

This contradiction shows that there does not exist any regular conditional probability function on $\Omega \times \mathfrak{A}$, given $\mathscr{D}$.

Proposition 6.1.9. Let $(\Omega, \mathscr{S}, P)$ be a probability space, and $\mathscr{D} \subset \mathscr{S}$ be a sub-σ-field. Let X be a random variable on $(\Omega, \mathscr{S}, P)$ such that $\mathscr{E}X$ exists. Suppose that there exists a regular conditional probability function $\hat{P}$ defined on $\Omega \times \mathscr{S}$. Then

$$\mathscr{E}\{X \mid \mathscr{D}\} = \int_\Omega X(\omega')\hat{P}(\omega, d\omega') \text{ a.s.} \tag{6.1.17}$$

Proof. First let $X = \chi_A$ with $A \in \mathscr{S}$. Then we have for $\omega \in \Omega$

$$\begin{aligned}\mathscr{E}\{\chi_A \mid \mathscr{D}\} &= P\{A \mid \mathscr{D}\} = \hat{P}(\omega, A)\\ &= \int_\Omega \chi_A(\omega')\hat{P}(\omega, d\omega') \text{ a.s.,}\end{aligned}$$

so that (6.1.17) holds. The proof for the general case can be carried out in the usual manner. ■

Remark 6.1.4. In Example 6.1.6 we saw that a regular conditional probability function does not always exist. By passing over to the space of values of a random variable and using the structure of the real line, we can overcome this difficulty.

6.1.4 Conditional Probability Distribution

Let $(\Omega, \mathscr{S}, P)$ be a probability space, and let $\mathscr{D} \subset \mathscr{S}$ be a σ-field. Let X be a random variable defined on Ω.

Definition 6.1.4. A function $\hat{P}_X$ defined on $\Omega \times \mathscr{B}$ is said to be a regular conditional probability distribution of X, given $\mathscr{D}$, if it satisfies the following conditions:

(i) For every fixed $\omega \in \Omega$, the set function $\hat{P}_X(\omega, \cdot)$ defined on $\mathscr{B}$ is a probability measure.
(ii) For every fixed $B \in \mathscr{B}$, the function $\hat{P}_X(\cdot, B)$ is a $\mathscr{D}$-measurable function on Ω.

(iii) For every $B \in \mathscr{B}$ and $E \in \mathscr{D}$, the relation

$$(6.1.18) \quad \int_E \hat{P}_X(\omega, X^{-1}(B))\, dP_{\mathscr{D}}(\omega) = P(E \cap X^{-1}(B))$$

holds.

Remark 6.1.5. In particular, if a regular conditional probability function $\hat{P}$, given $\mathscr{D}$, exists, we have

$$\hat{P}_X(\omega, B) = \hat{P}(\omega, X^{-1}(B)) \text{ a.s.}$$

for all $B \in \mathscr{B}$, $\omega \in \Omega$.

Definition 6.1.5. With the same notation as in Definition 6.1.4 set

$$(6.1.19) \quad F_X\{x|\mathscr{D}\} = F_X\{x|\mathscr{D}\}(\omega) = \hat{P}_X(\omega, (-\infty, x]) \text{ a.s.}, \qquad x \in \mathbb{R}.$$

The function F_X defined on $\mathbb{R} \times \Omega$ by (6.1.19) is called a conditional distribution function of X, given $\mathscr{D}$.

Remark 6.1.6. A function F_X defined on $\mathbb{R} \times \Omega$ is a conditional distribution function of X, given $\mathscr{D}$, if it satisfies the following conditions:

(i) For every fixed $\omega \in \Omega$, the function $F_X\{\cdot|\mathscr{D}\}$ is a distribution function on $\mathbb{R}$.
(ii) For every fixed $x \in \mathbb{R}$, the function $F_X\{x|\mathscr{D}\}(\cdot)$ is a $\mathscr{D}$-measurable function on Ω.
(iii) For every $x \in \mathbb{R}$ and $E \in \mathscr{D}$, the relation

$$\int_E F_X\{x|\mathscr{D}\}(\omega)\, dP_{\mathscr{D}}(\omega) = P(E \cap X^{-1}(-\infty, x])$$

holds.

We now prove the existence of a regular conditional probability distribution of a random variable X, given $\mathscr{D}$.

Proposition 6.1.10. Let $(\Omega, \mathscr{S}, P)$ be a probability space, and $\mathscr{D} \subset \mathscr{S}$ be a σ-field. Let X be a random variable defined on $(\Omega, \mathscr{S}, P)$. Then a regular conditional distribution $\hat{P}_X$ of X, given $\mathscr{D}$, always exists.

Proof. We first prove the existence of a conditional distribution function F_X of X, given $\mathscr{D}$. Let $\mathbb{Q} = \{r_n\}$ be the set of all rational numbers. For each $n \geq 1$, $\{X \leq r_n\} \in \mathscr{S}$, so that the conditional probability $P\{X \leq r_n|\mathscr{D}\}$ is well defined. For each fixed $r \in \mathbb{Q}$ select a version of $P\{X \le r|\mathscr{D}\}$. For $m, n \ge 1$ set

$$A_{m,n} = \{\omega: P\{X \le r_n|\mathscr{D}\} < P\{X \le r_m|\mathscr{D}\}\} \qquad \text{and} \qquad A = \bigcup_{r_n > r_m} A_{m,n}.$$

Since $r_m < r_n$ implies $P\{X \le r_m|\mathscr{D}\} \le P\{X \le r_n|\mathscr{D}\}$ a.s., it follows that $P(A) = 0$. Next set

$$B_n = \left\{\omega: \lim_{\substack{r_l \downarrow r_n \\ r_l > r_n}} P\{X \le r_l|\mathscr{D}\} \neq P\{X \le r_n|\mathscr{D}\}\right\} \qquad \text{and} \qquad B = \bigcup_{n=1}^{\infty} B_n.$$

In view of the conditional monotone convergence theorem we see that $P(B_n) = 0$ for every $n \ge 1$, so that $P(B) = 0$. Similarly, letting $r_n \uparrow \infty$ or $r_n \downarrow -\infty$, we see that

$$P\left\{\omega: \lim_{r_n \uparrow \infty} P\{X \le r_n|\mathscr{D}\} \neq 1\right\} = 0$$

and

$$P\left\{\omega: \lim_{r_n \downarrow -\infty} P\{X \le r_n|\mathscr{D}\} \neq 0\right\} = 0,$$

and let C be the union of these two null sets. Clearly $P(C) = 0$. Hence $P(A \cup B \cup C) = 0$.

It follows immediately that for every $\omega \in (A \cup B \cup C)^c$ the function $x \to P\{X \le x|\mathscr{D}\} = \lim_{r_n \downarrow x} P\{X \le r_n|\mathscr{D}\}$ is a nondecreasing, right-continuous function on $\mathbb{R}$ which tends to the limit 0 or 1 according as $x \to -\infty$ or $x \to \infty$. Let G be an arbitrary distribution function on $\mathbb{R}$, and define the function F_X on $\mathbb{R} \times \Omega$ by

$$(6.1.20) \qquad F_X\{x|\mathscr{D}\} = \begin{cases} G(x) & \text{for } \omega \in A \cup B \cup C, \\ \lim_{r_n \downarrow x} P\{X \le r_n|\mathscr{D}\} & \text{otherwise.} \end{cases}$$

We see easily that $F_X\{\cdot|\mathscr{D}\}$ satisfies conditions (i) through (iii) of Remark 6.1.6, so that it is a conditional distribution function of X, given $\mathscr{D}$. This proves the existence of a conditional distribution function of X, given $\mathscr{D}$.

Finally we construct a conditional probability distribution $\hat{P}_X$ of X, given $\mathscr{D}$, associated with the conditional distribution function $F_X\{\cdot|\mathscr{D}\}$. For every $\omega \in \Omega$ we define $P_X(\omega, \cdot)$ as the probability measure on $(\mathbb{R}, \mathscr{B})$ associated with the distribution function $F_X(\cdot|\mathscr{D})(\omega)$. To show that this $\hat{P}_X$ is a conditional probability distribution it is sufficient to show that $\hat{P}_X$ satisfies (6.1.18). For this purpose let $\mathscr{C}$ be the class of all sets $C \in \mathscr{B}$ such that for every $E \in \mathscr{D}$ the relation

$$\int_E \hat{P}_X(\omega, X^{-1}(C))\, dP_{\mathscr{D}}(\omega) = P(E \cap X^{-1}(C))$$

holds. Then we see easily that $\mathscr{C}$ contains all finite disjoint unions of left open and right closed intervals and, moreover, $\mathscr{C}$ is a monotone class of subsets of $\mathbb{R}$. Hence $\mathscr{C}$ is a σ-field containing $\mathscr{B}$ so that $\mathscr{C} = \mathscr{B}$. This completes the proof of Proposition 6.1.10. ∎

Proposition 6.1.11. Let X be a random variable defined on $(\Omega, \mathscr{S}, P)$. Let g be a Borel-measurable function on $\mathbb{R}$ such that $\mathscr{E}|g(X)| < \infty$. If $\hat{P}_X$ is a regular conditional probability distribution of X, given $\mathscr{D} \subset \mathscr{S}$, then

$$\mathscr{E}\{g(X)|\mathscr{D}\} = \int g(x)\hat{P}_X(\omega, dx) \text{ a.s.}$$

In practical applications the sub-σ-field $\mathscr{D}$ is generated by some random variable Y (or by a collection of random variables). It is therefore convenient to define and work with a regular conditional probability distribution of X, given $Y = y$. This is done in an analogous way as follows.

Let X and Y be random variables, and assume that $\mathscr{E}|X| < \infty$. We have defined $\mathscr{E}\{X|Y\} = \mathscr{E}\{X|\sigma(Y)\}$. For every $B \in \mathscr{B}$ define

$$Q(B) = \int_{Y^{-1}(B)} X\, dP.$$

Then $Q \ll P_Y$, where P_Y is the probability distribution on $\mathscr{B}$ induced by Y. In view of the Randon–Nikodym theorem there exists a Borel-measurable function, say $\mathscr{E}\{X|Y = y\}$, defined for each $y \in \mathbb{R}$ such that

$$\int_B \mathscr{E}\{X|Y = y\}\, dP_Y(y) = \int_{Y^{-1}(B)} X\, dP \tag{6.1.21}$$

for every $B \in \mathscr{B}$. We call $\mathscr{E}\{X|Y = y\}$ the conditional expectation of X, given $Y = y$. Clearly $\mathscr{E}\{X|Y = y\}$ is determined uniquely up to sets of P_Y-measure 0. It is easy to show that the function $\varphi(y) = \mathscr{E}\{X|Y = y\}$ has properties similar to those of $\mathscr{E}\{X|\mathscr{D}\}$. Moreover, $\mathscr{E}\{X|Y\}$ is the composition of φ, and Y: $\varphi \circ Y = \mathscr{E}\{X|Y\}$ a.s.

Next we define the conditional probability of an event $A \in \mathscr{S}$, given $Y = y$, by the relation

$$P\{A|Y = y\} = \mathscr{E}\{\chi_A|Y = y\}. \tag{6.1.22}$$

As before, it is easily verified that $P\{A|Y\}$ and $P\{A|Y = y\}$ are related by

$$P\{A|Y\} = \psi \circ Y \text{ a.s.},$$

where $\psi(y) = P\{A|Y = y\}$. We can also define a regular conditional probability distribution of X, given $Y = y$, and a regular conditional distribution function of X, given $Y = y$, and show, as in Proposition 6.1.10, that they always exist. Thus, for every rational number r and all $y \in \mathbb{R}$, let

$$F_{X|Y}(r|y) = P\{X \le r|Y = y\}, \tag{6.1.23}$$

where $P\{X \le r \mid Y = y\}$ is a fixed version of the conditional probability $P\{X^{-1}(-\infty, r] \mid Y = y\}$ defined above. If $x \in \mathbb{R}$, we proceed as in Proposition 6.1.10 and define

$$F_{X|Y}(x \mid y) = \lim_{r_n \downarrow x} P\{X \le r_n \mid Y = y\}, \tag{6.1.24}$$

where $r_n \in \mathbb{Q}$, $n \ge 1$, and so on.

The following results show the usefulness of this construction of the conditional distribution function.

Proposition 6.1.12. Let (X, Y) be a random vector with joint distribution function F. Then

$$F(x, y) = \int_{-\infty}^{y} F_{X|Y}(x \mid u)\, dF_2(u) \qquad \text{for } y \in \mathbb{R}, \tag{6.1.25}$$

where F_2 is the marginal distribution function of Y.

The random variables X and Y are independent if and only if for all $x \in \mathbb{R}$

$$F_{X|Y}(x \mid y) = F_1(x) \text{ a.s. } (P_Y), \tag{6.1.26}$$

where F_1 is the marginal distribution function of X.

Proof. First let $x \in \mathbb{Q}$ and $y \in \mathbb{R}$. Then

$$\begin{aligned} F(x, y) &= P\{X \le x, Y \le y\} \\ &= \int_{(-\infty, y]} P\{X \le x \mid Y = u\}\, dF_2(u) \end{aligned}$$

by (6.1.22) and (6.1.15), and since $x \in \mathbb{Q}$

$$F(x, y) = \int_{-\infty}^{y} F_{X|Y}(x \mid u)\, dF_2(u).$$

Thus (6.1.25) holds for all $x \in \mathbb{Q}$ and all $y \in \mathbb{R}$. To prove it for all $x \in \mathbb{R}$ and all $y \in \mathbb{R}$, we use (6.1.24) and the right continuity of a distribution function. If (6.1.26) holds, then, for all x and y,*

$$F(x, y) = \int_{-\infty}^{y} F_{X|Y}(x \mid u)\, dF_2(u) = \int_{-\infty}^{y} F_1(x)\, dF_2(u) = F_1(x) F_2(y),$$

* Define $F_{X|Y}(x \mid y) = 0$ for the values of $y \in N \subset \mathbb{R}$ where $P_Y(N) = 0$.

so that X and Y are independent. Conversely, if X and Y are independent, then

$$\int_{-\infty}^{y} F_1(x)\,dF_2(u) = F_1(x)F_2(y) = F(x, y) = \int_{-\infty}^{y} F_{X|Y}(x|u)\,dF_2(u),$$

so that $F_1(x) = F_{X|Y}(x|y)$ a.s. (P_Y) (Problem 6.6.16), and (6.1.26) holds. ∎

Finally, we define a regular conditional probability distribution of a random variable X, given $Y = y$. Let $F_{X|Y}(\cdot|y)$ be a regular conditional distribution function of X, given $Y = y$ as defined in (6.1.24). For every $y \in \mathbb{R}$ let $\hat{P}_{X|Y}(\cdot|y)$ be the probability measure on $\mathscr{B}$ associated with $F_{X|Y}(\cdot|y)$. Then $\hat{P}_{X|Y}(\cdot|y)$ is called a regular conditional probability distribution of X, given $Y = y$. It is easy to see that for every $A \in \mathscr{B}$ and $y \in \mathbb{R}$

$$\hat{P}_{X|Y}(A|y) = P\{X^{-1}(A)|\,Y = y\}.$$

Proposition 6.1.13. Let (X, Y) be a random vector with joint probability distribution $P_{X,Y}$, and let $\hat{P}_{X|Y}(\cdot|y)$ be a regular conditional probability distribution of X, given $Y = y$. Then, for every $A \in \mathscr{B}$, $B \in \mathscr{B}$, the relation

$$P_{X,Y}(A \times B) = \int_B \hat{P}_{X|Y}(A|y)\,dP_Y(y) \tag{6.1.27}$$

holds, where P_Y is the marginal probability distribution of Y.

The random variables X and Y are independent if and only if for every $A \in \mathscr{B}$

$$\hat{P}_{X|Y}(A|y) = P_X(A) \text{ a.s. } (P_Y), \tag{6.1.28}$$

where P_X is the marginal probability distribution of X.

Proof. Let $A \in \mathscr{B}$, $B \in \mathscr{B}$. Then

$$\begin{aligned}
P_{X,Y}(A \times B) &= P\{X^{-1}(A) \cap Y^{-1}(B)\} \\
&= \int_{Y^{-1}(B)} \chi_{X^{-1}(A)}\,dP \\
&= \int_B \mathscr{E}\{\chi_{X^{-1}(A)}|\,Y = y\}\,dP_Y(y) \qquad \text{[from (6.1.21)]} \\
&= \int_B P\{X^{-1}(A)|\,Y = y\}\,dP_Y(y) \qquad \text{[from (6.1.22)]} \\
&= \int_B \hat{P}_{X|Y}(A|y)\,dP_Y(y).
\end{aligned}$$

Suppose that (6.1.28) holds for all $A \in \mathscr{B}$ and a.s. (P_Y). Then for $y \in \mathbb{R}$

$$P_{X,Y}(A \times (-\infty, y]) = \int_{(-\infty, Y]} \hat{P}_{X|Y}(A \mid y)\, dP_Y(y)$$

$$= \int_{(-\infty, Y]} P_X(A)\, dP_Y(y) = P_X(A) P_Y(-\infty, y].$$

For $x \in \mathbb{R}$, set $A = (-\infty, x]$. Then

$$P_{X,Y}((-\infty, x] \times (-\infty, y]) = P_X(-\infty, x] P_Y(-\infty, y],$$

that is,

$$F(x, y) = F_1(x) F_2(y)$$

holds for all $x, y \in \mathbb{R}$. Hence X and Y are independent.

Conversely, if X and Y are independent, then for every $A, B \in \mathscr{B}$ the relation

$$\int_B \hat{P}_{X|Y}(A \mid y)\, dP_Y(y) = P_{X,Y}(A \times B)$$

$$= P_X(A) P_Y(B)$$

$$= \int_B P_X(A)\, dP_Y(y)$$

holds. It follows immediately that (6.1.28) holds. ∎

Next we prove the following two propositions.

Proposition 6.1.14. Let X and Y be two random variable on $(\Omega, \mathscr{S}, P)$, and g be a Borel-measurable function on $\mathbb{R}$ such that $\mathscr{E}|g(X)| < \infty$. Let $\hat{P}_{X|Y}(\cdot \mid y)$ be a regular conditional probability distribution of X, given $Y = y$. Then we have

(6.1.29) $$\mathscr{E}\{g(X) \mid Y = y\} = \int_{\mathbb{R}} g(x) \hat{P}_{X|Y}(dx \mid y) \text{ a.s. } (P_Y).$$

Proof. It is sufficient to prove (6.1.29) for indicator functions. Extension to arbitrary g can be carried out in the usual manner.

Let $g = \chi_A$ for $A \in \mathscr{B}$. Then for a.s. (P_Y)

$$\int_{\mathbb{R}} \chi_A(x) \hat{P}_{X|Y}(dx \mid y) = \int_A \hat{P}_{X|Y}(dx \mid y) = \hat{P}_{X|Y}(A \mid y)$$

$$= P\{X^{-1}(A) \mid Y = y\}$$

$$= \mathscr{E}\{\chi_{X^{-1}(A)} \mid Y = y\}$$

$$= \mathscr{E}\{\chi_A(X) \mid Y = y\},$$

which is (6.1.29). ∎

Proposition 6.1.15. Let (X, Y) be a random vector defined on $(\Omega, \mathscr{S}, P)$, and let g be a Borel-measurable function defined on $\mathbb{R}_2$ such that $\mathscr{E}|g(X, Y)| < \infty$. Let $\hat{P}_{X|Y}(\cdot|y)$ be a regular conditional probability distribution of X, given $Y = y$. Then

$$\mathscr{E}\{g(X, Y)|Y = y\} = \int_{\mathbb{R}} g(x, y)\hat{P}_{X|Y}(dx|y) \text{ a.s.} \tag{6.1.30}$$

In particular, if X and Y are independent, then

$$\mathscr{E}\{g(X, Y)|Y = y\} = \mathscr{E}g(X, y) \text{ a.s. } (P_Y).$$

Proof. It is sufficient to prove the result for the case where g is of the form

$$g(x, y) = \chi_A(x)\chi_B(y) \qquad \text{for } A, B \in \mathscr{B}, \quad x, y \in \mathbb{R}.$$

Proceeding exactly as in Proposition 6.1.14, we obtain a.s. (P_Y)

$$\begin{aligned}\int_{\mathbb{R}} \chi_A(x)\chi_B(y)\hat{P}_{X|Y}(dx|y) &= \chi_B(y)\mathscr{E}\{\chi_A(X)|Y = y\}\\ &= \mathscr{E}\{\chi_A(X)\chi_B(Y)|Y = y\},\end{aligned}$$

which is (6.1.30).

The proof of the remaining part can be easily carried out by using (6.1.28). ∎

Example 6.1.7. Let Y be a random variable with probability distribution

$$P\{Y = y_j\} = q_j > 0, \qquad \sum_{j=1}^{\infty} q_j = 1.$$

For $A \in \mathscr{S}$ define

$$\varphi(y_j) = \frac{P(A \cap \{Y = y_j\})}{P\{Y = y_j\}}, \qquad j = 1, 2, \ldots.$$

We show that $\varphi(y_j)$ is a version of $P\{A|Y = y_j\}$. Let $\mathscr{Y} = \{y_1, y_2, \ldots\} \subset \mathbb{R}$, and $\mathfrak{A}$ be the collection of all subsets of $\mathscr{Y}$. For $E \in \mathfrak{A}$

$$\begin{aligned}\int_E \varphi(y)\, dP_Y(y) &= \int_{\mathscr{Y}} \varphi(y)\chi_E(y)\, dP_Y(y)\\ &= \sum_{j=1}^{\infty} \varphi(y_j)\chi_E(y_j)P_Y\{y_j\}\\ &= \sum_{y_j \in E} \varphi(y_j)P\{Y = y_j\}\\ &= \sum_{y_j \in E} P(A \cap \{Y = y_j\})\\ &= P(A \cap \{Y \in E\}).\end{aligned}$$

Next suppose that (X, Y) is a random vector where Y has the probability distribution given above and $\mathscr{E}|X| < \infty$. We show that

$$\psi(y_j) = \frac{1}{P\{Y = y_j\}} \int_{\{Y = y_j\}} X \, dP$$

is a version of $\mathscr{E}\{X \mid Y = y_j\}$. Indeed, for $E \in \mathfrak{A}$ we have

$$\begin{aligned} \int_{\{Y \in E\}} X \, dP &= \sum_{y_j \in E} \int_{\{Y = y_j\}} X \, dP \\ &= \sum_{y_j \in E} P\{Y = y_j\} \psi(y_j) \\ &= \int_E \psi(y) \, dP_Y(y). \end{aligned}$$

If, in particular, X takes only a countable number of values $x_1, x_2, \ldots$, we see easily that

$$P\{X = x_i \mid Y = y_j\} = \frac{P\{X = x_i, Y = y_j\}}{P\{Y = y_j\}}, \qquad i = 1, 2, \ldots,$$

moreover, if $\mathscr{E}|X| < \infty$, then

$$\begin{aligned} \mathscr{E}\{X \mid Y = y_j\} &= \sum_{i=1}^{\infty} x_i \frac{P\{X = x_i, Y = y_j\}}{P\{Y = y_j\}} \\ &= \sum_{i=1}^{\infty} x_i P\{X = x_i \mid Y = y_j\}. \end{aligned}$$

Interchanging the roles of X and Y, and assuming that $P\{X = x_i\} > 0$ for all $i \geq 1$, we see also that

$$P\{Y = y_j \mid X = x_i\} = \frac{P\{X = x_i, Y = y_j\}}{P\{X = x_i\}}, \qquad j = 1, 2, \ldots,$$

and, moreover,

$$P\{Y = y_j \mid X = x_i\} = \frac{P\{Y = y_j\} P\{X = x_i \mid Y = y_j\}}{\sum_{j=1}^{\infty} P\{Y = y_j\} P\{X = x_i \mid Y = y_j\}},$$

which is a special case of the Bayes formula.

Example 6.1.8. Let (X, Y) be a random variable with an absolutely continuous distribution function given by probability density function f. Let f_1 and f_2 be the marginal probability density functions of X and Y, respectively. We show that for any $E \in \mathscr{B}$

$$P\{X \in E \mid Y = y\} = \int_E \frac{f(x, y)}{f_2(y)} \, dx \text{ a.s. } (P_Y),$$

that is, $\int_E g(x|y)\,dx$, where $g(x|y) = f(x, y)/f_2(y)$, is a version of

$$P\{X \in E | Y = y\}.$$

First note that $g(x|y)$ is defined only for $f_2(y) \neq 0$, but this set has probability 0. In fact, if $A = \{(x, y): f_2(y) = 0\}$, then

$$P\{(X, Y) \in A\} = \iint_A f(x, y)\,dx\,dy = \int_{\{y:\, f_2(y) = 0\}} \left[\int_{-\infty}^{\infty} f(x, y)\,dx\right] dy$$
$$= 0.$$

Clearly $\int_E g(x|y)\,dx$ is Borel-measurable. For $F \in \mathscr{B}$, we have

$$\begin{aligned} P\{X \in E, Y \in F\} &= \int_E \int_F f(x, y)\,dx\,dy \\ &= \int_F \left[\int_E f_2(y) \frac{f(x, y)}{f_2(y)}\,dx\right] dy \\ &= \int_F f_2(y)\left[\int_E g(x|y)\,dx\right] dy \\ &= \int_F \left[\int_E g(x|y)\,dx\right] dP_Y(y), \end{aligned}$$

and it follows that

$$P\{X \in E | Y = y\} = \int_E \frac{f(x, y)}{f_2(y)}\,dx \text{ a.s. } (P_Y).$$

Taking $E = (-\infty, x]$, $x \in \mathbb{R}$, we see that

$$P\{X \leq x | Y = y\} = \int_{-\infty}^{x} \frac{f(x, y)}{f_2(y)}\,dx \text{ a.s. } (P_Y),$$

so that $g(x|y)$ may be interpreted as the conditional probability density of X, given $Y = y$, and $P\{X \leq x | y = y\}$ as the conditional distribution function of X, given $Y = y$.

Interchanging the role of X and Y, putting

$$h(y|x) = \frac{f(x, y)}{f_1(x)} \qquad \text{if } f_1(x) > 0,$$

we see that

$$h(y|x) = \frac{f_2(y)g(x|y)}{f_1(x)} = \frac{f_2(y)g(x|y)}{\int_{-\infty}^{\infty} f_2(y)g(x|y)\,dy},$$

which is a variant of the well-known Bayes formula.

If X and Y are independent, then

$$g(x \mid y) = f_1(x) \qquad \text{and} \qquad h(y \mid x) = f_2(y).$$

6.1.5 Applications

We now consider some simple applications of conditional probability and conditional expectation. More applications will appear in Section 6.4. One of the most important applications of conditional probability distributions is in mathematical statistics. In problems of statistical inference it is frequently possible to obtain a reduction of the data by observing a sufficient statistic which contains all the relevant information regarding the unknown distribution. A statistic T is said to be *sufficient* for the family of distributions $\mathscr{P} = \{P_\theta : \theta \in \Omega\}$ defined on $(\Omega, \mathscr{S})$ if for each $A \in \mathscr{S}$ there exists a version of the conditional probability function $P_\theta\{A \mid T\}$ that is independent of θ. The concept of sufficiency plays a key role, in particular, in the theory of the testing of statistical hypotheses and minimum variance unbiased estimation.

The following result is quite useful in many statistical applications.

Proposition 6.1.16. Let Y be any random variable defined on $(\Omega, \mathscr{S}, P)$ with $\mathscr{E}Y^2 < \infty$, and let $\mathscr{D}$ be a sub-σ-field of $\mathscr{S}$. Then for every $\mathscr{D}$-measurable random variable Z with $\mathscr{E}Z^2 < \infty$ we have

$$\mathscr{E}(Y - Z)^2 \geq \mathscr{E}(Y - \mathscr{E}\{Y \mid \mathscr{D}\})^2 \tag{6.1.31}$$

with equality if and only if $Z = \mathscr{E}\{Y \mid \mathscr{D}\}$ a.s.

Proof. Clearly

$$\begin{aligned}\mathscr{E}(Y - Z)^2 = {} & \mathscr{E}(Y - \mathscr{E}\{Y \mid \mathscr{D}\})^2 + \mathscr{E}(Z - \mathscr{E}\{Y \mid \mathscr{D}\})^2 \\ & - 2\mathscr{E}[(Y - \mathscr{E}\{Y \mid \mathscr{D}\})(Z - \mathscr{E}\{Y \mid \mathscr{D}\})],\end{aligned}$$

and since

$$\mathscr{E}|Z - \mathscr{E}\{Y \mid \mathscr{D}\}| < \infty, \qquad \mathscr{E}|(Y - \mathscr{E}\{Y \mid \mathscr{D}\})(Z - \mathscr{E}\{Y \mid \mathscr{D}\})| < \infty,$$

and $(Z - \mathscr{E}\{Y \mid \mathscr{D}\})$ is $\mathscr{D}$-measurable, it follows that

$$\begin{aligned}&\mathscr{E}\{[(Y - \mathscr{E}\{Y \mid \mathscr{D}\})(Z - \mathscr{E}\{Y \mid \mathscr{D}\})] \mid \mathscr{D}\} \\ &\qquad = (Z - \mathscr{E}\{Y \mid \mathscr{D}\})\mathscr{E}\{(Y - \mathscr{E}\{Y \mid \mathscr{D}\}) \mid \mathscr{D}\} = 0 \text{ a.s.}\end{aligned}$$

Thus

$$\mathscr{E}(Y - Z)^2 = \mathscr{E}(Y - \mathscr{E}\{Y \mid \mathscr{D}\})^2 + \mathscr{E}(Z - \mathscr{E}\{Y \mid \mathscr{D}\})^2, \tag{6.1.32}$$

which proves (6.1.31). The equality in (6.1.31) holds if and only if

$$\mathscr{E}(Z - \mathscr{E}\{Y \mid \mathscr{D}\})^2 = 0,$$

that is, if and only if

$$Z = \mathscr{E}\{Y \mid \mathscr{D}\} \text{ a.s.} \qquad \blacksquare$$

Corollary 1. If $\mathscr{E}Y^2 < \infty$, then

$$\text{var}(Y) \geq \text{var}(\mathscr{E}\{Y \mid \mathscr{D}\}) \tag{6.1.33}$$

with equality if and only if $Y = \mathscr{E}\{Y \mid \mathscr{D}\}$ a.s.

Proof. For the proof we take $Z = \mathscr{E}Y$ in (6.1.32) and note that $\mathscr{E}Y = \mathscr{E}\{\mathscr{E}\{Y \mid \mathscr{D}\}\}$. It follows that

$$\begin{aligned} \text{var}(Y) &= \mathscr{E}(Y - \mathscr{E}\{Y \mid \mathscr{D}\})^2 + \mathscr{E}(\mathscr{E}Y - \mathscr{E}\{Y \mid \mathscr{D}\})^2 \\ &= \mathscr{E}(\text{var}\{Y \mid \mathscr{D}\}) + \text{var}(\mathscr{E}\{Y \mid \mathscr{D}\})^2, \end{aligned}$$

which yields (6.1.33) immediately. Equality in (6.1.33) holds if and only if

$$\mathscr{E}(\text{var}\{Y \mid \mathscr{D}\}) = \mathscr{E}(Y - \mathscr{E}\{Y \mid \mathscr{D}\})^2 = 0.$$

In particular, let $\mathbf{X} = (X_1, \ldots, X_n)$ be defined on $(\Omega, \mathscr{S})$, and let $\mathscr{P} = \{P_\theta : \theta \in \Theta\}$ be a family of probability measures on $\mathscr{S}$. Let $T = T(X_1, \ldots, X_n)$ be sufficient for θ (or for the family $\mathscr{P}$). The result in (6.1.33) can now be interpreted as follows. If Y is an unbiased estimate of θ with finite variance, then $\mathscr{E}\{Y \mid T\}$ is also an unbiased estimate and has smaller variance. In other words, if we are looking for an unbiased estimate of θ with minimum variance, we can restrict our search to estimates which depend on $\mathbf{X}$ only through T.

Proposition 6.1.16 is also useful in prediction. Frequently it is of interest to approximate Y by a $\mathscr{D}$-measurable random variable Z with $\mathscr{E}Z^2 < \infty$. Proposition 6.1.16 says that $Z = \mathscr{E}\{Y \mid \mathscr{D}\}$ a.s. best approximates Y in the sense of least squares (or mean square error), that is, $\mathscr{E}(Y - Z)^2$ is minimal if and only if

$$Z = \mathscr{E}\{Y \mid \mathscr{D}\} \text{ a.s.}$$

In particular, if

$$\mathbf{X} = (X_1, \ldots, X_n) \quad \text{and} \quad \mathscr{D} = \sigma(X_1, \ldots, X_n),$$

then $\mathscr{E}\{Y \mid X_1, \ldots, X_n\}$ is the best predictor of Y in the sense of least squares among all measurable functions $g: \mathbb{R}_n \to \mathbb{R}$.

Another application of Proposition 6.1.16 is in Bayesian estimation. Consider the problem of estimating some function $d: \Theta \to \mathbb{R}$ of the parameter

θ, $\theta \in \Theta$. In a Bayesian model θ itself is considered to be a realization of a random variable, say Θ, and P_θ is then interpreted as the conditional distribution of $\mathbf{X} = (X_1, \ldots, X_n)$, given $\Theta = \theta$. In this framework, if we want to find an estimate $\delta(X_1, \ldots, X_n)$ of d such that the risk $\mathscr{E}[\delta(\mathbf{X}) - d(\Theta)]^2$ is minimum, Proposition 6.1.16 immediately leads to the Bayes estimate

$$\delta^*(\mathbf{X}) = \mathscr{E}\{d(\Theta) \mid X_1, \ldots, X_n\}$$

as a solution of the problem. We can compute δ^* if we know the joint distribution of $(\mathbf{X}, d(\Theta))$.

The problem of the testing of hypotheses can be briefly stated as follows. Let $(\Omega, \mathscr{S})$ be a probability space, and Θ a set containing at least two elements. For each $\theta \in \Theta$ let Q_θ be a probability measure on $(\Omega, \mathscr{S})$. Let $\Theta_0 \subset \Theta$ and $\Theta_1 = \Theta - \Theta_0$. Let $\mathbf{X} = (X_1, \ldots, X_n)$ be the observation vector, and denote by P_θ the corresponding (induced) distribution of $\mathbf{X}$. The problem is to decide whether $\theta \in \Theta_0$ or $\theta \in \Theta_1$. The Neyman–Pearson approach to this problem consists of selecting a measurable function $\varphi\colon \mathbb{R}_n \to \mathbb{R}$, $0 \le \varphi \le 1$, which maximizes the power

$$\beta_\phi(\theta) = \int \varphi(\mathbf{x})\, dP_\theta(\mathbf{x}) \qquad \text{for all } \theta \in \Theta_1$$

subject to the condition

$$\int \varphi(\mathbf{x})\, dP_\theta(\mathbf{x}) \le \alpha \qquad \text{for all } \theta \in \Theta_0,$$

where α, $0 \le \alpha \le 1$, is a given constant. Let $T = T(\mathbf{X})$ be a sufficient statistic for θ, and φ be a test function satisfying

$$\int \varphi(\mathbf{x})\, dP_\theta(\mathbf{x}) \le \alpha \qquad \text{for all } \theta \in \Theta_0.$$

Then $\psi(t) = \mathscr{E}\{\varphi(\mathbf{X}) \mid t\}$ is also a test function (it is independent of θ), and, moreover,

$$\mathscr{E}_\theta\{\mathscr{E}\{\varphi(\mathbf{X}) \mid T\}\} = \mathscr{E}_\theta \varphi(\mathbf{X}) = \beta_\varphi(\theta),$$

so that ψ has the same power as φ. It is sufficient therefore to restrict attention to tests which are functions of the sufficient statistics T.

For further details we refer the reader to Zacks [92] and Lehmann [52]. Amongst other applications we mention conditional entropy in information theory (see Billingsley [7]) and characterizations of distributions (see Kagan, Linnik, and Rao [42]).

6.2 MARTINGALES

We now consider a special case of dependence which has its origin in gambling.

6.2.1 Definitions and Elementary Properties

Let $(\Omega, \mathscr{S}, P)$ be a probability space, and let $(T, \prec)$ be a partially ordered set. Let $\{\mathscr{D}_t, t \in T\}$ be a collection of sub-σ-fields of $\mathscr{S}$ such that $\mathscr{D}_s \subset \mathscr{D}_t$ for $s \prec t$, $s, t \in T$. Let $\mathfrak{X} = \{X_t, t \in T\}$ be a collection of random variables defined on Ω, each having a finite expectation.

Definition 6.2.1. The class $\mathfrak{X}$ is said to constitute a martingale with respect to $\{\mathscr{D}_t, t \in T\}$ if the following conditions hold:

(i) For every $t \in T$, X_t is $\mathscr{D}_t$-measurable.
(ii) For $s, t \in T$, $s \prec t$, the relation

$$\mathscr{E}\{X_t \mid \mathscr{D}_s\} = X_s \text{ a.s.} \tag{6.2.1}$$

holds.

The class $\mathfrak{X}$ is said to be a submartingale (respectively, supermartingale) if in (6.2.1) we replace the $=$ sign by the $\geq$ (respectively, $\leq$) sign.

Clearly, changing X_t into $-X_t$ interchanges "submartingale" and "supermartingale."

In the following, if $\{X_t\}$ is a martingale with respect to $\{\mathscr{D}_t\}$ (in the sense of Definition 6.2.1), we will say that $\{X_t, \mathscr{D}_t : t \in T\}$ or, simply, that $\{X_t, \mathscr{D}_t\}$ is a martingale. We will restrict attention mainly to the discrete parameter case where $T = \{1, 2, \ldots\}$ is the set of natural numbers. In particular, if $\mathscr{D}_n = \sigma(X_1, \ldots, X_n)$ for $n \geq 1$, we will simply say that $\{X_n\}$ is a martingale.

Proposition 6.2.1. $\{X_n, \mathscr{D}_n\}$ is a martingale (sub- or supermartingale) if and only if for every $n > 1$

$$\mathscr{E}\{X_n \mid \mathscr{D}_{n-1}\} = (\geq \text{ or } \leq)\ X_{n-1} \text{ a.s.} \tag{6.2.2}$$

Proof. Clearly (6.2.1) $\Rightarrow$ (6.2.2). Conversely, for $m < n$, since $\mathscr{D}_m \subset \mathscr{D}_{n-1}$,

$$\begin{aligned}
\mathscr{E}\{X_n \mid \mathscr{D}_m\} &= \mathscr{E}\{\mathscr{E}\{X_n \mid \mathscr{D}_{n-1}\} \mid \mathscr{D}_m\} \\
&\underset{(\geq \text{ or } \leq)}{=} \mathscr{E}\{X_{n-1} \mid \mathscr{D}_m\} \\
&\ \ \vdots \\
&\underset{(\geq \text{ or } \leq)}{=} \mathscr{E}\{X_{m+1} \mid \mathscr{D}_m\} \\
&\underset{(\geq \text{ or } \leq)}{=} X_m \text{ a.s.}
\end{aligned}$$

■

Example 6.2.1. Consider a sequence of independent, identically distributed random variables $\{X_n\}$ with common distribution given by $P\{X_1 = \pm 1\} = \frac{1}{2}$. We can interpret $X_i = 1$ as the event that a gambler who is playing a sequence of games in each of which he (or she) has probability $\frac{1}{2}$ of winning \$1 and probability $\frac{1}{2}$ of losing \$1 wins the ith game. Similarly, $X_i = -1$ is the event that he loses the ith game. If he wagers an amount $\$b_{i-1}$ on the ith game, his expected winnings on the ith game are

$$b_{i-1}P\{X_i = 1\} - b_{i-1}P\{X_i = -1\} = 0.$$

Suppose that the gambler's strategy is to bet $b_0(>0)$ on the first (trial) game and $b_n = b_n(X_1, \ldots, X_n)$ on the $(n + 1)$st game, $n \geq 1$, where b_n is $\mathscr{D}_n$-measurable. Let $S_0(>b_0)$ be his initial fortune, and let

$$S_n = S_n(b_0, b_1, \ldots, b_{n-1}, X_1, \ldots, X_n)$$

be his fortune after n trials. Clearly

$$S_{n+1} = S_0 + \sum_{k=0}^{n} b_k X_{k+1} = S_n + X_{n+1}b_n.$$

Let $\mathscr{D}_n = \sigma(X_1, \ldots, X_n)$. Since $\mathscr{E}X_n = 0$ for $n \geq 1$, we have

$$\begin{aligned}\mathscr{E}\{S_{n+1} | \mathscr{D}_n\} &= S_n + \mathscr{E}\{X_{n+1}b_n(X_1, \ldots, X_n) | \mathscr{D}_n\} \\ &= S_n + b_n\mathscr{E}(X_{n+1}) \\ &= S_n \text{ a.s.,}\end{aligned}$$

so that $\{S_n, \mathscr{D}_n\}$ is a martingale. In particular, if $b_n(X_1, \ldots, X_n) \equiv 1$ for $n \geq 0$ and $S_0 = 0$, then $S_n = \sum_{k=1}^{n} X_k$ for $n \geq 1$.

Consider now the following strategy for the gambler. He doubles his bet until he wins a game and then quits. Clearly the probability that he will win at least one game equals $\sum_{k=1}^{\infty} (\frac{1}{2})^k = 1$, so that he is certain to win at least one game. Let $b_0 = b > 0$ be a constant. Then

$$b_{n-1} = \begin{cases} b2^{n-1} & \text{if } X_i = -1 \quad \text{for } i = 1, 2, \ldots, n-1, \\ 0 & \text{otherwise.} \end{cases}$$

If the gambler wins the first time on the $(n + 1)$st game, he will have lost $\sum_{k=1}^{n} b2^{k-1} = b(2^n - 1)$ on the first n games, and since he (bets and) wins $b2^n$ on the $(n + 1)$st game, the probability is 1 that he will win b. The catch, of course, is that the gambler must have infinite initial fortune.

Finally, suppose that the gambler has the option of skipping individual games. Let $\delta_n = \delta_n(X_1, \ldots, X_{n-1})$ be a $\mathscr{D}_{n-1}$-measurable function taking the values

$$\delta_n = \begin{cases} 0 & \text{if he skips the } n\text{th game,} \\ 1 & \text{if he bets on the } n\text{th game.} \end{cases}$$

If S_n^* is his fortune after n trials, then

$$S_{n+1}^* = S_n^* + \delta_{n+1}(X_1, \ldots, X_n)b_n(X_1, \ldots, X_n)X_{n+1}.$$

It is easily seen that $\mathscr{E}|S_n^*| < \infty$ for all n, so that

$$\mathscr{E}\{S_{n+1}^* | \mathscr{D}_n\} = S_n^* + \delta_{n+1}b_n\mathscr{E}(X_{n+1}) = S_n^* \text{ a.s.,}$$

and $\{S_n^*, \mathscr{D}_n\}$ is a martingale.

Example 6.2.2. Let $\{X_n\}$ be a sequence of independent random variables such that $\mathscr{E}X_n = 0$ for all $n \geq 1$. Let $S_n = \sum_{k=1}^n X_k$, $n \geq 1$. Clearly $\mathscr{E}|S_n| \leq \sum_{k=1}^n \mathscr{E}|X_k| < \infty$ for all n. Then $\{S_n\}$ is a martingale. In fact, for every $n \geq 1$ we have a.s.

$$\begin{aligned}\mathscr{E}\{S_{n+1}|S_1, \ldots, S_n\} &= \mathscr{E}\{(S_n + X_{n+1})|X_1, \ldots, X_n\} \\ &= S_n + \mathscr{E}(X_{n+1}) \\ &= S_n.\end{aligned}$$

Clearly it is sufficient to assume that $\mathscr{E}\{X_{n+1}|X_1, \ldots, X_n\} = 0$ a.s. for all $n \geq 1$, and independence of the X_n is not required. Also, if $\{S_n\}$ is a martingale sequence, then, by setting $X_n = S_n - S_{n-1}$ for $n \geq 1$, $S_0 = 0$, we have

$$\begin{aligned}\mathscr{E}\{X_{n+1}|X_1, \ldots, X_n\} &= \mathscr{E}\{(S_{n+1} - S_n)|X_1, \ldots, X_n\} \\ &= S_n - S_n = 0 \text{ a.s.}\end{aligned}$$

It follows that the martingale property characterizes sums of random variables centered at conditional expectations, given the predecessors.

Example 6.2.3. Let $X_1, X_2, \ldots$ be independent random variables with $\mathscr{E}X_n = 1$ for all n. Let $Z_n = \prod_{i=1}^n X_i$, $n \geq 1$. Then $\mathscr{E}|Z_n| = \prod_{i=1}^n \mathscr{E}|X_i| < \infty$ for every n, and a.s.

$$\begin{aligned}\mathscr{E}\{Z_{n+1}|Z_1, \ldots, Z_n\} &= \mathscr{E}\{(X_{n+1}Z_n)|Z_1, \ldots, Z_n\} \\ &= Z_n\mathscr{E}\{X_{n+1}|Z_1, \ldots, Z_n\} \\ &= Z_n,\end{aligned}$$

so that $\{Z_n\}$ is a martingale.

Example 6.2.4. Consider an urn which contains $b \geq 1$ black and $w \geq 1$ white balls which are well mixed. Repeated drawings are made from the urn, and after each drawing the ball drawn is replaced, along with c balls of the same color. Here $c \geq 1$ is an integer. Let $X_0 = b/(b + w)$, and X_n be the proportion of black balls in the urn after the nth draw. We show that $\{X_n, n \geq 0\}$ is a martingale.

Let $Y_0 = 1$, and for $n \geq 1$ let $Y_n = 1$ if the nth ball drawn is black, and $Y_n = 0$ if the nth ball drawn is white. Let b_n and w_n denote the number of black and white balls, respectively, in the urn after the nth draw. Write $b_0 = b$ and $w_0 = w$. Then $X_n = b_n/(b_n + w_n)$, $n \geq 0$. Clearly, for $n \geq 0$

$$b_{n+1} = b_n + cY_{n+1}, \qquad w_{n+1} = w_n + c(1 - Y_{n+1}).$$

Now $P\{Y_{n+1} = 1 \mid Y_0, \ldots, Y_n\} = X_n$ for $n \geq 0$, so that a.s.

$$\begin{aligned}
\mathscr{E}\{X_{n+1} \mid Y_0, \ldots, Y_n\} &= \mathscr{E}\left\{\frac{b_{n+1}}{b_{n+1} + w_{n+1}} \,\middle|\, Y_0, \ldots, Y_n\right\} \\
&= \frac{b_n}{b_n + w_n + c} + \frac{c}{b_n + w_n + c}\mathscr{E}\{Y_{n+1} \mid Y_0, \ldots, Y_n\} \\
&= \frac{b_n}{b_n + w_n + c} + \frac{c}{b_n + w_n + c} X_n \\
&= \frac{b_n}{b_n + w_n} \\
&= X_n.
\end{aligned}$$

Since $\sigma(X_0, \ldots, X_n) \subset \sigma(Y_0, \ldots, Y_n)$, we see that

$$\begin{aligned}
\mathscr{E}\{X_{n+1} \mid X_0, \ldots, X_n\} &= \mathscr{E}\{\mathscr{E}\{X_{n+1} \mid Y_0, \ldots, Y_n\} \mid X_0, \ldots, X_n\} \\
&= X_n \text{ a.s.}
\end{aligned}$$

It follows that $\{X_n\}$ is a martingale. It also follows that

$$\begin{aligned}
P\{Y_{n+1} = 1\} &= \mathscr{E}\{P\{Y_{n+1} = 1 \mid Y_0, Y_2, \ldots, Y_n\}\} \\
&= \mathscr{E}X_n = \mathscr{E}X_0 = \frac{b}{b + w}.
\end{aligned}$$

Example 6.2.5. Let $\{X_n\}$ be a sequence of random variables, and suppose that the joint probability density of $(X_1, X_2, \ldots, X_n)$ is either p_n or q_n. In statistics the ratio

$$\lambda_n = \lambda_n(X_1, X_2, \ldots, X_n) = \frac{q_n(X_1, X_2, \ldots, X_n)}{p_n(X_1, X_2, \ldots, X_n)}$$

is known as a *likelihood ratio.* The ratio λ_n is likely to be small or large, according as the true probability density function is p_n or q_n, so that λ_n may be used to reach a decision.

For convenience let us assume that $p_n > 0$ and is continuous for all n. (It is sufficient to assume that $q_n = 0$ whenever $p_n = 0$.) If p_n is the true density, the conditional probability density function of X_{n+1}, given $X_1, X_2, \ldots, X_n$, is p_{n+1}/p_n, so that

$$\begin{aligned}\mathscr{E}\{\lambda_{n+1} \mid X_1 &= x_1, \ldots, X_n = x_n\} \\ &= \int_{-\infty}^{\infty} \lambda_{n+1}(x_1, \ldots, x_n, y) \frac{p_{n+1}(x_1, \ldots, x_n, y)}{p_n(x_1, \ldots, x_n)}\, dy \\ &= \int_{-\infty}^{\infty} \frac{q_{n+1}(x_1, \ldots, x_n, y)}{p_n(x_1, \ldots, x_n)}\, dy \\ &= \frac{q_n(x_1, \ldots, x_n)}{p_n(x_1, \ldots, x_n)} \text{ a.s.}\end{aligned}$$

It follows that $\mathscr{E}\{\lambda_{n+1} \mid X_1, \ldots, X_n\} = \lambda_n$ a.s., and hence

$$\mathscr{E}\{\mathscr{E}\{\lambda_{n+1} \mid X_1, \ldots, X_n\} \mid \lambda_1, \ldots, \lambda_n\} = \mathscr{E}\{\lambda_{n+1} \mid \lambda_1, \ldots, \lambda_n\} = \lambda_n \text{ a.s.}$$

in view of the fact that $\sigma(X_1, \ldots, X_n) \supset \sigma(\lambda_1, \ldots, \lambda_n)$. Thus $\{\lambda_n\}$ is a martingale.

The following proposition gives a procedure by which a martingale may be constructed.

Proposition 6.2.2. Let $\{\mathscr{D}_n\}$ be a nondecreasing sequence of sub-σ-fields of $\mathscr{S}$, and let X be a random variable such that $\mathscr{E}|X| < \infty$. Then the sequence $X_n = \mathscr{E}\{X \mid \mathscr{D}_n\}$ a.s. $n \geq 1$ is a martingale.

Proof. Clearly X_n is $\mathscr{D}_n$-measurable. We now show that (6.2.1) holds. We have a.s.

$$\begin{aligned}\mathscr{E}\{X_{n+1} \mid X_1, X_2, \ldots, X_n\} &= \mathscr{E}\{\mathscr{E}\{X \mid \mathscr{D}_{n+1}\} \mid X_1, \ldots, X_n\} \\ &= \mathscr{E}\{\mathscr{E}\{\mathscr{E}\{X \mid \mathscr{D}_{n+1}\} \mid \mathscr{D}_n\} \mid X_1, \ldots, X_n\},\end{aligned}$$

since $\mathscr{D}_n \supset \sigma(X_1, \ldots, X_n)$. Again $\mathscr{D}_n \subset \mathscr{D}_{n+1}$, so that

$$\begin{aligned}\mathscr{E}\{X_{n+1} \mid X_1, \ldots, X_n\} &= \mathscr{E}\{\mathscr{E}\{\mathscr{E}\{X \mid \mathscr{D}_n\} \mid \mathscr{D}_{n+1}\} \mid X_1, \ldots, X_n\} \\ &= \mathscr{E}\{\mathscr{E}\{X_n \mid \mathscr{D}_{n+1}\} \mid X_1, \ldots, X_n\} \\ &= \mathscr{E}\{X_n \mid X_1, \ldots, X_n\} \\ &= X_n \text{ a.s.,}\end{aligned}$$

since X_n is $\mathscr{D}_n$- and, hence, $\mathscr{D}_{n+1}$-measurable. ■

Remark 6.2.1. Given a martingale $\{X_n\}$, there does not necessarily exist a random variable X with $\mathscr{E}|X| < \infty$ and a sequence $\{\mathscr{D}_n\}$ of nondecreasing sub-σ-fields of $\mathscr{S}$ such that $X_n = \mathscr{E}\{X \mid \mathscr{D}_n\}$ a.s. Take $Z_1, Z_2, \ldots$ to be independent and identically distributed with distribution $P\{Z_i = 0\} = \frac{1}{2} = P\{Z_i = 2\}$. In Example 6.2.3 we showed that $X_n = \prod_{i=1}^n Z_i$ is a martingale sequence. Suppose there exists some random variable X with $\mathscr{E}|X| < \infty$ and sub-σ-fields $\mathscr{D}_n \subset \mathscr{D}_{n+1}$, $n \geq 1$, such that $X_n = \mathscr{E}\{X|\mathscr{D}_n\}$ a.s. Clearly X_n is $\mathscr{D}_n$-measurable, so that $A_n = \{X_n = 0\} \in \mathscr{D}_n$. By the definition of conditional probability

$$0 = \int_{A_n} X_n \, dP_{\mathscr{D}_n} = \int_{A_n} X \, dP = \mathscr{E}(X\chi_{A_n}).$$

Since $\chi_{A_n} \xrightarrow{\text{a.s.}} 1$, it follows from the dominated convergence theorem that as $n \to \infty$

$$0 = \int_{A_n} X_n \, dP_{\mathscr{D}_n} = \mathscr{E}(X\chi_{A_n}) \to \mathscr{E}(X) = \mathscr{E}(X_n) = 1,$$

which is a contradiction. (See Corollary 1 to Theorem 6.3.2.)

Proposition 6.2.3. Let $\{X_n\}$ be a sequence of random variables with $\mathscr{E}|X_n| < \infty$ for all n. Then $\{X_n\}$ is a martingale (submartingale, supermartingale) if and only if, for every $m \geq n$ and $A \in \sigma(X_1, \ldots, X_n)$,

$$\int_A X_m \, dP \underset{\substack{(\geq) \\ (\leq)}}{=} \int_A X_n \, dP. \tag{6.2.3}$$

Proof. By the definition of conditional expectation for every $m \geq n$ and all $A \in \sigma(X_1, \ldots, X_n)$

$$\int_A X_m \, dP = \int_A \mathscr{E}\{X_m | X_1, \ldots, X_n\} \, dP = \int_A X_n \, dP,$$

and the result follows. ∎

Proposition 6.2.4 (Decomposition of Submartingales). Let $\{X_n, \mathscr{D}_n : n \geq 1\}$ be a submartingale. Then X_n has a decomposition

$$X_n = X_n' + X_n'' \text{ a.s.},$$

where $\{X_n', \mathscr{D}_n\}$ is a martingale, and $\{X_n''\}$ is a nondecreasing sequence of a.s. nonnegative random variables such that X_n'' is $\mathscr{D}_{n-1}$-measurable, $n \geq 2$.

Proof. Note that

$$\begin{aligned} X_n &= X_1 + \sum_{j=1}^{n-1} (X_{j+1} - X_j) \\ &= X_1 + \sum_{j=1}^{n-1} \{[X_{j+1} - \mathscr{E}\{X_{j+1}|\mathscr{D}_j\}] + [\mathscr{E}\{X_{j+1}|\mathscr{D}_j\} - X_j]\}. \end{aligned}$$

Set $X'_1 = X_1$, and for $n \geq 2$

$$X'_n = X_1 + \sum_{j=1}^{n-1} [X_{j+1} - \mathscr{E}\{X_{j+1} | \mathscr{D}_j\}].$$

Also set $X''_1 = 0$, and for $n \geq 2$

$$X''_n = \sum_{j=1}^{n-1} [\mathscr{E}\{X_{j+1} | \mathscr{D}_j\} - X_j].$$

Since $\{X_n, \mathscr{D}_n\}$ is a submartingale, it follows that $X''_n \geq 0$ a.s., $\{X''_n\}$ nondecreasing and $\mathscr{D}_{n-1}$-measurable. Also

$$\mathscr{E}\{X'_{n+1} | \mathscr{D}_n\} = X'_n \text{ a.s.,}$$

so that $\{X'_n, \mathscr{D}_n\}$ is a martingale. ■

Remark 6.2.2. Let $X_n = X'_n + X''_n$, $n \geq 1$, where X'_n and X''_n are as in Proposition 6.2.4. Then for $n \geq 2$

$$\begin{aligned}\mathscr{E}\{X_n | \mathscr{D}_{n-1}\} &= \mathscr{E}\{X'_n | \mathscr{D}_{n-1}\} + \mathscr{E}\{X''_n | \mathscr{D}_{n-1}\} \\ &= X'_{n-1} + X''_n \geq X_{n-1} \text{ a.s.,}\end{aligned}$$

so that $\{X_n\}$ is a submartingale and the converse of Proposition 6.2.4 also holds.

Remark 6.2.3. Let $\{X_n\}$ be a submartingale. In view of Proposition 6.2.4 we have

$$\mathscr{E}X_n = \mathscr{E}X'_n + \mathscr{E}X''_n, \qquad \mathscr{E}|X'_n| \leq \mathscr{E}|X_n| + \mathscr{E}X''_n, \qquad \text{and} \qquad 0 \leq \mathscr{E}X''_n \uparrow.$$

If $\sup_{n \geq 1} \mathscr{E}|X_n| < \infty$, then $\mathscr{E}X'_n$ and $\mathscr{E}X''_n$ exist and $\sup_{n \geq 1} \mathscr{E}|X'_n| < \infty$, $\sup \mathscr{E}X''_n < \infty$. It follows that $0 \leq X''_n \uparrow X''$ a.s., and study of the convergence of the sequence $\{X_n\}$ reduces to that of the sequence $\{X'_n\}$, which is a martingale sequence with $\sup_{n \geq 1} \mathscr{E}|X'_n| < \infty$.

Proposition 6.2.5. Let $\{X_n\}$ be a martingale, and let g be a convex function on $\mathbb{R}$. Then $\{g(X_n)\}$ is a submartingale, provided that $\mathscr{E}|g(X_n)| < \infty$, for $n \geq 1$.

If $\{X_n\}$ is a submartingale and g is a convex, nondecreasing function such that $\mathscr{E}|g(X_n)| < \infty$, then $\{g(X_n)\}$ is a submartingale.

Proof. Let $\{X_n\}$ be a martingale. By Jensen's inequality (Proposition 6.1.8)

$$\begin{aligned}\mathscr{E}\{g(X_n) | X_1, \ldots, X_{n-1}\} &\geq g(\mathscr{E}\{X_n | X_1, \ldots, X_{n-1}\}) \\ &= g(X_{n-1}) \text{ a.s.}\end{aligned}$$

On the other hand, if $\{X_n\}$ is a submartingale, we use the nondecreasing property of g to obtain the same result. ■

Corollary. If $\{X_n\}$ is a submartingale, then so also is the sequence

$$\{\max(X_n, a): n \geq 1\}$$

for any $a \in \mathbb{R}$.

Proof. Clearly $\{X_n^+\}$ is a submartingale, and hence so also is $\{X_n^+ + a, n \geq 1\}$ for any $a \in \mathbb{R}$. Since $\max(X_n, a) = (X_n - a)^+ + a$, the result follows.

Definition 6.2.2. Let $\{X_n\}$ be a sequence of random variables such that $\mathscr{E}|X_n| < \infty$ for $n \geq 1$. Then $\{X_n\}$ is said to be a backward (or reverse) martingale if $\ldots, X_n, X_{n-1}, \ldots, X_1$ is a martingale, that is, if

$$\mathscr{E}\{X_n | X_{n+1}, X_{n+2}, \ldots\} = X_{n+1} \text{ a.s.}$$

We say that $\{X_n\}$ is a backward submartingale if

$$\mathscr{E}\{X_n | X_{n+1}, X_{n+2}, \ldots\} \geq X_{n+1} \text{ a.s.}$$

We note (cf. Proposition 6.2.1) that $\{X_n\}$ is a backward martingale if and only if for $m > n$

$$\mathscr{E}\{X_n | X_m, X_{m+1}, \ldots\} = X_m \text{ a.s.}$$

Also, if X is an integrable random variable and $\{\mathscr{D}_n\}$ a nonincreasing sequence of sub-σ-fields of $\mathscr{S}$, then $X_n = \mathscr{E}\{X | \mathscr{D}_n\}$ a.s. is a backward martingale (see Proposition 6.2.2). In fact, if $\{X_n\}$ is a backward martingale sequence, there exists a random variable X with $\mathscr{E}|X| < \infty$ and a sequence of sub-σ-fields $\{\mathscr{D}_n\}$ of $\mathscr{S}$, $\mathscr{D}_n \downarrow$, such that $X_n = \mathscr{E}\{X | \mathscr{D}_n\}$ a.s. Take $X = X_1$ and $\mathscr{D}_n = \sigma(X_n, X_{n+1}, \ldots)$. Then $\mathscr{D}_n \supset \mathscr{D}_{n+1}$ for all $n \geq 1$, and

$$\mathscr{E}\{X | \mathscr{D}_n\} = \mathscr{E}\{X_1 | X_n, X_{n+1}, \ldots\} = X_n \text{ a.s.}, \qquad n > 1.$$

(See Remark 6.2.1.)

6.2.2 Some Applications of Martingales

Let $\{X_n\}$ be a sequence of random variables with zero means, and let $S_n = \sum_{k=1}^n X_k$, $n = 1, 2, \ldots$. In the case where $X_1, X_2, \ldots$ are independent, we obtained many results on the convergence of normed sums $\{n^{-1}S_n\}$ by using only the orthogonality property of the X_n namely, that $\mathscr{E}X_jX_k = 0$ for $j \neq k$. Since this property holds for martingales that have increments with finite variance, it is possible to obtain extensions of some of these results.

Let $\{S_n, n \geq 1\}$ be a martingale with respect to the nondecreasing sequence of σ-fields $\{\mathscr{D}_n\}$ such that $\mathscr{E}S_n = 0$ for all n. Let us write $X_n = S_n - S_{n-1}$, $n \geq 1$ $(S_0 = 0)$, and assume that $\mathscr{E}X_n^2 < \infty$ for all n. Then the following result holds.

Proposition 6.2.6. The inequality

$$P\{|S_n| \geq \epsilon\} \leq \epsilon^{-2} \sum_{j=1}^{n} \mathscr{E} X_j^2 \tag{6.2.4}$$

holds for all $\epsilon > 0$.

Proof. Clearly

$$P\{|S_n| \geq \epsilon\} \leq \epsilon^{-2} \mathscr{E} S_n^2$$

and

$$\mathscr{E} S_n^2 = \sum_{j=1}^{n} \mathscr{E} X_j^2 + 2 \sum_{i>j} \mathscr{E}(X_i X_j).$$

But for $i > j$

$$\begin{aligned} \mathscr{E}(X_i X_j) &= \mathscr{E}\{X_j \mathscr{E}\{X_i | \mathscr{D}_{i-1}\}\} \\ &= \mathscr{E}\{X_j \mathscr{E}\{(S_i - S_{i-1}) | \mathscr{D}_{i-1}\}\} \\ &= 0 \text{ a.s.,} \end{aligned}$$

since $\{S_n\}$ is a martingale. Inequality (6.2.4) follows immediately. ∎

Corollary. In the above notation the weak law of large numbers holds if $n^{-2} \sum_{j=1}^{n} \mathscr{E} X_j^2 \to 0$.

Proposition 6.2.7. Let $\{X_n\}$ be a sequence of random variables, and let $S_n = \sum_{k=1}^{n} X_k$, $n = 1, 2, \ldots$ $(X_0 = 0)$. Then $n^{-1} S_n \xrightarrow{P} 0$ if the following three conditions are satisfied:

(i) $\sum_{j=1}^{n} P\{|X_j| \geq n\} \to 0$ as $n \to \infty$.
(ii) $n^{-1} \sum_{j=1}^{n} \mathscr{E}\{X_j^{(n)} | X_1, \ldots, X_{j-1}\} \xrightarrow{P} 0$ as $n \to \infty$.
(iii) $n^{-2} \sum_{j=1}^{n} \{\mathscr{E}[\mathscr{E}\{X_j^{(n)} | X_1, \ldots, X_{j-1}\}]^2 - [\mathscr{E}\{X_j^{(n)} | X_1, \ldots, X_{j-1}\}]^2\} \xrightarrow{P} 0$ as $n \to \infty$.

[Here $X_j^{(n)}$ is the random variable X_j truncated at n.]

Proof. This result extends Corollary 1 of Proposition 5.3.4. The proof in the independent case was obtained as a special case of the central convergence criterion for row-wise independent sequences of random variables.

A direct proof of the result (in the independent variable case) involves only truncation and Chebyshev's inequality, which has been shown to hold also for zero-mean martingale sequences.

For the proof we note that in view of Proposition 2.2.4 and condition (i) it suffices to prove that $\sum_{j=1}^{n} X_j^{(n)}/n \xrightarrow{P} 0$. Let

$$S_n^{(n)} = \sum_{j=1}^{n} [X_j^{(n)} - \mathscr{E}\{X_j^{(n)} | X_1, \ldots, X_{j-1}\}].$$

In view of (ii) it is sufficient to show that $n^{-1}S_n^{(n)} \xrightarrow{P} 0$. But this follows immediately from (6.2.4) and (iii), since $\{S_n^{(n)}\}$ is a martingale. ∎

Next we note that most of the sufficient conditions for a.s. convergence involve centering at expectations or medians. The same methods apply to the general case, provided that we center at conditional expectations or conditional medians so that the centering constants themselves become random variables. For example, we define a conditional median, $\text{med}\{X | \mathscr{D}\}$, of a random variable X, given a sub-σ-field $\mathscr{D} \subset \mathscr{S}$ as a $\mathscr{D}$-measurable function satisfying

$$P\{X - \text{med}\{X | \mathscr{D}\} \geq 0 | \mathscr{D}\} \geq \tfrac{1}{2} \text{ a.s.}$$

and

$$P\{X - \text{med}\{X | \mathscr{D}\} \leq 0 | \mathscr{D}\} \geq \tfrac{1}{2} \text{ a.s.}$$

Similarly, we say that a random variable X is centered at conditional expectation (given $\mathscr{D}$) if $\mathscr{E}\{X | \mathscr{D}\} = 0$. In practical applications X will be X_n, and $\mathscr{D}$ will be the σ-field generated by the predecessors $X_1, X_2, \ldots, X_{n-1}$, $n \geq 2$. More precisely, let $\{X_n\}$ be a sequence of random variables with $\mathscr{E}|X_n| < \infty$ for all $n \geq 1$. Let $\mu_1 = \mathscr{E}X_1$ a.s., and for $n \geq 2$ let

$$\mu_n = \mathscr{E}\{X_n | X_1, \ldots, X_{n-1}\} \text{ a.s.}$$

Then $\mathscr{E}\mu_n = \mathscr{E}X_n$ for all n, but $\{\mu_n\}$ is not a martingale. In fact,

$$\mathscr{E}\{\mu_n | X_1, \ldots, X_{n-1}\} = \mu_n \text{ a.s.}$$

For $m < n$

$$\mathscr{E}\{(X_m - \mu_m)(X_n - \mu_n) | X_1, \ldots, X_{n-1}\} = 0 \text{ a.s.,}$$

so that

$$\mathscr{E}\left\{\sum_{k=1}^{n} (X_k - \mu_k)\right\}^2 = \sum_{k=1}^{n} \mathscr{E}(X_k - \mu_k)^2.$$

Thus, if the random variables X_n are centered at μ_n and $S_n = \sum_{k=1}^{n} X_k$, then $\text{var}(S_n) = \sum_{k=1}^{n} \text{var}(X_k)$. This result immediately leads to the following extension of Kolmogorov's inequality.

Proposition 6.2.8. Let $\{X_n\}$ be a sequence of random variables with finite variances such that each X_n is centered at conditional expectation $\mu_n = \mathscr{E}\{X_n | X_1, \ldots, X_{n-1}\}$. Then for every $\epsilon > 0$ the following inequality holds:

$$P\left\{\max_{1 \le k \le n} |S_k| \ge \epsilon\right\} \le \epsilon^{-2} \sum_{k=1}^{n} \mathscr{E} X_k^2. \tag{6.2.5}$$

Proof. A re-examination of the proof of Proposition 2.3.1 shows that independence is needed only in two steps: first, to conclude that $\text{var}(S_k) = \sum_{k=1}^{n} \text{var}(X_k)$, and, second, to show that in (2.3.3) the product terms vanish. In view of the discussion preceding the statement of Proposition 6.2.8 we need only show that $S_k \chi_{E_k}$ and $S_n - S_k$ are orthogonal (in the notation of Proposition 2.3.1). Thus it suffices to show that

$$\mathscr{E}(S_k \chi_{E_k} X_{k+1}) = 0 \qquad \text{for every } k \ge 1.$$

Clearly $E_k \in \sigma(X_1, \ldots, X_k)$, so that

$$\begin{aligned} \mathscr{E}\{S_k \chi_{E_k} X_{k+1} | X_1, \ldots, X_k\} &= S_k \chi_{E_k} \mathscr{E}\{X_{k+1} | X_1, \ldots, X_k\} \\ &= 0 \text{ a.s.} \end{aligned}$$

as required. ∎

Corollary 1. Let $\{X_n\}$ be a sequence of random variables with $\mathscr{E} X_n^2 < \infty$ for $n \ge 1$. If $\sum_{n=1}^{\infty} \text{var}(X_n) < \infty$ and $\sum_{n=1}^{\infty} \mathscr{E}\{X_n | X_1, \ldots, X_{n-1}\}$ converges a.s., then $\sum_{n=1}^{\infty} X_n$ converges a.s.

Corollary 2. Let $\{X_n\}$ be a sequence of random variables with $\mathscr{E} X_n^2 < \infty$ for $n \ge 1$. If $\sum_{n=1}^{\infty} \text{var}(X_n)/n^2 < \infty$, then

$$n^{-1} \sum_{k=1}^{n} [X_k - \mathscr{E}\{X_k | X_1, \ldots, X_{k-1}\}] \xrightarrow{\text{a.s.}} 0.$$

Corollary 3. If for some constant $c > 0$ the following conditions hold:

(i) $\sum_{n=1}^{\infty} P\{|X_n| \ge c\} < \infty$,
(ii) $\sum_{n=1}^{\infty} \mathscr{E}\{X_n^c - \mathscr{E}\{X_n^c | X_1, \ldots, X_{n-1}\}\}^2 < \infty$,
(iii) $\sum_{n=1}^{\infty} \mathscr{E}\{X_n^c | X_1, \ldots, X_{n-1}\}$ converges a.s.,

the series $\sum_{n=1}^{\infty} X_n$ converges a.s. (Here X_n^c is the random variable X_n truncated at c.)

Similar extensions of the Lindeberg–Feller central limit theorem and the law of the iterated logarithm can be obtained, but we do not propose to carry them out here.

6.2.3 Martingale Convergence Theorem

We now study some limit properties of a martingale. In particular we will prove the martingale convergence theorem, which has many applications. We start with the following weaker version of the martingale convergence theorem.

Theorem 6.2.1. Let $\{S_n\}$ be a martingale with $\mathscr{E}S_n^2 < c < \infty$ for all $n \geq 1$. Then there exists a random variable S such that S_n converges a.s. and in mean square to S. Moreover, $\mathscr{E}S_n = \mathscr{E}S$ for all n.

Proof. First note that, since $\{S_n\}$ is a martingale for $n > m$, $\mathscr{E}\{S_n | S_m\} = S_m$ a.s., so that $\mathscr{E}S_n = \mathscr{E}S_m$. Also

$$\mathscr{E}\{S_m S_n | S_m\} = S_m^2 \text{ a.s.,}$$

so that $\mathscr{E}S_m S_n = \mathscr{E}S_m^2$. It follows that

$$\mathscr{E}(S_n - S_m)^2 = \operatorname{var}(S_n - S_m) = \mathscr{E}S_n^2 - \mathscr{E}S_m^2$$

for $n > m$. Since $\{S_n^2\}$ is a submartingale sequence (Proposition 6.2.5), $\{\mathscr{E}S_n^2\}$ is a bounded, nondecreasing sequence of real numbers. Hence it must have a finite limit so that $\lim_{m,n\to\infty} \mathscr{E}(S_n - S_m)^2 = 0$.

Let $\epsilon > 0$, and set

$$D(\epsilon) = \bigcap_{m=1}^{\infty} \bigcup_{n=1}^{\infty} \{|S_{m+n} - S_m| \geq \epsilon\} \qquad \text{and} \qquad D = \bigcup_{\epsilon>0} D(\epsilon).$$

We wish to show that $P(D) = 0$. Since $D(\epsilon) \uparrow$, it suffices to show that $P(D(\epsilon)) = 0$ for every $\epsilon > 0$.

Let

$$D_m(\epsilon) = \bigcup_{n=1}^{\infty} \{|S_{m+n} - S_m| \geq \epsilon\},$$

so that $D(\epsilon) = \bigcap_{m=1}^{\infty} D_m(\epsilon)$, and it is sufficient to prove that $\lim_{m\to\infty} P(D_m(\epsilon)) = 0$. Finally we set

$$D_{m,n}(\epsilon) = \left\{\max_{1\leq k\leq n} |S_{m+k} - S_m| \geq \epsilon\right\} = \bigcup_{k=1}^{n} \{|S_{m+k} - S_m| \geq \epsilon\},$$

so that $D_{m,n}(\epsilon) \uparrow D_m(\epsilon)$ as $n \to \infty$ for every $m \geq 1$. By Proposition 6.2.8

$$P(D_{m,n}(\epsilon)) \leq \epsilon^{-2}[\mathscr{E}S_{m+n}^2 - \mathscr{E}S_m^2],$$

so that

$$P(D_m(\epsilon)) \leq \lim_{n\to\infty} \epsilon^{-2}[\mathscr{E}S_{m+n}^2 - \mathscr{E}S_m^2].$$

It follows that $\lim_{m\to\infty} P(D_m(\epsilon)) = 0$ as required. Consequently the sequence $\{S_n\}$ is Cauchy a.s. and hence converges a.s. to a random variable S.

Next we show that $\mathscr{E}S^2 < \infty$ and, moreover, $S_n \xrightarrow{\mathscr{L}_2} S$. Since $S_n^2 \xrightarrow{\text{a.s.}} S^2$ as $n \to \infty$, $S_n^2 \ge 0$ a.s., and $\{\mathscr{E}S_n^2\}$ is a bounded, nondecreasing sequence, it follows from Fatou's lemma that $\mathscr{E}S^2 < \infty$. Clearly

$$\mathscr{E}(S_n - S)^2 \le \lim_{m\to\infty} \mathscr{E}(S_n - S_m)^2 \qquad \text{(Fatou's lemma)}$$

for every $n \ge 1$. Hence

$$\lim_{n\to\infty} \mathscr{E}(S_n - S)^2 \le \lim_{n\to\infty} \lim_{m\to\infty} \mathscr{E}(S_n - S_m)^2 = 0,$$

and $S_n \xrightarrow{\mathscr{L}_2} S$.

Finally we show that $\mathscr{E}S_n = S$. Clearly $\mathscr{E}S_n = \mathscr{E}S_1$ for all $n \ge 1$, so that $\mathscr{E}S_n$ is independent of n. Since $S_n \xrightarrow{\text{a.s.}} S$, $S_n \xrightarrow{L} S$, and since $\mathscr{E}S_n^2 \le c < \infty$ for all $n \ge 1$, it follows from Corollary 1 to Theorem 3.1.5 that $\mathscr{E}S_n = \mathscr{E}S$. ∎

Corollary 1. Let $\{X_n\}$ be a sequence of random variables such that

$$\mathscr{E}\{X_n | X_1, \ldots, X_{n-1}\} = 0 \text{ a.s.}$$

for all n. Then $\sum_{k=1}^{\infty} \mathscr{E}X_k^2/k^2 < \infty$ implies $n^{-1}S_n \xrightarrow{\text{a.s.}} 0$.

Proof. Let $Y_n = \sum_{k=1}^{n} k^{-1}X_k$. Then

$$\begin{aligned}\mathscr{E}\{Y_n | Y_1, \ldots, Y_{n-1}\} &= \mathscr{E}\{(Y_{n-1} + n^{-1}X_n) | Y_1, \ldots, Y_{n-1}\} \\ &= Y_{n-1} \text{ a.s.}\end{aligned}$$

Moreover,

$$\begin{aligned}\mathscr{E}Y_n^2 &= \sum_{k=1}^{n} k^{-2}\mathscr{E}X_k^2 + \sum_{j\neq k} j^{-1}k^{-1}\mathscr{E}X_jX_k \\ &= \sum_{k=1}^{n} k^{-2}\mathscr{E}X_k^2 < \sum_{k=1}^{\infty} k^{-2}\mathscr{E}X_k^2.\end{aligned}$$

It follows from Theorem 6.2.1 that there exists a random variable Y such that $Y_n \xrightarrow{\text{a.s.}} Y$. By Kronecker's lemma $n^{-1}S_n \xrightarrow{\text{a.s.}} 0$.

Corollary 2 (Teicher [79]). Let $\{X_n\}$ be a sequence of independent random variables with $\mathscr{E}X_n = 0$ and $\mathscr{E}X_n^2 = \sigma_n^2$, $n = 1, 2, \ldots$. If the following conditions hold:

(i) $\sum_{n=2}^{\infty} n^{-4}\sigma_n^2 \sum_{i=1}^{n-1} \sigma_i^2 < \infty$,
(ii) $n^{-2} \sum_{j=1}^{n} \sigma_j^2 \to 0$,
(iii) $\sum_{n=1}^{\infty} P\{|X_n| \ge a_n\} < \infty$ for some sequence of constants $a_n > 0$ with $\sum_{n=1}^{\infty} n^{-4}a_n^2\sigma_n^2 < \infty$,

then $n^{-1}S_n \xrightarrow{\text{a.s.}} 0$.

Proof. We have

$$(n^{-1}S_n)^2 = n^{-2}\sum_{k=1}^{n} X_k^2 + 2n^{-2}\sum_{j<k} X_jX_k.$$

Define $Y_n = X_n^2$ if $|X_n| < a_n$, and $= 0$ otherwise. Then

$$\sum_{n=1}^{\infty} P\{Y_n \neq X_n^2\} = \sum_{n=1}^{\infty} P\{|X_n| \geq a_n\} < \infty,$$

so that $n^{-2}\sum_{k=1}^{n} X_k^2$ and $n^{-2}\sum_{k=1}^{n} Y_k$ converge a.s. to the same limit. But

$$\text{var}(Y_n) \leq \mathscr{E}Y_n^2 = \int_{|X_n|<a_n} X_n^4\, dP < a_n^2\sigma_n^2,$$

so that $\sum_{n=1}^{\infty} n^{-4}\,\text{var}(Y_n) < \sum_{n=1}^{\infty} n^{-4}a_n^2\sigma_n^2 < \infty$, and Corollary 1 implies $n^{-2}\sum_{k=1}^{n}(Y_k - \mathscr{E}Y_k) \xrightarrow{\text{a.s.}} 0$. Since

$$n^{-2}\sum_{k=1}^{n} \mathscr{E}Y_k \leq n^{-2}\sum_{k=1}^{n} \sigma_k^2 \to 0$$

as $n \to \infty$, from condition (ii) it follows that $n^{-2}\sum_{i=1}^{n} Y_i \xrightarrow{\text{a.s.}} 0$.

Next write

$$W_n = \sum_{j=2}^{n} j^{-2}X_jS_{j-1} \qquad \text{for } n \geq 2.$$

It is easy to check that $\{W_n\}$ is a martingale sequence. Moreover,

$$\begin{aligned}\mathscr{E}W_n^2 &= \sum_{j=2}^{n} j^{-4}\mathscr{E}X_j^2\mathscr{E}S_{j-1}^2 \\ &= \sum_{j=2}^{n} j^{-4}\sigma_j^2\sum_{i=1}^{j-1}\sigma_i^2 < \infty \qquad \text{[in view of (i)]},\end{aligned}$$

so that from Theorem 6.2.1 W_n converges a.s. to some random variable. By Kronecker's lemma $n^{-2}\sum_{j=2}^{n} X_jS_{j-1} \xrightarrow{\text{a.s.}} 0$. Since

$$\sum_{j<k}^{n} X_jX_k = \sum_{j=2}^{n} X_jS_{j-1},$$

the proof is complete.

We next prove an important inequality due to Doob which is basic to the proof of our main result. Let $\{X_n\}$ be a sequence of random variables. The sequence $\{X_n\}$ has a limit, finite or infinite, if and only if the number of its oscillations between any two (rational) numbers $a, b, a < b$, is finite (depending on a, b, and $\omega \in \Omega$). We obtain an estimate of the expected number of such oscillations for a sub- or supermartingale sequence $\{X_n\}$.

Let $a, b \in \mathbb{R}$ with $a < b$. Let $\{X_n\}$ be any sequence of random variables. For each $\omega \in \Omega$, set

$$\begin{aligned} T_1(\omega) &= \min\{n: X_n(\omega) \le a\}, \\ T_2(\omega) &= \min\{n: n > T_1(\omega), X_n(\omega) \ge b\}, \\ &\vdots \\ T_{2k-1}(\omega) &= \min\{n: n > T_{2k-2}(\omega), X_n(\omega) \le a\}, \\ T_{2k}(\omega) &= \min\{n: n > T_{2k-1}(\omega), X_n(\omega) \ge b\}, \\ &\vdots \quad , \end{aligned}$$

with the convention that if any set on the right-hand side is empty the corresponding $T_s(\omega) = +\infty$. Clearly $T_1 < T_2 < \cdots$ a.s., and T_s is an extended† random variable for every s satisfying $T_s \ge s$ a.s.

For each $n \ge 1$ let us set $U_n(a, b)(\omega) = 0$ if $T_2(\omega) > n$, and

$$= \max\{s: T_{2s} \le n\}$$

otherwise, and call $U_n(a, b)$ the *number of upcrossings of the interval* $[a, b]$ by $X_1, \ldots, X_n$. Clearly $U_n(a, b) \le [n/2]$, where $[x]$ is the largest integer $\le x$. Moreover

$$U_n(a, b) = \sum_{s=1}^{[n/2]} s\chi_{\{T_{2s} \le n,\, T_{2s+2} > n\}},$$

so that $U_n(a, b)$ is a nonnegative, nondecreasing, and bounded random variable and hence is integrable.

Next, for $j \ge 2$ let

$$Y_j(\omega) = \begin{cases} 0 & \text{if for some } s,\ T_{2s+1}(\omega) < j \le T_{2s+2}(\omega), \\ 1 & \text{otherwise.} \end{cases}$$

Then

$$Y_2 = \chi_{\{X_1 > a\}},$$

and for $j > 2$

$$Y_j = \chi_{\{Y_{j-1}=0,\, X_{j-1} \ge b\} \cup \{Y_{j-1}=1,\, X_{j-1} > a\}}.$$

It follows immediately that Y_j is a random variable and that Y_j is

$$\sigma(X_1, \ldots, X_{j-1})\text{-measurable.}$$

Lemma 6.2.1. For any sequence of random variables $\{X_n\}$, the inequality

$$\sum_{s=2}^{n} Y_s(X_s - X_{s-1}) + (b - a)U_n(a, b) \le (X_n - a)^+ \tag{6.2.6}$$

holds a.s.

† Here T_s may take the value $+\infty$ with a positive probability.

Proof. CASE. $U_n(a, b) = 0$. In this case $T_2 > n$. If $T_1 = 1$, then $Y_s = 0$ for $s = 2, \ldots, n$, and (6.2.6) holds trivially. If $1 < T_1 \le n$, then

$$\sum_{s=2}^{n} Y_s(X_s - X_{s-1}) = X_{T_1} - X_1 < 0,$$

since $X_1 > a$ and $X_{T_1} \le a$, so (6.2.6) holds. Finally, if $T_1 > n$, then $X_1 > a$, and we have

$$\sum_{s=2}^{n} Y_s(X_s - X_{s-1}) = X_n - X_1 \le X_n - a \le (X_n - a)^+,$$

so that (6.2.6) again holds.

CASE. $U_n(a, b) > 0$. In this case $T_2 \le n$, so that $T_1 < n$. If $T_1 > 1$, then

$$\begin{aligned}\sum_{s=2}^{n} Y_s(X_s - X_{s-1}) &= (X_{T_1} - X_1) + \sum_{s=T_2+1}^{n} Y_s(X_s - X_{s-1})\\ &\le \sum_{s=T_2+1}^{n} Y_s(X_s - X_{s-1})\end{aligned}$$

with equality when $T_1 = 1$. There are two cases, according as $T_{2t+1} < n \le T_{2t+2}$ for some t or $T_{2t'} < n \le T_{2t'+1}$ for some t'. In the first case

$$\begin{aligned}\sum_{s=T_2+1}^{n} Y_s(X_s - X_{s-1}) &= (X_{T_3} - X_{T_2}) + \cdots + (X_{T_{2t+1}} - X_{2t})\\ &\le (a - b)t = (a - b)U_n(a, b).\end{aligned}$$

In the second case

$$\begin{aligned}\sum_{s=T_2+1}^{n} Y_s(X_s - X_{s-1}) &= (X_{T_3} - X_{T_2}) + \cdots + (X_{T_{2t-1}} - X_{T_{2t-2}})\\ &\quad + (a - X_{2t}) + (X_n - a)\\ &\le (a - b)t + (X_n - a)\\ &\le (a - b)U_n(a, b) + (X_n - a)^+.\end{aligned}$$

Hence in either case (6.2.6) holds, and the proof is complete. ■

Theorem 6.2.2 (Doob's Upcrossing Inequality). Let $\{X_1, X_2, \ldots, X_n\}$ be a submartingale with respect to $\mathscr{D}_1 \subset \mathscr{D}_2 \subset \cdots \subset \mathscr{D}_n \subset \mathscr{S}$. Then

$$\mathscr{E} U_n(a, b) \le \frac{1}{b - a} \mathscr{E}(X_n - a)^+. \tag{6.2.7}$$

Proof. We have, since Y_s is $\mathscr{D}_{s-1}$-measurable,

$$\begin{aligned}\mathscr{E}[Y_s(X_s - X_{s-1})] &= \mathscr{E}[\mathscr{E}\{Y_s(X_s - X_{s-1})|\mathscr{D}_{s-1}\}] \\ &= \mathscr{E}[Y_s\mathscr{E}\{(X_s - X_{s-1})|\mathscr{D}_{s-1}\}] \\ &\geq 0\end{aligned}$$

in view of the fact that $\{X_n, \mathscr{D}_n\}$ is a submartingale, so that $\mathscr{E}\{X_s|\mathscr{D}_{s-1}\} \geq X_{s-1}$ a.s. It follows therefore from Lemma 6.2.1 that

$$0 \leq \mathscr{E}\left\{\sum_{s=2}^{n} Y_s(X_s - X_{s-1})\right\} \leq (a - b)\mathscr{E}U_n(a, b) + \mathscr{E}(X_n - a)^+$$

as asserted. ∎

Inequality (6.2.7) immediately leads to the following important result.

Theorem 6.2.3 (Submartingale Convergence Theorem). Let $\{X_n, \mathscr{D}_n\}$ be a submartingale. Suppose that $\limsup_{n\to\infty} \mathscr{E}|X_n| < \infty$. Then there exists a random variable X which is $\sigma(\bigcup_{n=1}^{\infty} \mathscr{D}_n)$-measurable such that $X_n \xrightarrow{\text{a.s.}} X$. Moreover, the inequality

$$\mathscr{E}|X| \leq \limsup_{n\to\infty} \mathscr{E}|X_n| < \infty$$

holds.

Proof. Let $\mathbb{Q} \subset \mathbb{R}$ be the set of all rational numbers. For every pair (a, b), $a, b \in \mathbb{Q}$ with $a < b$, set

$$D(a, b) = \left\{\omega: \liminf_{n\to\infty} X_n(\omega) < a < b < \limsup_{n\to\infty} X_n(\omega)\right\}$$

and

$$D = \bigcup_{\substack{a, b \in \mathbb{Q} \\ a < b}} D(a, b).$$

Clearly D is the set of divergence of $\{X_n\}$. We show that $P(D) = 0$. Since $\mathbb{Q}$ is countable, it is sufficient to show that $P(D(a, b)) = 0$ for every pair (a, b).

Let $n \geq 1$ be a fixed integer, and let $U_n = U_n(a, b)$ be the number of upcrossings of $[a, b]$ by $X_1, X_2, \ldots, X_n$ as defined above. Since $\{U_n\}$ is a nondecreasing sequence of nonnegative random variables, it either converges a.s. to a finite limit or it diverges to $+\infty$ on a set of positive probability. For every $\omega \in \Omega$ we set $\lim_{n\to\infty} U_n(\omega) = U(\omega)$. We now show that

$$P\{\omega: U(\omega) = +\infty\} = 0.$$

In view of Lemma 6.2.1 we have

$$\begin{aligned}\lim_{n\to\infty} \mathscr{E}U_n &= \limsup_{n\to\infty} \mathscr{E}U_n \\ &\le \frac{1}{b-a} \limsup_{n\to\infty} \mathscr{E}(X_n - a)^+ \\ &\le \frac{1}{b-a} \limsup_{n\to\infty} \mathscr{E}\{|X_n| + |a|\} \\ &< \infty.\end{aligned}$$

From the monotone convergence theorem it follows that $\mathscr{E}(U)$ exists, and we have

$$\mathscr{E}U = \lim_{n\to\infty} \mathscr{E}U_n < \infty.$$

In particular, U is finite a.s. and hence $P\{U = \infty\} = 0$. Since $D(a, b) \subset \{\omega: U(\omega) = \infty\}$, it follows that $P(D(a, b)) = 0$ as asserted. Consequently X_n converges a.s. to some random variable X. Since X_n is $\mathscr{D}_n$-measurable for every $n \ge 1$, it is $\sigma(\bigcup_{n=1}^{\infty} \mathscr{D}_n)$-measurable. Therefore X is also $\sigma(\bigcup_{n=1}^{\infty} \mathscr{D}_n)$-measurable.

Finally, from Fatou's lemma we conclude that $\mathscr{E}|X| < \infty$ and, moreover, satisfies the inequality

$$\mathscr{E}|X| \le \liminf_{n\to\infty} \mathscr{E}|X_n| \le \limsup_{n\to\infty} \mathscr{E}|X| < \infty. \qquad \blacksquare$$

Remark 6.2.4. Theorem 6.2.3 holds if the condition $\limsup_{n\to\infty} \mathscr{E}|X_n| < \infty$ is replaced by the condition $\sup_{n\ge 1} \mathscr{E}|X_n| < \infty$. In this case

$$\mathscr{E}|X| \le \sup_{n\ge 1} \mathscr{E}|X_n| < \infty.$$

Remark 6.2.5. If $\{X_n\}$ is a supermartingale, $\{-X_n\}$ is a submartingale, so that Theorem 6.2.3 holds also when $\{X_n\}$ is a supermartingale sequence. Since a martingale is also a submartingale, the result also holds for martingale sequences. Similarly it holds also for a backward submartingale sequence.

Remark 6.2.6. Note that

$$\limsup_{n\to\infty} \mathscr{E}|X_n| < \infty \Leftrightarrow \limsup_{n\to\infty} \mathscr{E}X_n^+ < \infty$$

and

$$\sup_{n\ge 1} \mathscr{E}|X_n| < \infty \Leftrightarrow \sup_{n\ge 1} \mathscr{E}X_n^+ < \infty.$$

These conditions are satisfied, in particular, if $\{X_n\}$ converges in $\mathscr{L}_1$. In fact, since $\{X_n\}$ is a submartingale sequence,

$$\mathscr{E}|X_n| = 2\mathscr{E}X_n^+ - \mathscr{E}X_n \le 2\mathscr{E}X_n^+ - \mathscr{E}X_1.$$

It follows that

$$\limsup_{n\to\infty} \mathscr{E}|X_n| \le 2 \limsup_{n\to\infty} \mathscr{E}X_n^+ - \mathscr{E}X_1 < \infty$$

and

$$\sup_{n\ge 1} \mathscr{E}|X_n| \le 2 \sup_{n\ge 1} \mathscr{E}X_n^+ - \mathscr{E}X_1 < \infty.$$

Conversely, $\mathscr{E}X_n^+ \le \mathscr{E}|X_n|$, so that

$$\limsup_{n\to\infty} \mathscr{E}|X_n| < \infty \left(\sup_{n\ge 1} \mathscr{E}|X_n| < \infty\right)$$

implies $\limsup_{n\to\infty} \mathscr{E}X_n^+ < \infty (\sup_{n\ge 1} \mathscr{E}X_n^+ < \infty)$. Note that, since $\{X_n\}$ is a submartingale sequence, so is $\{X_n^+\}$, so that $\{\mathscr{E}X_n^+\}$ is a nondecreasing sequence of nonnegative real numbers and $\limsup_{n\to\infty} \mathscr{E}X_n^+$ $(\sup_{n\ge 1} \mathscr{E}X_n^+)$ is either finite or $+\infty$.

Remark 6.2.7. The condition $\limsup_{n\to\infty} \mathscr{E}X_n^+ < \infty$ $(\sup_n \mathscr{E}|X_n| < \infty)$ of Theorem 6.2.3 is not sufficient to ensure that $X_n \xrightarrow{\mathscr{L}_1} X$. Let $Z_1, Z_2, \ldots$ be independent, identically distributed random variables with $P\{Z_n = 0\} = P\{Z_n = 2\} = \frac{1}{2}$. Let $X_n = \prod_{i=1}^n Z_i$, $n = 1, 2, \ldots$. Then $\{X_n\}$ is a martingale sequence (Example 6.2.3) such that $\mathscr{E}X_n = 1$ for all n. It follows from Theorem 6.2.3 that $X_n \xrightarrow{\text{a.s.}} X$, say. Clearly $X = 0$ a.s. However, $\mathscr{E}|X_n - X| = \mathscr{E}X_n = 1$ for all n, so that $\mathscr{E}|X_n - X| \nrightarrow 0$. The same example shows that $\mathscr{E}X_n \nrightarrow \mathscr{E}X$ as $n \to \infty$ and that $\{X_1, X_2, \ldots, X\}$ is not a submartingale, since

$$\mathscr{E}\{X | X_1, \ldots, X_n\} = 0 \not\ge X_n \text{ a.s.}$$

The relationship between X and $\{X_n\}$ will be explored in Section 6.3.

Remark 6.2.8. Every uniformly bounded submartingale (or supermartingale) sequence converges a.s. Also every nonnegative martingale sequence converges a.s., since

$$\limsup_{n\to\infty} \mathscr{E}|X_n| = \limsup_{n\to\infty} \mathscr{E}X_n = \mathscr{E}X_1 < \infty.$$

In fact, every positive supermartingale and every negative submartingale converge a.s.

Example 6.2.6. In Example 6.2.4 we showed that $\{X_n\}$ is a martingale. Since $|X_n| \le 1$ for all n, Theorem 6.2.3 implies that there exists a random variable X such that $X_n \xrightarrow{\text{a.s.}} X$. Moreover, the Lebesgue dominated convergence theorem shows that $\mathscr{E}X = \lim_{n\to\infty} \mathscr{E}X_n = b/(b + w)$.

In Example 6.2.5 we showed that the sequence of likelihood ratios $\{\lambda_n\}$ defined there is a martingale. Clearly $\lambda_n \ge 0$ a.s., and $\mathscr{E}\lambda_n = 1$ for all n (if p_n is the true density). It follows from Theorem 6.2.3 that there exists a random variable λ such that $\mathscr{E}\lambda \le 1$ and $\lambda_n \xrightarrow{\text{a.s.}} \lambda$.

6.3 UNIFORM INTEGRABILITY

We now introduce the concept of uniform integrability and study its relationship to martingale theory.

Definition 6.3.1. Let $(\Omega, \mathscr{S}, P)$ be a probability space. Let $\{X_t, t \in T\}$ be a collection of random variables defined on Ω such that $\mathscr{E}|X_t| < \infty$ for all $t \in T$. We say that $\{X_t, t \in T\}$ is uniformly integrable (with respect to P) if

$$\sup_{t\in T} \int_{|X_t|\ge a} |X_t|\, dP \to 0 \qquad \text{as } a \to \infty.$$

Remark 6.3.1. Let T be a finite set. Then the family $\{X_t, t \in T\}$ is uniformly integrable. (See Corollary 1 to Proposition 6.3.1.)

Remark 6.3.2. In the example of Remark 6.2.6 the sequence $\{X_n\}$ is not uniformly integrable. In fact, $X_n = 2^n$ with probability 2^{-n}, and $X_n = 0$ with probability $(1 - 2^{-n})$, so that

$$\int_{|X_n|\ge a} |X_n|\, dP = \begin{cases} 0 & \text{if } a > 2^n, \\ 1 & \text{if } a \le 2^n. \end{cases}$$

It follows that $\int_{|X_n|\ge a} |X_n|\, dP \not\to 0$ (uniformly in n) as $a \to \infty$. It is for this reason that the limit X and the martingale sequence $\{X_n\}$ are not closely related in the sense that $\{X_1, X_2, \ldots, X\}$ is not always a martingale (or a submartingale).

The following propositions elucidate the concept of uniform integrability.

Proposition 6.3.1. Let $\{X_t, t \in T\}$ be uniformly integrable. Then the following assertions hold:

(a) $\sup_{t\in T} \mathscr{E}|X_t| < \infty$.
(b) $\sup_{t\in T} P\{|X_t| \ge a\} \to 0$ as $a \to \infty$.

(c) The collection of set functions $E \to \int_E |X_t|\, dP$, $t \in T$, $E \in \mathscr{S}$, is uniformly absolutely continuous (with respect to P).

Conversely (a) and (c) or (b) and (c) imply that $\{X_t, t \in T\}$ is uniformly integrable.

Proof. Let $\{X_t\}$ be uniformly integrable.

(a) There exists an $a_0 > 0$ such that

$$\sup_{t \in T} \int_{|X_t| \geq a_0} |X_t|\, dP < 1,$$

and hence for all $t \in T$ we have

$$\begin{aligned} \mathscr{E}|X_t| &\leq a_0 P\{|X_t| < a_0\} + 1 \\ &\leq a_0 + 1, \end{aligned}$$

which implies that

$$\sup_{t \in T} \mathscr{E}|X_t| \leq a_0 + 1 < \infty.$$

(b) In view of (a), for every $t \in T$ and $a > 0$ we have

$$\infty > c = \sup_{t \in T} E|X_t| \geq aP\{|X_t| \geq a\},$$

so that

$$\sup_{t \in T} P\{|X_t| \geq a\} \leq \frac{c}{a}.$$

Letting $a \to \infty$, we see that (a) $\Rightarrow$ (b).

(c) Let $E \in \mathscr{S}$. Then for every $a > 0$ and all $t \in T$

$$\int_E |X_t|\, dP = \int_{E \cap \{|X_t| \geq a\}} |X_t|\, dP + \int_{E \cap \{|X_t| < a\}} |X_t|\, dP.$$

For $\epsilon > 0$ choose $a_0 > 0$ such that $\int_{|X_t| \geq a_0} |X_t|\, dP < \epsilon/2$ for all $t \in T$. It follows that for sufficiently large $a_0 > 0$ and $E \in \mathscr{S}$

$$\int_E |X_t|\, dP \leq \frac{\epsilon}{2} + a_0 P(E) \qquad \text{for all } t \in T.$$

For given $\epsilon > 0$, set $\delta = \epsilon/3a_0$. Let $E \in \mathscr{S}$ be such that $P(E) < \delta$. Then $\sup_{t \in T} \int_E |X_t| < \epsilon$, which proves (c).

Conversely, we note that we need only show that (b) and (c) together imply that $\{X_t\}$ is uniformly integrable. Let $\epsilon > 0$. Then it follows from (c) that there exists a $\delta = \delta(\epsilon) > 0$ with the property that $\int_E |X_t|\, dP < \epsilon$ for $E \in \mathscr{S}$ such that $P(E) < \delta$ and for all $t \in T$. On the other hand, (b) implies that, for given $\delta > 0$, there exists an $a_0 > 0$ such that for all $a \geq a_0$ and for all $t \in T$

$$P\{|X_t| \geq a\} < \delta.$$

Setting $E = \{|X_t| \geq a\}$ above, we have for all $a \geq a_0$ and for all $t \in T$

$$\int_{|X_t| \geq a} |X_t| \, dP < \epsilon,$$

so that

$$\sup_{t \in T} \int_{|X_t| \geq a} |X_t| \, dP \to 0 \qquad \text{as } a \to \infty.$$

This completes the proof of Proposition 6.3.1. ∎

Corollary 1. Let $|X_t| \leq Y$ a.s. for all $t \in T$, and suppose that $\mathscr{E}(Y) < \infty$. Then $\{X_t\}$ is uniformly integrable. In particular, any finite collection of random variables with finite expectations is uniformly integrable.

Proof. Clearly $\sup_{t \in T} \mathscr{E}|X_t| < \infty$, and, moreover, for all $E \in \mathscr{S}$

$$\sup_{t \in T} \int_E |X_t| \, dP \leq \int_E Y \, dP.$$

Since $\mathscr{E}(Y) < \infty$, $\int_E Y \, dP \to 0$ as $P(E) \to 0$. In view of Proposition 6.3.1 $\{X_t\}$ is uniformly integrable.

In the special case when $T = \{1, 2, \ldots, n\}$, we take $Y = \sum_{k=1}^n |X_k|$.

Corollary 2. If $\{X_t, t \in T\}$ and $\{Y_t, t \in T\}$ are both uniformly integrable, then so is the family $\{X_t \pm Y_t, t \in T\}$.

Proposition 6.3.2. Let $\{X_n\}$ be a sequence of random variables such that $X_n \xrightarrow{P} X$ as $n \to \infty$.

(a) Suppose that $\{X_n\}$ is uniformly integrable. Then $\mathscr{E}|X| < \infty$ and $X_n \xrightarrow{\mathscr{L}_1} X$ as $n \to \infty$.

(b) Conversely, suppose that $\mathscr{E}|X_n| < \infty$ for all $n \geq 1$, $\mathscr{E}|X| < \infty$, and, moreover, $X_n \xrightarrow{\mathscr{L}_1} X$ as $n \to \infty$. Then $\{X_n\}$ is uniformly integrable.

Proof. Let $\{X_n\}$ be uniformly integrable. Then clearly $\sup_{n \geq 1} \mathscr{E}|X_n| < \infty$ in view of Proposition 6.3.1. Since $X_n \xrightarrow{P} X$, there exists a subsequence $\{X_{n_k}\} \subset \{X_n\}$ such that $X_{n_k} \xrightarrow{\text{a.s.}} X$ and hence $|X_{n_k}| \xrightarrow{\text{a.s.}} |X|$. From Fatou's lemma and Proposition 6.3.1

$$\begin{aligned}\mathscr{E}|X| &\leq \liminf_{k \to \infty} \mathscr{E}|X_{n_k}| \\ &\leq \sup_{k \geq 1} \mathscr{E}|X_{n_k}| \leq \sup_{n \geq 1} \mathscr{E}|X_n| < \infty.\end{aligned}$$

Next we show that $X_n \xrightarrow{\mathscr{L}_1} X$. Let $\epsilon > 0$. Then we have

$$\begin{aligned}\mathscr{E}|X_n - X| &\le \epsilon P\{|X_n - X| < \epsilon\} + \int_{|X_n - X| \ge \epsilon} |X_n - X| \, dP \\ &\le \epsilon + \sup_{n \ge 1} \int_{|X_n - X| \ge \epsilon} |X_n| \, dP + \int_{|X_n - X| \ge \epsilon} |X| \, dP.\end{aligned}$$

Since $X_n \xrightarrow{P} X$, $P\{|X_n - X| \ge \epsilon\} \to 0$, so that $\int_{|X_n - X| \ge \epsilon} |X| \, dP \to 0$ as $n \to \infty$ since $\mathscr{E}|X| < \infty$. On the other hand, $\int_{|X_n - X| \ge \epsilon} |X_n| \, dP \to 0$ as $n \to \infty$ in view of part (c) of Proposition 6.3.1. This completes the proof of (a).

For the proof of (b) first note that $\mathscr{E}|X_n - X| < \infty$ for all n. Since $\mathscr{E}|X_n - X| \to 0$ as $n \to \infty$, for $\epsilon > 0$ there exists an integer $n_0 = n_0(\epsilon)$ such that, for all $n \ge n_0$ and for all $a > 0$,

$$\int_{|X_n - X| \ge a} |X_n - X| \, dP \le \mathscr{E}|X_n - X| < \epsilon.$$

Now choose a large enough so that

$$\int_{|X_n - X| \ge a} |X_n - X| \, dP < \epsilon \qquad \text{for all } n < n_0.$$

This is possible, since $\mathscr{E}|X_n - X| < \infty$. It follows that for all $n \ge 1$ and sufficiently large $a > 0$

$$\int_{|X_n - X| \ge a} |X_n - X| \, dP < \epsilon.$$

Thus $\{X_n - X\}$ is uniformly integrable, and hence so is $\{X_n\}$. ∎

Proposition 6.3.3. Let φ be a nonnegative Borel-measurable function defined on $[0, \infty)$ such that $\varphi(x)/x \to \infty$ as $x \to \infty$. Let $\{X_t, t \in T\}$ be a collection of random variables with $\mathscr{E}|X_t| < \infty$ such that $\sup_{t \in T} \mathscr{E}\varphi(|X_t|) < \infty$. Then $\{X_t, t \in T\}$ is uniformly integrable.

Proof. Let $N > 0$ be a positive integer. Then there exists an $a_0 > 0$ so that

$$\frac{\varphi(x)}{x} \ge N \qquad \text{for all } x \ge a_0.$$

It follows that for every $t \in T$

$$\begin{aligned}\int_{|X_t| \ge a_0} |X_t| \, dP &\le N^{-1} \mathscr{E}\varphi(|X_t|) \\ &\le N^{-1} \sup_{t \in T} \mathscr{E}\varphi(|X_t|),\end{aligned}$$

so that for all $a \geq a_0$ we have

$$\sup_{t \in T} \int_{|X_t| \geq a} |X_t| \, dP \leq N^{-1} \sup_{t \in T} \mathscr{E}\varphi(|X_t|).$$

Taking the limit first as $a \to \infty$ and then as $N \to \infty$, we see that $\{X_t\}$ is uniformly integrable. ∎

Corollary. Let $\{X_n\}$ be a sequence of random variables such that

$$\sup_{n \geq 1} \mathscr{E}|X_n|^p < \infty \qquad \text{for some } p > 1.$$

Then $\{X_n\}$ is uniformly integrable.

Proof. Take $\varphi(x) = |x|^p$, $p > 1$. Then φ satisfies the conditions of Proposition 6.3.3, and it follows that $\{X_n\}$ is uniformly integrable.

We are now in a position to study the effect of uniform integrability on a submartingale.

Theorem 6.3.1. Let $\{X_n, \mathscr{D}_n\}$ be a uniformly integrable submartingale. Then there exists a random variable X with $\mathscr{E}|X| < \infty$ such that $X_n \xrightarrow{\text{a.s.}, \mathscr{L}_1} X$ as $n \to \infty$. Moreover, $\{X_1, X_2, \ldots, X\}$ is itself a submartingale.

In particular, let $\{X_n, \mathscr{D}_n\}$ be a martingale. Then $\{X_1, X_2, \ldots, X\}$ is a martingale.

Proof. Since $\{X_n\}$ is uniformly integrable, Proposition 6.3.1, Theorem 6.2.3, and Remark 6.2.4 immediately imply the existence of a random variable X with $\mathscr{E}|X| < \infty$ such that $X_n \xrightarrow{\text{a.s.}} X$. In view of Proposition 6.3.2(a) we see that $\mathscr{E}|X_n - X| \to 0$ as $n \to \infty$.

Let m be a positive integer, and let $E \in \mathscr{D}_m$. In view of Proposition 6.2.3 the sequence $\{\int_E X_n \, dP\}$ is nondecreasing for $n \geq m$. Also

$$\left| \int_E X_n \, dP - \int_E X \, dP \right| \leq \int_E |X_n - X| \, dP \leq \int_\Omega |X_n - X| \, dP \to 0 \quad \text{as } n \to \infty,$$

so that

$$\lim_{n \to \infty} \int_E X_n \, dP = \int_E X \, dP.$$

It follows that for every $E \in \mathscr{D}_m$ and $m \leq n$

$$\int_E X_m \, dP \leq \int_E X_n \, dP \leq \int_E X \, dP,$$

and hence

$$\mathscr{E}\{X \mid \mathscr{D}_m\} \geq X_m \text{ a.s.}$$

Consequently $\{X_1, X_2, \ldots, X\}$ is a submartingale. ■

Corollary. Let $\{X_n, \mathscr{D}_n\}$ be a martingale or a nonegative submartingale with $\sup_{n \geq 1} \mathscr{E}|X_n|^p < \infty$ for some $p > 1$. Then X_n converges a.s. and in $\mathscr{L}_p$ to a random variable X with $\mathscr{E}|X|^p < \infty$.

Proof. By the corollary to Proposition 6.3.3 $\{X_n\}$ is uniformly integrable, so that X_n converges a.s. and in $\mathscr{L}_1$ to a random variable X in view of Theorem 6.3.1. Moreover, $\{X_1, X_2, \ldots, X\}$ is a martingale and $\mathscr{E}|X|^p < \infty$.

Clearly $\{|X_n|^p, \mathscr{D}_n\}$ is a nonnegative submartingale which satisfies

$$\int_{|X_n| \geq a} |X_n|^p \, dP \leq \int_{|X_n| \geq a} |X|^p \, dP < \infty$$

for every $a > 0$ and every $n \geq 1$. Moreover, we note that

$$P\{|X_n| \geq a\} \leq \frac{\mathscr{E}|X_n|}{a} \leq \frac{\mathscr{E}|X|}{a} \to 0$$

uniformly in n as $a \to \infty$, which gives

$$\sup_{n \geq 1} \int_{|X_n| \geq a} |X_n|^p \, dP \to 0 \qquad \text{as } a \to \infty.$$

Therefore $\{|X_n|^p\}$ is uniformly integrable. By an argument similar to the one used in Proposition 6.3.2(a) it is easy to show that $X_n \xrightarrow{\mathscr{L}_p} X$.

The following result is usually referred to as the Lévy continuity theorem for conditional expectations.

Theorem 6.3.2. Let $(\Omega, \mathscr{S}, P)$ be a probability space, and let X be a random variable defined on Ω such that $\mathscr{E}|X| < \infty$. Let $\{\mathscr{D}_n\}$ be a sequence of sub-σ-fields of $\mathscr{S}$.

(a) Suppose that $\{\mathscr{D}_n\}$ is nondecreasing. Let $\mathscr{D}$ be the sub-σ-field generated by $\bigcup_{n=1}^{\infty} \mathscr{D}_n$. Then

$$\mathscr{E}\{X \mid \mathscr{D}_n\} \xrightarrow{\text{a.s.}, \mathscr{L}_1} \mathscr{E}\{X \mid \mathscr{D}\} \qquad \text{as } n \to \infty.$$

(b) Suppose that $\{\mathscr{D}_n\}$ is nonincreasing so that $\mathscr{D} = \bigcap_{n=1}^{\infty} \mathscr{D}_n$ is nonempty. Then

$$\mathscr{E}\{X \mid \mathscr{D}_n\} \xrightarrow{\text{a.s.}, \mathscr{L}_1} \mathscr{E}\{X \mid \mathscr{D}\} \qquad \text{as } n \to \infty.$$

Proof. (a) For simplicity in notation, we set $X_n = \mathscr{E}\{X \mid \mathscr{D}_n\}$ for $n \geq 1$. From Proposition 6.2.2 it follows that $\{X_n\}$ is a martingale. In order to apply Theorem 6.3.1 we first need to show that $\{X_n\}$ is uniformly integrable.

Let $a > 0$. Then we have by Markov's inequality

$$P\{|X_n| \geq a\} \leq P\{\mathscr{E}\{|X| \,|\, \mathscr{D}_n\} \geq a\}$$

$$\leq a^{-1}\mathscr{E}|X| \tag{6.3.1}$$

for all n. Therefore

$$\sup_{n \geq 1} P\{|X_n| \geq a\} \leq a^{-1}\mathscr{E}|X| \to 0 \qquad \text{as } a \to \infty.$$

Let $\epsilon > 0$. Since the set function $E \to \int_E |X| \, dP$ is absolutely continuous with respect to P, there exists a $\delta = \delta(\epsilon)$ such that

$$\int_E |X| \, dP < \epsilon \qquad \text{whenever } E \in \mathscr{S} \text{ such that } P(E) < \delta.$$

It follows from the above that, for given $\delta > 0$, there exists an $a_0 = a_0(\delta)$ such that, for all $a \geq a_0$ and all $n \geq 1$,

$$P\{|X_n| \geq a\} \leq a^{-1}\mathscr{E}|X| < \delta.$$

On the other hand, for all $a > 0$ and for all $n \geq 1$, we have

$$\int_{|X_n| \geq a} |X_n| \, dP \leq \int_{|X_n| \geq a} \mathscr{E}\{|X| \,|\, \mathscr{D}_n\} \, dP$$

$$= \int_{|X_n| \geq a} |X| \, dP,$$

where the last equality follows from the definition of conditional expectation. Since $P\{|X_n| \geq a\} < \delta$ for all $a \geq a_0$ and all $n \geq 1$, it follows that for all $a \geq a_0$

$$\sup_{n \geq 1} \int_{|X_n| \geq a} |X_n| \, dP \leq \int_{|X_n| \geq a} |X| \, dP < \epsilon.$$

Hence $\{\mathscr{E}\{X \mid \mathscr{D}_n\}\}$ is a uniformly integrable martingale. From Theorem 6.3.1 it follows that there exists a random variable, X_∞ say, with $\mathscr{E}|X_\infty| < \infty$ such that

$$\mathscr{E}\{X \mid \mathscr{D}_n\} \xrightarrow{\text{a.s.}, \mathscr{L}_1} X_\infty \qquad \text{as } n \to \infty.$$

Finally we show that $X_\infty = \mathscr{E}\{X \mid \mathscr{D}\}$ a.s. Let $E \in \bigcup_{n=1}^\infty \mathscr{D}_n$. Then there exists an $n_0 \geq 1$ such that $E \in \mathscr{D}_{n_0}$. Clearly $E \in \mathscr{D}_n$ for all $n \geq n_0$. For every positive integer $n \geq 1$ we set $X_n = \mathscr{E}\{X \mid \mathscr{D}_n\}$. Then for $n \geq 1$ we have

$$\left| \int_E X_\infty \, dP - \int_E \mathscr{E}\{X \mid \mathscr{D}\} \, dP \right| \leq \left| \int_E X_\infty \, dP - \int_E X_n \, dP \right| + \left| \int_E X_n \, dP - \int_E \mathscr{E}\{X \mid \mathscr{D}\} \, dP \right|$$

$$\leq \mathscr{E}|X_\infty - X_n| + \left| \int_E X_n \, dP - \int_E \mathscr{E}\{X \mid \mathscr{D}\} \, dP \right|. \tag{6.3.2}$$

For $n \geq n_0$

$$\int_E X_n \, dP = \int_E X \, dP = \int_E \mathscr{E}\{X \mid \mathscr{D}\} \, dP,$$

so that, on taking the limit as $n \to \infty$, (6.3.2) implies that

$$\int_E X_\infty \, dP = \int_E \mathscr{E}\{X \mid \mathscr{D}\} \, dP \qquad \text{for every } E \in \bigcup_{n=1}^{\infty} \mathscr{D}_n.$$

Since $\bigcup_{n=1}^{\infty} \mathscr{D}_n$ is a field and $\mathscr{D} = \sigma(\bigcup_{n=1}^{\infty} \mathscr{D}_n)$, it follows from the extension theorem that the relation

$$\int_E X_\infty \, dP = \int_E \mathscr{E}\{X \mid \mathscr{D}\} \, dP$$

holds for every $E \in \mathscr{D}$. Since both X_∞ and $\mathscr{E}\{X \mid \mathscr{D}\}$ are $\mathscr{D}$-measurable, it follows that $X_\infty = \mathscr{E}\{X \mid \mathscr{D}\}$ a.s. This completes the proof of (a).

For the proof of (b) we need only to note that when $\{\mathscr{D}_n\}$ is nonincreasing the sequence $\{\mathscr{E}\{X \mid \mathscr{D}_n\}\}$ is a backward martingale, so that we can apply the same argument as above. ■

Corollary 1. $\{X_n, \mathscr{D}_n\}$ is a uniformly integrable martingale if and only if there exists a random variable X with $\mathscr{E}|X| < \infty$ such that $X_n = \mathscr{E}\{X \mid \mathscr{D}_n\}$ a.s., $n = 1, 2, \ldots$. In this case $X_n \xrightarrow{\text{a.s., } \mathscr{L}_1} \mathscr{E}\{X \mid \mathscr{D}\}$, where $\mathscr{D}$ is the σ-field generated by $\bigcup_{n=1}^{\infty} \mathscr{D}_n$.

Proof. If X is a random variable with $\mathscr{E}|X| < \infty$ and $X_n = \mathscr{E}\{X \mid \mathscr{D}_n\}$ a.s., $n \geq 1$, then $\{X_n\}$ is a martingale in view of Proposition 6.2.2. Also we showed in Theorem 6.3.2 that $\{X_n\}$ is uniformly integrable. Therefore Theorem 6.3.2 implies that $X_n \to \mathscr{E}\{X \mid \mathscr{D}\}$ a.s. and in $\mathscr{L}_1$ as $n \to \infty$.

Conversely, if $\{X_n, \mathscr{D}_n\}$ is a uniformly integrable martingale, it follows from Theorem 6.3.1 that there exists a random variable X with $\mathscr{E}|X| < \infty$ such that $X_n \to X$ a.s. and in $\mathscr{L}_1$. Moreover, $\{X_1, X_2, \ldots, X\}$ is itself a uniformly integrable martingale. (Use the same argument as in Theorem 6.3.1 with the sign $\leq$ replaced by $=$.) Then $X_n = \mathscr{E}\{X \mid \mathscr{D}_n\}$ a.s. for $n = 1, 2, \ldots$.

Corollary 2. The sequence $\{X_n\}$ is a backward martingale if and only if

$$\mathscr{E}\{X_n | X_{n+k}, X_{n+k+1}, \ldots, X_{n+k+m}\} = X_{n+k} \text{ a.s.} \tag{6.3.3}$$

for all integers n, k and $m \geq 1$.

Proof. If $\{X_n\}$ is a backward martingale, then

$$\mathscr{E}\{X_n | X_{n+k}, X_{n+k+1}, \ldots\} = X_{n+k} \text{ a.s.},$$

so that

$$\mathscr{E}\{\mathscr{E}\{X_n | X_{n+k}, \ldots\} | X_{n+k}, \ldots, X_{n+k+m}\} = \mathscr{E}\{X_{n+k} | X_{n+k}, \ldots, X_{n+k+m}\}.$$

It follows that

$$\mathscr{E}\{X_n | X_{n+k}, \ldots, X_{n+k+m}\} = X_{n+k} \text{ a.s.},$$

and (6.3.3) holds. Conversely, if (6.3.3) holds, Theorem 6.3.2 implies that

$$\lim_{m \to \infty} \mathscr{E}\{X_n | X_{n+k}, \ldots, X_{n+k+m}\} = \mathscr{E}\{X_n | X_{n+k}, \ldots\} \text{ a.s.}$$

It follows that

$$\mathscr{E}\{X_n | X_{n+k}, \ldots\} = X_{n+k} \text{ a.s.},$$

so that $\{X_n\}$ is a backward martingale.

6.4 MORE APPLICATIONS OF MARTINGALE THEORY

We now consider some applications of martingale theory.

(a) Applications in Probability Theory. We first derive short proofs of the zero-one law and the strong law of large numbers for independent, identically distributed random variables.

Proposition 6.4.1 (The Zero-One Law). The tail σ-field of a sequence $\{X_n\}$ of independent random variables defined on $(\Omega, \mathscr{S}, P)$ is $\{\varnothing, \Omega\}$.

Proof. Let A be a tail event of the sequence of independent random variables $\{X_n\}$. Then, for each n, χ_A and $(X_1, X_2, \ldots, X_n)$ are independent, so that

$$\mathscr{E}\chi_A = \mathscr{E}\{\chi_A | X_1, \ldots, X_n\}.$$

Theorem 6.3.2 immediately implies that

$$\mathscr{E}\{\chi_A | X_1, \ldots, X_n\} \xrightarrow{\text{a.s.}} \mathscr{E}\{\chi_A | X_1, X_2, \ldots\},$$

and since χ_A is $\sigma(X_1, X_2, \ldots)$-measurable, it follows that

$$\mathscr{E}\chi_A = \chi_A \text{ a.s.} \qquad \blacksquare$$

Proposition 6.4.2 (The Strong Law of Large Numbers). Let $\{X_n\}$ be a sequence of independent, identically distributed random variables with $\mathscr{E}X_n = \mu$ for all n. Let $S_n = \sum_{k=1}^n X_k$, $n = 1, 2, \ldots$. Then $n^{-1}S_n$ converges a.s. and in $\mathscr{L}_1$ to μ.

Proof. In Example 6.1.4 we showed that

$$\mathscr{E}\{X_1 | S_n, S_{n+1}, \ldots\} = \mathscr{E}\{X_1 | S_n\} = \frac{S_n}{n} \text{ a.s.}$$

If $\mathscr{T}$ is the tail σ-field of $\{S_n\}$, it follows from Theorem 6.3.2 that

$$\frac{S_n}{n} \to \mathscr{E}\{X_1 | \mathscr{T}\} \text{ a.s.} \qquad \text{and in } \mathscr{L}_1.$$

Since $\lim_{n\to\infty} S_n/n$ is a tail function, it must be a constant a.s. In view of the convergence in $\mathscr{L}_1$ and the fact that $\mathscr{E}(S_n/n) = \mu$, it follows that $\mathscr{E}\{X_1 | \mathscr{T}\} = \mu$ a.s., that is, $S_n/n \xrightarrow{\text{a.s.}, \mathscr{L}_1} \mu$. This completes the proof. $\blacksquare$

We next extend the zero-one law and the Borel–Cantelli lemma to the case where the random variables are not necessarily independent.

Proposition 6.4.3. Let $X, X_1, X_2, \ldots$ be random variables defined on $(\Omega, \mathscr{S}, P)$ such that $\mathscr{E}|X| < \infty$. Then

$$\mathscr{E}\{X | X_1, X_2, \ldots, X_n\} \to \mathscr{E}\{X | X_1, X_2, \ldots\}$$

a.s. and in $\mathscr{L}_1$ as $n \to \infty$. If $\mathscr{E}|X|^p < \infty$ for some $p > 1$, then $\mathscr{E}\{X | X_1, \ldots, X_n\} \to \mathscr{E}\{X | X_1, X_2, \ldots\}$ a.s. and in $\mathscr{L}_p$ as $n \to \infty$.

In particular, if X is $\sigma(X_1, X_2, \ldots)$-measurable, then $\mathscr{E}\{X | X_1, X_2, \ldots\} = X$ a.s.

Proof. Suppose that $\mathscr{E}|X| < \infty$. Write $Y_n = \mathscr{E}\{X | X_1, X_2, \ldots, X_n\}$ for $n = 1, 2, \ldots$. In this case it follows immediately from Theorem 6.3.2 that

$$Y_n = \mathscr{E}\{X | X_1, X_2, \ldots, X_n\} \xrightarrow{\text{a.s.}, \mathscr{L}_1} \mathscr{E}\{X | X_1, X_2, \ldots\} \qquad \text{as } n \to \infty.$$

Next suppose that, for some $p > 1$, $\mathscr{E}|X|^p < \infty$. Clearly $\{Y_n\}$ is a martingale with

$$\mathscr{E}|Y_n|^p = \mathscr{E}|\mathscr{E}\{X|X_1, \ldots, X_n\}|^p \le \mathscr{E}|X|^p < \infty \qquad \text{for all } n.$$

It follows from Proposition 6.2.5 that $\{|Y_n|^p, n = 1, 2, \ldots\}$ is a nonnegative submartingale, and from the corollary to Theorem 6.3.1 that Y_n converges in $\mathscr{L}_p$ to $\mathscr{E}\{X|X_1, X_2, \ldots\}$. ∎

Corollary (Extended Zero-One Law). Let $E \in \sigma(X_1, X_2, \ldots)$. Then the sequence of conditional probabilities $P\{E|X_1, \ldots, X_n\}$ converges a.s. to χ_E as $n \to \infty$.

Proof. Let $X = \chi_E$. Then Proposition 6.4.3 immediately implies that

$$\mathscr{E}\{\chi_E|X_1, \ldots, X_n\} \xrightarrow{\text{a.s.}} \mathscr{E}\{\chi_E|X_1, X_2, \ldots\} = \chi_E.$$

Proposition 6.4.4. Let $\{X_n\}$ be a sequence of a.s. nonnegative, uniformly bounded random variables defined on $(\Omega, \mathscr{S}, P)$, and let $\{\mathscr{D}_n\}$ be a nondecreasing sequence of sub-σ-fields of $\mathscr{S}$ such that X_n is $\mathscr{D}_n$-measurable for each $n = 1, 2, \ldots$. Then the series

$$\sum_{n=1}^{\infty} X_n \qquad \text{and} \qquad \sum_{n=1}^{\infty} \mathscr{E}\{X_n|\mathscr{D}_{n-1}\}, \qquad \mathscr{D}_0 = \{\varnothing, \Omega\},$$

either both converge a.s. or both diverge a.s.

Proof. Let $Y_k = X_k - \mathscr{E}\{X_k|\mathscr{D}_{k-1}\}$ for $k \ge 1$, and write $S_n = \sum_{k=1}^n Y_k$, $n \ge 1$. Then $\{S_n, \mathscr{D}_n\}$ is a martingale with $\mathscr{E}(S_n) = 0$ for all $n \ge 1$.

By the submartingale convergence theorem it follows that $\lim_n S_n$ exists and is finite a.s., so that

$$P\left\{\lim_{n\to\infty} S_n = \infty\right\} = P\left\{\lim_{n\to\infty} S_n = -\infty\right\} = 0.$$ ∎

Corollary (Extended Borel–Cantelli Lemma). Let $\{\mathscr{D}_n\}$ be a nondecreasing sequence of sub-σ-fields of $\mathscr{S}$, and let $A_n \in \mathscr{D}_n$, $n = 1, 2, \ldots$. Write $p_1 = P(A_1)$, and for $n \ge 2$

$$p_n = P\{A_n|\mathscr{D}_{n-1}\}.$$

Then

$$P\left\{\left(\limsup_{n\to\infty} A_n\right) \Delta \left(\sum_{n=1}^{\infty} p_n = \infty\right)\right\} = 0.$$

In other words,

$$P\left(\limsup_n A_n\right) = 1 \qquad \text{if and only if } P\left\{\omega: \sum_{n=1}^{\infty} p_n(\omega) = \infty\right\} = 1,$$

that is, the A_n occur i.o. if and only if $\sum_{n=1}^{\infty} p_n = \infty$. [Here $A \,\Delta\, B = (A - B) \cup (B - A)$.]

Proof. Set $X_n = \chi_{A_n}$ for $n \geq 1$. Then $\{X_n\}$ is a sequence of nonnegative and uniformly bounded random variables, and the X_n are $\mathscr{D}_n$-measurable for each n. It follows immediately from Proposition 6.4.4 that the series $\sum_{n=1}^{\infty} X_n$ and $\sum_{n=1}^{\infty} p_n$ either both converge a.s. or both diverge a.s. On the other hand, we note that

$$\left\{\omega: \sum_{n=1}^{\infty} X_n(\omega) = \infty\right\} = \left\{\omega: \sum_{n=1}^{\infty} \chi_{A_n}(\omega) = \infty\right\}$$

$$= \{\omega: \omega \in A_n \text{ i.o.}\} = \limsup_n A_n.$$

It follows that

$$P\left\{\sum_{n=1}^{\infty} p_n = \infty\right\} = 1 \qquad \text{if and only if } P\left(\limsup_n A_n\right) = 1.$$

Hence we have

$$P\left\{\left(\limsup_{n\to\infty} A_n\right) \Delta \left(\sum_{n=1}^{\infty} p_n = \infty\right)\right\} = 0.$$

As the next application of martingale theory we show the equivalence of a.s. convergence and convergence in law of sums of independent random variables. (See also Corollary 2 to Proposition 3.9.4.)

Proposition 6.4.5. Let $\{X_n\}$ be a sequence of independent random variables, and write $S_n = \sum_{k=1}^{n} X_k$, $n \geq 1$. Then the sequence $\{S_n\}$ converges in law if and only if it converges a.s.

Proof. It suffices to show that, if $S_n \xrightarrow{L} S$, then $S_n \xrightarrow{\text{a.s.}} S$. Suppose that $\{S_n\}$ converges in law. Then the corresponding sequence of characteristic functions $\prod_{j=1}^{n} \varphi_j$, where φ_j is the characteristic function of X_j, $j = 1, 2, \ldots, n$, converges to a characteristic function, say φ. Clearly there exists a $t_0 > 0$ such that $\varphi(t) \neq 0$ for $|t| < t_0$. Then $\varphi_n(t) \neq 0$ for $|t| < t_0$ and all $n \geq 1$. For every $n \geq 1$ set

$$Z_n = \left[\prod_{j=1}^{n} \varphi_j(t)\right]^{-1} e^{itS_n}.$$

Clearly $\{Z_n\}$ is a sequence of uniformly bounded, complex-valued random variables, and, moreover, $\mathscr{E}Z_n = 1$ for all n. It is easily seen that

$$\{Z_n, \sigma(S_1, S_2, \ldots, S_n)\}$$

is a martingale, and by the submartingale convergence theorem there exists a random variable Z with $\mathscr{E}|Z| < \infty$ such that $Z_n \xrightarrow{\text{a.s.}} Z$, that is,

$$e^{itS_n} \xrightarrow{\text{a.s.}} \varphi(t)Z \qquad \text{as } n \to \infty.$$

Now consider the mapping $e^{itS_n(\omega)}: (-t_0, t_0) \times \Omega \to \mathbb{C}$ with product measure $\lambda \times P$, where λ is the Lebesgue measure on $(-t_0, t_0)$. Clearly this function is measurable for each $n \geq 1$, so that the set

$$A = \left\{(t, \omega): \lim_{n\to\infty} e^{itS_n(\omega)}\right\}$$

is measurable. By what has been shown above, each section of A by a fixed $t \in (-t_0, t_0)$ has probability 1. It follows from Fubini's theorem that almost every section of A by (fixed) $\omega \in \Omega$ must have Lebesgue measure $2t_0$. Consequently there exists a measurable subset of Ω, Ω^* say, with $P(\Omega^*) = 1$, and for $\omega \in \Omega^*$ there exists a subset E_ω of $(-t_0, t_0)$ with $\lambda(E_\omega) = 2t_0 > 0$ such that $\lim_{n\to\infty} e^{itS_n(\omega)}$ exists for $t \in E_\omega$. This implies that $S_n(\omega)$ converges for $\omega \in \Omega^*$, that is, that S_n converges a.s. (See Problem 6.6.42.) ∎

(b) Applications to Measure Theory. Let $(\Omega, \mathscr{S}, P)$ be a probability space, and suppose that $\mathscr{S}$ is countably generated by the sequence $\{A_n\}$, $A_n \subset \Omega$, for $n \geq 1$. Let $\mathscr{D}_n$ be the σ-field generated by $(A_1, A_2, \ldots, A_n)$. Then there exists a partition of Ω into a finite number of sets $A_{n,1}, A_{n,2}, \ldots, A_{n,k_n}$ of $\mathscr{S}$ which generates $\mathscr{D}_n$ such that every element of $\mathscr{D}_n$ is the union of some of these sets. Clearly $\mathscr{D}_n \subset \mathscr{D}_{n+1}$ for $n \geq 1$, and we can assume that the $(n+1)$th partition is a refinement of the nth in the sense that each $A_{n,j}$ is a union of sets in the $(n+1)$th partition. Also $\mathscr{S}$ is generated by $\bigcup_{n=1}^{\infty} \mathscr{D}_n$. Let Q be a finite measure on $\mathscr{S}$ which is absolutely continuous with respect to P. For each $\omega \in \Omega$ and $n \geq 1$ set

$$X_n(\omega) = \sum_{j=1}^{k_n} \frac{Q(A_{n,j})}{P(A_{n,j})} \chi_{A_{n,j}}(\omega)$$

with the convention that ratios of the form 0/0 are given the value 0. Clearly X_n is $\mathscr{D}_n$-measurable. Also

$$\int_{A_{n,j}} X_n(\omega)\, dP(\omega) = \begin{cases} Q(A_{n,j}) & \text{if } P(A_{n,j}) > 0, \\ 0 & \text{if } P(A_{n,j}) = 0, \end{cases}$$

and, since each $A_{n,i}$ is a union of certain $A_{n+1,j_{i'}}, i' = 1, 2, \ldots, l, \{j_1, \ldots, j_l\} \subset \{1, 2, \ldots, k_{n+1}\}$, we have

$$\int_{A_{n,i}} X_{n+1}(\omega)\, dP(\omega) = \sum_{i'=1}^{l} \frac{Q(A_{n+1,j_{i'}})}{P(A_{n+1,j_{i'}})} P(A_{n+1,j_{i'}})$$
$$= \sum_{P(A_{n+1,j_{i'}})>0} Q(A_{n+1,j_{i'}}) = Q(A_{n,i}).$$

The last equality holds since $Q \ll P$. It follows therefore that $\{X_n, \mathscr{D}_n\}$ is a martingale. Next note that the finiteness of Q implies that for every $\epsilon > 0$ there exists a $\delta > 0$ such that $P(A) < \delta \Rightarrow Q(A) < \epsilon$, $A \in \mathscr{S}$, for, if such a $\delta > 0$ does not exist, there must exist a sequence $\{A_n\}$ of sets of $\mathscr{S}$ such that $\sum P(A_n) < \infty$ and $Q(A_n) \geq \epsilon$ for every n. Set $\chi_A = \limsup_{n\to\infty} \chi_{A_n}$. Then $A \subset \bigcup_{n=p}^{\infty} A_n$ for every $p \geq 1$, and hence $P(A) \leq \sum_{n=p}^{\infty} P(A_n)$ for every p, so that $P(A) = 0$. On the other hand, it follows from Fatou's lemma applied to A_n^c that

$$Q(A) \geq \limsup_{n\to\infty} Q(A_n) \geq \epsilon,$$

which contradicts the absolute continuity of Q with respect to P.

Finally we show that $\{X_n, \mathscr{D}_n\}$ is uniformly integrable. In fact, for every $\alpha > 0$

$$\int_{X_n \geq \alpha} X_n\, dP = Q(X_n \geq \alpha)$$

and

$$P\{X_n \geq \alpha\} \leq \frac{\mathscr{E}X_n}{\alpha} = \frac{Q(\Omega)}{\alpha} \qquad \text{independently of } n.$$

Choosing α large enough, we can make $P\{X_n \geq \alpha\} < \delta$, so that $Q(X_n \geq \alpha) < \epsilon$ or $\int_{X_n \geq \alpha} X_n\, dP < \epsilon$ independently of n.

In view of the uniform integrability of the martingale $\{X_n, \mathscr{D}_n\}$ it follows from Theorem 6.3.1 that there exists a random variable X with $\mathscr{E}|X| < \infty$ such that $X_n \xrightarrow{\mathscr{L}_1} X$ and $\mathscr{E}\{X \mid \mathscr{D}_n\} = X_n$ a.s. Thus, for every $A \in \mathscr{D}_n$ and every $n \geq 1$,

$$\int_A X_n\, dP = \int_A X\, dP,$$

and, as before,

$$Q(A_j) = \int_{A_j} X_j\, dP \qquad \text{for every } j.$$

It follows that for every $j \geq 1$

$$Q(A_j) = \int_{A_j} X_j \, dP = \int_{A_j} X \, dP,$$

and hence for every $A \in \mathscr{S}$

$$Q(A) = \int_A X \, dP.$$

Thus X is a.s. (P) the Radon–Nikodym derivative of Q with respect to P.

In particular, let $\Omega = [0, 1)$, $\mathscr{S}$ be the Borel σ-field of subsets of Ω, and P be the Lebesgue measure on Ω. For each $n \geq 1$ let us choose a finite sequence $x_j^{(n)}$ of points

$$0 = x_1^{(n)} < \cdots < x_{k_n}^{(n)} = 1$$

such that all elements of $\{x_j^{(n)}\}$ occur in $\{x_j^{(n+1)}\}$ and such that

$$\lim_{n \to \infty} \max_{1 \leq j < k_n} (x_{j+1}^{(n)} - x_j^{(n)}) = 0.$$

Let $\mathscr{D}_n$ be the σ-field generated by the partition consisting of the intervals $[x_i^{(n)}, x_{i+1}^{(n)})$ for $i = 1, 2, \ldots, k_n - 1$ and each $n \geq 1$. Then $\mathscr{D}_n \subset \mathscr{D}_{n+1}$ and $\mathscr{S} = \sigma(\bigcup_{n=1}^{\infty} \mathscr{D}_n)$. Let f defined on Ω be measurable and Lebesgue-integrable. Set

$$f_n(x) = \int_{[x_i^{(n)}, x_{i+1}^{(n)})} \frac{f(y)}{x_{i+1}^{(n)} - x_i^{(n)}} \, dy$$

whenever $x \in [x_i^{(n)}, x_{i+1}^{(n)})$ for each $x \in [0, 1)$. It is easy to check that $\{f_n, \mathscr{D}_n\}$ is a martingale for which

$$\mathscr{E}|f_n| \leq \int_0^1 |f| \, dP < \infty.$$

It follows that $f_n \xrightarrow{\text{a.s.}} \mathscr{E}\{f \mid \mathscr{S}\}$ as $n \to \infty$, and since $f = \mathscr{E}\{f \mid \mathscr{S}\}$ a.s., we see that $f_n \xrightarrow{\text{a.s.}} f$. The limit f can be viewed as the derivative of $\int f$ with respect to the given partition.

(c) Branching Process. We consider a model for population growth of an organism. Suppose that the first generation of an organism has X_0 members initially. Assume that by the end of its lifetime each of these produces a random number of progeny. Let $Z_j^{(1)}$, $j = 1, 2, \ldots, X_0$, be the number of progeny produced by the jth of the original X_0 members. The number of progeny at the end of the first generation is $X_1 = \sum_{j=1}^{X_0} Z_j^{(1)}$. (It is assumed that all the original members die or leave the population.) In the later generations this process repeats itself. Thus, if there are X_n members in the nth generation, $n \geq 1$, and if the jth of these has $Z_j^{(n+1)}$ progeny ($j = 1, 2, \ldots, X_n$) during the $(n + 1)$st generation, the population size at the end of the $(n + 1)$st generation is $X_{n+1} = \sum_{j=1}^{X_n} Z_j^{(n+1)}$. We say that the population is *extinct* if $X_n = 0$ for some n. In that case $X_{n+k} = 0$ for $k \geq 0$.

We will assume that, for each n, $Z_1^{(n)}, Z_2^{(n)}, \ldots$ are conditionally independent, given $X_0, X_1, \ldots, X_{n-1}$ with common distribution given by

$$p_k = P\{Z_j^{(n)} = k \mid X_0 = k_0, X_1 = k_1, \ldots, X_{n-1} = k_{n-1}\}, \qquad k = 0, 1, 2, \ldots,$$

independently of $n, j, k_0, \ldots, k_{n-1}$. The sequence $\{X_n\}$ is known as a *branching process.*

To avoid trivialities we assume that $p_0 > 0$. Let $\mu = \sum_{j=0}^{\infty} jp_j$, $0 < \mu < \infty$. Note that $\mathscr{E}X_1 = X_0\mu$. Also $\mathscr{E}\{X_{n+1} \mid X_0, \ldots, X_n\} = \mu X_n$, so $\mathscr{E}X_{n+1} = \mu\mathscr{E}X_n$ and, by iteration, $\mathscr{E}X_{n+1} = \mu^n\mathscr{E}X_1 = \mu^{n+1}X_0$.

Clearly

$$\mathscr{E}\{\mu^{-(n+1)}X_{n+1} \mid X_0, \ldots, X_n\} = \mu^{-n}X_n,$$

so that $\{\mu^{-n}X_n\}$ is a martingale. We will show that if $0 < \mu < 1$ then the population becomes extinct with probability 1, while if $\mu > 1$ then X_n behaves like $\mu^n X_\infty$ for large n, where X_∞ is a random variable.

CASE 1. $0 < \mu < 1$. In this case $\mathscr{E}X_{n+1} = \mu\mathscr{E}X_n$, so since $\mu < 1$

$$\mathscr{E}\left(\sum_{n=0}^{\infty} X_n\right) = \sum_{n=0}^{\infty} \mu^{-n}X_0 < \infty.$$

It follows that $X_n \xrightarrow{\text{a.s.}} 0$, and since the X_n are integer-valued, with probability 1 X_n is zero for sufficiently large n. Hence $X_\infty = \lim_{n\to\infty} \mu^{-n}X_n = 0$ a.s. Note, however, that $X_0 = \mathscr{E}(\mu^{-n}X_n) \neq \mathscr{E}X_\infty = 0$, so that $\{\mu^{-n}X_n\}$ cannot be uniformly integrable when $\mu < 1$.

CASE 2. $1 < \mu < \infty$. Note that

$$\mathscr{E}\left(\frac{X_n}{\mu^n}\right) = X_0 < \infty,$$

so that, by the submartingale convergence theorem (Theorem 6.2.3), it follows that there exists a random variable X_∞ such that $\mu^{-n}X_n \xrightarrow{\text{a.s.}} X_\infty$ as $n \to \infty$. Since $\mu > 1$, it follows that, for large n, $X_n \sim X_\infty \mu^n$. Note that $\mathscr{E}X_\infty = \mathscr{E}(\mu^{-1}X_1) = X_0$.

Actually, with a little bit of effort one can show a lot more. Let $\alpha_n = P\{X_n = 0\}$. Since $\{X_n = 0\} \subset \{X_{n+1} = 0\}$, it follows that $\alpha_n \le \alpha_{n+1}$ for every $n \ge 1$ and hence the limit $\alpha = \lim_{n\to\infty} \alpha_n$ exists. The quantity α is known as the probability of *extinction*. One can show that if $\mu \le 1$ then $\alpha = 1$, while if $\mu > 1$ then, with probability β, X_n is eventually zero, and with probability $1 - \beta$, $X_n \xrightarrow{\text{a.s.}} \infty$, where β is the unique root in $[0, 1)$ of the equation $\beta = \varphi(\beta)$. Here $\varphi(s) = \mathscr{E}(s^{Z_j^{(n)}})$ is the *probability-generating function* of $Z_j^{(n)}$. (See Problem 6.6.49.)

6.5 STOPPING TIMES

In this section we briefly consider some discrete time stopping problems. The theory of optimal stopping has been developed in recent years and has been applied to a variety of problems. We also consider some applications.

6.5.1 Definitions and Elementary Properties

Let $\{X_n\}$ be any sequence of random variables defined on $(\Omega, \mathscr{S}, P)$. Suppose a gambler bets on the successive outcomes $X_1, X_2, \ldots$. Let S_n, $n \ge 1$, denote the gambler's fortune after n plays. Then S_n is a function of the outcomes $X_1, X_2, \ldots, X_n$. Suppose $\{\varphi_n\}$ is a $\mathscr{B}_n$-measurable function such that $S_n = \varphi_n(X_1, \ldots, X_n)$ and $\mathscr{E}|S_n| < \infty$. The sequence of games $\{X_n\}$ is called *fair* if $\{S_n\}$ is a martingale, *unfavorable* if $\{S_n\}$ is a supermartingale, and *favorable* if $\{S_n\}$ is a submartingale. Here S_n may be interpreted to signify the fortune of the gambler at time n. In this interpretation the gambler can decide to terminate the game at some random point of time N, depending on his or her mood or on the course of the game up to then.

Definition 6.5.1. Let $(\Omega, \mathscr{S}, P)$ be a probability space, and let $\{\mathscr{D}_n\}$ be a nondecreasing sequence of sub-σ-fields of $\mathscr{S}$. A stopping time N with respect to $\{\mathscr{D}_n\}$ is a random variable defined on Ω such that

$$P\{N < \infty\} = 1 \tag{6.5.1}$$

and

$$\{N = n\} \in \mathscr{D}_n \qquad \text{for every } n, \quad 1 \le n < \infty. \tag{6.5.2}$$

If, however, N is an extended random variable, that is, $P\{N = \infty\} > 0$ satisfying (6.5.2) for $1 \le n < \infty$, we call it a stopping rule.

Example 6.5.1. Every constant mapping $N: \Omega \to \{1, 2, \ldots\}$ is a stopping time. In this case $\{N = n\}$ is either $\varnothing$ or Ω for every n, $1 \le n < \infty$.

Example 6.5.2. Let $\{X_n\}$ be a sequence of random variables defined on $(\Omega, \mathcal{S}, P)$, and $\{\mathcal{D}_n\}$ be a nondecreasing sequence of sub-σ-fields of $\mathcal{S}$ such that X_n is $\mathcal{D}_n$-measurable for each n. Let $B \in \mathcal{B}$ be fixed. For each $\omega \in \Omega$ define

$$N_B(\omega) = \begin{cases} \inf\{n: X_n(\omega) \in B\} & \text{if such an } n \text{ exists,} \\ +\infty & \text{otherwise.} \end{cases}$$

Then N_B is a stopping rule, since for every finite $n \geq 1$

$$\{N_B = n\} = \{X_1 \notin B, X_2 \notin B, \ldots, X_{n-1} \notin B, X_n \in B\} \in \mathcal{D}_n,$$

and for $n = \infty$

$$\{N_B = +\infty\} = \Omega - \{N_B < \infty\} \in \mathcal{S}.$$

N_B is called the *first entry time* of B with respect to the sequence $\{X_n\}$.

Example 6.5.3. In Example 6.2.1 we considered the following "double your bet" strategy for a gambler. The gambler wins or loses each game with probability $\frac{1}{2}$. He (or she) starts with a bet of 1 and doubles his bet until he wins a game and then quits. Let S_n = net gain after n plays. Then

$$P\{S_1 = 1\} = \tfrac{1}{2} = P\{S_1 = -1\},$$

and for $n = 1, 2, \ldots$

$$P\{S_{n+1} = 1 | S_n = 1\} = 1$$

and

$$P\{S_{n+1} = 1 | S_n = 1 - 2^n\} = \tfrac{1}{2} = P\{S_{n+1} = 1 - 2^{n+1} | S_n = 1 - 2^n\}.$$

Clearly $\{S_n\}$ is a martingale. Define

$$N = \inf_{n \geq 1} \{n: S_n = 1\}.$$

Since $P\{N = n\} = (\frac{1}{2})^n$, $n = 1, 2, \ldots$, $P\{N < \infty\} = 1$ and N is a stopping time. Clearly $P\{S_N = 1\} = 1$ so that $\mathcal{E}S_N = 1 \geq \mathcal{E}S_1 = 0$. Therefore under this strategy the gambler cannot lose (provided that he has infinite capital and time to continue to play indefinitely). Note that the sequence $\{S_n\}$ is uniformly bounded above by 1 but is not bounded below. Moreover, $\{S_n\}$ is not uniformly integrable.

For each n, $1 \leq n < \infty$, let S_n be a $\mathcal{D}_n$-measurable random variable defined on Ω. Let N be a stopping time. Set

$$S_N = \sum_{n=1}^{\infty} S_n \chi_{\{N=n\}}.$$

Then for every $B \in \mathscr{B}$

$$\{S_N \in B\} = \bigcup_{n=1}^{\infty} \{S_n \in B, N = n\} \in \mathscr{S},$$

and it follows that S_N is a random variable.

With the stopping time N we associate a σ-field $\mathscr{D}_N$ by setting

$$\mathscr{D}_N = \{A \in \mathscr{S} : A \cap \{N = n\} \in \mathscr{D}_n \text{ for all } n\}.$$

Clearly $\Omega \in \mathscr{D}_N$, and $\mathscr{D}_N$ is closed under countable unions. Moreover, if $E \in \mathscr{D}_N$, then

$$E^c \cap \{N = n\} = \{\{N = n\} - E \cap \{N = n\}\} \in \mathscr{D}_n \qquad \text{for all } n,$$

so that $E^c \in \mathscr{D}_N$. It follows that $\mathscr{D}_N$ is a sub-σ-field of $\mathscr{S}$. In particular, if there exists an n_0 such that $P\{N = n_0\} = 1$, then $\mathscr{D}_N = \mathscr{D}_{n_0}$ (Example 6.5.1).

Proposition 6.5.1. The random variables N and S_N are $\mathscr{D}_N$-measurable.

Proof. Clearly N is $\mathscr{D}_N$-measurable. For every n, $1 \le n < \infty$, and every $B \in \mathscr{B}$ the set $\{S_n \in B, N = n\} \in \mathscr{D}_n$, so that $\{S_N \in B\} \cap \{N = n\} \in \mathscr{D}_n$ and hence $\{S_N \in B\} \in \mathscr{D}_N$. ∎

Let $\mathscr{D} = \sigma(\bigcup_{n=1}^{\infty} \mathscr{D}_n)$ be the sub-σ-field of $\mathscr{S}$ generated by $\{\mathscr{D}_n\}$. If N is a stopping time, $\mathscr{D}_N \subset \mathscr{D}$. If N is a stopping rule, $\{N < \infty\} \in \mathscr{D}$ and N is $\mathscr{D}$-measurable.

Remark 6.5.1. Since

$$\{N \le n\} = \bigcup_{k \le n} \{N = k\} \qquad \text{and} \qquad \{N = n\} = \{N \le n\} \cap \{N \le n - 1\}^c,$$

condition (6.5.2) is equivalent to the condition

$$\{N \le n\} \in \mathscr{D}_n \qquad \text{for all } n = 1, 2, \ldots. \tag{6.5.3}$$

Note that $\{N > n\} = \{N \le n\}^c \in \mathscr{D}_n$ for all n.

Remark 6.5.2. Let N_1, N_2 be stopping times with respect to $\{\mathscr{D}_n\}$. Then $N_1 + N_2$, $\min(N_1, N_2)$, and $\max(N_1, N_2)$ are all stopping times. Indeed,

$$\{N_1 + N_2 \le n\} = \bigcup_{\substack{k,l = 1, 2, \ldots \\ k + l \le n}} [\{N_1 \le k\} \cap \{N_2 \le l\}],$$

$$\{\min(N_1, N_2) \le n\} = \{N_1 \le n\} \cup \{N_2 \le n\},$$

and

$$\{\max(N_1, N_2) \le n\} = \{N_1 \le n\} \cap \{N_2 \le n\}.$$

In particular, if N is a stopping time, then so is $\min(n, N)$, for any fixed $n = 1, 2, \ldots$.

Remark 6.5.3. If N_1 and N_2 are stopping times such that $N_1 \le N_2$ a.s., then $\mathscr{D}_{N_1} \subset \mathscr{D}_{N_2}$. Indeed, $N_1 \le N_2 \Rightarrow \{N_2 \le n\} \subset \{N_1 \le n\}$ for all $n = 1, 2, \ldots$. Hence for every n and every $A \in \mathscr{S}$

$$A \cap \{N_2 = n\} = A \cap \{N_1 \le n\} \cap \{N_2 = n\}.$$

Therefore $A \in \mathscr{D}_{N_1} \Rightarrow A \in \mathscr{D}_{N_2}$.

Let $\{S_n, \mathscr{D}_n\}$ be a martingale. Interpreting S_n as a gambler's fortune after n games in a sequence of fair games, we see that a stopping time N is a strategy by which the gambler decides when to stop playing, and S_N is his or her terminal fortune. Requirement (6.5.2) serves just to ensure that the gambler's decision whether to stop at time n depends only on the past, namely, $X_1, X_2, \ldots, X_n$, and not on the future. In this gambling context it is therefore of interest to know whether the property of "fairness" of the sequence of games is preserved under any stopping time N, that is, whether $\mathscr{E}S_N = \mathscr{E}S_1$. In Example 6.5.3 this was not the case.

Theorem 6.5.1 (Doob-Optional Stopping). Let $\{S_n, \mathscr{D}_n\}$ be a martingale (submartingale) and N be a stopping time. If the following conditions hold:

(i) $\mathscr{E}|S_N| < \infty$,
(ii) $\liminf_{n\to\infty} \int_{\{N>n\}} |S_n|\, dP = 0$,

then

$$\mathscr{E}S_N \underset{(\ge)}{=} \mathscr{E}S_1.$$

Proof. In view of (i) we have

$$\mathscr{E}S_N = \sum_{j=1}^{\infty} \int_{\{N=j\}} S_j\, dP = \lim_{n\to\infty} \sum_{j=1}^{n} \int_{\{N=j\}} S_j\, dP$$
$$= \lim_{n\to\infty} \sum_{j=1}^{n} \left(\int_{\{N\ge j\}} S_j\, dP - \int_{\{N\ge j+1\}} S_j\, dP \right).$$

Write $A_{j-1} = \{\omega: N(\omega) \ge j\}$. Then $A_{j-1} \in \mathscr{D}_{j-1}$, and, moreover,

$$\mathscr{E}\{S_j | \mathscr{D}_{j-1}\} \underset{(\ge)}{=} S_{j-1} \text{ a.s.}$$

in view of the martingale (submartingale) property. It follows from Proposition 6.2.3 that for $j = 2, 3, \ldots$

$$\int_{A_{j-1}} S_j\, dP \underset{(\ge)}{=} \int_{A_{j-1}} S_{j-1}\, dP.$$

Thus, in view of (ii),

$$\mathscr{E}S_N \underset{(\geq)}{=} \limsup_{n\to\infty} \left(\mathscr{E}S_1 - \int_{\{N \geq n+1\}} S_n \, dP \right)$$

$$= \mathscr{E}S_1 - \liminf_{n\to\infty} \int_{\{N>n\}} S_n \, dP$$

$$= \mathscr{E}S_1. \qquad \blacksquare$$

Remark 6.5.4. Conditions (i) and (ii) of Theorem 6.5.1 hold if there exists a positive integer n_0 such that $P\{N \leq n_0\} = 1$.

Remark 6.5.5. Conditions (i) and (ii) hold if the sequence $\{S_n\}$ is uniformly bounded a.s. In fact, if $|S_n| \leq c$ a.s. for all $n = 1, 2, \ldots$, then clearly $|S_N| \leq c$ a.s., so that $\mathscr{E}|S_N| \leq c < \infty$ and, moreover,

$$\int_{\{N>n\}} |S_n| \, dP \leq cP\{N > n\} \to 0 \qquad \text{as } n \to \infty,$$

since $\{N > n\} \downarrow \varnothing$ as $n \to \infty$.

Remark 6.5.6. Condition (ii) holds if there exists a constant $c > 0$ such that $\mathscr{E}S_n^2 \leq nc$ for all $n \geq 1$ and $\mathscr{E}N < \infty$. Indeed, by the Cauchy–Schwartz inequality

$$\int_{\{N>n\}} |S_n| \, dP \leq [\mathscr{E}S_n^2 P\{N > n\}]^{1/2}$$

$$\leq [cnP\{N > n\}]^{1/2} \to 0 \qquad \text{as } n \to \infty,$$

since $\mathscr{E}N < \infty$.

Remark 6.5.7. Both (i) and (ii) hold if $\{S_n\}$ is uniformly integrable. First we note that condition (ii) follows since $\{N > n\} \downarrow \varnothing$ yields $P\{N > n\} \to 0$ as $n \to \infty$, so that uniform integrability implies that $\int_{N>n} |S_n| \, dP \to 0$ as $n \to \infty$. Since $\{S_n\}$ is uniformly integrable, there exists a $c > 0$ such that $\sup_{n\geq 1} \mathscr{E}|S_n| \leq c < \infty$. Then we see easily that for $n \geq 1$

$$\sum_{j=1}^{n} \int_{\{N=j\}} S_j \, dP \leq c - \int_{N>n} S_n \, dP,$$

so that

$$\mathscr{E}(S_n) = \lim_{n\to\infty} \sum_{j=1}^{n} \int_{\{N=j\}} S_N \, dP$$

exists and is finite.

The following result is quite useful in applications.

Proposition 6.5.2. Let $\{S_n\}$ be a martingale (submartingale), and let N be a stopping time. Suppose that $\mathscr{E}N < \infty$ and $\mathscr{E}\{|S_{n+1} - S_n| | S_1, \ldots, S_n\} \le c < \infty$ on $N \ge n$. Then $\mathscr{E}|S_N| < \infty$, and

$$\mathscr{E}S_N \underset{(\ge)}{=} \mathscr{E}S_1.$$

Proof. We need only verify conditions (i) and (ii) of Theorem 6.5.1. Clearly

$$\mathscr{E}|S_N| \le \mathscr{E}|S_1| + \sum_{j=2}^{N} |S_j - S_{j-1}|.$$

Set $S_0 = 0$. Then

$$\mathscr{E}\sum_{j=1}^{N} |S_j - S_{j-1}| = \sum_{n=1}^{\infty} \sum_{j=1}^{n} \int_{\{N=n\}} |S_j - S_{j-1}| \, dP.$$

On interchanging the order of summation, we have

$$\begin{aligned}\mathscr{E}|S_N| &\le \sum_{j=1}^{\infty} \sum_{n=j}^{\infty} \int_{\{N=n\}} |S_j - S_{j-1}| \, dP \\ &= \sum_{j=1}^{\infty} \int_{\{N \ge j\}} |S_j - S_{j-1}| \, dP.\end{aligned}$$

Now $\{N \ge j\} = \{N < j\}^c \in \sigma(S_1, \ldots, S_{j-1})$, so that

$$\begin{aligned}\int_{\{N \ge j\}} |S_j - S_{j-1}| \, dP &= \int_{\{N \ge j\}} \mathscr{E}\{|S_j - S_{j-1}| | S_1, \ldots, S_{j-1}\} \, dP \\ &\le cP\{N \ge j\}.\end{aligned}$$

It follows therefore that

$$\mathscr{E}|S_N| \le c \sum_{j=1}^{\infty} P\{N \ge j\} = c\mathscr{E}N < \infty,$$

which proves condition (i) of Theorem 6.5.1.

Again, for $N > n$,

$$\sum_{j=1}^{n} |S_j - S_{j-1}| \le \sum_{j=1}^{N} |S_j - S_{j-1}|,$$

so that

$$\int_{\{N > n\}} |S_n| \, dP \le \int_{\{N > n\}} \sum_{j=1}^{N} |S_j - S_{j-1}| \, dP.$$

Since $\mathscr{E}\sum_{j=1}^{N}|S_j - S_{j-1}| < \infty$ and $\{N > n\} \downarrow \varnothing$ as $n \to \infty$, a simple application of the Lebesgue dominated convergence theorem to the sequence $\{\chi_{\{N>n\}}\sum_{j=1}^{N}|S_j - S_{j-1}|\}$ shows that $\int_{N>n}|S_n|\,dP \to 0$ as $n \to \infty$, so that condition (ii) of Theorem 6.5.1 holds. ■

6.5.2 Applications

We now consider some applications of stopping times.

(a) Wald's Equation. Let $\{X_n\}$ be independent, identically distributed random variables with $\mathscr{E}X_1 = \mu$. Let $S_n = \sum_{i=1}^{n}(X_i - \mu)$, $n = 1, 2, \ldots$. If N is any stopping time with $\mathscr{E}N < \infty$, then $\mathscr{E}S_N$ exists and

$$\mathscr{E}\left(\sum_{k=1}^{N} X_k\right) = \mu\mathscr{E}N. \tag{6.5.4}$$

Proof. Clearly $\{S_n, \sigma(S_1, \ldots, S_n)\}$ is a martingale satisfying

$$\begin{aligned}\mathscr{E}\{|S_{n+1} - S_n|\,|S_1, \ldots, S_n\} &= \mathscr{E}\{|X_{n+1} - \mu|\,|X_1, \ldots, X_n\}\\ &= \mathscr{E}|X_{n+1} - \mu| = c < \infty \qquad \text{for all } n.\end{aligned}$$

It follows from Proposition 6.5.2 and Theorem 6.5.1 that

$$\mathscr{E}(S_N) = \mathscr{E}(S_1) = 0,$$

so that

$$\mathscr{E}\left(\sum_{k=1}^{N} X_k\right) = \mu\mathscr{E}N.$$

(b) Wald's Fundamental Identity. Let $\{X_n\}$ be a sequence of independent, identically distributed random variables, and let $S_n = \sum_{k=1}^{n} X_k$, $n = 1, 2, \ldots$. Suppose that, for some real $t \neq 0$, $M(t) = \mathscr{E}e^{tX_1}$ exists and that $M(t) \geq 1$. Let N be any stopping time for the sums $\{S_n\}$ such that $|S_n| < c$ a.s. for $n \leq N$ and $\mathscr{E}N < \infty$. Then

$$\mathscr{E}\{e^{tS_N}[M(t)]^{-N}\} = 1. \tag{6.5.5}$$

[Note that $M(t)$ is the moment-generating function of X_i if $\mathscr{E}e^{tX_i}$ exists in some open interval containing 0.]

Proof. For each n define

$$Z_n = e^{tS_n}[M(t)]^{-n}.$$

Then $\{Z_n\}$ is a martingale with $\mathscr{E}Z_n = 1$ for all $n \geq 1$. For $N \geq n$ we have

$$\mathscr{E}\{|Z_{n+1} - Z_n||S_1, \ldots, S_n\} = Z_n \mathscr{E}\left\{\left|\frac{e^{tX_{n+1}}}{M(t)} - 1\right||S_1, \ldots, S_n\right\}$$
$$= Z_n \mathscr{E}\{|e^{tX_1}[M(t)]^{-1} - 1|\},$$

and since $|S_n| < c$ a.s. for $N \geq n$, it follows that there exists a positive real number $\alpha > 0$ such that

$$\mathscr{E}\{|Z_{n+1} - Z_n||S_1, \ldots, S_n\} \leq \alpha < \infty \qquad \text{for } N \geq n.$$

From Proposition 6.5.2 we see immediately that

$$\mathscr{E}Z_N = \mathscr{E}Z_1 = 1$$

as asserted. [For $t = 0$ (6.5.5) holds trivially.]

A special case of (6.5.5) arises in the sequential probability ratio test, where N is defined by

$$N = \inf_{n \geq 1} \{n: S_n \leq -a \quad \text{or} \quad S_n \geq b\},$$

where $a > 0$, $b > 0$ are fixed real numbers, and the X_i are assumed to be nondegenerate. In this case we need only to check that $\mathscr{E}N < \infty$. Clearly there exists an integer $n_0 > 0$ and a $\delta > 0$ such that $P\{|S_{n_0}| > a + b\} > \delta$. Then for any integer $k \geq 0$

$$P\{N \geq kn_0\}$$
$$\leq P\{|S_{n_0}| \leq a + b, |S_{2n_0} - S_{n_0}| \leq a + b, \ldots, |S_{kn_0} - S_{(k-1)n_0}| \leq a + b\}$$
$$\leq (1 - \delta)^k.$$

It follows that

$$\mathscr{E}N = \sum_{n=1}^{\infty} P\{N \geq n\} \leq n_0 \sum_{k=0}^{\infty} P\{N \geq kn_0\}$$
$$\leq \frac{n_0}{\delta} < \infty,$$

and (6.5.5) holds for any $t \neq 0$, provided that $M(t)$ exists and ≥ 1.

(c) Gambler's Ruin. Let $\{X_n\}$ be a sequence of independent, identically distributed random variables with $P\{X_i = 1\} = \frac{1}{2} = P\{X_i = -1\}$. Let $S_n = \sum_{k=1}^{n} X_k$, $n = 1, 2, \ldots$. Then $\{S_n\}$ is a martingale. Let $N = \inf_{n \geq 1} \{n: S_n = 1\}$; then $P\{N < \infty\} = 1$ and $P\{S_N = 1\} = 1$. It follows that $\mathscr{E}S_N = 1$, but $\mathscr{E}S_1 = 0$ so $\mathscr{E}S_N \neq \mathscr{E}S_1$. Therefore from Section 6.5.2(a) it follows that $\mathscr{E}N \not< \infty$.

Let a, b be two positive integers, and let

$$N = \inf_{n \geq 1} \{n: S_n = -a \quad \text{or} \quad S_n = b\}.$$

If we regard X_i as a player's winnings on the ith game, S_n is her (or his) cumulative gain after n games. The numbers a and b may be regarded as the initial capitals of the player and her opponent, respectively. If the two continue playing until one of them is ruined (has lost all her capital), N denotes the duration of the game. Clearly $\mathscr{E}N < \infty$ [see Section 6.5.2(b)], and it follows from Wald's equation that

$$0 = \mathscr{E}S_N = p_{-a}(-a) + (1 - p_{-a})b,$$

where

$$\begin{aligned} p_{-a} &= \text{probability of absorption at } -a, \\ &= \text{probability that } S_n \text{ reaches } -a \text{ before reaching } b. \end{aligned}$$

Thus

$$p_{-a} = \frac{b}{a + b}.$$

Also $\{S_n^2 - n\}$ is a martingale which is easily shown to satisfy the conditions of Proposition 6.5.2. It follows that

$$\mathscr{E}(S_N^2 - N) = \mathscr{E}(S_1^2 - 1) = 0$$

and hence

$$\mathscr{E}N = \mathscr{E}S_N^2 = a^2 p_{-a} + (1 - p_{-a})b^2 = ab.$$

In the general case where $P\{X_i = 1\} = p$, $P\{X_i = -1\} = q = 1 - p$, suppose that $p > q$ and write $\mathscr{E}X_i = p - q = \mu > 0$. Then $\{S_n - n\mu\}$ and $\{(q/p)^{S_n}\}$ are both martingales. From Wald's equation $\mathscr{E}S_N = \mu\mathscr{E}N$ for any stopping time N with $\mathscr{E}N < \infty$. Also application of Proposition 6.5.2 to $\{(q/p)^{S_n}\}$ immediately gives

$$\mathscr{E}\left\{\left(\frac{q}{p}\right)^{S_N}\right\} = \mathscr{E}\left\{\left(\frac{q}{p}\right)^{S_1}\right\} = 1,$$

so that

$$1 = \left(\frac{q}{p}\right)^{-a} p_{-a} + \left(\frac{q}{p}\right)^{b} (1 - p_{-a}).$$

It follows that

$$p_{-a} = \left[1 - \left(\frac{q}{p}\right)^{b}\right]\left[\left(\frac{q}{p}\right)^{-a} - \left(\frac{q}{p}\right)^{b}\right]^{-1}.$$

Letting $b \to \infty$, we see that the probability of losing to an infinitely rich opponent is

$$\lim_{b \to \infty} p_{-a} = \left(\frac{q}{p}\right)^a.$$

A similar argument applies if $p < q$ and

$$\lim_{b \to \infty} p_{-a} = 1,$$

so that

$$P\{S_n = -a \quad \text{for some } n = 1, 2, \ldots\} = 1,$$

provided that $p < q$, that is, if $p < \frac{1}{2}$.

(d) Some Inequalities in Probability. Let $\{S_n, \mathscr{D}_n\}$ be a nonnegative submartingale, and let $\{\varphi_n\}$ be a nonincreasing sequence of (a.s.) nonnegative functions such that φ_n is $\mathscr{D}_{n-1}$-measurable for each n. Suppose further that, for each n, $\mathscr{E}\{\varphi_n|S_n - S_{n-1}|\} < \infty$, and let N be any stopping rule. Then

$$\text{(6.5.6)} \qquad \int_{N<\infty} \varphi_N S_N \, dP \le \sum_{k=1}^{\infty} \mathscr{E}\{\varphi_k \mathscr{E}\{(S_k - S_{k-1})|\mathscr{D}_{k-1}\}\}.$$

(Take $S_0 = 0$ a.s., and $\mathscr{D}_0 = \{\Omega, \phi\}$.)

Proof. Let $X_k = S_k - S_{k-1}$, $k = 1, 2, \ldots$, and set

$$Z_n = \varphi_n S_n - \sum_{k=1}^{n} \varphi_k \mathscr{E}\{X_k|\mathscr{D}_{k-1}\} + \sum_{k=1}^{n} (\varphi_{k-1} - \varphi_k) S_{k-1}.$$

Then Z_n is $\mathscr{D}_n$-measurable, and

$$Z_n - Z_{n-1} = \varphi_n(X_n - \mathscr{E}\{X_n|\mathscr{D}_{n-1}\}),$$

so that

$$\mathscr{E}\{Z_n|\mathscr{D}_{n-1}\} = Z_{n-1} \quad \text{a.s.}$$

If there exists an n_0, $0 < n_0 < \infty$, such that $P\{N \le n_0\} = 1$, it follows from Theorem 6.5.1 (see Remark 6.5.4) that $\mathscr{E}Z_N = \mathscr{E}Z_1 = 0$. Therefore

$$\mathscr{E}(\varphi_N S_N) = \mathscr{E}\left\{\sum_{k=1}^{N} \varphi_k \mathscr{E}\{X_k|\mathscr{D}_{k-1}\} - \sum_{k=1}^{N} (\varphi_{k-1} - \varphi_k) S_{k-1}\right\}$$

$$\le \mathscr{E}\left\{\sum_{k=1}^{n_0} \varphi_k \mathscr{E}\{X_k|\mathscr{D}_{k-1}\}\right\},$$

since $\varphi_{k-1} - \varphi_k \geq 0$ a.s. and $S_{k-1} \geq 0$ a.s. Thus

$$\mathscr{E}(\varphi_N S_N) \leq \mathscr{E}\left\{\sum_{k=1}^{\infty} \varphi_k \mathscr{E}\{X_k | \mathscr{D}_{k-1}\} \chi_{[N \geq k]}\right\}.$$

In the general case, set $N_0 = \min(N, n_0)$. Then

$$\mathscr{E}(\varphi_{N_0} S_{N_0}) \leq \mathscr{E}\left\{\sum_{k=1}^{\infty} \varphi_k \mathscr{E}\{X_k | \mathscr{D}_{k-1}\} \chi_{[N_0 \geq k]}\right\}.$$

Letting $n_0 \to \infty$, we get the result.

Corollary 1 (Chow). Let $\{S_n, \mathscr{D}_n\}$ be a submartingale, and let $\{c_n\}$ be a positive, nonincreasing sequence of real numbers. Then for every $\varepsilon > 0$

$$P\left\{\max_{1 \leq k \leq n} c_k S_k \geq \epsilon\right\} \leq \epsilon^{-1}\left[c_1 \mathscr{E} S_1^+ + \sum_{k=2}^{n} c_k \mathscr{E}(S_k^+ - S_{k-1}^+)\right]. \tag{6.5.7}$$

Proof. Let $N = \inf_{n \geq 1} \{n: c_n S_n^+ \geq \epsilon\}$. Now

$$\epsilon P\left\{\max_{1 \leq k \leq n} c_k S_k^+ \geq \epsilon\right\} \leq \int_{\{\max_{1 \leq k \leq n} c_k S_k^+ \geq \epsilon\}} c_N S_N^+ \, dP$$

$$= \int_{N \leq n} c_N S_N^+ \, dP.$$

Write $X_k = S_k^+ - S_{k-1}^+$, $k \geq 1$ $(S_0^+ = 0)$, and note that $\{S_n^+, \mathscr{D}_n\}$ is a nonnegative submartingale. It follows from (6.5.6) that

$$\int_{N \leq n} c_N S_N^+ \, dP \leq \mathscr{E}\left\{\sum_{k=1}^{n} c_k \mathscr{E}\{X_k | \mathscr{D}_{k-1}\}\right\}$$

$$= c_1 \mathscr{E} S_1^+ + \sum_{k=2}^{n} c_k \mathscr{E}(S_k^+ - S_{k-1}^+).$$

Corollary 2 (Hájek–Rényi). Let $\{X_n\}$ be a sequence of independent random variables with $\mathscr{E} X_n = 0$, $\mathscr{E} X_n^2 < \infty$ for all n, and let $\{c_n\}$ be a nonincreasing sequence of positive real numbers. Then for every $\epsilon > 0$

$$P\left\{\max_{1 \leq k \leq n} c_k |S_k| \geq \epsilon\right\} \leq \epsilon^{-2} \sum_{k=1}^{n} c_k^2 \mathscr{E} X_k^2,$$

where $S_n = \sum_{k=1}^{n} X_k$, $n = 1, 2, \ldots$.

Proof. For every $\epsilon > 0$ it follows from Corollary 1 that

$$P\left\{\max_{1\le k\le n} c_k|S_k| \ge \epsilon\right\} = P\left\{\max_{1\le k\le n} c_k^2 S_k^2 \ge \epsilon^2\right\}$$

$$\le \epsilon^{-2}\left[c_1^2 \mathscr{E} X_1^2 + \sum_{k=2}^{n} c_k^2 \mathscr{E}(S_k^2 - S_{k-1}^2)\right]$$

$$= \epsilon^{-2} \sum_{k=1}^{n} c_k^2 \mathscr{E} X_k^2.$$

Letting $c_n = 1$, $n \ge 1$, in Corollary 2, we get the Kolmogorov inequality (2.3.1).

6.6 PROBLEMS

SECTION 6.1

1. Let $\Omega = [-1, 1]$, $\mathscr{S}$ be the σ-field of Borel sets in Ω, and P = one-half the Lebesgue measure. Find $\mathscr{E}\{X \mid |\omega|\}$, $\omega \in \Omega$.

2. Let X_1, X_2 be independent, identically distributed random variables with common uniform distribution on (0, 1). Let $X_{(2)} = \max(X_1, X_2)$. Find $P\{X_1 \le x \mid X_{(2)} = y\}$ and $\mathscr{E}\{X_1 \mid X_{(2)} = y\}$.

3. Let $\Omega = [0, 1]$, $\mathscr{B}$ be the Borel σ-field of subsets of Ω, and P be the Lebesgue measure on $\mathscr{B}$. Let $\mathscr{D}$ be the σ-field generated by the class $\{[0, \frac{1}{3}], (\frac{1}{3}, \frac{2}{3}], (\frac{2}{3}, 1]\}$. Define X on Ω by $X(\omega) = \omega^2$. Find $\mathscr{E}\{X \mid \mathscr{D}\}$.

4. Let (X_1, X_2) be a random vector with joint probability mass function

$$P\{X_1 = x_1, X_2 = x_2\} = \frac{n!}{x_1!x_2!(n - x_1 - x_2)!} p_1^{x_1} p_2^{x_2} (1 - p_1 - p_2)^{n - x_1 - x_2},$$

where $0 < p_1 < 1$, $0 < p_2 < 1$, $p_1 + p_2 < 1$, and $x_1, x_2 = 0, 1, 2, \ldots, n$, $0 \le x_1 + x_2 \le n$. Find the conditional distribution of X_2, given X_1, and compute $\mathscr{E}\{X_2 \mid X_1\}$.

5. Let (X, Y) have a bivariate normal distribution with probability density function

$$f(x, y) = [2\pi\sigma_1\sigma_2(1 - \rho^2)^{1/2}]^{-1} \exp\left\{\frac{-Q(x, y)}{2}\right\}$$

where

$$Q(x, y) = (1 - \rho^2)^{-1}\left[\left(\frac{x - \mu_1}{\sigma_1}\right)^2 - 2\rho\left(\frac{x - \mu_1}{\sigma_1}\right)\left(\frac{y - \mu_2}{\sigma_2}\right) + \left(\frac{y - \mu_2}{\sigma_2}\right)^2\right],$$

$\mu_1 \in \mathbb{R}$, $\mu_2 \in \mathbb{R}$, $\sigma_1 > 0$, $\sigma_2 > 0$, $|\rho| < 1$, $(x, y) \in \mathbb{R}_2$. Find the conditional distribution of X, given Y, and compute $\mathscr{E}\{X \mid Y\}$.

6. Let X be a random variable that has normal distribution with mean μ and variance σ^2, and let Φ denote the standard normal distribution function. Show that

$$\mathscr{E}\Phi(X) = \Phi\left(\frac{\mu}{\sqrt{1+\sigma^2}}\right).$$

If

$$\Phi_2(x, y; \rho) = \frac{1}{2\pi(1-\rho^2)^{1/2}} \int_{-\infty}^{x} \int_{-\infty}^{y} \exp\left\{-\frac{u^2+v^2-2\rho uv}{2(1-\rho^2)}\right\} dv\, du,$$

show that

$$\mathscr{E}\Phi^2(X) = \Phi_2\left(\frac{\mu}{\sqrt{1+\sigma^2}}, \frac{\mu}{\sqrt{1+\sigma^2}}; \frac{\sigma^2}{1+\sigma^2}\right).$$

(*Hint*: Let U, V be independent standard normal random variables which are independent of X. Then

$$\Phi(X) = P\{U \le X \mid X\} \qquad \text{and} \qquad \Phi^2(X) = P\{U \le X, V \le X \mid X\}.)$$

7. Prove Propositions 6.1.3 and 6.1.4.

8. Let $(\Omega, \mathscr{S}, P)$ be a probability space, and $\mathscr{D} \subset \mathscr{S}$ be a σ-field. For some $p > 1$ let $X \in \mathscr{L}_p$. Show that for $1 \le p_1 < p$

$$[\mathscr{E}\{|X|^{p_1} \mid \mathscr{D}\}]^{1/p_1} \le [\mathscr{E}\{|X|^p \mid \mathscr{D}\}]^{1/p} \quad \text{a.s.}$$

In particular, $\mathscr{E}\{X \mid \mathscr{D}\} \in \mathscr{L}_p$; moreover,

$$\|\mathscr{E}\{X \mid \mathscr{D}\}\|_p \le \|X\|_p \qquad \text{for all } X \in \mathscr{L}_p.$$

If $X_n \xrightarrow{\mathscr{L}_p} X$, deduce that $\mathscr{E}\{X_n \mid \mathscr{D}\} \xrightarrow{\mathscr{L}_p} \mathscr{E}\{X \mid \mathscr{D}\}$.

9. (Conditional Hölder's Inequality) Let $0 < \alpha \le 1$, $0 < \beta \le 1$, $\alpha + \beta \le 1$. Let $(\Omega, \mathscr{S}, P)$ be a probability space, and let $\mathscr{D} \subset \mathscr{S}$ be a σ-field. Let $X, Y \in \mathscr{L}_1$. Show that

$$\mathscr{E}\{|X|^\alpha |Y|^\beta \mid \mathscr{D}\} \le [\mathscr{E}\{|X| \mid \mathscr{D}\}]^\alpha [\mathscr{E}\{|Y| \mid \mathscr{D}\}]^\beta \text{ a.s.}$$

In particular, show that

$$\mathscr{E}\{|XY| \mid \mathscr{D}\} \le [\mathscr{E}\{|X|^p \mid \mathscr{D}\}]^{1/p} [\mathscr{E}\{|Y|^q \mid \mathscr{D}\}]^{1/q} \text{ a.s.}$$

for $p \ge 1$, $q \ge 1$, $1/p + 1/q \le 1$, and $X \in \mathscr{L}_p$, $Y \in \mathscr{L}_q$.

10. (Jensen's Inequality) Let $(\Omega, \mathscr{S}, P)$ be a probability space, and $\mathscr{D} \subset \mathscr{S}$ be a sub-σ-field. Let O be an open convex set in $\mathbb{R}_n$, and let $\mathbf{X}\colon \Omega \to O$ be a random vector such that every component $X_i \in \mathscr{L}_1$. Suppose that

$$\mathscr{E}\{\mathbf{X} \mid \mathscr{D}\} = (\mathscr{E}\{X_1 \mid \mathscr{D}\}, \mathscr{E}\{X_2 \mid \mathscr{D}\}, \ldots, \mathscr{E}\{X_n \mid \mathscr{D}\}) \in O \text{ a.s.}$$

Let φ be a real-valued, twice differentiable function defined on O such that the matrix $((\partial^2\varphi/\partial t_i\, \partial t_j))$, $1 \le i, j \le n$, is positive definite and continuous at every $\mathbf{t} =$

$(t_1, t_2, \ldots, t_n) \in O$ and $\varphi(\mathbf{X}) \in \mathscr{L}_1$. Show that

$$\mathscr{E}\{\varphi(\mathbf{X}) | \mathscr{D}\} \geq \varphi(\mathscr{E}\{\mathbf{X} | \mathscr{D}\}) \text{ a.s.}$$

11. (Doob's Inequality) Let $\{\mathscr{D}_n\}$ be an increasing sequence of sub-σ-fields of $\mathscr{S}$ in the probability space $(\Omega, \mathscr{S}, P)$. Let $X \in \mathscr{L}_1$, and let $X_i = \mathscr{E}\{X | \mathscr{D}_i\}$, $i \geq 1$. Show that for any $\epsilon > 0$

$$P\left\{\max_{1 \leq k \leq n} |X_k| \geq \epsilon\right\} \leq \frac{\mathscr{E}|X|}{\epsilon}.$$

12. Let $(\Omega, \mathscr{S}, P)$ be a probability space, and $\{\mathscr{D}_n\}$ be an increasing sequence of sub-σ-fields of $\mathscr{S}$. Suppose that $\mathscr{D}_\infty = \bigvee_{n=1}^{\infty} \mathscr{D}_n$ is the σ-field generated by $\bigcup_{n=1}^{\infty} \mathscr{D}_n$, $X \in \mathscr{L}_p$, for some $p > 1$ and

$$X_n = \mathscr{E}\{X | \mathscr{D}_n\}, \qquad X^* = \sup_{n \geq 1} |X_n|.$$

Show that $X^* \in \mathscr{L}_p$, and, moreover, $X_n \xrightarrow{\mathscr{L}_p} \mathscr{E}\{X | \mathscr{D}_\infty\}$.

13. Construct an example to show that the result of Example 6.1.4 is not true if the X_i are dependent. In other words, construct random variables $X_1, X_2, \ldots, X_n$ which are identically distributed but for which

$$\mathscr{E}\{X_k | X_1 + X_2 + \cdots + X_n\} \neq \frac{X_1 + X_2 + \cdots + X_n}{n} \text{ a.s.}, \qquad 1 \leq k \leq n.$$

14. In the notation of Example 6.1.6 show that:

(a) The set function P defined on $\mathfrak{A}$ by

$$P\{(B_1 \cap A) \cup (B_2 \cap A^c)\} = \tfrac{1}{2}[\lambda(B_1) + \lambda(B_2)]$$

is well defined.

(b) The set function P defines a probability measure on $\mathfrak{A}$.

15. Let $\mathscr{L}_2$ be the space of all square-integrable random variables defined on $(\Omega, \mathscr{S}, P)$. Then $\mathscr{L}_2$ is a Hilbert space under the inner product $\langle X, Y \rangle = \mathscr{E}(XY)$, and norm $\|X\| = (\mathscr{E}X^2)^{1/2}$. Let $\mathscr{D}$ be a sub-σ-field of $\mathscr{S}$. Then show that the set $\mathscr{L}_2(\mathscr{D})$ of all $\mathscr{D}$-measurable random variables $X \in \mathscr{L}_2$ is a closed subspace of $\mathscr{L}_2$. For $X \in \mathscr{L}_2$, show that random variable $\mathscr{E}\{X | \mathscr{D}\}$ is the projection of X onto $\mathscr{L}_2(\mathscr{D})$.

16. Let F be a distribution function, and let P be the associated probability measure on $\mathscr{B}$. Let g_1, g_2 be two nonnegative Borel-measurable functions on $\mathbb{R}$ which are integrable with respect to P. Suppose that the relation

$$\int_{-\infty}^{x} g_1(y)\, dF(y) = \int_{-\infty}^{x} g_2(y)\, dF(y)$$

holds for all $x \in \mathbb{R}$. Show that

$$g_1 = g_2 \text{ a.s. } (P).$$

(*Hint*: Show that the given condition implies the relation $\int_E g_1\, dP = \int_E g_2\, dP$ for all $E \in \mathscr{B}$.)

17. Let (X, Y) be a random vector with joint probability density function f. Suppose that $\mathscr{E}|X| < \infty$. Show that for $z \in \mathbb{R}$

$$\frac{\iint_{\{x \in \mathbb{R}, y \in \mathbb{R}: x+y=z\}} x f(x, y)\, dx\, dy}{\iint_{\{x \in \mathbb{R}, y \in \mathbb{R}: x+y=z\}} f(x, y)\, dx\, dy}$$

is a version of $\mathscr{E}\{X|X + Y = z\}$.

18. Let $g: I \to \mathbb{R}$ be a Borel-measurable function, where I is a finite or an infinite open interval on $\mathbb{R}$. Show that, if g is convex on I, then g is continuous.

19. If λ and μ are two σ-finite measures on a measurable space $(\Omega, \mathscr{S})$ such that λ is absolutely continuous with respect to $\mu(\lambda \ll \mu)$, we say that λ is dominated by μ. By the Radon–Nikodym theorem there exists a nonnegative Borel-measurable function f on Ω such that

$$\lambda(E) = \int_E f\, d\mu \qquad \text{for all } E \in \mathscr{S}.$$

We write $f = d\lambda/d\mu$ and call it the (generalized) density of λ with respect to μ. We note that f is determined uniquely a.s. (μ). If $\lambda \ll \mu$ and $\mu \ll \lambda$, we write $\lambda \equiv \mu$ and say that λ and μ are equivalent.

(a) Let P_1, P_2 be two probability measures on $(\Omega, \mathscr{S})$ such that $P_1 \ll P_2$, and let $\{\mathscr{D}_n\}$ be an increasing sequence of sub-σ-fields of $\mathscr{S}$ such that $\mathscr{D} = \bigvee_{n=1}^{\infty} \mathscr{D}_n$ is the σ-field generated by $\bigcup_{n=1}^{\infty} \mathscr{D}_n$. Let $f = dP_1/dP_2$ a.s. (P_2), and let f_n be the Radon–Nikodym derivative of P_1 with respect to P_2 in $(\Omega, \mathscr{D}_n)$. Show that

$$f_n \to f \text{ a.s. } (P_2)$$

and

$$f_n \to f \text{ in } \mathscr{L}_1(P_2).$$

(b) Let $P_1 \equiv P_2$ be two probability measures on $(\Omega, \mathscr{S})$, and let g be the Radon–Nikodym derivative of P_1 with respect to P_2. Show that

$$\mathscr{E}_{P_1}\{X|g\} = \mathscr{E}_{P_2}\{X|g\} \text{ a.s. } (P_1)$$

for every bounded random variable X on $(\Omega, \mathscr{S})$. In particular, the conditional probability distributions, given dP_1/dP_2 (with respect to P_1 and P_2), agree a.s. [This says, in view of the definition in Section 6.1.5, that (the likelihood ratio) dP_1/dP_2 is sufficient for the family $\{P_1, P_2\}$.]

20. Let $\mathbf{X} = (X_1, X_2, \ldots, X_n)$ be a random vector defined on $(\Omega, \mathscr{S}, P)$. Suppose that $\mathbf{X}$ has a continuous joint distribution function with continuous marginal distribution functions. Let

$$T(x_1, x_2, \ldots, x_n) = (x_{(1)}, x_{(2)}, \ldots, x_{(n)}),$$

where $x_{(1)} \leq x_{(2)} \leq \cdots \leq x_{(n)}$ denote the ordered x_i. The statistic

$$T(\mathbf{X}) = (X_{(1)}, X_{(2)}, \ldots, X_{(n)})$$

is called the *ordered statistic* of X. Clearly $X_{(1)} < X_{(2)} < \cdots < X_{(n)}$ with probability 1. Let $\mathfrak{X}$ be the set of all n-tuples with distinct coordinates, $\mathfrak{T}$ the set of all ordered n-tuples, and $\mathfrak{A}$ and $\mathfrak{A}^*$ the classes of Borel subsets of $\mathfrak{X}$ and $\mathfrak{T}$, respectively. Let $\mathscr{D} \subset \mathfrak{A}$ be the class of all sets that are symmetric; that is, if $\mathbf{x} \in D \in \mathscr{D}$, then $(x_{i_1}, x_{i_2}, \ldots, x_{i_n}) \in D$ for all permutations $\{i_1, i_2, \ldots, i_n\}$ of $\{1, 2, \ldots, n\}$. Let $f(\mathbf{X})$ be an integrable function, and suppose that $\mathbf{X}$ has joint density h which is symmetric in the x_i. Show that

$$\mathscr{E}\{f(X_1, X_2, \ldots, X_n) \mid T(\mathbf{X}) = T(\mathbf{x})\} = \frac{\sum f(x_{i_1}, \ldots, x_{i_n}) h(x_{i_1}, \ldots, x_{i_n})}{\sum h(x_{i_1}, \ldots, x_{i_n})},$$

where the summation on the right-hand side extends over all $n!$ permutations of $(x_1, \ldots, x_n)$.

In particular, if $X_1, X_2, \ldots, X_n$ are independent and identically distributed with a common continuous distribution function, show that

$$\mathscr{E}\{f(X_1, X_2, \ldots, X_n) \mid T(\mathbf{x})\} = (n!)^{-1} \sum_{\substack{\text{all} \\ \text{permutations}}} f(x_{i_1}, \ldots, x_{i_n}).$$

Conclude that the set of order statistics $T(\mathbf{X}) = (X_{(1)}, X_{(2)}, \ldots, X_{(n)})$ is sufficient for the family of joint distributions of $\mathbf{X}$. (The conditional expectation $\mathscr{E}\{f \mid T\}$ is called the U-statistic corresponding to f and has a wide variety of applications in nonparametric statistical inference. See, for example, Lehmann [53].)

21. We recall that two measures μ and ν are equivalent, $\mu \equiv \nu$, if and only if $\mu \ll \nu$ and $\nu \ll \mu$. Two families of measures $\mathfrak{M}$ and $\mathfrak{N}$ are said to be equivalent if $\mu(A) = 0$ for all $\mu \in \mathfrak{M} \Rightarrow \nu(A) = 0$ for all $\nu \in \mathfrak{N}$, and conversely.

(a) Show that a family $\mathscr{P}$ of probability measures is dominated by a σ-finite measure if and only if $\mathscr{P}$ has a countable equivalent subset. As a dominating measure for $\mathscr{P}$ we may take $\lambda = \sum_{k=1}^{\infty} c_k P_k$, where $c_k > 0$, $\sum_{k=1}^{\infty} c_k = 1$, and $\{P_n\}$ is a suitably chosen, countable equivalent subset of $\mathscr{P}$. Then λ is equivalent to $\mathscr{P}$.

(b) (Factorization Criterion) Let $\mathscr{P} = \{P_\theta : \theta \in \Theta\}$ be a family of probability measures on $(\mathbb{R}_n, \mathscr{B}_n)$ which is dominated by a σ-finite measure, and let $\lambda = \sum_{i=1}^{\infty} c_i P_{\theta_i}$ be equivalent to $\mathscr{P}$ [as in part (a)]. Let $T : (\mathbb{R}_n, \mathscr{B}_n) \to (\mathfrak{T}, \mathscr{B}_{\mathfrak{T}})$ be a statistic. (Here $\mathfrak{T}$ is some Euclidean space, and $\mathscr{B}_{\mathfrak{T}}$ is the Borel σ-field of subsets of $\mathfrak{T}$.) Show that the statistic T is sufficient for $\mathscr{P}$ if and only if there exist nonnegative $\mathscr{B}$-measurable functions $g : \mathfrak{T} \times \Theta \to \mathbb{R}$ such that

$$dP_\theta(x_1, \ldots, x_n) = g(T(x_1, \ldots, x_n), \theta)\, d\lambda(x_1, \ldots, x_n)$$

for all $\theta \in \Theta$, $(x_1, \ldots, x_n) \in \mathbb{R}_n$.

(c) Suppose that the probability measures $P_\theta \in \mathscr{P}$ have probability densities $p_\theta = dP_\theta/d\mu$ with respect to some σ-finite measure μ. Then show that a statistic T is sufficient for $\mathscr{P}$ if and only if there exist nonnegative $\mathscr{B}_{\mathfrak{T}}$-measurable functions $g : \mathfrak{T} \times \Theta \to \mathbb{R}$ and a nonnegative $\mathscr{B}_n$-measurable function $h : \mathbb{R}_n \to \mathbb{R}$ such that the relation

$$p_\theta(x_1, \ldots, x_n) = g(T(x_1, \ldots, x_n), \theta) h(x_1, \ldots, x_n)$$

holds for almost all $(x_1, \ldots, x_n)$ (with respect to the measure μ).

(Halmos and Savage [30])

22. Let p and q be the probability densities of probability measures P and Q, respectively, with respect to a measure μ, and let p' and q' be the densities of $P' = PT^{-1}$ and $Q' = QT^{-1}$, respectively, with respect to a measure ν. Show that, if $q/(p + q) = g \circ T$, then $q'/(p' + q') = g$.

SECTION 6.2

23. Let $\{Y_n\}$ be a sequence of random variables, and for each $n \geq 0$ let $g_n: \mathbb{R}_{n+1} \to \mathbb{R}$ be a Borel-measurable function. Set $Z_n = g_n(Y_0, Y_1, \ldots, Y_n)$, $n \geq 0$. Let f be a Borel-measurable function on $\mathbb{R}$ such that $\mathscr{E}|f(Z_n)| < \infty$ for all $n \geq 0$. Let $a_n: \mathbb{R}_n \to \mathbb{R}$ be a bounded function for each $n \geq 1$. Show that the sequence

$$X_n = \sum_{k=1}^{n} [f(Z_k) - \mathscr{E}\{f(Z_k) | Y_0, Y_1, \ldots, Y_{k-1}\}] a_k(Y_0, Y_1, \ldots, Y_{k-1}),$$

defined for $n \geq 1$ ($X_0 = 0$), defines a martingale sequence relative to $\{Y_n\}$.

24. Let U be a random variable which has uniform distribution on $[0, 1)$, and for $n \geq 0$ set $Y_n = k2^{-n}$, where k is the unique integer (depending on n and U) which satisfies $k2^{-n} \leq U < (k + 1)2^{-n}$. Let f be a bounded, real-valued function defined on $[0, 1]$, and set $X_n = 2^n[f(Y_n + 2^{-n}) - f(Y_n)]$, $n \geq 0$. Show that $\{X_n\}$ is a martingale sequence with respect to $\{Y_n\}$.

25. Let $\{R_n\}$ be the sequence of Rademacher functions, and $\{c_n\}$ be a sequence of constants. Define

$$L_n(x) = \prod_{k=1}^{n} [1 + c_k R_k(x)], \qquad 0 \leq x \leq 1.$$

Show that $\{L_n, \mathscr{D}_n\}$ is a martingale sequence, where $\mathscr{D}_n$, $n = 0, 1, \ldots$, is the σ-field of subsets of $[0, 1]$ generated by the partition sets

$$\left(\frac{k}{2^{n+1}}, \frac{k+1}{2^{n+1}}\right), \qquad k = 0, 1, 2, \ldots, 2^{n+1}.$$

26. Let $\{X_n\}$ be a sequence of independent random variables with $\mathscr{E}X_n = 0$ and $\mathscr{E}X_n^2 = 1$ for $n = 1, 2, \ldots$. Let $((a_{jk}))$ be a real matrix such that $\sum_{j=1}^{\infty} \sum_{k=1}^{\infty} a_{jk}^2 < \infty$. Write $S_n = \sum_{j=1}^{n} \sum_{k=1}^{n} a_{jk} X_j X_k$, $n \geq 1$. Show that, if $\sum_{k=1}^{n} |a_{kk}| < \infty$, then S_n converges a.s. to a random variable S. (Set

$$K_n = \sum_{j=1}^{n} X_j \sum_{k=1}^{j-1} a_{jk} X_k, \qquad L_n = \sum_{k=1}^{n} X_k \sum_{j=1}^{k-1} a_{jk} X_j, \qquad \text{and} \qquad M_n = \sum_{k=1}^{n} a_{kk} X_k^2$$

Then $S_n = K_n + L_n + M_n$. Show that $\{L_n\}$ and $\{K_n\}$ are square-integrable martingale sequences, and M_n is the sum of an absolutely integrable martingale sequence and a sequence of constants.) (Varberg [84])

27. Let $\{\mathbf{X}_n = (X_{n1}, X_{n2}, \ldots, X_{np}), n \geq 1\}$ be a sequence of independent, identically distributed random variables defined on $(\Omega, \mathscr{S}, P)$, where $\mathbf{X}_n$ has a continuous distribution function F defined on $\mathbb{R}_p$. Let F_n be the sample distribution function of $\mathbf{X}_1, \mathbf{X}_2, \ldots, \mathbf{X}_n$ In other words, for $\mathbf{x} \in \mathbb{R}_p$

$$F_n(\mathbf{x}) = n^{-1} \sum_{j=1}^{n} \epsilon(\mathbf{x} - \mathbf{X}_j), \qquad \mathbf{x} = (x_1, \ldots, x_p), \quad n \geq 1,$$

where $\epsilon(\mathbf{x}) = 1$ if all $x_1, x_2, \ldots, x_p \geq 0$, and $\epsilon(\mathbf{x}) = 0$ otherwise. Then the statistics

$$D_n^+ = \sup_{\mathbf{x} \in \mathbb{R}_p} \{F_n(\mathbf{x}) - F(\mathbf{x})\}, \qquad D_n^- = \sup_{\mathbf{x} \in \mathbb{R}_p} \{F(\mathbf{x}) - F_n(\mathbf{x})\}$$

and

$$D_n = \max\{D_n^+, D_n^-\}, \qquad n \geq 1,$$

are called the general p-variate Kolmogorov-Smirnov statistics. For each $n \geq 1$ let $\mathscr{D}_n$ be the σ-field generated by $(\mathbf{X}_1, \mathbf{X}_2, \ldots, \mathbf{X}_n)$ and $(\mathbf{X}_{n+1}, \mathbf{X}_{n+2}, \ldots)$.

(a) Show that $\mathscr{D}_n \downarrow$.
(b) Show that both $\{D_n^+, \mathscr{D}_n\}$ and $\{D_n^-, \mathscr{D}_n\}$ are nonnegative reverse submartingales for every $p \geq 1$.

28. Let $X_1, X_2, \ldots, X_n$ be independent random variables, and suppose that X_i has a continuous distribution function F_i, $i = 1, 2, \ldots, n$. For testing the hypothesis

$$H_0\colon F_1(x) = F_2(x) = \cdots = F_n(x) = F(x), \qquad x \in \mathbb{R},$$

where F is continuous but unknown, a linear rank statistic of the form

$$L_n = \sum_{i=1}^{n} (c_i - \bar{c}) a_n(R_{ni})$$

is often used, where c_i are known constants, $\bar{c} = n^{-1} \sum_{i=1}^{n} c_i$, $R_{ni} = \sum_{j=1}^{n} \epsilon(X_i - X_j)$, $1 \leq i \leq n$ [$\epsilon(x) = 1$ if $x \geq 0$ and $=0$ otherwise], and $a_n(i)$, $1 \leq i \leq n$, are suitable scores. Let φ be an integrable function on (0, 1), and $U_{n1} < U_{n2} < \cdots < U_{nn}$ be the order statistic from the uniform distribution on (0, 1). Let $\mathscr{D}_n = \sigma(R_{n1}, R_{n2}, \ldots, R_{nn})$. Then $\{\mathscr{D}_n\}$ is a nondecreasing sequence of sub-σ-fields. Choose $a_n^0(i) = \mathscr{E}\varphi(U_{ni})$, $1 \leq i \leq n$, and set

$$L_n^0 = \sum_{i=1}^{n} (c_i - \bar{c}) a_n^0(R_{ni}), \qquad n \geq 1.$$

Show that $(L_n^0, \mathscr{D}_n)$ is a martingale sequence under H_0.

29. Let $X_1, X_2, \ldots, X_n$ be independent, identically distributed random variables with common continuous distribution function F. For testing the hypothesis of symmetry that

$$H_0: F(x) + F(-x) = 1 \qquad \text{for all } x \geq 0,$$

one uses a signed rank statistic of the form

$$S_n = \sum_{j=1}^{n} \operatorname{sgn} X_j a_n(R_{nj}^+),$$

where a_n is a score function and $R_{nj}^+ = \sum_{i=1}^n \epsilon(|X_j| - |X_i|)$ is the rank of $|X_j|$ among $|X_1|, |X_2|, \ldots, |X_n|$. [Here $\epsilon(x) = 1$ if $x \geq 0$ and $=0$ otherwise.) Let φ be an integrable function on (0, 1), and let $U_{n1}, \ldots, U_{nn}$ be the order statistic of a sample of size n from the uniform distribution on (0, 1). Choose $a_n(i) = \mathscr{E}\varphi(U_{ni})$, $i = 1, 2, \ldots, n$. Let $\mathscr{D}_n$ be the σ-field generated by $(\operatorname{sgn} X_1, \operatorname{sgn} X_2, \ldots, \operatorname{sgn} X_n)$ and $(R_{n1}^+, R_{n2}^+, \ldots, R_{nn}^+)$.

(a) Show that $\mathscr{D}_n$ is nondecreasing.
(b) Under H_0 the sequence $\{S_n, \mathscr{D}_n\}$ is a martingale, where S_n is as defined above with scores $a_n(R_{nj}^+) = \mathscr{E}\varphi(U_{nR_{nj}^+})$, $j = 1, 2, \ldots, n$.

30. Let $\{X_n\}$ be a sequence of random variables, and let $\mathscr{D}_{nm}$ be the σ-field generated by $X_n, X_{n+1}, \ldots, X_m$, $m \geq n$. We say that the sequence $\{X_n\}$ is **-mixing* if there exist a positive integer N and a nondecreasing function ψ defined on integers $n \geq N$ with $\lim_{n\to\infty} \psi(n) = 0$, such that for $n \geq N$, $A \in \mathscr{D}_{1m}$, and $B \in \mathscr{D}_{m+n,m+n}$ the relation

$$|P(A \cap B) - P(A)P(B)| \leq \psi(n)P(A)P(B)$$

holds for any integer $m \geq 1$.

(a) Show that the *-mixing condition is equivalent to the condition

$$|P\{B|\mathscr{D}_{1m}\} - P(B)| \leq \psi(n)P(B) \text{ a.s.}$$

for $B \in \mathscr{D}_{m+n,m+n}$ and $m \geq 1$, and implies

$$|\mathscr{E}\{X_{m+n}|X_1, X_2, \ldots, X_m\} - \mathscr{E}(X_{m+n})| \leq \psi(n)\mathscr{E}|X_{m+n}|$$

with probability 1.

(b) (Blum, Hanson and Koopman) Let $\{X_n\}$ be a *-mixing sequence such that $\mathscr{E}(X_n) = 0$ and $\mathscr{E}(X_n)^2 < \infty$, $n \geq 1$. Suppose that $\mathscr{E}|X_n| \leq c$, $n \geq 1$, where $c > 0$ is a constant and $\sum_{n=1}^{\infty} \mathscr{E}(X_n^2)/n^2 < \infty$. Show that $S_n/n \xrightarrow{\text{a.s.}} 1$. (*Hint*: Use part (a) above and Corollary 2 to Proposition 6.2.8.)

31. Let $\{X_n, \mathscr{D}_n\}$ be a submartingale. Show that for every $\epsilon > 0$ the following inequalities hold:

(a) $$\epsilon P\left\{\max_{1\leq k\leq n} X_k \geq \epsilon\right\} \leq \int_{\{\max_{1\leq k\leq n} X_k \geq \epsilon\}} X_n \, dP \leq \mathscr{E}(X_n^+).$$

(b) $$\epsilon P\left\{\min_{1\leq k\leq n} X_k \leq -\epsilon\right\} \leq \mathscr{E}(X_n - X_1) - \int_{\{\min_{1\leq k\leq n} X_k \leq -\epsilon\}} X_n \, dP \leq \mathscr{E}(X_n^+) - \mathscr{E}(X_1).$$

Deduce that, if $\{X_n\}$ is a martingale sequence, then for every $\epsilon > 0$

$$\epsilon P\left\{\max_{1 \le k \le n} |X_k| \ge \epsilon\right\} \le \int_{\{\max_{1 \le k \le n} |X_k| \ge \epsilon\}} |X_n| \, dP.$$

32. Let $\{X_n\}$ be a backward submartingale sequence. Show that there exists an extended random variable X such that $X_n \xrightarrow{\text{a.s.}} X$.

SECTION 6.3

33. Let $\{X_n\}$ be a sequence of uniformly integrable random variables. Show that

$$\int_\Omega \left(\liminf_{n\to\infty} X_n\right) dP \le \liminf_{n\to\infty} \int_\Omega X_n \, dP$$
$$\le \limsup_{n\to\infty} \int_\Omega X_n \, dP \le \int_\Omega \left(\limsup_{n\to\infty} X_n\right) dP.$$

Deduce that, if $X_n \to X$ in probability or a.s., then $X \in \mathscr{L}_1$ and $\mathscr{E}X_n \to \mathscr{E}X$ as $n \to \infty$.

34. Let $\{X_n\}$ be a sequence of independent, identically distributed random variables such that $\mathscr{E}|X_1| < \infty$. Show that the sequence $\{\sum_{k=1}^n X_k/n\}$ is uniformly integrable.

35. (a) Let $\{X_1, X_2, \ldots, X\}$ be a submartingale. Show that for every $\epsilon > 0$

$$\epsilon P\left\{\sup_{n \ge 1} X_n \ge \epsilon\right\} \le \int_{\{\sup_{n \ge 1} X_n \ge \epsilon\}} X \, dP.$$

In particular, if $X_1, X_2, \ldots, X$ is a martingale, then for every $\epsilon > 0$

$$P\left\{\sup_{n \ge 1} |X_n| \ge \epsilon\right\} \le \epsilon^{-1} \int_{\{\sup_{n \ge 1} |X_n| \ge \epsilon\}} |X| \, dP.$$

(b) If $X_1, X_2, \ldots$ is a backward martingale, show that for every $\epsilon > 0$

$$P\left\{\sup_{n \ge 1} |X_n| \ge \epsilon\right\} \le \epsilon^{-1} \int_{\{\sup_{n \ge 1} |X_n| \ge \epsilon\}} |X| \, dP.$$

36. Let $\{X_1, X_2, \ldots, X\}$ be a submartingale. Show that for any $\epsilon \in \mathbb{R}$ the sequence $\{\max(X_n, \epsilon)\}$ is uniformly integrable. In particular, if $\{X_1, X_2, \ldots, X\}$ is a martingale and for some $p \ge 1$ the random variables $|X_1|^p, |X_2|^p, \ldots, |X|^p$ are integrable, show that $\{|X_n|^p\}$ is uniformly integrable.

37. If $\{X_n\}$ is a reverse martingale, show that $\{X_n\}$ is uniformly integrable. Show also that there exists a random variable X with $\mathscr{E}|X| < \infty$ such that $X_n \xrightarrow{\text{a.s.}, \mathscr{L}_1} X$. [Use the inequality in part (b) of Problem 35 and the result in Problem 32. Show that in view of the uniform integrability of $\{X_n\}$ the limit X has to be finite a.s.]

38. Let $\{X_n, \mathscr{D}_n\}$ be a submartingale. Then show that $\{X_n^+\}$ is uniformly integrable if and only if there exists an integrable random variable X such that $X_n \xrightarrow{\text{a.s.}} X$ and such that $\{X_1, X_2, \ldots, X\}$ is a submartingale with respect to $\{\mathscr{D}_1, \mathscr{D}_2, \ldots, \mathscr{D}\}$, where $\mathscr{D} = \sigma(\bigcup_{n=1}^{\infty} \mathscr{D}_n)$.

39. Let $\{X_n\}$ be a sequence of random variables such that $\{|X_n|^p\}$ is uniformly integrable for some $0 < p \leq 1$. Let $S_n = \sum_{k=1}^{n} X_k, n \geq 1$. Show that $\mathscr{E}|S_n - a_n|^p = o(n)$ as $n \to \infty$, where $a_n = 0$ if $0 < p < 1$, and $a_n = \sum_{k=1}^{n} \mathscr{E}\{X_k | X_1, \ldots, X_{k-1}\}$ if $p = 1$.

SECTION 6.4

40. (a) Let R_n denote the nth Rademacher function (Problem 1.5.48), and $\{c_n\}$ be a sequence of constants such that $\sum_{n=1}^{\infty} c_n^2 = \infty$. Show that the series $\sum_{n=1}^{\infty} c_n R_n$ diverges a.s.

(b) Let R_n denote the nth Rademacher function, and let $c_{n,k}$ be constants such that $\sum_{n=2}^{\infty} \sum_{k=1}^{n-1} c_{n,k}^2 < \infty$. Consider the series

$$\sum_{n=2}^{\infty} \sum_{k=1}^{n-1} c_{n,k} R_n(x) R_k(x),$$

and for $N = 2, 3, \ldots, r = 1, 2, \ldots, N$ set

$$S_{\{N(N-1)/2\}+r}(x) = \sum_{n=2}^{N} \sum_{k=1}^{n-1} c_{n,k} R_n(x) R_k(x) + \sum_{k=1}^{r} c_{N+1,r} R_{N+1}(x) R_k(x).$$

Show that $\{S_{N(N-1)/2}\}$ is a martingale sequence such that $\mathscr{E}S^2_{N(N-1)/2} < \infty$. Also show that S_n converges a.s. in (0, 1).

41. Consider the Haar orthonormal system $\{\text{H}_n\}$ defined on [0, 1] as follows:

$$H_0(x) = 1 \qquad \text{for } 0 \leq x \leq 1$$

and for $n \geq 1$

$$H_n(x) = \begin{cases} 2^{k/2} & \text{for } 0 \leq x < 2^{-k-1}, \\ -2^{k/2} & \text{for } 2^{-k-1} \leq x < 2^{-k}, \\ 0 & \text{for } 2^{-k} \leq x < 1, \end{cases}$$

if $n = 2^k$, $k = 0, 1, 2, \ldots,$ and

$$H_n(x) = \begin{cases} H_{2^k}(x - 2^{-k}j) & \text{if } 2^{-k}j \leq x < 2^{-k}(j+1), \\ 0 & \text{otherwise,} \end{cases}$$

if $n = 2^k + j$, $k = 1, 2, \ldots$ and $1 \leq j \leq 2^k - 1$.

Consider the probability space $([0, 1), \mathscr{B}_{[0,1)}, \lambda)$, where λ is the Lebesgue measure on [0, 1]. Let f be a measurable function on [0, 1] such that $\int_0^1 f^2(x)\, dx < \infty$. Consider the expansion $\sum_{n=0}^{\infty} c_n H_n(x)$ of x, where

$$c_n = \int_0^1 f(t)H_n(t)\,dt, \qquad n \geq 0,$$

and set $S_N(x) = \sum_{n=0}^{N} c_n H_n(x)$, $N \geq 0$.

(a) Show that $\{S_N(x), N \geq 0\}$ is a martingale.
(b) Let $\mathscr{H}_N$ denote the σ-field of subsets of $[0, 1)$ generated by $H_1, H_2, \ldots, H_N$. Show that a.s.

$$\mathscr{E}\{f(x)\,|\,\mathscr{H}_N\}(x) = S_N(x), \qquad N \geq 0.$$

(c) Show that $S_N(x)$ converges a.s. and in $\mathscr{L}_2$ to $f(x)$.

46. In the notation of Proposition 6.4.5 show that $e^{itS_n} \xrightarrow{\text{a.s.}} \varphi(t)Z$ as $n \to \infty$ implies that S_n converges a.s., that is, supply the details of the proof.

43. Let p be a continuous probability density function on $\mathbb{R}$. Show that the only bounded and continuous solutions to the functional equation

$$f(y) = \int_{-\infty}^{\infty} f(y + x)p(x)\,dx, \; y \in \mathbb{R},$$

are constant functions.
(*Hint*: Let $\{X_n\}$ be independent and identically distributed with probability density function p, and set $S_n = \sum_{k=1}^{n} X_k$. Then, for every $x \in \mathbb{R}$, $\{f(x + S_n)\}$ is a martingale sequence which converges a.s. and in mean square to a constant (independent of x).)

44. Let f be any Borel-measurable function on $[0, 1)$ such that $\int_0^1 |f(x)|\,dx < \infty$. For $\omega \in [0, 1)$ define the step function

$$f_n(\omega) = 2^n \int_{k2^{-n}}^{(k+1)2^{-n}} f(y)\,dy \qquad \text{if } k2^{-n} \leq \omega < (k+1)2^{-n}.$$

Show that $f_n \xrightarrow{\text{a.s.}} f$ on $[0, 1)$.
(*Hint*: In the notation of Problem 24 set $Z = f(U)$. Then $\mathscr{E}\{Z\,|\,Y_0, Y_1, \ldots, Y_n\} = f_n(U)$ a.s.)

45. (Stochastic Approximation) Let F be a distribution function on $\mathbb{R}$, $\alpha \in (0, 1)$, and suppose there exists a $\theta \in \mathbb{R}$ such that $F(\theta) = \alpha$. Let F be differentiable at θ, and $F'(\theta) > 0$. Let

$$X_{n+1} = X_n - \frac{1}{n}(Y_n - \alpha),$$

where Y_n is a random variable such that

$$P\{Y_n = y\,|\,X_1, X_2, \ldots, X_n, Y_1, \ldots, Y_{n-1}\} = \begin{cases} F(X_n) & \text{if } y = 1, \\ 1 - F(X_n) & \text{if } y = 0. \end{cases}$$

Show that X_n converges in $\mathscr{L}_2$ to θ.

[In many problems one observes whether or not a subject responds to a certain dose, and the problem is to determine a critical dose for a given quantitative response. Let Z be a random variable with distribution function F. If $x \in \mathbb{R}$, and $Y(x)$ is the random variable defined by $Y(x) = 1$ if $Z \le x$ (response) and $Y(x) = 0$ if $Z > x$ (nonresponse), then $P\{Y(x) = y\} = F(x)$ if $y = 1$ and $P\{Y(x) = y\} = 1 - F(x)$ if $y = 0$ and $\mathscr{E}Y(x) = F(x)$. Let $\alpha \in (0, 1)$. Then the problem is to determine x for a given response α. The procedure suggested here is a simple case of the Robbins-Monro method.]

46. Consider an infinite sequence $\mathbf{X} = (X_1, X_2, \ldots)$ of exchangeable random variables [that is, for each $n \ge 2$ let $(X_1, X_2, \ldots, X_n)$ be exchangeable]. A Borel-measurable function f of $\mathbf{X}$ is said to be *n-symmetric* if it remains invariant under permutations of $X_1, X_2, \ldots, X_n$. Let $\mathscr{D}_n$ be the σ-field generated by the class of all n-symmetric functions of $\mathbf{X}$. Then $\mathscr{D}_n \supset \mathscr{D}_{n+1}, n \ge 1$.

(a) Let f be a Borel-measurable function on $\mathbb{R}$ for which $\mathscr{E}|f(X_1)| < \infty$, and let $Y = g(\mathbf{X})$ be a bounded n-symmetric random variable. Show that for $1 \le j \le n$

$$\mathscr{E}\left\{n^{-1}\sum_{j=1}^{n} Yf(X_j)\right\} = \mathscr{E}\{Yf(X_1)\}.$$

Conclude that for $n \ge 1$

$$\mathscr{E}\{f(X_1)|\mathscr{D}_n\} = n^{-1}\sum_{j=1}^{n} f(X_j) \text{ a.s.}$$

(b) (Strong Law of Large Numbers) Show that

$$n^{-1}\sum_{j=1}^{n} f(X_j) \xrightarrow{\text{a.s.}} \mathscr{E}\{f(X_1)|\mathscr{D}_\infty\},$$

where $\mathscr{D}_\infty = \bigcap_{n=1}^{\infty} \mathscr{D}_n$.

(c) (De Finetti) For $x \in \mathbb{R}$ set $F(x) = P\{X_1 \le x | \mathscr{D}_\infty\}$. Show that the X_n are independent and identically distributed conditional on the σ-field $\mathscr{D}_\infty$. In other words, show that

$$P\{X_1 \le x_1, X_2 \le x_2, \ldots, X_k \le x_k | \mathscr{D}_\infty\} = F(x_1)F(x_2)\cdots F(x_k) \text{ a.s.}$$

holds for every positive integer $k \ge 1$ and $x_1, x_2, \ldots, x_k \in \mathbb{R}$.

47. Let $\{X_n\}$ be a sequence of random variables such that $\mathscr{E}(X_n^2) < \infty$ for all $n \ge 1$ and

$$\mathscr{E}(X_m X_n) = 0 \qquad \text{for all } m \ne n.$$

Let $S_n = \sum_{k=1}^{n} X_k$, and suppose that $\{S_n\}$ is a submartingale sequence. Show that if, for some $0 < p \le 2$, $\sum_{n=1}^{\infty} \mathscr{E}|X_n|^p < \infty$ then S_n converges a.s.

48. In the notation of Section 6.4(c), let $0 < \mu < \infty$. Set $\varphi(s) = \mathscr{E}(s^{Z_j(n)})$ and $\varphi_n(s) = \mathscr{E}(s^{X_n})$, $n \ge 1$, for $|s| \le 1$. Show that for $n \ge 2$ and $|s| \le 1$

$$\varphi_n(s) = \varphi_{n-1}(\varphi(s)),$$

and conclude that

$$\mathscr{E}(X_n) = \varphi_n'(1) \quad \text{and} \quad \operatorname{var}(X_n) = \varphi_n''(1) + \varphi_n'(1) - [\varphi_n'(1)]^2.$$

Setting $\varphi''(1) = \mathscr{E}(X_1)^2 - \mathscr{E}(X_1) = \sigma^2 + \mu^2 - \mu = K$, where $\sigma^2 = \operatorname{var}(X_1)$, show that

$$\operatorname{var}(X_n) = \begin{cases} nX_0\sigma^2 + X_0(1 - X_0) & \text{if } \mu = 1, \\ X_0\sigma^2\mu^{n-1}\dfrac{(\mu^n - 1)}{\mu - 1} + \mu^{2n}X_0(1 - X_0) & \text{if } \mu \neq 1. \end{cases}$$

In particular, show that $\mathscr{E}(X_n/\mu^n)^2 \leq c$, where $c > 0$ is a constant (independent of n).

49. In the notation of Section 6.4(c) show that for $\mu > 1$ the equation $\varphi(\beta) = \beta$ has a unique root in $[0, 1)$. If $0 < \beta < 1$, show that, with probability β, X_n is eventually zero (that is, extinction occurs with probability β), and with probability $1 - \beta$, $X_n \xrightarrow{\text{a.s.}} \infty$. In the case $\mu = 1$, show that extinction occurs with probability 1.

SECTION 6.5

50. Let N_1 be a stopping time with respect to a nondecreasing sequence $\{\mathscr{D}_n\}$ of σ-fields, and let N_2 be an integer-valued, $\mathscr{D}_{N_1}$-measurable random variable such that $N_2 \geq N_1$ a.s. Show that N_2 is also a stopping time.

51. (Optional Sampling) Let $\{X_n\}$ be a sequence of random variables, and let $\{T_n\}$ be an increasing sequence of stopping times. Then the sequence defined by $Y_n = X_{T_n}$, $n = 1, 2, \ldots$, is known as the sequence derived by *optional sampling* from the sequence $\{X_n\}$.

(a) Let $\{X_n\}$ be a submartingale (martingale) sequence, $\{T_n\}$ an increasing sequence of stopping times, and $\{X_{T_n}\}$ the optional sampling sequence derived from $\{X_n\}$. Show that if the following conditions hold:

(i) $\mathscr{E}|X_{T_n}| < \infty$ for all n,
(ii) $\liminf_{N\to\infty} \int_{\{T_n > N\}} |X_N|\, dP = 0$ for all n,

then $\{X_{T_n}\}$ is a submartingale (martingale).

(b) Show that conditions (i) and (ii) in part (a) hold in each of the following cases:
(1) For each $n \geq 1$, there exists a positive constant c_n such that $T_n \leq c_n$ a.s.
(2) $\mathscr{E}(\sup_{n\geq 1} |X_n|) < \infty$.

(c) Suppose that conditions (i) and (ii) in part (a) hold, and in addition

$$\limsup_{n\to\infty} \mathscr{E}|X_n| < \infty.$$

Show that:
(1) $\mathscr{E}(X_1) \leq \mathscr{E}X_{T_n} \leq \limsup_{n\to\infty} \mathscr{E}(X_n)$.
(2) $\mathscr{E}|X_{T_n}| \leq 2 \limsup_{n\geq 1} \mathscr{E}|X_n| - \mathscr{E}(X_1)$.

(d) Suppose that in part (a) condition (ii) holds, and in addition $\limsup_{n\to\infty} \mathscr{E}|X_n| < \infty$. Show that $\sup_{n\geq 1} \mathscr{E}|X_{T_n}| < \infty$. In particular, it follows that the conclusion in part (a) holds.

52. Let $\{X_1, X_2, \ldots, X\}$ be a submartingale (martingale) sequence, and let $\{T_n\}$ be an increasing sequence of stopping times for $\{X_n\}$. Show that $\{X_{T_n}\}$ is a submartingale (martingale) sequence.

53. Let $(\Omega, \mathscr{S}, P)$ be a probability space, and $\{\mathscr{D}_n\}$ an increasing sequence of sub-σ-fields of $\mathscr{S}$. Let $\{X_n\}$ be a sequence of integrable random variables on $(\Omega, \mathscr{S}, P)$ such that X_n is $\mathscr{D}_n$-measurable, $n \geq 1$. Let N be a stopping time with respect to $\{\mathscr{D}_n\}$ such that $\mathscr{E}(X_N)$ exists and is finite. The *value* V of $\{X_n, \mathscr{D}_n\}$ is defined to be $\sup \mathscr{E}(X_N)$, where the supremum is taken over all stopping times N such that $\mathscr{E}(X_N)$ is finite. Then $-\infty < \mathscr{E}(X_1) \leq V \leq \infty$. If there exists a stopping time N such that $\mathscr{E}(X_N)$ exists and equals V, we call it an *optimal* stopping time.

(a) Let X_n be degenerate at $1 - (1/n)$, $n \geq 1$, that is, let $P\{X_n = 1 - (1/n)\} = 1$. Find the value V. Does there exist an optimal stopping time?

(b) Let $\{Y_n\}$ be independent and identically distributed with common probability distribution $P\{Y_1 = \pm 1\} = \frac{1}{2}$, and let

$$X_n = \frac{n2^n}{n+1} \prod_{j=1}^{n} \left(\frac{Y_j + 1}{2}\right), \qquad n \geq 1.$$

Show that

$$\mathscr{E}\left\{X_{n+1} \,|\, X_n = \frac{n2^n}{n+1}\right\} > X_n \text{ a.s.}$$

Consider the class of stopping times $\{N_k, k \geq 1\}$, where $N_k = k$. Show that $V = 1$ but no optimal stopping time exists in this class.

(c) Let $\{Y_n\}$ be as in part (b), but

$$X_n = \min\left(1, \sum_{j=1}^{n} Y_j\right) - \frac{n}{n+1}, \qquad n \geq 1.$$

Define

$$N = \text{first } n \geq 1 \text{ such that } \sum_{j=1}^{n} Y_j = 1.$$

Show that N is a stopping time,

$$\mathscr{E}(X_N) = 1 - \mathscr{E}\left(\frac{N}{N+1}\right) > 0,$$

and N is optimal, so that $V = \mathscr{E}(X_N)$.

NOTES AND COMMENTS

As pointed out earlier, the concept of conditional probability is basic to the study of mathematical statistics and probability theory. The most rigorous treatment of conditional probability (and conditional expectation) was first given by Kolmogorov [48].

The theory of martingales was systematically developed by Doob [17]. For applications to optimal stopping we refer the reader to Chow, Robbins, and Siegmund [12].

CHAPTER 7

Random Variables Taking Values in a Normed Linear Space

In preceding chapters we studied, for the most part, the classical probability theory on the real line. All the well-known results like the laws of large numbers, the central limit theorem, and the law of the iterated logarithm concerned real-valued random variables. This theory can be extended to finite-dimensional random variables without much difficulty. In the case of infinite-dimensional random variables, however, some problems arise that need special consideration. Recent consideration of a stochastic process as a random element in a function space has inspired the study of laws of large numbers, the central limit theorem, and the law of the iterated logarithm for random elements in abstract spaces. In this chapter we derive some of these results for random variables taking values in a normed linear space.

In Sections 7.1 through 7.3 we derive analogues of the weak and the strong laws of large numbers for independent random variables taking values in a Banach space. In Section 7.4 we introduce the concept of the characteristic functional and use it to establish the convergence equivalence of a series of independent random variables taking values in a Banach space.

In Sections 7.5 through 7.8 we restrict our investigation to random variables taking values in a Hilbert space. In particular, we derive the Lévy–Khintchine representation of an infinitely divisible probability measure on a Hilbert space and study in detail the general central limit problem.

7.1 DEFINITIONS AND PRELIMINARY RESULTS

Let $(\Omega, \mathcal{S}, P)$ be a probability space, and B be a real Banach space with norm $\|\cdot\|$. Clearly B is a topological vector space with respect to the metric topology induced by $\|\cdot\|$. Let $\mathcal{B}$ be the σ-field generated by the class of all open subsets of B. Then $\mathcal{B}$ is known as the *Borel σ-field* on B, and elements of $\mathcal{B}$ are called *Borel sets.*

Definition 7.1.1. A mapping $X: \Omega \to B$ is called a B-valued random variable if X is $\mathcal{B}$-measurable, that is, for every $E \in \mathcal{B}$

$$X^{-1}(E) = \{\omega: X(\omega) \in E\} \in \mathcal{S}.$$

Remark 7.1.1. Let $X: (\Omega, \mathcal{S}, P) \to (B, \mathcal{B})$ be a random variable, and let B_0 be another Banach space with Borel σ-field $\mathcal{B}_0$. Let T be a measurable mapping of $(B, \mathcal{B})$ into $(B_0, \mathcal{B}_0)$. Then $T(X)$ is a B_0-valued random variable.

Remark 7.1.2. Let B^* be the dual (or conjugate) of B, that is, B^* is the Banach space consisting of all bounded (continuous) linear functionals on B. Then it follows immediately from Remark 7.1.1 that, for every $l \in B^*$, $l(X)$ is a real-valued random variable. In particular, if B is separable and $l(X)$ is a random variable for every $l \in B^*$, then X is a B-valued random variable. (See, for example, Padgett and Taylor [63], p. 20.)

Proposition 7.1.1. Let B be a separable Banach space, and X a B-valued random variable. Then $\|X\|$ is a random variable.

Proof. Since B is separable, there exists a sequence $\{x_n \in B: n \geq 1\}$ which is everywhere dense in B. Using the Hahn–Banach theorem (Royden [72], p. 187) for every $n \geq 1$, we can construct a bounded linear functional $l_n \in B^*$ such that $l_n(x_n) = \|x_n\|$ and $|l_n(x)| \leq \|x\|$ for every $x \in B$. Set $\theta(x) = \sup_{n \geq 1} l_n(x)$ for $x \in B$. Then for every $x \in B$

$$l_n(x) \leq |l_n(x)| \leq \|x\|,$$

so that

$$\theta(x) \leq \|x\|, \qquad x \in B.$$

On the other hand, for every $n \geq 1$

$$\begin{aligned}\theta(x) \geq l_n(x) &= l_n(x_n) + l_n(x - x_n)\\ &\geq \|x_n\| - \|x - x_n\|\\ &\geq \|x\| - 2\|x - x_n\|.\end{aligned}$$

Let $\epsilon > 0$. Since $\{x_n\}$ is dense in B, there exists an $n_0 = n_0(\epsilon)$ such that

$$\|x - x_n\| < \frac{\epsilon}{2} \qquad \text{for } n \geq n_0,$$

so that for $n \geq n_0$

$$\theta(x) > \|x\| - \epsilon.$$

Since ϵ is arbitrary, it follows that

$$\theta(x) = \|x\| \qquad \text{for } x \in B.$$

Let $\omega \in \Omega$ be fixed. Then we have

$$\|X\|(\omega) = \|X(\omega)\| = \theta(X(\omega)) = \sup_{n \geq 1} l_n(X(\omega)),$$

so that $\|X\| = \sup_{n\geq 1} l_n(X)$, and since $l_n(X)$ is a random variable for every $n \geq 1$, so is $\|X\|$. ■

Let P_X be the set function on $\mathscr{B}$ defined by

$$P_X(E) = P(X^{-1}(E)), \qquad E \in \mathscr{B}.$$

Clearly P_X is a probability measure on $\mathscr{B}$ induced by X and is known as the *probability distribution* of X.

Definition 7.1.2. Let X and Y be two B-valued random variables defined on $(\Omega, \mathscr{S}, P)$. Then X and Y are said to be identically distributed if $P_X = P_Y$. A collection of B-valued random variables is said to be identically distributed if every pair has the same probability distribution. In particular, X is said to be symmetric if X and $-X$ are identically distributed. Here $-X$ is the B-valued random variable defined by $(-X)(\omega) = -X(\omega)$ for all $\omega \in \Omega$.

Remark 7.1.3. Let $\{X_\alpha, \alpha \in A\}$ be a collection of identically distributed B-valued random variables, and let T be a measurable mapping of $B \to B_0$, where B_0 is a Banach space. Then $\{T(X_\alpha), \alpha \in A\}$ is a collection of identically distributed B_0-valued random variables. In particular, if B is separable and, for every $l \in B^*$, $\{l(X_\alpha), \alpha \in A\}$ is a collection of identically distributed random variables, then $\{X_\alpha, \alpha \in A\}$ are identically distributed.

Definition 7.1.3. A (finite) set of B-valued random variables $\{X_1, X_2, \ldots, X_n\}$ is said to be independent if for every $E_1, E_2, \ldots, E_n \in \mathscr{B}$ the relation

$$P\{X_1 \in E_1, \ldots, X_n \in E_n\} = \prod_{j=1}^{n} P\{X_j \in E_j\} \tag{7.1.1}$$

holds. More generally, a collection of B-valued random variables is said to be independent if every finite subset is independent in the sense of (7.1.1).

Remark 7.1.4. Let $\{X_\alpha, \alpha \in A\}$ be a collection of independent B-valued random variables, and let $\{T_\alpha, \alpha \in A\}$ be a collection of measurable mappings from $B \to B_0$, where B_0 is also a Banach space. Then $\{T_\alpha(X_\alpha), \alpha \in A\}$ is a collection of independent B_0-valued random variables. In particular, if B is separable and $\{l_\alpha(X_\alpha), \alpha \in A, l_\alpha \in B^*\}$ is a collection of independent random variables, $\{X_\alpha, \alpha \in A\}$ is also independent.

The expected value of a B-valued random variable, where B is separable, is defined in terms of a Pettis integral.

Definition 7.1.4. Let B be a separable Banach space, and let X be a B-valued random variable. We say that the expected value of X exists if the following conditions are met:

(i) $\mathscr{E}|l(X)| < \infty$ for all $l \in B^*$.
(ii) There exists an element $\mathscr{E}X \in B$ such that the relation

$$l(\mathscr{E}X) = \mathscr{E}l(X) = \int_\Omega l(X)\, dP \tag{7.1.2}$$

holds for all $l \in B^*$. Here $\mathscr{E}X$ is called the expected value of X and is unique.

We note that if $\mathscr{E}\|X\| < \infty$ then $\mathscr{E}X$ exists. The following proposition gives some properties of $\mathscr{E}X$.

Proposition 7.1.2. Let B be a separable Banach space, and let X be a B-valued random variable such that $\mathscr{E}X$ exists. Then the following conditions hold:

(i) For every $\alpha \in \mathbb{R}$, $\mathscr{E}(\alpha X) = \alpha\mathscr{E}X$.
(ii) If $P(X = x) = 1$, then $\mathscr{E}X = x$; moreover, if in addition ζ is a real-valued random variable such that $\mathscr{E}|\zeta| < \infty$, then $\mathscr{E}(\zeta X) = x\mathscr{E}(\zeta)$.
(iii) Let T be a bounded (continuous) linear operator from $B \to B_0$, where B_0 is a Banach space. Then $\mathscr{E}T(X)$ exists, and the relation

$$\mathscr{E}T(X) = T(\mathscr{E}X)$$

holds.
(iv) The relation

$$\|\mathscr{E}X\| \leq \mathscr{E}\|X\|$$

holds where $\mathscr{E}\|X\|$ may be infinite.

(v) Let Y be a B-valued random variable such that $\mathscr{E}Y$ exists. Then $\mathscr{E}(X + Y)$ also exists, and the relation $\mathscr{E}(X + Y) = \mathscr{E}X + \mathscr{E}Y$ holds.

Proof. The proof follows immediately from (7.1.2) and the properties of linear operators. ∎

Example 7.1.1. Let $B = c$ be the set of all convergent sequences $x = (x_1, x_2, \ldots)$ of real numbers with norm $\|x\| = \sup_{n \geq 1} |x_n|$. Then B is a Banach space. Let X be a B-valued random variable defined on some probability space $(\Omega, \mathscr{S}, P)$. For $\omega \in \Omega$ let $X(\omega) = (X_1(\omega), X_2(\omega), \ldots)$. Then $\{X_n\{\omega\}\}$ is a convergent sequence of real numbers for each $\omega \in \Omega$. Suppose that

$$\mathscr{E}\|X\| = \mathscr{E}\sup_{n \geq 1} |X_n| < \infty.$$

Then $\mathscr{E}X = (\mathscr{E}X_1, \mathscr{E}X_2, \ldots,)$. In fact, note that $\lim_{n \to \infty} X_n$ exists a.s. and $\sup_{n \geq 1} X_n$ has a finite expectation. It follows from the Lebesgue dominated convergence theorem that $\lim_{n \to \infty} \mathscr{E}X_n$ exists. Hence $(\mathscr{E}X_1, \mathscr{E}X_2, \ldots) \in B$. Next note that the dual B^* consists of all summable sequences $l = (l_0, l_1, l_2, \ldots)$, $\sum_{n=0}^{\infty} |l_n| < \infty$. Hence for any $l \in B^*$ we have

$$\begin{aligned} l(\mathscr{E}X) &= l_0 \lim_{n \to \infty} \mathscr{E}X_n + \sum_{n=1}^{\infty} l_n \mathscr{E}X_n \\ &= \mathscr{E}\left\{l_0 \lim X_n + \sum_{n=1}^{\infty} l_n X_n\right\} \\ &= \mathscr{E}l(X), \end{aligned}$$

so that (7.1.2) holds. It follows that $\mathscr{E}X = (\mathscr{E}X_1, \mathscr{E}X_2, \ldots)$.

Example 7.1.2. Let $B = \mathscr{C}[0, 1]$, the set of all continuous real-valued functions on the interval $[0, 1]$ with norm

$$\|x\| = \sup_{0 \leq t \leq 1} |x(t)|,$$

where $x = \{x(t), 0 \leq t \leq 1\} \in B$. Then B is a separable Banach space. Let $X = \{X_t, 0 \leq t \leq 1\}$ be a B-valued random variable. Suppose that

$$\mathscr{E}\|X\| = \mathscr{E} \sup_{0 \leq t \leq 1} |X_t| < \infty.$$

We show that $\mathscr{E}X = \{\mathscr{E}X_t, 0 \leq t \leq 1\}$. First note that

$$\lim_{h \to 0} |\mathscr{E}X_{t+h} - \mathscr{E}X_t| \leq \lim_{h \to 0} \mathscr{E}|X_{t+h} - X_t| = 0$$

in view of the Lebesgue dominated convergence theorem. It follows that $\{\mathscr{E}X_t, 0 \leq t \leq 1\} \in B$. Now B^* consists of finite signed measures μ on $[0, 1]$, so for any $\mu \in B^*$ we have

$$\mu(\mathscr{E}X) = \int_0^1 \mathscr{E}X_t \, d\mu(t) = \mathscr{E} \int_0^1 X_t \, d\mu(t) = \mathscr{E}\mu(X),$$

and it follows that $\mathscr{E}X = \{\mathscr{E}X_t, 0 \leq t \leq 1\}$.

Definition 7.1.5. Let B be a separable Banach space, and let X be a B-valued random variable. We say that the variance of X exists if $\mathscr{E}X$ exists and

$$\operatorname{var}(X) = \int_\Omega \|X - \mathscr{E}X\|^2 \, dP < \infty.$$

Here $\operatorname{var}(X)$ is known as the variance of X, and its nonnegative square root $\sigma(X)$ is called the standard deviation of X.

Remark 7.1.5. Let X be a B-valued random variable such that $\operatorname{var}(X)$ exists. Then the following inequality (Chebyshev's inequality):

$$P\{\|X - \mathscr{E}X\| \geq \epsilon\} < \epsilon^{-2} \operatorname{var}(X) \tag{7.1.3}$$

holds for every $\epsilon > 0$. The proof of (7.1.3) is an immediate consequence of Proposition 7.1.1, the Chebyshev inequality for real-valued random variables, and the definition of $\operatorname{var}(X)$.

Remark 7.1.6. Moments of any order p, $1 \leq p < \infty$, can be defined in a similar manner.

We now define various types of convergence for sequences of B-valued random variables. In what follows we assume that B is separable to ensure that $\|X\|$ is a random variable.

Definition 7.1.6. Let $\{X_n\}$ be a sequence of B-valued random variables.

(a) The sequence $\{X_n\}$ is said to converge in probability to a B-valued random variable X if, for every $\epsilon > 0$, $\lim_{n\to\infty} P\{\|X_n - X\| \geq \epsilon\} = 0$. In this case we write $X_n \xrightarrow{P} X$ as $n \to \infty$.

(b) The sequence $\{X_n\}$ is said to converge almost surely (a.s.) or with probability 1 to a B-valued random variable X if

$$P\left\{\lim_{n\to\infty} \|X_n - X\| = 0\right\} = 1.$$

In this case we write $X_n \xrightarrow{\text{a.s.}} X$ as $n \to \infty$.

(c) We say that the sequence $\{X_n\}$ converges in law to a B-valued random variable X if the sequence of probability measures $\{P_n\}$ induced by $\{X_n\}$ on $\mathscr{B}$ converges weakly to the probability measure induced by X on $\mathscr{B}$.

(d) The sequence $\{X_n\}$ is said to converge in the pth mean $(1 \leq p < \infty)$ to a B-valued random variable X if $\lim_{n\to\infty} \mathscr{E}\|X_n - X\|^p = 0$, and in this case we write $X_n \xrightarrow{\mathscr{L}_p} X$ as $n \to \infty$.

Remark 7.1.7. It can be easily verified that (b) ⇒ (a), (a) ⇒ (c), and (d) ⇒ (a), as in the case of real-valued random variables.

7.2 STRONG LAW OF LARGE NUMBERS

Let $(\Omega, \mathscr{S}, P)$ be a probability space, and let B be a Banach space with Borel σ-field $\mathscr{B}$. We now consider some analogues of the strong law of large numbers (Theorem 2.3.1) for sequences $\{X_n\}$ of B-valued random variables. In the following we assume that B is separable.

Theorem 7.2.1 (Mourier). Let $\{X_n\}$ be a sequence of independent, identically distributed B-valued random variables. Suppose that $\mathscr{E}X_n$ exists. Set $S_n = \sum_{k=1}^n X_k$, $n = 1, 2, \ldots$. Then the strong law of large numbers holds, that is,

$$P\left\{\lim_{n\to\infty} \frac{\|S_n - \mathscr{E}S_n\|}{n} = 0\right\} = 1. \tag{7.2.1}$$

Proof. Without any loss of generality we assume that $\mathscr{E}X_n = 0$ for $n = 1, 2, \ldots$.

First we note that, when B is finite-dimensional, Theorem 7.2.1 is a straightforward generalization of Theorem 2.3.1. We now consider the case when B is infinite-dimensional. Since B is separable, there exists a sequence $\{x_n \in B, n \geq 1\}$ which is everywhere dense in B. Let $\epsilon > 0$ be fixed. Set

$$O_j = \{x \in B: \|x - x_j\| < \epsilon\}, \qquad j = 1, 2, \ldots,$$

and consider the sequence

$$E_1 = O_1$$

and

$$E_j = O_j - \bigcup_{i=1}^{j-1} O_i, \qquad j \geq 2.$$

Note that $\{E_j, j \geq 1\}$ is a disjoint sequence of Borel sets in B such that

$$\bigcup_{j=1}^{\infty} E_j = \bigcup_{j=1}^{\infty} O_j.$$

Define the sequence X_i^* as follows:

$$X_i^*(\omega) = \begin{cases} x_j & \text{if } X_i(\omega) \in E_j \text{ for } i \geq 1, j \geq 1, \\ 0 & \text{otherwise.} \end{cases}$$

Then $\{X_i^*\}$ is also a sequence of independent B-valued random variables satisfying

$$\|X_i^*(\omega) - X_i(\omega)\| < \epsilon \qquad \text{for all } \omega \in \Omega, \tag{7.2.2}$$

so that

$$\begin{aligned} \sum_{j=1}^{\infty} \|x_j\| P\{X_i \in E_j\} &= \mathscr{E}\|X_i^*\| \\ &\leq \mathscr{E}\|X_i\| + \epsilon < \infty \end{aligned}$$

for all $i \geq 1$. Hence there exists an integer $N = N(\epsilon)$ such that

$$\sum_{j=N}^{\infty} \|x_j\| P\{X_i \in E_j\} < \epsilon \qquad \text{for all } i \geq 1.$$

Let N be fixed, and set

$$X_i^{**} = \begin{cases} x_j & \text{if } X_i \in E_j, j \geq N, \\ 0 & \text{if } j < N, \end{cases}$$

and

$$Y_i = X_i^* - X_i^{**}, \qquad Z_i = X_i - Y_i, \quad i \geq 1.$$

Clearly $\{Y_i\}$ and $\{Z_i\}$ are sequences of independent B-valued random variables. We have for all $i \geq 1$

$$\|\mathscr{E}X_i^{**}\| = \sum_{j=N}^{\infty} \|x_j\| P\{X_i \in E_j\} < \epsilon,$$

and since $\mathscr{E}X_i = 0$,

$$\begin{aligned} \|\mathscr{E}Y_i\| &\leq \|\mathscr{E}X_i^*\| + \|\mathscr{E}X_i^{**}\| \\ &= \|\mathscr{E}(X_i^* - X_i) + \mathscr{E}X_i\| + \|\mathscr{E}X_i^{**}\| \\ &= \|\mathscr{E}(X_i - X_i^*)\| + \|\mathscr{E}X_i^{**}\| \\ &< \epsilon + \epsilon = 2\epsilon \end{aligned}$$

in view of (7.2.2).

We note also that

$$\begin{aligned} Z_i = X_i - Y_i &= (X_i - X_i^*) + (X_i^* - Y_i) \\ &= (X_i - X_i^*) + X_i^{**}, \end{aligned}$$

so that

$$\begin{aligned} \mathscr{E}\|Z_i\| &\leq \mathscr{E}\|X_i - X_i^*\| + \mathscr{E}\|X_i^{**}\| \\ &< 2\epsilon. \end{aligned}$$

We can also write for $i \geq 1$

$$\begin{aligned} X_i &= X_i - \mathscr{E}X_i \\ &= (Y_i - \mathscr{E}Y_i) + (Z_i - \mathscr{E}Z_i). \end{aligned} \tag{7.2.3}$$

Clearly $\{Y_i - \mathscr{E}Y_i\}$ is a sequence of independent, identically distributed B-valued random variables taking only $N - 1$ distinct values. In view of our remark above, the strong law of large numbers holds for the sequence $\{Y_i - \mathscr{E}Y_i\}$, that is,

$$P\left\{\lim_{n\to\infty} \left\|\frac{1}{n}\sum_{i=1}^{n}(Y_i - \mathscr{E}Y_i)\right\| = 0\right\} = 1. \tag{7.2.4}$$

On the other hand,

$$\begin{aligned} \left\|\frac{1}{n}\sum_{i=1}^{n}(Z_i - \mathscr{E}Z_i)\right\| &\leq \frac{1}{n}\sum_{i=1}^{n}\|Z_i\| + \frac{1}{n}\sum_{i=1}^{n}\|\mathscr{E}Z_i\| \\ &\leq \frac{1}{n}\sum_{i=1}^{n}\|Z_i\| + \frac{1}{n}\sum_{i=1}^{n}\mathscr{E}\|Z_i\| \\ &< \frac{1}{n}\sum_{i=1}^{n}\|Z_i\| + 2\epsilon. \end{aligned}$$

Since $\{\|Z_i\|\}$ is a sequence of independent, identically distributed random variables, Theorem 2.3.1 implies that $n^{-1}\sum_{i=1}^{n}\|Z_i\| \xrightarrow{\text{a.s.}} 0$. In view of (7.2.3) and (7.2.4) the proof of (7.2.1) is now complete. ■

Remark 7.2.1. For the case of nonidentically distributed, independent B-valued random variables the strong law of large numbers holds when some additional restrictions are imposed on the Banach space B. These include, for example, the Beck convexity condition (Problem 7.9.5) and the g_α-conditions due to Woyczynski [90]. We will not pursue this line of investigation here.

We next consider an extension of Proposition 2.3.7 for the special case of Hilbert-space-valued random variables.

Theorem 7.2.2. Let $\mathscr{H}$ be a real separable Hilbert space, and let $\{X_n\}$ be a sequence of independent $\mathscr{H}$-valued random variables, each having a finite variance. Suppose that $\sum_{n=1}^{\infty} \operatorname{var}(X_n)/n^2 < \infty$. Then $\{X_n\}$ satisfies the strong law of large numbers.

Proof. The method of proof parallels the proof of Proposition 2.3.7. First note that the following extension of Kolmogorov's inequality:

$$P\left\{\max_{1 \le k \le n} \left\| \sum_{i=1}^{k} (X_i - \mathscr{E}X_i) \right\| \ge \epsilon \right\} \le \epsilon^{-2} \sum_{i=1}^{n} \operatorname{var}(X_i) \tag{7.2.5}$$

holds for all $\epsilon > 0$.

It then follows immediately from (7.2.5) that, if $\sum_{n=1}^{\infty} \operatorname{var}(X_n)/n^2 < \infty$, then $\sum_{n=1}^{\infty} (X_n - \mathscr{E}X_n)/n$ converges a.s. The result now follows from an extension of Kronecker's lemma for real separable Hilbert spaces. ∎

7.3 WEAK LAW OF LARGE NUMBERS

We now consider some analogues of the weak law of large numbers (Theorem 2.1.3) for B-valued random variables. For this purpose we need to impose some additional restrictions on the Banach space B.

Definition 7.3.1. Let B be a real, separable Banach space. A sequence $\{x_n\}$ of elements of B is said to form a Schauder basis for B if for every $x \in B$ there exists a sequence $\{a_n\}$ of real numbers determined uniquely by x such that

$$\lim_{n \to \infty} \left\| x - \sum_{i=1}^{n} a_i x_i \right\| = 0. \tag{7.3.1}$$

Example 7.3.1. In a real, separable Hilbert space $\mathscr{H}$ every complete orthonormal system of elements forms a Schauder basis for $\mathscr{H}$.

Example 7.3.2. Let $\delta^1 = (1, 0, 0, \ldots)$, $\delta^2 = (0, 1, 0, \ldots)$, ... Then the space c of all real convergent sequences $x = (x_1, x_2, \ldots)$ has $\{1, \delta^1, \delta^2, \ldots\}$ as a Schauder basis. The subspace $c_0 \subset c$ consisting of all null convergent sequences has basis $\{\delta^n\}$. The space $\mathscr{C}[0, 1]$ of continuous real-valued functions on $[0, 1]$ has Schauder basis $x_n(t) = 0$ if $t \notin [0, 1]$; for $t \in [0, 1]$, $x_0(t) = t$, $x_1(t) = 1 - t$, $x_2(t) = 2t$ if $0 \le t \le \frac{1}{2}$ and $= 2(1 - t)$ if $\frac{1}{2} \le t \le 1$, $x_3(t) = x_2(2t)$, $x_4(t) = x_2(2t - 1)$, etc. In general, $x_{2^n + j}(t) = x_2(2^n t - j + 1)$, $j = 1, 2, \ldots, 2^n$.

When B has a Schauder basis $\{x_n\}$, every $x \in B$ has the expansion

$$x = \sum_{n=1}^{\infty} l_n(x) x_n$$

where the series on the right-hand side converges and $l_n(x) = a_n$ for every $n \geq 1$. It can be shown (Goffman and Pedrick [29], p. 102) that, for every n, l_n is a bounded linear functional on B. Also note that a sequence of bounded linear operators $\{T_n\}$ can be defined on B by setting

$$T_n(x) = \sum_{i=1}^{n} l_i(x) x_i, \qquad x \in B, \quad n \geq 1.$$

Here $\{l_n\}$ are called the *coordinate functionals*, and $\{T_n\}$ the *partial sum operators* for the Schauder basis $\{x_n\}$. We need yet another concept.

Definition 7.3.2. Let B be a real, separable Banach space with a Schauder basis $\{x_n\}$. We say that $\{x_n\}$ is a monotone basis if, for each $x \in B$ of the form $x = \sum_{n=1}^{\infty} a_n x_n$, the sequence $\{\|\sum_{k=1}^{n} a_k x_k\|, n \geq 1\}$, is nondecreasing.

Example 7.3.3. The spaces c_0 and l_p, $p \geq 1$, have monotone basis $\{\delta^n\}$. The space c has $\{1, \delta^1, \delta^2, \ldots\}$ as the monotone basis.

For further discussion we refer the reader to Wilansky [89], p. 86.

Proposition 7.3.1. Let B be a real, separable Banach space with a Schauder basis $\{x_n\}$, and let $\{T_n\}$ be the partial sum operators for the basis $\{x_n\}$. Then there exists a $\gamma > 0$ such that

$$\|T_n\| \leq \gamma \qquad \text{for all } n \geq 1. \tag{7.3.2}$$

Proof. For every $x \in B$ we set

$$p(x) = \sup_{n \geq 1} \|T_n(x)\| = \sup_{n \geq 1} \left\{ \left\| \sum_{k=1}^{n} l_k(x) x_k \right\| \right\}. \tag{7.3.3}$$

Then we can verify easily that p is a norm on B. We now show that $\{x_n\}$ is a monotone basis with respect to p. In fact, let $\alpha_1, \alpha_2, \ldots, \alpha_N$ be real numbers, and set $x = \sum_{k=1}^{N-1} \alpha_k x_k$ and $y = \sum_{k=1}^{N} \alpha_k x_k$. Then

$$\begin{aligned} p(x) &= \sup_{n \geq 1} \left\{ \left\| \sum_{k=1}^{n} l_k(x) x_k \right\| \right\} \\ &= \sup_{1 \leq n \leq N-1} \left\{ \left\| \sum_{k=1}^{n} \alpha_k x_k \right\| \right\} \\ &\leq \sup_{1 \leq n \leq N} \left\{ \left\| \sum_{k=1}^{n} \alpha_k x_k \right\| \right\} = p(y), \end{aligned}$$

which proves the assertion. Using this fact, we can show that B is complete with respect to p. Moreover,

$$\begin{aligned} p(x) &= \sup_{n \geq 1} \|T_n(x)\| \\ &\geq \|T_n(x)\| \end{aligned}$$

for every $n \geq 1$, so that

$$p(x) \geq \|x\|$$

for every $x \in B$. It follows immediately from the open-mapping theorem (Royden [71], p. 195) that the two norms are equivalent, so that there exists a positive number $\gamma > 0$ such that $p(x) < \gamma\|x\|$ for all $x \in B$. Hence we have

$$\|T_n(x)\| \leq p(x) \leq \gamma\|x\|,$$

so that $\|T_n\| \leq \gamma$ for $n \geq 1$. ∎

Theorem 7.3.1 (Taylor). Let B be a real, separable Banach space with a Schauder basis $\{x_n\}$. Let X_n be a sequence of identically (not necessarily independent) B-valued random variables such that $\mathscr{E}\|X_1\| < \infty$. Then for each coordinate functional l_k the weak law of large numbers holds for the B-valued random variables $\{l_k(X_n): n \geq 1\}$ if and only if

$$n^{-1} \sum_{k=1}^{n} X_k \xrightarrow{P} \mathscr{E}X_1. \tag{7.3.4}$$

Proof. First suppose that the weak law holds for $\{l_k(X_n): n \geq 1\}$, that is,

$$n^{-1} \sum_{j=1}^{n} l_k(X_j) \xrightarrow{P} \mathscr{E}l_k(X_1). \tag{7.3.5}$$

We show that (7.3.4) holds. Clearly $\mathscr{E}X_1$ exists. Without loss of generality we let $\mathscr{E}X_1 = 0$. Let $\epsilon > 0$, and $\delta > 0$ be given. It is sufficient to show that there exists a positive integer $N_0 = N_0(\epsilon, \delta)$ such that for all $n \geq N_0$ the inequality

$$P\left\{\left\|n^{-1} \sum_{k=1}^{n} X_k\right\| \geq \epsilon\right\} < \delta \tag{7.3.6}$$

holds. Let $\gamma > 0$ be a positive constant such that $\|T_N\| \leq \gamma$ for all $N \geq 1$ (Proposition 7.3.1). For every $N \geq 1$ let W_N be the operator on B, defined by $W_N(x) = x - T_N(x)$ for $x \in B$. Clearly W_N is a bounded linear operator satisfying

$$\|W_N\| \leq \gamma + 1 \qquad \text{for all } N \geq 1.$$

Then, for each $n \geq 1$ and $N \geq 1$,

$$n^{-1} \sum_{k=1}^{n} X_k = n^{-1} \sum_{k=1}^{n} T_N(X_k) + n^{-1} \sum_{k=1}^{n} W_N(X_k). \tag{7.3.7}$$

Now, for every fixed $N \geq 1$,

$$P\left\{\left\| n^{-1} \sum_{k=1}^{n} W_N(X_k) \right\| > \frac{\epsilon}{2}\right\} \leq P\left\{n^{-1} \sum_{k=1}^{n} \| W_N(X_k) \| \geq \frac{\epsilon}{2}\right\}$$

$$\leq \frac{2}{\epsilon} \mathscr{E} \| W_N(X_1) \|,$$

so that

$$\sup_{n \geq 1} \mathrm{P}\left\{\left\| n^{-1} \sum_{k=1}^{n} W_N(X_k) \right\| \geq \frac{\epsilon}{2}\right\} \leq \frac{2}{\epsilon} \mathscr{E} \| W_N(X_1) \| \tag{7.3.8}$$

by Markov's inequality and the fact that the X_i are identically distributed. Since $\| W_N(X_1) \| \to 0$ pointwise as $N \to \infty$ and $\| W_N(X_1) \| \leq (\gamma + 1) \| X_1 \|$, we conclude that $\mathscr{E} \| W_N(X_1) \| \to 0$ as $N \to \infty$. Consequently for given $\delta > 0$ we can choose $N_1(\epsilon, \delta) = N_1$ (independently of n) such that for all $N \geq N_1$

$$\sup_{n \geq 1} P\left\{\left\| n^{-1} \sum_{k=1}^{n} W_N(X_k) \right\| \geq \frac{\epsilon}{2}\right\} < \frac{\delta}{2}. \tag{7.3.9}$$

On the other hand, it follows from the construction of T_N that for $x \in B$

$$T_N(x) = \sum_{i=1}^{N} l_i(x) x_i,$$

so that for all $n \geq 1$, $N \geq 1$, and $\epsilon > 0$

$$P\left\{\left\| n^{-1} \sum_{k=1}^{n} T_N(X_k) \right\| \geq \frac{\epsilon}{2}\right\}$$

$$= P\left\{\left\| \sum_{i=1}^{N} l_i\left(n^{-1} \sum_{k=1}^{n} X_k\right) x_i \right\| \geq \frac{\epsilon}{2}\right\}$$

$$\leq P\left\{\sum_{i=1}^{N} \left| l_i\left(n^{-1} \sum_{k=1}^{n} X_k\right) \right| \| x_i \| \geq \frac{\epsilon}{2}\right\}$$

$$\leq \sum_{i=1}^{N} P\left\{\left| n^{-1} \sum_{k=1}^{n} l_i(X_k) \right| \geq \frac{\epsilon}{2N \| x_i \|}\right\}. \tag{7.3.10}$$

By hypothesis the weak law of large numbers holds for every sequence $\{l_i(X_k) : k \geq 1\}$; moreover, $\mathscr{E} l_i(X_1) = 0$ for each i, so that

$$\lim_{n \to \infty} P\left\{\left| n^{-1} \sum_{k=1}^{n} l_i(X_k) \right| \geq \frac{\epsilon}{2N \| x_i \|}\right\} = 0$$

for each $i \geq 1$. Hence with N_1 as defined in (7.3.8) we can choose a positive integer $N_0(\epsilon, \delta)$ such that for all $n \geq N_0(\epsilon, \delta)$

$$\sum_{i=1}^{N_1} P\left\{\left|n^{-1}\sum_{k=1}^{n} l_i(X_k)\right| \geq \frac{\epsilon}{2N_1\|x_i\|}\right\} < \frac{\epsilon}{2}. \tag{7.3.11}$$

Using (7.3.7), (7.3.9), (7.3.10) and (7.3.11), we get (7.3.4).

Conversely, suppose that (7.3.4) holds, that is, suppose that

$$P\left\{\left\|n^{-1}\sum_{k=1}^{n} X_k - \mathscr{E}X_1\right\| \geq \epsilon\right\} \to 0 \qquad \text{as } n \to \infty.$$

Since the convergence in norm topology implies the convergence in weak topology, it follows that (7.3.5) holds. This completes the proof of Theorem 7.3.1. ■

We now give some results which follow as corollaries to Theorem 7.3.1.

Corollary 1. Let B be a real, separable Banach space with a Schauder basis $\{x_n\}$, and let $\{X_n\}$ be a sequence of identically distributed (not necessarily independent), B-valued random variables such that $\mathscr{E}\|X_1\| < \infty$. Then for every $l \in B^*$ the weak law of large numbers holds for the sequence $\{l(x_n)\}$ if and only if (7.3.4) holds.

Proof. The proof follows immediately from Theorem 7.3.1.

For Corollary 2 we need the following definition.

Definition 7.3.3. Let B be a real, separable Banach space, and let $\{X_n\}$ be a sequence of B-valued random variables. The sequence $\{X_n\}$ is said to be coordinate-wise uncorrelated if there exists a Schauder basis $\{x_n\}$ for B such that for every coordinate functional l_k the following conditions hold:

(i) $\mathscr{E}\{l_k(X_n)\}^2 < \infty$ for every $n \geq 1$.
(ii) $\mathscr{E}\{l_k(X_m)l_k(X_n)\} = \mathscr{E}\{l_k(X_m)\}\mathscr{E}\{l_k(X_n)\}$ for every pair $m, n, m \neq n$.

We note that, if the X_n are independent B-valued random variables with finite variances, $\{X_n\}$ is coordinate-wise uncorrelated.

Corollary 2 to Theorem 7.3.1. Let B be a real, separable Banach space, and let $\{X_n\}$ be a sequence of identically distributed, coordinate-wise uncorrelated B-valued random variables such that $\mathscr{E}\|X_1\| < \infty$. Then (7.3.4) holds.

7.4 SUMS OF INDEPENDENT RANDOM VARIABLES

Let $(\Omega, \mathscr{S}, P)$ be a probability space, and let B be a real, separable Banach space with Borel σ-field $\mathscr{B}$. Let $\{X_n\}$ be a sequence of independent B-valued random variables, and set $S_n = \sum_{k=1}^{n} X_k$, $n = 1, 2, \ldots$. In this section we discuss the convergence in law of the sequence of partial sums $\{S_n\}$.

7.4.1 Characteristic Functionals

To imitate the development in the real-valued case we need an analogue of characteristic functions. In the following, B is a real, separable Banach space with dual B^*, and $\mathscr{B}$ is the Borel σ-field of subsets of B.

Definition 7.4.1. Let $\mathscr{P}$ be the class of all probability measures on $(B, \mathscr{B})$. Let $P \in \mathscr{P}$. Then the Fourier transform $\hat{P}$ of P is a function defined on B^* by

$$\hat{P}(l) = \int_B e^{il(x)}\, dP(x), \qquad l \in B^*. \tag{7.4.1}$$

The Fourier transform of a finite measure on $(B, \mathscr{B})$ can be defined in a similar manner.

Let X be a B-valued random variable, and let $P_X \in \mathscr{P}$ be the probability measure induced by X on $\mathscr{B}$. Then the *characteristic functional* of the random variable X is the mathematical expectation of $e^{il(X)}$, $l \in B^*$, so that

$$\begin{aligned} \mathscr{E} e^{il(X)} &= \int_B e^{il(x)}\, dP_X(x) \\ &= \hat{P}_X(l), \qquad l \in B^*. \end{aligned} \tag{7.4.2}$$

We also need the following definition.

Definition 7.4.2. For $n \geq 1$ let $\mathscr{B}_n$ be the Borel σ-field of subsets of $\mathbb{R}_n$. A subset E of B is said to be a Borel cylinder set if it is of the form

$$E = \{x \in B : (l_1(x), l_2(x), \ldots, l_n(x)) \in M\} \tag{7.4.3}$$

for some $n \geq 1$, $l_1, l_2, \ldots, l_n \in B^*$ (arbitrary) and some $M \in \mathscr{B}_n$.

The following result is useful in the subsequent development.

Lemma 7.4.1. Let $\mathscr{F}$ be the class of all cylinder sets of B. Then $\mathscr{F}$ is a field, and $\mathscr{F} \subset \mathscr{B}$. Moreover, let $\sigma(\mathscr{F})$ be the σ-field generated by $\mathscr{F}$. Then

$$\sigma(\mathscr{F}) = \mathscr{B}.$$

Proof. Clearly every Borel cylinder set E defined by (7.4.3) belongs to $\mathscr{B}$. Moreover, the class $\mathscr{F}$ is a field, so that $\sigma(\mathscr{F}) \subset \mathscr{B}$. We show that $\mathscr{B} \subset \sigma(\mathscr{F})$. Since B is separable, there exists a countable set of elements $x_n \in B$ which is everywhere dense in B. By the Hahn–Banach theorem for every $n \geq 1$ we construct an $l_n \in B^*$ such that $\|l_n\| = 1$ and $l_n(x_n) = \|x_n\|$. Next we show show that for every $a > 0$

$$\{x \in B: \|x\| \leq a\} = \bigcap_{n=1}^{\infty} \{x \in B: l_n(x) \leq a\}. \tag{7.4.4}$$

It follows from the definition of l_n that

$$l_n(x) \leq \|l_n\| \, \|x\| = \|x\|,$$

so that

$$\{x \in B: \|x\| \leq a\} \subset \bigcap_{n=1}^{\infty} \{x \in B: l_n(x) \leq a\}.$$

We now show that

$$\{x \in B: \|x\| > a\} \subset \bigcup_{n=1}^{\infty} \{x \in B: l_n(x) > a\}.$$

Let $x \in B$ with $\|x\| > a$. Since $\{x_n\}$ is everywhere dense in B, there exists an n_0 such that $\|x - x_{n_0}\| < \frac{1}{2}(\|x\| - a)$. We have

$$\begin{aligned} \|x_{n_0}\| &\geq \|x\| - \|x_{n_0} - x\| \\ &> \|x\| - \tfrac{1}{2}(\|x\| - a) = \tfrac{1}{2}(\|x\| + a). \end{aligned}$$

On the other hand,

$$\begin{aligned} |l_{n_0}(x) - \|x_{n_0}\|| &= |l_{n_0}(x) - l_{n_0}(x_{n_0})| \\ &\leq \|l_{n_0}\| \, \|x - x_{n_0}\| \\ &< \tfrac{1}{2}(\|x\| - a). \end{aligned}$$

Hence

$$\begin{aligned} l_{n_0}(x) &> \|x_{n_0}\| - \tfrac{1}{2}(\|x\| - a) \\ &> \tfrac{1}{2}(\|x\| + a) - \tfrac{1}{2}(\|x\| - a) = a. \end{aligned}$$

It follows that (7.4.4) holds. Consequently for every $a > 0$ the set

$$\{x \in B: \|x\| \leq a\} \in \sigma(\mathscr{F}).$$

Hence, for every $a > 0$, $\{\|x\| > a\} \in \sigma(\mathscr{F})$. Since $\sigma(\mathscr{F})$ is translation invariant, we conclude that every open sphere $\in \sigma(\mathscr{F})$. Since $\mathscr{B}$ is the σ-field generated by the class of all open spheres in B, it follows that $\mathscr{B} \subset \sigma(\mathscr{F})$. ■

We are now ready to study some elementary properties of the characteristic functional $\hat{P}$ of a probability measure $P \in \mathscr{P}$.

Proposition 7.4.1. The Fourier transform $\hat{P}$ of $P \in \mathscr{P}$ has the following properties:

(a) $\hat{P}(0) = 1$.
(b) $|\hat{P}(l)| \leq 1$ for all $l \in B^*$.
(c) (Hermitian property) $\overline{\hat{P}(l)} = \hat{P}(-l)$ for all $l \in B^*$, where $\bar{z}$ denotes the complex conjugate of z.
(d) $\hat{P}$ is positive definite† on B^*.
(e) $\hat{P}$ is uniformly continuous on B^* with respect to the usual norm topology.
(f) (Uniqueness) Let $P_1, P_2 \in \mathscr{P}$. Then $\hat{P}_1 = \hat{P}_2 \Leftrightarrow P_1 = P_2$.
(g) (Convolution) $\widehat{P_1 * P_2}(l) = \hat{P}_1(l)\hat{P}_2(l)$, $l \in B^*$.

Proof. The proofs of (a), (b), (c), and (d) being simple, we need prove only (e), (f), and (g).

(e) Let $l_1, l_2 \in B^*$ be arbitrary. Then

$$\hat{P}(l_1) - \hat{P}(l_2) = \int_B [e^{il_1(x)} - e^{il_2(x)}]\, dP(x),$$

so that

$$|\hat{P}(l_1) - \hat{P}(l_2)| \leq \int_B |1 - e^{i(l_1 - l_2)(x)}|\, dP(x)$$

$$= 2\int_B |\sin \tfrac{1}{2}(l_1 - l_2)(x)|\, dP(x).$$

Let $\epsilon > 0$ be arbitrary. Then we select $\gamma = \gamma(\epsilon) > 0$ such that

$$P\{x \in B: \|x\| \geq \gamma\} < \epsilon.$$

Then

$$|\hat{P}(l_1) - \hat{P}(l_2)| \leq 2\left(\int_{\|x\| \geq \gamma} + \int_{\|x\| < \gamma}\right)|\sin \tfrac{1}{2}(l_1 - l_2)(x)|\, dP(x)$$

$$< 2\epsilon + \gamma\|l_1 - l_2\|.$$

† Let φ be a complex-valued function defined on B^*. Then φ is said to be positive definite if for every positive integer N, for every finite set of complex numbers $\omega_1, \omega_2, \ldots, \omega_N$, and for every finite set of elements $l_1, l_2, \ldots, l_N$ in B^*,

$$\sum_{j=1}^{N}\sum_{k=1}^{N} \omega_j \bar{\omega}_k \varphi(l_j - l_k) \geq 0.$$

Let $l_1, l_2 \in B^*$ be such that $\|l_1 - l_2\| < \epsilon/\gamma$. Then we have

$$|\hat{P}(l_1) - \hat{P}(l_2)| < 3\epsilon$$

whenever $\|l_1 - l_2\| < \epsilon/\gamma$. This proves that $\hat{P}$ is uniformly continuous.

(f) It suffices to show that $\hat{P}_1 = \hat{P}_2 \Rightarrow P_1 = P_2$, since the converse is trivial. Let $n \geq 1$, and let $l_1, l_2, \ldots, l_n \in B^*$ be fixed. Set

$$l = \sum_{j=1}^{n} t_j l_j,$$

where $t_j \in \mathbb{R}, j = 1, 2, \ldots, n$. Clearly $l \in B^*$, and since $\hat{P}_1(l) = \hat{P}_2(l)$, we verify easily, using the uniqueness of the Fourier transform of probability measures on $(\mathbb{R}_n, \mathscr{B}_n)$, that $P_1 = P_2$ on $\mathscr{F}$. In view of the extension theorem, $P_1 = P_2$ on $\sigma(\mathscr{F})$ and hence on $\mathscr{B}$ from Lemma 7.4.1.

(g) In view of the uniqueness established above, the proof follows in the usual manner. ■

Corollary. Let $P \in \mathscr{P}$ be such that $\hat{P}(l) = 1$ for all $l \in B^*$ with $\|l\| < \epsilon$ for some $\epsilon > 0$. Then P is the degenerate probability measure on $\mathscr{B}$.

Proof. Let $l \in B^*$, $l \neq 0$, be fixed. Define the function φ on $\mathbb{R}$ be setting

$$\varphi(t) = \hat{P}(tl), \qquad t \in \mathbb{R}.$$

Then φ is the Fourier transform of a probability measure on $(\mathbb{R}, \mathscr{B}_1)$, where $\mathscr{B}_1$ is the Borel σ-field on $\mathbb{R}$. Moreover, $\varphi(t) = 1$ for $|t| < \epsilon/\|l\|$. Consequently the probability measure on $\mathscr{B}_1$ corresponding to φ is degenerate, so that $\varphi(t) \equiv 1$ for all $t \in \mathbb{R}$. In particular, for $t = 1$ we obtain

$$\hat{P}(l) = \varphi(1) = 1,$$

so that $\hat{P} \equiv 1$. Let P_0 be the degenerate probability measure on $\mathscr{B}$. Then $\hat{P}_0 \equiv 1$. It follows from uniqueness that $P = P_0$.

We now prove an extension of the Lévy continuity theorem (Theorem 3.5.1) which will be useful in our subsequent investigation.

Proposition 7.4.2.

(a) Let $P_n \in \mathscr{P}$, $n \geq 1$, be a sequence of probability measures such that $P_n \Rightarrow P$, where $P \in \mathscr{P}$, as $n \to \infty$. Then $\hat{P}_n \to \hat{P}$ as $n \to \infty$ on B^*.

(b) Conversely, let $P_n \in \mathscr{P}$, $n \geq 1$, be a tight sequence such that $\hat{P}_n$ converges to some function θ on B^*. Then there exists a $P \in \mathscr{P}$ such that $P_n \Rightarrow P$. Moreover, in this case $\theta = \hat{P}$. (For the definition of weak convergence of probability measures on metric spaces we refer to Section 3.9.)

Proof. Part (a) follows easily from the definition of weak convergence of probability measures.

(b) Since $\{P_n\}$ is tight, we conclude from Theorem 3.9.2 that $\{P_n\}$ is relatively compact, so that there exists a subsequence $\{P_{n_k}\} \subset \{P_n\}$ which converges weakly to some probability measure P as $k \to \infty$. It follows from the definition of weak convergence that $\hat{P}_{n_k} \to \hat{P}$ as $k \to \infty$. According to our hypothesis, $\theta = \hat{P}$. Suppose that the sequence $\{P_n\}$ contains another subsequence which converges weakly to some other probability measure Q. Proceeding as above, we conclude that $\theta = \hat{Q}$. Hence $\hat{P} = \hat{Q} = \theta$. By the uniqueness of the characteristic functional, it follows that $P = Q$. Consequently $P_n \Rightarrow P$ as $n \to \infty$, and, moreover, $\theta = \hat{P}$. ■

Remark 7.4.1. It should be noted that $\hat{P}_n \to \hat{P}$ as $n \to \infty$ does not imply $P_n \Rightarrow P$. As an example, let $\{x_n\}$ be a sequence of elements in B which converges weakly to some element $x_0 \in B$, but not strongly. More precisely, this means that

$$l(x_n) \to l(x_0) \qquad \text{as } n \to \infty$$

for every $l \in B^*$, but

$$\|x_n - x_0\| \not\to 0 \qquad \text{as } n \to \infty.$$

For every $n \geq 1$ let P_n be the probability measure degenerate at x_n. Let P_0 be the measure degenerate at x_0. Clearly $\hat{P}_n \to \hat{P}_0$ as $n \to \infty$. However, $P_n \not\Rightarrow P_0$. In fact, since $\|x_n - x_0\| \not\to 0$ as $n \to \infty$, there exists an $\epsilon > 0$ and a subsequence $\{x_{n_k}\} \subset \{x_n\}$ such that $\|x_{n_k} - x_0\| \geq \epsilon$ for all $k \geq 1$. Let f be a bounded, continuous function on B such that $f(x_0) = 1$, while f vanishes outside the open sphere with center x_0 and radius ϵ. Clearly

$$\int_B f\, dP_0 = 1,$$

while for all $k \geq 1$

$$\int_B f\, dP_{n_k} = 0,$$

so that $\int_B f\, dP_n \not\to \int_B f\, dP_0$.

7.4.2 Convergence Equivalence of Sequences of Sums of Independent Random Variables

Let B be a separable Banach space, and $\mathscr{B}$ be the Borel σ-field of subsets of B. Let $\{X_n\}$ be a sequence of independent B-valued random variables. Set $S_n = \sum_{k=1}^{n} X_k$, $n \geq 1$. We now consider the convergence properties of the sequence $\{S_n\}$. In the case of real-valued random variables we showed

(Corollary 2 to Proposition 3.9.4) the equivalence of convergence of S_n in law, in probability, and almost surely. We now show that this result also holds when the X_n are B-valued random variables.

First we prove the following analogue of Lévy's inequality (Lemma 2.4.1) for B-valued random variables, which was obtained independently by Lai and Buldygin. See also Kahane [43], p. 12.

Lemma 7.4.2 (Levy's Inequality). Let $X_1, X_2, \ldots, X_n$ be independent, symmetric, B-valued random variables. Then for every $\epsilon > 0$

$$P\left\{\max_{1 \le j \le n} \|S_j\| \ge \epsilon\right\} \le 2P\{\|S_n\| \ge \epsilon\}, \tag{7.4.5}$$

where $S_j = \sum_{k=1}^{j} X_k, j = 1, 2, \ldots, n$.

Proof. Set

$$E = \left\{\max_{1 \le j \le n} \|S_j\| \ge \epsilon\right\},$$

and partition E into sets:

$$\begin{aligned} E_1 &= \{\|S_1\| \ge \epsilon\}, \\ E_2 &= \{\|S_1\| < \epsilon, \|S_2\| \ge \epsilon\}, \\ E_3 &= \{\|S_1\| < \epsilon, \|S_2\| < \epsilon, \|S_3\| \ge \epsilon\}, \end{aligned}$$

etc.

Let m $(1 \le m \le n)$ be a fixed integer, and set

$$S_{m,n} = S_m - \sum_{j=m+1}^{n} X_j = S_m - (S_n - S_m).$$

In view of our hypothesis, $S_{m,n}$ and S_n are identically distributed. Clearly

$$\{\|S_m\| \ge \epsilon\} \subset \{\|S_{m,n}\| \ge \epsilon\} \cup \{\|S_n\| \ge \epsilon\},$$

and since $E_m \subset \{\|S_m\| \ge \epsilon\}$ we have

$$\begin{aligned} P(E_m) &\le P\{E_m \cap \{\|S_{m,n}\| \ge \epsilon\}\} + P\{E_m \cap \{\|S_n\| \ge \epsilon\}\} \\ &= 2P\{E_m \cap \{\|S_n\| \ge \epsilon\}\} \qquad \text{for all } m, \quad 1 \le m \le n. \end{aligned}$$

Summing over m, we get

$$P(E) \le 2P\{\|S_n\| \ge \epsilon\}$$

as asserted. ■

We now prove the following result, which is due to Ito and Nisio [40].

Theorem 7.4.1. Let $\{X_n\}$ be a sequence of independent B-valued random variables, and set $S_n = \sum_{k=1}^n X_k$, $n = 1, 2, \ldots$. Then the following assertions are equivalent:

(a) S_n converges a.s.
(b) S_n converges in probability.
(c) S_n converges in law.

Proof. Clearly (a) $\Rightarrow$ (b) $\Rightarrow$ (c). It is therefore sufficient to show that (c) $\Rightarrow$ (b) $\Rightarrow$ (a). First we show that (c) $\Rightarrow$ (b).

Let m and n ($m < n$) be two positive integers. For every $n \geq 1$ let P_n be the probability measure on $\mathscr{B}$ induced by S_n, and let $P_{m,n}$ be the probability measure on $\mathscr{B}$ induced by $S_n - S_m$. According to our hypothesis, P_n converges weakly to a probability measure Q on $\mathscr{B}$ so that $\{P_n\}$ is relatively compact and hence tight (Theorem 3.9.2). Hence, given $\epsilon > 0$, there exists a compact set $K = K_\epsilon \subset B$ such that

$$P_n(K) \geq 1 - \epsilon \qquad \text{for all } n.$$

We set $K_0 = \{x - y: x, y \in K\}$. Then K_0 is a compact subset of B. Since $S_n, S_m \in K \Rightarrow S_n - S_m \in K_0$, we have

$$\begin{aligned} P_{m,n}(K_0) &\geq P(S_n \in K, S_m \in K) \\ &\geq 1 - P(S_n \in K^c) - P(S_m \in K^c) \\ &= 1 - P_n(K^c) - P_m(K^c) \\ &> 1 - 2\epsilon, \end{aligned}$$

so that the sequence $\{P_{m,n}, m < n, n = 1, 2, \ldots\}$ is tight and hence relatively compact. To complete the proof of (b) it is sufficient to show that the sequence $\{P_{m,n}\}$ converges weakly to the measure P_0 degenerate at zero, which is equivalent to the following. For a given $\epsilon > 0$ and a given neighborhood U_ϵ of zero in B there exists a positive integer $N_0 = N_0(\epsilon)$ such that for all $m < n < N_0$

$$P_{m,n}(U_\epsilon) > 1 - \epsilon.$$

Suppose this is not the case. Then, for some $\epsilon_0 > 0$ and some neighborhood $U_0 = U_{\epsilon_0}$ of zero in B and for all $N \geq 1$, there exist positive integers $m(N)$, $n(N)$ with $n(N) > m(N) > N$ such that

$$P_{m(N),\,n(N)}(U_0) \leq 1 - \epsilon_0.$$

Since $\{P_{m,n}\}$ is relatively compact, it contains an infinite subsequence which converges weakly to a probability measure Q_0 on $\mathscr{B}$. Without loss of generality we may assume that the sequence $\{P_{m(N),n(N)}, N \geq 1\}$ itself converges weakly to Q_0. Then we have from Theorem 3.9.1

$$Q_0(U_0) \leq \liminf_{N\to\infty} P_{m(N),\,n(N)}(U_0) \leq 1 - \epsilon_0. \tag{7.4.6}$$

On the other hand, using the independence of the sequence $\{X_n\}$, we have

$$\mathscr{E}e^{il(S_{n(N)})} = \mathscr{E}e^{il(S_{m(N)})} \cdot \mathscr{E}e^{il(S_{n(N)} - S_{m(N)})} \qquad \text{for every } l \in B^*.$$

Rewriting, we have

$$\hat{P}_{n(N)} = \hat{P}_{m(N)} \cdot \hat{P}_{m(N),\, n(N)} \qquad \text{for all } N \geq 1.$$

Taking the limits as $N \to \infty$ and using Proposition 7.4.2, we have

$$\hat{Q} = \hat{Q}\hat{Q}_0.$$

Since $\hat{Q}(0) = 1$ and $\hat{Q}$ is continuous, there exists an $r > 0$ such that $\hat{Q}(l) \neq 0$ for all $l \in B^*$ with $\|l\| < r$. Hence we conclude that $\hat{Q}_0(l) = 1$ for all $l \in B^*$ with $\|l\| < r$. In view of the corollary to Proposition 7.4.1 we conclude that $Q_0 = P_0$. This contradicts (7.4.6). Hence $\{P_{m,n}\}$ converges weakly to the measure P_0 degenerate at 0. This proves (c) $\Rightarrow$ (b).

Finally we need to show that (b) $\Rightarrow$ (a). This is done in the same way as in the real case (see Proposition 3.9.4 and its corollaries), using either the inequality

$$P\left\{\max_{m<k\leq n} \|S_k - S_m\| > 2\epsilon\right\} \leq \frac{P\{\|S_n - S_m\| > \epsilon\}}{1 - \max_{m+1\leq k\leq n} P\{\|S_n - S_k\| > \epsilon\}}, \tag{7.4.7}$$

which holds for every $\epsilon > 0$, or Lévy's inequality (7.4.5).

To prove (7.4.7) we set

$$E_k = \{\|S_{m+1} - S_m\| \leq 2\epsilon, \ldots, \|S_{k-1} - S_m\| < 2\epsilon, \|S_k - S_m\| > 2\epsilon\}$$

and

$$F_k = \{\|S_n - S_k\| \leq \epsilon\}$$

for $k = m + 1, \ldots, n$. Clearly, the sets E_k and F_k $(m < k \leq n)$ are independent. Moreover, the sets are disjoint and satisfy

$$\bigcup_{k=m+1}^{n} E_k = \left\{\max_{m<k\leq n} \|S_k - S_m\| > 2\epsilon\right\}.$$

Therefore

$$\begin{aligned} P\{\|S_n - S_m\| > \epsilon\} &\geq P\left\{\bigcup_{k=m+1}^{n} (E_k \cap F_k)\right\} \\ &= \sum_{k=m+1}^{n} P(E_k)P(F_k) \\ &\geq \min_{m+1\leq k\leq n} P(F_k)P\left\{\max_{m+1\leq k\leq n} \|S_k - S_m\| > 2\epsilon\right\}, \end{aligned}$$

which yields (7.4.7) immediately. ∎

Corollary. Let $\{X_n\}$ be a sequence of independent B-valued random variables, and let $S_n = \sum_{k=1}^n X_k$, $n \geq 1$. Let P_n be the probability measure on $\mathscr{B}$ induced by S_n for every $n \geq 1$. Suppose that $\{P_n\}$ is a tight sequence. Then there exists a sequence of elements $\{a_n \in B, n \geq 1\}$ such that $S_n - a_n$ converges a.s.

Proof. Let $\{Y_n\}$ be another sequence of independent B-valued random variables with the following properties:

(a) For every $n \geq 1$, X_n and Y_n are identically distributed.
(b) The sequence $\{X_n, Y_n, n \geq 1\}$ consists of independent random variables.

Clearly $\{X_n - Y_n\}$ is an independent sequence. Set $T_n = \sum_{k=1}^n Y_k$ and $W_n = S_n - T_n$, for $n \geq 1$. Let Q_n be the induced measure of W_n.

According to our hypothesis, for a given $\epsilon > 0$ there exists a $K = K_\epsilon$ such that

$$P_n(K) \geq 1 - \epsilon \qquad \text{for every } n \geq 1.$$

Set $K_0 = \{x - y : x, y \in K\}$. Then $K_0 \subset B$ is a compact set, and we have for every $n \geq 1$

$$\begin{aligned} Q_n(K_0) &= P\{S_n - T_n \in K_0\} \geq \{S_n \in K, T_n \in K\} \\ &\geq 1 - P(S_n \in K^c) - P(T_n \in K^c) \\ &= 1 - 2P_n(K^c) \\ &> 1 - 2\epsilon. \end{aligned}$$

This shows that $\{Q_n\}$ is tight and hence relatively compact.

On the other hand, for all $l \in B^*$

$$\begin{aligned} \hat{Q}_n(l) = \mathscr{E}e^{il(W_n)} &= \mathscr{E}e^{il(S_n - T_n)} \\ &= \mathscr{E}e^{il(S_n)} \cdot \mathscr{E}e^{-il(T_n)} \\ &= \prod_{k=1}^n \{\mathscr{E}e^{il(X_k)} \mathscr{E}e^{-il(Y_k)}\} \\ &= \prod_{k=1}^n |\mathscr{E}e^{il(X_k)}|^2. \end{aligned}$$

Since for every $k \geq 1$

$$0 \leq |\mathscr{E}e^{il(X_k)}|^2 \leq 1,$$

we conclude that $\lim_{n\to\infty} \hat{Q}_n(l)$ exists and is finite for every $l \in B^*$. Since the sequence $\{Q_n\}$ is relatively compact, we conclude from Proposition 7.4.2 that the sequence $\{Q_n\}$ converges weakly to a probability measure. In view of Theorem 7.4.1 we conclude that $\{W_n\}$ converges a.s., that is, $\{S_n - T_n\}$ converges a.s. Since $\{S_n\}$ and $\{T_n\}$ are independent, we see from Fubini's theorem that the sequence $\{S_n - a_n\}$ converges a.s. for almost every realization $a_n \in B$ of T_n, $n \geq 1$. This completes the proof.

7.5 PROBABILITY MEASURES ON A HILBERT SPACE

Let $\mathscr{H}$ be a real, separable Hilbert space with inner product $\langle \cdot, \cdot \rangle$. Let $\mathscr{B}$ be the σ-field generated by the class of all open subsets of $\mathscr{H}$. Since $\mathscr{H}$ is separable, we note that $\mathscr{B}$ coincides with the σ-field generated by the class of all open spheres in $\mathscr{H}$. We now study briefly some properties of probability measures on $(\mathscr{H}, \mathscr{B})$. In particular, we shall prove an analogue of the Lévy central limit theorem.

Definition 7.5.1. Let P be a probability measure on $\mathscr{B}$. Then the Fourier transform $\hat{P}$ of P is a complex-valued function defined on $\mathscr{H}$ by the formula

$$\hat{P}(y) = \int_{\mathscr{H}} \exp(i\langle y, x \rangle)\, dP(x), \qquad y \in \mathscr{H}. \tag{7.5.1}$$

In particular, let X be an $\mathscr{H}$-valued random variable, and let P_X be the probability measure induced by X on $\mathscr{B}$. Then the characteristic functional of X is defined by

$$\mathscr{E} \exp(i\langle y, X \rangle) = \hat{P}_X(y), \qquad y \in \mathscr{H}, \tag{7.5.2}$$

where $\hat{P}_X$ is the Fourier transform of P_X as defined in (7.5.1).

Remark 7.5.1. Propositions 7.4.1 and 7.4.2 hold for the Fourier transforms of probability measures on $(\mathscr{H}, \mathscr{B})$ in view of the Riesz representation theorem, which states that every bounded linear functional l on $\mathscr{H}$ can be represented as

$$l(x) = \langle y, x \rangle \qquad x \in \mathscr{H},$$

where $y \in \mathscr{H}$ is determined uniquely by l. Therefore the dual space $\mathscr{H}^*$ of $\mathscr{H}$ can be identified with the space $\mathscr{H}$ itself.

We now give a simple sufficient condition for the tightness of a set of probability measures on $\mathscr{B}$ which is due to Prokhorov. For this purpose we choose and fix a complete orthonormal system (orthonormal basis) of elements in $\mathscr{H}$, say e_ν, $\nu = 1, 2, \ldots$. For every $x \in \mathscr{H}$ we denote by x_ν the νth coordinate of x, defined by $x_\nu = \langle x, e_\nu \rangle$, $\nu \geq 1$. For every $x \in \mathscr{H}$ and every $N \geq 1$ we set

$$r_N^2(x) = \sum_{\nu=N}^{\infty} x_\nu^2 = \sum_{\nu=N}^{\infty} \langle x, e_\nu \rangle^2. \tag{7.5.3}$$

Proposition 7.5.1. Let $\mathscr{P}_0$ be a class of probability measures on $(\mathscr{H}, \mathscr{B})$ satisfying the following condition:

$$\lim_{N\to\infty} \sup_{P\in\mathscr{P}_0} \int_{\mathscr{H}} r_N^2(x)\, dP(x) = 0. \tag{7.5.4}$$

Then $\mathscr{P}_0$ is tight.

Proof. For simplicity in notation we set

$$\psi(N) = \sup_{P\in\mathscr{P}_0} \int_{\mathscr{H}} r_N^2(x)\, dP(x), \qquad N \geq 1.$$

Then, according to (7.5.4), $\psi(N) \downarrow 0$ as $N \to \infty$. Hence for given $\epsilon > 0$ we can choose a sequence $0 < \alpha_N \to \infty$ as $N \to \infty$ such that $\sum_{N=1}^{\infty} \alpha_N \psi(N) < \epsilon$. For every $N \geq 1$ we set $K_N = \{x \in \mathscr{H} : r_N^2(x) \leq \alpha_N^{-1}\}$ and $K = \bigcap_{N=1}^{\infty} K_N$. Clearly K is a closed subset of $\mathscr{H}$. We now show that K is compact.

First we note that $K \subset K_1 = \{x \in \mathscr{H} : \|x\|^2 \leq \alpha_1^{-1}\}$, so that K is bounded and hence is weakly (sequentially) compact (see Theorem 1, p. 46, of Akhiezer and Glazman [3]). Let $x^{(n)} \in K$, $n \geq 1$. Then $\{x^{(n)}\}$ contains a subsequence $\{x^{(n_k)}\}$ such that, for every $y \in \mathscr{H}$, $\langle x^{(n_k)}, y\rangle \to \langle x, y\rangle$ as $k \to \infty$ for some $x \in K$. We show that $\|x^{(n_k)} - x\| \to 0$ as $k \to \infty$. In fact, for $N > 1$ and $k \geq 1$ we have

$$\begin{aligned}
\|x^{(n_k)} - x\|^2 &= \sum_{\nu=1}^{\infty} \langle x^{(n_k)} - x, e_\nu\rangle^2 \\
&= \sum_{\nu=1}^{N-1} \langle x^{(n_k)} - x, e_\nu\rangle^2 + \sum_{\nu=N}^{\infty} \langle x^{(n_k)} - x, e_\nu\rangle^2 \\
&\leq \sum_{\nu=1}^{N-1} \langle x^{(n_k)} - x, e_\nu\rangle^2 + \sum_{\nu=N}^{\infty} \langle x^{(n_k)}, e_\nu\rangle^2 \\
&\quad + \sum_{\nu=N}^{\infty} \langle x, e_\nu\rangle^2 + 2 \sum_{\nu=N}^{\infty} |\langle x^{(n_k)}, e_\nu\rangle\langle x, e_\nu\rangle| \\
&\leq \sum_{\nu=1}^{N-1} \langle x^{(n_k)} - x, e_\nu\rangle^2 + 2r_N^2(x^{(n_k)}) + 2r_N^2(x) \\
&\leq \sum_{\nu=1}^{N-1} \langle x^{(n_k)} - x, e_\nu\rangle^2 + \frac{4}{\alpha_N}.
\end{aligned}$$

Let $\epsilon > 0$. Choose $N = N(\epsilon)$ sufficiently large so that $4/\alpha_N < \epsilon/2$. On the other hand, since $\langle x^{(n_k)}, e_\nu\rangle \to \langle x, e_\nu\rangle$ as $k \to \infty$ for every $\nu \geq 1$, we can choose k sufficiently large so that

$$\sum_{\nu=1}^{N-1} \langle x^{(n_k)} - x, e_\nu\rangle^2 < \frac{\epsilon}{2}.$$

It follows that $\|x^{(n_k)} - x\| \to 0$ as $k \to \infty$, so that K is a compact subset of $\mathscr{H}$.

For every $P \in \mathscr{P}_0$ we have from Chebyshev's inequality

$$\begin{aligned} P(K^c) &\leq \sum_{N=1}^{\infty} P(K_N^c) \\ &\leq \sum_{N=1}^{\infty} \alpha_N \psi(N) \\ &< \epsilon, \end{aligned}$$

so that $P(K) > 1 - \epsilon$ for every $P \in \mathscr{P}_0$. This proves that $\mathscr{P}_0$ is tight. ∎

In the following we restrict our investigation to probability measures P on $(\mathscr{H}, \mathscr{B})$ such that $\int_{\mathscr{H}} \|x\|^2 \, dP(x) < \infty$. We denote this class of probability measures by $\mathscr{P}_2$. Note that, for every $P \in \mathscr{P}_2$, $\int_{\mathscr{H}} \|x\| \, dP(x) < \infty$.

Let $P \in \mathscr{P}_2$. Since $\int_{\mathscr{H}} \|x\| \, dP(x) < \infty$, we see easily that for every $y \in \mathscr{H}$ the integral $\int_{\mathscr{H}} \langle x, y \rangle \, dP(x)$ exists and is finite. Moreover, the mapping $y \to \int_{\mathscr{H}} \langle x, y \rangle \, dP(x)$ is a bounded linear functional on $\mathscr{H}$. Hence in view of the Riesz representation theorem there exists an element $\mu_P \in \mathscr{H}$ determined uniquely by P such that the relation

$$\langle \mu_P, y \rangle = \int_{\mathscr{H}} \langle x, y \rangle \, dP(x) \tag{7.5.5}$$

holds for all $y \in \mathscr{H}$. We also note that for $P \in \mathscr{P}_2$ the integral

$$\int_{\mathscr{H}} \langle x, y \rangle \langle x, z \rangle \, dP(x)$$

exists and is finite for all $y, z \in \mathscr{H}$. Moreover, it can be verified that the mapping

$$\begin{aligned} (y, z) \to & \int_{\mathscr{H}} \langle x, y \rangle \langle x, z \rangle \, dP(x) - \langle \mu_P, y \rangle \langle \mu_P, z \rangle \\ &= \int_{\mathscr{H}} \langle x - \mu_P, y \rangle \langle x - \mu_P, z \rangle \, dP(x) \end{aligned}$$

for $y, z \in \mathscr{H}$ of $\mathscr{H} \times \mathscr{H}$ into $\mathbb{R}$ defines a bounded bilinear form on $\mathscr{H}$. Hence it follows (see Akhiezer and Glazman [3], p. 42) that there exists a bounded linear operator $\mathbf{S}_P$ on $\mathscr{H}$ determined uniquely by P by the relation

$$\begin{aligned} \langle \mathbf{S}_P y, z \rangle &= \int_{\mathscr{H}} \langle x, y \rangle \langle x, z \rangle \, dP(x) - \langle \mu_P, y \rangle \langle \mu_P, z \rangle \\ &= \int_{\mathscr{H}} \langle x - \mu_P, y \rangle \langle x - \mu_P, z \rangle \, dP(x) \end{aligned} \tag{7.5.6}$$

for all $y, z \in \mathscr{H}$. In particular, the relation

$$(7.5.7) \quad \langle \mathbf{S}_P y, y \rangle = \int_{\mathscr{H}} \langle x, y \rangle^2 \, dP(x) - \langle \mu_P, y \rangle^2 = \int_{\mathscr{H}} \langle x - \mu_P, y \rangle^2 \, dP(x)$$

holds for all $y \in \mathscr{H}$.

Definition 7.5.2. The element $\mu_P \in \mathscr{H}$ defined by (7.5.5) is called the mean vector of P, and the operator $\mathbf{S}_P$ defined by (7.5.7) is called the covariance operator of P.

Proposition 7.5.2. The covariance operator $\mathbf{S}_P$ associated with the probability measure $P \in \mathscr{P}_2$ defined by (7.5.7) has the following properties:

(a) $\mathbf{S}_P$ is compact, that is, $\mathbf{S}_P$ maps every bounded set in $\mathscr{H}$ onto a relatively compact set in $\mathscr{H}$.
(b) $\mathbf{S}_P$ is self-adjoint, that is, $\mathbf{S}_P = \mathbf{S}_P^*$, where $\mathbf{S}_P^*$ is the adjoint of $\mathbf{S}_P$.
(c) $\mathbf{S}_P$ is positive, that is, $\langle \mathbf{S}_P y, y \rangle \geq 0$ for all $y \in \mathscr{H}$.
(d) $\mathbf{S}_P$ has a finite trace.

Moreover, in this case the trace of $\mathbf{S}_P$ is given by

$$\operatorname{tr} \mathbf{S}_P = \int_{\mathscr{H}} \|x\|^2 \, dP(x) - \|\mu_P\|^2 = \int_{\mathscr{H}} \|x - \mu_P\|^2 \, dP(x).$$

Proof. Clearly $\mathbf{S}_P$ is bounded, linear, and (b) and (c) follow immediately from the definition of $\mathbf{S}_P$.

Next we note that a bounded, positive, self-adjoint operator $\mathbf{S}$ has finite trace if for some orthonormal basis $\{e_\nu\}$ in $\mathscr{H}$

$$\sum_{\nu=1}^{\infty} \langle \mathbf{S} e_\nu, e_\nu \rangle < \infty.$$

In this case it can be shown that $\mathbf{S}$ is compact and, moreover, if $\{e'_\nu\}$ is another orthonormal basis, then

$$\sum_{\nu=1}^{\infty} \langle \mathbf{S} e_\nu, e_\nu \rangle = \sum_{\nu=1}^{\infty} \langle \mathbf{S} e'_\nu, e'_\nu \rangle < \infty,$$

and this common value is called the trace of $\mathbf{S}$.

For the proof of (a) and (d) it is then sufficient to show that

$$\sum_{\nu=1}^{\infty} \langle \mathbf{S}_P e_\nu, e_\nu \rangle = \int \|x\|^2 \, dP(x) - \|\mu_P\|^2,$$

where $\{e_\nu\}$ is an orthonormal basis in $\mathscr{H}$. In fact, for every $N \geq 1$ we have

$$\sum_{\nu=1}^{N} \langle \mathbf{S}_P e_\nu, e_\nu \rangle = \int_{\mathscr{H}} \sum_{\nu=1}^{N} \langle x, e_\nu \rangle^2 \, dP(x) - \sum_{\nu=1}^{N} \langle \mu_P, e_\nu \rangle^2.$$

In view of the monotone convergence theorem we take the limits as $N \to \infty$ on both sides and obtain

$$\sum_{\nu=1}^{\infty} \langle \mathbf{S}_P e_\nu, e_\nu \rangle = \int_{\mathscr{H}} \sum_{\nu=1}^{\infty} \langle x, e_\nu \rangle^2 \, dP(x) - \sum_{\nu=1}^{\infty} \langle \mu_P, e_\nu \rangle^2$$
$$= \int_{\mathscr{H}} \|x\|^2 \, dP(x) - \|\mu_P\|^2,$$

where we have used the Parseval relation

$$\sum_{\nu=1}^{\infty} \langle x, e_\nu \rangle \langle y, e_\nu \rangle = \langle x, y \rangle, \qquad x, y \in \mathscr{H}.$$

The proof is now complete. ■

Definition 7.5.3. An operator on $\mathscr{H}$ satisfying (a) through (d) of Proposition 7.5.2 is known as an **S**-operator.

Clearly every covariance operator is an **S**-operator.

Remark 7.5.2. Let $\mathscr{P}_0 \subset \mathscr{P}_2$ be a class of probability measures on $(\mathscr{H}, \mathscr{B})$ such that $\mu_P = 0$ for all $P \in \mathscr{P}_0$. Then it can be verified easily that condition (7.5.4) can be written in the form

$$\lim_{N\to\infty} \sup_{P \in \mathscr{P}_0} \sum_{\nu=N}^{\infty} \langle \mathbf{S}_P e_\nu, e_\nu \rangle = 0, \tag{7.5.8}$$

where $\mathbf{S}_P$ is the covariance operator associated with $P \in \mathscr{P}_0$, and $\{e_\nu\}$ is an orthonormal basis for $\mathscr{H}$.

We now define an important probability measure belonging to the class $\mathscr{P}_2$.

Definition 7.5.4. A probability measure ν on $(\mathscr{H}, \mathscr{B})$ is said to be a Gaussian measure if its Fourier transform $\hat{\nu}$ is of the form

$$\hat{\nu}(y) = \exp(i\langle \mu, y \rangle - \tfrac{1}{2}\langle \mathbf{S}y, y \rangle), \qquad y \in \mathscr{H}, \tag{7.5.9}$$

where $\mu \in \mathscr{H}$ is fixed, and **S** is an **S**-operator on $\mathscr{H}$ as defined above. An $\mathscr{H}$-valued random variable X is said to be Gaussian if its characteristic functional is of the form (7.5.9).

Clearly the measure ν is uniquely determined by $\hat{\nu}$ and hence by the element $\mu \in \mathscr{H}$ and the operator **S**.

Proposition 7.5.3. Let ν be a Gaussian measure on $(\mathscr{H}, \mathscr{B})$ with Fourier transform $\hat{\nu}$ given by (7.5.9). Then $\nu \in \mathscr{P}_2$, and, moreover, μ is the mean vector and **S** the covariance operator of ν.

Proof. The proof will be given in Section 7.6 (Proposition 7.6.4). ■

Proposition 7.5.4.

(a) Let X be an $\mathscr{H}$-valued Gaussian random variable with mean vector μ and covariance operator **S** (that is, μ is the mean vector and **S** the covariance operator of the probability measure induced by X). Let A be a bounded linear operator on $\mathscr{H}$. Then the $\mathscr{H}$-valued random variable $Y = AX$ is also Gaussian with mean vector $A\mu$ and covariance operator $A\mathbf{S}A^*$, where A^* is the adjoint of A.

(b) Let X and Y be two independent $\mathscr{H}$-valued Gaussian random variables with mean vectors μ_X and μ_Y and covariance operators $\mathbf{S}_X$ and $\mathbf{S}_Y$, respectively. Then the sum $X + Y$ is a Gaussian random variable with mean vector $\mu_X + \mu_Y$ and covariance operator $\mathbf{S}_X + \mathbf{S}_Y$.

Proof. The proof is straightforward. ■

Some other properties of Gaussian measures on a Hilbert space are discussed in Section 7.6. Finally we prove an analogue of the Lévy central limit theorem (Corollary 1 to Theorem 5.1.1) for $\mathscr{H}$-valued random variables.

Theorem 7.5.1. Let $\{X_n\}$ be a sequence of independent, identically distributed $\mathscr{H}$-valued random variables with

$$\mathscr{E}X_n = 0 \qquad \text{and} \qquad \mathscr{E}\|X_n\|^2 < \infty, \qquad n \geq 1.$$

Set

$$Z_n = n^{-1/2} \sum_{k=1}^{n} X_k, \qquad n \geq 1.$$

Then $\{Z_n\}$ converges in law to a Gaussian random variable with mean vector zero and covariance operator **S**, where **S** is the common covariance operator of each X_n.

Proof. Let $y \in \mathscr{H}$ be fixed. Clearly $\{\langle y, X_n \rangle\}$ is a sequence of independent, identically distributed, real-valued random variables each with zero mean and finite variance. For every $n \geq 1$ we have

$$\langle y, Z_n \rangle = n^{-1/2} \sum_{k=1}^{n} \langle y, X_k \rangle.$$

Let P_n be the probability measure on $\mathscr{B}$ induced by Z_n. Then the characteristic function of $\langle y, Z_n \rangle$ is given by $t \to \hat{P}_n(ty)$, $t \in \mathbb{R}$. In view of the Lévy central limit theorem for the real-valued case (Corollary 1 to Theorem 5.1.1), the sequence $\{\hat{P}_n(ty)\}$ converges (pointwise) for every $t \in \mathbb{R}$ to the normal characteristic function $\exp\{-(t^2/2)\mathscr{E}\langle y, X_n\rangle^2\}$ as $n \to \infty$. On the other hand, we see easily that

$$\exp\left(-\frac{t^2}{2}\mathscr{E}\langle y, X_n\rangle^2\right) = \exp\left(-\frac{t^2}{2}\langle \mathbf{S}y, y\rangle\right). \tag{7.5.10}$$

Setting $t = 1$, we see that $\hat{P}_n(y) \to \exp(-\frac{1}{2}\langle \mathbf{S}y, y\rangle)$ for every $y \in \mathscr{H}$ as $n \to \infty$.

It follows from (7.5.10) that $\mathbf{S}$ is an S-operator, so (7.5.8) is satisfied. Hence $\{P_n\}$ is a tight sequence (Proposition 7.5.1). We conclude at once from Proposition 7.4.2 that $\{P_n\}$ converges weakly to the Gaussian measure with mean vector zero and covariance operator $\mathbf{S}$. The proof is now complete. ■

7.6 THE MINLOS–SAZONOV THEOREM

Let $\mathscr{H}$ be a real, separable Hilbert space with inner product $\langle \cdot, \cdot \rangle$, and let $\mathscr{B}$ be the σ-field generated by the class of all open subsets of $\mathscr{H}$. Let φ be a complex-valued function defined on $\mathscr{H}$. In this section we derive a necessary and sufficient condition for φ to be the Fourier transform of a probability measure P on $\mathscr{B}$. This result generalizes Bochner's theorem on positive definite functions (Theorem 3.6.1) to the case of (infinite-dimensional) Hilbert spaces. For this purpose we need some preliminaries.

7.6.1 Preliminaries

Let $\mathfrak{U}$ be a finite-dimensional subspace of $\mathscr{H}$, and let $\mathscr{B}_{\mathfrak{U}}$ be the σ-field of Borel subsets of $\mathfrak{U}$. Let $\pi_{\mathfrak{U}}$ be the projection of $\mathscr{H}$ onto $\mathfrak{U}$.

A set $E \in \mathscr{H}$ is said to be a *Borel cylinder set with base* in $\mathfrak{U}$ if E can be written in the form

$$E = \pi_{\mathfrak{U}}^{-1}(B) = \{x \in \mathscr{H} : \pi_{\mathfrak{U}}(x) \in B\}, \qquad B \in \mathscr{B}_{\mathfrak{U}}.$$

Let $\mathscr{B}^{\mathfrak{U}}$ denote the class of all Borel cylinder sets in $\mathscr{H}$ with base in $\mathfrak{U}$. Then $\mathscr{B}^{\mathfrak{U}}$ is a σ-field, and, moreover, $\mathscr{B}^{\mathfrak{U}} \subset \mathscr{B}$. Set $\mathscr{B}^0 = \bigcup_{\mathfrak{U}} \mathscr{B}^{\mathfrak{U}}$, where the union is taken over all finite-dimensional subspaces $\mathfrak{U}$ of $\mathscr{H}$. Then $\mathscr{B}^0$ is a field, and, moreover, $\mathscr{B}^0 \subset \mathscr{B}$.

Proposition 7.6.1. $\mathscr{B}$ coincides with $\sigma(\mathscr{B}^0)$.

Proof. Clearly $\sigma(\mathscr{B}^0) \subset \mathscr{B}$. We now show that $\mathscr{B} \subset \sigma(\mathscr{B}^0)$. For every $n \geq 1$ let $\mathfrak{U}_n$ be an n-dimensional subspace of $\mathscr{H}$ such that $\mathfrak{U}_n \subset \mathfrak{U}_{n+1}$ for all n, and, moreover, $\bigcup_{n\geq 1} \mathfrak{U}_n$ is everywhere dense in $\mathscr{H}$. Set $\mathscr{B}^1 = \bigcup_{n \geq 1} \mathscr{B}^{\mathfrak{U}_n}$. Then clearly $\mathscr{B}^1$ is a field, and $\mathscr{B}^1 \subset \mathscr{B}^0$. Hence it is sufficient to show that $\mathscr{B} \subset \sigma(\mathscr{B}^1)$. Since $\mathscr{B}$ is generated by the class of all open spheres in $\mathscr{H}$ and every open sphere in $\mathscr{H}$ can be written as a countable union of an increasing sequence of closed spheres, it is sufficient to show that every closed sphere in $\mathscr{H}$ belongs to $\sigma(\mathscr{B}^1)$. Let $S^* = \{x\colon \|x - x_0\| \leq \rho\}$ be an arbitrary closed sphere in $\mathscr{H}$. We show that $S^* \in \sigma(\mathscr{B}^1)$. In fact, for every $n \geq 1$ let π_n denote the projection of $\mathscr{H}$ onto $\mathfrak{U}_n$. Set

$$S_n^* = \{x \in \mathscr{H}\colon \|\pi_n(x) - \pi_n(x_0)\| \leq \rho\}.$$

Then clearly, for every $n \geq 1$, $S_n^* \in \mathscr{B}^{\mathfrak{U}_n} \subset \mathscr{B}^1$, and, moreover, $S^* \subset S_n^*$. We now show that $S^* = \bigcap_{n=1}^{\infty} S_n^*$. It suffices to show that $\bigcap_{n=1}^{\infty} S_n^* \subset S^*$. For this purpose suppose $x \notin S^*$. Then for some $\epsilon > 0$

$$\|x - x_0\| = \rho + \epsilon.$$

On the other hand, we note that $\lim_{n\to\infty} \pi_n(x - x_0) = x - x_0$, so that

$$\lim_{n\to\infty} \|\pi_n(x - x_0)\| = \|x - x_0\|.$$

It follows that for sufficiently large n

$$\|\pi_n x - \pi_n x_0\| > \rho,$$

so that $x \notin S_n^*$. Hence $x \notin \bigcap_{n=1}^{\infty} S_n^*$. This completes the proof. ■

Remark 7.6.1. We note that the field $\mathscr{B}^0$ of Borel cylinder sets in $\mathscr{H}$ coincides with the field $\mathscr{F}$ of cylinder sets as defined in Lemma 7.4.1.

Let P be a probability measure on $\mathscr{B}$. Let $\mathfrak{U}$ be an arbitrary finite-dimensional subspace of $\mathscr{H}$, and let $\mathscr{B}_{\mathfrak{U}}$ be the σ-field of Borel subsets of $\mathfrak{U}$. Define the set function $P_{\mathfrak{U}}$ on $(\mathfrak{U}, \mathscr{B}_{\mathfrak{U}})$ by the relation

(7.6.1) $$P_{\mathfrak{U}}(E) = P(\pi_{\mathfrak{U}}^{-1}(E)), \qquad E \in \mathscr{B}_{\mathfrak{U}}.$$

Then $P_{\mathfrak{U}}$ is a probability measure on $\mathscr{B}_{\mathfrak{U}}$ and is known as the projection of P onto $(\mathfrak{U}, \mathscr{B}_{\mathfrak{U}})$.

Definition 7.6.1. Let $\{Q_{\mathfrak{U}}\}$ be a family of finite-dimensional probability distributions defined on $\{(\mathfrak{U}, \mathscr{B}_{\mathfrak{U}})\}$, where $\mathfrak{U}$ runs through all the finite-dimensional subspaces of $\mathscr{H}$. Then $\{Q_{\mathfrak{U}}\}$ is said to be a compatible family if for every $\mathfrak{U}_1 \subset \mathfrak{U}_2$ and for every $E \in \mathscr{B}_{\mathfrak{U}_1}$ the relation

(7.6.2) $$Q_{\mathfrak{U}_1}(E) = Q_{\mathfrak{U}_2}(\pi_{\mathfrak{U}_1}^{-1}(E) \cap \mathfrak{U}_2)$$

holds.

Remark 7.6.2. We note that the family $\{P_{\mathfrak{U}}\}$ associated with the probability measure P on $\mathscr{B}$ as defined in (7.6.1) is a compatible family. In fact, let $\mathfrak{U}_1 \subset \mathfrak{U}_2$ and $E \in \mathscr{B}_{\mathfrak{U}_1}$. Then

$$\pi_{\mathfrak{U}_1}^{-1}(E) = \{x \in \mathscr{H} : \pi_{\mathfrak{U}_1}(x) \in E\} = \pi_{\mathfrak{U}_2}^{-1}(E_2),$$

where $E_2 \in \mathscr{B}_{\mathfrak{U}_2}$ is defined by

$$E_2 = \{x \in \mathfrak{U}_2 : \pi_{\mathfrak{U}_1}(x) \in E\} = \pi_{\mathfrak{U}_1}^{-1}(E) \cap \mathfrak{U}_2.$$

Clearly for $E \in \mathscr{B}_{\mathfrak{U}_1}$

$$\begin{aligned} P_{\mathfrak{U}_1}(E) = P(\pi_{\mathfrak{U}_1}^{-1}(E)) &= P(\pi_{\mathfrak{U}_2}^{-1}(E_2)) = P_{\mathfrak{U}_2}(E_2) \\ &= P_{\mathfrak{U}_2}(\pi_{\mathfrak{U}_1}^{-1}(E) \cap \mathfrak{U}_2), \end{aligned}$$

which proves (7.6.2).

We now derive a necessary and sufficient condition under which there exists a probability measure on $\mathscr{B}$ associated with a given compatible family of finite-dimensional probability distributions.

Proposition 7.6.2. Let $\{P_{\mathfrak{U}}\}$ be a compatible family of finite-dimensional probability distributions defined on $\{(\mathfrak{U}, \mathscr{B}_{\mathfrak{U}})\}$. Then there exists a probability measure P on $(\mathscr{H}, \mathscr{B})$ associated with $\{P_{\mathfrak{U}}\}$ if and only if for given $\epsilon > 0$ there exists an $N_0 = N_0(\epsilon) \geq 1$ such that for all $N \geq N_0$

(7.6.3) $$\sup P_{\mathfrak{U}}(\{x \in \mathscr{H} : \|x\| \geq N\} \cap \mathfrak{U}) < \epsilon,$$

where the supremum is taken over all finite-dimensional subspaces $\mathfrak{U}$ of $\mathscr{H}$.

Proof. First suppose that there exists a probability measure P on $\mathscr{B}$ which is associated with the family $\{P_{\mathfrak{U}}\}$. Then we show that (7.6.3) holds. Since

$$\lim_{N \to \infty} P\{x \in \mathscr{H} : \|x\| \geq N\} = 0,$$

it follows that for $\epsilon > 0$ we can choose an $N_0 = N_0(\epsilon)$ such that for all $N \geq N_0$

$$P\{\|x\| \geq N\} < \epsilon.$$

Hence for all $N \geq N_0$ and every finite-dimensional subspace $\mathfrak{U}$ of $\mathscr{H}$ we have

$$P_{\mathfrak{U}}(\{\|x\| \geq N\} \cap \mathfrak{U}) = P(\pi_{\mathfrak{U}}^{-1}[\{\|x\| \geq N\} \cap \mathfrak{U}]) \\ \leq P\{\|x\| \geq N\} < \epsilon.$$

Conversely, let $\{P_{\mathfrak{U}}\}$ be a compatible family of finite-dimensional probability distributions which satisfies (7.6.3). We show that there exists a probability measure P on $(\mathscr{H}, \mathscr{B})$ associated with $\{P_{\mathfrak{U}}\}$. On the field $\mathscr{B}^0 = \bigcup_{\mathfrak{U}} \mathscr{B}^{\mathfrak{U}}$ we define a set function P as follows. Let $E \in \mathscr{B}^0$. Then $E \in \mathscr{B}^{\mathfrak{U}}$ for some $\mathfrak{U}$, so that there exists a $B \in \mathscr{B}_{\mathfrak{U}}$ such that $E = \pi_{\mathfrak{U}}^{-1}(B)$. Set

$$P(E) = P_{\mathfrak{U}}(B). \tag{7.6.4}$$

We see easily that P is determined uniquely on $\mathscr{B}^0$ by (7.6.4), and, moreover, P is a finite, finitely additive, and nonnegative set function on $\mathscr{B}^0$ with $P(\mathscr{H}) = 1$. Next we show that P is a probability measure on $\mathscr{B}^0$ and can be uniquely extended to a probability measure on $\mathscr{B}$. Since $\mathscr{B} = \sigma(\mathscr{B}^0)$, it is sufficient to show that P is countably additive on $\mathscr{B}^0$. In other words, it is sufficient to show that P is continuous from above at $\varnothing$ on $\mathscr{B}^0$. Let $E_n \in \mathscr{B}^0$, $n \geq 1$, such that $E_n \downarrow \varnothing$. We show that $P(E_n) \downarrow 0$. For every $n \geq 1$ let $\mathfrak{U}_n$ be an n-dimensional subspace of $\mathscr{H}$ such that $\mathfrak{U}_n \subset \mathfrak{U}_{n+1}, n \geq 1$. Without loss of generality we may assume that, for every $n \geq 1$, E_n is a Borel cylinder set with base in $\mathfrak{U}_n$. Let $B_n \in \mathscr{B}_{\mathfrak{U}_n}$ be such that $E_n = \pi_{\mathfrak{U}_n}^{-1}(B_n)$. Then

$$P(E_n) = P_{\mathfrak{U}_n}(B_n), \qquad n \geq 1.$$

First we consider the case where, for every $n \geq 1$, B_n is a compact, and hence a closed, bounded set in $\mathfrak{U}_n$. Since $\pi_{\mathfrak{U}_n}$ is continuous, E_n is a closed set in $\mathscr{H}$. Moreover, it can be verified that E_n is weakly closed for every $n \geq 1$, that is, $x_k \in E_n$ and $\langle x_k, y\rangle \to \langle x, y\rangle$ for every $y \in \mathscr{H}$ as $k \to \infty$ implies that $x \in E_n$. For every $N \geq 1$ the closed sphere $S_N^* = \{x \in \mathscr{H} : \|x\| \leq N\}$ is weakly closed and weakly compact. Consequently for fixed $N \geq 1$ the sets $\{S_N^* \cap E_n,\ n \geq 1\}$ are weakly closed and weakly compact. Moreover, $\{S_N^* \cap E_n\} \downarrow \varnothing$ as $n \to \infty$. Therefore for given $N \geq 1$ there exists an $n_0 = n_0(N) \geq 1$ such that $S_N^* \cap E_{n_0} = \varnothing$ and consequently $S_N^* \cap E_n = \varnothing$ for all $n \geq n_0$. Hence, for $n \geq n_0$, $E_n \subset \{\|x\| > N\}$. We now show that, for all $n \geq n_0$, $B_n \subset \mathfrak{U}_n \cap \{x \in \mathscr{H} : \|x\| \geq N\}$. In fact, let $y \in B_n$. Then $y \in \mathfrak{U}_n$, so that $\pi_{\mathfrak{U}_n}(y) = y$ and $\pi_{\mathfrak{U}_n}(y) \in B_n$. Since $E_n = \pi_{\mathfrak{U}_n}^{-1}(B_n)$, it follows that $y \in E_n$ so that $y \in \mathfrak{U}_n \cap \{x \in \mathscr{H} : \|x\| > N\}$. Hence for $n \geq n_0$

$$B_n \subset \mathfrak{U}_n \cap \{\|x\| > N\} \subset \mathfrak{U}_n \cap \{\|x\| \geq N\},$$

so that for $n \geq n_0$

$$P(E_n) = P_{\mathfrak{U}_n}(B_n) \leq P_{\mathfrak{U}_n}(\mathfrak{U}_n \cap \{\|x\| \geq N\}).$$

It follows from (7.6.3) that for a given $\epsilon > 0$ there exists an $N_0 = N_0(\epsilon)$ such that

$$P_{\mathfrak{U}_n}(\mathfrak{U}_n \cap \{\|x\| \geq N_0\}) < \epsilon \qquad \text{for all } n \geq 1.$$

Hence for given $\epsilon > 0$ there exists an $n_0 = n_0(N_0) = n_0(\epsilon)$ such that for all $n \geq n_0$

$$P(E_n) < \epsilon,$$

so that $\lim_{n\to\infty} P(E_n) = 0$.

Finally we consider the case where $B_n \in \mathscr{B}_{\mathfrak{U}_n}$ is arbitrary. Let $\epsilon > 0$. Then for every $n \geq 1$ there exists a compact set $C_n \subset \mathfrak{U}_n$ such that $C_n \subset B_n$ and, moreover, $P_{\mathfrak{U}_n}(B_n - C_n) < \epsilon/2^n$, since $P_{\mathfrak{U}_n}$ on $\mathscr{B}_{\mathfrak{U}_n}$ is tight. Let $F_n \in \mathscr{B}^0$ be such that $F_n = \pi_{\mathfrak{U}_n}^{-1}(C_n)$. Clearly $F_n \subset E_n$, and since P is finitely additive on $\mathscr{B}^0$ we have

$$P(E_n - F_n) = P_{\mathfrak{U}_n}(B_n - C_n) < \frac{\epsilon}{2^n}, \qquad n \geq 1.$$

For $n \geq 1$ set $G_n = F_1 \cap F_2 \cap \cdots \cap F_n$. Then $G_n \in \mathscr{B}^0$, and $G_n \subset F_n \subset E_n$. Since E_n is decreasing, we have

$$E_n - G_n = E_n - \bigcap_{j=1}^{n} F_j = \bigcup_{j=1}^{n} (E_n - F_j) \subset \bigcup_{j=1}^{n} (E_j - F_j),$$

so that

$$P(E_n - G_n) \leq \sum_{j=1}^{n} P(E_j - F_j) < \epsilon.$$

Hence $P(E_n) < P(G_n) + \epsilon$ for every $n \geq 1$. We note that $G_n \downarrow$ and $G_n \subset E_n$, so that $G_n \downarrow \varnothing$.

We now show that G_n is of the form $G_n = \pi_{\mathfrak{U}_n}^{-1}(D_n)$, where D_n is a compact set in $\mathfrak{U}_n$. Set $D_n = (\bigcap_{j=1}^{n} \pi_{\mathfrak{U}_j}^{-1}(C_j)) \cap \mathfrak{U}_n$, $n \geq 1$. Clearly D_n is a closed set in $\mathfrak{U}_n$, and moreover $D_n \subset C_n$, so that D_n is a compact set in $\mathfrak{U}_n$. Now

$$G_n = \bigcap_{j=1}^{n} F_j = \bigcap_{j=1}^{n} \pi_{\mathfrak{U}_j}^{-1}(C_j) = \pi_{\mathfrak{U}_n}^{-1}(D_n).$$

It follows from what has been shown above that

$$\lim_{n\to\infty} P(G_n) = 0.$$

Hence

$$0 \leq \lim_{n\to\infty} P(E_n) \leq \epsilon,$$

and since ϵ is arbitrary we have $\lim_{n\to\infty} P(E_n) = 0$. Consequently P is countably additive and hence is a probability measure on $\mathscr{B}^0$. Finally, by the extension theorem P can be uniquely extended to a probability P on $\mathscr{B} = \sigma(\mathscr{B}^0)$. Clearly P is the probability measure on $(\mathscr{H}, \mathscr{B})$ associated with the compatible family $\{P_{\mathfrak{U}}\}$. This completes the proof of Proposition 7.6.2. ■

As an application of Propositions 7.6.1 and 7.6.2 we now give a simple proof of the uniqueness property of the Fourier transforms.

Proposition 7.6.3. Let P and Q be two probability measures on $(\mathscr{H}, \mathscr{B})$, and let $\hat{P}$ and $\hat{Q}$, respectively, be their Fourier transforms. Suppose that $\hat{P} = \hat{Q}$ on $\mathscr{H}$. Then $P = Q$ on $\mathscr{B}$.

Proof. Let $\mathfrak{U}$ be an arbitrary finite-dimensional subspace of $\mathscr{H}$, and let $\mathscr{B}_{\mathfrak{U}}$ be the σ-field of Borel subsets of $\mathfrak{U}$. Let $P_{\mathfrak{U}}$ and $Q_{\mathfrak{U}}$ be the projections of P and Q, respectively, onto $(\mathfrak{U}, \mathscr{B}_{\mathfrak{U}})$ as defined in (7.6.1). First we show that $P_{\mathfrak{U}} = Q_{\mathfrak{U}}$ on $\mathscr{B}_{\mathfrak{U}}$. Let $x \in \mathscr{H}$. Then

$$\hat{P}(\pi_{\mathfrak{U}}(x)) = \int_{\mathscr{H}} \exp\{i\langle \pi_{\mathfrak{U}}(x), y\rangle\}\, dP(y).$$

For $y \in \mathscr{H}$ we write

$$y = \pi_{\mathfrak{U}}(y) + (y - \pi_{\mathfrak{U}}(y)),$$

where $y - \pi_{\mathfrak{U}}(y)$ is orthogonal to $\mathfrak{U}$, so that

$$\hat{P}(\pi_{\mathfrak{U}}(x)) = \int_{\mathscr{H}} \exp\{i\langle \pi_{\mathfrak{U}}(x), \pi_{\mathfrak{U}}(y)\rangle\}\, dP(y).$$

In view of the transformation theorem (Halmos [31], p. 163) we have

$$\hat{P}(\pi_{\mathfrak{U}}(x)) = \int_{\mathscr{H}} \exp\{i\langle \pi_{\mathfrak{U}}(x), y\rangle\}\, dP_{\mathfrak{U}}(y) \qquad \text{for } x \in \mathscr{H}.$$

Similarly

$$\hat{Q}(\pi_{\mathfrak{U}}(x)) = \int_{\mathfrak{U}} \exp\{i\langle \pi_{\mathfrak{U}}(x), y\rangle\}\, dQ_{\mathfrak{U}}(y).$$

Since $\hat{P} = \hat{Q}$, we conclude from the uniqueness of Fourier transforms of probability distributions on $(\mathbb{R}_n, \mathscr{B}_n)$ (Corollary 1 to Theorem 3.3.3) that

$$P_{\mathfrak{U}} = Q_{\mathfrak{U}} \qquad \text{on } \mathscr{B}_{\mathfrak{U}}.$$

Let $\mathscr{B}^0$ be the field of all Borel cylinder sets of $\mathscr{H}$, and let $E \in \mathscr{B}^0$. Then for some finite-dimensional subspace $\mathfrak{U}$ of $\mathscr{H}$ there exists a $B \in \mathscr{B}_{\mathfrak{U}}$ such that

$$E = \pi_{\mathfrak{U}}^{-1}(B).$$

Clearly

$$P(E) = P_{\mathrm{u}}(B), \qquad Q(E) = Q_{\mathrm{u}}(B),$$

so that

$$P(E) = Q(E) \qquad \text{for all } E \in \mathscr{B}^0.$$

Since $\mathscr{B} = \sigma(\mathscr{B}^0)$ from Proposition 7.6.1, it follows immediately from the extension theorem that $P = Q$ on $\mathscr{B}$. This completes the proof. ■

7.6.2 The Minlos–Sazonov Theorem

We are now in a position to prove the main result of this section.

Theorem 7.6.1. Let $\mathscr{H}$ be a real, separable Hilbert space, and let $\mathscr{B}$ be the σ-field generated by the class of all open sets of $\mathscr{H}$. Let φ be a complex-valued, continuous, positive definite function on $\mathscr{H}$ such that $\varphi(0) = 1$. Then φ is the Fourier transform of a probability measure P on $\mathscr{B}$ if and only if the following condition is satisfied: For given $\epsilon > 0$ there exists an **S**-operator, $\mathbf{S}_\epsilon$ on $\mathscr{H}$, such that

$$1 - \operatorname{Re} \varphi(x) < \epsilon \qquad \text{whenever } \langle \mathbf{S}_\epsilon x, x \rangle < 1. \tag{7.6.5}$$

Proof. First suppose that φ is the Fourier transform of a probability measure P on $\mathscr{B}$. Then in view of Proposition 7.4.1 φ is a continuous, positive definite function on $\mathscr{H}$ such that $\varphi(0) = 1$. Next we show that φ satisfies (7.6.5). Let $x \in \mathscr{H}$ and $\gamma > 0$. Then we have

$$\begin{aligned} 0 \le 1 - \operatorname{Re} \varphi(x) &= \int_{\mathscr{H}} (1 - \cos\langle x, y \rangle)\, dP(y) \\ &\le 2 \int_{\|x\| < \gamma} \sin^2 \frac{\langle x, y \rangle}{2}\, dP(y) + 2 \int_{\|x\| \ge \gamma} dP(y) \\ &\le \frac{1}{2} \int_{\|x\| < \gamma} \langle x, y \rangle^2\, dP(y) + 2P\{x \in \mathscr{H} : \|x\| \ge \gamma\}. \end{aligned} \tag{7.6.6}$$

We note that for $\gamma > 0$

$$\int_{\|x\| < \gamma} \|x\|^2\, dP(x) \le \gamma^2 < \infty.$$

Consequently, proceeding exactly as in Section 7.5 (discussion following Proposition 7.5.1), we see easily that $\mathbf{S}'_\gamma$, defined on $\mathscr{H}$ by

$$\langle \mathbf{S}'_\gamma y, z \rangle = \int_{\|x\| < \gamma} \langle x, y \rangle \langle x, z \rangle\, dP(x), \qquad y, z \in \mathscr{H},$$

is an S-operator. Let $\epsilon > 0$, and choose $\gamma = \gamma(\epsilon) > 0$ such that

$$P\{x \in \mathscr{H} : \|x\| \geq \gamma\} < \frac{\epsilon}{4}.$$

Then we have from (7.6.6)

$$0 \leq 1 - \text{Re } \varphi(x) \leq \tfrac{1}{2}\langle \mathbf{S}'_\gamma x, x\rangle + \tfrac{1}{2}\epsilon, \qquad x \in \mathscr{H}.$$

Now set $\mathbf{S}_\epsilon = \epsilon^{-1}\mathbf{S}'_\gamma$. Clearly $\mathbf{S}_\epsilon$ is also an S-operator on $\mathscr{H}$. Let $x \in \mathscr{H}$ be such that $\langle \mathbf{S}_\epsilon x, x\rangle < 1$. Then we have

$$0 \leq 1 - \text{Re } \varphi(x) \leq \frac{\epsilon}{2}\langle \mathbf{S}_\epsilon x, x\rangle + \frac{\epsilon}{2} < \epsilon.$$

This proves (7.6.5).

Conversely, let φ be a (complex-valued) continuous, positive definite function on $\mathscr{H}$ such that $\varphi(0) = 1$ and φ satisfies (7.6.5). We now show that there exists a probability measure P on $\mathscr{B}$ such that φ is the Fourier transform of P.

Let $\mathfrak{U}$ be an arbitrary, finite-dimensional subspace of $\mathscr{H}$, and let $\mathscr{B}_\mathfrak{U}$ be the σ-field of Borel sets of $\mathfrak{U}$. Clearly φ, considered as a function on $\mathfrak{U}$, is also continuous and positive definite, satisfying $\varphi(0) = 1$. In view of the multidimensional version of Bochner's theorem (Problem 3.10.38) we conclude that there exists a probability measure $P_\mathfrak{U}$ on $\mathscr{B}_\mathfrak{U}$ such that φ is the Fourier transform of $P_\mathfrak{U}$, that is, the integral representation

$$\varphi(x) = \int_\mathfrak{U} \exp(i\langle x, y\rangle)\, dP_\mathfrak{U}(y) \tag{7.6.7}$$

holds for all $x \in \mathscr{H}$.

We now show that the family $\{P_\mathfrak{U}\}$ (where $\mathfrak{U}$ runs through all finite-dimensional subspaces of $\mathscr{H}$) is a compatible family of finite-dimensional probability distributions. In fact, let $\mathfrak{U}_1 \subset \mathfrak{U}_2$, and let $x \in \mathfrak{U}_1$. Then we have

$$\int_{\mathfrak{U}_1} \exp(i\langle x, y\rangle)\, dP_{\mathfrak{U}_1}(y) = \varphi(x) = \int_{\mathfrak{U}_2} \exp(i\langle x, y\rangle)\, dP_{\mathfrak{U}_2}(y). \tag{7.6.8}$$

For $y \in \mathfrak{U}_2$ we write $y = \pi_{\mathfrak{U}_1}(y) + (y - \pi_{\mathfrak{U}_1}(y))$, where $y - \pi_{\mathfrak{U}_1}(y)$ is orthogonal to $\mathfrak{U}_1$. Then for $x \in \mathfrak{U}_1$ we have

$$\langle x, y\rangle = \langle x, \pi_{\mathfrak{U}_1}(y)\rangle,$$

so that

$$\int_{\mathfrak{U}_2} \exp(i\langle x, y\rangle)\, dP_{\mathfrak{U}_2}(y) = \int_{\mathfrak{U}_2} \exp\{i\langle x, \pi_{\mathfrak{U}_1}(y)\rangle\}\, dP_{\mathfrak{U}_2}(y).$$

Hence it follows from (7.6.8) and the transformation theorem that the relation

$$\int_{\mathfrak{U}_1} \exp(i\langle x, y\rangle)\, dP_{\mathfrak{U}_1}(y) = \int_{\mathfrak{U}_2} \exp\{i\langle x, \pi_{\mathfrak{U}_1}(y)\rangle\}\, dP_{\mathfrak{U}_2}(y)$$

$$= \int_{\mathfrak{U}_1} \exp(i\langle x, y\rangle)\, dP_{\mathfrak{U}_2}\pi_{\mathfrak{U}_1}^{-1}(y)$$

holds for all $x \in \mathfrak{U}_1$. Consequently, from the uniqueness of the Fourier transforms, we conclude that for every $\mathfrak{U}_1 \subset \mathfrak{U}_2$ and $E \in \mathscr{B}_{\mathfrak{U}_1}$ the relation

$$P_{\mathfrak{U}_1}(E) = P_{\mathfrak{U}_2}(\pi_{\mathfrak{U}_1}^{-1}(E) \cap \mathfrak{U}_2)$$

holds. This proves that the family $\{P_{\mathfrak{U}}\}$ is compatible.

Next we show that there exists a probability measure P on $\mathscr{B}$ associated with $\{P_{\mathfrak{U}}\}$. For this purpose it is sufficient to verify that condition (7.6.3) is satisfied. In fact, let $\mathfrak{U}$ be a finite-dimensional subspace of $\mathscr{H}$ with dimension $n_{\mathfrak{U}}$. First we note that for every positive integer N

$$\{x \in \mathscr{H} : \|x\| \geq N\} \cap \mathfrak{U} = \{x \in \mathfrak{U} : \|x\| \geq N\}.$$

Let $\alpha > 0$ and $N \geq 1$. Then we note that

$$\int_{\mathfrak{U}} \left[1 - \exp\left(-\frac{\alpha}{2}\|x\|^2\right)\right] dP_{\mathfrak{U}}(x) \geq \int_{\|x\| \geq N} \left[1 - \exp\left(-\frac{\alpha}{2}\|x\|^2\right)\right] dP_{\mathfrak{U}}(x)$$

$$\geq \left[1 - \exp\left(-\frac{\alpha}{2}N^2\right)\right] P_{\mathfrak{U}}(\|x\| \geq N),$$

so that we have

$$\left[1 - \exp\left(-\frac{\alpha}{2}N^2\right)\right] P_{\mathfrak{U}}(\|x\| \geq N)$$

$$\leq \int_{\mathfrak{U}} \left[1 - \exp\left(-\frac{\alpha}{2}\|x\|^2\right)\right] dP_{\mathfrak{U}}(x)$$

$$= \int_{\mathfrak{U}} (2\pi\alpha)^{-n_{\mathfrak{U}}/2} \int_{\mathfrak{U}} (1 - \cos\langle x, u\rangle) \exp\left(-\frac{1}{2\alpha}\|y\|^2\right) d\lambda_{\mathfrak{U}}(y)\, dP_{\mathfrak{U}}(x),$$

where $\lambda_{\mathfrak{U}}$ is the Lebesgue measure on $(\mathfrak{U}, \mathscr{B}_{\mathfrak{U}})$. Using Fubini's theorem and interchanging the order of integration, we obtain

$$\left[1 - \exp\left(-\frac{\alpha}{2}N^2\right)\right] P_{\mathfrak{U}}(\|x\| \geq N) \leq (2\pi\alpha)^{-n_{\mathfrak{U}}/2}$$

$$\times \int_{\mathfrak{U}} [1 - \operatorname{Re} \varphi(y)] \exp\left(-\frac{1}{2\alpha}\|y\|^2\right) d\lambda_{\mathfrak{U}}(y).$$

Let $\epsilon > 0$. Then in view of (7.6.5) there exists an S-operator, $\mathbf{S}_\epsilon$ on $\mathscr{H}$, such that $1 - \operatorname{Re} \varphi(y) < \epsilon/2$ whenever $\langle \mathbf{S}_\epsilon y, y \rangle < 1$. Then for all $y \in \mathscr{H}$

$$0 \leq 1 - \operatorname{Re} \varphi(y) \leq \frac{\epsilon}{2} + 2\langle \mathbf{S}_\epsilon y, y \rangle,$$

so that

$$\begin{aligned}
&\left[1 - \exp\left(-\frac{\alpha}{2} N^2\right)\right] P_{\mathfrak{U}}(\|x\| \geq N) \\
&\quad \leq (2\pi\alpha)^{-n_{\mathfrak{U}}/2} \int_{\mathfrak{U}} \left(\frac{\epsilon}{2} + 2\langle \mathbf{S}_\epsilon y, y \rangle\right) \exp\left(-\frac{1}{2\alpha} \|y\|^2\right) d\lambda_{\mathfrak{U}}(y) \\
&\quad \leq \frac{\epsilon}{2} + 2(2\pi\alpha)^{-n_{\mathfrak{U}}/2} \int_{\mathfrak{U}} \langle \mathbf{S}_\epsilon y, y \rangle \exp\left(-\frac{1}{2\alpha} \|y\|^2\right) d\lambda_{\mathfrak{U}}(y).
\end{aligned}$$

We now show that the inequality

$$(2\pi\alpha)^{-n_{\mathfrak{U}}/2} \int_{\mathfrak{U}} \langle \mathbf{S}_\epsilon y, y \rangle \exp\left(-\frac{1}{2\alpha} \|y\|^2\right) d\lambda_{\mathfrak{U}}(y) \leq \alpha \operatorname{tr} \mathbf{S}_\epsilon$$

holds. In fact, we choose an orthonormal basis $\{e_\nu\}$ in $\mathscr{H}$ such that the subspace $\mathfrak{U}$ is spanned by $\{e_1, e_2, \ldots, e_{n_{\mathfrak{U}}}\}$. Then we have

$$\begin{aligned}
&(2\pi\alpha)^{-n_{\mathfrak{U}}/2} \int_{\mathfrak{U}} \langle \mathbf{S}_\epsilon y, y \rangle \exp\left(-\frac{1}{2\alpha} \|y\|^2\right) d\lambda_{\mathfrak{U}}(y) \\
&= (2\pi\alpha)^{-n_{\mathfrak{U}}/2} \sum_{j,k=1}^{n_{\mathfrak{U}}} \int_{\mathfrak{U}} \langle \mathbf{S}_\epsilon e_j, e_k \rangle \langle y, e_j \rangle \langle y, e_k \rangle \left[\prod_{j=1}^{n_{\mathfrak{U}}} \exp\left(-\frac{\langle y, e_j \rangle^2}{2\alpha}\right)\right] d\lambda_{\mathfrak{U}}(y) \\
&= \sum_{j=1}^{n_{\mathfrak{U}}} \langle \mathbf{S}_\epsilon e_j, e_j \rangle \left[(2\pi\alpha)^{-1/2} \int_{-\infty}^{\infty} u^2 \exp\left(-\frac{u^2}{2\alpha}\right) du\right] \\
&= \alpha \sum_{j=1}^{n_{\mathfrak{U}}} \langle \mathbf{S}_\epsilon e_j, e_j \rangle = \alpha \operatorname{tr} \mathbf{S}_\epsilon,
\end{aligned}$$

as asserted. Hence for every $N \geq 1$ and for any $\alpha > 0$ we have

$$P_{\mathfrak{U}}(\|x\| \geq N) \leq \left(\frac{\epsilon}{2} + 2\alpha \operatorname{tr} \mathbf{S}_\epsilon\right)\left[1 - \exp\left(-\frac{\alpha}{2} N^2\right)\right]^{-1}. \tag{7.6.9}$$

Clearly for given $\epsilon > 0$ we can select an $\alpha = \alpha(\epsilon) > 0$ and a positive integer $N_0 = N_0(\epsilon)$ such that (7.6.3) holds for all $N \geq N_0$. Consequently it follows from Proposition 7.6.2 that there exists a probability measure P on $\mathscr{B}$ associated with the family $\{P_{\mathfrak{U}}\}$.

Finally we show that φ is the Fourier transform of P. Let $x \in \mathscr{H}$. Then from the transformation theorem we have

$$\begin{aligned}\varphi(\pi_{\mathfrak{U}}(x)) &= \int_{\mathfrak{U}} \exp\{i\langle \pi_{\mathfrak{U}}(x), y\rangle\}\, dP_{\mathfrak{U}}(y)\\ &= \int_{\mathscr{H}} \exp\{i\langle \pi_{\mathfrak{U}}(x), \pi_{\mathfrak{U}}(y)\rangle\}\, dP(y).\end{aligned}$$

Since, for every $y \in \mathscr{H}$, $y - \pi_{\mathfrak{U}}(y)$ is orthogonal to $\mathfrak{U}$, it follows that

$$\langle \pi_{\mathfrak{U}}(x), \pi_{\mathfrak{U}}(y)\rangle = \langle \pi_{\mathfrak{U}}(x), y\rangle,$$

so that

$$\varphi(\pi_{\mathfrak{U}}(x)) = \int_{\mathscr{H}} \exp\{i\langle \pi_{\mathfrak{U}}(x), y\rangle\}\, dP(y) \qquad \text{for all } x \in \mathscr{H}.$$

Let $\{\mathfrak{U}_n, n \geq 1\}$ be an increasing sequence of finite-dimensional subspaces of $\mathscr{H}$ such that $\bigcup_{n\geq 1} \mathfrak{U}_n$ is everywhere dense in $\mathscr{H}$. Then for every $n \geq 1$ and $x \in \mathscr{H}$ the relation

$$\varphi(\pi_{\mathfrak{U}_n}(x)) = \int_{\mathscr{H}} \exp\{i\langle \pi_{\mathfrak{U}_n}(x), y\rangle\}\, dP(y) \tag{7.6.10}$$

holds. Clearly $\pi_{\mathfrak{U}_n}(x) \to x$ as $n \to \infty$. We now take the limit of both sides as $n \to \infty$. Using the continuity of φ on the left-hand side and the Lebesgue dominated convergence theorem on the right-hand side of (7.6.10), we obtain

$$\varphi(x) = \int_{\mathscr{H}} \exp(i\langle x, y\rangle)\, dP(y), \qquad x \in \mathscr{H}.$$

Consequently φ is the Fourier transform of P. This completes the proof of Theorem 7.6.1. ■

Remark 7.6.3. In view of the uniqueness property of the Fourier transforms we note that the probability measure P on $\mathscr{B}$ in Theorem 7.6.1 is determined uniquely by φ.

As an application of the Minlos–Sazonov theorem we derive some basic properties of a Gaussian measure on a Hilbert space. First we complete the proof of Proposition 7.5.3.

Proposition 7.6.4. Let $(\mathscr{H}, \mathscr{B})$ be a real, separable Hilbert space. Let φ be a complex-valued function defined by the formula

$$\varphi(y) = \exp(i\langle \mu, y\rangle - \tfrac{1}{2}\langle \mathbf{S}y, y\rangle), \qquad y \in \mathscr{H}, \tag{7.6.11}$$

where $\mu \in \mathscr{H}$ and $\mathbf{S}$ is an S-operator on $\mathscr{H}$. Then φ is the Fourier transform of a probability measure P on $\mathscr{B}$ satisfying the condition

$$\int_{\mathscr{H}} \|x\|^2 \, dP(x) < \infty.$$

Moreover, in this case μ is the mean vector and $\mathbf{S}$ is the covariance operator associated with the probability measure P. (We recall from Definition 7.5.4 that in this case P is a Gaussian measure on $\mathscr{B}$ with mean vector μ and covariance operator $\mathbf{S}$.)

Proof. First we show that there exists a probability measure P on $\mathscr{B}$ such that φ is the Fourier transform of P. Clearly φ is continuous on $\mathscr{H}$ and $\varphi(0) = 1$. Since $\mathbf{S}$ is a bounded, positive, self-adjoint operator on $\mathscr{H}$, φ considered as a function on any finite-dimensional subspace $\mathfrak{U}$ of $\mathscr{H}$ is the Fourier transform of a finite-dimensional Gaussian (that is, multivariate normal) probability distribution. Hence, in particular, φ is positive definite on $\mathfrak{U}$. It follows from the continuity property of φ that φ is positive definite on $\mathscr{H}$. We now verify that condition (7.6.5) of Theorem 7.6.1 is satisfied. In fact, for $y \in \mathscr{H}$

$$\begin{aligned} 0 \leq 1 - \operatorname{Re} \varphi(y) &= 1 - \cos\langle \mu, y\rangle \exp(-\tfrac{1}{2}\langle \mathbf{S}y, y\rangle) \\ &= 1 - \exp(-\tfrac{1}{2}\langle \mathbf{S}y, y\rangle) \\ &\quad + (1 - \cos\langle \mu, y\rangle) \exp(-\tfrac{1}{2}\langle \mathbf{S}y, y\rangle) \\ &\leq \tfrac{1}{2}\{\langle \mathbf{S}y, y\rangle + \langle \mu, y\rangle^2\}. \end{aligned}$$

Define the operator $\mathbf{T}_\mu$ on $\mathscr{H}$ by setting $\mathbf{T}_\mu y = \langle \mu, y\rangle \mu$. Then we see easily that $\mathbf{T}_\mu$ is an S-operator on $\mathscr{H}$ such that $\operatorname{tr} \mathbf{T}_\mu = \|\mu\|^2$. Let $\epsilon > 0$. Define the operator $\mathbf{S}_\epsilon$ on $\mathscr{H}$ by the relation $\mathbf{S}_\epsilon = (1/2\epsilon)(\mathbf{S} + \mathbf{T}_\mu)$. Clearly $\mathbf{S}_\epsilon$ is an S-operator such that

$$\operatorname{tr} \mathbf{S}_\epsilon = \frac{1}{2\epsilon}(\operatorname{tr} \mathbf{S} + \|\mu\|^2) < \infty.$$

Let $y \in \mathscr{H}$ be such that $\langle \mathbf{S}_\epsilon y, y\rangle < 1$. Then

$$\frac{1}{2\epsilon}\{\langle \mathbf{S}y, y\rangle + \langle \mu, y\rangle^2\} < 1,$$

which implies that

$$0 \leq 1 - \operatorname{Re} \varphi(y) < \epsilon.$$

It follows from the Minlos–Sazonov theorem (Theorem 7.6.1) that there exists a probability measure P on $\mathscr{B}$ such that $\varphi = \hat{P}$.

Next we show that $\int_{\mathscr{H}} \|x\|^2 \, dP(x) < \infty$. Without loss of generality we assume that $\mu = 0$. Then for $y \in \mathscr{H}$ we have

$$\int_{\mathscr{H}} \cos\langle x, y\rangle \, dP(x) = \operatorname{Re} \varphi(y) = \exp(-\tfrac{1}{2}\langle \mathbf{S}y, y\rangle),$$

so that

$$\begin{aligned} \exp(-\tfrac{1}{2}\langle \mathbf{S}y, y\rangle) &= 1 - 2\int_{\mathscr{H}} \sin^2 \frac{\langle x, y\rangle}{2} \, dP(x) \\ &\le \exp\left\{-2\int_{\mathscr{H}} \sin^2 \frac{\langle x, y\rangle}{2} \, dP(x)\right\}. \end{aligned}$$

It follows that

$$2\int_{\mathscr{H}} \sin^2 \frac{\langle x, y\rangle}{2} \, dP(x) \le \tfrac{1}{2}\langle \mathbf{S}y, y\rangle.$$

Let $\{e_\nu\}$ be an orthonormal basis in $\mathscr{H}$. Then for every $t \in \mathbb{R}$ and $\nu \ge 1$ we have

$$2\int_{\mathscr{H}} \sin^2 \frac{t\langle x, e_\nu\rangle}{2} \, dP(x) \le \tfrac{1}{2}t^2\langle \mathbf{S}e_\nu, e_\nu\rangle,$$

so that

$$\liminf_{t\to 0} \frac{2}{t^2}\int_{\mathscr{H}} \sin^2 \frac{t\langle x, e_\nu\rangle}{2} \, dP(x) \le \tfrac{1}{2}\langle \mathbf{S}e_\nu, e_\nu\rangle < \infty.$$

It follows immediately from Fatou's lemma that for all $\nu \ge 1$ we have

$$\int_{\mathscr{H}} \langle x, e_\nu\rangle^2 \, dP(x) \le \langle \mathbf{S}e_\nu, e_\nu\rangle,$$

so that for $N \ge 1$

$$\sum_{\nu=1}^{N} \int_{\mathscr{H}} \langle x, e_\nu\rangle^2 \, dP(x) \le \sum_{\nu=1}^{N} \langle \mathbf{S}e_\nu, e_\nu\rangle \le \operatorname{tr} \mathbf{S} < \infty.$$

Taking the limit on both sides as $N \to \infty$, and using the monotone convergence theorem, we get

$$\begin{aligned} \int_{\mathscr{H}} \|x\|^2 \, dP(x) &= \sum_{\nu=1}^{\infty} \int_{\mathscr{H}} \langle x, e_\nu\rangle^2 \, dP(x) \\ &\le \operatorname{tr} \mathbf{S} \\ &< \infty. \end{aligned}$$

Finally we note that for any $t \in \mathbb{R}$ and any $y \in \mathscr{H}$

$$\varphi(ty) = \int_{\mathscr{H}} \exp(it\langle x, y\rangle)\, dP(x)$$
$$= \exp(it\langle \mu, y\rangle - \tfrac{1}{2}t^2\langle \mathbf{S}y, y\rangle).$$

Differentiating both sides with respect to t and evaluating at $t = 0$, we obtain

$$\int_{\mathscr{H}} \langle x, y\rangle\, dP(x) = \langle \mu, y\rangle$$

and

$$\int_{\mathscr{H}} \langle x, y\rangle^2\, dP(x) - \langle \mu, y\rangle^2 = \langle \mathbf{S}y, y\rangle$$

for all $y \in \mathscr{H}$. Hence μ is the mean vector and $\mathbf{S}$ is the covariance operator associated with P. This completes the proof. ■

Remark 7.6.4. Let ν be a Gaussian measure on $(\mathscr{H}, \mathscr{B})$, and $\mathfrak{U}$ be any finite-dimensional subspace of $\mathscr{H}$ with σ-field $\mathscr{B}_{\mathfrak{U}}$ of its Borel subsets. Then the projection $\nu_{\mathfrak{U}}$ of ν onto $\mathscr{B}_{\mathfrak{U}}$ as defined in (7.6.1) is a finite-dimensional normal probability distribution on $\mathscr{B}_{\mathfrak{U}}$.

Proposition 7.6.5. Let $(\mathscr{H}, \mathscr{B})$ be a real, separable Hilbert space. An $\mathscr{H}$-valued random variable X is Gaussian if and only if for every $y \in \mathscr{H}$ the real-valued random variable $\langle X, y\rangle$ is a normal random variable.

Proof. First suppose that X is Gaussian. Then its characteristic functional φ is of the form

$$\varphi(y) = \mathscr{E} \exp(i\langle X, y\rangle) = \exp(i\langle \mu, y\rangle - \tfrac{1}{2}\langle \mathbf{S}y, y\rangle) \qquad \text{for } y \in \mathscr{H},$$

where $\mu \in \mathscr{H}$ is the mean vector, and $\mathbf{S}$ defined on $\mathscr{H}$ is the covariance operator associated with the probability distribution P_X of X. Then for $t \in \mathbb{R}$ we have

$$\mathscr{E} \exp(it\langle X, y\rangle) = \varphi(ty)$$
$$= \exp(it\langle \mu, y\rangle - \tfrac{1}{2}t^2\langle \mathbf{S}y, y\rangle),$$

so that $\langle X, y\rangle$ is a normal random variable with mean $\langle \mu, y\rangle$ and variance $\langle \mathbf{S}y, y\rangle$.

Conversely, let X be an $\mathscr{H}$-valued random variable such that for every $y \in \mathscr{H}$ the real-valued random variable $\langle X, y\rangle$ is normally distributed. Let μ_y and σ_y^2 be the mean and the variance, respectively, of $\langle X, y\rangle$. Let P_X

be the probability distribution of X. Then for every $y \in \mathscr{H}$ we have

$$\mu_y = \mathscr{E}\langle X, y\rangle = \int_{\mathscr{H}} \langle x, y\rangle \, dP_X(x)$$

and

$$\sigma_y^2 = \mathscr{E}\langle X, y\rangle^2 - \{\mathscr{E}\langle X, y\rangle\}^2$$
$$= \int_{\mathscr{H}} \langle x, y\rangle^2 \, dP_X(x) - \mu_y^2.$$

We see easily that $y \to \mu_y$, $y \in \mathscr{H}$, is a linear functional on $\mathscr{H}$. We now show that this functional is also bounded. For this purpose it is sufficient to show that there exists a $\gamma > 0$ such that

$$\sup_{\|y\| \le 1} \int_{\mathscr{H}} |\langle x, y\rangle| \, dP_X(x) \le \gamma < \infty.$$

For $y \in \mathscr{H}$ we set

$$\theta(y) = \int_{\mathscr{H}} |\langle x, y\rangle| \, dP_X(x)$$

and for every $n \ge 1$

$$\theta_n(y) = \int_{\mathscr{H}} \frac{n|\langle x, y\rangle|}{n + \|x\|} \, dP_X(x).$$

Note that $0 \le \theta_n \uparrow \theta$ as $n \to \infty$. Moreover, in view of the Lebesgue dominated convergence theorem, for every $n \ge 1$, θ_n is weakly continuous on $\mathscr{H}$. For every $n \ge 1$ and $N \ge 1$ we set

$$K_{n,N} = \{y \in \mathscr{H} : \theta_n(y) \ge N\} \cap \{y \in \mathscr{H} : \|y\| \le 1\}.$$

Then $K_{n,N}$ is weakly closed because of the weak continuity of θ_n, and since $K_{n,N}$ is bounded it is also weakly compact. For $N \ge 1$ set

$$K_N = \bigcap_{n=1}^{\infty} K_{n,N}.$$

Then

$$K_N = \{y \in \mathscr{H} : \theta(y) \ge N\} \cap \{y \in \mathscr{H} : \|y\| \le 1\}.$$

Clearly K_N is also weakly closed and weakly compact. We now show that there exists an $N_0 \ge 1$ such that $K_{N_0} = \varnothing$. Suppose this is not the case. Then $K_N \ne \varnothing$ for all $N \ge 1$. This implies that

$$\bigcap_{N=1}^{\infty} K_N \ne \varnothing,$$

so that there exists a $y_0 \in \bigcap_{N=1}^{\infty} K_N$. Thus $\theta(y_0) = \infty$, which is a contradiction. Hence there exists an N_0 such that

$$K_{N_0} = \{y \in \mathscr{H} : \theta(y) \geq N_0\} \cap \{y \in \mathscr{H} : \|y\| \leq 1\} = \emptyset,$$

so that

$$\sup_{\|y\| \leq 1} \theta(y) = \sup_{\|y\| \leq 1} \int_{\mathscr{H}} |\langle x, y \rangle| \, dP_X(x) < N_0 < \infty.$$

Consequently there exists a vector $\mu \in \mathscr{H}$ such that the relation

$$\int_{\mathscr{H}} \langle x, y \rangle \, dP_X(x) = \mu_y = \langle \mu, y \rangle$$

holds for every $y \in \mathscr{H}$.

Next we show that

$$\sup_{\|y\| \leq 1} \int \langle x, y \rangle^2 \, dP_X(x) < \infty.$$

For this purpose we set

$$\psi(y) = \int_{\mathscr{H}} \langle x, y \rangle^2 \, dP_X(x), \qquad y \in \mathscr{H},$$

and for every $n \geq 1$ and $y \in \mathscr{H}$

$$\psi_n(y) = \int_{\mathscr{H}} \frac{n\langle x, y \rangle^2}{n + \|x\|^2} \, dP_X(x),$$

and complete the proof in the same manner as above. Then we see easily that the mapping

$$\begin{aligned}(y, z) \to \int_{\mathscr{H}} \langle x, y \rangle \langle x, z \rangle \, dP_X(x) - \mu_y \mu_z \\ = \int_{\mathscr{H}} \langle x - \mu, y \rangle \langle x - \mu, z \rangle \, dP_X(x)\end{aligned}$$

is a bounded bilinear form on $\mathscr{H}$, so that there exists bounded linear operator $\mathbf{S}$ on $\mathscr{H}$ such that the relation

$$\langle \mathbf{S}y, z \rangle = \int_{\mathscr{H}} \langle x - \mu, y \rangle \langle x - \mu, z \rangle \, dP_X(x)$$

holds for all $y, z \in \mathscr{H}$. Clearly

$$\begin{aligned}\langle \mathbf{S}y, y\rangle &= \int_{\mathscr{H}} \langle x - \mu, y\rangle^2 \, dP_X(x)\\ &= \int_{\mathscr{H}} \langle x, y\rangle^2 \, dP_X(x) - \mu_y^2\\ &= \sigma_y^2 \geq 0,\end{aligned}$$

so that $\mathbf{S}$ is a bounded, positive, self-adjoint operator on $\mathscr{H}$.

For every $t \in \mathbb{R}$, $y \in \mathscr{H}$ we have

$$\begin{aligned}\mathscr{E} \exp(it\langle X, y\rangle) &= \exp(it\mu_y - \tfrac{1}{2}t^2\sigma_y^2)\\ &= \exp(it\langle \mu, y\rangle - \tfrac{1}{2}t^2\langle \mathbf{S}y, y\rangle),\end{aligned}$$

and, in particular, by setting $t = 1$, we have for all $y \in \mathscr{H}$

$$\varphi(y) = \mathscr{E} \exp(i\langle X, y\rangle) = \exp(i\langle \mu, y\rangle - \tfrac{1}{2}\langle \mathbf{S}y, y\rangle).$$

Finally we show that $\mathbf{S}$ is an S-operator on $\mathscr{H}$. Since φ is the Fourier transform of P_X it follows from Theorem 7.6.1 that for given ϵ, $0 < \epsilon < 1$, there exists an S-operator $\mathbf{S}_\epsilon$ on $\mathscr{H}$ such that

$$1 - \operatorname{Re} \varphi(y) < \epsilon \qquad \text{whenever } \langle \mathbf{S}_\epsilon y, y\rangle < 1.$$

Let $y \in \mathscr{H}$ be such that $\langle \mathbf{S}_\epsilon y, y\rangle < 1$. Then

$$\begin{aligned}\tfrac{1}{2}\langle \mathbf{S}y, y\rangle &< \exp(\tfrac{1}{2}\langle \mathbf{S}y, y\rangle) - 1\\ &= \frac{1 - \exp(-\frac{1}{2}\langle \mathbf{S}y, y\rangle)}{1 - [1 - \exp(-\frac{1}{2}\langle \mathbf{S}y, y\rangle)]}\\ &< \frac{1 - \exp(-\frac{1}{2}\langle \mathbf{S}y, y\rangle)\cos\langle \mu, y\rangle}{1 - [1 - \exp(-\frac{1}{2}\langle \mathbf{S}y, y\rangle)\cos\langle \mu, y\rangle]}\\ &< \frac{\epsilon}{1 - \epsilon}\end{aligned}$$

since $1 - \operatorname{Re} \varphi(y) = 1 - \cos\langle \mu, y\rangle \exp(-\frac{1}{2}\langle \mathbf{S}y, y\rangle) < \epsilon$. (Note that for $\epsilon < 1$, $\cos\langle \mu, y\rangle > 0$.) Therefore

$$\langle \mathbf{S}y, y\rangle < \frac{2\epsilon}{1 - \epsilon} \qquad \text{whenever } \langle \mathbf{S}_\epsilon y, y\rangle < 1.$$

Hence for all $y \in \mathscr{H}$

$$\langle \mathbf{S}y, y\rangle < \frac{2\epsilon}{1 - \epsilon} \langle \mathbf{S}_\epsilon y, y\rangle,$$

so that **S** is an S-operator on $\mathscr{H}$ such that

$$\operatorname{tr} \mathbf{S} < \frac{2\epsilon}{1-\epsilon} \operatorname{tr} \mathbf{S}_\epsilon < \infty.$$

It follows that X is a Gaussian random variable with mean vector μ and covariance operator **S**. This completes the proof. ■

7.7 INFINITELY DIVISIBLE PROBABILITY MEASURES ON A HILBERT SPACE

Let $\mathscr{H}$ be a real, separable Hilbert space with inner product $\langle \cdot, \cdot \rangle$, and let $\mathscr{B}$ be the σ-field generated by the class of all open subsets of $\mathscr{H}$. In this section we study the concept of infinite divisibility of probability measures on $\mathscr{B}$. In particular, we derive an analogue of the Lévy–Khintchine representation (Theorem 4.1.1) for the Fourier transform of an infinitely divisible (i.d.) probability measure on $\mathscr{B}$.

7.7.1 A Necessary and Sufficient Condition for Tightness

First we obtain a necessary and sufficient condition for tightness of a family of probability measures on $\mathscr{B}$. This will be used in the derivation of the Lévy–Khintchine representation of an i.d. probability measure on $\mathscr{B}$.

Theorem 7.7.1. Let $\mathscr{P}$ be a family of probability measures on $\mathscr{B}$. For every $P \in \mathscr{P}$, let $\hat{P}$ denote the Fourier transform of P. Then the family $\mathscr{P}$ is tight if and only if the following conditions are satisfied:

(i) For given $\epsilon > 0$ and for every $P \in \mathscr{P}$ there exists an S-operator $\mathbf{S}_{P,\epsilon}$ on $\mathscr{H}$ such that

$$0 \leq 1 - \operatorname{Re} \hat{P}(y) < \epsilon \qquad \text{whenever } \langle \mathbf{S}_{P,\epsilon} y, y \rangle < 1, \qquad y \in \mathscr{H}.$$

(ii) $\sup_{P \in \mathscr{P}} \operatorname{tr} \mathbf{S}_{P,\epsilon} < \infty$.

(iii) For some orthonormal basis $\{e_\nu\}$ in $\mathscr{H}$ the series

$$\operatorname{tr} \mathbf{S}_{P,\epsilon} = \sum_{\nu=1}^{\infty} \langle \mathbf{S}_{P,\epsilon} e_\nu, e_\nu \rangle$$

converges uniformly in $P \in \mathscr{P}$, that is,

$$\lim_{N \to \infty} \sup_{P \in \mathscr{P}} \sum_{\nu=N}^{\infty} \langle S_{P,\epsilon} e_\nu, e_\nu \rangle = 0.$$

Proof. First suppose that $\mathscr{P}$ is tight. We show that (i), (ii), and (iii) hold. Let $\epsilon > 0$. Then there exists a compact set $K = K_\epsilon \subset \mathscr{H}$ such that

$$\sup_{P \in \mathscr{P}} P(\mathscr{H} - K) < \frac{\epsilon}{4}.$$

Let $P \in \mathscr{P}$ and $y \in \mathscr{H}$. Then we have

$$\begin{aligned} 0 \leq 1 - \operatorname{Re} \hat{P}(y) &= \int_{\mathscr{H}} (1 - \cos\langle x, y\rangle)\, dP(x) \\ &= \left(\int_K + \int_{\mathscr{H} - K}\right)(1 - \cos\langle x, y\rangle)\, dP(x) \\ &\leq 2\int_K \sin^2 \frac{\langle x, y\rangle}{2}\, dP(x) + 2P(\mathscr{H} - K) \\ &\leq \tfrac{1}{2}\int_K \langle x, y\rangle^2\, dP(x) + \frac{\epsilon}{2}. \end{aligned}$$

Since K is compact and the function $x \to \|x\|$ is continuous on $\mathscr{H}$, it is bounded on K. Set $\gamma = \sup_{x \in K} \|x\|$. Clearly

$$\sup_{P \in \mathscr{P}} \int_K \|x\|^2\, dP(x) \leq \gamma^2.$$

Proceeding exactly as in Section 7.5, we now define the operator $\mathbf{S}_{P,\epsilon}$ on $\mathscr{H}$ by the relation

$$\langle \mathbf{S}_{P,\epsilon} y, y\rangle = \frac{1}{\epsilon}\int_K \langle x, y\rangle^2\, dP(x).$$

Then we see easily that $\mathbf{S}_{P,\epsilon}$ is an S-operator on $\mathscr{H}$ such that

$$\operatorname{tr} \mathbf{S}_{P,\epsilon} = \frac{1}{\epsilon}\int_K \|x\|^2\, dP(x) \leq \frac{\gamma^2}{\epsilon} < \infty.$$

Let $y \in \mathscr{H}$ be such that $\langle \mathbf{S}_{P,\epsilon} y, y\rangle < 1$. Then

$$0 \leq 1 - \operatorname{Re} \hat{P}(y) \leq \frac{\epsilon}{2}\langle \mathbf{S}_{P,\epsilon} y, y\rangle + \frac{\epsilon}{2} < \epsilon,$$

so that (i) holds.

Clearly $\sup_{P \in \mathscr{P}} \operatorname{tr} \mathbf{S}_{P,\epsilon} \leq \gamma^2\epsilon < \infty$, so that (ii) holds.

Next let $\{e_\nu\}$ be an orthonormal basis in $\mathscr{H}$. For every $N \geq 1$ and $x \in \mathscr{H}$ set

$$r_N^2(x) = \sum_{\nu=N}^{\infty} \langle x, e_\nu\rangle^2.$$

We note that the function r_N^2 is continuous and, moreover, $r_N^2 \downarrow 0$ as $N \to \infty$. Now set

$$\psi_N = \sup_{x \in K} r_N^2(x).$$

Since K is compact, we see easily that $0 \le \psi_N \downarrow 0$ as $N \to \infty$. For $N \ge 1$ we have

$$\sum_{\nu=N}^{\infty} \langle \mathbf{S}_{P,\epsilon} e_\nu, e_\nu \rangle = \frac{1}{\epsilon} \sum_{\nu=N}^{\infty} \int_K \langle x, e_\nu \rangle^2 \, dP(x) \le \frac{1}{\epsilon} \psi_N,$$

so that

$$\sup_{P \in \mathscr{P}} \sum_{\nu=N}^{\infty} \langle \mathbf{S}_{P,\epsilon} e_\nu, e_\nu \rangle \le \frac{1}{\epsilon} \psi_N \to 0 \qquad \text{as } N \to \infty.$$

This proves (iii).

Conversely, suppose that (i), (ii), and (iii) hold. We show that $\mathscr{P}$ is tight. For this purpose we need the following result.

Lemma 7.7.1. Let $\mathscr{P}$ be a family of probability measures on $\mathscr{B}$. Suppose there exists a family $\{\mathbf{S}_P, P \in \mathscr{P}\}$ of $\mathbf{S}$-operators on $\mathscr{H}$ satisfying these conditions:

(i) $\sup_{P \in \mathscr{P}} \operatorname{tr} \mathbf{S}_P < \infty$.
(ii) For some orthonormal basis $\{e_\nu\}$ in $\mathscr{H}$

$$\lim_{N \to \infty} \sup_{P \in \mathscr{P}} \sum_{\nu=N}^{\infty} \langle \mathbf{S}_P e_\nu, e_\nu \rangle = 0.$$

Then every $\mathbf{S}_P$ can be represented in the form

$$\mathbf{S}_P = B \mathbf{T}_P B,$$

where B is a compact, positive, self-adjoint operator on $\mathscr{H}$, and $\mathbf{T}_P$ is an $\mathbf{S}$-operator on $\mathscr{H}$ such that $\operatorname{tr} \mathbf{T}_P \le 1$ for every $P \in \mathscr{P}$.

Proof. For every $N \ge 1$ we set

$$\beta_N = \sup_{P \in \mathscr{P}} \sum_{\nu=N}^{\infty} \langle \mathbf{S}_P e_\nu, e_\nu \rangle.$$

It then follows from (i) and (ii) that $\beta_N \downarrow 0$ as $N \to \infty$. We now choose a sequence $\{\alpha_N, N \ge 1\}$ of positive real numbers such that $\sum_{N=1}^{\infty} \alpha_N = \infty$ and $\sum_{N=1}^{\infty} \alpha_N \beta_N < \infty$. Then for every $P \in \mathscr{P}$

$$\sum_{n=1}^{\infty} \sum_{k=1}^{n} \alpha_k \langle \mathbf{S}_P e_n, e_n \rangle = \sum_{k=1}^{\infty} \alpha_k \sum_{n=k}^{\infty} \langle \mathbf{S}_P e_n, e_n \rangle \le \sum_{k=1}^{\infty} \alpha_k \beta_k < \infty.$$

For every $k \ge 1$ we set

$$\lambda_k = \left[\frac{\sum_{n=1}^{\infty} \alpha_n \beta_n}{\sum_{n=1}^{k} \alpha_n} \right]^{1/2}.$$

Then $0 < \lambda_k \downarrow 0$ as $k \to \infty$. We define a linear operator B on $\mathscr{H}$ with eigenvalues λ_k, $k \geq 1$, such that e_k is the eigenvector corresponding to λ_k, that is, B is defined by the relation $Be_k = \lambda_k e_k$ for $k \geq 1$. Clearly B is a bounded, positive, self-adjoint operator on $\mathscr{H}$; moreover, B is compact, since $\lambda_k \to 0$ as $k \to \infty$.

For every $N \geq 1$, let $\mathfrak{U}_N$ be the N-dimensional subspace of $\mathscr{H}$ spanned by $\{e_1, e_2, \ldots, e_N\}$. Then $\mathfrak{U}_N \subset \mathfrak{U}_{N+1}$, $N \geq 1$, and $\bigcup_{N=1}^{\infty} \mathfrak{U}_N$ is everywhere dense in $\mathscr{H}$. Let π_N be the projection of $\mathscr{H}$ onto $\mathfrak{U}_N$, and set $B_N = B\pi_N$, $N \geq 1$. Then clearly B_N leaves $\mathfrak{U}_N$ invariant; moreover, B_N considered as an operator on $\mathfrak{U}_N$ is invertible. In this case for $P \in \mathscr{P}$ we have

$$
\begin{aligned}
\sum_{k=1}^{N} \langle B_N^{-1} \mathbf{S}_P B_N^{-1} e_k, e_k \rangle &= \sum_{k=1}^{N} \langle \mathbf{S}_P B_N^{-1} e_k, B_N^{-1} e_k \rangle \\
&= \sum_{k=1}^{N} \frac{1}{\lambda_k^2} \langle \mathbf{S}_P e_k, e_k \rangle \\
&= \sum_{k=1}^{N} \frac{\sum_{n=1}^{k} \alpha_n}{\sum_{n=1}^{\infty} \alpha_n \beta_n} \langle \mathbf{S}_P e_k, e_k \rangle \\
&= \frac{\sum_{k=1}^{N} \langle \mathbf{S}_P e_k, e_k \rangle \sum_{n=1}^{k} \alpha_n}{\sum_{n=1}^{\infty} \alpha_n \beta_n} \\
&\leq \frac{\sum_{k=1}^{\infty} \sum_{n=1}^{k} \alpha_n \langle \mathbf{S}_P e_k, e_k \rangle}{\sum_{n=1}^{\infty} \alpha_n \beta_n} \\
&\leq 1. \qquad (7.7.1)
\end{aligned}
$$

For every $N \geq 1$ and $P \in \mathscr{P}$ we define a linear operator $T_{P,N}$ on $\mathfrak{U}_N$ by setting $T_{P,N} e_k = B_N^{-1} \mathbf{S}_P B_N^{-1} e_k$, $1 \leq k \leq N$. It follows from (7.7.1) that

$$
\operatorname{tr} T_{P,N} = \sum_{k=1}^{N} \langle T_{P,N} e_k, e_k \rangle \leq 1. \qquad (7.7.2)
$$

Since $\mathfrak{U}_N$ is increasing and $\bigcup_{N=1}^{\infty} \mathfrak{U}_N$ is everywhere dense in $\mathscr{H}$, we can verify easily that for every $P \in \mathscr{P}$ there exists a bounded linear operator $\mathbf{T}_P$ on $\mathscr{H}$ such that $\mathbf{T}_P$ coincides with $T_{P,N}$ on $\mathfrak{U}_N$ and, moreover, for every $x \in \mathscr{H}$, $T_{P,N} x \to \mathbf{T}_P x$ as $N \to \infty$. Then it follows from (7.7.2) that $\mathbf{T}_P$ is an S-operator on $\mathscr{H}$ satisfying the condition

$$
\operatorname{tr} \mathbf{T}_P \leq 1, \qquad P \in \mathscr{P}.
$$

On the other hand, for every $P \in \mathscr{P}$ and $N \geq 1$ we have the relation

$$
\mathbf{S}_P = B_N T_{P,N} B_N = B\pi_N T_{P,N} \pi_N B.
$$

We now take the limit as $N \to \infty$. Since, for $x \in \mathscr{H}$, $\pi_N(x) \to x$, and $T_{P,N}x \to \mathbf{T}_P x$ as $N \to \infty$, we conclude that

$$\mathbf{S}_P = B\mathbf{T}_P B.$$

This completes the proof of the lemma. ■

We now return to the proof of Theorem 7.7.1. Let $\epsilon > 0$. In view of Lemma 7.7.1 the operator $\mathbf{S}_{P,\epsilon}$ can be represented in the form

$$\mathbf{S}_{P,\epsilon} = B_\epsilon \mathbf{T}_{P,\epsilon} B_\epsilon,$$

where B_ϵ is a compact, positive, self-adjoint operator in $\mathscr{H}$, and $\mathbf{T}_{P,\epsilon}$ is an S-operator on $\mathscr{H}$ with the property that $\operatorname{tr} \mathbf{T}_{P,\epsilon} \leq 1$ for all $P \in \mathscr{P}$. In view of condition (i) of the theorem we have for every $P \in \mathscr{P}$ and $y \in \mathscr{H}$

$$0 \leq 1 - \operatorname{Re} \hat{P}(y) < \frac{\epsilon}{2} + 2\langle B_\epsilon \mathbf{T}_{P,\epsilon} B_\epsilon y, y\rangle. \tag{7.7.3}$$

Let $\{e_\nu\}$ be the orthonormal basis as in Lemma 7.7.1. For every $k \geq 1$ let λ_k be the eigenvalue of B_ϵ associated with the eigenvector e_k for $k \geq 1$. For every $N \geq 1$ let $\mathfrak{U}_N$, π_N, and $B_N = B\pi_N$ be as defined in the proof of Lemma 7.7.1. Let P_N be the projection of P onto $\mathfrak{U}_N$ for $P \in \mathscr{P}$ and $N \geq 1$. Let $\alpha > 0$, $N \geq 1$, and $P \in \mathscr{P}$ be fixed. We now estimate the integral

$$\int_{\mathfrak{U}_N} \left[1 - \exp\left(-\frac{\alpha}{2}\langle B_N^{-2}x, x\rangle\right)\right] dP_N(x).$$

For $x \in \mathscr{H}$ we write $\langle x, e_\nu\rangle = x_\nu$, $\nu \geq 1$. Then for $x \in \mathfrak{U}_N$ we have

$$x = \sum_{k=1}^{N} x_k e_k,$$

$$\|x\|^2 = \sum_{k=1}^{N} x_k^2,$$

and

$$\langle B_N^{-2}x, x\rangle = \sum_{k=1}^{N} \frac{x_k^2}{\lambda_k^2}.$$

We note that

$$\exp\left(-\frac{\alpha}{2}\sum_{k=1}^{N}\frac{x_k^2}{\lambda_k^2}\right) = \frac{1}{(2\pi\alpha)^{N/2}} \prod_{k=1}^{N} \lambda_k \int_{-\infty}^{\infty} \cdots \int_{-\infty}^{\infty} \exp\left(i \sum_{k=1}^{N} x_k y_k - \frac{1}{2\alpha}\sum_{k=1}^{N} \lambda_k^2 y_k^2\right) dy_1 \cdots dy_N,$$

where $y_k = \langle y, e_k \rangle$, $1 \leq k \leq N$. Hence we have

$$\int_{\mathfrak{U}_N} \left[1 - \exp\left(-\frac{\alpha}{2}\langle B_N^{-2}x, x\rangle\right)\right] dP_N(x)$$

$$= \int_{\mathfrak{U}_N} \frac{1}{(2\pi\alpha)^{N/2}} \left\{\prod_{k=1}^{N} \lambda_k \int_{-\infty}^{\infty} \cdots \int_{-\infty}^{\infty} \left[1 - \cos \sum_{k=1}^{n} x_k y_k\right]\right.$$

$$\left. \times \exp\left(-\frac{1}{2\alpha}\sum_{k=1}^{N} \lambda_k^2 y_k^2\right) dy_1 \cdots dy_N \right\} dP_N(x)$$

$$= \frac{1}{(2\pi\alpha)^{N/2}} \prod_{k=1}^{N} \lambda_k \int_{-\infty}^{\infty} \cdots \int_{-\infty}^{\infty} \left[1 - \operatorname{Re} \hat{P}\left(\sum_{k=1}^{N} y_k e_k\right)\right]$$

$$\times \exp\left(-\frac{1}{2\alpha}\sum_{k=1}^{N} \lambda_k^2 y_k^2\right) dy_1 \cdots dy_N$$

in view of Fubini's theorem. Then, using (7.7.3), we have

$$\int_{\mathfrak{U}_N} \left[1 - \exp\left(-\frac{\alpha}{2}\langle B_N^{-2}x, x\rangle\right)\right] dP_N(x)$$

$$\leq \frac{1}{(2\pi\alpha)^{N/2}} \prod_{k=1}^{N} \lambda_k \int_{-\infty}^{\infty} \cdots \int_{-\infty}^{\infty} \left(\frac{\epsilon}{2} + 2\left\langle B_\epsilon \mathbf{T}_{P,\epsilon} B_\epsilon \sum_{k=1}^{N} y_k e_k, \sum_{k=1}^{N} y_k e_k\right\rangle\right)$$

$$\times \exp\left(-\frac{1}{2\alpha}\sum_{k=1}^{N} \lambda_k^2 y_k^2\right) dy_1 \ldots dy_N$$

$$\leq \frac{\epsilon}{2} + 2\alpha \operatorname{tr} \mathbf{T}_{P,\epsilon} \leq \frac{\epsilon}{2} + 2\alpha,$$

where the computation of the last integral is carried out exactly as in the proof of Theorem 7.6.1. On the other hand, we have for $\gamma > 0$

$$\int_{\mathfrak{U}_N} \left[1 - \exp\left(-\frac{\alpha}{2}\langle B_N^{-2}x, x\rangle\right)\right] dP_N(x)$$

$$\geq \left[1 - \exp\left(-\frac{\alpha}{2}\gamma^2\right)\right] P_N\{x \in \mathfrak{U}_N : \langle B_N^{-2}x, x\rangle > \gamma^2\}.$$

Hence for $\alpha > 0$, $\gamma > 0$, $N \geq 1$, and $P \in \mathscr{P}$ we have

$$\text{(7.7.4)} \quad P_N\{x \in \mathfrak{U}_N : \|B_N^{-1}x\| > \gamma\} \leq \left(\frac{\epsilon}{2} + 2\alpha\right)\left[1 - \exp\left(-\frac{\alpha}{2}\gamma^2\right)\right]^{-1}.$$

For $N \geq 1$ we now set

$$K_N = \{x \in \mathfrak{U}_N : \|B_N^{-1}x\| \leq \gamma\}.$$

Then K_N is a compact set in $\mathfrak{U}_N$, and we have

$$P_N(\mathfrak{U}_N - K_N) \leq \left(\frac{\epsilon}{2} + 2\alpha\right)\left[1 - \exp\left(-\frac{\alpha}{2}\gamma^2\right)\right]^{-1}.$$

For given $\epsilon > 0$ we select $\alpha = \alpha(\epsilon) > 0$ and $\gamma = \gamma(\epsilon) > 0$ such that

$$\left(\frac{\epsilon}{2} + 2\alpha\right)\left[1 - \exp\left(-\frac{\alpha}{2}\gamma^2\right)\right]^{-1} < \epsilon.$$

Then we have

$$\sup_{N \geq 1} P_N(\mathfrak{U}_N - K_N) < \epsilon,$$

where $K_N = \{B_N y\colon \|y\| \leq \gamma\}$ is a compact set in $\mathfrak{U}_N$. Let $K \subset \mathscr{H}$ be such that $K = \pi_N^{-1}(K_N)$, $N \geq 1$. Clearly

$$K = \{x \in \mathscr{H} : \pi_N(x) \in K_N \text{ for all } N \geq 1\} \subset \overline{\{By \in \mathscr{H} : \|y\| \leq \gamma\}},$$

since $\pi_N(x) \to x$ as $N \to \infty$, where $\bar{A}$ is the closure of A. Since B is compact, the set $\{By \in \mathscr{H} : \|y\| \leq \gamma\}$ is relatively compact. On the other hand, since K is closed, K is a compact set in $\mathscr{H}$. Moreover, for every $P \in \mathscr{P}$,

$$P(\mathscr{H} - K) = P_N(\mathfrak{U}_N - K_N) < \epsilon,$$

which proves that $\mathscr{P}$ is tight. This completes the proof of Theorem 7.7.1. ■

Remark 7.7.1. Theorem 7.7.1 holds if $\mathscr{P} = \mathscr{M}$ is a family of uniformly bounded, finite measures on $\mathscr{B}$, that is,

$$\sup_{\mu \in \mathscr{M}} \mu(\mathscr{H}) < \infty.$$

In this case in condition (i) we need only to replace $1 - \operatorname{Re} \hat{P}(y)$ by $\hat{\mu}(0) - \operatorname{Re} \hat{\mu}(y)$.

Finally we derive a sufficient condition for tightness of a family of finite measures on $\mathscr{B}$.

Proposition 7.7.1. Let $\mathscr{M}$ be a family of finite measures on $(\mathscr{H}, \mathscr{B})$. Suppose that the following two conditions are satisfied:

(i) $\sup_{\mu \in \mathscr{M}} \hat{\mu}(0) < \infty$.

(ii) For given $\epsilon > 0$ there exists an S-operator $\mathbf{S}_\epsilon$ on $\mathscr{H}$ such that

$$\sup_{\mu \in \mathscr{M}} [\hat{\mu}(0) - \operatorname{Re} \hat{\mu}(y)] < \epsilon \qquad \text{whenever } \langle \mathbf{S}_\epsilon y, y\rangle < 1, \qquad y \in \mathscr{H}.$$

Then $\mathscr{M}$ is tight.

Proof. In Theorem 7.7.1 we set $\mathbf{S}_{P,\epsilon} = \mathbf{S}_{\mu,\epsilon} = \mathbf{S}_\epsilon$ for every $\mu \in \mathscr{M}$. Then conditions (i), (ii), and (iii) of Theorem 7.7.1 hold. ■

7.7.2 Lévy–Khintchine Representation

We first define an i.d. probability measure on $\mathscr{B}$, and then derive the Lévy–Khintchine representation of its Fourier transform.

Definition 7.7.1. Let $\mathbf{P}$ be a probability measure on $(\mathscr{H}, \mathscr{B})$. Then $\mathbf{P}$ is said to be i.d. if for every positive integer n there exists a probability measure $\mathbf{P}_n$ on $\mathscr{B}$ such that $\mathbf{P}$ can be written as

$$\mathbf{P} = \underbrace{\mathbf{P}_n * \mathbf{P}_n * \cdots * \mathbf{P}_n}_{n\text{-fold}}.$$

Equivalently, $\mathbf{P}$ is i.d. if its Fourier transform $\hat{\mathbf{P}}$ satisfies the following condition: For every positive integer n there exists a Fourier transform $\hat{\mathbf{P}}_n$ of a probability measure $\mathbf{P}_n$ on $\mathscr{B}$ such that

$$\hat{\mathbf{P}}(y) = [\hat{\mathbf{P}}_n(y)]^n, \qquad y \in \mathscr{H}.$$

If $\mathbf{P}$ is i.d., we say that $\hat{\mathbf{P}}$ is i.d.

We first state some elementary properties of the Fourier transform $\hat{\mathbf{P}}$ of an i.d. probability measure $\mathbf{P}$ on $\mathscr{B}$.

Proposition 7.7.2.

(a) Let $\mathbf{P}$ be an i.d. probability measure on $(\mathscr{H}, \mathscr{B})$. Then, for every $x \in \mathscr{H}$, $\hat{\mathbf{P}}(x) \neq 0$.

(b) Let $\mathbf{P}_1$ and $\mathbf{P}_2$ be two i.d. probability measures on $\mathscr{B}$. Then their convolution $\mathbf{P}_1 * \mathbf{P}_2$ is also i.d.

(e) Let $\{\mathbf{P}_n\}$ be a sequence of i.d. probability measures on $\mathscr{B}$, and suppose that $\mathbf{P}_n \Rightarrow \mathbf{P}$, where $\mathbf{P}$ is a probability measure on $\mathscr{B}$. Then $\mathbf{P}$ is i.d.

Proof. The proofs of (a), (b), and (c) can be carried out by proceeding exactly in the same manner as in the proofs of Propositions 4.1.1, 4.1.2, and 4.1.3, respectively. ■

We next derive the following result for the representation of the Fourier transform of an i.d. probability measure on $(\mathscr{H}, \mathscr{B})$.

Theorem 7.7.2. Let $(\mathscr{H}, \mathscr{B})$ be a real, separable Hilbert space, and let φ be a complex-valued function defined on $\mathscr{H}$. Then φ is the Fourier transform of an i.d. probability measure $\mathbf{P}$ on $\mathscr{B}$ if and only if $\ln \varphi$ can be represented in the form

$$\begin{aligned}\ln \varphi(x) \\ &= i\langle a, x\rangle - \tfrac{1}{2}\langle \mathbf{S}x, x\rangle \\ &\quad + \int_{\mathscr{H}} \left\{\exp(i\langle x, y\rangle) - 1 - \frac{i\langle x, y\rangle}{1 + \|y\|^2}\right\} \frac{1 + \|y\|^2}{\|y\|^2} d\mu(y), \qquad x \in \mathscr{H},\end{aligned} \tag{7.7.5}$$

where μ is a finite measure on $\mathscr{B}$ with $\mu\{0\} = 0$, $a \in \mathscr{H}$ is a fixed vector, and $\mathbf{S}$ is an S-operator on $\mathscr{H}$. Moreover, in this case a, $\mathbf{S}$, and μ are determined uniquely by φ.

Proof. First suppose that φ is the Fourier transform of an i.d. probability measure $\mathbf{P}$ on $(\mathscr{H}, \mathscr{B})$. We show that $\ln \varphi$ can be represented in the form (7.7.5). For this purpose we proceed in the following manner.

Let X be an $\mathscr{H}$-valued random variable defined on $(\Omega, \mathscr{S}, P)$ with probability distribution $P_X = \mathbf{P}$. Then the characteristic functional of the random variable X is given by

$$\begin{aligned}\mathscr{E} \exp(i\langle y, X\rangle) &= \int_{\mathscr{H}} \exp(i\langle y, x\rangle)\, d\mathbf{P}(x) \\ &= \hat{\mathbf{P}}(y) = \varphi(y), \qquad y \in \mathscr{H}.\end{aligned}$$

For every $n \geq 1$ let $X_{n1}, X_{n2}, \ldots, X_{nn}$ be n independent, identically distributed $\mathscr{H}$-valued random variables defined on $(\Omega, \mathscr{S}, P)$, each having probability distribution $P_{X_n} = \mathbf{P}_n$ such that $X = \sum_{k=1}^n X_{nk}$, $n \geq 1$. Clearly

$$\mathscr{E} \exp(i\langle y, X_{nk}\rangle) = \hat{\mathbf{P}}_n(y), \qquad y \in \mathscr{H}, \quad k = 1, 2, \ldots, n;$$

moreover, for every $n \geq 1$

$$\varphi = \hat{\mathbf{P}} = (\hat{\mathbf{P}}_n)^n.$$

We need the following lemmas.

Lemma 7.7.2. Let $0 < \epsilon < \frac{1}{4}$ and $n \geq 1$. Then there exists a $\delta = \delta(\epsilon) > 0$ such that for all $1 \leq k \leq n$ and $n \geq 1$

$$P\left\{\left\| \sum_{j=k}^{n} X_{nj} \right\| \geq \delta\right\} < \epsilon. \tag{7.7.6}$$

Proof. Let $\epsilon > 0$. In view of the Minlos–Sazonov theorem (Theorem 7.6.1) there exists an S-operator $\mathbf{S}_\epsilon$ on $\mathscr{H}$ such that

$$1 - \operatorname{Re} \varphi(y) < \frac{\epsilon}{2} \qquad \text{whenever } \langle \mathbf{S}_\epsilon y, y\rangle < 1, \quad y \in \mathscr{H}.$$

Clearly

$$|\text{Im } \varphi(y)| \leq [1 - (\text{Re } \varphi(y))^2]^{1/2} < \epsilon^{1/2} \qquad \text{whenever } \langle \mathbf{S}_\epsilon y, y \rangle < 1.$$

Now choose $\epsilon < \frac{1}{4}$. Then

$$|\arg \varphi(y)| \leq \arc \tan \frac{\epsilon^{1/2}}{1 - \epsilon/2} < \frac{\pi}{4}.$$

Consequently, for all $1 \leq k \leq n$ and $n \geq 1$, we have

$$\begin{aligned} 1 - \text{Re}\{\hat{\mathbf{P}}_n(y)\}^{n-k+1} &= 1 - |\varphi(y)|^{(n-k+1)/n} \cos\left\{\frac{n-k+1}{n} \arg \varphi(y)\right\} \\ &\leq 1 - |\varphi(y)| \cos\{\arg \varphi(y)\} \\ &= 1 - \text{Re } \varphi(y) < \frac{\epsilon}{2} \qquad \text{whenever } \langle \mathbf{S}_\epsilon y, y \rangle < 1. \end{aligned}$$

Using (7.6.9), we can verify that for every $\delta > 0, \alpha > 0$ we have, for $1 \leq k \leq n$, $n \geq 1$,

$$P\left\{\left\| \sum_{j=k}^{n} X_{nj} \right\| \geq \delta\right\} \leq \left(\frac{\epsilon}{2} + 2\alpha \text{ tr } \mathbf{S}_\epsilon\right)\left[1 - \exp\left(\frac{-\alpha\delta^2}{2}\right)\right]^{-1}. \tag{7.7.7}$$

For given $0 < \epsilon < \frac{1}{4}$ we now choose $\alpha = \alpha(\epsilon) > 0$ and $\delta = \delta(\epsilon) > 0$ such that the right-hand side of (7.7.7) is $<\epsilon$. This completes the proof of Lemma 7.7.2. ■

Lemma 7.7.3. Let $\delta > 0$ be sufficiently large. Then

$$\sup_{n \geq 1} nP\{\|\mathbf{X}_{n1}\| \geq \delta\} < \infty \tag{7.7.8}$$

and

$$\lim_{\delta \to \infty} \sup_{n \geq 1} nP\{\|X_{n1}\| \geq \delta\} = 0. \tag{7.7.9}$$

Proof. In view of (7.4.7) and Lemma 7.7.2 we obtain for all $n \geq 1$ the inequality

$$\begin{aligned} P\left\{\max_{1 \leq k \leq n} \left\| \sum_{j=1}^{k} X_{nj} \right\| \geq 2\delta\right\} &\leq \frac{P\{\|\sum_{j=1}^{n} X_{nj}\| \geq \delta\}}{1 - \max_{1 \leq k \leq n} P\{\|\sum_{j=k+1}^{n} X_{nj}\| \geq \delta\}} \\ &< \frac{\epsilon}{1 - \epsilon}. \end{aligned} \tag{7.7.10}$$

We see easily from (7.7.10) that

$$P\left\{\max_{1\le k\le n}\|X_{nk}\|\ge 4\delta\right\}\le P\left\{\max_{1\le k\le n}\left\|\sum_{j=1}^{k}X_{nj}\right\|\ge 2\delta\right\}<\frac{\epsilon}{1-\epsilon},$$

so that for all $n \ge 1$

$$\prod_{k=1}^{n}P\{\|X_{nk}\|<4\delta\}\ge 1-\frac{\epsilon}{1-\epsilon}=\frac{1-2\epsilon}{1-\epsilon}.$$

This implies

$$\exp\left\{-\sum_{k=1}^{n}P\{\|X_{nk}\|\ge 4\delta\}\right\}\ge\prod_{k=1}^{n}P\{\|X_{nk}\|<4\delta\}$$
$$\ge\frac{1-2\epsilon}{1-\epsilon},$$

so that for all $n \ge 1$

$$\sum_{k=1}^{n}P\{\|X_{nk}\|\ge 4\delta\}\le\ln\frac{1-\epsilon}{1-2\epsilon}.$$

Thus we obtain

$$\sup_{n\ge 1} nP\{\|X_{n1}\|\ge 4\delta\}\le\ln\frac{1-\epsilon}{1-2\epsilon}<\infty. \tag{7.7.11}$$

Then for $\delta' > 4\delta$ we have

$$\sup_{n\ge 1} nP\{\|X_{n1}\|>\delta'\}\le\ln\frac{1-\epsilon}{1-2\epsilon}<\infty.$$

This proves (7.7.8). Taking the limit as $\delta' \to \infty$ and $\epsilon \to 0$, we obtain (7.7.9). This completes the proof of Lemma 7.7.3. ■

Lemma 7.7.4. Let $X_1, X_2, \ldots, X_n$ be independent, $\mathscr{H}$-valued random variables defined on $(\Omega, \mathscr{S}, P)$. Suppose there exists a constant $\delta > 0$ such that $\|X_i\| < \delta$ a.s., $1 \le i \le n$. Set $S_k = \sum_{j=1}^{k} X_j$, $1 \le k \le n$. Then for every positive integer N and every $\gamma > 0$ the inequality

$$P\left\{\max_{1\le k\le n}\|S_k\|\ge N\gamma+(N-1)\delta\right\}\le\left[P\left\{\max_{1\le k\le n}\|S_k\|\ge\frac{\gamma}{2}\right\}\right]^N \tag{7.7.12}$$

holds.

Proof. For $1 \le k \le n$ define the sets $E_k \subset \Omega$ as follows:

$$E_1=\{\|S_1\|\ge(N-1)\gamma+(N-2)\delta\},$$

and for $1 < k \le n$,

$$E_k = \{\|S_k\| \ge (N-1)\gamma + (N-2)\delta$$

and

$$\|S_i\| < (N-1)\gamma + (N-2)\delta \quad \text{for} \quad 1 \le i < k\}.$$

Then $E_1, E_2, \ldots, E_n$ are disjoint sets in $\mathscr{S}$, and, moreover,

$$\bigcup_{k=1}^{n} E_k = \left\{\max_{1 \le k \le n} \|S_k\| \ge (N-1)\gamma + (N-2)\delta\right\}.$$

Now set $E = \{\max_{1 \le k \le n}\|S_k\| \ge N\gamma + (N-1)\delta\}$. Since $E \subset \bigcup_{j=1}^{n} E_j$, it follows that

$$\begin{aligned}
P(E) &= P\left\{\max_{1 \le k \le n} \|S_k\| \ge N\gamma + (N-1)\delta\right\} = P\left(E \cap \bigcup_{j=1}^{n} E_j\right) \\
&= \sum_{j=1}^{n} P(E \cap E_j) \\
&\le \sum_{j=1}^{n} P\left\{\max_{j < k \le n} \|S_k - S_j\| \ge \gamma\right\} P(E_j) \\
&\le \max_{1 \le j \le n} P\left\{\max_{j < k \le n} \|S_k - S_j\| \ge \gamma\right\} \sum_{j=1}^{n} P(E_j) \\
&\le P\left\{\max_{1 \le j < k \le n} \|S_k - S_j\| \ge \gamma\right\} \sum_{j=1}^{n} P(E_j) \\
&\le P\left\{\max_{1 \le k \le n} \|S_k\| \ge \frac{\gamma}{2}\right\} P\left\{\max_{1 \le k \le n} \|S_k\| \ge (N-1)\gamma + (N-2)\delta\right\}.
\end{aligned}$$

Iterating this last inequality, we see immediately that (7.7.12) holds. ∎

We return now to the proof of Theorem 7.7.2. In view of Lemma 7.7.3 we choose and fix $\delta_0 > 0$ sufficiently large so that

$$\sup_{n \ge 1} nP\{\|\mathrm{X}_{n1}\| \ge \delta_0\} < \tfrac{1}{6}.$$

We define truncated random variables

$$X_{nk}^* = \begin{cases} X_{nk}, & \|X_{nk}\| < \delta_0, \\ 0, & \text{otherwise}, \end{cases}$$

for $1 \le k \le n$, $n \ge 1$. Then we have for $\alpha > 0$

$$P\left\{\max_{1 \le k \le n} \left\| \sum_{j=1}^{k} X_{nj}^* \right\| \ge \alpha\right\} \le P\left\{\max_{1 \le k \le n} \left\| \sum_{j=1}^{n} X_{nj} \right\| \ge \alpha\right\} + nP\{\|X_{n1}\| \ge \delta_0\},$$

so that for $\alpha_0 > 0$ sufficiently large and the above choice of δ_0 we have, from (7.7.10) and the fact that $0 < \epsilon < \frac{1}{4}$,

$$P\left\{\max_{1 \le k \le n} \left\| \sum_{j=1}^{k} X_{nj}^* \right\| \ge \alpha_0\right\} < \frac{\epsilon}{1-\epsilon} + \frac{1}{6} < \frac{1}{2}, \tag{7.7.13}$$

which holds for all $n \ge 1$. (We note that $0 < \epsilon < \frac{1}{4}$.)

In view of Lemma 7.7.4 we see that for all $N \ge 1$

$$\begin{aligned} P\left\{\max_{1 \le k \le n} \left\| \sum_{j=1}^{k} X_{nj}^* \right\| \ge N(2\alpha_0 + \delta_0)\right\} &\le P\left\{\max_{1 \le k \le n} \left\| \sum_{j=1}^{k} X_{nj}^* \right\| \ge 2N\alpha_0 + (N-1)\delta_0\right\} \\ &\le \left[P\left\{\max_{1 \le k \le n} \left\| \sum_{j=1}^{k} X_{nj}^* \right\| \ge \alpha_0\right\}\right]^N. \end{aligned}$$

In view of (7.7.13) we have for all $N \ge 1$

$$P\left\{\max_{1 \le k \le n} \left\| \sum_{j=1}^{k} X_{nj}^* \right\| \ge N(2\alpha_0 + \delta_0)\right\} < (\tfrac{1}{2})^N$$

holding for all $n \ge 1$. Writing $t = N(2\alpha_0 + \delta_0)$, we obtain for $n \ge 1$

$$P\left\{\max_{1 \le k \le n} \left\| \sum_{j=1}^{k} X_{nj}^* \right\| \ge t\right\} < e^{-\lambda t},$$

where $\lambda = \ln 2/(\delta_0 + 2\alpha_0) > 0$. After some elementary computation we see that, for all $r > 0$, $\mathscr{E}\|\sum_{j=1}^{n} X_{nj}^*\|^r$ is uniformly bounded in n. In particular, the quantities

$$\mathscr{E}\left\| \sum_{j=1}^{n} X_{nj}^* \right\|, \qquad \left\| \mathscr{E}\sum_{j=1}^{n} X_{nj}^* \right\|, \qquad \text{and} \qquad \mathscr{E}\left\| \sum_{j=1}^{n} X_{nj}^* \right\|^2$$

are uniformly bounded in n. Moreover, since

$$\left\| \mathscr{E}\left(\sum_{j=1}^{n} X_{nj}^*\right) \right\| = n\|\mathscr{E}(X_{n1}^*)\|,$$

we see that

$$\|\mathscr{E}X_{n1}^*\| = O(n^{-1}) \qquad \text{as } n \to \infty. \tag{7.7.14}$$

For every $n \ge 1$ we now define the set function μ_n on $\mathscr{B}$ by the relation

$$\mu_n(E) = n \int_E \frac{\|y\|^2}{1 + \|y\|^2}\, d\mathbf{P}_n(y), \qquad E \in \mathscr{B}, \tag{7.7.15}$$

where $\mathbf{P}_n$ is the probability distribution of each X_{nk}, $1 \le k \le n$. Clearly μ_n is a finite measure on $\mathscr{B}$ such that $\mu_n(\mathscr{H}) \le n$ for $n \ge 1$.

Lemma 7.7.5. The sequence $\{\mu_n\}$ of finite measures as defined in (7.7.15) has the following properties:

(a) It is uniformly bounded.
(b) It satisfies the relation

$$\lim_{\delta\to\infty}\sup_{n\geq 1}\mu_n\{y\in\mathscr{H}: \|y\|\geq\delta\}=0. \tag{7.7.16}$$

(c) It is tight.

Proof. (a) and (b). Let $n \geq 1$, and let $\delta_0 > 0$ be as in the definition of X_{nk}^*. Then we have

$$\begin{aligned}
\mu_n(\mathscr{H}) &= n\int_{\mathscr{H}}\frac{\|y\|^2}{1+\|y\|^2}\,d\mathbf{P}_n(y)\\
&= n\left(\int_{\|y\|<\delta_0}+\int_{\|y\|\geq\delta_0}\right)\frac{\|y\|^2}{1+\|y\|^2}\,d\mathbf{P}_n(y)\\
&\leq n\int_{\|y\|<\delta_0}\|y\|^2\,d\mathbf{P}_n(y)+nP\{\|X_{n1}\|\geq\delta_0\}\\
&\leq \mathscr{E}\left\|\sum_{j=1}^{n}X_{nj}^*\right\|^2+nP\{\|X_{n1}\|\geq\delta_0\}
\end{aligned}$$

since $\{X_{nj}^*,\ 1\leq j\leq n\}$ is independent and identically distributed. Hence

$$\sup_{n\geq 1}\mu_n(\mathscr{H})\leq\gamma<\infty$$

in view of Lemma 7.7.3 and the fact that $\mathscr{E}\|\sum_{j=1}^n X_{nj}^*\|^2$ is uniformly bounded in n. Moreover, (7.7.16) follows immediately from (7.7.9).

(c) In view of Proposition 7.7.1 it is sufficient to show that for given $\epsilon > 0$ there exists an S-operator $\mathbf{S}_\epsilon$ on $\mathscr{H}$ such that

$$\sup_{n\geq 1}\left\{\mu_n(\mathscr{H})-\operatorname{Re}\int_{\mathscr{H}}\exp(i\langle x, y\rangle)\,d\mu_n(y)\right\}<\epsilon \qquad \text{whenever } \langle\mathbf{S}_\epsilon x, x\rangle<1.$$

In fact, for $n \geq 1$ and $x \in \mathscr{H}$

$$\begin{aligned}
\mu_n(\mathscr{H})-\operatorname{Re}\int_{\mathscr{H}}\exp(i\langle x, y\rangle)\,d\mu_n(y) &= n\int_{\mathscr{H}}[1-\cos\langle x, y\rangle]\frac{\|y\|^2}{1+\|y\|^2}\,d\mathbf{P}_n(y)\\
&\leq n[1-\operatorname{Re}\hat{\mathbf{P}}_n(x)]\\
&= n\{1-\operatorname{Re}[\varphi(x)]^{1/n}\}
\end{aligned}$$

since φ, being an i.d. characteristic functional, has no zeros on $\mathscr{H}$ (Proposition 7.7.2). Hence

$$\begin{aligned}\mu_n(\mathscr{H}) - \operatorname{Re} \int_{\mathscr{H}} \exp(i\langle x, y\rangle)\, d\mu_n(y) &\le n\left[1 - |\varphi(x)|^{1/n} \cos\left(\frac{1}{n} \arg \varphi(x)\right)\right] \\ &= n(1 - |\varphi(x)|^{1/n}) + n|\varphi(x)|^{1/n}\left[1 - \cos\left(\frac{1}{n} \arg \varphi(x)\right)\right] \\ &\le \frac{1 - |\varphi(x)|}{|\varphi(x)|} + \frac{1}{2n} [\arg \varphi(x)]^2.\end{aligned}$$

For given ϵ $(0 < \epsilon < \frac{1}{4})$, we now choose $x \in \mathscr{H}$ such that $1 - \operatorname{Re} \varphi(x) < \epsilon/2$. Then

$$|\operatorname{Im} \varphi(x)| \le \sqrt{1 - (\operatorname{Re} \varphi(x))^2} < \epsilon^{1/2},$$

so that

$$|\arg \varphi(x)| < \operatorname{arc\,tan} \frac{\epsilon^{1/2}}{1 - \epsilon/2} < \frac{\pi}{4}$$

and

$$1 - |\varphi(x)| < \frac{\epsilon}{2}.$$

Since φ is the Fourier transform of a probability measure, it follows from the Minlos–Sazonov theorem that for given $\epsilon > 0$ there exists an S-operator $\mathbf{S}_\epsilon$ on $\mathscr{H}$ such that

$$1 - \operatorname{Re} \varphi(x) < \frac{\epsilon}{2} \qquad \text{whenever } \langle \mathbf{S}_\epsilon x, x\rangle < 1.$$

Then for such an $x \in \mathscr{H}$ and all $n \ge 1$ we have

$$\mu_n(\mathscr{H}) - \operatorname{Re} \int_{\mathscr{H}} \exp(i\langle x, y\rangle)\, d\mu_n(y)$$
$$< \frac{\epsilon/2}{1 - \epsilon/2} + \frac{1}{2n} \frac{\epsilon}{(1 - \epsilon/2)^2} \le \frac{\epsilon}{2 - \epsilon} + \frac{\epsilon}{2n(1 - \epsilon/2)^2}.$$

For $n > 1$ we can choose a sufficiently small $\epsilon > 0$ (independent of n) such that the right-hand side of the above inequality is $< \epsilon$. This proves that $\{\mu_n\}$ is tight. ■

We again return to the proof of the theorem and note that for $n \geq 1$ the mapping

$$x \to n \int_{\mathscr{H}} \frac{\langle x, y\rangle}{1 + \|y\|^2} \, d\mathbf{P}_n(y)$$

is a bounded linear functional on $\mathscr{H}$. Hence there exists a vector $a_n \in \mathscr{H}$ satisfying the relation

$$\langle a_n, x\rangle = n \int_{\mathscr{H}} \frac{\langle x, y\rangle}{1 + \|y\|^2} \, d\mathbf{P}_n(y) = n\mathscr{E}\left(\frac{\langle x, X_{n1}\rangle}{1 + \|X_{n1}\|^2}\right), \qquad x \in \mathscr{H}. \tag{7.7.17}$$

We show that $\{a_n\}$ is uniformly bounded. In fact, for $x \in \mathscr{H}$

$$\mathscr{E}\left(\frac{\langle x, X_{n1}\rangle}{1 + \|X_{n1}\|^2}\right) = \mathscr{E}\left(\frac{\langle x, X_{n1}^*\rangle}{1 + \|X_{n1}^*\|^2}\right) + \int_{\|y\| \geq \delta_0} \frac{\langle x, y\rangle}{1 + \|y\|^2} \, d\mathbf{P}_n(y).$$

Then for $x \in \mathscr{H}$

$$\begin{aligned} |\langle a_n, x\rangle| &\leq n|\langle x, \mathscr{E}(X_{n1}^*)\rangle| + \frac{n\|x\|}{\delta_0} P\{\|X_{n1}\| \geq \delta_0\} \\ &\leq \|x\|\left[n\|\mathscr{E}(X_{n1}^*)\| + \frac{n}{\delta_0} P\{\|X_{n1}\| \geq \delta_0\}\right] \\ &\leq c\|x\|, \end{aligned}$$

where $c > 0$ is a constant independent of n in view of (7.7.14) and Lemma 7.7.3. We see easily that $\|a_n\| \leq c$ for all $n \geq 1$.

Next note that for every $n \geq 1$ the mapping

$$\langle x, z\rangle \to n \int_{\mathscr{H}} \frac{\langle x, y\rangle\langle z, y\rangle}{1 + \|y\|^2} \, d\mathbf{P}_n(y), \qquad x, z \in \mathscr{H},$$

is a bounded bilinear form on $\mathscr{H}$, so that there exists a bounded linear operator $\mathbf{S}_n$ on $\mathscr{H}$ such that the relation

$$\langle \mathbf{S}_n x, z\rangle = n \int_{\mathscr{H}} \frac{\langle x, y\rangle\langle z, y\rangle}{1 + \|y\|^2} \, d\mathbf{P}_n(y)$$

holds for all $x, z \in \mathscr{H}$ and $n \geq 1$. In particular, for $x \in \mathscr{H}$ and $n \geq 1$ we have

$$\langle \mathbf{S}_n x, x\rangle = n \int_{\mathscr{H}} \frac{\langle x, y\rangle^2}{1 + \|y\|^2} \, d\mathbf{P}_n(y). \tag{7.7.18}$$

It is easily verified that $\mathbf{S}_n$ is an S-operator and, moreover,

$$\operatorname{tr} \mathbf{S}_n = n \int_{\mathscr{H}} \frac{\|y\|^2}{1 + \|y\|^2} \, d\mathbf{P}_n(y) = \mu_n(\mathscr{H}).$$

It follows immediately from Lemma 7.7.5 that $\sup_{n\geq 1} \operatorname{tr} \mathbf{S}_n < \infty$. Moreover, it follows from (7.7.18) that

$$\langle \mathbf{S}_n x, x\rangle \leq \|x\|^2 \operatorname{tr} \mathbf{S}_n,$$

so that $\|\mathbf{S}_n\| = \sup_{\|x\|=1} \langle \mathbf{S}_n x, x\rangle \leq \operatorname{tr} \mathbf{S}_n < \infty$. Hence

$$\sup_{n\geq 1} \|\mathbf{S}_n\| \leq \sup_{n\geq 1} \operatorname{tr} \mathbf{S}_n < \infty.$$

Since $\{\mu_n\}$ is tight and $\{a_n\}$ and $\{\|\mathbf{S}_n\|\}$ are uniformly bounded, we can select a subsequence $\{n_k\}$ with $n_k \to \infty$ as $k \to \infty$ such that the following conditions hold:

(i) $\{\mu_{n_k}\}$ converges weakly to a finite measure μ' on $\mathscr{B}$.
(ii) $\{a_{n_k}\}$ converges weakly to a vector $a' \in \mathscr{H}$.
(iii) For every $x \in D$, where D is a countable subset of $\mathscr{H}$ which is everywhere dense in $\mathscr{H}$, the sequence $\{\mathbf{S}_{n_k} x\}$ converges weakly to an element in $\mathscr{H}$, that is, the sequence $\{\langle \mathbf{S}_{n_k} x, y\rangle\}$ converges to a real number, say $\theta(x, y)$, for every $y \in \mathscr{H}$. This is possible in view of Helly's theorem (Theorem 3.1.2) and the fact that every bounded set in $\mathscr{H}$ is weakly compact.

Clearly θ is bilinear, and moreover

$$|\theta(x, y)| \leq \rho \|x\| \, \|y\|, \qquad x \in D \quad \text{and} \quad y \in \mathscr{H},$$

where $\rho = \sup_{n\geq 1} \|\mathbf{S}_n\| < \infty$. Consequently, θ can be uniquely extended by continuity to a bounded bilinear form on $\mathscr{H}$ so that there exists a bounded linear operator $\mathbf{S}'$ on $\mathscr{H}$ such that the relation

$$\langle \mathbf{S}' x, y\rangle = \theta(x, y) = \lim_{k\to\infty} \langle \mathbf{S}_{n_k} x, y\rangle$$

holds for all $x, y \in \mathscr{H}$. In particular, for every $x \in \mathscr{H}$ we have

$$\langle \mathbf{S}' x, x\rangle = \lim_{k\to\infty} \langle \mathbf{S}_{n_k} x, x\rangle \geq 0,$$

so that $\mathbf{S}'$ is a positive, bounded, self-adjoint operator on $\mathscr{H}$. We see easily that $\mathbf{S}'$ is also an S-operator on $\mathscr{H}$ such that

$$\operatorname{tr} \mathbf{S}' \leq \liminf_{k\to\infty} \operatorname{tr} \mathbf{S}_{n_k} \leq \sup_{n\geq 1} \operatorname{tr} \mathbf{S}_n < \infty.$$

Now we note that for $x \in \mathscr{H}$ and $k \geq 1$

$$n_k[\hat{\mathbf{P}}_{n_k}(x) - 1]$$

$$= n_k \int_{\mathscr{H}} [\exp(i\langle x, y\rangle) - 1]\, d\mathbf{P}_{n_k}(y)$$

$$= i\langle a_{n_k}, x\rangle - \tfrac{1}{2}\langle \mathbf{S}_{n_k} x, x\rangle$$

$$+ n_k \int_{\mathscr{H}} \left[\exp(i\langle x, y\rangle) - 1 - i\frac{\langle x, y\rangle}{1 + \|y\|^2} + \frac{1}{2}\frac{\langle x, y\rangle^2}{1 + \|y\|^2}\right] d\mathbf{P}_{n_k}(y)$$

$$= i\langle a_{n_k}, x\rangle - \tfrac{1}{2}\langle \mathbf{S}_{n_k} x, x\rangle$$

$$+ \int_{\mathscr{H}} \left[\exp(i\langle x, y\rangle) - 1 - i\frac{\langle x, y\rangle}{1 + \|y\|^2} + \frac{1}{2}\frac{\langle x, y\rangle^2}{1 + \|y\|^2}\right] \frac{1 + \|y\|^2}{\|y\|^2}\, d\mu_{n_k}(y).$$

On the other hand, we note that for $x \in \mathscr{H}$ and $k \geq 1$

$$\varphi(x) = \hat{\mathbf{P}}(x) = [\hat{\mathbf{P}}_{n_k}(x)]^{n_k}.$$

Since φ has no zeros on $\mathscr{H}$, we can write

$$\hat{\mathbf{P}}_{n_k}(x) = \exp\left\{\frac{1}{n_k} \ln \varphi(x)\right\} = 1 + \frac{1}{n_k} \ln \varphi(x) + o(n_k^{-1}) \qquad \text{as } k \to \infty.$$

Hence for $x \in \mathscr{H}$

$$\ln \varphi(x) = \lim_{k\to\infty} n_k[\hat{\mathbf{P}}_{n_k}(x) - 1]$$

$$= \lim_{k\to\infty} \Bigg\{ i\langle a_{n_k}, x\rangle - \tfrac{1}{2}\langle \mathbf{S}_{n_k} x, x\rangle$$

$$+ \int_{\mathscr{H}} \left[\exp(i\langle x, y\rangle) - 1 - i\frac{\langle x, y\rangle}{1 + \|y\|^2} + \frac{1}{2}\frac{\langle x, y\rangle^2}{1 + \|y\|^2}\right]$$

$$\times \frac{1 + \|y\|^2}{\|y\|^2}\, d\mu_{n_k}(y)\Bigg\}$$

$$= i\langle a', x\rangle - \tfrac{1}{2}\langle \mathbf{S}' x, x\rangle$$

$$+ \int_{\mathscr{H}} \left[\exp(i\langle x, y\rangle) - 1 - \frac{i\langle x, y\rangle}{1 + \|y\|^2} + \frac{1}{2}\frac{\langle x, y\rangle^2}{1 + \|y\|^2}\right]$$

$$\times \frac{1 + \|y\|^2}{\|y\|^2}\, d\mu'(y).$$

We note that the integrand on the right-hand side is defined to be zero at $x = 0$ by continuity. Next we define a set function μ on $\mathscr{B}$ by setting $\mu(\{0\}) = 0$

and $\mu(E) = \mu'(E)$ for every $E \in \mathscr{B}$ with $\{0\} \not\subset E$. Then for $x \in \mathscr{H}$

$$\ln \varphi(x) = i\langle a', x\rangle - \tfrac{1}{2}\langle \mathbf{S}'x, x\rangle + \int_{\mathscr{H}} \left[\exp(i\langle x, y\rangle) - 1 - \frac{i\langle x, y\rangle}{1 + \|y\|^2} + \frac{1}{2}\frac{\langle x, y\rangle^2}{1 + \|y\|^2}\right] \times \frac{1 + \|y\|^2}{\|y\|^2}\, d\mu(y).$$

Now define the operator $\mathbf{S}$ on $\mathscr{H}$ by setting

$$\langle \mathbf{S}x, x\rangle = \langle \mathbf{S}'x, x\rangle - \int_{\mathscr{H}} \frac{\langle x, y\rangle^2}{\|y\|^2}\, d\mu(y), \qquad x \in \mathscr{H}. \tag{7.7.19}$$

Then for $x \in \mathscr{H}$ we have

$$\ln \varphi(x) = i\langle a', x\rangle - \tfrac{1}{2}\langle \mathbf{S}x, x\rangle + \int_{\mathscr{H}} \left[\exp(i\langle x, y\rangle) - 1 - \frac{i\langle x, y\rangle}{1 + \|y\|^2}\right] \times \frac{1 + \|y\|^2}{\|y\|^2}\, d\mu(y). \tag{7.7.20}$$

Finally we show that the operator $\mathbf{S}$ defined by (7.7.19) is an $\mathbf{S}$-operator on $\mathscr{H}$. For this purpose it is sufficient to show that $\langle \mathbf{S}x, x\rangle \geq 0$ for every $x \in \mathscr{H}$. In fact, for $x \in \mathscr{H}$

$$\begin{aligned}
\langle \mathbf{S}x, x\rangle &= \lim_{k\to\infty} \langle \mathbf{S}_{n_k} x, x\rangle - \int_{\mathscr{H}} \frac{\langle x, y\rangle^2}{\|y\|^2}\, d\mu(y) \\
&= \lim_{k\to\infty} n_k \int_{\mathscr{H}} \frac{\langle x, y\rangle^2}{1 + \|y\|^2}\, d\mathbf{P}_{n_k}(y) - \int_{\mathscr{H}} \frac{\langle x, y\rangle^2}{\|y\|^2}\, d\mu(y) \\
&= \lim_{k\to\infty} \int_{\mathscr{H}} \frac{\langle x, y\rangle^2}{\|y\|^2}\, d\mu_{n_k}(y) - \int_{\mathscr{H}} \frac{\langle x, y\rangle^2}{\|y\|^2}\, d\mu(y) \\
&= \lim_{k\to\infty} \left\{\int_{\|y\| \geq \epsilon} \frac{\langle x, y\rangle^2}{\|y\|^2}\, d\mu_{n_k}(y) - \int_{\|y\| \geq \epsilon} \frac{\langle x, y\rangle^2}{\|y\|^2}\, d\mu(y)\right\} \\
&\quad + \lim_{k\to\infty} \left\{\int_{\|y\| < \epsilon} \frac{\langle x, y\rangle^2}{\|y\|^2}\, d\mu_{n_k}(y) - \int_{\|y\| < \epsilon} \frac{\langle x, y\rangle^2}{\|y\|^2}\, d\mu(y)\right\}.
\end{aligned} \tag{7.7.21}$$

Since $\mu_{n_k} \Rightarrow \mu'$ (as $k \to \infty$) and $\mu'(E) = \mu(E)$ for $E \in \mathscr{B}$ with $\{0\} \not\subset E$, the first term on the right-hand side of (7.7.21) is zero. On the other hand, since $\mu(\{0\}) = 0$, using the continuity property of the finite measure μ for fixed $x \in \mathscr{H}$, we can choose an $\epsilon = \epsilon(x) > 0$ such that

$$\int_{\|y\| < \epsilon} \frac{\langle x, y\rangle^2}{\|y\|^2}\, d\mu(y) \leq \|x\|^2 \mu\{y \in \mathscr{H} : \|y\| < \epsilon\}$$

can be made sufficiently small. In particular, we can choose $\epsilon > 0$ such that

$$\int_{\|y\|<\epsilon} \frac{\langle x, y\rangle^2}{\|y\|^2}\, d\mu(y) \le \lim_{k\to\infty} \int_{\|y\|<\epsilon} \frac{\langle x, y\rangle^2}{\|y\|^2}\, d\mu_{n_k}(y)$$
$$= \int_{\|y\|<\epsilon} \frac{\langle x, y\rangle^2}{\|y\|^2}\, d\mu'(y).$$

This proves that $\langle \mathbf{S}x, x\rangle \ge 0$ for all $x \in \mathscr{H}$, and hence $\mathbf{S}$ is an S-operator on $\mathscr{H}$. Setting $a' = a$ in (7.7.20), we conclude that, if φ is the Fourier transform of an i.d. probability measure $\mathbf{P}$ on $\mathscr{B}$, then φ has a representation of the form (7.7.5).

Next we show that $a \in \mathscr{H}$, $\mathbf{S}$, and μ in representation (7.7.5) are uniquely determined by φ. First we note that for every $x \in \mathscr{H}$

$$\langle \mathbf{S}x, x\rangle = -2 \lim_{t\to\infty} \frac{1}{t^2} \ln \varphi(tx),$$

so that $\mathbf{S}$ is determined uniquely by φ. We may therefore assume, without loss of generality, that $\mathbf{S} = 0$.

Choose an orthonormal basis $\{e_\nu\}$ for $\mathscr{H}$. Let $\alpha_\nu > 0$ be such that

$$\sum_{\nu=1}^{\infty} \alpha_\nu < \infty.$$

Then for every $t \in \mathbb{R}$ the sequence

$$\sum_{\nu=1}^{n} \alpha_\nu \left[1 - \frac{\exp(it\langle x, e_\nu\rangle) + \exp(-it\langle x, e_\nu\rangle)}{2}\right] = \sum_{\nu=1}^{n} \alpha_\nu [1 - \cos(t\langle x, e_\nu\rangle)]$$

is a uniformly bounded, nondecreasing sequence of nonnegative real numbers. In view of the Lebesgue dominated convergence theorem we have for $t \in \mathbb{R}$ and $x \in \mathscr{H}$

$$\sum_{\nu=1}^{\infty} \alpha_\nu \left[\ln \varphi(x) - \frac{\ln \varphi(x + te_\nu) + \ln \varphi(x - te_\nu)}{2}\right]$$

$$\text{(7.7.22)} \qquad = \int_{\mathscr{H}} \exp(i\langle x, y\rangle) \left\{\sum_{\nu=1}^{\infty} \alpha_\nu [1 - \cos(t\langle y, e_\nu\rangle)]\right\} \frac{1 + \|y\|^2}{\|y\|^2}\, d\mu(y),$$

so that the series on the left-hand side of (7.7.22) converges. Let $h > 0$. Integrating both sides of (7.7.22) with respect to dt over the interval $[-h, h]$ and interchanging the order of integration, we obtain

$$\frac{1}{2h} \int_{-h}^{h} \left\{\sum_{\nu=1}^{\infty} \alpha_\nu \left[\ln \varphi(x) - \frac{\ln \varphi(x + te_\nu) + \ln \varphi(x - te_\nu)}{2}\right]\right\} dt$$
$$= \int_{\mathscr{H}} \exp(i\langle x, y\rangle) \left\{\sum_{\nu=1}^{\infty} \alpha_\nu \left[1 - \frac{\sin h\langle y, e_\nu\rangle}{h\langle y, e_\nu\rangle}\right]\right\}$$
$$\text{(7.7.23)} \qquad \times \frac{1 + \|y\|^2}{\|y\|^2}\, d\mu(y) = \psi(x),$$

say. Now define the set function $\tilde{\mu}$ on $\mathscr{B}$ by the relation

$$\tilde{\mu}(E) = \int_E \left\{ \sum_{v=1}^{\infty} \alpha_v \left[1 - \frac{\sin h\langle y, e_v \rangle}{h\langle y, e_v \rangle} \right] \right\} \frac{1 + \|y\|^2}{\|y\|^2}\, d\mu(y), \qquad E \in \mathscr{B}.$$

We can verify easily that $\tilde{\mu}$ is a finite measure on $\mathscr{B}$, and, moreover, ψ is the Fourier transform of $\tilde{\mu}$. Consequently $\tilde{\mu}$ is uniquely determined by ψ and hence by φ. On the other hand, we note that μ is uniquely determined from $\tilde{\mu}$ by the relations

$$\mu(\{0\}) = 0$$

and, for $E \in \mathscr{B}$ such that $\{0\} \not\subset E$.

$$\mu(E) = \int_E \left\{ \sum_{v=1}^{\infty} \alpha_v \left[1 - \frac{\sin h\langle y, e_v \rangle}{h\langle y, e_v \rangle} \right] \right\}^{-1} \frac{\|y\|^2}{1 + \|y\|^2}\, d\tilde{\mu}(y).$$

Hence the measure μ is uniquely determined by φ, and consequently the vector $a \in \mathscr{H}$ is also uniquely determined.

Conversely, suppose that $\ln \varphi$ has a representation of the form (7.7.5). We show that φ is the Fourier transform of an i.d. probability measure P on $(\mathscr{H}, \mathscr{B})$. First we note that φ is continuous on $\mathscr{H}$ and $\varphi(0) = 1$. Next we show that φ is positive definite on $\mathscr{H}$. Let $\mathfrak{U}$ be an arbitrary finite-dimensional subspace of $\mathscr{H}$. Let $\pi_{\mathfrak{U}}$ be the projection of $\mathscr{H}$ onto $\mathfrak{U}$. Define $\varphi_{\mathfrak{U}}$ on $\mathscr{H}$ by the relation

$$\varphi_{\mathfrak{U}}(\pi_{\mathfrak{U}}(x)) = \varphi(\pi_{\mathfrak{U}}(x)), \qquad x \in \mathscr{H}. \tag{7.7.24}$$

Then from representation (7.7.5) of $\ln \varphi$ we see easily that $\varphi_{\mathfrak{U}}$ is the Fourier transform of an i.d. probability measure on $(\mathfrak{U}, \mathscr{B}_{\mathfrak{U}})$, where $\mathscr{B}_{\mathfrak{U}}$ is the σ-field of Borel sets of $\mathfrak{U}$. In particular, $\varphi_{\mathfrak{U}}$ is positive definite on $\mathfrak{U}$ for every finite-dimensional subspace $\mathfrak{U}$ of $\mathscr{H}$. Then, using the continuity property of φ and relation (7.7.24), we see easily that φ is positive definite on $\mathscr{H}$. To show that φ is the Fourier transform of a probability measure on $\mathscr{B}$ it is sufficient to show (Minlos–Sazonov theorem) that for given $\epsilon > 0$ there exists an S-operator, $\mathbf{S}_\epsilon$ on $\mathscr{H}$, such that

$$1 - \operatorname{Re} \varphi(x) < \epsilon \qquad \text{whenever } \langle \mathbf{S}_\epsilon x, x \rangle < 1.$$

Let $x \in \mathscr{H}$. Then it follows from (7.7.5) that

$$1 - |\varphi(x)| \le \tfrac{1}{2}\langle \mathbf{S}x, x \rangle + \int_{\mathscr{H}} (1 - \cos\langle x, y \rangle) \frac{1 + \|y\|^2}{\|y\|^2}\, d\mu(y)$$

$$\le \tfrac{1}{2}\langle \mathbf{S}x, x \rangle + \int_{\mathscr{H}} (1 - \cos\langle x, y \rangle)\, d\mu(y) + \frac{1}{2} \int_{\mathscr{H}} \frac{\langle x, y \rangle^2}{\|y\|^2}\, d\mu(y) \tag{7.7.25}$$

and

$$\arg \varphi(x) = \langle a, x \rangle + \int_{\mathscr{H}} \sin\langle x, y \rangle \, d\mu(y)$$

(7.7.26)
$$+ \int_{\mathscr{H}} \frac{\sin\langle x, y \rangle - \langle x, y \rangle}{\|y\|^2} \, d\mu(y).$$

Using the elementary inequalities

$$|\sin t - t| \leq \frac{t^2}{2}, \qquad \sin^2 t \leq 2(1 - \cos t), \qquad t \in \mathbb{R},$$

and

$$t^2 + u^2 \geq 2tu, \qquad u, t \in \mathbb{R},$$

we see that for $x \in \mathscr{H}$

$$\begin{aligned}
1 - \operatorname{Re} \varphi(x) &= 1 - |\varphi(x)| \cos \arg \varphi(x) \\
&= 1 - |\varphi(x)| + |\varphi(x)|(1 - \cos \arg \varphi(x)) \\
&\leq 1 - |\varphi(x)| + \tfrac{1}{2}(\arg \varphi(x))^2 \\
&\leq \tfrac{1}{2}\langle \mathbf{S}x, x \rangle + \int_{\mathscr{H}} (1 - \cos\langle x, y \rangle) \frac{1 + \|y\|^2}{\|y\|^2} \, d\mu(y) \\
&\quad + \frac{1}{2}\Bigg\{\langle a, x \rangle + \int_{\mathscr{H}} \sin\langle x, y \rangle \, d\mu(y) \\
&\quad + \int_{\mathscr{H}} \frac{\sin\langle x, y \rangle - \langle x, y \rangle}{\|y\|^2} \, d\mu(y)\Bigg\}^2 \\
&\leq \tfrac{1}{2}\langle \mathbf{S}x, x \rangle + \int_{\mathscr{H}} (1 - \cos\langle x, y \rangle) \, d\mu(y) + \frac{1}{2} \int_{\mathscr{H}} \frac{\langle x, y \rangle^2}{\|y\|^2} \, d\mu(y) \\
&\quad + \frac{3}{2}\Bigg\{\langle a, x \rangle^2 + \mu(\mathscr{H}) \int_{\mathscr{H}} \sin^2\langle x, y \rangle \, d\mu(y) \\
&\quad + \left[\int_{\mathscr{H}} \frac{\sin\langle x, y \rangle - \langle x, y \rangle}{\|y\|^2} \, d\mu(y)\right]^2\Bigg\} \\
&\leq \tfrac{1}{2}\langle \mathbf{S}x, x \rangle + \int_{\mathscr{H}} (1 - \cos\langle x, y \rangle) \, d\mu(y) + \frac{1}{2} \int_{\mathscr{H}} \frac{\langle x, y \rangle^2}{\|y\|^2} \, d\mu(y) \\
&\quad + \tfrac{3}{2}\langle a, x \rangle^2 + 3\mu(\mathscr{H}) \int_{\mathscr{H}} (1 - \cos\langle x, y \rangle) \, d\mu(y) \\
&\quad + \frac{3}{4}\left[\int_{\mathscr{H}} \frac{\langle x, y \rangle^2}{\|y\|^2} \, d\mu(y)\right]^2 \\
&\leq C\Bigg\{\langle a, x \rangle^2 + \langle \mathbf{S}x, x \rangle + \int_{\mathscr{H}} (1 - \cos\langle x, y \rangle) \, d\mu(y)
\end{aligned}$$

(7.7.27)
$$+ \int_{\mathscr{H}} \frac{\langle x, y \rangle^2}{\|y\|^2} \, d\mu(y) + \left[\int_{\mathscr{H}} \frac{\langle x, y \rangle^2}{\|y\|^2} \, d\mu(y)\right]^2\Bigg\},$$

where $C = \max\{\frac{3}{2}, 3\mu(\mathscr{H})\}$ is a positive constant. Since μ is a finite measure, it follows from the Minlos–Sazonov theorem that for given $\epsilon > 0$ we can find an S-operator, $\mathbf{S}'_\epsilon$ on $\mathscr{H}$, such that

$$\int_{\mathscr{H}} (1 - \cos\langle x, y\rangle)\, d\mu(y) < \frac{\epsilon}{2C} \qquad \text{whenever } \langle \mathbf{S}'_\epsilon x, x\rangle < 1.$$

Define operators $\mathbf{S}'$ and $\mathbf{S}''$ on $\mathscr{H}$ by the relations

$$\langle \mathbf{S}'x, x\rangle = \langle a, x\rangle^2, \qquad x \in \mathscr{H},$$

and

$$\langle \mathbf{S}''x, x\rangle = \int_{\mathscr{H}} \frac{\langle x, y\rangle^2}{\|y\|^2}\, d\mu(y), \qquad x \in \mathscr{H}.$$

Then we see easily that $\mathbf{S}'$ and $\mathbf{S}''$ are S-operators on $\mathscr{H}$, and, moreover,

$$tr\, \mathbf{S}' = \|a\|^2 < \infty, \qquad \text{tr}\, \mathbf{S}'' = \mu(\mathscr{H}) < \infty.$$

Finally define $\mathbf{S}_\epsilon$ on $\mathscr{H}$ by setting

$$\mathbf{S}_\epsilon = \frac{2C}{\epsilon - \epsilon^2}(\mathbf{S} + \mathbf{S}' + \mathbf{S}'') + \mathbf{S}'_\epsilon \qquad (0 < \epsilon < 1).$$

Then $\mathbf{S}_\epsilon$ is an S-operator on $\mathscr{H}$. Let $x \in \mathscr{H}$ such that $\langle \mathbf{S}_\epsilon x, x\rangle < 1$. Then $\langle \mathbf{S}'_\epsilon x, x\rangle < 1$, and, moreover,

$$\langle \mathbf{S}x, x\rangle + \langle \mathbf{S}'x, x\rangle + \langle \mathbf{S}''x, x\rangle < \frac{\epsilon - \epsilon^2}{2C}.$$

In particular, $\langle \mathbf{S}''x, x\rangle < \epsilon/2C$, so that $\langle \mathbf{S}''x, x\rangle^2 < \epsilon^2/4C^2$. Then it follows from (7.7.27) that

$$1 - \text{Re}\, \varphi(x) \le C\left\{\langle \mathbf{S}'x, x\rangle + \langle \mathbf{S}x, x\rangle + \frac{\epsilon}{2C} + \langle \mathbf{S}''x, x\rangle + \langle \mathbf{S}''x, x\rangle^2\right\}$$

$$< \frac{\epsilon}{2} + \frac{\epsilon - \epsilon^2}{2} + \frac{\epsilon^2}{4C} < \epsilon,$$

since $C \ge \frac{3}{2}$. Consequently φ is the Fourier transform of a probability measure $\mathbf{P}$ on $(\mathscr{H}, \mathscr{B})$.

To complete the proof of Theorem 7.7.2 we show $\mathbf{P}$ is i.d. In fact, for every $n \ge 1$, we see that $(1/n) \ln \varphi$ has the representation given by (7.7.5) with a, $\mathbf{S}$, and μ replaced by a/n, $\mathbf{S}/n$, and μ/n, respectively, so that $\varphi^{1/n}$ is the Fourier transform of a probability measure on $\mathscr{B}$ for every $n \ge 1$. Hence $\mathbf{P}$ is i.d. ■

7.8 THE GENERAL CENTRAL LIMIT PROBLEM IN A HILBERT SPACE

Let $(\mathscr{H}, \mathscr{B})$ be a real, separable Hilbert space with inner product $\langle \cdot, \cdot \rangle$ and norm $\|\cdot\|$. Let $\{X_{nk}, 1 \leq k \leq k_n, n \geq 1\}$ be a sequence of $\mathscr{H}$-valued, row-wise independent random variables defined on $(\Omega, \mathscr{S}, P)$. Let $S_n = \sum_{k=1}^{k_n} X_{nk}$, $n \geq 1$. We say that the random variables X_{nk} satisfy the *u.a.n. condition* if for every $\epsilon > 0$

$$\lim_{n \to \infty} \max_{1 \leq k \leq k_n} P\{\|X_{nk}\| \geq \epsilon\} = 0. \tag{7.8.1}$$

In the following we assume that the X_{nk} satisfy condition (7.8.1).

We now investigate the conditions under which the sequence $\{S_n\}$ converges in law to a certain random variable. Denote by P_{nk}, $1 \leq k \leq k_n$, the probability distribution of X_{nk}, and by P_n the probability distribution of S_n, $n \geq 1$. We note that

$$P_n = P_{n1} * P_{n2} * \cdots * P_{nk_n}, \qquad n \geq 1.$$

Proposition 7.8.1. Let $0 < \delta < 1$ be fixed. Then for sufficiently large n and $1 \leq k \leq k_n$ there exist $a_{nk} \in \mathscr{H}$ satisfying the relations

$$\langle a_{nk}, y \rangle = \int_{\mathscr{H}} \frac{\langle x, y \rangle}{1 + \|x - a_{nk}\|^2} \, dP_{nk}(y) \qquad \text{for all } y \in \mathscr{H} \tag{7.8.2}$$

and

$$\|a_{nk}\| \leq \delta. \tag{7.8.3}$$

Moreover, in this case the a_{nk} are determined uniquely by P_{nk} by relations (7.8.2) and (7.8.3).

Proof. Let $0 < \delta < 1$ be fixed, and let $S_\delta^* = \{x \in \mathscr{H}: \|x\| \leq \delta\}$. Let $a \in S_\delta^*$. Then for given $\epsilon > 0$ we have

$$\begin{aligned}
\int_{\mathscr{H}} \frac{\|x\|}{1 + \|x - a\|^2} \, dP_{nk}(x) &= \left(\int_{\|x\| < \epsilon} + \int_{\|x\| \geq \epsilon} \right) \frac{\|x\|}{1 + \|x - a\|^2} \, dP_{nk}(x) \\
&\leq \epsilon + (\|a\| + \tfrac{1}{2}) P_{nk}\{x \in \mathscr{H}: \|x\| \geq \epsilon\} \\
&\leq \epsilon + (\delta + \tfrac{1}{2}) P_{nk}\{\|x\| \geq \epsilon\}.
\end{aligned}$$

We see easily that the mapping

$$y \to \int_{\mathscr{H}} \frac{\langle y, x \rangle}{1 + \|x\|^2} \, dP_{nk}(x) \qquad \text{for } y \in \mathscr{H}$$

defines a bounded linear functional on $\mathscr{H}$, so that in view of the Riesz representation theorem there exists an $a' \in \mathscr{H}$ determined uniquely by a such that the relation

$$\langle a', y\rangle = \int_{\mathscr{H}} \frac{\langle y, x\rangle}{1 + \|x\|^2}\, dP_{nk}(x)$$

holds for all $y \in \mathscr{H}$. We now set

$$Ta = a' \qquad \text{for every } a \in S_\delta^*.$$

Then we have

$$\|Ta\| = \|a'\| \leq \int_{\mathscr{H}} \frac{\|x\|}{1 + \|x - a\|^2}\, dP_{nk}(x)$$

$$\leq \epsilon + (\delta + \tfrac{1}{2})P_{nk}\{\|x\| \geq \epsilon\}. \tag{7.8.4}$$

For given δ, $0 < \delta < 1$, we select $0 < \epsilon \leq \delta/2$, and in view of condition (7.8.1) we choose n sufficiently large so that

$$P_{nk}\{\|x\| \geq \epsilon\} < \frac{\delta}{2\delta + 1} \qquad \text{for } 1 \leq k \leq k_n.$$

Then, for every $a \in S_\delta^*$ and n sufficiently large, $\|Ta\| \leq \delta$, so that T maps S_δ^* into itself.

Next we show that, for sufficiently large n, T is a contraction mapping. In fact, let $a, b \in S_\delta^*$. Then for $y \in \mathscr{H}$

$$\langle Ta - Tb, y\rangle = \int_{\mathscr{H}} \left\{\frac{\|x - b\|^2 - \|x - a\|^2}{(1 + \|x - a\|^2)(1 + \|x - b\|^2)}\right\}\langle y, x\rangle\, dP_{nk}(x),$$

so that

$$\|Ta - Tb\| \leq \int_{\mathscr{H}} \|x\| \frac{|\,\|x - b\|^2 - \|x - a\|^2|}{(1 + \|x - a\|^2)(1 + \|x - b\|^2)}\, dP_{nk}(x)$$

$$\leq \|a - b\| \int_{\mathscr{H}} \frac{(2\|x\| + \|a\| + \|b\|)\|x\|}{(1 + \|x - a\|^2)(1 + \|x - b\|^2)}\, dP_{nk}(x).$$

Let $\epsilon > 0$, and set

$$\gamma = \sup\left\{\frac{(2\|x\| + \|a\| + \|b\|)\|x\|}{(1 + \|x - a\|^2)(1 + \|x - b\|^2)};\ a, b \in S_\delta^*,\ x \in \mathscr{H}\right\}.$$

We note that

$$\gamma \leq 2\sup\left\{\left(\frac{\|x\|}{1 + \|x - a\|^2}\right)\left(\frac{\|x\| + \delta}{1 + \|x - b\|^2}\right);\ a, b \in S_\delta^*,\ x \in \mathscr{H}\right\}$$

$$\leq 2(\delta + \tfrac{1}{2})(2\delta + \tfrac{1}{2}).$$

Then we see easily that

$$\begin{aligned}\|Ta - Tb\| &\le \|a - b\|[2\epsilon^2 + 2\delta\epsilon + \gamma P_{nk}\{\|x\| \ge \epsilon\}]\\ &\le \|a - b\|[2\epsilon^2 + 2\delta\epsilon + (2\delta + 1)^2 P_{nk}\{\|x\| \ge \epsilon\}].\end{aligned}$$

Then in view of (7.8.1) for given δ, $0 < \delta < 1$, we choose ϵ sufficiently small and n sufficiently large so that for all $1 \le k \le k_n$

$$2\epsilon^2 + 2\delta\epsilon + (2\delta + 1)^2 P_{nk}\{\|x\| \ge \epsilon\} \le \rho < 1.$$

Therefore, for sufficiently large n, T is a contraction mapping of S_δ^* into itself. Since S_δ^* is closed, it is complete. In view of the contraction mapping theorem (Kolmogorov and Fomin [49]), p. 43) there exists an element $a_{nk} \in S_\delta^*$ satisfying the relation

$$Ta_{nk} = a_{nk}$$

for n sufficiently large and for $1 \le k \le k_n$, where the a_{nk} are determined uniquely by T and hence by P_{nk}. This completes the proof of the proposition. ∎

Corollary. In the notation of Proposition 7.8.1 the relation

$$\lim_{n\to\infty} \max_{1 \le k \le k_n} \|a_{nk}\| = 0 \tag{7.8.5}$$

holds.

Proof. Since $a_{nk} = Ta_{nk}$, it follows, in view of (7.8.4), that for $\epsilon > 0$

$$\|a_{nk}\| \le \epsilon + (\delta + \tfrac{1}{2}) P_{nk}\{\|x\| \ge \epsilon\},$$

where $0 < \delta < 1$ is fixed as above. Hence

$$\max_{1 \le k \le k_n} \|a_{nk}\| \le \epsilon + (\delta + \tfrac{1}{2}) \max_{1 \le k \le k_n} P_{nk}\{\|x\| \ge \epsilon\}.$$

We now take the limit as $n \to \infty$ and then as $\epsilon \to 0$. In view of the u.a.n. condition (7.8.1) the result follows.

In the following we will assume that $0 < \delta < 1$ is fixed and n is sufficiently large so that $a_{nk} \in \mathscr{H}$ as defined by relations (7.8.2) and (7.8.3) exist and are uniquely determined.

For $1 \le k \le k_n$ we now define the operator $\mathbf{S}_{nk}$ on $\mathscr{H}$ by the relation

$$\langle \mathbf{S}_{nk} y, y \rangle = \int_{\mathscr{H}} \frac{\langle x - a_{nk}, y\rangle^2}{1 + \|x - a_{nk}\|^2}\, dP_{nk}(x) \qquad \text{for all } y \in \mathscr{H}. \tag{7.8.6}$$

Then it is easy to verify that $\mathbf{S}_{nk}$ is an S-operator on $\mathscr{H}$ such that

$$\operatorname{tr} \mathbf{S}_{nk} = \int_{\mathscr{H}} \frac{\|x - a_{nk}\|^2}{1 + \|x - a_{nk}\|^2}\, dP_{nk}(x). \tag{7.8.7}$$

Remark 7.8.1. We note that

(7.8.8) $$\lim_{n\to\infty} \max_{1\le k\le k_n} \operatorname{tr} \mathbf{S}_{nk} = 0.$$

In fact, for $\epsilon > 0$ we have from (7.8.7) and (7.8.3)

$$\operatorname{tr} \mathbf{S}_{nk} = \left(\int_{\|x\|<\epsilon} + \int_{\|x\|\ge\epsilon}\right) \frac{\|x - a_{nk}\|^2}{1 + \|x - a_{nk}\|^2} dP_{nk}(x)$$
$$\le \epsilon(\epsilon + 2) + \|a_{nk}\|^2 + P_{nk}\{\|x\| \ge \epsilon\},$$

so that

$$\max_{1\le k\le k_n} \operatorname{tr} \mathbf{S}_{nk} \le \epsilon(\epsilon + 2) + \max_{1\le k\le k_n} \|a_{nk}\|^2 + \max_{1\le k\le k_n} P_{nk}\{\|x\| \ge \epsilon\}.$$

Taking the limit first as $n \to \infty$ and then as $\epsilon \to 0$ and using (7.8.1) and (7.8.5), we obtain (7.8.8).

We define $a_n \in \mathscr{H}$ by the relation

(7.8.9) $$\langle a_n, y\rangle = \sum_{k=1}^{k_n} \left[\langle a_{nk}, y\rangle + \int_{\mathscr{H}} \frac{\langle y, x - a_{nk}\rangle}{1 + \|x - a_{nk}\|^2} dP_{nk}(x)\right], \qquad y \in \mathscr{H}$$

and the operators $\mathbf{S}_n$ on $\mathscr{H}$ by

(7.8.10) $$\langle \mathbf{S}_n y, y\rangle = \sum_{k=1}^{k_n} \langle \mathbf{S}_{nk} y, y\rangle, \qquad y \in \mathscr{H}.$$

Clearly $\mathbf{S}_n$ is also an S-operator on $\mathscr{H}$, and

(7.8.11) $$\operatorname{tr} \mathbf{S}_n = \sum_{k=1}^{k_n} \operatorname{tr} \mathbf{S}_{nk} = \sum_{k=1}^{k_n} \int_{\mathscr{H}} \frac{\|x - a_{nk}\|^2}{1 + \|x - a_{nk}\|^2} dP_{nk}(x).$$

Finally we define the finite measures μ_n on $\mathscr{B}$ satisfying the relation

(7.8.12) $$\int_{\mathscr{H}} g(x)\, d\mu_n(x) = \sum_{k=1}^{k_n} \int_{\mathscr{H}} g(x - a_{nk}) \frac{\|x - a_{nk}\|^2}{1 + \|x - a_{nk}\|^2} dP_{nk}(x)$$

for every bounded measurable function g on $\mathscr{H}$. We note that the measure μ_n is determined uniquely by relation (7.8.12). We now prove the main result of this section.

Theorem 7.8.1. The sequence $\{P_n\}$ of probability measures converges weakly to some probability measure $\mathbf{P}$ on $\mathscr{B}$ if and only if the following three conditions are satisfied:

(i) The limit $\lim_{n\to\infty} a_n = a$ exists.
(ii) The sequence $\{\mu_n\}$ converges weakly to a finite measure μ' on $\mathscr{B}$.
(iii) The sequence of S-operators $\mathbf{S}_n$ satisfies:
(a) $\sup_n \operatorname{tr} \mathbf{S}_n < \infty$.
(b) The series $\sum_{\nu=1}^{\infty} \langle \mathbf{S}_n e_\nu, e_\nu \rangle$ converges uniformly in n for some orthonormal basis $\{e_\nu\}$ in $\mathscr{H}$, that is,

$$\lim_{N\to\infty} \sup_n \sum_{\nu=N}^{\infty} \langle \mathbf{S}_n e_\nu, e_\nu \rangle = 0.$$

(c) For every $y \in \mathscr{H}$, $\lim_{n\to\infty} \langle \mathbf{S}_n y, y \rangle = \langle \mathbf{S}' y, y \rangle$ exists, where $\mathbf{S}'$ is an S-operator in $\mathscr{H}$.

Moreover, in this case the Fourier transform $\hat{\mathbf{P}}$ of the limiting measure $\mathbf{P}$ is given by

$$\hat{\mathbf{P}}(y) = \exp\Bigg\{ i\langle a, y\rangle - \tfrac{1}{2}\langle \mathbf{S}y, y\rangle$$

$$+ \int_{\mathscr{H}} \left[\exp(i\langle x, y\rangle) - 1 - \frac{i\langle x, y\rangle}{1 + \|x\|^2} \right] \frac{1 + \|x\|^2}{\|x\|^2}\, d\mu(x) \Bigg\}, \qquad y \in \mathscr{H}. \tag{7.8.13}$$

Here $\mathbf{S}$ is defined by

$$\langle \mathbf{S}y, y\rangle = \langle \mathbf{S}'y, y\rangle - \int_{\mathscr{H}} \frac{\langle y, x\rangle^2}{\|x\|^2}\, d\mu(x), \qquad y \in \mathscr{H}, \tag{7.8.14}$$

and μ is defined by

$$\mu\{0\} = 0 \qquad \text{and} \qquad \mu(E) = \mu'(E) \quad \text{for } E \in \mathscr{B},\ \{0\} \not\subset E. \tag{7.8.15}$$

Remark 7.8.2. It follows as an immediate consequence of Theorems 7.8.1 and 7.7.1 that the class of limit distributions of the sequence of partial sums $\{S_n\}$ coincides with the class of all i.d. probability measures on $(\mathscr{H}, \mathscr{B})$.

Proof of Theorem 7.8.1. First suppose that conditions (i), (ii), and (iii) are satisfied. We show that there exists a probability measure $\mathbf{P}$ on $\mathscr{B}$ such that $P_n \Rightarrow \mathbf{P}$ and, moreover, $\hat{\mathbf{P}}$ is given by (7.8.13).

Let $\hat{P}_{nk}$ and $\hat{P}_n$ be the Fourier transforms of P_{nk} and P_n, respectively. First we prove the following lemmas.

Lemma 7.8.1

$$\lim_{n\to\infty} \max_{1\le k\le k_n} |\hat{P}_{nk}(y) - 1| = 0 \tag{7.8.16}$$

uniformly in every closed sphere $\{\|y\| \le c\}$ for $c > 0$.

Proof. Let $c > 0$ and $y \in \mathscr{H}$ be such that $\|y\| \leq c$. Let $\epsilon > 0$. Then

$$\begin{aligned}|\hat{P}_{nk}(y) - 1| &\leq \int_{\mathscr{H}} |\exp(i\langle y, x\rangle) - 1| \, dP_{nk}(x) \\ &= \left(\int_{\|x\| < \epsilon} + \int_{\|x\| \geq \epsilon}\right) |\exp(i\langle y, x\rangle) - 1| \, dP_{nk}(x) \\ &- \int_{\|x\| < \epsilon} |\langle y, x\rangle| \, dP_{nk}(x) + 2P_{nk}\{\|x\| \geq \epsilon\} \\ &\leq c\epsilon + 2P_{nk}\{\|x\| \geq \epsilon\},\end{aligned}$$

so that

$$\sup_{\|y\| \leq c} \max_{1 \leq k \leq k_n} |\hat{P}_{nk}(y) - 1| \leq c\epsilon + 2 \max_{1 \leq k \leq k_n} P_{nk}\{\|x\| \geq \epsilon\}.$$

We take the limit first as $n \to \infty$ and then as $\epsilon \to 0$ and obtain (7.8.16). ■

Let Q_{nk} be the probability distribution of the random variable $Y_{nk} = X_{nk} - a_{nk}$, and let $\hat{Q}_{nk}$ be its Fourier transform. Clearly, for $y \in \mathscr{H}$,

$$\begin{aligned}\hat{Q}_{nk}(y) &= \exp(-i\langle y, a_{nk}\rangle)\hat{P}_{nk}(y) \\ &= \int_{\mathscr{H}} \exp(i\langle y, x - a_{nk}\rangle) \, dP_{nk}(x).\end{aligned}$$

Remark 7.8.3. In view of (7.8.1) and (7.8.5) we note that for every $\epsilon > 0$

$$\lim_{n \to \infty} \max_{1 \leq k \leq k_n} P\{\|Y_{nk}\| \geq \epsilon\} = 0,$$

so that the Y_{nk} satisfy the u.a.n. condition. It follows immediately from Lemma 7.8.1 that

$$\lim_{n \to \infty} \max_{1 \leq k \leq k_n} |\hat{Q}_{nk}(y) - 1| = 0 \tag{7.8.17}$$

uniformly in every closed sphere $\{\|y\| \leq c\}$ for $c > 0$.

Lemma 7.8.2

$$\sup_n \sum_{k=1}^{k_n} |\hat{Q}_{nk}(y) - 1| < \infty \tag{7.8.18}$$

uniformly in every closed sphere $\{\|y\| \leq c\}$.

Proof. For $y \in \mathscr{H}$ we have

$$\hat{Q}_{nk}(y) - 1 = \int_{\mathscr{H}} [\exp(i\langle y, x - a_{nk}\rangle) - 1]\, dP_{nk}(x)$$

$$= \int_{\mathscr{H}} \frac{\|x - a_{nk}\|^2}{1 + \|x - a_{nk}\|^2} [\exp(i\langle y, x - a_{nk}\rangle) - 1]\, dP_{nk}(x)$$

$$+ \int_{\mathscr{H}} \frac{1}{1 + \|x - a_{nk}\|^2}$$

$$\times [\exp(i\langle y, x - a_{nk}\rangle) - 1 - i\langle y, x - a_{nk}\rangle]\, dP_{nk}(x)$$

$$+ i \int_{\mathscr{H}} \frac{\langle y, x - a_{nk}\rangle}{1 + \|x - a_{nk}\|^2}\, dP_{nk}(x),$$

so that

$$|\hat{Q}_{nk}(y) - 1| \leq 2 \int_{\mathscr{H}} \frac{\|x - a_{nk}\|^2}{1 + \|x - a_{nk}\|^2}\, dP_{nk}(x)$$

$$+ \frac{1}{2} \int_{\mathscr{H}} \frac{\langle y, x - a_{nk}\rangle^2}{1 + \|x - a_{nk}\|^2}\, dP_{nk}(x)$$

$$+ \left| \int_{\mathscr{H}} \frac{\langle y, x - a_{nk}\rangle}{1 + \|x - a_{nk}\|^2}\, dP_{nk}(x) \right|.$$

In view of (7.8.2) we see easily that

$$\left| \int_{\mathscr{H}} \frac{\langle y, x - a_{nk}\rangle}{1 + \|x - a_{nk}\|^2}\, dP_{nk}(x) \right| = |\langle a_{nk}, y\rangle| \int_{\mathscr{H}} \frac{\|x - a_{nk}\|^2}{1 + \|x - a_{nk}\|^2}\, dP_{nk}(x),$$

so that in view of (7.8.11)

$$\sum_{k=1}^{k_n} |\hat{Q}_{nk}(y) - 1| \leq 2 \operatorname{tr} \mathbf{S}_n + \frac{c^2}{2} \operatorname{tr} \mathbf{S}_n + c\delta \operatorname{tr} \mathbf{S}_n$$

for $\|y\| \leq c$. It follows from condition (iii)(a) of Theorem 7.8.1 that

$$\sup_{\|y\| \leq c} \sup_n \sum_{k=1}^{k_n} |\hat{Q}_{nk}(y) - 1| \leq \gamma \sup_n \operatorname{tr} \mathbf{S}_n < \infty,$$

where $\gamma = 2 + c\delta + c^2/2$. This proves Lemma 7.8.2. ∎

Lemma 7.8.3. Set $\hat{Q}_n = \prod_{k=1}^{k_n} \hat{Q}_{nk}$. Then

(7.8.19)
$$\lim_{n \to \infty} \left\{ \ln \hat{Q}_n(y) - \sum_{k=1}^{k_n} [\hat{Q}_{nk}(y) - 1] \right\} = 0$$

uniformly in every closed sphere $\{y \in \mathscr{H} : \|y\| \leq c\}$ for $c > 0$.

Proof. Let $c > 0$ be fixed, and let $y \in \mathscr{H}$ be such that $\|y\| \le c$. Then it follows from (7.8.17) that for sufficiently large n the inequality

$$|\hat{Q}_{nk}(y) - 1| < \tfrac{1}{2}$$

holds uniformly in $1 \le k \le k_n$. Then we can write

$$\begin{aligned}
\ln \hat{Q}_n(y) &= \sum_{k=1}^{k_n} \ln \hat{Q}_{nk}(y) \\
&= \sum_{k=1}^{k_n} \ln\{1 + [\hat{Q}_{nk}(y) - 1]\} \\
&= \sum_{k=1}^{k_n} [\hat{Q}_{nk}(y) - 1] + R_n(y),
\end{aligned}$$

where

$$R_n(y) = \sum_{k=1}^{k_n} \sum_{s=2}^{\infty} \frac{(-1)^{s-1}}{s} [\hat{Q}_{nk}(y) - 1]^s.$$

Then

$$\begin{aligned}
|R_n(y)| &\le \sum_{k=1}^{k_n} \sum_{s=2}^{\infty} \frac{|\hat{Q}_{nk}(y) - 1|^s}{s} \\
&\le \frac{1}{2} \sum_{k=1}^{k_n} \frac{|\hat{Q}_{nk}(y) - 1|^2}{1 - |\hat{Q}_{nk}(y) - 1|} \le \sum_{k=1}^{k_n} |\hat{Q}_{nk}(y) - 1|^2 \\
&\le \max_{1 \le k \le k_n} |\hat{Q}_{nk}(y) - 1| \sup_n \sum_{k=1}^{k_n} |\hat{Q}_{nk}(y) - 1|.
\end{aligned}$$

We conclude at once from (7.8.17) and (7.8.18) that

$$\sup_{\|y\| \le c} |R_n(y)| \to 0 \qquad \text{as } n \to \infty.$$

This proves (7.8.19). ■

We now return to the proof of Theorem 7.8.1, and note that for every $y \in \mathscr{H}$

$$\hat{Q}_n(y) = \exp\left(-i \sum_{k=1}^{k_n} \langle a_{nk}, y \rangle \right) \hat{P}_n(y),$$

so that for n sufficiently large we have, in view of Lemma 7.8.3,

$$\begin{aligned}
\ln \hat{P}_n(y) &= i \sum_{k=1}^{k_n} \langle a_{nk}, y\rangle + \ln \hat{Q}_n(y) \\
&= i\langle a_n, y\rangle - i \sum_{k=1}^{k_n} \int_{\mathscr{H}} \frac{\langle y, x - a_{nk}\rangle}{1 + \|x - a_{nk}\|^2} \, dP_{nk}(x) \\
&\quad + \sum_{k=1}^{k_n} [\hat{Q}_{nk}(y) - 1] + R_n(y),
\end{aligned}$$

where $R_n(y) = o(1)$ as $n \to \infty$. Thus

$$\begin{aligned}
\ln \hat{P}_n(y) &= i\langle a_n, y\rangle - \tfrac{1}{2}\langle \mathbf{S}_n y, y\rangle \\
&\quad + \sum_{k=1}^{k_n} \int_{\mathscr{H}} \left[\exp(i\langle y, x - a_{nk}\rangle) - 1 - i\frac{\langle y, x - a_{nk}\rangle}{1 + \|x - a_{nk}\|^2}\right. \\
&\quad \left. + \frac{1}{2}\frac{\langle y, x - a_{nk}\rangle^2}{1 + \|x - a_{nk}\|^2}\right] dP_{nk}(x) + R_n(y) \\
&= i\langle a_n, y\rangle - \tfrac{1}{2}\langle \mathbf{S}_n y, y\rangle \\
&\quad + \int_{\mathscr{H}} \left[\exp(i\langle y, x\rangle) - 1 - \frac{i\langle y, x\rangle}{1 + \|x\|^2} + \frac{1}{2}\frac{\langle y, x\rangle^2}{1 + \|x\|^2}\right] \\
&\quad \times \frac{1 + \|x\|^2}{\|x\|^2} \, d\mu_n(x) + R_n(y)
\end{aligned} \tag{7.8.20}$$

in view of (7.8.12). We now take the limit on both sides of (7.8.20) as $n \to \infty$. Using conditions (i), (ii), and (iii) of the theorem, we obtain

$$\begin{aligned}
\lim_{n\to\infty} \ln \hat{P}_n(y) &= i\langle a, y\rangle - \tfrac{1}{2}\langle \mathbf{S}' y, y\rangle \\
&\quad + \int_{\mathscr{H}} \left[\exp(i\langle y, x\rangle) - 1 - \frac{i\langle y, x\rangle}{1 + \|x\|^2} + \frac{1}{2}\frac{\langle y, x\rangle^2}{1 + \|x\|^2}\right] \\
&\quad \times \frac{1 + \|x\|^2}{\|x\|^2} \, d\mu'(x).
\end{aligned} \tag{7.8.21}$$

We note that the integrand on the right-hand side of (7.8.21) is defined by continuity to be zero at $x = 0$. Using expressions (7.8.14) and (7.8.15) for $\mathbf{S}$

and μ, respectively, we can rewrite (7.8.21) in the form

$$\lim_{n\to\infty} \hat{P}_n(y) = \exp\left\{ i\langle a, y\rangle - \tfrac{1}{2}\langle \mathbf{S}y, y\rangle \right.$$

$$\left. + \int_{\mathcal{H}} \left[\exp(i\langle y, x\rangle) - 1 - \frac{i\langle y, x\rangle}{1 + \|x\|^2} \right] \frac{1 + \|x\|^2}{\|x\|^2}\, d\mu(x) \right\}. \tag{7.8.22}$$

In view of Theorem 7.7.1 we note that the expression on the right-hand side of (7.8.22) is the Fourier transform $\hat{\mathbf{P}}$ of an i.d. probability measure $\mathbf{P}$ on $\mathcal{B}$. To show that $P_n \Rightarrow \mathbf{P}$ it is sufficient, in view of Proposition 7.4.2, to prove that the sequence $\{P_n\}$ is tight. This is carried out in Lemma 7.8.4.

Lemma 7.8.4. The sequence $\{P_n\}$ of probability measures on $\mathcal{B}$ is tight.

Proof. Let $y \in \mathcal{H}$. Then we have

$$|1 - \hat{P}_n(y)| = \left| 1 - \exp\left(i \sum_{k=1}^{k_n} \langle a_{nk}, y\rangle \right) \hat{Q}_n(y) \right|$$

$$= |1 - \exp(i\langle a_n, y\rangle)\theta_n(y)|,$$

where

$$\theta_n(y) = \hat{Q}_n(y) \exp\left\{ -i \sum_{k=1}^{k_n} \int_{\mathcal{H}} \frac{\langle y, x - a_{nk}\rangle}{1 + \|x - a_{nk}\|^2}\, dP_{nk}(x) \right\}.$$

We can rewrite as

$$|1 - \hat{P}_n(y)| = |1 - \exp(i\langle a_n, y\rangle) + \exp(i\langle a_n, y\rangle)(1 - \theta_n(y))|$$

$$\leq |\langle a_n, y\rangle| + |1 - \theta_n(y)|$$

$$= |\langle a_n, y\rangle|$$

$$+ \left| 1 - \prod_{k=1}^{k_n} \left\{ \hat{Q}_{nk}(y) \exp\left[-i \int_{\mathcal{H}} \frac{\langle y, x - a_{nk}\rangle}{1 + \|x - a_{nk}\|^2}\, dP_{nk}(x) \right] \right\} \right|.$$

Since for any two complex numbers z_1, z_2 such that $|z_1| \leq 1$, $|z_2| \leq 1$ we have

$$|1 - z_1 z_2| \leq |1 - z_1| + |1 - z_2|,$$

we obtain

$$
\begin{aligned}
|1 - \hat{P}_n(y)| &\le |\langle a_n, y\rangle| + \sum_{k=1}^{k_n} \left| 1 - \hat{Q}_{nk}(y) \exp\left[-i \int_{\mathscr{H}} \frac{\langle y, x - a_{nk}\rangle}{1 + \|x - a_{nk}\|^2} \, dP_{nk}(x) \right] \right| \\
&= |\langle a_n, y\rangle| + \sum_{k=1}^{k_n} \left| \hat{Q}_{nk}(y) - \exp\left[i \int_{\mathscr{H}} \frac{\langle y, x - a_{nk}\rangle}{1 + \|x - a_{nk}\|^2} \, dP_{nk}(x) \right] \right| \\
&= |\langle a_n, y\rangle| + \sum_{k=1}^{k_n} \left| \hat{Q}_{nk}(y) - 1 - i \int_{\mathscr{H}} \frac{\langle y, x - a_{nk}\rangle}{1 + \|x - a_{nk}\|^2} \, dP_{nk}(x) \right. \\
&\quad - \left\{ \exp\left[i \int_{\mathscr{H}} \frac{\langle y, x - a_{nk}\rangle}{1 + \|x - a_{nk}\|^2} \, dP_{nk}(x) \right] - 1 \right. \\
&\quad \left. \left. - i \int_{\mathscr{H}} \frac{\langle y, x - a_{nk}\rangle}{1 + \|x - a_{nk}\|^2} \, dP_{nk}(x) \right\} \right| \\
&\le |\langle a_n, y\rangle| + \sum_{k=1}^{k_n} \left| \hat{Q}_{nk}(y) - 1 - i \int_{\mathscr{H}} \frac{\langle y, x - a_{nk}\rangle}{1 + \|x - a_{nk}\|^2} \, dP_{nk}(x) \right| \\
&\quad + \frac{1}{2} \sum_{k=1}^{k_n} \left(\int_{\mathscr{H}} \frac{\langle y, x - a_{nk}\rangle}{1 + \|x - a_{nk}\|^2} \, dP_{nk}(x) \right)^2 \\
&\le |\langle a_n, y\rangle| + \sum_{k=1}^{k_n} \int_{\mathscr{H}} (1 - \cos\langle y, x - a_{nk}\rangle) \, dP_{nk}(x) \\
&\quad + \sum_{k=1}^{k_n} \left| \int_{\mathscr{H}} \left[\sin\langle y, x - a_{nk}\rangle - \frac{\langle y, x - a_{nk}\rangle}{1 + \|x - a_{nk}\|^2} \right] dP_{nk}(x) \right| \\
&\quad + \frac{1}{2} \sum_{k=1}^{k_n} \left(\int_{\mathscr{H}} \frac{\langle y, x - a_{nk}\rangle}{1 + \|x - a_{nk}\|^2} \, dP_{nk}(x) \right)^2. \qquad (7.8.23)
\end{aligned}
$$

Let $\alpha > 0$ and $\gamma > 0$, to be chosen later. Let $y \in \mathscr{H}$. Then we can write

$$
|\langle a_n, y\rangle| \le \alpha + \frac{1}{\alpha} \langle a_n, y\rangle^2. \qquad (7.8.24)
$$

Also

$$
\begin{aligned}
&\int_{\mathscr{H}} (1 - \cos\langle y, x - a_{nk}\rangle) \, dP_{nk}(x) \\
&\quad = \left(\int_{\|x - a_{nk}\| < \gamma} + \int_{\|x - a_{nk}\| \ge \gamma} \right) (1 - \cos\langle y, x - a_{nk}\rangle) \, dP_{nk}(x) \\
&\quad \le \frac{1}{2} \int_{\|x - a_{nk}\| < \gamma} \langle y, x - a_{nk}\rangle^2 \, dP_{nk}(x) + 2 \int_{\|x - a_{nk}\| \ge \gamma} dP_{nk}(x) \\
&\quad \le \tfrac{1}{2}(1 + \gamma^2) \int_{\|x - a_{nk}\| < \gamma} \frac{\langle y, x - a_{nk}\rangle^2}{1 + \|x - a_{nk}\|^2} \, dP_{nk}(x) \\
&\qquad + 2 \, \frac{1 + \gamma^2}{\gamma^2} \int_{\|x - a_{nk}\| \ge \gamma} \frac{\langle x - a_{nk}\rangle^2}{1 + \|x - a_{nk}\|^2} \, dP_{nk}(x),
\end{aligned}
$$

so that

$$\sum_{k=1}^{k_n} \int_{\mathscr{H}} (1 - \cos\langle y, x - a_{nk}\rangle)\, dP_{nk}(x)$$
$$\le \frac{1+\gamma^2}{2} \sum_{k=1}^{k_n} \langle \mathbf{S}_{nk} y, y\rangle$$
$$+ 2\frac{1+\gamma^2}{\gamma^2} \sum_{k=1}^{k_n} \int_{\|x-a_{nk}\| \ge \gamma} \frac{\|x - a_{nk}\|^2}{1 + \|x - a_{nk}\|^2}\, dP_{nk}(x)$$
$$= \frac{1+\gamma^2}{2} \langle \mathbf{S}_n y, y\rangle + 2\frac{1+\gamma^2}{\gamma^2} \mu_n\{x \in \mathscr{H} : \|x\| \ge \gamma\}. \tag{7.8.25}$$

Next we note that

$$\int_{\mathscr{H}} \left[\sin\langle y, x - a_{nk}\rangle - \frac{\langle y, x - a_{nk}\rangle}{1 + \|x - a_{nk}\|^2}\right] dP_{nk}(x)$$
$$= \int_{\mathscr{H}} \frac{\|x - a_{nk}\|^2}{1 + \|x - a_{nk}\|^2} \sin\langle y, x - a_{nk}\rangle\, dP_{nk}(x)$$
$$+ \int_{\mathscr{H}} \frac{\sin\langle y, x - a_{nk}\rangle - \langle y, x - a_{nk}\rangle}{1 + \|x - a_{nk}\|^2}\, dP_{nk}(x),$$

so that

$$\int_{\mathscr{H}} \left[\sin\langle y, x - a_{nk}\rangle - \frac{\langle y, x - a_{nk}\rangle}{1 + \|x - a_{nk}\|^2}\right] dP_{nk}(x)$$
$$\le \int_{\mathscr{H}} \frac{\|x - a_{nk}\|^2}{1 + \|x - a_{nk}\|^2} |\sin\langle y, x - a_{nk}\rangle|\, dP_{nk}(x)$$
$$+ \frac{1}{2} \int_{\mathscr{H}} \frac{\langle y, x - a_{nk}\rangle^2}{1 + \|x - a_{nk}\|^2}\, dP_{nk}(x)$$
$$\le \alpha \int_{\mathscr{H}} \frac{\|x - a_{nk}\|^2}{1 + \|x - a_{nk}\|^2}\, dP_{nk}(x) + \frac{1}{\alpha} \int_{\mathscr{H}} \sin^2\langle y, x - a_{nk}\rangle\, dP_{nk}(x)$$
$$+ \tfrac{1}{2}\langle \mathbf{S}_{nk} y, y\rangle$$
$$\le \alpha \operatorname{tr} \mathbf{S}_{nk} + \frac{1}{\alpha} \int_{\|x-a_{nk}\| < \gamma} \langle y, x - a_{nk}\rangle^2\, dP_{nk}(x)$$
$$+ \frac{1}{\alpha} \int_{\|x-a_{nk}\| \ge \gamma} dP_{nk}(x) + \frac{1}{2} \langle \mathbf{S}_{nk} y, y\rangle.$$

Hence

$$\sum_{k=1}^{k_n} \left| \int_{\mathscr{H}} \left[\sin\langle y, x - a_{nk}\rangle - \frac{\langle y, x - a_{nk}\rangle}{1 + \|x - a_{nk}\|^2} \right] dP_{nk}(x) \right|$$

$$\leq \alpha \operatorname{tr} \mathbf{S}_n + \frac{1}{\alpha}(1 + \gamma^2)\langle \mathbf{S}_n y, y\rangle + \frac{1}{\alpha}\frac{1 + \gamma^2}{\gamma^2}\mu_n\{x \in \mathscr{H} : \|x\| \geq \gamma\}$$

$$+ \tfrac{1}{2}\langle \mathbf{S}_n y, y\rangle. \tag{7.8.26}$$

Finally, in view of the Schwartz inequality we have

$$\left(\int_{\mathscr{H}} \frac{\langle y, x - a_{nk}\rangle}{1 + \|x - a_{nk}\|^2} dP_{nk}(x) \right)^2$$

$$\leq \int_{\mathscr{H}} \frac{\langle y, x - a_{nk}\rangle^2}{(1 + \|x - a_{nk}\|^2)^2} dP_{nk}(x)$$

$$\leq \int_{\mathscr{H}} \frac{\langle y, x - a_{nk}\rangle^2}{1 + \|x - a_{nk}\|^2} dP_{nk}(x)$$

$$= \langle \mathbf{S}_{nk} y, y\rangle,$$

so that

$$\frac{1}{2}\sum_{k=1}^{k_n} \left(\int_{\mathscr{H}} \frac{\langle y, x - a_{nk}\rangle}{1 + \|x - a_{nk}\|^2} dP_{nk}(x) \right)^2 \leq \tfrac{1}{2}\langle \mathbf{S}_n y, y\rangle. \tag{7.8.27}$$

Using estimates (7.8.24) through (7.8.27) on the right-hand side of (7.8.23), we obtain for $y \in \mathscr{H}$

$$|1 - \hat{P}_n(y)| \leq \alpha + \frac{1}{\alpha}\langle a_n, y\rangle^2 + \frac{1 + \gamma^2}{2}\langle \mathbf{S}_n y, y\rangle$$

$$+ 2\frac{1 + \gamma^2}{\gamma^2}\mu_n\{\|x\| \geq \gamma\} + \alpha \operatorname{tr} \mathbf{S}_n + \frac{1}{\alpha}(1 + \gamma^2)\langle \mathbf{S}_n y, y\rangle$$

$$+ \frac{1}{\alpha}\frac{1 + \gamma^2}{\gamma^2}\mu_n\{\|x\| \geq \gamma\} + \langle \mathbf{S}_n y, y\rangle,$$

so that for $y \in \mathscr{H}$ we have

$$0 \leq 1 - \operatorname{Re} \hat{P}_n(y) \leq |1 - \hat{P}_n(y)|$$

$$\leq \alpha\left(1 + \sup_n \operatorname{tr} \mathbf{S}_n\right) + \left(2 + \frac{1}{\alpha}\right)\frac{1 + \gamma^2}{\gamma^2}\sup_n \mu_n\{\|x\| \geq \gamma\}$$

$$+ \frac{1}{\alpha}\langle a_n, y\rangle^2 + \left[(1 + \gamma^2)\left(\frac{1}{2} + \frac{1}{\alpha}\right) + 1\right]\langle \mathbf{S}_n y, y\rangle. \tag{7.8.28}$$

Let $\epsilon > 0$. Since $\sup_n \operatorname{tr} \mathbf{S}_n < \infty$ and $\mu_n \Rightarrow \mu'$, where μ' is a finite measure on $\mathscr{B}$, for given $\epsilon > 0$ we first choose $\alpha = \alpha(\epsilon) > 0$ sufficiently small and then $\gamma = \gamma(\alpha, \epsilon) > 0$ sufficiently large so that

$$\alpha\left(1 + \sup_n \operatorname{tr} \mathbf{S}_n\right) + \left(2 + \frac{1}{\alpha}\right)\frac{1 + \gamma^2}{\gamma^2} \sup_n \mu_n\{\|x\| \geq \gamma\} < \frac{\epsilon}{2}.$$

Set $C = (1 + \gamma^2)(\frac{1}{2} + 1/\alpha) + 1$. Then we can rewrite (7.8.28) in the form

$$0 \leq 1 - \operatorname{Re} \hat{P}_n(y) < C\{\langle \mathbf{S}_n y, y\rangle + \langle a_n, y\rangle^2\} + \frac{\epsilon}{2}. \tag{7.8.29}$$

Now define the operator $\mathbf{T}_n$ on $\mathscr{H}$ by setting

$$\mathbf{T}_n y = \langle a_n, y\rangle a_n, \qquad y \in \mathscr{H}.$$

Then $\mathbf{T}_n$ is an S-operator on $\mathscr{H}$ such that $\operatorname{tr} \mathbf{T}_n = \|a_n\|^2$. Next define the operator $\mathbf{S}_{n,\epsilon}$ on $\mathscr{H}$ by setting

$$\mathbf{S}_{n,\epsilon} = \frac{2C}{\epsilon}(\mathbf{S}_n + \mathbf{T}_n). \tag{7.8.30}$$

Then $\mathbf{S}_{n,\epsilon}$ is also an S-operator on $\mathscr{H}$ such that

$$\operatorname{tr} \mathbf{S}_{n,\epsilon} = \frac{2C}{\epsilon}(\operatorname{tr} \mathbf{S}_n + \|a_n\|^2) < \infty.$$

It follows easily from (7.8.29) and (7.8.30) that

$$1 - \operatorname{Re} \hat{P}_n(y) < \frac{\epsilon}{2}\langle \mathbf{S}_{n,\epsilon} y, y\rangle + \frac{\epsilon}{2}. \tag{7.8.31}$$

Let $y \in \mathscr{H}$ such that $\langle \mathbf{S}_{n,\epsilon} y, y\rangle < 1$. Then it follows from (7.8.31) that

$$0 \leq 1 - \operatorname{Re} \hat{P}_n(y) < \epsilon.$$

Since $a_n \to a$, $\|a_n\| \to \|a\|$ so that

$$\sup_n \operatorname{tr} \mathbf{T}_n = \sup_n \|a_n\|^2 < \infty.$$

Hence

$$\sup_n \operatorname{tr} \mathbf{S}_{n,\epsilon} \leq \frac{2C}{\epsilon}\left(\sup_n \operatorname{tr} \mathbf{S}_n + \sup_n \operatorname{tr} \mathbf{T}_n\right)$$
$$< \infty.$$

Finally we show that the series $\sum_{\nu=1}^{\infty} \langle \mathbf{S}_{n,\epsilon} e_\nu, e_\nu\rangle$ converges uniformly in n for some orthonormal basis $\{e_\nu\}$ in $\mathscr{H}$. In view of condition (iii)(b) of Theorem

7.8.1 there exists an orthonormal basis $\{e_\nu\}$ in $\mathscr{H}$ such that

$$\sup_n \sum_{\nu=N}^{\infty} \langle \mathbf{S}_n e_\nu, e_\nu \rangle \to 0 \qquad \text{as } N \to \infty.$$

On the other hand, since $a_n \to a$ we note that

$$\sum_{\nu=1}^{\infty} [\langle a_n, e_\nu \rangle^2 - \langle a, e_\nu \rangle^2] = \|a_n\|^2 - \|a\|^2 \to 0 \qquad \text{as } n \to \infty.$$

This implies that

$$\sup_n \sum_{\nu=N}^{\infty} \langle \mathbf{T}_n e_\nu, e_\nu \rangle = \sup_n \sum_{\nu=N}^{\infty} \langle a_n, e_\nu \rangle^2 \to 0 \qquad \text{as } N \to \infty.$$

Consequently

$$\sup_n \sum_{\nu=N}^{\infty} \langle \mathbf{S}_{n,\epsilon} e_\nu, e_\nu \rangle \le \frac{2C}{\epsilon} \left\{ \sup_n \sum_{\nu=N}^{\infty} \langle \mathbf{S}_n e_\nu, e_\nu \rangle + \sup_n \sum_{\nu=N}^{\infty} \langle \mathbf{T}_n e_\nu, e_\nu \rangle \right\}$$
$$\to 0 \qquad \text{as } N \to \infty.$$

We have thus shown that $\{P_n\}$ satisfies the conditions of Proposition 7.7.1 and hence the sequence $\{P_n\}$ is tight. This completes the proof of Lemma 7.8.4. ■

In view of Lemma 7.8.4 we conclude that $P_n \Rightarrow \mathbf{P}$, where $\mathbf{P}$ is an i.d. probability measure on $\mathscr{B}$ with Fourier transform $\hat{\mathbf{P}}$ given by (7.8.13). This completes the proof of the sufficiency part of Theorem 7.8.1.

Next we suppose that the sequence $\{P_n\}$ converges weakly to some probability measure $\mathbf{P}$ on $\mathscr{B}$. We show that conditions (i), (ii), and (iii) of the theorem are satisfied and, moreover, the Fourier transform $\hat{\mathbf{P}}$ of $\mathbf{P}$ is of the form (7.8.13). First we prove the following lemmas.

Lemma 7.8.5. For sufficiently large $\alpha > 0$ we have

$$\sup_n \sum_{k=1}^{k_n} P\{\|X_{nk}\| \ge \alpha\} < \infty. \tag{7.8.32}$$

Proof. For every $n \ge 1$ let $X'_{n1}, X'_{n2}, \ldots, X'_{nk_n}$ be independent $\mathscr{H}$-valued random variables defined on $(\Omega, \mathscr{S}, P)$ such that the random vectors

$$(X_{n1}, X_{n2}, \ldots, X_{nk_n}) \qquad \text{and} \qquad (X'_{n1}, X'_{n2}, \ldots, X'_{nk_n})$$

are independent and, moreover, for every k, $1 \le k \le k_n$, X_{nk} and X'_{nk} are identically distributed. Let $X^s_{nk} = X_{nk} - X'_{nk}$ be the symmetrized random

$\beta = 4\alpha$, where α is defined above, and obtain for every $\gamma > 0$

$$\sup_n \sum_{k=1}^{k_n} P\{\|X_{nk}\| < \gamma\}P\{\|X_{nk}\| \geq 4\alpha + \gamma\} \leq \sup_n \sum_{k=1}^{k_n} P\{\|X_{nk}\| \geq 4\alpha\}$$
$$\leq -\ln(1-\epsilon) < \infty.$$

Since $\{X_{nk}\}$ satisfies the u.a.n. condition, for sufficiently large n and for given $0 < \epsilon < 1$ we have

$$P\{\|X_{nk}\| \geq \gamma\} < \epsilon$$

uniformly in k, $1 \leq k \leq k_n$, so that

$$\min_{1 \leq k \leq k_n} P\{\|X_{nk}\| < \gamma\} > 1 - \epsilon$$

for sufficiently large n. It follows that for all $\gamma > 0$ and for $\alpha > 0$ sufficiently large we have

$$\sup_n \sum_{k=1}^{k_n} P\{\|X_{nk}\| \geq 4\alpha + \gamma\} < \frac{-\ln(1-\epsilon)}{1-\epsilon} < \infty.$$

This completes the proof of the lemma. ■

Let $\alpha > 0$, and let χ_α be the indicator function of the open sphere $\{\|x\| < \alpha\}$.

Lemma 7.8.6. For $\alpha > 0$ sufficiently large we have

$$\sup_n \mathscr{E}\left(\left\| \sum_{k=1}^{k_n} \chi_\alpha(X_{nk}^s)X_{nk}^s \right\|^2\right) < \infty. \tag{7.8.33}$$

Proof. Let $\gamma > 0$. Then in view of Lemma 7.4.2 we have

$$P\left\{\max_{1 \leq j \leq k_n} \left\| \sum_{k=1}^{j} \chi_\alpha(X_{nk}^s)X_{nk}^s \right\| \geq \gamma\right\}$$
$$\leq 2P\left\{\left\| \sum_{k=1}^{k_n} \chi_\alpha(X_{nk}^s)X_{nk}^s \right\| \geq \gamma\right\}$$
$$\leq 2P\left\{\left\| \sum_{k=1}^{k_n} X_{nk}^s \right\| \geq \gamma\right\} + 2P\left\{\max_{1 \leq k \leq k_n} \|X_{nk}^s\| \geq \alpha\right\}$$
$$\leq 2P\left\{\left\| \sum_{k=1}^{k_n} X_{nk}^s \right\| \geq \gamma\right\} + 2P\left\{\max_{1 \leq j \leq k_n} \left\| \sum_{k=1}^{j} X_{nk}^s \right\| \geq \frac{\alpha}{2}\right\}$$
$$\leq 2P\left\{\left\| \sum_{k=1}^{k_n} X_{nk}^s \right\| \geq \gamma\right\} + 4P\left\{\left\| \sum_{k=1}^{k_n} X_{nk}^s \right\| \geq \frac{\alpha}{2}\right\}$$
$$\leq 4P\left\{\|S_n\| \geq \frac{\gamma}{2}\right\} + 8P\left\{\|S_n\| \geq \frac{\alpha}{2}\right\}$$
$$= 4P_n\left\{\|x\| \geq \frac{\gamma}{2}\right\} + 8P_n\left\{\|x\| \geq \frac{\alpha}{2}\right\}.$$

Since $P_n \Rightarrow \mathbf{P}$ for given ϵ, $0 < \epsilon < \frac{1}{2}$, we can choose $\gamma = \gamma(\epsilon) > 0$ and $\alpha = \alpha(\epsilon) > 0$ sufficiently large so that

$$\sup_n P\left\{ \max_{1 \le j \le k_n} \left\| \sum_{k=1}^{j} \chi_\alpha(X_{nk}^s) X_{nk}^s \right\| \ge \gamma \right\} < \epsilon < \tfrac{1}{2}. \tag{7.8.34}$$

Equation (7.8.34) is similar to (7.7.10). Using Lemma 7.7.4 and a similar argument, we obtain (7.8.33). ■

In the following we choose and fix an $\alpha > 0$ sufficiently large so that both (7.8.32) and (7.8.33) hold.

Lemma 7.8.7. Let the a_{nk} be as in Proposition 7.8.1. Then we have

$$\sup_n \sum_{k=1}^{k_n} \int_{\mathscr{H}} \frac{\|x - a_{nk}\|^2}{1 + \|x - a_{nk}\|^2} \, dP_{nk}(x) < \infty. \tag{7.8.35}$$

Proof. Let $\epsilon > 0$. Let $\alpha > 0$ be fixed as above. Then for $1 \le k \le k_n$ and $n \ge 1$ we have

$$\begin{aligned}
\mathscr{E}(\chi_\alpha(X_{nk}^s)\|X_{nk}^s\|^2) &= \int_{\|x-y\|<\alpha} \|x - y\|^2 \, dP_{nk}(x) \, dP_{nk}(y) \\
&\ge \int_{\{\|y\|<\epsilon,\ \|x\|<\alpha-\epsilon\}} \|x - y\|^2 \, dP_{nk}(x) \, dP_{nk}(y) \\
&\ge \int_{\|y\|<\epsilon} dP_{nk}(y) \int_{\|x\|<\alpha-\epsilon} \|x - y\|^2 \, dP_{nk}(x). \\
&\ge P_{nk}\{\|y\| < \epsilon\} \inf_{\|a\| \le \epsilon} \int_{\|x\|<\alpha-\epsilon} \|x - a\|^2 \, dP_{nk}(x). \qquad (7.8.36)
\end{aligned}$$

We note that, for sufficiently large n, $P_{nk}\{\|x\| < \alpha - \epsilon\} > 0$ uniformly in $1 \le k \le k_n$. Clearly the mapping

$$y \to \frac{1}{P_{nk}\{\|x\| < \alpha - \epsilon\}} \int_{\|x\|<\alpha-\epsilon} \langle y, x \rangle \, dP_{nk}(x)$$

is a bounded linear functional on $\mathscr{H}$, so that there exists a vector $\tilde{a}_{nk} \in \mathscr{H}$ determined uniquely by the relation

$$P_{nk}\{\|x\| < \alpha - \epsilon\} \ \langle \tilde{a}_{nk}, y \rangle = \int_{\|x\|<\alpha-\epsilon} \langle y, x \rangle \, dP_{nk}(x).$$

Clearly

$$\|\tilde{a}_{nk}\| \le \alpha - \epsilon.$$

Let $0 < \epsilon < \alpha/2$, and let $\|a\| \le \epsilon$. Then we note that

$$\|x - a\|^2 = \|x - \tilde{a}_{nk}\|^2 + \|a - \tilde{a}_{nk}\|^2 + 2\langle x - \tilde{a}_{nk}\rangle\langle \tilde{a}_{nk} - a\rangle,$$

so that, using the definition of $\tilde{a}_{nk}$, we obtain

$$\int_{\|x\| < \alpha - \epsilon} \|x - a\|^2 \, dP_{nk}(x) = \int_{\|x\| < \alpha - \epsilon} \|x - \tilde{a}_{nk}\|^2 \, dP_{nk}(x) + \|a - \tilde{a}_{nk}\|^2 P_{nk}\{\|x\| < \alpha - \epsilon\}.$$

Hence

$$\inf_{\|a\| \le \epsilon} \int_{\|x\| < \alpha - \epsilon} \|x - a\|^2 \, dP_{nk}(x) = \int_{\|x\| < \alpha - \epsilon} \|x - \tilde{a}_{nk}\|^2 \, dP_{nk}(x),$$

and it follows from (7.8.36) that

$$\mathscr{E}(\chi_\alpha(X^s_{nk})\|X^s_{nk}\|^2) \ge P_{nk}\{\|y\| < \epsilon\} \int_{\|x\| < \alpha - \epsilon} \|x - \tilde{a}_{nk}\|^2 \, dP_{nk}(x). \tag{7.8.37}$$

Since the X^s_{nk} are symmetric and independent, it follows from (7.8.37) and (7.8.33) that

$$\sup_n \sum_{k=1}^{k_n} P_{nk}\{\|y\| < \epsilon\} \int_{\|x\| < \alpha - \epsilon} \|x - \tilde{a}_{nk}\|^2 \, dP_{nk}(x) < \infty.$$

In view of the u.a.n. condition we have

$$\inf_{1 \le k \le n} P_{nk}\{\|y\| < \epsilon\} \to 1 \qquad \text{as } n \to \infty,$$

so that

$$\sup_n \sum_{k=1}^{k_n} \int_{\|x\| < \alpha - \epsilon} \|x - \tilde{a}_{nk}\|^2 \, dP_{nk}(x) < \infty. \tag{7.8.38}$$

Therefore

$$\sup_n \sum_{k=1}^{k_n} \int_{\|x\| < \alpha - \epsilon} \frac{\|x - \tilde{a}_{nk}\|^2}{1 + \|x - \tilde{a}_{nk}\|^2} \, dP_{nk}(x) < \infty.$$

On the other hand, it follows from (7.8.32) that for sufficiently large $\alpha > 0$ and $0 < \epsilon < \alpha/2$ we have

$$\sup_n \sum_{k=1}^{k_n} \int_{\|x\| \ge \alpha - \epsilon} \frac{\|x - \tilde{a}_{nk}\|^2}{1 + \|x - \tilde{a}_{nk}\|^2} \, dP_{nk}(x) < \infty,$$

so that

$$\sup_n \sum_{k=1}^{k_n} \int_{\mathscr{H}} \frac{\|x - \tilde{a}_{nk}\|^2}{1 + \|x - \tilde{a}_{nk}\|^2} \, dP_{nk}(x) < \infty. \tag{7.8.39}$$

Next we show that

$$\lim_{n\to\infty} \max_{1\le k\le k_n} \|\tilde{a}_{nk}\| = 0. \tag{7.8.40}$$

In fact, for $0 < \epsilon < \alpha/2$ we have

$$\begin{aligned}
\|\tilde{a}_{nk}\| &\le \frac{\int_{\|x\|<\alpha-\epsilon} \|x\|\, dP_{nk}(x)}{P_{nk}\{\|x\| < \alpha - \epsilon\}} \\
&= \frac{1}{P_{nk}\{\|x\| < \alpha - \epsilon\}} \left(\int_{\|x\|<\epsilon} + \int_{\epsilon\le\|x\|<\alpha-\epsilon} \right) \|x\|\, dP_{nk}(x) \\
&\le \epsilon + (\alpha - \epsilon) \frac{P_{nk}\{\|x\| \ge \epsilon\}}{P_{nk}\{\|x\| < \alpha - \epsilon\}},
\end{aligned}$$

so that

$$\max_{1\le k\le k_n} \|\tilde{a}_{nk}\| \le \epsilon + (\alpha - \epsilon) \frac{\max_{1\le k\le k_n} P_{nk}\{\|x\| \ge \epsilon\}}{\min_{1\le k\le k_n} P_{nk}\{\|x\| < \alpha - \epsilon\}}.$$

We now take the limit, first as $n \to \infty$ and then as $\epsilon \to 0$. In view of the u.a.n. condition we obtain (7.8.40).

Using (7.8.40), (7.8.5), and (7.8.39), we see easily that

$$\begin{aligned}
&\sup_n \sum_{k=1}^{k_n} \int_{\mathscr{H}} \frac{\|x - \tilde{a}_{nk}\|^2}{1 + \|x - a_{nk}\|^2}\, dP_{nk}(x) \\
&\qquad \le C' \sup_n \sum_{k=1}^{k_n} \int_{\mathscr{H}} \frac{\|x - \tilde{a}_{nk}\|^2}{1 + \|x - \tilde{a}_{nk}\|^2}\, dP_{nk}(x) \\
&\qquad < \infty.
\end{aligned} \tag{7.8.41}$$

Finally, using the relation

$$\|x - \tilde{a}_{nk}\|^2 = \|x - a_{nk}\|^2 + \|a_{nk} - \tilde{a}_{nk}\|^2 + 2\langle x - a_{nk}, a_{nk} - \tilde{a}_{nk}\rangle,$$

we can verify easily that

$$\begin{aligned}
&\int_{\mathscr{H}} \frac{\|x - \tilde{a}_{nk}\|^2}{1 + \|x - a_{nk}\|^2}\, dP_{nk}(x) \\
&\quad = \|a_{nk} - \tilde{a}_{nk}\|^2 + (1 + \|a_{nk}\|^2 - \|\tilde{a}_{nk}\|^2) \int_{\mathscr{H}} \frac{\|x - a_{nk}\|^2}{1 + \|x - a_{nk}\|^2}\, dP_{nk}(x).
\end{aligned}$$

Therefore (7.8.41) implies that

$$\sup_n \sum_{k=1}^{k_n} (1 + \|a_{nk}\|^2 - \|\tilde{a}_{nk}\|^2) \int_{\mathscr{H}} \frac{\|x - a_{nk}\|^2}{1 + \|x - a_{nk}\|^2}\, dP_{nk}(x) < \infty. \tag{7.8.42}$$

On the other hand, in view of (7.8.5) and (7.8.40), for sufficiently large n there exists a constant $C'' > 0$ (independent of n) such that

$$1 + \|a_{nk}\|^2 - \|\tilde{a}_{nk}\|^2 \geq C''$$

uniformly in k, $1 \leq k \leq n$, and uniformly in n. It follows immediately from (7.8.42) that (7.8.35) holds. ∎

Lemma 7.8.8. The sequence $\{\mu_n\}$ of finite measures as defined in (7.8.12) is tight.

Proof. It follows from (7.8.11) and (7.8.12) that

$$\text{tr } \mathbf{S}_n = \mu_n(\mathscr{H}) = \sum_{k=1}^{k_n} \int_{\mathscr{H}} \frac{\|x - a_{nk}\|^2}{1 + \|x - a_{nk}\|^2} \, dP_{nk}(x),$$

so that in view of (7.8.35) we have

$$\sup_n \text{tr } \mathbf{S}_n = \sup_n \mu_n(\mathscr{H}) < \infty. \tag{7.8.43}$$

Consequently Lemmas 7.8.2 and 7.8.3 hold. Therefore (7.8.20) also holds for all $y \in \mathscr{H}$. We now equate the real parts on both sides of (7.8.20), and obtain

$$\begin{aligned} 0 \leq -\ln|\hat{P}_n(y)| &= -\text{Re} \ln \hat{P}_n(y) \\ &= \tfrac{1}{2}\langle \mathbf{S}_n y, y\rangle + \sum_{k=1}^{k_n} \int_{\mathscr{H}} \Bigg(1 - \cos\langle y, x - a_{nk}\rangle \\ &\qquad - \frac{1}{2} \frac{\langle y, x - a_{nk}\rangle^2}{1 + \|x - a_{nk}\|^2}\Bigg) \, dP_{nk}(x) + \text{Re } R_n(y), \end{aligned}$$

where $R_n(y) = o(1)$ as $n \to \infty$. Therefore

$$-\ln|\hat{P}_n(y)| = \sum_{k=1}^{k_n} \int_{\mathscr{H}} (1 - \cos\langle y, x - a_{nk}\rangle) \, dP_n(x) + o(1) \tag{7.8.44}$$

as $n \to \infty$. Let $0 < \epsilon < \frac{1}{2}$. Since $P_n \Rightarrow \mathbf{P}$, the sequence $\{P_n\}$ is tight, so that it follows from Proposition 7.7.1 that for every $n \geq 1$ there exist S-operators $\mathbf{S}_{n,\epsilon}$ on $\mathscr{H}$ satisfying these conditions:

(i) $1 - \text{Re } \hat{P}_n(y) < \epsilon$ whenever $\langle \mathbf{S}_{n,\epsilon} y, y\rangle < 1$.
(ii) $\sup_n \text{tr } \mathbf{S}_{n,\epsilon} < \infty$.
(iii) There exists an orthonormal basis $\{e_\nu\}$ in $\mathscr{H}$ such that

$$\lim_{N\to\infty} \sup_n \sum_{\nu=N}^{\infty} \langle \mathbf{S}_{n,\epsilon} e_\nu, e_\nu\rangle = 0.$$

Let $y \in \mathscr{H}$ be such that $\langle \mathbf{S}_{n,\epsilon} y, y\rangle < 1$. Then it follows from (i) that $0 \leq 1 - |\hat{P}_n(y)| \leq 1 - \operatorname{Re} \hat{P}_n(y) < \epsilon/2$, so that after some elementary computations we obtain

$$-\ln|\hat{P}_n(y)| = -\ln\{1 - [1 - |\hat{P}_n(y)|]\} < \frac{\epsilon}{2}\left(\frac{1}{2} + \ln 2\right) < \epsilon.$$

It follows from (7.8.44) that

$$\sum_{k=1}^{k_n} \int_{\mathscr{H}} (1 - \cos\langle y, x - a_{nk}\rangle)\, dP_{nk}(x)$$

$$< \epsilon \qquad \text{whenever } \langle \mathbf{S}_{n,\epsilon} y, y\rangle < 1. \tag{7.8.45}$$

In view of (7.8.12) and (7.8.45) we obtain

$$\begin{aligned}
\mu_n(\mathscr{H}) - \operatorname{Re} \hat{\mu}_n(y) &= \int_{\mathscr{H}} (1 - \cos\langle y, x\rangle)\, d\mu_n(x) \\
&= \sum_{k=1}^{k_n} \int_{\mathscr{H}} (1 - \cos\langle y, x - a_{nk}\rangle) \\
&\quad \times \frac{\|x - a_{nk}\|^2}{1 + \|x - a_{nk}\|^2}\, dP_{nk}(x) \\
&\leq \sum_{k=1}^{k_n} \int_{\mathscr{H}} (1 - \cos\langle y, x - a_{nk}\rangle)\, dP_{nk}(x) \\
&< \epsilon \qquad \text{whenever } \langle \mathbf{S}_{n,\epsilon} y, y\rangle < 1.
\end{aligned} \tag{7.8.46}$$

Thus the conditions of Proposition 7.7.1 are satisfied, and it follows that $\{\mu_n\}$ is tight. This completes the proof of Lemma 7.8.8. ■

Lemma 7.8.9

(a) The sequence $\{\mu_n\}$ converges weakly to a finite measure μ' on $\mathscr{B}$.
(b) The sequence $\{\mathbf{S}_n\}$ of S-operators on $\mathscr{H}$ satisfies (a), (b), (c) of condition (iii) in the statement of Theorem 7.8.1.
(c) The Fourier transform $\hat{\mathbf{P}}$ of the limiting probability measure $\mathbf{P}$ is of the form (7.8.13).

Proof. First we observe that (iii)(a) in the statement of Theorem 7.8.1 holds in view of (7.8.43). Next we note that for every $y \in \mathscr{H}$

$$\begin{aligned}
\langle \mathbf{S}_n y, y\rangle &= \sum_{k=1}^{k_n} \int_{\mathscr{H}} \frac{\langle y, x - a_{nk}\rangle^2}{1 + \|x - a_{nk}\|^2}\, dP_{nk}(x) \\
&= \int_{\mathscr{H}} \frac{\langle y, x\rangle^2}{1 + \|x\|^2} \frac{1 + \|x\|^2}{\|x\|^2}\, d\mu_n(x) \\
&= \int_{\mathscr{H}} \frac{\langle y, x\rangle^2}{\|x\|^2}\, d\mu_n(x).
\end{aligned} \tag{7.8.47}$$

It follows from the definition of μ_n in (7.8.12) that

$$\int_{\{0\}} \frac{\langle y, x\rangle^2}{\|x\|^2}\, d\mu_n(x) = 0 \qquad \text{for all } y \in \mathscr{H} \text{ and } n \geq 1.$$

Let $\epsilon > 0$. Since $\{\mu_n\}$ is tight, there exists a compact set $K = K(\epsilon) \subset \mathscr{H}$ such that $\mu_n(\mathscr{H} - K) < \epsilon$ for all $n \geq 1$. Let $\{e_\nu\}$ be an orthonormal basis in $\mathscr{H}$. Then, in view of (7.8.47), we have for every $N \geq 1$

$$\begin{aligned}\sum_{\nu=N}^{\infty} \langle \mathbf{S}_n e_\nu, e_\nu\rangle &= \sum_{\nu=N}^{\infty} \int_K \frac{\langle x, e_\nu\rangle^2}{\|x\|^2}\, d\mu_n(x) \\ &\quad + \sum_{\nu=N}^{\infty} \int_{\mathscr{H}-K} \frac{\langle x, e_\nu\rangle^2}{\|x\|^2}\, d\mu_n(x) \\ &\leq \sum_{\nu=N}^{\infty} \int_K \frac{\langle x, e_\nu\rangle^2}{\|x\|^2}\, d\mu_n(x) + \epsilon. \end{aligned} \tag{7.8.48}$$

For every $N \geq 1$ and $x \in \mathscr{H}$, $x \neq 0$, we set

$$\psi_N(x) = \frac{1}{\|x\|^2} \sum_{\nu=N}^{\infty} \langle x, e_\nu\rangle^2, \qquad \psi_N(0) = 0, \qquad \text{and} \qquad \theta_N = \sup_{x \in K} \psi_N(x).$$

Since K is compact, we see easily that $0 \leq \theta_N \downarrow 0$ as $N \to \infty$. It follows from (7.8.48) that

$$\sup_n \sum_{\nu=N}^{\infty} \langle \mathbf{S}_n e_\nu, e_\nu\rangle \leq \theta_N \sup_n \mu_n(\mathscr{H}) + \epsilon.$$

Taking the limit on both sides, first as $N \to \infty$ and then as $\epsilon \to 0$, and noting that $\sup_n \mu_n(\mathscr{H}) < \infty$, we obtain (iii)(b) in the statement of Theorem 7.8.1.

It follows easily from (7.8.47) that $\|\mathbf{S}_n\| \leq \operatorname{tr} \mathbf{S}_n$ so that $\sup_n \|\mathbf{S}_n\| \leq \sup_n \operatorname{tr} \mathbf{S}_n < \infty$. We now proceed exactly as in the proof of Theorem 7.7.2 (see the development following the proof of Lemma 7.7.5), and select a subsequence $n_k \to \infty$ as $k \to \infty$ such that:

(a) $\{\mu_{n_k}\}$ converges weakly to a finite measure μ' on $\mathscr{B}$.
(b) For every $y \in \mathscr{H}$

$$\lim_{k\to\infty} \langle \mathbf{S}_{n_k} y, y\rangle = \langle \mathbf{S}' y, y\rangle$$

exists, where $\mathbf{S}'$ is an S-operator on $\mathscr{H}$.

On the other hand, we obtain from (7.8.20) the result that, for every $y \in \mathscr{H}$,

$$\lim_{n\to\infty} \hat{P}_n(y) = \lim_{n\to\infty} \exp\left\{i\langle a_n, y\rangle - \frac{1}{2}\langle \mathbf{S}_n y, y\rangle \right.$$
$$+ \int_{\mathscr{H}} \left[\exp(i\langle y, x\rangle) - 1 - \frac{i\langle y, x\rangle}{1 + \|x\|^2} + \frac{1}{2}\frac{\langle y, x\rangle^2}{1 + \|x\|^2}\right]$$
$$\left. \times \frac{1 + \|x\|^2}{\|x\|^2}\, d\mu_n(x)\right\}. \tag{7.8.49}$$

Since $P_n \Rightarrow \mathbf{P}$, it follows that $\hat{P}_n \to \hat{\mathbf{P}}$ on $\mathscr{H}$, so that the limit $\lim_{k\to\infty} \langle a_{n_k}, y\rangle$ also exists and is finite for all $y \in \mathscr{H}$. Therefore the sequence $\{a_{n_k}\}$ is bounded, so there exists a vector $a \in \mathscr{H}$ such that the relation

$$\lim_{k\to\infty} \langle a_{n_k}, y\rangle = \langle a, y\rangle$$

holds for all $y \in \mathscr{H}$. Then it follows from (7.8.49) that for $y \in \mathscr{H}$

$$\hat{\mathbf{P}}(y) = \lim_{k\to\infty} \hat{P}_{n_k}(y) = \exp\left\{i\langle a, y\rangle - \tfrac{1}{2}\langle \mathbf{S}'y, y\rangle \right.$$
$$+ \int_{\mathscr{H}} \left[\exp(i\langle y, x\rangle) - 1 - \frac{i\langle y, x\rangle}{1 + \|x\|^2} + \frac{1}{2}\frac{\langle y, x\rangle^2}{1 + \|x\|^2}\right]$$
$$\left. \times \frac{1 + \|x\|^2}{\|x\|^2}\, d\mu'(x)\right\} = \exp\left\{i\langle a, y\rangle - \tfrac{1}{2}\langle \mathbf{S}y, y\rangle \right.$$
$$\left. + \int_{\mathscr{H}} \left[\exp(i\langle y, x\rangle) - 1 - \frac{i\langle y, x\rangle}{1 + \|x\|^2}\right] \frac{1 + \|x\|^2}{\|x\|^2}\, d\mu(x)\right\}, \tag{7.8.50}$$

where $\mathbf{S}$ and μ are as defined in (7.8.14) and (7.8.15), respectively. This proves that $\mathbf{P}$ is an i.d. probability measure on $\mathscr{B}$ with Fourier transform $\hat{\mathbf{P}}$ of the form (7.8.13).

From the uniqueness of the representation in (7.8.50) we conclude that $\mu_n \Rightarrow \mu'$, $\langle \mathbf{S}_n y, y\rangle \to \langle \mathbf{S}'y, y\rangle$ for every $y \in \mathscr{H}$, and $\langle a_n, y\rangle \to \langle a, y\rangle$ for every $y \in \mathscr{H}$ as $n \to \infty$. This completes the proof of Lemma 7.8.9. ■

The proof of the necessity part of Theorem 7.8.1 will be complete if we show that

$$\lim_{n\to\infty} a_n = a.$$

This is carried out in the following lemma.

Lemma 7.8.10. $\lim_{n\to\infty} a_n = a$.

Proof. We showed in Lemma 7.8.9 that $\{a_n\}$ converges weakly to a. Consequently $\{a_n\}$ is bounded. Since $P_n \Rightarrow \mathbf{P}$, we note that $\hat{P}_n \to \hat{\mathbf{P}}$ uniformly in every closed sphere $\|y\| \le c$ (Problem 7.9.20) and also $\langle \mathbf{S}_n y, y\rangle \to \langle \mathbf{S}'y, y\rangle$ uniformly in every closed sphere $\|y\| \le c$ (Problem 7.9.28). Moreover, since $\mu_n \Rightarrow \mu'$, it can be shown (Problem 7.9.29) that

$$\int_{\mathscr{H}} \left[\exp(i\langle y, x\rangle) - 1 - \frac{i\langle y, x\rangle}{1 + \|x\|^2} + \frac{1}{2}\frac{\langle y, x\rangle^2}{1 + \|x\|^2}\right]\frac{1 + \|x\|^2}{\|x\|^2}\, d\mu_n(x)$$

$$\to \int_{\mathscr{H}} \left[\exp(i\langle y, x\rangle) - 1 - \frac{i\langle y, x\rangle}{1 + \|x\|^2} + \frac{1}{2}\frac{\langle y, x\rangle^2}{1 + \|x\|^2}\right]\frac{1 + \|x\|^2}{\|x\|^2}\, d\mu'(x)$$

uniformly in every closed sphere $\|y\| \le c$. Hence $\langle a_n, y\rangle \to \langle a, y\rangle$ uniformly in every closed sphere $\|y\| \le c$. Since $\{a_n\}$ is bounded, we see easily that

$$\lim_{n\to\infty} \|a_n\|^2 = \lim_{n\to\infty} \langle a_n, a_n\rangle = \lim_{n\to\infty} \langle a_n, a\rangle$$
$$= \langle a, a\rangle = \|a\|^2,$$

which implies that $\|a_n - a\|^2 \to 0$ as $n \to \infty$. This completes the proof of Lemma 7.8.10. ■

The proof of Theorem 7.8.1 is now complete. ■

7.9 PROBLEMS

SECTION 7.1

1. Show that $\mathscr{E}\|X\| < \infty$ entails the existence of $\mathscr{E}(X)$.

2. Prove Proposition 7.1.2.

3. Let $\mathscr{H}$ be a real, separable Hilbert space with inner product $\langle\cdot, \cdot\rangle$, and let $\mathscr{B}$ be the σ-field generated by the class of all open subsets of $\mathscr{H}$. Let X and Y be $\mathscr{H}$-valued random variables such that $\mathscr{E}\|X\|^2$ and $\mathscr{E}\|Y\|^2 < \infty$. Then X and Y are said to be uncorrelated if $\mathscr{E}\langle X, Y\rangle = \langle \mathscr{E}X, \mathscr{E}Y\rangle$. If X and Y are independent, show that they are uncorrelated.

SECTION 7.2

4. Show with the help of an example that the assumption of identical distribution in Theorem 7.2.1 cannot be relaxed by imposing bounds on the moments of $\{\|X_n\|\}$.

5. A normed linear space $\mathfrak{X}$ is said to be B-convex if there is an integer $n > 0$ and an $\epsilon > 0$ such that, for all $x_1, x_2, \ldots, x_n \in \mathfrak{X}$ with $\|x_k\| \le 1$, $1 \le k \le n$,

$$\|\pm x_1 \pm x_2 \pm \cdots \pm x_n\| < n(1 - \epsilon)$$

for some choice of $+$ and $-$ signs. Show that the spaces $\mathscr{L}_p$ and l_p, $1 < p < \infty$, and inner product spaces are B-convex. The spaces l_1, l_∞, and c_0 are not B-convex.

6. Show with the help of an example that Theorem 7.2.2 does not extend to Banach spaces even when Beck's convexity condition is satisfied.

7. Let $\mathscr{H}$ be a real, separable Hilbert space, and let $\{X_n\}$ be a sequence of independent $\mathscr{H}$-valued random variables, each with finite variance. Show that Kolmogorov's inequality (7.2.5) holds.

8. (a) Let B be a real, separable Banach space with norm $\|\cdot\|$, and let $X_1, X_2, \ldots, X_n$ be independent, symmetric, B-valued random variables. Show that for $\epsilon > 0$

$$P\left\{\max_{1 \le j \le k} \|X_j\| \ge \epsilon\right\} \le 2P\left\{\|S_n\| \ge \frac{\epsilon}{2}\right\}.$$

(b) Let φ be a strictly increasing function defined on $[0, \infty)$ such that $\varphi(0) = 0$ and $\varphi(3x) \le C\varphi(x)$ for some finite constant C and $x > 0$. Let X_1, X_2 be independent, B-valued random variables such that, for some $a > 0$, $P\{\|X_2\| < a\} > \frac{1}{2}$. Show that for $\epsilon > \varphi(a)$

$$P\{\varphi(\|X_1 - X_2\|) \ge \epsilon\} \ge \tfrac{1}{2}P\{\varphi(\|X_1\|) \ge C\epsilon\}.$$

9. Let φ and ψ be strictly increasing, nonnegative functions defined on $[0, \infty)$. Set $\theta = \varphi \circ \psi$, and $\beta(j) = \theta(j+1) - \theta(j), j = 1, 2, \ldots$. Suppose that, for some $C_1, C_2 > 0$, $C_1 \le C_2, \beta(j+1) \le \beta(j), j \ge 1$. Show that for a B-valued random variable X

$$\mathscr{E}\varphi(\|X\|) < \infty \Leftrightarrow \sum_{j=1}^{\infty} \beta(j)P\{\|X\| \ge \psi(j)\} < \infty.$$

In particular, if $r > 0$ and $t > 0$, then

$$\mathscr{E}\|X\|^t < \infty \Leftrightarrow \sum_{j=1}^{\infty} j^{r-1}P\{\|X\| \ge j^{r/t}\} < \infty.$$

(Jain [41])

SECTION 7.3

10. Construct examples to show that no implications exist between coordinate-wise uncorrelated random variables (Definition 7.3.3) and uncorrelated random variables (Problem 3).

11. Construct an example to show that the condition of identical distribution on the X_n in Theorem 7.3.1 (and hence also in the two corollaries following it) cannot be replaced by a uniform bound in norm on the X_n.

12. Let $\mathscr{H}$ be a real, separable Hilbert space. Let $\{X_n\}$ be a sequence of $\mathscr{H}$-valued, uncorrelated random variables such that $n^{-2}\sum_{k=1}^{n} \operatorname{var}(X_k) \to 0$ as $n \to \infty$. Show that

$$\left\| n^{-1} \sum_{k=1}^{n} (X_k - \mathscr{E}X_k) \right\| \xrightarrow{P} 0.$$

13. Let $\mathscr{H}$ be a real, separable Hilbert space with an orthonormal basis $\{e_\nu\}$, and let X and Y be $\mathscr{H}$-valued random variables with $\mathscr{E}\|X\|^2 < \infty$ and $\mathscr{E}\|Y\|^2 < \infty$. Show that, if X and Y are coordinate-wise uncorrelated with respect to the basis $\{e_\nu\}$, then X and Y are uncorrelated.

SECTION 7.4

14. Let X be a B-valued random variable, and let $a \in B$. Let $X^s = X - X'$, where X and X' are independent and identically distributed. Then for every $\epsilon > 0$ show that:

(a) $P\{\|X^s\| \geq \epsilon\} \leq 2P\{\|X - a\| \geq \epsilon/2\}$.
(b) $P\{\|X\| < \epsilon\}P\{\|X\| \geq \epsilon + \eta\} \leq P\{\|X^s\| \geq \eta\}$ for $\eta > 0$.

SECTION 7.5

15. Let $\{Y_{n1}, Y_{n2}, \ldots, Y_{nk_n}\}$ be a sequence of independent, real-valued random variables which are independent for each n. Suppose that $\mathscr{E}(Y_{nk}) = 0$, $\text{var}(Y_{nk}) > 0$, and

$$\sum_{k=1}^{k_n} \text{var}(Y_{nk}) = 1.$$

Set $t_{n0} = 0$ and $t_{nk} = \sum_{j=1}^{k} \text{var}(Y_{nj})$, for $1 \leq k \leq k_n$, and define

$$X_{nk}(t) = \begin{cases} 0 & \text{if } t \leq t_{n,k-1}, \\ Y_{nk} & \text{if } t > t_{n,k-1}, \end{cases}$$

$$S_n = \sum_{k=1}^{k_n} X_{nk}.$$

Then the X_{nk} are $\mathscr{L}_2(0, 1)$-valued variables. Let P_n be the distribution of S_n, $n \geq 1$. Show that the sequence $\{P_n\}$ is tight.

SECTION 7.6

16. Let $\mathscr{H}$ be a real, separable Hilbert space, and $\mathscr{B}$ the σ-field generated by open subsets of $\mathscr{H}$. Let $\mathfrak{U}$ be a finite-dimensional subspace of $\mathscr{H}$, and let $\lambda_{\mathfrak{U}}$ denote the Lebesgue measure on the Borel σ-field $\mathscr{B}_{\mathfrak{U}}$ of subsets of $\mathfrak{U}$. Let $2n$ be the dimension of $\mathfrak{U}$. Define

$$P_{\mathfrak{U}}(E) = (2\pi)^{-n} \int_E \exp(-\tfrac{1}{2}\|x\|^2)\, d\lambda_{\mathfrak{U}}(x), \qquad E \in \mathscr{B}_{\mathfrak{U}}.$$

Show that $\{P_{\mathfrak{U}}\}$ is a compatible family of probability measures.

17. Let φ be the Fourier transform of a finite measure on $\mathscr{H}$. Show that there exists an S-operator $\mathbf{S}$ on $\mathscr{H}$ such that

$$|\varphi(x_1) - \varphi(x_2)| \to 0 \qquad \text{if } \langle \mathbf{S}(x_1 - x_2), x_1 - x_2 \rangle \to 0$$

SECTION 7.7

18. Let $\mathscr{P}$ be a family of probability measures on $(\mathscr{H}, \mathscr{B})$. Show that the following two conditions are sufficient in order that $\mathscr{P}$ be tight:

(i) For every $N \geq 1$

$$\lim_{c \to \infty} \inf_{P \in \mathscr{P}} P\left\{x \in \mathscr{H} : \max_{1 \leq \nu < N} |\langle x, e_\nu \rangle| \leq c\right\} = 1.$$

(ii) $\lim_{N\to\infty} \sup_{P\in\mathscr{P}} J_N(P) = 0$,
where

$$J_N(P) = \int_{\mathscr{H}} [1 - \exp\{-\tfrac{1}{2}r_N^2(x)\}]\, dP(x)$$

and $\{e_\nu\}$ is a fixed orthonormal basis in $\mathscr{H}$.

19. Let $\mathscr{P}$ be a family of probability measures on $(\mathscr{H}, \mathscr{B})$. In order that $\mathscr{P}$ be tight show that the following two conditions are sufficient:

(i) For every $N \geq 1$ the family of functions ψ defined on $\mathbb{R}_N$ by

$$\psi_P(y_1, y_2, \ldots, y_N) = \hat{P}(y_1e_1 + y_2e_2 + \cdots + y_Ne_N),$$

for $P \in \mathscr{P}$ and $y_1, y_2, \ldots, y_N \in \mathbb{R}$, is equicontinuous at the origin.

(ii) $\lim_{N\to\infty} \sup_{P\in\mathscr{P}} \lim_{p\to\infty} J_{N,p}[1 - \operatorname{Re} \hat{P}(y_Ne_N + \cdots + y_{N+p-1}e_{N+p-1})] = 0$,
where

$$J_{N,p}(g) = (2\pi)^{-p/2} \int \cdots \int g \exp\{-\tfrac{1}{2}r_{N,p}^2(y)\}\, dy_N \cdots dy_{N+p-1}$$

and

$$r_{N,p}^2(y) = \sum_{i=N}^{N+p-1} y_i^2.$$

20. Let $\{P_n\}$, P be probability measures on $(\mathscr{H}, \mathscr{B})$. Let $\hat{P}_n$, $\hat{P}$ be the corresponding Fourier transforms. Suppose that $P_n \Rightarrow P$. Then show that $\hat{P}_n \to \hat{P}$ uniformly in every closed sphere $\{x \in \mathscr{H}: \|x\| \leq c\}$, where $c > 0$.

21. (a) (Kolmogorov's Three-Series Criterion) Let $\{X_n\}$ be a sequence of independent $\mathscr{H}$-valued random variables. Show that, in order that the series $\sum_{n=1}^{\infty} X_n$ converge, it is necessary that for any $\epsilon > 0$ the three series

$$\sum_{n=1}^{\infty} \int_{\|x\|<\epsilon} x\, dP_{X_n}(x), \qquad \sum_{n=1}^{\infty} \int_{\|x\|<\epsilon} \left\| x - \int_{\|y\|<\epsilon} y\, dP_{X_n}(y) \right\| dP_{X_n}(x),$$

and

$$\sum_{n=1}^{\infty} P\{\|X_n\| \geq \epsilon\}$$

be convergent, and it is sufficient that the series converge for at least one $\epsilon > 0$.

(b) Deduce the following: In order that the series $\sum_{n=1}^{\infty} X_n$ converge, it is sufficient that the series

$$\sum_{n=1}^{\infty} \mathscr{E}X_n \qquad \text{and} \qquad \sum_{n=1}^{\infty} \mathscr{E}\|X_n - \mathscr{E}X_n\|^2$$

converge, where $\mathscr{E}X_n = \int_{\mathscr{H}} x\, dP_{X_n}(x)$ is a vector in $\mathscr{H}$ such that, for all $y \in \mathscr{H}$, $\langle \mathscr{E}X_n, y\rangle = \int_{\mathscr{H}} \langle x, y\rangle\, dP_{X_n}(x)$.

22. Let $\{X_n\}$ be a sequence of independent $\mathscr{H}$-valued random variables, and let φ_n be the characteristic functional of X_n, $n \geq 1$. Show that $\sum_{n=1}^{\infty} X_n$ converges a.s. if and only if the product $\prod_{n=1}^{\infty} \varphi_n$ converges uniformly in each closed sphere $\{x \in \mathscr{H}: \|x\| \leq \epsilon\}$ to a characteristic functional φ.

23. Let P be a probability measure on $\mathscr{B}$, the Borel σ-field of subsets of a real, separable Hilbert space $\mathscr{H}$. Then the concentration function Q_P of P is defined as

$$Q_P(\epsilon) = \sup_{x \in \mathscr{H}} P(S_\epsilon + x), \qquad \epsilon > 0,$$

where $S_\epsilon = \{x \in \mathscr{H}: \|x\| \leq \epsilon\}$ and $S_\epsilon + x$ denotes the translate of S_ϵ by the element $x \in \mathscr{H}$. Show that:

(a) Q_P is a nondecreasing function of ϵ satisfying $\lim_{\epsilon \to \infty} Q_P(\epsilon) = 1$.
(b) If $P_1 * P_2 = P$, then for every $\epsilon > 0$, $Q_P(\epsilon) \leq \min\{Q_{P_1}(\epsilon), Q_{P_2}(\epsilon)\}$.

24. Let $X_1, X_2, \ldots, X_n$ be independent, symmetric, $\mathscr{H}$-valued random variables, and let $S_j = X_1 + \cdots + X_j$, $j \geq 1$. Let Q be the concentration function of S_n. Set $T = \sup_{1 \leq j \leq n} \|S_j\|$. Show that for $\epsilon > 0$

$$P\{T > 4\epsilon\} \leq 2\{1 - Q(\epsilon)\}.$$

(Varadhan [85])

25. Let $X_1, X_2, \ldots, X_n$ be independent $\mathscr{H}$-valued random variables with $\mathscr{E}X_i = 0$, $i = 1, 2, \ldots, n$, and $\|X_i\| \leq c$ for $i = 1, 2, \ldots, n$. Let $d > 0$. Suppose that

$$P\left\{\sup_{1 \leq j \leq n} \left\| \sum_{k=1}^{j} X_k \right\| \leq d\right\} \geq \epsilon > 0.$$

Show that

$$\mathscr{E} \left\| \sum_{j=1}^{n} X_j \right\|^2 \leq \frac{d^2 + (c + d)^2}{\epsilon}.$$

(Varadhan [85])

26. Let μ be a Gaussian probability measure on $(\mathscr{H}, \mathscr{B})$ with mean vector zero and covariance operator $\mathbf{S}$. Let λ be Gaussian on $(\mathscr{H}, \mathscr{B})$ with mean $a \in \mathscr{H}$ and the same covariance operator $\mathbf{S}$. Then show that λ and μ are either equivalent or orthogonal, depending on whether or not a is in the range of $\mathbf{S}^{1/2}$.

SECTION 7.8

27. Let $c > 0$ be fixed, and let the a_{nk} be defined [instead of by (7.8.2) and (7.8.3)] by the relation

$$\langle a_{nk}, y \rangle = \int_{\|x\| \leq c} \langle y, x \rangle \, dP_{nk}(x), \qquad y \in \mathscr{H}.$$

Let μ_n, $\mathbf{S}_{nk}$, and $\mathbf{S}_n$ be as defined in Theorem 7.8.1. Show that the sequence $\{P_n\}$ converges weakly to some probability measure $\mathbf{P}$ on $\mathscr{B}$ if and only if conditions (i), (ii), and

(iii) of Theorem 7.8.1 are satisfied. Moreover, in this case $\hat{\mathbf{P}}$ has the representation

$$\ln \hat{\mathbf{P}}(y) = i\langle a, y\rangle - \tfrac{1}{2}\langle \mathbf{S}y, y\rangle + \int_{\|x\|>c} [\exp(i\langle y, x\rangle - 1] \frac{1 + \|x\|^2}{\|x\|^2}\, d\mu(x)$$

$$+ \int_{\|x\|\le c} [\exp(i\langle y, x\rangle) - 1 - i\langle y, x\rangle] \frac{1 + \|x\|^2}{\|x\|^2}\, d\mu(x)$$

for $y \in \mathscr{H}$, provided that c is chosen such that $\mu\{x \in \mathscr{H}: \|x\| = c\} = 0$.

28. By using condition (iii) of Theorem 7.8.1, show that $\langle \mathbf{S}_n, y\rangle \to \langle \mathbf{S}'y, y\rangle$ uniformly in every closed sphere $\|y\| \le c$.

29. Show that $\mu_n \Rightarrow \mu'$ implies

$$\int_{\mathscr{H}} \left[\exp(i\langle y, x\rangle) - 1 - \frac{i\langle y, x\rangle}{1 + \|x\|^2} + \frac{1}{2}\frac{\langle y, x\rangle^2}{1 + \|x\|^2}\right] \frac{1 + \|x\|^2}{\|x\|^2}\, d\mu_n(x)$$

$$\to \int_{\mathscr{H}} \left[\exp(i\langle y, x\rangle) - 1 - \frac{i\langle y, x\rangle}{1 + \|x\|^2} + \frac{1}{2}\frac{\langle y, x\rangle^2}{1 + \|x\|^2}\right] \frac{1 + \|x\|^2}{\|x\|^2}\, d\mu'(x)$$

uniformly in every closed sphere $\|y\| \le c$.

NOTES AND COMMENTS

The current interest in the general theory of probability measures on abstract spaces can be traced to the fundamental paper by Prokorov [67]. A detailed account of this work and a report on subsequent investigation are given in Parthasarathy [64] and Billingsley [8, 9]. The laws of large numbers for random variables taking values in normed linear spaces are discussed in Padgett and Taylor [63]. An exhaustive account of the theory of probability measures on a Hilbert space is given in Parthasarathy [64], Gihman and Skorohod [27], and Skorohod [76]. In Sections 7.6 through 7.8 we followed closely the development in Gihman and Skorohod [27]. In the special case where the Hilbert space is finite-dimensional we immediately obtain the multidimensional analogues of the results in Sections 7.5 through 7.8. We note, however, that the multidimensional analogue of the general central limit problem is due to Rvačeva [72] and is also discussed in Cuppens [14].

References

1. Aczel, J., *Lectures on Functional Equations and Their Applications*, Academic Press, New York, 1966.
2. Akhiezer, N. I., *The Classical Moment Problem*, Oliver and Boyd, Edinburgh, 1965.
3. Akhiezer, N. I. and I. M. Glazman, *Theory of Linear Operators in a Hilbert Space*, Vol. I, Frederick Ungar, New York, 1966.
4. Andersen, G. R., Large deviation probabilities for positive random variables, *Proc. Amer. Math. Soc.*, **24** (1970), 382–384.
5. Berry, A. C., The accuracy of the Gaussian approximation to the sum of independent variates, *Trans. Amer. Math. Soc.*, **48** (1941), 122–136.
6. Bhattacharya, R. N. and R. Ranga Rao, *Normal Approximation and Asymptotic Expansions*, John Wiley, New York, 1976.
7. Billingsley, P., *Ergodic Theory and Information*, John Wiley, New York, 1965.
8. Billingsley, P., *Convergence of Probability Measures*, John Wiley, New York, 1968.
9. Billingsley, P., *Weak Convergence of Measures: Applications in Probability*, Society for Industrial and Applied Mathematics, Philadelphia, Pa., 1971.
10. Broughton, A. and B. W. Huff, A comment on unions of sigma-fields, *Amer. Math. Monthly*, **84** (1977), 553–554.
11. Carleman, T., *Sur les Équations Intégrales Singulières*, Uppsala, 1923.
12. Chow, Y. S., H. Robbins, and D. Siegmund, *Great Expectations: The Theory of Optimal Stopping*, Houghton Mifflin, Boston, 1971.
13. Cramèr, H., *Mathematical Methods of Statistics*, Princeton University Press, Princeton, N.J., 1946.
14. Cuppens, R., *Decomposition of Multivariate Probability*, Academic Press, New York, 1975.
15. Darling, D. A. and H. Robbins, Iterated logarithm inequalities, *Proc. Nat. Acad. Sci.*, **57** (1967), 1188–1192.
16. De Finetti, B., *Theory of Probability*, Vol. 2, John Wiley, New York, 1975.
17. Doob, J. L., *Stochastic Processes*, John Wiley, New York, 1953.
18. Dubins, L. E. and L. J. Savage, *Inequalities for Stochastic Processes* (*How to Gamble If You Must*), Dover, New York, 1976.

19. Dunford, N. and J. T. Schwartz, *Linear Operators*, Part I: *General Theory*, Wiley-Interscience, New York, 1957.
20. Dynkin, E. B., *Markov Processes*, Vol. 1, Springer-Verlag, Berlin, 1965.
21. Esseen, C. G., Fourier analysis of distribution functions: A mathematical study of the Laplace–Gaussian law, *Acta Math.*, **77** (1945), 1–125.
22. Feller, W., A limit theorem for random variables with infinite moments, *Amer. J. Math.*, **68** (1946), 257–262.
23. Feller, W., An extension of the law of the iterated logarithm to variables without variance, *J. Math. Mech.*, **18** (1968), 343–355.
24. Feller, W., *An Introduction to Probability Theory and Its Applications*, Vol. 1, 3rd ed., John Wiley, New York, 1968.
25. Feller, W., *An Introduction to Probability Theory and Its Applications*, Vol. 2, 2nd ed., John Wiley, New York, 1971.
26. Fisz, M., *Probability Theory and Mathematical Statistics*, 3rd ed., John Wiley, New York, 1963.
27. Gihman, I. I. and A. V. Skorohod, *The Theory of Stochastic Processes*, Vol. I, Springer-Verlag, Berlin, 1974.
28. Gnedenko, B. V. and A. N. Kolmogorov, *Limit Distributions for Sums of Independent Random Variables*, Addison-Wesley, Reading, Mass., 1954.
29. Goffman, C. and G. Pedrick, *First Course in Functional Analysis*, Prentice-Hall, Englewood Cliffs, N.J., 1965.
30. Halmos, P. R. and L. J. Savage, Application of the Radon–Nikodym theorem to the theory of sufficient statistics, *Ann. Math. Stat.*, **20** (1949), 225–241.
31. Halmos, P. R., *Measure Theory*, D. Van Nostrand, New York, 1950.
32. Hartman, P. and A. Wintner, On the law of the iterated logarithm, *Amer. J. Math.*, **63** (1941), 169–176.
33. Hewitt, E. and K. Stromberg, *Real and Abstract Analysis*, Springer-Verlag, New York, 1965.
34. Heyde, C. C., On the converse to the iterated logarithm law, *J. Appl. Probability*, **5** (1968), 210–215.
35. Heyde, C. C., A note concerning behavior of iterated logarithm type, *Proc. Amer. Math. Soc.*, **23** (1969), 85–90.
36. Hille, E. and R. S. Phillips, *Functional Analysis and Semigroups*, American Mathematical Society, Providence, R.I., 1957.
37. Hille, E., *Analytic Function Theory*, Vol. 1, Ginn, Boston, 1959.
38. Hsu, P. L. and H. Robbins, Complete convergence and the law of large numbers, *Proc. Nat. Acad. Sci.*, **33** (1947), 25–31.
39. Ibragimov, I. A. and Yu. V. Linnik, *Independent and Stationary Sequences of Random Variables*, Wolters-Noordhoff, Groningen, 1971.
40. Ito, K. and M. Nisio, On the convergence of sums of independent Banach space valued random variables, *Osaka J. Math.*, **5** (1968), 35–48.
41. Jain, N. C., Tail probabilities for sums of independent Banach space valued random variables, *Z. Wahrscheinlichkeitstheorie verw. Geb.*, **33** (1975), 155–166.
42. Kagan, A. M., Yu. V. Linnik, and C. R. Rao, *Characterization Problems in Mathematical Statistics*, Wiley-Interscience, New York, 1973.

43. Kahane, Jean-Pierre, *Some Random Series of Functions*, Heath, Lexington, Mass., 1968.

44. Kawata, T., *Fourier Analysis in Probability Theory*, Academic Press, New York, 1972.

45. Keilson, J. and F. W. Steutel, Families of Infinitely divisible distributions closed under mixing and convolution, *Center for System Science, University of Rochester, Technical Report* CSS 70-18, 1970.

46. Kelley, J. L., *General Topology*, D. Van Nostrand, Princeton, N.J., 1963.

47. Kolmogorov, A. N., Über das Gesetz iterierten Logarithmus, *Math. Ann.*, **101** (1929), 126–135.

48. Kolmogorov, A. N., *Foundations of the Theory of Probability*, Chelsea, New York, 1956.

49. Kolmogorov, A. N. and S. V. Fomin, *Elements of the Theory of Functions and Functional Analysis*, Vol. 1, Graylock, Rochester, N.Y., 1957.

50. Kolmogorov, A. N. and S. V. Fomin, *Elements of the Theory of Functions and Functional Analysis*, Vol. 2, Graylock, Rochester, N.Y., 1965.

51. Krasnoselsky, M. A. and Y. B. Rutitsky, *Convex Functions and Orlicz Spaces*, Hindustan Publishing, Delhi, 1962.

52. Lehman, E. L., *Testing Statistical Hypotheses*, John Wiley, New York, 1959.

53. Lehmann, E. L., *Nonparametrics: Statistical Methods Based on Ranks*, Holden-Day, San Francisco, 1975.

54. Lévy, P., *Théorie de l'Addition des Variables Aléatoires*, 2nd ed., Gauthier-Villars, Paris, 1954.

55. Linnik, Yu. V., On the probability of large deviations for the sums of independent random variables, *Proc. 4th Berkeley Symposium*, **2** (1961), 289–306.

56. Linnik, Yu. V., *Decomposition of Probability Distributions*, Oliver and Boyd, London, 1964.

57. Linnik, Yu. V. and I. V. Ostrovskii, *Decomposition of Random Variables and Random Vectors*, American Mathematical Society, Providence, R.I., 1977.

58. Lukacs, E., A characterization of the normal distribution, *Ann. Math. Stat.*, **13** (1942), 91–93.

59. Lukacs, E. and R. G. Laha, *Applications of Characteristic Functions*, Griffin, London, 1964.

60. Lukacs, E., *Characteristic Functions*, 2nd ed., Hafner, New York, 1970.

61. Marcinkiewicz, J. and A. Zygmund, Remarque sur la loi du logarithme itere, *Fund. Math.*, **29** (1937), 215–222.

62. Natanson, I. P., *Theory of Functions of a Real Variable*, Frederick Ungar, New York, 1955.

63. Padgett, W. J. and R. L. Taylor, *Laws of Large Numbers for Normed Linear Spaces and Certain Fréchet Spaces*, Lecture Notes No. 360, Springer-Verlag, New York, 1970.

64. Parthasarathy, K. R., *Probability Measures on Metric Spaces*, Academic Press, New York, 1967.

65. Petrov, V. V., On a relation between an estimate of the remainder term in the central limit theorem and the law of iterated logarithm, *Theory Probability Appl.*, **11** (1966), 454–458.

66. Petrov, V. V., *Sums of Independent Random Variables*, Springer-Verlag, New York, 1975.

67. Prokhorov, Yu. V., Convergence of random processes and limit theorems in probability theory, *Theory Probability Appl.*, **1** (1956), 157–214.

68. Rényi, A., *Foundations of Probability*, Holden-Day, San Francisco, 1970.

69. Révész, P., *The Laws of Large Numbers*, Academic Press, New York, 1968.

70. Rohatgi, V. K., *An Introduction to Probability Theory and Mathematical Statistics*, John Wiley, New York, 1976.

71. Royden, H. L., *Real Analysis*, 2nd ed., Macmillan, New York, 1968.

72. Rvačeva, E. L., On domains of attraction of multidimensional distributions, in *Selected Translations in Mathematics and Statistics*, Vol. 2, American Mathematical Society, Providence, R.I., 1962, pp. 183–205.

73. Schmetterer, L., *Introduction to Mathematical Statistics*, Springer-Verlag, New York, 1974.

74. Shapiro, J., Domains of attraction for reciprocals of powers of random variables, *SIAM J. Appl. Math.*, **29** (1975), 734–739.

75. Shohat, J. A. and J. D. Tamarkin, *The Problems of Moments*, American Mathematical Society, Providence, R.I., 1943.

76. Skorohod, A. V., *Integration in Hilbert Space*, Springer-Verlag, Berlin, 1974.

77. Stout, W. F., *Almost Sure Convergence*, Academic Press, New York, 1974.

78. Strassen V., A converse to the law of the iterated logarithm, *Z. Wahrscheinlichkeitstheorie verw. Geb.*, **4** (1966), 265–268.

79. Teicher, H., Some new conditions for the strong law, *Proc. Nat. Acad. Soc.*, **59** (1968), 705–707.

80. Titchmarsh, E. C., *Introduction to the Theory of Fourier Integrals*, Clarendon Press, Oxford, 1937.

81. Titchmarsh, E. C., *The Theory of Functions*, Oxford University Press, London, 1961.

82. Trotter, H., An elementary proof of the central limit theorem, *Arch. Math.*, **10** (1959), 226–234.

83. Tucker, H., On moments of distribution functions attracted to stable laws, *Houston, J. Math.*, **1** (1975), 149–152.

84. Varberg, D. E., Almost sure convergence of quadratic forms in independent random variables, *Ann. Math. Stat.*, **39** (1968), 1502–1506.

85. Varadhan, S. R. S., Limit theorems for sums of independent random variables with values in a Hilbert space, *Sankhyā*, **24** (1962), 213–238.

86. Von Mises, R., *Probability, Statistics and Truth*, 2nd rev. English ed., Allen and Unwin, London, 1957.

87. Whittle, P., *Probability*, Penguin, Baltimore, Md., 1970.

88. Widder, D. V., *The Laplace Transform*, Princeton University Press, Princeton, N.J., 1946.

89. Wilansky, A., *Functional Analysis*, Blaisdell, New York, 1964.

90. Woyczynski, W. A., Random series and laws of large numbers in some Banach spaces, *Theory Probability Appl.*, **18** (1973), 350–355.

91. Yosida, K. *Functional Analysis*, 2nd ed., Springer-Verlag, New York, 1968.

92. Zacks, S., *The Theory of Statistical Inference*, John Wiley, New York, 1971.

Some Frequently Used Symbols and Abbreviations

Page numbers refer to the first occurrence of these symbols or abbreviations.

$\mathbb{R}$, 3. *see also* p. 19
$\mathbb{R}_n$, 19
$\mathcal{B}$, 3. *see also* pp. 19, 200, 447
$\mathcal{B}_n$, 19
$\overline{\mathbb{R}}$, 181
$\mathbb{Q}$, 114
$\mathbb{C}$, 123
$\overline{C}_0$, 152
$\bar{\varphi}_0$, 181
$\bar{\varphi}_\infty$, 183
$\mathbb{R}^\times_+$, 187
$\mathbb{C}^\times$, 215
$\alpha_n \beta_\lambda$, μ_n, 30
χ_E, 29
$\epsilon(x)$, 12
$\xrightarrow{\text{a.s.}}$, 43
$\xrightarrow{P}$, 45
$\mathcal{L}_1$, $\xrightarrow{\mathcal{L}_1}$, 49
$\mathcal{L}_p$, $\xrightarrow{\mathcal{L}_p}$, 52
$\xrightarrow{\omega}$, 131
$\xrightarrow{c}$, 131
$\xrightarrow{L}$, 132
$\xrightarrow{s}$, 184
C_F, 131
$\Rightarrow$, 158. *see also* pp. 160, 203
$\sigma(x)$, 60
$\sigma(x_1, x_2, \ldots, x_n)$, 75
$\mathcal{T}$, 76
i.o., 73
med(X), 81
$*$, 150. *see also* pp. 153, 181
A_F, 184
i.d., 232
u.a.n., 295
$\hat{P}, \hat{\mu}$, 460, 469
$\underset{\sim}{S}, \underset{\sim}{S}_p$, 473

variable associated with X_{nk}, $1 \leq k \leq k_n$, $n \geq 1$. Let $\alpha > 0$. Clearly

$$P\left\{\left\|\sum_{k=1}^{n} X_{nk}^s\right\| \geq 2\alpha\right\} \leq 2P\{\|S_n\| \geq \alpha\},$$

so that, in view of Lemma 7.4.2, we have

$$\begin{aligned} P\left\{\max_{1 \leq j \leq k_n}\left\|\sum_{k=1}^{j} X_{nk}^s\right\| \geq 2\alpha\right\} &\leq 2P\left\{\left\|\sum_{k=1}^{k_n} X_{nk}^s\right\| \geq 2\alpha\right\} \\ &\leq 4P\{\|S_n\| \geq \alpha\}. \end{aligned}$$

This implies that

$$\begin{aligned} P\left\{\max_{1 \leq k \leq k_n}\|X_{nk}^s\| \geq 4\alpha\right\} &\leq P\left\{\max_{1 \leq j \leq k_n}\left\|\sum_{k=1}^{n} X_{nk}^s\right\| \geq 2\alpha\right\} \\ &\leq 4P\{\|S_n\| \geq \alpha\} \\ &= 4P_n\{x \in \mathscr{H}: \|x\| \geq \alpha\}. \end{aligned}$$

Since $P_n \Rightarrow \mathbf{P}$, for given $0 < \epsilon < 1$ we can choose an $\alpha_0 = \alpha_0(\epsilon)$ such that for all $\alpha > \alpha_0$

$$\sup_n P_n\{\|x\| \geq \alpha\} < \frac{\epsilon}{4},$$

so that for $\alpha > \alpha_0$

$$\sup_n P\left\{\max_{1 \leq k \leq k_n}\|X_{nk}^s\| \geq 4\alpha\right\} < \epsilon.$$

Then, using the inequality $t < -\ln(1 - t)$ for $0 < t < 1$, we have

$$\begin{aligned} \sum_{k=1}^{k_n} P\{\|X_{nk}^s\| \geq 4\alpha\} &\leq -\sum_{k=1}^{k_n} \ln P\{\|X_{nk}^s\| < 4\alpha\} \\ &= -\ln P\left\{\max_{1 \leq k \leq k_n}\|X_{nk}^s\| < 4\alpha\right\} \\ &\leq -\ln(1 - \epsilon), \end{aligned}$$

so that

$$\sup_{n \geq 1} \sum_{k=1}^{k_n} P\{\|X_{nk}^s\| \geq 4\alpha\} \leq -\ln(1 - \epsilon) < \infty.$$

We now use the elementary inequality

$$P\{\|X^s\| \geq \beta\} \geq P\{\|X\| < \gamma\}P\{\|X\| \geq \beta + \gamma\},$$

where $\beta > 0$, $\gamma > 0$, and X^s is the symmetrized random variable associated with an $\mathscr{H}$-valued random variable X. (See Problem 7.9.14.) We choose

Author Index

Subject Index